INTRODUCTORY MATHEMATICAL ANALYSIS

4TH EDITION

INTRODUCTORY MATHEMATICAL ANALYSIS

For Students of Business and Economics

Ernest F. Haeussler, Jr.

Richard S. Paul

Department of Mathematics
The Pennsylvania State University

RESTON PUBLISHING COMPANY, INC.

A Prentice-Hall Company
Reston, Virginia

Library of Congress Cataloging in Publication Data

Haeussler, Ernest F.
Introductory mathematical analysis for students of
business and economics.

  Includes index.
  1. Mathematical analysis. 2. Economics, Mathematical.
3. Business mathematics. I. Paul, Richard S. II. Title.
QA300.H33 1983   515   82-25038
ISBN 0-8359-3274-5

Cover photograph compliments of Intel Corporation

10 9 8 7 6 5 4 3 2 1

Printed in the United States of America.

CONTENTS

v

PREFACE

The fourth edition of *Introductory Mathematical Analysis* continues to provide a mathematical foundation, including calculus, matrix algebra, and linear programming, for students enrolled in business and economics curricula. A chapter on basic probability theory now appears.

In order to make the material meaningful and relevant to the student, we have included many topics in this text that show how mathematical concepts are applied to describe business and economic phenomena. Among these applications are supply and demand, equilibrium, break-even point, marginal revenue, marginal cost, marginal propensity to consume, marginal propensity to save, marginal revenue product, maximization of profit under monopoly, elasticity, consumers' surplus, producers' surplus, and budget equations. However, the text is virtually self-contained in the sense that it *assumes no prior exposure to the concepts on which the applications are based.*

As an aid to planning a course outline, we point out that the following sections are marked optional: 11-1, 11-2, 11-6, 12-4, 12-6, 12-9, 12-10, 15-3, and 15-5. Also, 2-4 may be omitted. Section 14-9 may be omitted if 14-10 is not covered.

Answers to odd-numbered problems appear at the end of the book. For many of the differentiation problems in Chapter 8, the answers appear in both unsimplified and simplified forms.

Available from the publisher is an extensive instructor's manual that includes answers to all problems and detailed solutions to a great many of them, including all applied problems. Included also are problems suitable for examinations.

We express our appreciation to the following colleagues who contributed comments and suggestions that were valuable to us in developing the manuscript:

R. M. Alliston *(Pennsylvania State University)*, M. N. de Arce *(University of Puerto Rico)*, G. R. Bates *(Western Illinois University)*, C. Bernett *(Harper College)*, A. Bishop *(Western Illinois University)*, S. A. Book *(California State University)*, A. Brink *(St. Cloud State University)*, R. Brown *(York University)*, R. W. Brown *(University of Alaska)*, S. D. Bulman-Fleming *(Wilfrid Laurier University)*, K. S. Chung *(Kapiolani Community College)*, E. L. Cohen *(University of Ottawa)*, J. Dawson *(Pennsylvania State University)*, A. Dollins *(Pennsylvania State University)*, G. A. Earles *(St. Cloud State University)*, J. R. Elliott *(Wilfrid Laurier University)*, J. Fitzpatrick *(University of Texas at El Paso)*, M. J. Flynn *(Rhode Island Junior College)*, G. J. Fuentes *(University of Maine)*, G. Goff *(Oklahoma State University)*, J. Goldman *(DePaul University)*, L. Griff *(Pennsylvania State University)*, F. H. Hall *(Pennsylvania State University)*, V. E. Hanks *(Western Kentucky University)*, J. N. Henry *(California State University)*, W. U. Hodgson *(West Chester State College)*, B. C. Horne, Jr. *(Virginia Polytechnic Institute and State University)*, J. Hradnansky *(Pennsylvania State University)*, C. Hurd *(Pennsylvania State University)*, W. C. Jones *(Western Kentucky University)*, R. M. King *(Gettysburg College)*, M. M. Kostreva *(University of Maine)*, M. R. Latina *(Rhode Island Junior College)*, I. Marshak *(Loyola University of Chicago)*, F. B. Mayer *(Mt. San Antonio College)*, P. McDougle *(University of Miami)*, F. Miles *(California State University)*, E. Mohnike *(Mt. San Antonio College)*, C. Monk *(University of Richmond)*, J. G. Morris *(University of Wisconsin—Madison)*, D. Mullin *(Pennsylvania State University)*, E. Nelson *(Pennsylvania State University)*, R. H. Oehmke *(University of Iowa)*, Y. Y. Oh *(Pennsylvania State University)*, N. B. Patterson *(Pennsylvania State University)*, E. Pemberton *(Wilfrid Laurier University)*, M. Perkel *(Wright State University)*, D. B. Priest *(Harding College)*, J. R. Provencio *(University of Texas)*, L. R. Pulsinelli *(Western Kentucky University)*, M. Racine *(University of Ottawa)*, N. M. Rice *(Queen's University)*, A. Santiago *(University of Puerto Rico)*, W. H. Seybold, Jr. *(West Chester State College)*, J. R. Schaefer *(University of Wisconsin—Milwaukee)*, S. Singh *(Pennsylvania State University)*, E. Smet *(Huron College)*, M. Stoll *(University of South Carolina)*, B. Toole *(University of Maine)*, J. W. Toole *(University of Maine)*, D. H. Trahan *(Naval Postgraduate School)*, J. P. Tull *(Ohio State University)*, L. O. Vaughan, Jr. *(University of Alabama in Birmingham)*, L. A. Vercoe *(Pennsyl-*

vania State University), B. K. Waits (Ohio State University), A. Walton (Virginia Polytechnic Institute and State University), H. Walum (Ohio State University), A. J. Weidner (Pennsylvania State University), L. Weiss (Pennsylvania State University), N. A. Weigmann (California State University), C. R. B. Wright (University of Oregon), C. Wu (University of Wisconsin—Milwaukee).

Finally, we express thanks to our editor Sherry Goldbecker for her cooperation and efficiency.

Ernest F. Haeussler, Jr.
Richard S. Paul

0-1 PURPOSE

This chapter is designed to give you a brief review of some terms and methods of manipulative mathematics. No doubt you have been exposed to much of this material before. However, because these topics are important in handling the mathematics that comes later, perhaps an immediate second exposure to them would be beneficial. Devote whatever time is necessary to those sections in which you need review.

0-2 SETS AND REAL NUMBERS

In simplest terms, a *set* is a collection of objects. For example, we can speak of the set of even numbers between 5 and 11, namely 6, 8, and 10. An object in a set is called a *member* or *element* of that set.

One way to specify a set is by listing its members, in any order, inside braces. For example, the previous set is $\{6, 8, 10\}$, which we can denote by a letter such as A. A set A is said to be a **subset** of a set B if and only if every element of A is an element of B. For example, if $A = \{6, 8, 10\}$ and $B = \{6, 8, 10, 12\}$, then A is a subset of B.

Certain sets of numbers have special names. The numbers 1, 2, 3, etc. form the set of **positive integers** (or **natural numbers**):

$$\begin{array}{c} set\ of \\ positive\ integers \end{array} = \{1, 2, 3, \ldots\}.$$

The three dots mean that the listing of elements is unending, although we know what the elements are.

The positive integers together with 0 and the **negative integers** $-1, -2, -3, \ldots$ form the set of **integers**:

$$\begin{array}{c} set\ of \\ integers \end{array} = \{\ldots, -3, -2, -1, 0, 1, 2, 3, \ldots\}.$$

The set of **rational numbers** consists of numbers, such as $\frac{1}{2}$ and $\frac{5}{3}$, which can be written as a ratio (quotient) of two integers. That is, a rational number is one that can be written as p/q, where p and q are integers and $q \neq 0$. (The symbol "$\neq$" is read "is not equal to.") **We cannot divide by zero.** The numbers $\frac{19}{20}$, $\frac{-2}{7}$, and $\frac{-6}{-2}$ are rational. The integer 2 is rational since $2 = \frac{2}{1}$. In fact, every integer is rational. We point out that $\frac{2}{4}$, $\frac{1}{2}$, $\frac{3}{6}$, $\frac{-4}{-8}$, and .5 all represent the same rational number.

All rational numbers can be represented by decimal numbers that *terminate*, such as $\frac{3}{4} = .75$ and $\frac{3}{2} = 1.5$, or by *nonterminating repeating* decimals (a group of digits repeats without end), such as $\frac{2}{3} = .666\ldots$, $\frac{-4}{11} = -.3636\ldots$, and $\frac{2}{15} = .1333\ldots$. Numbers represented by *nonterminating nonrepeating* decimals are called **irrational numbers**. An irrational number cannot be written as an integer divided by an integer. The numbers π (pi) and $\sqrt{2}$ are irrational.

Together, the rational numbers and irrational numbers form the set of **real numbers**. Real numbers can be represented by points on a line. To do this, we first choose a point on the line to represent zero. This point is called the *origin* (see Fig. 0-1). Then a standard measure of distance, called a "unit distance," is

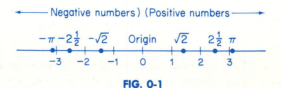

FIG. 0-1

chosen and is successively marked off both to the right and to the left of the origin. With each point on the line we associate a directed distance, or *signed number*, which depends on the position of the point with respect to the origin. Positions to the right of the origin are considered positive (+), and positions to the left are negative (−). Thus, with the point one unit to the right of the origin there corresponds the signed number +1 (or 1), positive one; with the point two and one-half units to the left of the origin there corresponds the signed number −2½, negative two and one-half. In Fig. 0-1 some points and their associated real numbers are identified. To each point on the line there corresponds a unique real number, and to each real number there corresponds a unique point on the line. For this reason we say that there is a *one-to-one correspondence* between points on the line and real numbers. We call this line the **real number line**.

EXERCISE 0-2

In Problems **1–12,** *classify the statement as either true or false. If false, give a reason.*

1. −7 is an integer.

2. $\frac{1}{6}$ is rational.

3. −3 is a natural number.

4. 0 is not rational.

5. 5 is rational.

6. $\frac{7}{0}$ is a rational number.

7. $\frac{4}{2}$ is not a positive integer.

8. π is a real number.

9. $\frac{0}{6}$ is rational.

10. 0 is a natural number.

11. −3 is to the right of −4 on the real number line.

12. Every integer is positive or negative.

0-3 SOME PROPERTIES OF REAL NUMBERS

In this section we shall give a few important properties of the real numbers.

I. The Transitive Property.
Let a, b, and c be real numbers.

$$If \quad a = b \quad and \quad b = c, \quad then \cdot a = c.$$

Thus, two numbers that are both equal to a third number are equal to each other. For example, if $x = y$ and $y = 7$, then $x = 7$.

II. The Commutative Properties.
If a and b are real numbers, then

$$a + b = b + a \quad and \quad ab = ba.$$

This means that we can add or multiply two real numbers in any order. For example, $3 + 4 = 4 + 3$ and $7(-4) = (-4)(7)$.

III. The Associative Properties.
If a, b, and c are real numbers, then

$$a + (b + c) = (a + b) + c \quad and \quad a(bc) = (ab)c.$$

This means that in addition or multiplication, numbers can be grouped in any order. For example, $2 + (3 + 4) = (2 + 3) + 4$. Also, $6(\frac{1}{3} \cdot 5) = (6 \cdot \frac{1}{3}) \cdot 5$ and $2x + (x + y) = (2x + x) + y$.

IV. The Inverse Properties.
a. *For each real number a, there is a unique real number denoted $-a$ such that*

$$a + (-a) = 0.$$

*The number $-a$ is called the **additive inverse**, or **negative**, of a.*

For example, since $6 + (-6) = 0$, the additive inverse of 6 is -6. The additive inverse of a number is not necessarily a negative number. For example, the additive inverse of -6 is 6, since $(-6) + (6) = 0$. That is, the negative of -6 is $-(-6) = 6$.

b. *For each real number a, except 0, there is a unique real number a^{-1} such that*

$$a \cdot a^{-1} = 1.$$

*The number a^{-1} is called the **multiplicative inverse** of a.*

Thus all numbers except 0 have a multiplicative inverse. You may recall that a^{-1} can be written $\frac{1}{a}$ and is also called the *reciprocal* of a. For example, the multiplicative inverse of 3 is $\frac{1}{3}$, since $3(\frac{1}{3}) = 1$. Thus $\frac{1}{3}$ is the reciprocal of 3. The reciprocal of $\frac{1}{3}$ is 3, since $(\frac{1}{3})(3) = 1$. We emphasize that **the reciprocal of 0 is not defined**.

V. The Distributive Properties.
If a, b and c are real numbers, then

$$a(b + c) = ab + ac \quad and \quad (b + c)a = ba + ca.$$

For example,

$$2(3 + 4) = 2(3) + 2(4) = 6 + 8 = 14,$$
$$(2 + 3)(4) = 2(4) + 3(4) = 8 + 12 = 20,$$
$$x(z + 4) = x(z) + x(4) = xz + 4x.$$

The distributive property can be extended to the form $a(b + c + d) = ab + ac + ad$. In fact, it can be extended to sums involving any other number of terms.

It is by the additive inverse property that we formally define *subtraction:* $a - b$ means $a + (-b)$, where $-b$ is the additive inverse of b. Thus $6 - 8$ means $6 + (-8)$. Subtraction is therefore defined in terms of addition.

In a similar way we define *division* in terms of multiplication. If $b \neq 0$. then $a \div b$, or $\dfrac{a}{b}$, is defined by

$$\frac{a}{b} = a(b^{-1}).$$

Since $b^{-1} = \dfrac{1}{b}$,

$$\frac{a}{b} = a(b^{-1}) = a\left(\frac{1}{b}\right).$$

Thus $\frac{3}{5}$ means 3 times $\frac{1}{5}$, where $\frac{1}{5}$ is the multiplicative inverse of 5. Sometimes we refer to $a \div b$ or $\dfrac{a}{b}$ as the *ratio* of a to b.

The following examples show some manipulations involving the above properties.

EXAMPLE 1

a. $x(y - 3z + 2w) = (y - 3z + 2w)x$, by the commutative property of multiplication.

b. By the associative property of multiplication, $3(4 \cdot 5) = (3 \cdot 4)5$. Thus, the result of multiplying 3 by the product of 4 and 5 is the same as the result of multiplying the product of 3 and 4 by 5. In either case the result is 60.

c. By the definition of subtraction, $2 - \sqrt{2} = 2 + (-\sqrt{2})$. However, by the commutative property, $2 + (-\sqrt{2}) = -\sqrt{2} + 2$. Thus, by the transitive property, $2 - \sqrt{2} = -\sqrt{2} + 2$. More concisely we can write

$$2 - \sqrt{2} = 2 + (-\sqrt{2}) = -\sqrt{2} + 2.$$

d.
$$(8 + x) - y = (8 + x) + (-y) \qquad \text{(definition of subtraction)}$$
$$= 8 + [x + (-y)] \qquad \text{(associative property)}$$
$$= 8 + (x - y) \qquad \text{(definition of subtraction)}.$$

Hence, by the transitive property,

$$(8 + x) - y = 8 + (x - y).$$

e. By the definition of division,

$$\frac{ab}{c} = (ab) \cdot \frac{1}{c} \quad \text{for} \quad c \neq 0.$$

But by the associative property,

$$(ab) \cdot \frac{1}{c} = a \left(b \cdot \frac{1}{c} \right).$$

However, by the definition of division, $b \cdot \dfrac{1}{c} = \dfrac{b}{c}$. Thus

$$\frac{ab}{c} = a \left(\frac{b}{c} \right).$$

We can also show that $\dfrac{ab}{c} = \left(\dfrac{a}{c} \right) b.$

EXAMPLE 2

a. *Show that* $3(4x + 2y + 8) = 12x + 6y + 24.$

By the distributive property,

$$3(4x + 2y + 8) = 3(4x) + 3(2y) + 3 \cdot 8.$$

But by the associative property of multiplication,

$$3(4x) = (3 \cdot 4)x = 12x \quad \text{and similarly} \quad 3(2y) = 6y.$$

Thus $3(4x + 2y + 8) = 12x + 6y + 24.$

b. *Show that* $x(y - z) = xy - xz.$

By the definition of subtraction and the distributive property,

$$x(y - z) = x[y + (-z)]$$
$$= xy + x(-z).$$

Recalling that $-z = (-1)z$, we can then say that $x(-z) = x[(-1)z]$. By the associative and commutative properties, $x[(-1)z] = [x(-1)]z = [(-1)x]z = (-1)(xz)$. Hence,

$$x(y - z) = xy + x(-z)$$
$$= xy + (-1)xz.$$

Referring again to the definition of subtraction, we have $x(y - z) = xy - xz$.

c. *Show that if $c \neq 0$, then* $\dfrac{a + b}{c} = \dfrac{a}{c} + \dfrac{b}{c}$.

By the definition of division and the distributive property,

$$\frac{a + b}{c} = (a + b)\frac{1}{c} = a \cdot \frac{1}{c} + b \cdot \frac{1}{c}.$$

However,

$$a \cdot \frac{1}{c} + b \cdot \frac{1}{c} = \frac{a}{c} + \frac{b}{c}.$$

Hence,

$$\frac{a + b}{c} = \frac{a}{c} + \frac{b}{c}.$$

This important result **does not** mean that $\dfrac{a}{b + c} = \dfrac{a}{b} + \dfrac{a}{c}$, a very common error. For example,

$$\frac{3}{2 + 1} \neq \frac{3}{2} + \frac{3}{1}.$$

Finding the product of several numbers can be done only by considering products of numbers taken two at a time. For example, to find the product of x, y, and z we could first multiply x by y and then multiply that product by z, or alternatively we could multiply x by the product of y and z. The associative property of multiplication says that both results are identical, regardless of how the numbers are grouped. Thus it is not ambiguous to write xyz. This concept can be extended to more than three numbers and applies equally well to addition.

One final comment before we end this section. Not only should you be aware of the manipulative aspects of the properties of the real numbers, but you should also be aware of and familiar with the terminology involved.

EXERCISE 0-3

*In Problems **1–10**, classify the statements as either true or false.*

1. Every real number has a reciprocal.

2. The reciprocal of $\frac{2}{5}$ is $\frac{5}{2}$.

3. The additive inverse of 5 is $\frac{1}{5}$.

4. $2(3 \cdot 4) = (2 \cdot 3)(2 \cdot 4)$.

5. $-x + y = y - x$.

6. $(x + 2)(4) = 4x + 8$.

7. $\dfrac{x + 2}{2} = \dfrac{x}{2} + 1$.

8. $3\left(\dfrac{x}{4}\right) = \dfrac{3x}{4}$.

9. $x + (y + 5) = (x + y) + (x + 5)$.

10. $8(9x) = 72x$.

In Problems 11–20, tell which properties of the real numbers are being used.

11. $2(x + y) = 2x + 2y$.

12. $(x + 5) + y = y + (x + 5)$.

13. $2(3y) = (2 \cdot 3)y$.

14. $\frac{6}{7} = 6 \cdot \frac{1}{7}$.

15. $2(x - y) = (x - y)(2)$.

16. $x + (x + y) = (x + x) + y$.

17. $8 - y = 8 + (-y)$.

18. $5(4 + 7) = 5(7 + 4)$.

19. $(7 + x)y = 7y + xy$.

20. $(-1)[-3 + 4] = (-1)(-3) + (-1)(4)$.

In Problems 21–26, show that the statements are true by using properties of the real numbers.

21. $5a(x + 3) = 5ax + 15a$.

22. $(2 - x) + y = 2 + (y - x)$.

23. $(x - y)(2) = 2x - 2y$.

24. $2[27 + (x + y)] = 2[(y + 27) + x]$.

25. $x[(2y + 1) + 3] = 2xy + 4x$.

26. $(x + 1)(y + 1) = xy + x + y + 1$.

27. Show that $a(b + c + d) = ab + ac + ad$. Hint: $b + c + d = (b + c) + d$.

0-4 OPERATIONS WITH SIGNED NUMBERS

Listed below are important properties of signed numbers which you should study thoroughly. Being able to manipulate signed numbers is essential to your success in mathematics. A numerical example follows each property. All denominators are different from zero and a knowledge of addition and subtraction of signed numbers is assumed.

Property	*Example*
1. $a - b = a + (-b)$.	$2 - 7 = 2 + (-7) = -5$.
2. $a - (-b) = a + b$.	$2 - (-7) = 2 + 7 = 9$.
3. $-a = (-1)(a)$.	$-7 = (-1)(7)$.
4. $a(b + c) = ab + ac$.	$6(7 + 2) = 6 \cdot 7 + 6 \cdot 2 = 54$.
5. $a(b - c) = ab - ac$.	$6(7 - 2) = 6 \cdot 7 - 6 \cdot 2 = 30$.
6. $-(a + b) = -a - b$.	$-(7 + 2) = -7 - 2 = -9$.

Property	*Example*
7. $-(a - b) = -a + b.$	$-(2 - 7) = -2 + 7 = 5.$
8. $-(-a) = a.$	$-(-2) = 2.$
9. $a(0) = (-a)(0) = 0.$	$2(0) = (-2)(0) = 0.$
10. $(-a)(b) = -(ab) = a(-b).$	$(-2)(7) = -(2 \cdot 7) = 2(-7).$
11. $(-a)(-b) = ab.$	$(-2)(-7) = 2 \cdot 7 = 14.$
12. $\dfrac{a}{1} = a.$	$\dfrac{7}{1} = 7, \quad \dfrac{-2}{1} = -2.$
13. $\dfrac{a}{b} = a\left(\dfrac{1}{b}\right).$	$\dfrac{2}{7} = 2\left(\dfrac{1}{7}\right).$
14. $\dfrac{1}{-a} = -\dfrac{1}{a} = \dfrac{-1}{a}.$	$\dfrac{1}{-4} = -\dfrac{1}{4} = \dfrac{-1}{4}.$
15. $\dfrac{a}{-b} = -\dfrac{a}{b} = \dfrac{-a}{b}.$	$\dfrac{2}{-7} = -\dfrac{2}{7} = \dfrac{-2}{7}.$
16. $\dfrac{-a}{-b} = \dfrac{a}{b}.$	$\dfrac{-2}{-7} = \dfrac{2}{7}.$
17. $\dfrac{0}{a} = 0.$	$\dfrac{0}{7} = 0.$
18. $\dfrac{a}{a} = 1.$	$\dfrac{2}{2} = 1, \quad \dfrac{-5}{-5} = 1.$
19. $a\left(\dfrac{b}{a}\right) = b.$	$2\left(\dfrac{7}{2}\right) = 7.$
20. $a \cdot \dfrac{1}{a} = 1.$	$2 \cdot \dfrac{1}{2} = 1.$
21. $\dfrac{1}{a} \cdot \dfrac{1}{b} = \dfrac{1}{ab}.$	$\dfrac{1}{2} \cdot \dfrac{1}{7} = \dfrac{1}{2 \cdot 7} = \dfrac{1}{14}.$
22. $\dfrac{ab}{c} = \left(\dfrac{a}{c}\right)b = a\left(\dfrac{b}{c}\right).$	$\dfrac{2 \cdot 7}{3} = \dfrac{2}{3} \cdot 7 = 2 \cdot \dfrac{7}{3}.$
23. $\dfrac{a}{bc} = \left(\dfrac{a}{b}\right)\left(\dfrac{1}{c}\right) = \left(\dfrac{1}{b}\right)\left(\dfrac{a}{c}\right).$	$\dfrac{2}{3 \cdot 7} = \dfrac{2}{3} \cdot \dfrac{1}{7} = \dfrac{1}{3} \cdot \dfrac{2}{7}.$
24. $\dfrac{a}{b} = \left(\dfrac{a}{b}\right)\left(\dfrac{c}{c}\right) = \dfrac{ac}{bc}.$	$\dfrac{2}{7} = \left(\dfrac{2}{7}\right)\left(\dfrac{5}{5}\right) = \dfrac{2 \cdot 5}{7 \cdot 5}.$
25. $\dfrac{a}{b(-c)} = \dfrac{a}{(-b)(c)} = \dfrac{-a}{bc}$ $= \dfrac{-a}{(-b)(-c)} = -\dfrac{a}{bc}.$	$\dfrac{2}{3(-5)} = \dfrac{2}{(-3)(5)} = \dfrac{-2}{3(5)}$ $= \dfrac{-2}{(-3)(-5)} = -\dfrac{2}{3(5)} = -\dfrac{2}{15}.$

	Property		*Example*

26. $\dfrac{a}{c} + \dfrac{b}{c} = \dfrac{a+b}{c}.$ $\dfrac{2}{9} + \dfrac{3}{9} = \dfrac{2+3}{9} = \dfrac{5}{9}.$

27. $\dfrac{a}{c} - \dfrac{b}{c} = \dfrac{a-b}{c}.$ $\dfrac{2}{9} - \dfrac{3}{9} = \dfrac{2-3}{9} = \dfrac{-1}{9}.$

28. $\dfrac{a}{b} + \dfrac{c}{d} = \dfrac{ad+bc}{bd}.$ $\dfrac{4}{5} + \dfrac{2}{3} = \dfrac{4\cdot3 + 5\cdot2}{5\cdot3} = \dfrac{22}{15}.$

29. $\dfrac{a}{b} - \dfrac{c}{d} = \dfrac{ad-bc}{bd}.$ $\dfrac{4}{5} - \dfrac{2}{3} = \dfrac{4\cdot3 - 5\cdot2}{5\cdot3} = \dfrac{2}{15}.$

30. $\dfrac{a}{b} \cdot \dfrac{c}{d} = \dfrac{ac}{bd}.$ $\dfrac{2}{3} \cdot \dfrac{4}{5} = \dfrac{2\cdot4}{3\cdot5} = \dfrac{8}{15}.$

31. $\dfrac{a}{\frac{b}{c}} = a \div \dfrac{b}{c} = \dfrac{ac}{b}.$ $\dfrac{2}{\frac{3}{5}} = 2 \div \dfrac{3}{5} = \dfrac{2\cdot5}{3} = \dfrac{10}{3}.$

32. $\dfrac{\frac{a}{b}}{c} = \dfrac{a}{b} \div c = \dfrac{a}{bc}.$ $\dfrac{\frac{2}{3}}{5} = \dfrac{2}{3} \div 5 = \dfrac{2}{3\cdot5} = \dfrac{2}{15}.$

33. $\dfrac{\frac{a}{b}}{\frac{c}{d}} = \dfrac{a}{b} \div \dfrac{c}{d} = \dfrac{a}{b} \cdot \dfrac{d}{c} = \dfrac{ad}{bc}.$ $\dfrac{\frac{2}{3}}{\frac{7}{5}} = \dfrac{2}{3} \div \dfrac{7}{5} = \dfrac{2}{3} \cdot \dfrac{5}{7} = \dfrac{10}{21}.$

Property 24 is essentially the **fundamental principle of fractions**, which states that *multiplying or dividing both the numerator and denominator of a fraction by the same number, except* 0, *results in a fraction which is equivalent to* (*that is, it has the same value as*) *the original fraction.* Thus,

$$\frac{7}{\frac{1}{8}} = \frac{7\cdot8}{\frac{1}{8}\cdot8} = \frac{56}{1} = 56.$$

By properties 28 and 24 we have

$$\frac{2}{5} + \frac{4}{15} = \frac{2\cdot15 + 5\cdot4}{5\cdot15} = \frac{50}{75} = \frac{2\cdot25}{3\cdot25} = \frac{2}{3}.$$

We can do this problem another way. The fractions $\frac{2}{5}$ and $\frac{4}{15}$ can be written with a common denominator of $5\cdot15$: $\dfrac{2}{5} = \dfrac{2\cdot15}{5\cdot15}$ and $\dfrac{4}{15} = \dfrac{4\cdot5}{15\cdot5}$. However, 15 is the *least* such common denominator and is called the *least common denominator*

(L.C.D.) of $\frac{2}{5}$ and $\frac{4}{15}$. Thus,

$$\frac{2}{5} + \frac{4}{15} = \frac{2 \cdot 3}{5 \cdot 3} + \frac{4}{15} = \frac{6}{15} + \frac{4}{15} = \frac{10}{15} = \frac{2}{3}.$$

Similarly,

$$\frac{3}{8} - \frac{5}{12} = \frac{3 \cdot 3}{8 \cdot 3} - \frac{5 \cdot 2}{12 \cdot 2}$$

$$= \frac{9}{24} - \frac{10}{24} = \frac{9 - 10}{24} = -\frac{1}{24}.$$

EXERCISE 0-4

Simplify each of the following if possible.

1. $-2 + (-4)$.

2. $-6 + 2$.

3. $6 + (-4)$.

4. $7 - 2$.

5. $7 - (-4)$.

6. $-7 - (-4)$.

7. $-8 - (-6)$.

8. $(-2)(9)$.

9. $7(-9)$.

10. $(-2)(-12)$.

11. $(-1)6$.

12. $-(-9)$.

13. $-(-6 + x)$.

14. $-7(x)$.

15. $-12(x - y)$.

16. $-[-6 + (-y)]$.

17. $-2 \div 6$.

18. $-2 \div (-4)$.

19. $4 \div (-2)$.

20. $2(-6 + 2)$.

21. $3[-2(3) + 6(2)]$.

22. $(-2)(-4)(-1)$.

23. $(-5)(-5)$.

24. $x(0)$.

25. $3(x - 4)$.

26. $4(5 + x)$.

27. $-(x - 2)$.

28. $0(-x)$.

29. $8\left(\dfrac{1}{11}\right)$.

30. $\dfrac{7}{1}$.

31. $\dfrac{-5x}{7y}$.

32. $\dfrac{3}{-2x}$.

33. $\dfrac{2}{3} \cdot \dfrac{1}{x}$.

34. $\dfrac{x}{y}(2z)$.

35. $(2x)\left(\dfrac{3}{2x}\right)$.

36. $\dfrac{-15x}{-3y}$.

37. $\dfrac{7}{y} \cdot \dfrac{1}{x}$.

38. $\dfrac{2}{x} \cdot \dfrac{5}{y}$.

39. $\dfrac{1}{2} + \dfrac{1}{3}$.

40. $\dfrac{5}{12} + \dfrac{3}{4}$.

41. $\dfrac{3}{10} - \dfrac{7}{15}$.

42. $\dfrac{2}{3} + \dfrac{7}{3}$.

43. $\dfrac{x}{9} - \dfrac{y}{9}$.

44. $\dfrac{3}{2} - \dfrac{1}{4} + \dfrac{1}{6}$.

45. $\dfrac{2}{3} - \dfrac{5}{8}$.

46. $\dfrac{6}{\frac{x}{y}}.$

47. $\dfrac{\frac{x}{6}}{y}.$

48. $\dfrac{\frac{-7}{2}}{\frac{5}{8}}.$

49. $\dfrac{7}{0}.$

50. $\dfrac{0}{7}.$

51. $\dfrac{0}{0}.$

52. $0 \cdot 0.$

0-5 EXPONENTS AND RADICALS

The product

$$x \cdot x \cdot x$$

is abbreviated x^3. In general, for n a positive integer, x^n is the abbreviation for the product of n x's. The letter n in x^n is called the *exponent* and x is called the *base*. More specifically, if n is a positive integer we have:

1. $x^n = \underbrace{x \cdot x \cdot x \cdot \ldots \cdot x}_{n \text{ factors}}.$

2. $x^{-n} = \dfrac{1}{x^n} = \dfrac{1}{\underbrace{x \cdot x \cdot x \cdot \ldots \cdot x}_{n \text{ factors}}}.$

3. $\dfrac{1}{x^{-n}} = x^n.$

4. $x^0 = 1$ if $x \neq 0$. 0^0 is not defined.

EXAMPLE 1

a. $\left(\dfrac{1}{2}\right)^4 = \left(\dfrac{1}{2}\right)\left(\dfrac{1}{2}\right)\left(\dfrac{1}{2}\right)\left(\dfrac{1}{2}\right) = \dfrac{1}{16}.$

b. $3^{-5} = \dfrac{1}{3^5} = \dfrac{1}{3 \cdot 3 \cdot 3 \cdot 3 \cdot 3} = \dfrac{1}{243}.$

c. $\dfrac{1}{3^{-5}} = 3^5 = 243.$

d. $2^0 = 1, \quad \pi^0 = 1, \quad (-5)^0 = 1.$

e. $x^1 = x.$

If $r^n = x$ where n is a positive integer, then r is an *nth root* of x. For example, $3^2 = 9$ and so 3 is a second root (usually called a *square root*) of 9. Since $(-3)^2 = 9$, -3 is also a square root of 9. Similarly, -2 is a *cube root* of -8 since $(-2)^3 = -8$.

Some numbers do not have an *nth* root that is a real number. For example, since the square of any real number is nonnegative, there is no real number that is a square root of -4.

The **principal *nth* root** of x is that *nth* root of x which is positive if x is positive, and is negative if x is negative and n is odd. We denote it by $\sqrt[n]{x}$. Thus,

$$\sqrt[n]{x} \text{ is } \begin{cases} \text{positive if } x \text{ is positive,} \\ \text{negative if } x \text{ is negative and } n \text{ is odd.} \end{cases}$$

For example, $\sqrt[2]{9} = 3$, $\sqrt[3]{-8} = -2$, and $\sqrt[3]{\frac{1}{27}} = \frac{1}{3}$. We define $\sqrt[n]{0} = 0$.

The symbol $\sqrt[n]{x}$ is called a **radical**. Here n is the *index*, x is the *radicand*, and $\sqrt{}$ is the *radical sign*. With principal square roots we usually omit the index and write $\sqrt{x}$ instead of $\sqrt[2]{x}$. Thus $\sqrt{9} = 3$.

PITFALL. *Although 2 and -2 are square roots of 4, the **principal** square root of 4 is 2, not -2. Hence* $\sqrt{4} = 2$.

If x is positive, the expression $x^{p/q}$ where p and q are integers and q is positive is defined to be $\sqrt[q]{x^p}$. Thus,

$$x^{3/4} = \sqrt[4]{x^3} \; ; \qquad 8^{2/3} = \sqrt[3]{8^2} = \sqrt[3]{64} = 4;$$
$$4^{-1/2} = \sqrt[2]{4^{-1}} = \sqrt{\frac{1}{4}} = \frac{1}{2}.$$

Below are the basic laws of exponents and radicals.[†]

Law	*Example*
1. $x^m \cdot x^n = x^{m+n}$.	$2^3 \cdot 2^5 = 2^8 = 256; \quad x^2 \cdot x^3 = x^5.$
2. $x^0 = 1$ if $x \neq 0$.	$2^0 = 1.$
3. $x^{-n} = \dfrac{1}{x^n}$.	$2^{-3} = \dfrac{1}{2^3} = \dfrac{1}{8}.$

[†]Although some laws involve restrictions, they are not vital to our discussion.

Law	*Example*

4. $\dfrac{1}{x^{-n}} = x^n.$ $\qquad$ $\dfrac{1}{2^{-3}} = 2^3 = 8; \quad \dfrac{1}{x^{-5}} = x^5.$

5. $\dfrac{x^m}{x^n} = x^{m-n} = \dfrac{1}{x^{n-m}}.$ $\qquad$ $\dfrac{2^{12}}{2^8} = 2^4 = 16; \quad \dfrac{x^8}{x^{12}} = \dfrac{1}{x^4}.$

6. $\dfrac{x^m}{x^m} = 1$ if $x \neq 0.$ $\qquad$ $\dfrac{2^4}{2^4} = 1.$

7. $(x^m)^n = x^{mn}.$ $\qquad$ $(2^3)^5 = 2^{15}; \quad (x^2)^3 = x^6.$

8. $(xy)^n = x^n y^n.$ $\qquad$ $(2 \cdot 4)^3 = 2^3 \cdot 4^3 = 8 \cdot 64.$

9. $\left(\dfrac{x}{y}\right)^n = \dfrac{x^n}{y^n}.$ $\qquad$ $\left(\dfrac{2}{3}\right)^3 = \dfrac{2^3}{3^3}; \left(\dfrac{1}{3}\right)^5 = \dfrac{1^5}{3^5} = \dfrac{1}{3^5} = 3^{-5}.$

10. $\left(\dfrac{x}{y}\right)^{-n} = \left(\dfrac{y}{x}\right)^n.$ $\qquad$ $\left(\dfrac{3}{4}\right)^{-2} = \left(\dfrac{4}{3}\right)^2 = \dfrac{16}{9}.$

11. $x^{1/n} = \sqrt[n]{x}.$ $\qquad$ $3^{1/5} = \sqrt[5]{3}.$

12. $x^{-1/n} = \dfrac{1}{x^{1/n}} = \dfrac{1}{\sqrt[n]{x}}.$ $\qquad$ $4^{-1/2} = \dfrac{1}{4^{1/2}} = \dfrac{1}{\sqrt{4}} = \dfrac{1}{2}.$

13. $\sqrt[n]{x}\,\sqrt[n]{y} = \sqrt[n]{xy}.$ $\qquad$ $\sqrt[3]{9}\,\sqrt[3]{2} = \sqrt[3]{18}.$

14. $\dfrac{\sqrt[n]{x}}{\sqrt[n]{y}} = \sqrt[n]{\dfrac{x}{y}}.$ $\qquad$ $\dfrac{\sqrt[3]{90}}{\sqrt[3]{10}} = \sqrt[3]{\dfrac{90}{10}} = \sqrt[3]{9}.$

15. $\sqrt[m]{\sqrt[n]{x}} = \sqrt[mn]{x}.$ $\qquad$ $\sqrt[3]{\sqrt[4]{2}} = \sqrt[12]{2}.$

16. $x^{m/n} = \sqrt[n]{x^m} = (\sqrt[n]{x})^m.$ $\qquad$ $8^{2/3} = \sqrt[3]{8^2} = (\sqrt[3]{8})^2 = 2^2 = 4.$

17. $(\sqrt[m]{x})^m = x.$ $\qquad$ $(\sqrt[8]{7})^8 = 7.$

EXAMPLE 2

a. By Law 1,

$$x^6 x^8 = x^{6+8} = x^{14},$$

$$a^3 b^2 a^5 b = a^3 a^5 b^2 b = a^8 b^3,$$

$$x^{11} x^{-5} = x^{11-5} = x^6,$$

$$z^{2/5} z^{3/5} = z^1 = z,$$

$$x x^{1/2} = x^1 x^{1/2} = x^{3/2}.$$

b. By Law 16,

$$\left(\frac{1}{4}\right)^{3/2} = \left(\sqrt{\frac{1}{4}}\right)^3 = \left(\frac{1}{2}\right)^3 = \frac{1}{8}.$$

c. $\left(-\frac{8}{27}\right)^{4/3} = \left(\sqrt[3]{\frac{-8}{27}}\right)^4 = \left(\frac{\sqrt[3]{-8}}{\sqrt[3]{27}}\right)^4$ (Laws 16 and 14)

$$= \left(\frac{-2}{3}\right)^4$$

$$= \frac{(-2)^4}{3^4} = \frac{16}{81}$$ (Law 9).

d. $(64a^3)^{2/3} = 64^{2/3}(a^3)^{2/3}$ (Law 8)

$$= (\sqrt[3]{64})^2 a^2$$ (Laws 16 and 7)

$$= (4)^2 a^2 = 16a^2.$$

The following examples illustrate various applications of the laws of exponents and radicals.

EXAMPLE 3

a. *Simplify* $(x^5y^8)^5$.

$$(x^5y^8)^5 = x^{25}y^{40}.$$

b. *Eliminate negative exponents in* $\dfrac{x^{-2}y^3}{z^{-2}}$.

$$\frac{x^{-2}y^3}{z^{-2}} = \frac{y^3z^2}{x^2}.$$

Thus we can bring a factor of the numerator down to the denominator by changing the sign of the exponent, and vice versa.

c. *Simplify* $\dfrac{x^2y^7}{x^3y^5}$.

$$\frac{x^2y^7}{x^3y^5} = \frac{y^{7-5}}{x^{3-2}} = \frac{y^2}{x}.$$

d. *Simplify* $\dfrac{x^3}{y^2} \div \dfrac{x^6}{y^5}$.

$$\frac{x^3}{y^2} \div \frac{x^6}{y^5} = \frac{x^3}{y^2} \cdot \frac{y^5}{x^6} = \frac{x^3y^5}{y^2x^6} = \frac{y^{5-2}}{x^{6-3}} = \frac{y^3}{x^3}.$$

e. *Eliminate negative exponents in* $x^{-1} + y^{-1}$ *and simplify.*

$$x^{-1} + y^{-1} = \frac{1}{x} + \frac{1}{y} = \frac{y + x}{xy}. \qquad \left(\text{Note: } x^{-1} + y^{-1} \neq \frac{1}{x + y}.\right)$$

f. *Eliminate negative exponents in* $7x^{-2} + (7x)^{-2}$.

$$7x^{-2} + (7x)^{-2} = \frac{7}{x^2} + \frac{1}{(7x)^2} = \frac{7}{x^2} + \frac{1}{49x^2}.$$

g. *Eliminate negative exponents in* $(x^{-1} - y^{-1})^{-2}$.

$$(x^{-1} - y^{-1})^{-2} = \left(\frac{1}{x} - \frac{1}{y}\right)^{-2} = \left(\frac{y - x}{xy}\right)^{-2}$$

$$= \left(\frac{xy}{y - x}\right)^{2} = \frac{x^2 y^2}{(y - x)^2}.$$

EXAMPLE 4

a. *Simplify* $(x^{5/9}y^{4/3})^{18}$.

$$(x^{5/9}y^{4/3})^{18} = (x^{5/9})^{18}(y^{4/3})^{18} = x^{10}y^{24}.$$

b. *Simplify* $\left(\dfrac{x^{1/5}y^{6/5}}{z^{2/5}}\right)^{5}$.

$$\left(\frac{x^{1/5}y^{6/5}}{z^{2/5}}\right)^{5} = \frac{(x^{1/5}y^{6/5})^{5}}{(z^{2/5})^{5}} = \frac{xy^6}{z^2}.$$

c. *Apply the distributive law to* $x^{2/5}(y^{1/2} + 2z^{6/5})$.

$$x^{2/5}(y^{1/2} + 2z^{6/5}) = x^{2/5}y^{1/2} + 2x^{2/5}z^{6/5}.$$

d. *Simplify* $x^{3/2} - x^{1/2}$ *by using the distributive property.*

$$x^{3/2} - x^{1/2} = x^{1/2}(x - 1).$$

EXAMPLE 5

a. *Use exponents to rewrite* $\sqrt{2 + 5x}$.

$$\sqrt{2 + 5x} = (2 + 5x)^{1/2}.$$

b. *Simplify* $\sqrt[3]{x^6 y^4}$.

$$\sqrt[3]{x^6 y^4} = \sqrt[3]{(x^2)^3 y^3 y} = \sqrt[3]{(x^2)^3} \cdot \sqrt[3]{y^3} \cdot \sqrt[3]{y}$$
$$= x^2 y \sqrt[3]{y} .$$

c. *Simplify* $\sqrt[4]{48}$.

$$\sqrt[4]{48} = \sqrt[4]{16 \cdot 3} = \sqrt[4]{16} \, \sqrt[4]{3} = 2\sqrt[4]{3} .$$

d. *Simplify* $\sqrt{250} - \sqrt{50} + 15\sqrt{2}$.

$$\sqrt{250} - \sqrt{50} + 15\sqrt{2} = \sqrt{25 \cdot 10} - \sqrt{25 \cdot 2} + 15\sqrt{2}$$
$$= 5\sqrt{10} - 5\sqrt{2} + 15\sqrt{2}$$
$$= 5\sqrt{10} + 10\sqrt{2} .$$

e. *Simplify* $\sqrt{\dfrac{2}{7}}$.

$$\sqrt{\frac{2}{7}} = \sqrt{\frac{2}{7} \cdot \frac{7}{7}} = \sqrt{\frac{14}{7^2}} = \frac{\sqrt{14}}{\sqrt{7^2}} = \frac{\sqrt{14}}{7} .$$

f. $\sqrt{x^2}$ is defined to be

$$\sqrt{x^2} = \begin{cases} x & \text{if } x \text{ is positive,} \\ -x & \text{if } x \text{ is negative,} \\ 0 & \text{if } x = 0. \end{cases}$$

Thus, $\sqrt{2^2} = 2$ and $\sqrt{(-3)^2} = -(-3) = 3$.

Rationalizing the denominator of a fraction is a procedure in which a fraction having a radical in its denominator is expressed as an equivalent fraction without a radical in its denominator. We use the fundamental principle of fractions.

EXAMPLE 6

Rationalize the denominators.

a. $\dfrac{2}{\sqrt{5}} = \dfrac{2}{5^{1/2}} = \dfrac{2 \cdot 5^{1/2}}{5^{1/2} \cdot 5^{1/2}} = \dfrac{2 \cdot 5^{1/2}}{5^1} = \dfrac{2\sqrt{5}}{5}$.

b. $\dfrac{2}{\sqrt[6]{3x^5}} = \dfrac{2}{\sqrt[6]{3}\cdot\sqrt[6]{x^5}} = \dfrac{2}{3^{1/6}x^{5/6}} = \dfrac{2\cdot3^{5/6}x^{1/6}}{3^{1/6}x^{5/6}\cdot3^{5/6}x^{1/6}} = \dfrac{2(3^5x)^{1/6}}{3x} = \dfrac{2\sqrt[6]{3^5x}}{3x}.$

c. $\dfrac{\sqrt[5]{2}}{\sqrt[3]{6}} = \dfrac{2^{1/5}\cdot6^{2/3}}{6^{1/3}\cdot6^{2/3}} = \dfrac{2^{3/15}6^{10/15}}{6} = \dfrac{(2^3 6^{10})^{1/15}}{6} = \dfrac{\sqrt[15]{2^3 6^{10}}}{6}.$

d. $\dfrac{\sqrt{20}}{\sqrt{5}} = \sqrt{\dfrac{20}{5}} = \sqrt{4} = 2.$

EXERCISE 0-5

*In Problems **1–36**, simplify and express all answers in terms of positive exponents.*

1. $(2^3)(2^2).$

2. $(3^4)(3^2).$

3. $x^6 x^9.$

4. $x^{20} x^{15}.$

5. $w^4 w^8.$

6. $y^9 y^{91}.$

7. $x^6 x^4 x^3.$

8. $x^2 x^3 x^4.$

9. $\dfrac{x^2 x^6}{y^7 y^{10}}.$

10. $\dfrac{x^{100} x^{200}}{z^9 z^{90}}.$

11. $(x^{12})^4.$

12. $(x^6)^6.$

13. $\dfrac{(x^2)^5}{(y^5)^{10}}.$

14. $(2x^3)^3.$

15. $\left(\dfrac{x^2}{y^3}\right)^5.$

16. $\left(\dfrac{2}{x}\right)^4.$

17. $(2x^2 y^3)^3.$

18. $(xy^2)^4.$

19. $\left(\dfrac{w^2 s^3}{y^2}\right)^2$

20. $\dfrac{(xy^2)^2}{(z^2 w)^3}.$

21. $\dfrac{x^8}{x^2}.$

22. $\dfrac{y^4}{y^9}.$

23. $\left(\dfrac{2x^2}{4x^4}\right)^3.$

24. $\dfrac{2^6\cdot2^9}{(2^6)^3}.$

25. $\dfrac{(x^3)^6}{x(x^3)}$

26. $\left(\dfrac{2}{3}x^2 y\right)^2.$

27. $\dfrac{(x^2)^3(x^3)^2}{(x^3)^4}.$

28. $(-1)^5(-1)^8.$

29. $\dfrac{x^{-2}y^{-6}z^2}{xy^{-1}}.$

30. $-\dfrac{2x^{-3}}{4x^{-5}}.$

31. $\dfrac{x^{-2}yzw}{x^{-3}y^5}.$

32. $\dfrac{x^3 y^{-5}}{(x^{-8}y^6)^{-3}}.$

33. $\dfrac{2(x^2 y)^2}{3y^{13}z^{-2}}.$

34. $\dfrac{(xyz^{-1})^{-2}}{(x^2)^{-4}}.$

35. $\dfrac{3^0}{(2^{-3}x^{-4}y^6)^3}.$

36. $\left[\left(\dfrac{x}{y}\right)^{-2}\right]^{-4}.$

In Problems 37–42, write the exponential forms in equivalent forms involving radicals.

37. $(8x - y)^{4/5}$.

38. $(ab^2c^3)^{3/4}$.

39. $x^{-4/5}$.

40. $2x^{1/2} - (2y)^{1/2}$.

41. $2x^{-2/5} - (2x)^{-2/5}$.

42. $[(x^{-4})^{1/5}]^{1/6}$.

In Problems 43–54, write the expressions in terms of positive exponents only. Avoid all radicals in the final form. For example, $y^{-1}\sqrt{x} = \dfrac{x^{1/2}}{y}$.

43. $\dfrac{x^3y^{-2}}{z^2}$.

44. $\sqrt[5]{x^2y^3z^{-10}}$.

45. $2x^{-1}x^{-3}$.

46. $x + y^{-1}$.

47. $(3t)^{-2}$.

48. $(3 - z)^{-4}$.

49. $\sqrt[3]{7s^2}$.

50. $(x^{-2}y^2)^{-2}$.

51. $\sqrt{x} - \sqrt{y}$.

52. $\dfrac{x^{-2}y^{-6}z^2}{xy^{-1}}$.

53. $x^2\sqrt[4]{xy^{-2}z^3}$.

54. $(\sqrt[5]{xy^{-3}})x^{-1}y^{-2}$.

In Problems 55–78, simplify the expressions.

55. $\sqrt{25}$.

56. $\sqrt[3]{64}$.

57. $\sqrt[5]{-32}$.

58. $\sqrt{.04}$.

59. $\sqrt[4]{\frac{1}{16}}$.

60. $\sqrt[3]{-\frac{8}{27}}$.

61. $(100)^{1/2}$.

62. $(64)^{1/3}$.

63. $4^{3/2}$.

64. $(25)^{-3/2}$.

65. $(32)^{-2/5}$.

66. $(.09)^{-1/2}$.

67. $\left(\frac{1}{16}\right)^{5/4}$.

68. $\left(-\frac{27}{64}\right)^{2/3}$.

69. $\sqrt{32}$.

70. $\sqrt[3]{24}$.

71. $\sqrt[3]{2x^3}$.

72. $\sqrt{4x}$.

73. $\sqrt{16x^4}$.

74. $\sqrt[4]{x/16}$.

75. $(9z^4)^{1/2}$.

76. $(16y^8)^{3/4}$.

77. $\left(\frac{27t^3}{8}\right)^{2/3}$.

78. $\left(\frac{1000}{a^9}\right)^{-2/3}$.

In Problems 79–88, rationalize the denominators.

79. $\dfrac{3}{\sqrt{7}}$.

80. $\dfrac{5}{\sqrt{11}}$.

81. $\dfrac{4}{\sqrt{2x}}$.

82. $\dfrac{y}{\sqrt{2y}}$.

83. $\dfrac{1}{\sqrt[3]{3x}}$.

84. $\dfrac{4}{3\sqrt[3]{x^2}}$.

85. $\dfrac{\sqrt{32}}{\sqrt{2}}$.

86. $\dfrac{\sqrt{18}}{\sqrt{2}}$.

87. $\dfrac{\sqrt[4]{2}}{\sqrt[3]{xy^2}}$.

88. $\dfrac{\sqrt{2}}{\sqrt[3]{3}}$.

In Problems **89–110**, *simplify the expressions. Express all answers in terms of positive exponents. Rationalize the denominator where necessary to avoid fractional exponents in the denominator.*

89. $2x^2y^{-3}x^4$.

90. $\dfrac{2}{x^{3/2}y^{1/3}}$.

91. $\sqrt{\sqrt[3]{t^4}}$.

92. $\{[(2x^2)^3]^{-4}\}^{-1}$.

93. $\dfrac{2^0}{(2^{-2}x^{1/2}y^{-2})^3}$.

94. $\dfrac{\sqrt{s^5}}{\sqrt[3]{s^2}}$.

95. $\sqrt[3]{x^2yz^3}\,\sqrt[3]{xy^2}$.

96. $(\sqrt[5]{2}\,)^{10}$.

97. $3^2(27)^{-4/3}$.

98. $(\sqrt[5]{x^3y}\,)^{2/5}$.

99. $(2x^{-1}y^2)^2$.

100. $\dfrac{3}{\sqrt[3]{y}\,\sqrt[4]{x}}$.

101. $\sqrt{x}\,\sqrt{x^2y^3}\,\sqrt{xy^2}$.

102. $\sqrt{75k^4}$.

103. $\dfrac{(x^2y^{-1}z)^{-2}}{(xy^2)^{-4}}$.

104. $\sqrt{6(6)}$.

105. $\dfrac{(x^2)^3}{x^4} \div \left[\dfrac{x^3}{(x^3)^2}\right]^{-2}$.

106. $\sqrt{(-6)(-6)}$.

107. $-\dfrac{8s^{-2}}{2s^3}$.

108. $(x^{-1}y^{-2}\sqrt{z}\,)^4$.

109. $(2x^2y \div 3y^3z^{-2})^2$.

110. $\dfrac{1}{\left(\dfrac{\sqrt{2}\,x^{-2}}{\sqrt{16}\,x^3}\right)^2}$.

111. Given that $\sqrt{2}$ is approximately 1.4142, find $\dfrac{1}{\sqrt{2}}$ by long division. Then compute $\dfrac{\sqrt{2}}{2}$ by long division. At what conclusion do you arrive about the use of rationalizing the denominator in the approximation of certain expressions?

0-6 OPERATIONS WITH ALGEBRAIC EXPRESSIONS

If numbers, represented by symbols, are combined by the operations of addition, subtraction, multiplication, division, or extraction of roots, then the resulting expression is called an *algebraic expression*.

EXAMPLE 1

a. $\sqrt[3]{\dfrac{3x^3 - 5x - 2}{10 - x}}$ is an algebraic expression in the variable x.

b. $10 - 3\sqrt{y} + \dfrac{5}{7 + y^2}$ is an algebraic expression in the variable y.

c. $\dfrac{(x + y)^3 - xy}{y} + 2$ is an algebraic expression in the variables x and y.

The algebraic expression $5ax^3 - 2bx + 3$ consists of three *terms:* $+5ax^3$, $-2bx$, and $+3$. Some of the *factors* of the first term $5ax^3$ are 5, a, x, x^2, x^3, $5ax$, and ax^2. Also, $5a$ is the *coefficient* of x^3 and 5 is the *numerical coefficient* of ax^3. If a and b represent fixed numbers throughout a discussion, then a and b are called *constants*.

Algebraic expressions with exactly one term are called *monomials*. Those having exactly two terms are *binomials*, and those with exactly three terms are *trinomials*. Algebraic expressions with more than one term are called *multinomials*. Thus the multinomial $2x - 5$ is a binomial; the multinomial $3\sqrt{y} + 2y - 4y^2$ is a trinomial.

A *polynomial in x* is an algebraic expression of the form

$$c_n x^n + c_{n-1} x^{n-1} + \ldots + c_1 x + c_0{}^\dagger$$

where n is a positive integer and $c_0, c_1, \ldots, c_n$ are real numbers with $c_n \neq 0$. We call n the *degree* of the polynomial. Hence $4x^3 - 5x^2 + x - 2$ is a polynomial in x of degree 3, and $y^5 - 2$ is a polynomial in y of degree 5. A nonzero constant is treated as a polynomial of degree zero; thus 5 is a polynomial of degree zero. The constant 0 is also a polynomial; however, no degree is assigned to it.

EXAMPLE 2

Simplify $(3x^2y - 2x + 1) + (4x^2y + 6x - 3)$.

We shall first remove the parentheses. Next, using the commutative property of addition, we gather all similar terms together. *Similar terms* are those terms which differ only by their numerical coefficients. In our case, $3x^2y$ and $4x^2y$ are similar, as are the pairs $-2x$

---

†The three dots indicate the terms that are understood to be included in the sum.

and $6x$, and 1 and -3. Thus,

$$(3x^2y - 2x + 1) + (4x^2y + 6x - 3)$$

$$= 3x^2y - 2x + 1 + 4x^2y + 6x - 3$$

$$= 3x^2y + 4x^2y - 2x + 6x + 1 - 3.$$

By the distributive property,

$$3x^2y + 4x^2y = (3 + 4)x^2y = 7x^2y$$

and $\qquad -2x + 6x = (-2 + 6)x = 4x.$

Hence,

$$(3x^2y - 2x + 1) + (4x^2y + 6x - 3) = 7x^2y + 4x - 2.$$

EXAMPLE 3

Simplify $(3x^2y - 2x + 1) - (4x^2y + 6x - 3).$

Here we apply the definition of subtraction and the distributive property:

$$(3x^2y - 2x + 1) - (4x^2y + 6x - 3)$$

$$= (3x^2y - 2x + 1) + (-1)(4x^2y + 6x - 3)$$

$$= (3x^2y - 2x + 1) + (-4x^2y - 6x + 3)$$

$$= 3x^2y - 2x + 1 - 4x^2y - 6x + 3$$

$$= 3x^2y - 4x^2y - 2x - 6x + 1 + 3$$

$$= (3 - 4)x^2y + (-2 - 6)x + 1 + 3$$

$$= -x^2y - 8x + 4.$$

EXAMPLE 4

Simplify $3\{2x[2x + 3] + 5[4x^2 - (3 - 4x)]\}.$

We shall first remove the innermost grouping symbols (parentheses) by using the distributive property. Then we repeat the process until all grouping symbols are removed —combining similar terms whenever possible.

$$3\{2x[2x + 3] + 5[4x^2 - (3 - 4x)]\}$$

$$= 3\{2x[2x + 3] + 5[4x^2 - 3 + 4x]\}$$

$$= 3\{4x^2 + 6x + 20x^2 - 15 + 20x\}$$

$$= 3\{24x^2 + 26x - 15\}$$

$$= 72x^2 + 78x - 45.$$

The distributive property is the key tool in multiplying expressions. For example, to multiply $ax + c$ by $bx + d$ we can consider $ax + c$ as a single number and then use the distributive property.

$$(ax + c)(bx + d) = (ax + c)bx + (ax + c)d.$$

Using the distributive property again, we have

$$(ax + c)bx + (ax + c)d = abx^2 + cbx + adx + cd$$
$$= abx^2 + (ad + cb)x + cd.$$

Thus, $(ax + c)(bx + d) = abx^2 + (ad + cb)x + cd$. In particular, if $a = 2$, $b = 1$, $c = 3$, and $d = -2$, then

$$(2x + 3)(x - 2) = 2(1)x^2 + [2(-2) + 3(1)]x + 3(-2)$$
$$= 2x^2 - x - 6.$$

EXAMPLE 5

Multiply $(2t - 3)(5t^2 + 3t - 1)$.

We treat $2t - 3$ as a single number and apply the distributive property.

$$(2t - 3)(5t^2 + 3t - 1) = (2t - 3)5t^2 + (2t - 3)3t - (2t - 3)1$$
$$= 10t^3 - 15t^2 + 6t^2 - 9t - 2t + 3$$
$$= 10t^3 - 9t^2 - 11t + 3.$$

Below is a list of special products that can be obtained from the distributive property and are useful in multiplying algebraic expressions.

SPECIAL PRODUCTS

(I) $x(y + z) = xy + xz$ (Distributive property).

(II) $(x + a)(x + b) = x^2 + (a + b)x + ab$.

(III) $(ax + c)(bx + d) = abx^2 + (ad + cb)x + cd$.

(IV) $(x + a)^2 = x^2 + 2ax + a^2$ (Square of a binomial).

(V) $(x - a)^2 = x^2 - 2ax + a^2$ (Square of a binomial).

(VI) $(x + a)(x - a) = x^2 - a^2$ (Product of sum and difference).

(VII) $(x + a)^3 = x^3 + 3ax^2 + 3a^2x + a^3$ (Cube of a binomial).

(VIII) $(x - a)^3 = x^3 - 3ax^2 + 3a^2x - a^3$ (Cube of a binomial).

EXAMPLE 6

a. By II, $(x + 2)(x - 5) = [x + 2][x + (-5)]$

$$= x^2 + (2 - 5)x + 2(-5)$$

$$= x^2 - 3x - 10.$$

b. By III, $(3z + 5)(7z + 4) = 3 \cdot 7z^2 + (3 \cdot 4 + 5 \cdot 7)z + 5 \cdot 4$

$$= 21z^2 + 47z + 20.$$

c. By V, $(x - 4)^2 = x^2 - 2(4)x + 4^2$

$$= x^2 - 8x + 16.$$

d. By VI, $(\sqrt{y^2 + 1} + 3)(\sqrt{y^2 + 1} - 3) = [(y^2 + 1)^{1/2} + 3][(y^2 + 1)^{1/2} - 3]$

$$= [(y^2 + 1)^{1/2}]^2 - 3^2$$

$$= (y^2 + 1) - 9$$

$$= y^2 - 8.$$

e. By VII, $(3x + 2)^3 = (3x)^3 + 3(2)(3x)^2 + 3(2)^2(3x) + (2)^3$

$$= 27x^3 + 54x^2 + 36x + 8.$$

In Example 2(c) of Sec. 0-3 we showed that $\dfrac{a + b}{c} = \dfrac{a}{c} + \dfrac{b}{c}$. Similarly, $\dfrac{a - b}{c} = \dfrac{a}{c} - \dfrac{b}{c}$. Using these results, we can divide a multinomial by a monomial by dividing each term in the multinomial by the monomial.

EXAMPLE 7

a. $\dfrac{x^3 + 3x}{x} = \dfrac{x^3}{x} + \dfrac{3x}{x} = x^2 + 3.$

b. $\dfrac{4z^3 - 8z^2 + 3z - 6}{2z} = \dfrac{4z^3}{2z} - \dfrac{8z^2}{2z} + \dfrac{3z}{2z} - \dfrac{6}{2z} = 2z^2 - 4z + \dfrac{3}{2} - \dfrac{3}{z}.$

To divide a polynomial by a polynomial, we use so-called "long division" when the degree of the divisor is less than or equal to the degree of the dividend, as the next example shows.

EXAMPLE 8

Divide $2x^3 - 14x - 5$ by $x - 3$.

Here $2x^3 - 14x - 5$ is the *dividend* and $x - 3$ is the *divisor*. To avoid errors it is best to write the dividend as $2x^3 + 0x^2 - 14x - 5$. Note that the powers of x are in decreasing order.

$$
\begin{array}{r}
2x^2 + 6x + 4 \quad \leftarrow \text{Quotient} \\
x - 3 \overline{)\,2x^3 + 0x^2 - 14x - 5\,} \\
\underline{2x^3 - 6x^2} \\
6x^2 - 14x \\
\underline{6x^2 - 18x } \\
4x - 5 \\
\underline{4x - 12} \\
7 \leftarrow \text{Remainder}
\end{array}
$$

Here we divided x into $2x^3$ and obtained $2x^2$. Then we multiplied $2x^2$ by $x - 3$, getting $2x^3 - 6x^2$. After subtracting $2x^3 - 6x^2$ from $2x^3 + 0x^2$, we obtained $6x^2$ and then "brought down" the term $-14x$. This process is continued until we arrive at 7, the *remainder*. We always stop when the remainder is zero or a polynomial whose degree is less than the degree of the divisor. Our answer may be written as

$$2x^2 + 6x + 4 + \frac{7}{x - 3}.$$

A way of checking a division is to verify that

$$(\text{Quotient})(\text{Divisor}) + \text{Remainder} = \text{Dividend}.$$

Notice that if the remainder is 0, then the divisor is a factor of the dividend. By using this equation you should verify the result of the example.

EXERCISE 0-6

Perform the indicated operations and simplify.

1. $(8x - 4y + 2) + (3x + 2y - 5)$.

2. $(6x^2 - 10xy + 2) + (2z - xy + 4)$.

3. $(8t^2 - 6s^2) + (4s^2 - 2t^2 + 6)$.

4. $(\sqrt{x} + 2\sqrt{x}) + (\sqrt{x} + 3\sqrt{x})$.

5. $(\sqrt{x} + \sqrt{2y}) + (\sqrt{x} + \sqrt{3z})$.

6. $(3x + 2y - 5) - (8x - 4y + 2)$.

7. $(6x^2 - 10xy + \sqrt{2}) - (2z - xy + 4)$.

8. $(\sqrt{x} + 2\sqrt{x}) - (\sqrt{x} + 3\sqrt{x})$.

9. $(\sqrt{x} + \sqrt{2y}) - (\sqrt{x} + \sqrt{3z})$.

10. $4(2z - w) - 3(w - 2z)$.

11. $3(3x + 2y - 5) - 2(8x - 4y + 2)$.   **12.** $(2s + t) - 3(s - 6) + 4(1 - t)$.

13. $3(x^2 + y^2) - x(y + 2x) + 2y(x + 3y)$. **14.** $2 - [3 + 4(s - 3)]$.

15. $2\{3[3(x^2 + 2) - 2(x^2 - 5)]\}$.   **16.** $4\{3(t + 5) - t[1 - (t + 1)]\}$.

17. $-3\{4x(x + 2) - 2[x^2 - (3 - x)]\}$.

18. $-\{-2[2a + 3b - 1] + 4[a - 2b] - a[2(b - 3)]\}$.

19. $(x + 4)(x + 5)$.   **20.** $(x + 3)(x + 2)$.

21. $(x + 3)(x - 2)$.   **22.** $(z - 7)(z - 3)$.

23. $(2x + 3)(5x + 2)$.   **24.** $(y - 4)(2y + 3)$.

25. $(x + 3)^2$.   **26.** $(2x - 1)^2$.

27. $(x - 5)^2$.   **28.** $(\sqrt{x} - 1)(2\sqrt{x} + 5)$.

29. $(\sqrt{2y} + 3)^2$.   **30.** $(y - 3)(y + 3)$.

31. $(2s - 1)(2s + 1)$.   **32.** $(z^2 - 3w)(z^2 + 3w)$.

33. $(x^2 - 3)(x + 4)$.   **34.** $(x + 1)(x^2 + x + 3)$.

35. $(x^2 - 1)(2x^2 + 2x - 3)$.   **36.** $(2x - 1)(3x^3 + 7x^2 - 5)$.

37. $x\{3(x - 1)(x - 2) + 2[x(x + 7)]\}$.   **38.** $[(2z + 1)(2z - 1)](4z^2 + 1)$.

39. $(x + y + 2)(3x + 2y - 4)$.   **40.** $(x^2 + x + 1)^2$.

41. $(x + 5)^3$.   **42.** $(x - 2)^3$.

43. $(2x - 3)^3$.   **44.** $(x + 2y)^3$.

45. $\dfrac{z^2 - 4z}{z}$.   **46.** $\dfrac{2x^3 - 7x + 4}{x}$.

47. $\dfrac{6x^5 + 4x^3 - 1}{2x^2}$.   **48.** $\dfrac{(3x - 4) - (x + 8)}{4x}$.

49. $(x^2 + 3x - 1) \div (x + 3)$.   **50.** $(x^2 - 5x + 4) \div (x - 4)$.

51. $(3x^3 - 2x^2 + x - 3) \div (x + 2)$.   **52.** $(x^4 + 2x^2 + 1) \div (x - 1)$.

53. $t^2 \div (t - 8)$.   **54.** $(4x^2 + 6x + 1) \div (2x - 1)$.

55. $(3x^2 - 4x + 3) \div (3x + 2)$.   **56.** $(z^3 + z^2 + z) \div (z^2 - z + 1)$.

0-7 FACTORING

If two or more expressions are multiplied together, the expressions are called
factors of the product. Thus if $c = ab$, then a and b are both factors of the
product c. The process by which an expression is written as a product of its
factors is called *factoring*.

Listed below are factorization rules, most of which arise from the special products discussed in Section 0-6. The right side of each identity is the factored form of the left side.

(I) $xy + xz = x(y + z)$ (Common factor).

(II) $x^2 + (a + b)x + ab = (x + a)(x + b)$.

(III) $abx^2 + (ad + cb)x + cd = (ax + c)(bx + d)$.

(IV) $x^2 + 2ax + a^2 = (x + a)^2$ (Perfect-square trinomial).

(V) $x^2 - 2ax + a^2 = (x - a)^2$ (Perfect-square trinomial).

(VI) $x^2 - a^2 = (x + a)(x - a)$ (Difference of two squares).

(VII) $x^3 + a^3 = (x + a)(x^2 - ax + a^2)$ (Sum of two cubes).

(VIII) $x^3 - a^3 = (x - a)(x^2 + ax + a^2)$ (Difference of two cubes).

When factoring a polynomial we usually choose factors which themselves are polynomials. For example, $x^2 - 4 = (x + 2)(x - 2)$. We shall not write $x - 4$ as $(\sqrt{x} + 2)(\sqrt{x} - 2)$.

Always factor completely. For example,

$$2x^2 - 8 = 2(x^2 - 4) = 2(x + 2)(x - 2).$$

EXAMPLE 1

a. *Completely factor* $3k^2x^2 + 9k^3x$.

Since $3k^2x^2 = (3k^2x)(x)$ and $9k^3x = (3k^2x)(3k)$, each term of the original expression contains the common factor $3k^2x$. Thus by Rule I, $3k^2x^2 + 9k^3x = 3k^2x(x + 3k)$. Note that although $3k^2x^2 + 9k^3x = 3(k^2x^2 + 3k^3x)$, we do not say that the expression is completely factored, since $k^2x^2 + 3k^3x$ can yet be factored.

b. *Completely factor* $8a^5x^2y^3 - 6a^2b^3yz - 2a^4b^4xy^2z^2$.

$$8a^5x^2y^3 - 6a^2b^3yz - 2a^4b^4xy^2z^2 = 2a^2y(4a^3x^2y^2 - 3b^3z - a^2b^4xyz^2).$$

c. *Completely factor* $3x^2 + 6x + 3$.

$$3x^2 + 6x + 3 = 3(x^2 + 2x + 1)$$

$$= 3(x + 1)^2 \qquad \text{(Rule IV)}.$$

EXAMPLE 2

a. *Completely factor $x^2 - x - 6$.*

If this trinomial factors into the form $(x + a)(x + b)$, which is a product of two binomials, we must determine a and b. Since $(x + a)(x + b) = x^2 + (a + b)x + ab$, then

$$x^2 + (-1)x + (-6) = x^2 + (a + b)x + ab.$$

By equating corresponding coefficients, we want

$$a + b = -1 \quad \text{and} \quad ab = -6.$$

If $a = -3$ and $b = 2$, then both conditions are met and hence

$$x^2 - x - 6 = (x - 3)(x + 2).$$

b. *Completely factor $x^2 - 7x + 12$.*

$$x^2 - 7x + 12 = (x - 3)(x - 4).$$

EXAMPLE 3

Listed below are expressions that are completely factored. The numbers in parentheses refer to the rules used.

a. $x^2 + 8x + 16 = (x + 4)^2$ (IV).

b. $9x^2 + 9x + 2 = (3x + 1)(3x + 2)$ (III).

c. $6y^3 + 3y^2 - 18y = 3y(2y^2 + y - 6)$ (I)

                $= 3y(2y - 3)(y + 2)$ (III).

d. $x^2 - 6x + 9 = (x - 3)^2$ (V).

e. $z^{1/4} + z^{5/4} = z^{1/4}(1 + z)$ (I).

f. $x^4 - 1 = (x^2 + 1)(x^2 - 1)$ (VI)

        $= (x^2 + 1)(x + 1)(x - 1)$ (VI).

g. $x^{2/3} - 5x^{1/3} + 4 = (x^{1/3} - 1)(x^{1/3} - 4)$ (II).

h. $8 - x^3 = (2)^3 - (x)^3 = (2 - x)(4 + 2x + x^2)$ (VIII).

i. $x^6 - y^6 = (x^3)^2 - (y^3)^2 = (x^3 + y^3)(x^3 - y^3)$ (VI)

     $= (x + y)(x^2 - xy + y^2)(x - y)(x^2 + xy + y^2)$ (VII), (VIII).

j. $ax^2 - ay^2 + bx^2 - by^2 = (ax^2 - ay^2) + (bx^2 - by^2)$

$$= a(x^2 - y^2) + b(x^2 - y^2) \qquad \text{(I)}$$

$$= (a + b)(x^2 - y^2) \qquad \text{(I)}$$

$$= (a + b)(x + y)(x - y) \qquad \text{(VI)}.$$

Note in Example 3(f) that $x^2 - 1$ is factorable but $x^2 + 1$ is not. In 3(j) we factored by making use of grouping.

EXERCISE 0-7

Completely factor the expressions.

1. $6x + 4$.

2. $6y^2 - 4y$.

3. $10xy + 5xz$.

4. $3x^2y - 9x^3y^3$.

5. $8a^3bc - 12ab^3cd + 4b^4c^2d^2$.

6. $6z^2t^3 + 3zst^4 - 12z^2t^3$.

7. $x^2 - 25$.

8. $x^2 + 3x - 4$.

9. $p^2 + 4p + 3$.

10. $s^2 - 6s + 8$.

11. $16x^2 - 9$.

12. $x^2 + 5x - 24$.

13. $z^2 + 6z + 8$.

14. $4t^2 - 9s^2$.

15. $x^2 + 6x + 9$.

16. $y^2 - 15y + 50$.

17. $2x^2 + 12x + 16$.

18. $2x^2 + 7x - 15$.

19. $3x^2 - 3$.

20. $4y^2 - 8y + 3$.

21. $6y^2 + 13y + 2$.

22. $4x^2 - x - 3$.

23. $12s^3 + 10s^2 - 8s$.

24. $9z^2 + 24z + 16$.

25. $x^{2/3}y - 4x^{8/3}y^3$.

26. $9x^{4/7} - 1$.

27. $2x^3 + 2x^2 - 12x$.

28. $x^2y^2 - 4xy + 4$.

29. $(4x + 2)^2$.

30. $3s^2(3s - 9s^2)^2$.

31. $x^3y^2 - 10x^2y + 25x$.

32. $(3x^2 + x) + (6x + 2)$.

33. $(x^3 - 4x) + (8 - 2x^2)$.

34. $(x^2 - 1) + (x^2 - x - 2)$.

35. $(y^{10} + 8y^6 + 16y^2) - (y^8 + 8y^4 + 16)$.

36. $x^3y - xy + z^2x^2 - z^2$.

37. $x^3 + 8$.

38. $x^3 - 1$.

39. $x^6 - 1$.

40. $27 + 8x^3$.

41. $(x + 3)^3(x - 1) + (x + 3)^2(x - 1)^2.$ **42.** $(x + 5)^2(x + 1)^3 + (x + 5)^3(x + 1)^2.$

43. $(x + 4)(2x + 1) + (x + 4).$ **44.** $(x - 3)(2x + 3) - (2x + 3)(x + 5).$

45. $x^4 - 16.$ **46.** $81x^4 - y^4.$

47. $y^8 - 1.$ **48.** $t^4 - 4.$

49. $x^4 + x^2 - 2.$ **50.** $x^4 - 5x^2 + 4.$

51. $x^5 - 2x^3 + x.$ **52.** $4x^3 - 6x^2 - 4x.$

0-8 FRACTIONS

By using the fundamental principle of fractions (Sec. 0-4), we may be able to simplify fractions. That principle allows us to multiply or divide both the numerator and the denominator of a fraction by the same nonzero quantity. The resulting fraction will be equivalent to the original one. The fractions that we shall consider are assumed to have nonzero denominators.

EXAMPLE 1

a. *Simplify* $\dfrac{x^2 - x - 6}{x^2 - 7x + 12}.$

First, we completely factor the numerator and denominator:

$$\frac{x^2 - x - 6}{x^2 - 7x + 12} = \frac{(x - 3)(x + 2)}{(x - 3)(x - 4)}.$$

Since both numerator and denominator have the common factor $x - 3$, we shall multiply each of them by the multiplicative inverse of $x - 3$, namely $\dfrac{1}{x - 3}$, and simplify.

$$\frac{(x - 3)(x + 2)}{(x - 3)(x - 4)} = \frac{\dfrac{1}{x - 3}(x - 3)(x + 2)}{\dfrac{1}{x - 3}(x - 3)(x - 4)}$$

$$= \frac{1(x + 2)}{1(x - 4)} = \frac{x + 2}{x - 4}.$$

Usually we just write

$$\frac{x^2 - x - 6}{x^2 - 7x + 12} = \frac{\overset{1}{\cancel{(x - 3)}}(x + 2)}{\underset{1}{\cancel{(x - 3)}}(x - 4)} = \frac{x + 2}{x - 4}.$$

or
$$\frac{x^2 - x - 6}{x^2 - 7x + 12} = \frac{(x-3)(x+2)}{(x-3)(x-4)} = \frac{x+2}{x-4}.$$

The process we have used here is commonly referred to as "cancellation."

b. *Simplify* $\dfrac{2x^2 + 6x - 8}{8 - 4x - 4x^2}.$

$$\frac{2x^2 + 6x - 8}{8 - 4x - 4x^2} = \frac{2(x-1)(x+4)}{4(1-x)(2+x)}$$

$$= \frac{2(x-1)(x+4)}{2(2)[(-1)(x-1)](2+x)}$$

$$= \frac{x+4}{-2(x+2)} = -\frac{x+4}{2(x+2)}.$$

Recall that the product of $\dfrac{a}{b}$ and $\dfrac{c}{d}$ is

$$\frac{a}{b} \cdot \frac{c}{d} = \frac{ac}{bd},$$

and that $\dfrac{a}{b}$ divided by $\dfrac{c}{d}$, where $c \neq 0$, is

$$\frac{d}{b} \div \frac{c}{d} = \frac{\dfrac{a}{b}}{\dfrac{c}{d}} = \frac{a}{b} \cdot \frac{d}{c}.$$

EXAMPLE 2

a. $\dfrac{x}{x+2} \cdot \dfrac{x+3}{x-5} = \dfrac{x(x+3)}{(x+2)(x-5)}.$

b. $\dfrac{x^2 - 4x + 4}{x^2 + 2x - 3} \cdot \dfrac{6x^2 - 6}{x^2 + 2x - 8} = \dfrac{[(x-2)^2][6(x+1)(x-1)]}{[(x+3)(x-1)][(x+4)(x-2)]}$

$$= \frac{6(x-2)(x+1)}{(x+3)(x+4)}.$$

c. $\dfrac{x}{x+2} \div \dfrac{x+3}{x-5} = \dfrac{x}{x+2} \cdot \dfrac{x-5}{x+3} = \dfrac{x(x-5)}{(x+2)(x+3)}.$

d. $\dfrac{\dfrac{x-5}{x-3}}{2x} = \dfrac{\dfrac{x-5}{x-3}}{\dfrac{2x}{1}} = \dfrac{x-5}{x-3} \cdot \dfrac{1}{2x} = \dfrac{x-5}{2x(x-3)}.$

e. $\dfrac{\dfrac{4x}{x^2 - 1}}{\dfrac{2x^2 + 8x}{x - 1}} = \dfrac{4x}{x^2 - 1} \cdot \dfrac{x - 1}{2x^2 + 8x} = \dfrac{4x(x - 1)}{[(x + 1)(x - 1)][2x(x + 4)]}$

$$= \dfrac{2}{(x + 1)(x + 4)}.$$

Sometimes a denominator of a fraction has two terms and involves square roots, such as $2 - \sqrt{3}$ or $\sqrt{5} + \sqrt{2}$. The denominator may be rationalized by multiplying by an expression that makes the denominator a difference of two squares. For example,

$$\dfrac{4}{\sqrt{5} + \sqrt{2}} = \dfrac{4}{\sqrt{5} + \sqrt{2}} \cdot \dfrac{\sqrt{5} - \sqrt{2}}{\sqrt{5} - \sqrt{2}}$$

$$= \dfrac{4(\sqrt{5} - \sqrt{2})}{(\sqrt{5})^2 - (\sqrt{2})^2} = \dfrac{4(\sqrt{5} - \sqrt{2})}{5 - 2}$$

$$= \dfrac{4(\sqrt{5} - \sqrt{2})}{3}.$$

EXAMPLE 3

Rationalize the denominators.

a. $\dfrac{x}{\sqrt{2} - 6} = \dfrac{x}{\sqrt{2} - 6} \cdot \dfrac{\sqrt{2} + 6}{\sqrt{2} + 6} = \dfrac{x(\sqrt{2} + 6)}{(\sqrt{2})^2 - 6^2}$

$$= \dfrac{x(\sqrt{2} + 6)}{2 - 36} = -\dfrac{x(\sqrt{2} + 6)}{34}.$$

b. $\dfrac{\sqrt{5} - \sqrt{2}}{\sqrt{5} + \sqrt{2}} = \dfrac{\sqrt{5} - \sqrt{2}}{\sqrt{5} + \sqrt{2}} \cdot \dfrac{\sqrt{5} - \sqrt{2}}{\sqrt{5} - \sqrt{2}}$

$$= \dfrac{(\sqrt{5} - \sqrt{2})^2}{5 - 2} = \dfrac{5 - 2\sqrt{5}\sqrt{2} + 2}{3} = \dfrac{7 - 2\sqrt{10}}{3}.$$

In Example 2(c) of Sec. 0-3, it was shown that $\dfrac{a}{c} + \dfrac{b}{c} = \dfrac{a + b}{c}$. That is, if we add two fractions having a common denominator, then the result is a fraction whose denominator is the common denominator. The numerator is the sum of the numerators of the original fractions. Similarly, $\dfrac{a}{c} - \dfrac{b}{c} = \dfrac{a - b}{c}$.

EXAMPLE 4

a. $\dfrac{p^2 - 5}{p - 2} + \dfrac{3p + 2}{p - 2} = \dfrac{(p^2 - 5) + (3p + 2)}{p - 2}$

$\qquad\qquad\qquad = \dfrac{p^2 + 3p - 3}{p - 2}.$

b. $\dfrac{x^2 - 5x + 4}{x^2 + 2x - 3} - \dfrac{x^2 + 2x}{x^2 + 5x + 6} = \dfrac{(x - 1)(x - 4)}{(x - 1)(x + 3)} - \dfrac{x(x + 2)}{(x + 2)(x + 3)}$

$\qquad\qquad\qquad = \dfrac{x - 4}{x + 3} - \dfrac{x}{x + 3} = \dfrac{(x - 4) - x}{x + 3} = -\dfrac{4}{x + 3}.$

c. $\dfrac{x^2 + x - 5}{x - 7} - \dfrac{x^2 - 2}{x - 7} + \dfrac{-4x + 8}{x^2 - 9x + 14} = \dfrac{x^2 + x - 5}{x - 7} - \dfrac{x^2 - 2}{x - 7} + \dfrac{-4(x - 2)}{(x - 2)(x - 7)}$

$\qquad\qquad\qquad = \dfrac{(x^2 + x - 5) - (x^2 - 2) + (-4)}{x - 7}$

$\qquad\qquad\qquad = \dfrac{x - 7}{x - 7} = 1.$

To add (or subtract) two fractions with different denominators, use the fundamental principle of fractions to transform the fractions into equivalent fractions that have the same denominator. Then proceed with the addition (or subtraction) by the method described above.

For example, to find

$$\frac{2}{x^3(x - 3)} + \frac{3}{x(x - 3)^2},$$

we can convert the first fraction into the equivalent fraction

$$\frac{2(x - 3)}{x^3(x - 3)^2},$$

and we can convert the second fraction into

$$\frac{3x^2}{x^3(x - 3)^2}.$$

These fractions have the same denominator. Hence,

$$\frac{2}{x^3(x - 3)} + \frac{3}{x(x - 3)^2} = \frac{2(x - 3)}{x^3(x - 3)^2} + \frac{3x^2}{x^3(x - 3)^2}$$

$$= \frac{3x^2 + 2x - 6}{x^3(x - 3)^2}.$$

We could have converted the original fractions into equivalent fractions with any common denominator. However, we chose to convert them into fractions with the denominator $x^3(x - 3)^2$. This is the **least common denominator (L.C.D.)** of the fractions $2/[x^3(x - 3)]$ and $3/[x(x - 3)^2]$.

In general, to find the L.C.D. of two or more fractions, first factor each denominator completely. *The L.C.D. is the product of each of the distinct factors appearing in the denominators, each raised to the highest power to which it occurs in any one denominator.*

EXAMPLE 5

a. *Subtract:* $\dfrac{t}{3t + 2} - \dfrac{4}{t - 1}.$

Here the denominators are already factored. The L.C.D. is $(3t + 2)(t - 1)$.

$$\frac{t}{3t + 2} - \frac{4}{t - 1} = \frac{t(t - 1)}{(3t + 2)(t - 1)} - \frac{4(3t + 2)}{(3t + 2)(t - 1)}$$

$$= \frac{t(t - 1) - 4(3t + 2)}{(3t + 2)(t - 1)} = \frac{t^2 - t - 12t - 8}{(3t + 2)(t - 1)}$$

$$= \frac{t^2 - 13t - 8}{(3t + 2)(t - 1)}.$$

b. $\dfrac{4}{q - 1} + 3 = \dfrac{4}{q - 1} + \dfrac{3(q - 1)}{q - 1} = \dfrac{4 + 3(q - 1)}{q - 1} = \dfrac{3q + 1}{q - 1}.$

EXAMPLE 6

$$\frac{x - 2}{x^2 + 6x + 9} - \frac{x + 2}{2(x^2 - 9)}$$

$$= \frac{x - 2}{(x + 3)^2} - \frac{x + 2}{2(x + 3)(x - 3)} \qquad \left[\text{L.C.D.} = 2(x + 3)^2(x - 3) \right]$$

$$= \frac{(x - 2)(2)(x - 3)}{(x + 3)^2(2)(x - 3)} - \frac{(x + 2)(x + 3)}{2(x + 3)(x - 3)(x + 3)}$$

$$= \frac{(x - 2)(2)(x - 3) - (x + 2)(x + 3)}{2(x + 3)^2(x - 3)}$$

$$= \frac{2(x^2 - 5x + 6) - [x^2 + 5x + 6]}{2(x + 3)^2(x - 3)}$$

$$= \frac{2x^2 - 10x + 12 - x^2 - 5x - 6}{2(x + 3)^2(x - 3)}$$

$$= \frac{x^2 - 15x + 6}{2(x + 3)^2(x - 3)}.$$

EXAMPLE 7

$$\frac{\dfrac{1}{x+h}-\dfrac{1}{x}}{h}=\frac{\dfrac{x}{x(x+h)}-\dfrac{x+h}{x(x+h)}}{h}=\frac{\dfrac{x-(x+h)}{x(x+h)}}{h}$$

$$=\frac{\dfrac{-h}{x(x+h)}}{\dfrac{h}{1}}=\frac{-h}{x(x+h)h}=-\frac{1}{x(x+h)}.$$

The original fraction can also be simplified by multiplying the numerator and denominator by $x(x+h)$:

$$\frac{\dfrac{1}{x+h}-\dfrac{1}{x}}{h}=\frac{x(x+h)\left[\dfrac{1}{x+h}-\dfrac{1}{x}\right]}{x(x+h)h}$$

$$=\frac{x-(x+h)}{x(x+h)h}=\frac{-h}{x(x+h)h}=-\frac{1}{x(x+h)}.$$

EXERCISE 0-8

In Problems **1–42**, perform the operations and simplify as much as possible.

1. $\dfrac{y^2}{y-3}\cdot\dfrac{-1}{y+2}.$

2. $\dfrac{z^2-4}{z^2+2z}\cdot\dfrac{z^2}{z-2}.$

3. $\dfrac{2x-3}{x-2}\cdot\dfrac{2-x}{2x+3}.$

4. $\dfrac{x^2-y^2}{x+y}\cdot\dfrac{x^2+2xy+y^2}{y-x}.$

5. $\dfrac{2x-2}{x^2-2x-8}\div\dfrac{x^2-1}{x^2+5x+4}.$

6. $\dfrac{x^2+2x}{3x^2-18x+24}\div\dfrac{x^2-x-6}{x^2-4x+4}.$

7. $\dfrac{\dfrac{x^2}{6}}{\dfrac{x}{3}}.$

8. $\dfrac{\dfrac{4x^3}{9x}}{\dfrac{x}{18}}.$

9. $\dfrac{\dfrac{2m}{n^3}}{\dfrac{4m}{n^2}}.$

10. $\dfrac{\dfrac{c+d}{c}}{\dfrac{c-d}{2c}}.$

11. $\dfrac{\dfrac{4x}{3}}{2x}.$

12. $\dfrac{4x}{\dfrac{3}{2x}}.$

13. $\dfrac{-9x^3}{\dfrac{x}{3}}.$

14. $\dfrac{\dfrac{-9x^3}{x}}{3}.$

15. $\dfrac{\dfrac{x-5}{x^2-7x+10}}{x-2}.$

16. $\dfrac{\dfrac{x^2 + 6x + 9}{x}}{x + 3}$.

17. $\dfrac{\dfrac{10x^3}{x^2 - 1}}{\dfrac{5x}{x + 1}}$.

18. $\dfrac{\dfrac{x^2 - 4}{x^2 + 2x - 3}}{\dfrac{x^2 - x - 6}{x^2 - 9}}$.

19. $\dfrac{\dfrac{x^2 + 7x + 10}{x^2 - 2x - 8}}{\dfrac{x^2 + 6x + 5}{x^2 - 3x - 4}}$.

20. $\dfrac{\dfrac{(x + 2)^2}{3x - 2}}{\dfrac{9x + 18}{4 - 9x^2}}$.

21. $\dfrac{\dfrac{4x^2 - 9}{x^2 + 3x - 4}}{\dfrac{2x - 3}{1 - x^2}}$.

22. $\dfrac{\dfrac{6x^2y + 7xy - 3y}{xy - x + 5y - 5}}{\dfrac{x^3y + 4x^2y}{xy - x + 4y - 4}}$.

23. $\dfrac{x^2}{x + 3} + \dfrac{5x + 6}{x + 3}$.

24. $\dfrac{2}{x + 2} + \dfrac{x}{x + 2}$.

25. $\dfrac{1}{t} + \dfrac{2}{3t}$.

26. $\dfrac{4}{x^2} - \dfrac{1}{x}$.

27. $1 - \dfrac{p^2}{p^2 - 1}$.

28. $\dfrac{4}{s + 4} + s$.

29. $\dfrac{4}{2x - 1} + \dfrac{x}{x + 3}$.

30. $\dfrac{x + 1}{x - 1} - \dfrac{x - 1}{x + 1}$.

31. $\dfrac{1}{x^2 - x - 2} + \dfrac{1}{x^2 - 1}$.

32. $\dfrac{y}{3y^2 - 5y - 2} - \dfrac{2}{3y^2 - 7y + 2}$.

33. $\dfrac{4}{x - 1} - 3 + \dfrac{-3x^2}{5 - 4x - x^2}$.

34. $\dfrac{2x - 3}{2x^2 + 11x - 6} - \dfrac{3x + 1}{3x^2 + 16x - 12} + \dfrac{1}{3x - 2}$.

35. $(1 + x^{-1})^2$.

36. $(x^{-1} + y^{-1})^2$.

37. $(x^{-1} - y)^{-1}$.

38. $(x - y^{-1})^{-2}$.

39. $\dfrac{1 + \dfrac{1}{x}}{3}$.

40. $\dfrac{\dfrac{x + 3}{x}}{x - \dfrac{9}{x}}$.

41. $\dfrac{3 - \dfrac{1}{2x}}{x + \dfrac{x}{x + 2}}$.

42. $\dfrac{\dfrac{x - 1}{x^2 + 5x + 6} - \dfrac{1}{x + 2}}{3 + \dfrac{x - 7}{3}}$.

In Problems **43** *and* **44**, *perform the indicated operations, but do not rationalize the denominators.*

43. $\dfrac{2}{\sqrt{x+h}} - \dfrac{2}{\sqrt{x}}$.

44. $\dfrac{x\sqrt{x}}{\sqrt{1+x}} + \dfrac{1}{\sqrt{x}}$.

In Problems **45–54** *simplify and express your answer in a form that is free of radicals in the denominator.*

45. $\dfrac{1}{2+\sqrt{3}}$.

46. $\dfrac{1}{1-\sqrt{2}}$.

47. $\dfrac{\sqrt{2}}{\sqrt{3}-\sqrt{6}}$.

48. $\dfrac{5}{\sqrt{6}+\sqrt{7}}$.

49. $\dfrac{2\sqrt{2}}{\sqrt{2}-\sqrt{3}}$.

50. $\dfrac{2\sqrt{3}}{\sqrt{5}-\sqrt{2}}$.

51. $\dfrac{1}{x+\sqrt{5}}$.

52. $\dfrac{x-3}{\sqrt{x}-1} + \dfrac{4}{\sqrt{x}-1}$.

53. $\dfrac{5}{1+\sqrt{3}} - \dfrac{4}{2-\sqrt{2}}$.

54. $\dfrac{4}{\sqrt{x}+2} \cdot \dfrac{x^2}{3}$.

EQUATIONS

Even a beginning student of economics or business is faced with solving elementary equations. In this chapter we shall develop techniques to accomplish this task. These methods will be applied in the next chapter to some practical situations.

1-1 LINEAR EQUATIONS

An **equation** is a statement that two expressions are equal. The two expressions that make up an equation are called its **sides** or **members**. They are separated by the **equality sign** "$=$."

EXAMPLE 1

The following are equations.

a. $x + 2 = 3$,   b. $x^2 + 3x + 2 = 0$,

c. $\dfrac{y}{y-5} = 7,$ d. $w = 7 - z.$

In Example 1 each equation contains at least one variable. A **variable** is a symbol that can be replaced by any one of a set of different numbers. The most popular symbols for variables are letters from the latter part of the alphabet, such as x, y, z, w, and t. Hence equations (a) and (c) are said to be in the variables x and y, respectively. Equation (d) is in the variables w and z. In the equation $x + 2 = 3$, the numbers 2 and 3 are called *constants*. They are fixed numbers.

We never allow a variable to have a value for which any expression in the equation is undefined. Thus in $y/(y - 5) = 7$, y cannot be 5 since this would result in division by 0.

To *solve* an equation means to find all values of its variables for which the equation is true. These values are called *solutions* of the equation and are said to *satisfy* the equation. When only one variable is involved, a solution is also called a **root**. The set of all solutions is called the **solution set** of the equation. Sometimes a letter representing an unknown quantity in an equation is simply called an *unknown*. Let us illustrate these terms.

EXAMPLE 2

a. In the equation $x + 2 = 3$, the variable x is the unknown. The only value of x which satisfies the equation is 1. Hence 1 is a root and the solution set is $\{1\}$.

b. $w = 7 - z$ is an equation in two unknowns. One solution is the pair of values $w = 4$ and $z = 3$. However, there are infinitely many solutions. Can you think of another?

c. -2 is a root of $x^2 + 3x + 2 = 0$ because by substitution we have $(-2)^2 + 3(-2) + 2 = 0$.

In solving an equation we want any operation on it to result in another equation having exactly the same solutions as the given equation. When this occurs, the equations are said to be **equivalent**. There are three operations that guarantee equivalence:

(1) *Adding (subtracting) the same polynomial*[†] *to (from) both sides of an equation, where the polynomial is in the same variable as that occurring in the equation.*

For example, if $3x = 5 - 6x$, then adding $6x$ to both sides gives the equivalent equation $3x + 6x = 5 - 6x + 6x$.

----

†See p. 21 for a definition of a polynomial.

(2) *Multiplying (dividing) both sides of an equation by the same constant, except by zero.*

For example, if $10x = 5$, then dividing both sides by 10 gives the equivalent equation $\dfrac{10x}{10} = \dfrac{5}{10}$.

(3) *Replacing either side of an equation by an equal expression.*

For example, if $x(x + 2) = 3$, then replacing the left side by the equal expression $x^2 + 2x$ gives the equivalent equation $x^2 + 2x = 3$.

We repeat: Applying operations 1–3 guarantees that the resulting equation is equivalent to the given one.

Sometimes in solving an equation we have to apply operations other than 1–3. These operations may not necessarily result in equivalent equations. They include:

(4) *Multiplying both sides of an equation by an expression involving the variable;*

(5) *Dividing both sides of an equation by an expression involving the variable;*

(6) *Raising both sides of an equation to equal powers.*

Let us illustrate the last three operations. For example, by inspection the only root of $x - 1 = 0$ is 1. Multiplying each side by x (operation 4) gives $x^2 - x = 0$, which is satisfied if x is 0 or 1 (you should check this by substitution). But 0 *does not* satisfy the *original* equation. Thus the equations are not equivalent.

Continuing, you may check that $(x - 4)(x - 3) = 0$ is satisfied when x is 3 or 4. Dividing both sides by $x - 4$ (operation 5) gives $x - 3 = 0$, whose only root is 3. Again, we do not have equivalence since, in this case, a root has been "lost." Note that when x is 4, division by $x - 4$ is actually division by 0, an invalid operation.

Finally, squaring each side of the equation $x = 2$ (operation 6) gives $x^2 = 4$, which is true if $x = 2$ or -2. But -2 is not a root of the given equation.

From our discussion it is clear that when operations 4–6 are performed, we must be careful about drawing conclusions concerning the roots of a given equation. Operations 4 and 6 *can* produce an equation with more roots. Thus you should check whether or not each "solution" obtained by these operations satisfies the *original* equation. Operation 5 *can* produce an equation with fewer roots. In this case, any "lost" roots may never be determined. Thus, avoid operation 5 whenever possible.

In summary, an equation may be thought of as a set of restrictions on any variable in the equation. Operations 4–6 may increase or decrease the restrictions, giving different solutions from the original equation. However, operations 1–3 never affect the restrictions.

The principles presented so far will now be demonstrated in the solution of a *linear equation*.

DEFINITION.   *A **linear equation** in the variable x is an equation that can be written in the form*

$$ax + b = 0 \tag{1}$$

where a and b are constants and $a \neq 0$.

A linear equation is also called a *first-degree equation* or an *equation of degree one*, since the highest power of the variable that occurs in Eq. (1) is one. To solve a linear equation we perform operations on it until the roots are obvious, as the following examples show.

EXAMPLE 3

Solve the following equations.

a. $5x - 6 = 3x$.

$$
\begin{aligned}
5x - 6 &= 3x, & \\
5x - 6 + (-3x) &= 3x + (-3x) & \text{(adding } -3x \text{ to both sides),} \\
2x - 6 &= 0 & \text{(simplifying, that is, operation 3),} \\
2x - 6 + 6 &= 0 + 6 & \text{(adding 6 to both sides),} \\
2x &= 6 & \text{(simplifying),} \\
\frac{2x}{2} &= \frac{6}{2} & \text{(dividing both sides by 2),} \\
x &= 3.
\end{aligned}
$$

Clearly 3 is the only root of the last equation. Since each equation is equivalent to the one before it, we conclude that 3 must be the only root of $5x - 6 = 3x$. That is, the solution set is {3}. We can describe the first step in the solution as moving a term from one side of an equation to the other while changing its sign; this is commonly called *transposing*. Note that since the original equation can be put in the form $2x + (-6) = 0$, it is a linear equation.

b. $2(p + 4) = 7p + 2$.

$$
\begin{aligned}
2(p + 4) &= 7p + 2, & \\
2p + 8 &= 7p + 2 & \text{(distributive property),} \\
2p &= 7p - 6 & \text{(subtracting 8 from both sides),} \\
-5p &= -6 & \text{(subtracting } 7p \text{ from both sides),} \\
p &= \frac{-6}{-5} & \text{(dividing both sides by } -5\text{),} \\
p &= \frac{6}{5}.
\end{aligned}
$$

c. $\dfrac{7x + 3}{2} - \dfrac{9x - 8}{4} = 6.$

We first clear the equation of fractions by multiplying *both* sides by the least common denominator (L.C.D.), which is 4.

$$4\left(\frac{7x + 3}{2} - \frac{9x - 8}{4}\right) = 4(6),$$

$$4 \cdot \frac{7x + 3}{2} - 4 \cdot \frac{9x - 8}{4} = 24,$$

$$2(7x + 3) - (9x - 8) = 24,$$

$$14x + 6 - 9x + 8 = 24,$$

$$5x + 14 = 24,$$

$$5x = 10,$$

$$x = 2.$$

Each equation in Example 3 has one and only one root. This is typical of every linear equation in one variable.

Equations in which some of the constants are represented by letters are called **literal equations**. For example, in the literal equation $x + a = 4b$ we consider a and b to be unspecified constants. Formulas, such as $I = Prt$, which express a relationship between certain quantities may be regarded as literal equations. If we want to express a particular letter in a formula in terms of the others, this letter is considered the unknown.

EXAMPLE 4

a. *The equation $I = Prt$ is the formula for the simple interest I on a principal of P dollars at the annual interest rate of r for a period of t years. Express r in terms of I, P, and t.*

Here we consider r to be the unknown and then proceed to solve for r.

$$I = Prt,$$

$$\frac{I}{Pt} = \frac{Prt}{Pt},$$

$$\frac{I}{Pt} = r \quad \text{or} \quad r = \frac{I}{Pt}.$$

When we divided both sides by Pt, we assumed $Pt \neq 0$ to avoid division by 0. Similar assumptions will be made in solving other literal equations.

b. *The equation $S = P + Prt$ is the formula for the value S of an investment of a principal of P dollars at a simple annual interest rate of r for a period of t years. Solve for P.*

$$S = P + Prt,$$

$$S = P(1 + rt),$$

$$\frac{S}{1 + rt} = P.$$

c. *Solve $(a + c)x + x^2 = (x + a)^2$ for x.*

We shall first simplify the equation and get all terms involving x on one side.

$$(a + c)x + x^2 = (x + a)^2,$$

$$ax + cx + x^2 = x^2 + 2ax + a^2,$$

$$cx - ax = a^2,$$

$$x(c - a) = a^2,$$

$$x = \frac{a^2}{c - a}.$$

EXERCISE 1-1

In Problems 1–6, determine by substitution which of the given numbers, if any, satisfy the given equation.

1. $9x - x^2 = 0$;   1, 0.   2. $20 - 9x = -x^2$;   5, 4.

3. $y + 2(y - 3) = 4$; $\frac{10}{3}$, 1.   4. $2x + x^2 - 8 = 0$;   2, −4.

5. $x(7 + x) - 2(x + 1) - 3x = -2$;   −3.   6. $x(x + 1)^2(x + 2) = 0$;   0, −1, 2.

*In Problems 7–16, determine what operations were applied to the first equation to obtain the second. State whether or not the operations **guarantee** that the equations are equivalent. Do not solve the equations.*

7. $x - 5 = 4x + 10$; $x = 4x + 15$.

8. $8x - 4 = 16$; $x - \frac{1}{2} = 2$.

9. $x = 4$; $x^2 = 16$.

10. $\frac{1}{2}x^2 + 3 = x - 9$; $x^2 + 6 = 2x - 18$.

11. $x^2 - 2x = 0$; $x - 2 = 0$.

12. $\frac{2}{x - 2} + x = x^2$; $2 + x(x - 2) = x^2(x - 2)$.

13. $\dfrac{x^2 - 1}{x - 1} = 3$; $x^2 - 1 = 3(x - 1)$.

14. $x(x + 5)(x + 9) = x(x + 1)$; $(x + 5)(x + 9) = x + 1$.

15. $\dfrac{x(x + 1)}{x - 5} = x(x + 9)$; $x + 1 = (x + 9)(x - 5)$.

16. $2x^2 - 9 = x$; $x^2 - \frac{1}{2}x = \frac{9}{2}$.

In Problems **17–46**, *solve the equations.*

17. $4x = 10$.   **18.** $.2x = 5$.

19. $3y = 0$.   **20.** $2x - 4x = -5$.

21. $-5x = 10 - 15$.   **22.** $3 - 2x = 4$.

23. $5x - 3 = 9$.   **24.** $\sqrt{2}\,x + 3 = 8$.

25. $7x + 7 = 2(x + 1)$.   **26.** $6z + 5z - 3 = 41$.

27. $2(p - 1) - 3(p - 4) = 4p$.   **28.** $t = 2 - 2[2t - 3(1 - t)]$.

29. $\dfrac{x}{5} = 2x - 6$.   **30.** $\dfrac{5y}{7} - \dfrac{6}{7} = 2 - 4y$.

31. $5 + \dfrac{4x}{9} = \dfrac{x}{2}$.   **32.** $\dfrac{x}{3} - 4 = \dfrac{x}{5}$.

33. $q = \dfrac{3}{2}q - 4$.   **34.** $\dfrac{x}{2} + \dfrac{x}{3} = 7$.

35. $3x + \dfrac{x}{5} - 5 = \dfrac{1}{5} + 5x$.   **36.** $y - \dfrac{y}{2} + \dfrac{y}{3} - \dfrac{y}{4} = \dfrac{y}{5}$.

37. $\dfrac{2y - 3}{4} = \dfrac{6y + 7}{3}$.   **38.** $\dfrac{p}{3} + \dfrac{3}{4}p = \dfrac{9}{2}(p - 1)$.

39. $w + \dfrac{w}{2} - \dfrac{w}{3} + \dfrac{w}{4} = 5$.   **40.** $\dfrac{7 + 2(x + 1)}{3} = \dfrac{8x}{5}$.

41. $\dfrac{x + 2}{3} - \dfrac{2 - x}{6} = x - 2$.   **42.** $\dfrac{x}{5} + \dfrac{2(x - 4)}{10} = 7$.

43. $\dfrac{9}{5}(3 - x) = \dfrac{3}{4}(x - 3)$.   **44.** $\dfrac{2y - 7}{3} + \dfrac{8y - 9}{14} = \dfrac{3y - 5}{21}$.

45. $\frac{3}{2}(4x - 3) = 2[x - (4x - 3)]$.

46. $(3x - 1)^2 - (5x - 3)^2 = -(4x - 2)^2$.

In Problems **47–54**, *express the indicated symbol in terms of the remaining symbols.*

47. $I = Prt$; P.   **48.** $ax + b = 0$; x.

49. $p = 8q - 1$; q.   **50.** $p = -3q + 6$; q.

51. $S = P(1 + rt);$ $r.$ **52.** $r = \dfrac{2mI}{B(n + 1)};$ $m.$

53. $S = \dfrac{n}{2}(a_1 + a_n);$ $a_1.$ **54.** $S = \dfrac{R[(1 + i)^n - 1]}{i};$ $R.$

1-2 EQUATIONS LEADING TO LINEAR EQUATIONS

Some equations that are not linear do not have any solutions. In this case we say that the solution set is the **empty set** or **null set**, which we denote by { } or ∅. The following examples illustrate that solving a nonlinear equation may lead to a linear equation.

EXAMPLE 1

Solve the following equations.

a. $\dfrac{5}{x - 4} = \dfrac{6}{x - 3}.$

This is called a *fractional equation*, since it involves a variable in a denominator. To solve it, we shall first write it in a form that is free of fractions. Multiplying both sides by the L.C.D., $(x - 4)(x - 3)$, we have

$$(x - 4)(x - 3)\left(\frac{5}{x - 4}\right) = (x - 4)(x - 3)\left(\frac{6}{x - 3}\right),$$

$$5(x - 3) = 6(x - 4),$$

$$5x - 15 = 6x - 24,$$

$$9 = x.$$

In the first step we multiplied each side by an expression involving the variable x. As we mentioned in Sec. 1-1, this means that we must check whether or not 9 satisfies the *original* equation. If 9 is substituted for x in that equation, the left side is

$$\frac{5}{9 - 4} = \frac{5}{5} = 1$$

and the right side is

$$\frac{6}{9 - 3} = \frac{6}{6} = 1.$$

Since both sides are equal, 9 is a root.

b. $\dfrac{3x + 4}{x + 2} - \dfrac{3x - 5}{x - 4} = \dfrac{12}{x^2 - 2x - 8}.$

Notice that $x^2 - 2x - 8 = (x + 2)(x - 4)$. Thus the L.C.D. of the fractions is clearly $(x + 2)(x - 4)$. Multiplying both sides by the L.C.D., we have

$$(x - 4)(3x + 4) - (x + 2)(3x - 5) = 12,$$

$$3x^2 - 8x - 16 - (3x^2 + x - 10) = 12,$$

$$3x^2 - 8x - 16 - 3x^2 - x + 10 = 12,$$

$$-9x - 6 = 12,$$

$$-9x = 18,$$

$$x = -2.$$

However, the *original* equation is not defined for $x = -2$ (division by zero), and so there are no roots. The solution set is $\varnothing$.

c. $\dfrac{4}{x - 5} = 0.$

The only way a fraction can equal zero is if the numerator is 0 and the denominator is different from 0. Since the numerator, 4, is never 0, the solution set is $\varnothing$.

EXAMPLE 2

Solve the following equations.

a. $\sqrt{x^2 + 33} = x + 3.$

This is called a *radical equation*, since a variable occurs in a radicand. To solve the given equation we raise both sides to the same power so as to eliminate the radical. With this operation we must check any resulting "solutions."

$$\sqrt{x^2 + 33} = x + 3,$$

$$x^2 + 33 = (x + 3)^2 \quad \text{(squaring both sides)},$$

$$x^2 + 33 = x^2 + 6x + 9,$$

$$24 = 6x,$$

$$4 = x.$$

You should show by substitution that 4 is indeed a root.

b. $\sqrt{y - 3} - \sqrt{y} = -3.$

When an equation contains two terms involving radical expressions, first write the equation so that one radical is on each side, if possible.

$$\sqrt{y-3} = \sqrt{y} - 3,$$
$$y - 3 = y - 6\sqrt{y} + 9 \quad \text{(squaring both sides)},$$
$$6\sqrt{y} = 12,$$
$$\sqrt{y} = 2,$$
$$y = 4 \quad \text{(squaring both sides)}.$$

Substituting 4 into the left side of the *original* equation gives $\sqrt{1} - \sqrt{4}$, which is -1. Since this does not equal the right side, -3, there is no solution. That is, the solution set is $\varnothing$.

EXERCISE 1-2

Solve the equations in Problems **1–34**.

1. $\dfrac{5}{x} = 25.$ **2.** $\dfrac{4}{x-1} = 2.$

3. $\dfrac{3}{7-x} = 0.$ **4.** $\dfrac{5x-2}{x+1} = 0.$

5. $\dfrac{4}{8-x} = \dfrac{3}{4}.$ **6.** $\dfrac{x+3}{x} = \dfrac{2}{5}.$

7. $\dfrac{q}{3q-4} = 3.$ **8.** $\dfrac{4p}{7-p} = 1.$

9. $\dfrac{1}{p-1} = \dfrac{2}{p-2}.$ **10.** $\dfrac{2x-3}{4x-5} = 6.$

11. $\dfrac{1}{x} + \dfrac{1}{5} = \dfrac{4}{5}.$ **12.** $\dfrac{4}{t-3} = \dfrac{3}{t-4}.$

13. $\dfrac{3x-2}{2x+3} = \dfrac{3x-1}{2x+1}.$ **14.** $\dfrac{x+2}{x-1} + \dfrac{x+1}{2-x} = 0.$

15. $\dfrac{y-6}{y} - \dfrac{6}{y} = \dfrac{y+6}{y-6}.$ **16.** $\dfrac{y-3}{y+3} = \dfrac{y-3}{y+2}.$

17. $\dfrac{-4}{x-1} = \dfrac{7}{2-x} + \dfrac{3}{x+1}.$ **18.** $\dfrac{1}{x-3} - \dfrac{3}{x-2} = \dfrac{4}{1-2x}.$

19. $\dfrac{9}{x-3} = \dfrac{3x}{x-3}.$ **20.** $\dfrac{x}{x+3} - \dfrac{x}{x-3} = \dfrac{3x-4}{x^2-9}.$

21. $\sqrt{x+6} = 3.$ **22.** $\sqrt{z-2} = 3.$

23. $\sqrt{5x-6} - 16 = 0.$ **24.** $6 - \sqrt{2x+5} = 0.$

25. $\sqrt{\dfrac{x}{2} + 1} = \dfrac{2}{3}$.

26. $(x + 6)^{1/2} = 7$.

27. $\sqrt{4x - 6} = \sqrt{x}$.

28. $\sqrt{7 - 2x} = \sqrt{x - 1}$.

29. $(x - 3)^{3/2} = 8$.

30. $\sqrt{y^2 - 9} = 9 - y$.

31. $\sqrt{y} + \sqrt{y + 2} = 3$.

32. $\sqrt{x} - \sqrt{x + 1} = 1$.

33. $\sqrt{z^2 + 2z} = 3 + z$.

34. $\sqrt{\dfrac{1}{w}} - \sqrt{\dfrac{2}{5w - 2}} = 0$.

In Problems 35–37, express the indicated letter in terms of the remaining letters.

35. $r = \dfrac{d}{1 - dt}$;　d.

36. $\dfrac{x - a}{b - x} = \dfrac{x - b}{a - x}$;　x.

37. $r = \dfrac{2mI}{B(n + 1)}$;　n.

1-3 QUADRATIC EQUATIONS

To learn how to solve more complicated problems, we turn to methods of solving *quadratic equations*.

> **DEFINITION.** *A **quadratic equation** in the variable x is an equation that can be written in the form*
>
> $$ax^2 + bx + c = 0$$
>
> *where a, b, and c are constants and a $\neq$ 0.*

A quadratic equation is also called a *second-degree equation* or an *equation of degree two*, since the highest power of the variable that occurs is the second. Whereas a linear equation has only one root, some quadratic equations have two different roots.

A useful method of solving quadratic equations is based on factoring $ax^2 + bx + c$, as the following example shows.

EXAMPLE 1

Solve the following quadratic equations.

a. $x^2 + x - 12 = 0$.

The left side factors easily:

$$(x - 3)(x + 4) = 0.$$

Think of this as two quantities, $x - 3$ and $x + 4$, whose product is zero. **Whenever a product of two or more quantities is zero, at least one of the quantities *must* be zero.** This means that either $x - 3 = 0$ or $x + 4 = 0$. Solving these gives $x = 3$ and $x = -4$. The roots are 3 and -4 and the solution set is $\{3, -4\}$.

b. $6w^2 = 5w$.

We *do not* divide both sides by w (a variable) since equivalence is not guaranteed and we may "lose" a root. Instead, we write the equation as

$$6w^2 - 5w = 0.$$

Factoring gives

$$w(6w - 5) = 0.$$

Setting each factor equal to 0, we have

$$w = 0 \qquad \text{and} \qquad 6w - 5 = 0.$$

Thus the roots are $w = 0$ and $w = \frac{5}{6}$. Note that if we had divided both sides of $6w^2 = 5w$ by w and obtained $6w = 5$, our only solution would be $w = \frac{5}{6}$. That is, we would lose the root $w = 0$. This confirms our discussion of operation 5 in Sec. 1–1.

Some equations that are not quadratic may be solved by the method of factoring, as Example 2 shows.

EXAMPLE 2

a. *Solve* $4x - 4x^3 = 0$.

$$4x - 4x^3 = 0,$$
$$4x(1 - x^2) = 0,$$
$$4x(1 - x)(1 + x) = 0.$$

Setting each factor equal to 0 gives

$$x = 0, 1, -1,$$

which we can write as $x = 0, \pm 1$.

b. *Solve* $x(x + 2)^2(x + 5) + x(x + 2)^3 = 0$.

Since the factor $x(x + 2)^2$ is common to both terms in the left side, we have

$$x(x + 2)^2[(x + 5) + (x + 2)] = 0,$$
$$x(x + 2)^2(2x + 7) = 0.$$

Hence $x = 0$, $x + 2 = 0$, or $2x + 7 = 0$, from which we conclude that $x = 0, -2, -\frac{7}{2}$.

EXAMPLE 3

a. *Solve* $(3x - 4)(x + 1) = -2.$

> **PITFALL.** *You should approach a problem like this with caution. If the product of two quantities is equal to* −2, *it is not true that at least one of the quantities must be* −2. *Why?*

We first multiply the factors in the left side:

$$3x^2 - x - 4 = -2.$$

Rewriting so that 0 appears on one side, we have

$$3x^2 - x - 2 = 0.$$

Factoring gives

$$(3x + 2)(x - 1) = 0.$$

Thus,

$$x = -\tfrac{2}{3}, 1.$$

b. *Solve* $\dfrac{y + 1}{y + 3} + \dfrac{y + 5}{y - 2} = \dfrac{7(2y + 1)}{y^2 + y - 6}.$ (1)

Multiplying both sides by the L.C.D., $(y + 3)(y - 2)$, we get

$$(y - 2)(y + 1) + (y + 3)(y + 5) = 7(2y + 1). \qquad (2)$$

Since Eq. (1) was multiplied by an expression involving the variable y, remember (from Sec. 1-1) that Eq. (2) is not necessarily equivalent to Eq. (1). After simplifying Eq. (2) we have

$$2y^2 - 7y + 6 = 0,$$
$$(2y - 3)(y - 2) = 0.$$

Thus $\tfrac{3}{2}$ and 2 are *possible* roots of the given equation. But 2 cannot be a root of Eq. (1) since substitution leads to division by 0. However, you should check that $\tfrac{3}{2}$ does indeed satisfy the *original* equation. Thus its only root is $\tfrac{3}{2}$.

EXAMPLE 4

Solve $u^2 = 3.$

This equation is equivalent to

$$u^2 - 3 = 0.$$

Factoring, we obtain

$$(u - \sqrt{3})(u + \sqrt{3}) = 0.$$

Thus $u - \sqrt{3} = 0$ or $u + \sqrt{3} = 0$. The roots are $\pm \sqrt{3}$. More generally,

$$\text{if } u^2 = k, \quad \text{then} \quad u = \pm \sqrt{k}.$$

Solving quadratic equations by factoring can be quite difficult, as is evident by trying that method on $.7x^2 - \sqrt{2}x - 8\sqrt{5} = 0$. However, there is a formula called the *quadratic formula*[†] that gives the roots of any quadratic equation:

QUADRATIC FORMULA

If $ax^2 + bx + c = 0$, where a, b, and c are constants and $a \neq 0$, then

$$x = \frac{-b \pm \sqrt{b^2 - 4ac}}{2a}.$$

These values of x are the roots of the quadratic equation above.

EXAMPLE 5

a. *Solve $4x^2 - 17x + 15 = 0$ by the quadratic formula.*

Here $a = 4$, $b = -17$, and $c = 15$.

$$x = \frac{-b \pm \sqrt{b^2 - 4ac}}{2a} = \frac{-(-17) \pm \sqrt{(-17)^2 - 4(4)(15)}}{2(4)}$$

$$= \frac{17 \pm \sqrt{49}}{8} = \frac{17 \pm 7}{8}.$$

The roots are $\dfrac{17 + 7}{8} = \dfrac{24}{8} = 3$ and $\dfrac{17 - 7}{8} = \dfrac{10}{8} = \dfrac{5}{4}$.

b. *Solve $2 + 6\sqrt{2}y + 9y^2 = 0$ by the quadratic formula.*

Look at the arrangement of the terms. Here $a = 9$, $b = 6\sqrt{2}$, and $c = 2$.

$$y = \frac{-b \pm \sqrt{b^2 - 4ac}}{2a} = \frac{-6\sqrt{2} \pm \sqrt{0}}{2(9)}.$$

[†]A derivation of the quadratic formula appears in the next section as a supplement.

Thus, $y = \dfrac{-6\sqrt{2} + 0}{18} = -\dfrac{\sqrt{2}}{3}$ or $y = \dfrac{-6\sqrt{2} - 0}{18} = -\dfrac{\sqrt{2}}{3}$. Therefore, the only value for the root is $-\dfrac{\sqrt{2}}{3}$.

c. *Solve $z^2 + z + 1 = 0$ by the quadratic formula.*

Here $a = 1$, $b = 1$, and $c = 1$. The roots are

$$\frac{-b \pm \sqrt{b^2 - 4ac}}{2a} = \frac{-1 \pm \sqrt{-3}}{2}.$$

Now, $\sqrt{-3}$ denotes a number whose square is -3. However, no such real number exists since the square of any real number is nonnegative. Hence, the equation has no real roots.[†]

PITFALL. *Be certain that you use the quadratic formula correctly. Do not write*
$$x = -b \pm \frac{\sqrt{b^2 - 4ac}}{2a}.$$

From Example 5 you can see that a quadratic equation has two different real roots, exactly one real root, or no real roots, depending on whether $b^2 - 4ac$ is > 0, $= 0$, or < 0.

EXERCISE 1-3

In Problems **1–30**, solve by factoring.

1. $x^2 - 4x + 4 = 0$.   **2.** $t^2 + 3t + 2 = 0$.

3. $y^2 - 7y + 12 = 0$.   **4.** $x^2 + x - 12 = 0$.

5. $x^2 - 2x - 3 = 0$.   **6.** $x^2 - 16 = 0$.

7. $x^2 - 12x = -36$.   **8.** $3w^2 - 12w + 12 = 0$.

9. $x^2 - 4 = 0$.   **10.** $2x^2 + 4x = 0$.

11. $z^2 - 8z = 0$.   **12.** $x^2 + 9x = -14$.

13. $4x^2 + 1 = 4x$.   **14.** $2z^2 + 7z = 4$.

15. $y(2y + 3) = 5$.   **16.** $8 + 2x - 3x^2 = 0$.

17. $-x^2 + 3x + 10 = 0$.   **18.** $\frac{1}{7}y^2 = \frac{3}{7}y$.

[†] $\dfrac{-1 \pm \sqrt{-3}}{2}$ can be expressed as $\dfrac{-1 \pm i\sqrt{3}}{2}$, where $i \ (= \sqrt{-1}\,)$ is called the imaginary unit.

19. $2p^2 = 3p$.

20. $-r^2 - r + 12 = 0$.

21. $x(x - 1)(x + 2) = 0$.

22. $(x - 2)^2(x + 1)^2 = 0$.

23. $x^3 - 64x = 0$.

24. $x^3 - 4x^2 - 5x = 0$.

25. $6x^3 + 5x^2 - 4x = 0$.

26. $(x + 1)^2 - 5x + 1 = 0$.

27. $(x + 3)(x^2 - x - 2) = 0$.

28. $3(x^2 + 2x - 8)(x - 5) = 0$.

29. $p(p - 3)^2 - 4(p - 3)^3 = 0$.

30. $x^4 - 3x^2 + 2 = 0$.

*In Problems **31–44**, find all real roots by using the quadratic formula.*

31. $x^2 + 2x - 24 = 0$.

32. $x^2 - 2x - 15 = 0$.

33. $4x^2 - 12x + 9 = 0$.

34. $p^2 + 2p = 0$.

35. $p^2 - 5p + 3 = 0$.

36. $2 - 2x + x^2 = 0$.

37. $4 - 2n + n^2 = 0$.

38. $2x^2 + x = 5$.

39. $6x^2 + 7x - 5 = 0$.

40. $w^2 - 2\sqrt{2}w + 2 = 0$.

41. $2x^2 - 3x = 20$.

42. $.01x^2 + .2x - .6 = 0$.

43. $2x^2 + 4x = 5$.

44. $-2x^2 - 6x + 5 = 0$.

*In Problems **45–66**, solve by any method.*

45. $x^2 = \dfrac{x + 3}{2}$.

46. $\dfrac{x}{3} = \dfrac{6}{x} - 1$.

47. $\dfrac{3}{x - 4} + \dfrac{x - 3}{x} = 2$.

48. $\dfrac{2}{x - 1} - \dfrac{6}{2x + 1} = 5$.

49. $\dfrac{6x + 7}{2x + 1} - \dfrac{6x + 1}{2x} = 1$.

50. $\dfrac{6(w + 1)}{2 - w} + \dfrac{w}{w - 1} = 3$.

51. $\dfrac{2}{r - 2} - \dfrac{r + 1}{r + 4} = 0$.

52. $\dfrac{2x - 3}{2x + 5} + \dfrac{2x}{3x + 1} = 1$.

53. $\dfrac{y + 1}{y + 3} + \dfrac{y + 5}{y - 2} = \dfrac{14y + 7}{y^2 + y - 6}$.

54. $\dfrac{3}{t + 1} + \dfrac{4}{t} = \dfrac{12}{t + 2}$.

55. $\dfrac{2}{x^2 - 1} - \dfrac{1}{x(x - 1)} = \dfrac{2}{x^2}$.

56. $5 - \dfrac{3(x + 3)}{x^2 + 3x} = \dfrac{1 - x}{x}$.

57. $\sqrt{x + 2} = x - 4$.

58. $3\sqrt{x + 4} = x - 6$.

59. $q + 2 = 2\sqrt{4q - 7}$.

60. $x + \sqrt{x} - 2 = 0$.

61. $\sqrt{x + 7} - \sqrt{2x} - 1 = 0$.

62. $\sqrt{x} - \sqrt{2x - 8} - 2 = 0$.

63. $\sqrt{x} - \sqrt{2x + 1} + 1 = 0$.

64. $\sqrt{y - 2} + 2 = \sqrt{2y + 3}$.

65. $\sqrt{x + 5} + 1 = 2\sqrt{x}$.

66. $\sqrt{\sqrt{x} + 2} = \sqrt{2x - 4}$.

67. On page 354 of Samuelson's *Economics*[†] it is stated that one root of the equation

$$\overline{M} = \frac{Q(Q + 10)}{44}$$

is $-5 + \sqrt{25 + 44\overline{M}}$. Verify this by using the quadratic formula to solve for Q in terms of $\overline{M}$. Here Q is real income and $\overline{M}$ is level of money supply.

1-4 SUPPLEMENT

The following is a derivation of the quadratic formula.

Suppose $ax^2 + bx + c = 0$ is a quadratic equation. Since $a \neq 0$, we can divide both sides by a:

$$x^2 + \frac{b}{a}x + \frac{c}{a} = 0,$$

$$x^2 + \frac{b}{a}x = -\frac{c}{a}.$$

If we add $\left(\dfrac{b}{2a}\right)^2$ to both sides,

$$x^2 + \frac{b}{a}x + \left(\frac{b}{2a}\right)^2 = \left(\frac{b}{2a}\right)^2 - \frac{c}{a},$$

then the left side factors into $\left(x + \dfrac{b}{2a}\right)^2$ and the right side simplifies into $\dfrac{b^2 - 4ac}{4a^2}$. Thus,

$$\left(x + \frac{b}{2a}\right)^2 = \frac{b^2 - 4ac}{4a^2}.$$

This equation has the form $u^2 = k$, where $u = x + \dfrac{b}{2a}$ and $k = \dfrac{b^2 - 4ac}{4a^2}$. By Example 4 of Sec. 1-3 we have

$$x + \frac{b}{2a} = \pm\sqrt{\frac{b^2 - 4ac}{4a^2}} = \pm\frac{\sqrt{b^2 - 4ac}}{2a}.$$

Solving for x gives

$$x = -\frac{b}{2a} \pm \frac{\sqrt{b^2 - 4ac}}{2a} = \frac{-b \pm \sqrt{b^2 - 4ac}}{2a}.$$

[†]From Paul A. Samuelson, *Economics*, 10th ed. (New York: McGraw-Hill, Inc., 1976).

It can be shown that the two values $\dfrac{-b + \sqrt{b^2 - 4ac}}{2a}$ and $\dfrac{-b - \sqrt{b^2 - 4ac}}{2a}$ do indeed satisfy $ax^2 + bx + c = 0$. In summary, the roots of the quadratic equation $ax^2 + bx + c = 0$ are given by the **quadratic formula**:

$$x = \frac{-b \pm \sqrt{b^2 - 4ac}}{2a}.$$

1-5 REVIEW

IMPORTANT TERMS AND SYMBOLS IN CHAPTER 1

equation   *(p. 38)*   side of equation   *(p. 38)*

variable   *(p. 39)*   root of equation   *(p. 39)*

solution set   *(p. 39)*   equivalent equations   *(p. 39)*

linear equation   *(p. 41)*   literal equation   *(p. 42)*

empty set   *(p. 45)*   Ø   *(p. 45)*

quadratic equation   *(p. 48)*   quadratic formula   *(p. 51)*

REVIEW SECTION

1. In the equation $x^3 + 7x^2 + 5 = x + 7$, we call $x^3 + 7x^2 + 5$ the ___(a)___ side and $x + 7$ the ___(b)___ side.

   Ans.   (a) left,   (b) right.

2. The number -3 _(is) (is not)_ a root of $-x^2 + 2x = 15$.

   Ans.   is not.

3. The equation $2x + 5 = 7$ is of the ___(a)___ degree and its only root is ___(b)___ .

   Ans.   (a) first,   (b) 1.

4. The equation $x + 2 = 3$ _(is) (is not) (a)_ equivalent to $x + 4 = 5$. The equation $x + 2 = 3$ _(is) (is not) (b)_ equivalent to $2x + 4 = 3$.

   Ans.   (a) is,   (b) is not.

5. The roots of $x^2 - 4 = 0$ are _____ .

   Ans. ± 2.

6. The roots of $x^2(x - 2)(2x + 1)$ are _____ .

　　　　　　　　　　　　　　　　　　　　　　　　Ans. $0, 2, -\frac{1}{2}$.

7. If $I = Prt$, then $t =$ _____ .

　　　　　　　　　　　　　　　　　　　　　　　　Ans. $\dfrac{I}{Pr}$.

8. True or false: Every quadratic equation has two different roots. _____ .

　　　　　　　　　　　　　　　　　　　　　　　　Ans. false.

9. A quadratic equation is of the ___(a)___ degree and can be written in the form ___(b)___ .

　　　　　　　Ans. (a) second,　(b) $ax^2 + bx + c = 0, a \neq 0$.

10. The roots of $x^2 + 50x = 0$ are _____ .

　　　　　　　　　　　　　　　　　　　　　　　　Ans. $0, -50$.

11. The number of roots of a linear equation is _____ .

　　　　　　　　　　　　　　　　　　　　　　　　Ans. one.

REVIEW PROBLEMS

Solve the following equations.

1. $4 - 3x = 2 + 5x$.

2. $\frac{5}{7}x - \frac{2}{3}x = \frac{3}{21}x$.

3. $3[2 - 4(1 + x)] = 5 - 3(3 - x)$.

4. $3(x + 4)^2 + 6x = 3x^2 + 7$.

5. $2 - w = 3 + w$.

6. $x = 2x$.

7. $x = 2x - (7 + x)$.

8. $3x - 8 = 4(x - 2)$.

9. $2(4 - \frac{3}{5}p) = 5$.

10. $\frac{5}{7}x - \frac{2}{3}x = \frac{3}{21}$.

11. $\dfrac{3x - 1}{x + 4} = 0$.

12. $\dfrac{5}{p + 3} - \dfrac{2}{p + 3} = 0$.

13. $\dfrac{2x}{x - 3} - \dfrac{x + 1}{x + 2} = 1$.

14. $\dfrac{t + 3t + 4}{7 - t} = 14$.

15. $3x^2 + 2x - 5 = 0$.

16. $x^2 - 2x - 2 = 0$.

17. $5q^2 = 7q$.

18. $2x^2 - x = 0$.

19. $x^2 - 10x + 25 = 0$.

20. $r^2 + 10r - 25 = 0$.

21. $3x^2 - 5 = 0$.

22. $x(x - 9) = 0$.

23. $(8t - 5)(2t + 6) = 0$.

24. $2(x^2 - 1) + 2x = x^2 - 6x + 1$.

25. $-3x^2 + 5x - 1 = 0$.

26. $y^2 = 6$.

27. $x(x^2 - 9) = 4(x^2 - 9)$.

28. $4x^2(x - 5) - 9(x - 5) = 0$.

29. $\dfrac{6w+7}{2w+1} - \dfrac{6w+1}{2w} = 1.$

30. $\dfrac{3}{x+1} + \dfrac{4}{x} - \dfrac{12}{x+2} = 0.$

31. $\dfrac{2}{x^2-9} - \dfrac{3x}{x+3} = \dfrac{1}{x-3}.$

32. $\dfrac{3}{x^2-4} + \dfrac{2}{x^2+4x+4} - \dfrac{4}{x+2} = 0.$

33. $x + 2 = 2\sqrt{4x-7}.$

34. $\sqrt{3z} - \sqrt{5z+1} + 1 = 0.$

2

APPLICATIONS OF EQUATIONS AND INEQUALITIES

2-1 APPLICATIONS OF EQUATIONS

In most cases, to solve practical problems you must translate the relationships stated in the problems into mathematical symbols. This is called *modeling*. The following examples illustrate basic techniques and concepts. Examine each of them carefully before going to the exercises.

In the first example we shall refer to some business terms relative to a manufacturing firm. **Fixed cost** (or *overhead*) is the sum of all costs that are independent of the level of production, such as rent, insurance, etc. This cost must be paid whether or not output is produced. **Variable cost** is the sum of all costs that are dependent on the level of output, such as labor and material. **Total cost** is the sum of variable cost and fixed cost:

$$\text{Total Cost} = \text{Variable Cost} + \text{Fixed Cost}.$$

Total revenue is the price per unit of output times the number of units sold:

$$\text{Total Revenue} = (\text{Price per unit})(\text{Number of units sold}).$$

Profit is *total* revenue minus total cost:

$$\text{Profit} = \text{Total Revenue} - \text{Total Cost.}$$

EXAMPLE 1

The Anderson Company produces Product A for which the variable cost per unit is $6 and fixed cost is $80,000. Each unit has a selling price of $10. Determine the number of units that must be sold for the company to earn a profit of $60,000.

If the number of units which must be sold is denoted by n, then the variable cost (in dollars) is $6n$. The *total cost* for the business is therefore $6n + 80,000$. The total revenue from the sale of n units is $10n$. Since

$$\text{Profit} = \text{Total Revenue} - \text{Total Cost,}$$

our model for this problem is

$$60,000 = 10n - (6n + 80,000).$$

Solving gives

$$60,000 = 10n - 6n - 80,000,$$
$$140,000 = 4n,$$
$$35,000 = n.$$

Thus 35,000 units must be sold to earn a profit of $60,000.

EXAMPLE 2

A company manufactures women's sportswear and is planning to sell its new line of slacks sets to retail outlets. The cost to the retailer will be $33 per set. As a convenience to the retailer, the manufacturer will attach a price tag to each set. What amount should be marked on the price tag so that the retailer may reduce this price by 20 percent during a sale and still make a profit of 15 percent on the cost?

Here we use the fact that

$$\text{Selling Price} = \text{Cost Per Set} + \text{Profit Per Set.}$$

Let p be the tag price per set in dollars. During the sale the retailer receives $p - .2p$. This must equal his cost, 33, plus his profit, $(.15)(33)$. Hence,

$$\text{Selling Price} = \text{Cost} + \text{Profit.}$$
$$p - .2p = 33 + (.15)(33),$$
$$.8p = 37.95,$$
$$p = 47.4375.$$

From a practical point of view, the company should mark the price tag at $47.44.

EXAMPLE 3

A total of $10,000 was invested in two business ventures, A and B. At the end of the first year, A and B yielded returns of 6 percent and $5\frac{3}{4}$ percent, respectively, on the original investments. How was the original amount allocated if the total amount earned was $588.75?

Let x be the amount (in dollars) invested at 6 percent. Then $10,000 - x$ was invested at $5\frac{3}{4}$ percent. The interest earned was $(.06)(x)$ and $(.0575)(10,000 - x)$, which total 588.75. Hence,

$$(.06)x + (.0575)(10,000 - x) = 588.75,$$

$$.06x + 575 - .0575x = 588.75,$$

$$.0025x = 13.75,$$

$$x = 5500.$$

Thus $5500 was invested at 6 percent, and $10,000 - \$5500 = \$4500$ was invested at $5\frac{3}{4}$ percent.

EXAMPLE 4

The board of directors of a corporation agrees to redeem some of its bonds in two years. At that time $1,102,500 will be required. Suppose they presently set aside $1,000,000. At what annual rate of interest, compounded annually, will this money have to be invested in order that its future value be sufficient to redeem the bonds?

Let r be the required annual rate of interest. At the end of the first year, the accumulated amount will be $1,000,000 plus the interest $1,000,000r$, for a total of

$$1,000,000 + 1,000,000r = 1,000,000(1 + r).$$

Under compound interest, at the end of the second year the accumulated amount will be $1,000,000(1 + r)$ plus the interest on this which is $[1,000,000(1 + r)]r$. Thus the total value at the end of the second year will be $1,000,000(1 + r) + 1,000,000(1 + r)r$. This must equal $1,102,500:

$$1,000,000(1 + r) + 1,000,000(1 + r)r = 1,102,500. \qquad (1)$$

Since $1,000,000(1 + r)$ is a common factor of both terms on the left side, we have

$$1,000,000(1 + r)[1 + r] = 1,102,500,$$

$$1,000,000(1 + r)^2 = 1,102,500,$$

$$(1 + r)^2 = \frac{1,102,500}{1,000,000} = \frac{11,025}{10,000} = \frac{441}{400},$$

$$1 + r = \pm\sqrt{\frac{441}{400}} = \pm\frac{21}{20},$$

$$r = -1 \pm \frac{21}{20}.$$

Thus, $r = -1 + (21/20) = .05$ or $r = -1 - (21/20) = -2.05$. Although .05 and -2.05 are roots of Eq. (1), we reject -2.05 since we want r to be nonnegative. Hence $r = .05$, and the desired rate is 5 percent.

At times there may be more than one way to model a verbal problem, as Example 5 shows.

EXAMPLE 5

A real estate firm owns the Parklane Garden Apartments, which consist of 70 apartments. At $250 per month every apartment can be rented. However, for each $10 per month increase there will be two vacancies with no possibility of filling them. The firm wants to receive $17,980 per month from rents. What rent should be charged for each apartment?

Method I. Suppose r is the rent (in dollars) to be charged per apartment. Then the increase over the $250 level is $r - 250$. Thus the number of $10 increases is $\dfrac{r - 250}{10}$. Since each $10 increase results in two vacancies, the total number of vacancies will be $2\left(\dfrac{r - 250}{10}\right)$. Hence the total number of apartments rented will be $70 - 2\left(\dfrac{r - 250}{10}\right)$. Since

$$\text{Total rent} = (\text{rent per apartment})(\text{number of apartments rented}),$$

we have

$$17,980 = r\left[70 - \frac{2(r - 250)}{10}\right],$$

$$17,980 = r\left[70 - \frac{r - 50}{5}\right],$$

$$17,980 = r\left[\frac{350 - r + 250}{5}\right],$$

$$89,900 = r[600 - r],$$

$$r^2 - 600r + 89,900 = 0.$$

By the quadratic formula,

$$r = \frac{600 \pm \sqrt{(600)^2 - 4(1)(89,900)}}{2(1)}$$

$$= \frac{600 \pm \sqrt{400}}{2} = \frac{600 \pm 20}{2}.$$

Thus the rent for each apartment should be $310 or $290.

Method II. Suppose n is the number of $10 increases. Then the increase in rent per apartment will be $10n$ and there will be $2n$ vacancies. Since

$$\text{Total rent} = (\text{rent per apartment})(\text{number of apartments rented}),$$

we have

$$17,980 = (250 + 10n)(70 - 2n),$$
$$17,980 = 17,500 + 200n - 20n^2,$$
$$20n^2 - 200n + 480 = 0,$$
$$n^2 - 10n + 24 = 0,$$
$$(n - 6)(n - 4) = 0.$$

Thus $n = 6$ or $n = 4$. The rent charged should be $250 + 10(6) = \$310$ or $250 + 10(4) = \$290$.

EXERCISE 2-1

1. The Geometric Products Company produces Product Z at a variable cost per unit of $2.20. If fixed costs are $95,000 and each unit sells for $3, how many units must be sold for the company to have a profit of $50,000?

2. The Clark Company management would like to know the total sales units that are required for the company to earn a profit of $100,000. The following data are available: unit selling price of $20; variable cost per unit of $15; total fixed cost of $600,000. From these data determine the required sales units.

3. A person wishes to invest $20,000 in two enterprises so that the total income per year will be $1440. One enterprise pays 6 percent annually; the other has more risk and pays $7\frac{1}{2}$ percent annually. How much must be invested in each?

4. A person invests $20,000: part at an interest rate of 6 percent annually and the remainder at 7 percent annually. The total interest at the end of one year is equivalent to an annual $6\frac{3}{4}$ percent rate on the entire $20,000. How much was invested at each rate?

5. The cost of a product to a retailer is $3.40. If he wishes to make a profit of 20 percent on the selling price, at what price should the product be sold?

6. In two years a company will require $1,123,600 in order to retire some bonds. If the company now invests $1,000,000 for this purpose, what annual rate of interest, compounded annually, must it receive on this amount in order to retire the bonds?

7. In two years a company will begin an expansion program. It has decided to invest $2,000,000 now so that in two years the total value of the investment will be $2,163,200, the amount required for the expansion. What is the annual rate of interest, compounded annually, that the company must receive to achieve its purpose?

8. A company finds that if it produces and sells q units of a product, its total sales revenue in dollars is $100\sqrt{q}$. If the variable cost per unit is $2 and the fixed cost is $1200, find the values of q for which:

$$\text{Total Sales Revenue} = \text{Variable Cost} + \text{Fixed Cost}$$

(that is, profit is zero).

9. Suppose that consumers will purchase q units of a product when the price is $(80 - q)/4$ dollars *each*. How many units must be sold in order that sales revenue be $400?

10. How long would it take to double an investment at simple interest with a rate of 5 percent per year? (*Hint*: See Example 4(a) of Sec. 1-1 and express 5 percent as .05.)

11. The inventor of a new toy offers the Kiddy Toy Company exclusive rights to manufacture and sell his product for a lump-sum payment of $25,000. After estimating that future sales possibilities beyond one year are nonexistent, the company management is reviewing the following alternate proposal: to give him a lump-sum payment of $2000 plus a royalty of $0.50 for each unit sold. How many units must be sold the first year to make this alternative as economically attractive to the inventor as his original request? (*Hint*: Determine when his incomes under both proposals are the same.)

12. A company parking lot is 120 ft long and 80 ft wide. Due to an increase in personnel, it is decided to double the area of the lot by adding strips of equal width to one end and one side. Find the width of one such strip.

13. You are the chief financial advisor to a corporation which owns an office complex consisting of 50 suites. At $400 per month every suite can be rented. However, for each $20 per month increase there will be two vacancies with no possibility of filling them. The corporation wants to receive a total of $20,240 per month from rents in the complex. You are asked to determine the rent that should be charged for each suite. What is your reply?

14. Six months ago an investment company had a $3,000,000 portfolio consisting of blue chip and glamour stocks. Since then, the value of the blue chip investment increased by $1/10$, while the value of the glamour stocks decreased by $1/10$. The current value of the portfolio is $3,140,000. What is the *current* value of the blue chip investment?

15. The monthly revenue R of a certain company is given by $R = 800p - 7p^2$, where p is the price in dollars of the product they manufacture. At what price will the revenue be $10,000 if the price must be greater than $50?

16. The *price-earning ratio* (P/E ratio) of a company is the ratio of the market value of one share of its outstanding common stock to the earnings per share. If the P/E ratio increases by 10 percent and the earnings per share increase by 20 percent, determine the percentage increase in the market value per share of the common stock.

17. When the price of a product is p dollars each, suppose that a manufacturer will supply $2p - 8$ units of the product to the market and that consumers will demand to buy $300 - 2p$ units. At the value of p for which supply equals demand, the market is said to be in equilibrium. Find this value of p.

18. Repeat Problem 17 for the following conditions: at a price of p dollars each, the supply is $3p^2 - 4p$ and the demand is $24 - p^2$.

19. For security reasons a company will enclose a rectangular area of 11,200 ft² in the rear of its plant. One side will be bounded by the building and the other three sides

by fencing. See Fig. 2-1. If 300 ft of fencing will be used, what will be the dimensions of the rectangular area?

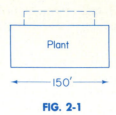

FIG. 2-1

20. A company is designing a package for its product. One part of the package is to be an open box made from a square piece of aluminum by cutting out a 3-in. square from each corner and folding up the sides. See Fig. 2-2. The box is to contain 75 cubic inches. What are the dimensions of the square piece of aluminum that must be used?

FIG. 2-2

21. A candy company makes the popular Dandy Bar. The rectangular-shaped bar is 10 centimeters (cm) long, 5 cm wide, and 2 cm thick. See Fig. 2-3. Because of increasing costs, the company has decided to cut the volume of the bar by a drastic 28 percent. The thickness will be the same, but the length and width will be reduced by equal amounts. What will be the length and width of the new bar?

FIG. 2-3

22. A candy company makes a washer-shaped candy (a candy with a hole in it). See Fig. 2-4. Because of increasing costs, the company will cut the volume of candy in each piece by 20 percent. To do this they will keep the same thickness and outer radius, but will make the inner radius larger. At present the thickness is 2 millimeters (mm), the inner radius is 2 mm, and the outer radius is 7 mm. Find the inner radius of the new-style candy. *Hint*: The volume V of a solid disc is $\pi r^2 h$, where r is the radius and h is the thickness.

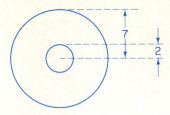

FIG. 2-4

23. A *compensating balance* refers to that practice wherein a bank requires a borrower to maintain on deposit a certain portion of a loan during the term of the loan. For example, if a firm makes a $100,000 loan which requires a compensating balance of 20 percent, it would have to leave $20,000 on deposit and would have the use of $80,000.

To meet the expenses of retooling, the Victor Manufacturing Company finds it must borrow $95,000. The Third National Bank, with whom they have had no prior association, requires a compensating balance of 15 percent. To the nearest thousand dollars, what must be the amount of the loan to obtain the needed funds?

24. A machine company has an incentive plan for its salespeople. For each machine that a salesperson sells, the commission is $20. The commission for *every* machine sold will increase by $0.02 for each machine sold over 600. For example, the commission on each of 602 machines sold is $20.04 How many machines must a salesperson sell in order to earn $15,400?

25. A land investment company purchased a parcel of land for $7200. After having sold all but 20 acres at a profit of $30 per acre, the entire cost of the parcel had been regained. How many acres were sold?

26. The *margin of profit* of a company is the net income divided by the total sales. A company's margin of profit increased by .02 from last year. In that year the company sold its product at $3.00 each and had a net income of $4500. This year it increased the price of its product by $0.50 each, sold 2000 more, and had a net income of $7140. The company never has had a margin of profit greater than .15. How many of its product were sold last year and how many were sold this year?

27. A company manufactures products A and B. The cost of producing each unit of A is $2 more than that of B. The costs of production of A and B are $1500 and $1000, respectively, and 25 more units of A are produced than of B. How many of each are produced?

2-2 LINEAR INEQUALITIES

Suppose a and b are two points on the real number line. Then either a and b coincide, or a lies to the left of b, or a lies to the right of b (see Fig. 2-5).

If a and b coincide, then $a = b$. If a lies to the left of b, we say a is less than b and write $a < b$, where the *inequality symbol* "$<$" is read "is less than."

FIG. 2-5

On the other hand, if a lies to the right of b, we say a is greater than b, written $a > b$. To write $a < b$ is equivalent to writing $b > a$.

Another inequality symbol "$\leqslant$" is read "is less than or equal to" and is defined: $a \leqslant b$ if and only if $a < b$ or $a = b$. Similarly, the symbol "$\geqslant$" is defined: $a \geqslant b$ if and only if $a > b$ or $a = b$. In this case we say "a is greater than or equal to b."

We shall use the words *real numbers* and *points* interchangeably, since there is a one-to-one correspondence between real numbers and points on a line. Thus we can speak of the points -5, -2, 0, 7, and 9, and can write $7 < 9$, $-2 > -5$, $7 \leqslant 7$, and $7 \geqslant 0$ (see Fig. 2-6). Clearly if $a > 0$, then a is positive; if $a < 0$, then a is negative.

FIG. 2-6

Suppose that $a < b$ and x is between a and b (see Fig. 2-7). Then not only is $a < x$, but $x < b$. We indicate this by writing $a < x < b$. For example, $0 < 7 < 9$ (see Fig. 2-6).

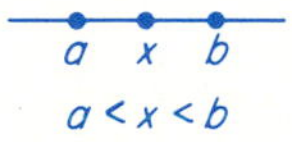

FIG. 2-7

In defining an inequality below we shall use the less than relation ($<$), but the others ($>$, $\geqslant$, $\leqslant$) would also apply.

DEFINITION. *An **inequality** is a statement that one number is less than another number.*

Of course we represent inequalities by means of inequality symbols. If two inequalities have their inequality symbols pointing in the same direction, then the inequalities are said to have the *same sense*. If not, they are said to be *opposite in sense* or one is said to have the *reverse sense* of the other. Hence, $a < b$ and $c < d$ have the same sense, but $a < b$ has the reverse sense of $c > d$.

Solving an inequality, such as $2(x - 3) < 4$, means to find all values of the variable for which the inequality is true. This involves the application of certain rules which we now state.

(1) *If the same number is added to or subtracted from both sides of an inequality, the resulting inequality has the same sense as the original inequality.*

Symbolically, if $a < b$, then $a + c < b + c$ and $a - c < b - c$.

For example, $7 < 10$ and $7 + 3 < 10 + 3$.

(2) *If both sides of an inequality are multiplied or divided by the same **positive** number, the resulting inequality has the same sense as the original inequality.*

Symbolically, if $a < b$ and $c > 0$, then $ac < bc$ and $\dfrac{a}{c} < \dfrac{b}{c}$.

For example, since $3 < 7$ and $2 > 0$, then $3(2) < 7(2)$. Also, $\frac{3}{2} < \frac{7}{2}$.

(3) *If both sides of an inequality are multiplied or divided by the same **negative** number, then the resulting inequality has the **reverse** sense of the original inequality.*

Symbolically, if $a < b$ and $c > 0$, then $a(-c) > b(-c)$ and $\dfrac{a}{-c} > \dfrac{b}{-c}$.

For example, $4 < 7$ but $4(-2) > 7(-2)$. Also $\frac{4}{-2} > \frac{7}{-2}$.

(4) *Any side of an inequality can be replaced by an expression equal to it.*

Symbolically, if $a < b$ and $a = c$, then $c < b$.

For example, if $x < 2$ and $x = y + 4$, then $y + 4 < 2$.

(5) *If the sides of an inequality are either both positive or both negative, then their respective reciprocals[†] are unequal in the **reverse** sense.*

For example, $2 < 4$ but $\frac{1}{2} > \frac{1}{4}$.

(6) *If both sides of an inequality are positive and we raise each side to the same positive power, then the resulting inequality has the same sense as the original inequality.*

Thus if $a > b > 0$ and $n > 0$, then

$$a^n > b^n$$
$$\text{and } \sqrt[n]{a} > \sqrt[n]{b}.$$

For example, $9 > 4$ and so $9^2 > 4^2$ and $\sqrt{9} > \sqrt{4}$.

The result of applying rules 1-4 to an inequality is called an *equivalent inequality*. It is an inequality whose solution is exactly the same as that of the original inequality. We shall apply these rules to solve a *linear inequality*.

[†]The *reciprocal* of a nonzero number a is defined to be $\dfrac{1}{a}$.

DEFINITION.   *A **linear inequality** in the variable x is an inequality which can be written in the form*

$$ax + b < 0 \quad or \quad ax + b \leqslant 0,$$

where a and b are constants and a $\neq$ 0.

In the following examples of solving linear inequalities, the property used is indicated to the right. In each step the given inequality will be replaced by an equivalent one until the solution is evident.

EXAMPLE 1

a. *Solve* $2(x - 3) < 4$.

$$2(x - 3) < 4,$$
$$2x - 6 < 4 \qquad (4),$$
$$2x - 6 + 6 < 4 + 6 \qquad (1),$$
$$2x < 10 \qquad (4),$$
$$\frac{2x}{2} < \frac{10}{2} \qquad (2),$$
$$x < 5.$$

All of the inequalities are equivalent. Thus the original inequality is true for *all* real numbers x such that $x < 5$. We shall write our solution simply as $x < 5$. Geometrically, we may represent this by the bold line segment in Fig. 2-8. The parenthesis indicates that 5 *is not included* in the solution.

$$x < 5$$

FIG. 2-8

b. *Solve* $3 - 2x \leqslant 6$.

$$3 - 2x \leqslant 6,$$
$$-2x \leqslant 3 \qquad (1),$$
$$x \geqslant -\frac{3}{2} \qquad (3).$$

The solution is $x \geqslant -3/2$. This is represented geometrically in Fig. 2-9. The square bracket indicates that $-3/2$ *is included* in the solution.

$$x \geq -\tfrac{3}{2}$$

FIG. 2-9

c. *Solve $\tfrac{3}{2}(s - 2) + 1 > -2(s - 4)$.*

$$\tfrac{3}{2}(s - 2) + 1 > -2(s - 4),$$

$$2\left[\tfrac{3}{2}(s - 2) + 1\right] > 2[-2(s - 4)] \qquad (2),$$

$$3(s - 2) + 2 > -4(s - 4),$$

$$3s - 4 > -4s + 16,$$

$$7s > 20 \qquad (1),$$

$$s > \frac{20}{7} \qquad (2).$$

See Fig. 2-10.

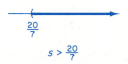

$$s > \tfrac{20}{7}$$

FIG. 2-10

EXAMPLE 2

a. *Solve $2(x - 4) - 3 > 2x - 1$.*

$$2(x - 4) - 3 > 2x - 1,$$

$$2x - 8 - 3 > 2x - 1,$$

$$-11 > -1 \qquad (1).$$

Since $-11 > -1$ is never true, there is no solution and the solution set is $\emptyset$.

b. *Solve $2(x - 4) - 3 < 2x - 1$.*

Proceeding in the same manner as in (a), we obtain $-11 < -1$. This inequality is true for all real numbers x. We write our solution as $-\infty < x < \infty$. See Fig. 2-11. The symbols $-\infty$ and ∞ are not numbers, but are merely a convenience for indicating that the solution is all real numbers.

$$-\infty < x < \infty$$

FIG. 2-11

Frequently we shall use the term *interval* to describe certain sets of real numbers. For example, the set of all numbers x for which $a \leqslant x \leqslant b$ is called a **closed interval** because it includes the *endpoints* a and b. We denote it by $[a, b]$. The set of all x for which $a < x < b$ is called an **open interval** and is denoted by (a, b). The endpoints are *not* part of this set. See Fig. 2-12.

Closed interval $[a,b]$ Open interval (a,b)

FIG. 2-12

Extending these concepts, we have the intervals shown in Fig. 2-13, where the symbols ∞ and $-\infty$ are not numbers but merely a convenience for indicating that an interval extends indefinitely in some direction.

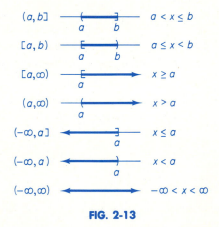

FIG. 2-13

In Problems 1–34, solve the inequalities and indicate your answers geometrically on the real number line.

1. $3x > 12$.

2. $4x < -2$.

3. $4x - 13 \leqslant 7$.

4. $3x \geqslant 0$.

5. $-4x \geqslant 2$.

6. $2y + 1 > 0$.

7. $3 - 5s > 5$.

8. $4s - 1 < -5$.

9. $3 < 2y + 3$.

10. $6 \leqslant 5 - 3y$.

11. $2x - 3 \leqslant 4 + 7x$.

12. $-3 \geqslant 8(2 - x)$.

13. $3(2 - 3x) > 4(1 - 4x)$.

14. $8(x + 1) + 1 < 3(2x) + 1$.

15. $2(3x - 2) > 3(2x - 1)$.

16. $3 - 2(x - 1) \leqslant 2(4 + x)$.

17. $x + 2 < \sqrt{3} - x$.

18. $\sqrt{2}\,(x + 2) > \sqrt{8}\,(3 - x)$.

19. $\frac{5}{3}x < 10$.

20. $-\frac{1}{2}x > 6$.

21. $\dfrac{9y + 1}{4} \leqslant 2y - 1$.

22. $\dfrac{4y - 3}{2} \geqslant \dfrac{1}{3}$.

23. $4x - 1 \geqslant 4(x - 2) + 7$.

24. $0x \leqslant 0$.

25. $\dfrac{1 - t}{2} < \dfrac{3t - 7}{3}$.

26. $\dfrac{3(2t - 2)}{2} > \dfrac{6t - 3}{5} + \dfrac{t}{10}$.

27. $2x + 3 \geqslant \frac{1}{2}x - 4$.

28. $4x - \frac{1}{2} \leqslant \frac{3}{2}x$.

29. $\frac{2}{3}r < \frac{5}{6}r$.

30. $\frac{7}{4}t > -\frac{2}{3}t$.

31. $\dfrac{y}{2} + \dfrac{y}{3} > y + \dfrac{y}{5}$.

32. $9 - .1x \leqslant \dfrac{2 - .01x}{.2}$.

33. $.1(.03x + 4) \geqslant .02x + .434$.

34. $\dfrac{5y - 1}{-3} < \dfrac{7(y + 1)}{-2}$.

35. Each month last year, a company had earnings that were greater than $37,000 but less than $53,000. If S represents the total earnings for the year, describe S by using inequalities.

36. Using inequalities, symbolize the statement: The number of man-hours x to produce a product is not less than $2\frac{1}{2}$ nor more than 4.

2-3 APPLICATIONS OF INEQUALITIES

Solving verbal problems may sometimes involve inequalities, as the following examples illustrate.

EXAMPLE 1

For a manufacturer of thermostats, the combined cost for labor and material is $4 per thermostat. Fixed costs (costs incurred in a given time period, regardless of output) are $60,000. If the selling price of a thermostat is $7, how many must be sold for the company to earn a profit?

Let n be the number of thermostats that must be sold. Then their cost is $4n$. The total cost for the company is therefore $4n + 60,000$. The total revenue from the sale of n

thermostats will be $7n$. Now,

$$\text{Profit} = \text{Total Revenue} - \text{Total Cost},$$

and we want Profit > 0. Thus,

$$\text{Total Revenue} - \text{Total Cost} > 0.$$
$$7n - (4n + 60{,}000) > 0,$$
$$3n > 60{,}000,$$
$$n > 20{,}000.$$

Therefore, at least 20,001 thermostats must be sold for the company to earn a profit.

EXAMPLE 2

Gene Fierro is president of the Fierro Construction Company. He is interested in comparing costs involved in purchasing or renting a piece of machinery needed for excavation. If he were to purchase it, his fixed annual cost would be $4000 and daily operation and maintenance costs would be $80 for each day it is used. In "Digger," a local trade journal, he finds that he can rent the same machinery for $600 per month (on a yearly basis). If the machinery were rented, the daily cost (gas, oil, driver) would be $60 for each day it is used. Neglecting any other considerations, determine the least number of days he would have to use the machinery each year to justify renting the equipment rather than purchasing it.

Let d be the number of days the machinery is used. By renting the machine, the total yearly cost consists of rental fees which are $(12)(600)$ and daily charges of $60d$. By purchasing the machine, the cost per year is $4000 + 80d$. We want

$$\text{Cost}_{\text{Rent}} < \text{Cost}_{\text{Purchase}}.$$
$$12(600) + 60d < 4000 + 80d,$$
$$3200 < 20d,$$
$$160 < d.$$

Thus he must use the machine at least 161 days to justify renting it.

EXAMPLE 3

The *current ratio* of a business is the ratio of its current assets (such as cash, merchandise inventory, and accounts receivable) to its current liabilities (such as short-term loans and taxes payable).

After consulting with the comptroller, the president of the Ace Sports Equipment Company decides to make a short-term loan to build up its inventory. The company has current assets of $350,000 and current liabilities of $80,000. How much can they borrow if they want their current ratio to be no less than 2.5? (Note: The funds they receive are considered as current assets and the loan as a current liability.)

Let x denote the amount which the company can borrow. Then their current assets will be $350,000 + x$, and their current liabilities will be $80,000 + x$. Thus,

$$\text{Current Ratio} = \frac{\text{Current Assets}}{\text{Current Liabilities}} = \frac{350,000 + x}{80,000 + x}.$$

We want

$$\frac{350,000 + x}{80,000 + x} \geqslant 2.5.$$

Since x is positive, so is $80,000 + x$. Thus we can multiply both sides of the inequality by $80,000 + x$ and the sense of the inequality will remain the same.

$$350,000 + x \geqslant 2.5(80,000 + x),$$

$$150,000 \geqslant 1.5x,$$

$$100,000 \geqslant x.$$

Hence they may borrow as much as $100,000 and yet maintain a current ratio of no less than 2.5.

EXAMPLE 4

A publishing company finds that the cost of publishing each copy of a certain magazine is $0.38. The revenue from dealers is $0.35 per copy. The advertising revenue is 10 percent of the revenue received from dealers for all copies sold beyond 10,000. What is the least number of copies which must be sold so as to have a profit for the company?

Let x be the number of copies that are sold. The revenue from dealers is $.35x$ and the revenue from advertising is $(.10)[(.35)(x - 10,000)]$. The total cost of publication is $.38x$. Since Profit = Total Revenue − Total Cost, we want

$$\text{Total Revenue} - \text{Total Cost} > 0.$$

$$.35x + (.10)[(.35)(x - 10,000)] - .38x > 0,$$

$$.35x + .035x - 350 - .38x > 0,$$

$$.005x - 350 > 0,$$

$$.005x > 350,$$

$$x > 70,000.$$

Thus the total number of copies must be greater than 70,000. That is, at least 70,001 copies must be sold to guarantee a profit.

EXERCISE 2-3

1. The Davis Company manufactures a product that has a unit selling price of $20 and a unit cost of $15. If fixed costs are $600,000, determine the least number of units that must be sold for the company to have a profit.

2. To produce one unit of a new product, a company determines that the cost for material is $2.50 and the cost of labor is $4. The constant overhead, regardless of sales volume, is $5000. If the cost to a wholesaler is $7.40 per unit, determine the least number of units that must be sold by the company to realize a profit.

3. For business purposes Mr. Michael Joseph wants to determine the difference between the costs of owning and renting an automobile. He can rent a compact for $135 per month (on an annual basis). Under this plan his cost per mile (gas and oil) is $0.05. If he were to purchase the car, his fixed annual expense would be $1000 and other costs would amount to $0.10 per mile. What is the least number of miles he would have to drive per year to make renting no more expensive than purchasing?

4. A shirt manufacturer produces N shirts at a total labor cost (in dollars) of $1.2N$ and a total material cost of $.3N$. The constant overhead for the plant is $6000. If each shirt sells for $3, how many must be sold by the company to realize a profit?

5. The cost of publication of each copy of a magazine is $0.65. It is sold to dealers for $0.60 each, and the amount received for advertising is 10 percent of the amount received for all magazines issued beyond 10,000. Find the least number of magazines that can be published without loss, that is, such that profit $\geqslant 0$. (Assume that all issues will be sold.)

6. A company produces alarm clocks. During the regular work week the labor cost for producing one clock is $2.00. However, if a clock is produced in overtime the labor cost is $3.00. Management has decided to spend no more than a total of $25,000 per week for labor. The company must produce 11,000 clocks this week. What is the minimum number of clocks that must be produced during the regular work week?

7. A company invests a total of $30,000 of surplus funds at two annual rates of interest: 5 percent and $6\frac{3}{4}$ percent. It wishes an annual yield of no less than $6\frac{1}{2}$ percent. What is the least amount of money that it must invest at the $6\frac{3}{4}$ percent rate?

8. The current ratio of Precision Machine Products is 3.8. If their current assets are $570,000, what are their current liabilities? To raise additional funds, what is the maximum amount they can borrow on a short-term basis if they want their current ratio to be no less than 2.6? (See Example 3 for an explanation of current ratio.)

9. A manufacturer presently has 2500 units of his product in stock. The product is now selling at $4 per unit. Next month the unit price will increase by $0.50. The manufacturer wants the total revenue received from the sale of the 2500 units to be no less than $10,750. What is the maximum number of units that can be sold this month?

10. Suppose that consumers will purchase x units of a product at a price of $\dfrac{100}{x} + 1$ dollars per unit. What is the minimum number of units that must be sold in order that sales revenue be greater than $5000?

2-4 ABSOLUTE VALUE

Sometimes it is useful to consider, on the real number line, the distance between a number x and 0. We call this distance the **absolute value** of x and denote it by $|x|$. For example, $|5| = 5$ and $|-5| = 5$ because both 5 and -5 are five units from 0 (see Fig. 2-14). Similarly, $|0| = 0$.

FIG. 2-14

If x is positive, clearly $|x| = x$. Just as $|-5| = 5 = -(-5)$, it should not be difficult to convince yourself that if x is any negative number, then $|x|$ is the positive number $-x$. The minus sign indicates that we have changed the sign of x. Thus, aside from its geometrical interpretation, absolute value can be defined as follows:

DEFINITION.   *The **absolute value** of a real number x, written $|x|$, is*

$$|x| = \begin{cases} x, \text{ if } x > 0, \\ 0, \text{ if } x = 0, \\ -x, \text{ if } x < 0. \end{cases}$$

Applying the definition, we have $|3| = 3$; $|-8| = -(-8) = 8$; $|\frac{1}{2}| = \frac{1}{2}$; $-|2| = -2$; and $-|-2| = -2$. Notice that $|x|$ is always positive or zero; that is, $|x| \geqslant 0$.

PITFALL. $\sqrt{x^2}$ *is not necessarily x, but $\sqrt{x^2} = |x|$. For example, $\sqrt{(-2)^2} = |-2| = 2$, not -2. This agrees with the fact that $\sqrt{(-2)^2} = \sqrt{4} = 2$. Also, it is usually the case that $|-x| \neq x$ and $|-x - 1| \neq x + 1$. For example, if $x = -3$, then $|-(-3)| \neq -3$ and $|-(-3) - 1| \neq -3 + 1$.*

EXAMPLE 1

a. *Solve $|x - 3| = 2$.*

This equation states that $x - 3$ is a number two units from 0. Thus, either

$$x - 3 = 2 \quad \text{or} \quad x - 3 = -2.$$

Solving these gives $x = 5$ and $x = 1$.

b. *Solve $|7 - 3x| = 5$.*

The equation is true if $7 - 3x = 5$ or if $7 - 3x = -5$. Solving these gives $x = \frac{2}{3}$ and $x = 4$.

c. *Solve* $|x - 4| = -3$.

The absolute value of a number is never negative. Thus the solution set is $\varnothing$.

The numbers 5 and 9 are 4 units apart. Also

$$|9 - 5| = |4| = 4,$$

$$|5 - 9| = |-4| = 4.$$

In general, we may interpret $|a - b|$ or $|b - a|$ as the distance between a and b.

For example, the equation $|x - 3| = 2$ states that the distance between x and 3 is 2 units. Thus x can be 1 or 5, as shown in Example 1(a) and Fig. 2-15.

FIG. 2-15

Let us turn now to inequalities. If $|x| < 3$, then x is less than 3 units from 0. Thus x must lie between -3 and 3. That is, $-3 < x < 3$ [see Fig. 2-16(a)]. On the other hand, if $|x| > 3$, then x must be greater than 3 units from 0. Thus there are two cases: either $x > 3$ or $x < -3$ [see Fig. 2-16(b)]. We can extend these ideas. If $|x| \leqslant 3$, then $-3 \leqslant x \leqslant 3$. If $|x| \geqslant 3$, then $x \geqslant 3$ or $x \leqslant -3$.

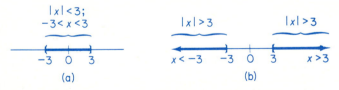

FIG. 2-16

In general, the solution of $|x| < d$ or $|x| \leqslant d$, where d is a positive number, consists of one interval, namely $-d < x < d$ or $-d \leqslant x \leqslant d$. However, when $|x| > d$ or $|x| \geqslant d$ there are two intervals in the solution, namely $x < -d$ and $x > d$, or $x \leqslant -d$ and $x \geqslant d$.

EXAMPLE 2

a. *Solve* $|x - 2| < 4$.

The number $x - 2$ must be less than 4 units from 0. From our discussion above this means that $-4 < x - 2 < 4$. We may set up the procedure for solving this inequality as follows:

$$-4 < x - 2 < 4,$$
$$-4 + 2 < x < 4 + 2 \qquad \text{(adding 2 to each member)},$$
$$-2 < x < 6.$$

Thus the solution is $-2 < x < 6$. This means that all numbers between -2 and 6 satisfy the original inequality. See Fig. 2-17.

FIG. 2-17

b. *Solve* $|3 - 2x| \leqslant 5$.

$$-5 \leqslant 3 - 2x \leqslant 5,$$
$$-5 - 3 \leqslant -2x \leqslant 5 - 3 \qquad \text{(subtracting 3 from each member)},$$
$$-8 \leqslant -2x \leqslant 2,$$
$$4 \geqslant x \geqslant -1 \qquad \text{(dividing each member by } -2),$$
$$-1 \leqslant x \leqslant 4 \qquad \text{(rewriting)}.$$

Note that the sense of the original inequality was reversed when we divided by a negative number.

EXAMPLE 3

a. *Solve* $|x + 5| \geqslant 7$.

The number $x + 5$ must be *at least* 7 units from 0. Thus, either $x + 5 \leqslant -7$ or $x + 5 \geqslant 7$. This means that either $x \leqslant -12$ or $x \geqslant 2$. See Fig. 2-18.

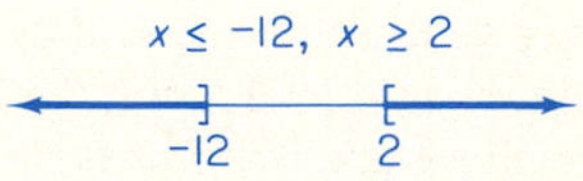

$$x \leq -12, \ x \geq 2$$

FIG. 2-18

b. *Solve* $|3x - 4| > 1$.

Either $3x - 4 < -1$ or $3x - 4 > 1$. Thus, either $3x < 3$ or $3x > 5$. Therefore, $x < 1$ or $x > \frac{5}{3}$.

EXAMPLE 4

Using absolute value notation, express the following statements:

a. *x is less than 3 units from 5.*

$$|x - 5| < 3.$$

b. *x differs from 6 by at least 7.*

$$|x - 6| \geqslant 7.$$

c. *x < 3 and x > -3 simultaneously.*

$$|x| < 3.$$

d. *x is more than 1 unit from -2.*

$$|x - (-2)| > 1,$$
$$|x + 2| > 1.$$

e. *x is strictly within σ (a Greek letter read "sigma") units of μ (a Greek letter read "mu").*

$$|x - \mu| < \sigma.$$

Three basic properties of absolute value are

1. $|ab| = |a| \cdot |b|.$

2. $\left|\dfrac{a}{b}\right| = \dfrac{|a|}{|b|}.$

3. $|a - b| = |b - a|.$

EXAMPLE 5

a. $|(-7) \cdot 3| = |-7| \cdot |3| = 21;$ $|(-7)(-3)| = |-7| \cdot |-3| = 21.$

b. $|4 - 2| = |2 - 4| = 2.$

c. $|7 - x| = |x - 7|.$

d. $\left|\dfrac{-7}{3}\right| = \dfrac{|-7|}{|3|} = \dfrac{7}{3};$ $\left|\dfrac{-7}{-3}\right| = \dfrac{|-7|}{|-3|} = \dfrac{7}{3}.$

e. $\left|\dfrac{x - 3}{-5}\right| = \dfrac{|x - 3|}{|-5|} = \dfrac{|x - 3|}{5}.$

EXERCISE 2-4

In Problems **1–10**, *write an equivalent form without the absolute value symbol.*

1. $|-13|$.

2. $|2^{-1}|$.

3. $|8 - 2|$.

4. $|(-4 - 6)/2|$.

5. $|3(-\frac{5}{3})|$.

6. $|2 - 7| - |7 - 2|$.

7. $|x| < 3$.

8. $|x| < 10$.

9. $|2 - \sqrt{5}|$.

10. $|\sqrt{5} - 2|$.

11. Using the absolute value symbol, express the fact that
  a. x is strictly within 3 units of 7,
  b. x differs from 2 by less than 3,
  c. x is no more than 5 units from 7,
  d. the distance between 7 and x is 4,
  e. $x + 4$ is strictly within 2 units of 0,
  f. x is strictly between -3 and 3,
  g. $x < -6$ or $x > 6$,
  h. $x - 6 > 4$ or $x - 6 < -4$,
  i. the number x of hours that a machine will operate efficiently differs from 105 by less than 3,
  j. the average monthly income x (in dollars) of a family differs from 850 by less than 100.

12. Use absolute value notation to indicate that x and μ differ by no more than σ.

13. Use absolute value notation to indicate that the prices p_1 and p_2 of two products may differ by no more than 2 (dollars).

14. Find all values of x such that $|x - \mu| \leqslant 2\sigma$.

In Problems **15–36**, *solve the given equation or inequality.*

15. $|x| = 7$.

16. $|-x| = 2$.

17. $\left|\dfrac{x}{3}\right| = 2$.

18. $\left|\dfrac{4}{x}\right| = 8$.

19. $|x - 5| = 8$.

20. $|4 + 3x| = 2$.

21. $|5x - 2| = 0$.

22. $|7x + 3| = x$.

23. $|7 - 4x| = 5$.

24. $|1 - 2x| = 1$.

25. $|x| < 4$.

26. $|-x| < 3$.

27. $\left|\dfrac{x}{4}\right| > 2$.

28. $\left|\dfrac{x}{3}\right| > \dfrac{1}{2}$.

29. $|x + 7| < 2.$

30. $|5x - 1| < -6.$

31. $|x - \frac{1}{2}| > \frac{1}{2}.$

32. $|1 - 3x| > 2.$

33. $|5 - 2x| \leqslant 1.$

34. $|4x - 1| \geqslant 0.$

35. $\left|\dfrac{3x - 8}{2}\right| \geqslant 4.$

36. $\left|\dfrac{x - 8}{4}\right| \leqslant 2.$

37. In statistical analysis, the Chebyshev inequality asserts that if x is a random variable, μ its mean, and σ its standard deviation, then

$$(\text{Probability that } |x - \mu| > h\sigma) \leqslant \frac{1}{h^2}.$$

Find those values of x such that $|x - \mu| > h\sigma$.

38. In the manufacture of widgets, the average dimension of a part is .01 cm. Using the absolute value symbol, express the fact that an individual measurement x of a part does not differ from the average by more than .005 cm.

2-5 REVIEW

IMPORTANT TERMS AND SYMBOLS IN CHAPTER 2

fixed cost   *(p. 58)*

overhead   *(p. 58)*

variable cost   *(p. 58)*

total cost   *(p. 58)*

total revenue   *(p. 58)*

profit   *(p. 59)*

$a < b, a \leqslant b$ *(p. 65)*

$a > b, a \geqslant b$ *(p. 66)*

$a < x < b$ *(p. 66)*

inequality   *(p. 66)*

linear inequality   *(p. 68)*

$-\infty < x < \infty$ *(p. 69)*

intervals   *(p. 70)*

absolute value, $|x|$ *(p. 75)*

REVIEW SECTION

1. True or false: If $x > -3$, then $-x > 3.$ __(a)__

If $x > 4$, then $2x > 8.$ __(b)__

If $2 < x < 3$, then both $x > 2$ and $x < 3.$ __(c)__

Ans.   (a) false,   (b) true,   (c) true.

2. The solution of $4x \leqslant 12$ is _____.

Ans. $x \leqslant 3.$

3. The solution of $-5x > 15$ is _____.

Ans. $x < -3.$

4. The solution of $x + 1 < x + 2$ is _____ .

Ans. $-\infty < x < \infty$.

5. The solution set of $x - 2 \geqslant x + 1$ is _____ .

Ans. $\emptyset$.

6. The solution of $2x - 6 < 0$ is _____ .

Ans. $x < 3$.

7. If $x = 0$, then $|x| = $ __(a)__ . Otherwise, $|x|$ is always (positive)(negative)(b).

Ans. (a) 0, (b) positive.

8. The solution of $|x - 2| = 0$ is _____ .

Ans. 2.

9. If $-x > 0$, then $|x| = $ _____ .

Ans. $-x$.

10. In absolute value notation, the fact that $2x$ is strictly within 5 units of 6 would be written _____ .

Ans. $|2x - 6| < 5$.

11. The solution of $|x - 4| \leqslant 2$ is _____ .

Ans. $2 \leqslant x \leqslant 6$.

REVIEW PROBLEMS

In Problems **1–15**, *solve the equation or inequality*.

1. $3x - 8 \geqslant 4(x - 2)$.

2. $2x - (7 + x) \leqslant x$.

3. $-(5x + 2) < -(2x + 4)$.

4. $-2(x + 6) > x + 4$.

5. $3p(1 - p) > 3(2 + p) - 3p^2$.

6. $2(4 - \frac{3}{5}q) < 5$.

7. $\dfrac{x + 1}{3} - \dfrac{1}{2} \leqslant 2$.

8. $\dfrac{x}{2} + \dfrac{x}{3} < \dfrac{x}{4}$.

9. $\dfrac{1}{4}s - 3 \leqslant \dfrac{1}{8}(3 + 2s)$.

10. $\dfrac{1}{3}(t + 2) \geqslant \dfrac{1}{4}t + 4$.

11. $|3 - 2x| = 7$.

12. $\left|\dfrac{5x - 8}{13}\right| = 0$.

13. $|4t - 1| < 1$.

14. $4 < \left|\dfrac{2}{3}x + 5\right|$.

15. $|3 - 2x| \geqslant 4$.

16. A profit of 40 percent on the selling price of a product is equivalent to what percent profit on the cost?

17. On a certain day, there were 1132 different issues traded on the New York Stock Exchange. There were 48 more issues showing an increase than showing a decline and no issues remained the same. How many issues suffered a decline?

18. The sales tax in a certain state is 6 percent. If a total of $3017.29 in purchases, including tax, is made in the course of a year, how much of it is tax?

19. A company will manufacture a total of 10,000 units of its product at plants A and B. Available data is shown in the table below.

	PLANT A	PLANT B
Unit cost for labor and material	$5	$5.50
Fixed costs	$30,000	$35,000

Between the two plants the company has decided to allot no more than $117,000 for total costs. What is the minimum number of units that must be produced at plant A?

20. A company is replacing two cylindrical oil-storage tanks with one new tank. The old tanks are each 16 ft high. One has a radius of 15 ft and the other a radius of 20 ft. The new tank will also be 16 ft high. Find its radius if it is to have the same volume as the old tanks combined. *Hint:* The volume V of a cylindrical tank is $V = \pi r^2 h$, where r is the radius and h is the height.

FUNCTIONS AND GRAPHS

3-1 FUNCTIONS

In 1694 Gottfried Wilhelm Leibniz, one of the developers of calculus, introduced the word *function* into the mathematical vocabulary. The concept of a "function" is no doubt one of the most basic in all of mathematics.

To introduce it let us consider the equation

$$y = x + 2.$$

Replacing x by various numbers, we get corresponding values of y. For example,

$$\text{if } x = 0, \quad \text{then} \quad y = 0 + 2 = 2;$$
$$\text{if } x = 1, \quad \text{then} \quad y = 1 + 2 = 3.$$

Notice that for each value of x that "goes into" the equation, only *one* value of y "comes out."

Think of the equation $y = x + 2$ as defining a rule: add 2 to x. This rule assigns to each *input number x* exactly one *output number y*:

$$x \xrightarrow{\qquad} y \quad (= x + 2).$$

input   output
number   number

We call this rule a *function* in the following sense:

> **DEFINITION.** *A **function** is a rule that assigns to each input number exactly one output number. The set of all input numbers to which the rule applies is called the **domain** of the function. The set of all output numbers is called the **range**.*

In general, the inputs or outputs need not be numbers. For example, a table of U. S. cities and their populations assigns to each city (which is not a number) its population (exactly one output). However, for now we shall use the word "function" in the restricted sense of having a domain and range consisting of only numbers.

Unless otherwise stated, the domain of a function consists of all real numbers for which the function is defined. For the function given by $y = x + 2$, if the input x is any real number, then $x + 2$ is defined, that is, $x + 2$ is a real number. Thus the domain is all real numbers. To the input number 0 is assigned the output number 2:

$$x \to y \, (= x + 2),$$
$$0 \to 2 \, (= 0 + 2).$$

Thus 2 is in the range.

A variable that represents input numbers for a function is called an **independent variable**. One that represents output numbers is a **dependent variable** because its value *depends* on the value of the independent variable. We say that the dependent variable is a *function of* the independent variable. Thus, in the equation $y = x + 2$ the independent variable is x, the dependent variable is y, and y is a function of x.

Not all equations define y as a function of x, as Example 1 shows.

EXAMPLE 1

Let $y^2 = x$.

a. Suppose x is an input number, say $x = 9$. Then $y^2 = 9$ and so $y = \pm 3$. Thus, with the input number 9 there are assigned not one but *two* output numbers, $+3$ and -3. Hence y is *not a function of x*.

b. Now, suppose y is an input number. Then the rule

$$y \to x \text{ where } x = y^2, \text{ as above,}$$

determines exactly one output number, x. For example, if $y = 3$, then $x = y^2 = 3^2 = 9$. Thus x is a function of y. Since y can be any real number, the domain is all real numbers. The independent variable is y and the dependent variable is x.

In some cases the domain of a function is restricted for physical or economic reasons, as Example 2 will show.

EXAMPLE 2

Suppose that the equation $p = 100/q$ describes the relationship between the price per unit, p, of a certain product and the number of units, q, of that product that consumers will buy (that is, demand) per week. This equation is called a *demand equation* for the product. If q is an input number, then to each value of q there is assigned exactly one output number p:

$$q \to \frac{100}{q} = p.$$

For example,

$$20 \to \frac{100}{20} = 5.$$

Thus price p is a function of quantity demanded, q. Here q is the independent variable and p is the dependent variable. Since q cannot be 0 (division by 0 is not defined) and cannot be negative (q represents quantity), the domain is all values of q such that $q > 0$. This function is called a **demand function.**

Usually, letters such as f, g, h, F, G, etc. are used to name functions. Suppose we let f represent the function defined by $y = x + 2$. Then the notation

$$f(x), \text{ which is read "}f \text{ of } x,"$$

is used to represent the *output number* corresponding to the input number x.

input
$$\downarrow$$
$$f(x)$$
output

Thus $f(x)$ is the same as y. But since $y = x + 2$ we may write

$$f(x) = x + 2. \tag{1}$$

To find $f(3)$, the output that corresponds to the input 3, we replace x in Eq. (1) by 3:

$$f(3) = 3 + 2 = 5.$$

Similarly,

$$f(8) = 8 + 2 = 10,$$
$$f(-4) = -4 + 2 = -2,$$
$$f(0) = 0 + 2 = 2.$$

Output numbers, such as $f(3)$, $f(8)$, etc., are called *functional values*. They are in the range of f.

PITFALL. $f(x)$ *does not mean "f times x."*

Quite often, functions are defined by "functional notation." For example, $g(x) = x^3 + x^2$ defines the function g which assigns to an input number x the output number $x^3 + x^2$. This is represented by the following notation:

$$g: x \to x^3 + x^2.$$

Some functional values are

$$g(0) = 0^3 + 0^2 = 0,$$
$$g(2) = 2^3 + 2^2 = 12,$$
$$g(-1) = (-1)^3 + (-1)^2 = -1 + 1 = 0,$$
$$g(t) = t^3 + t^2,$$
$$g(x + 1) = (x + 1)^3 + (x + 1)^2.$$

Note that $g(x + 1)$ was found by replacing each x in $x^3 + x^2$ by the input $x + 1$. That is, g adds the cube and the square of an input number.

When referring to the function g defined by $g(x) = x^3 + x^2$, we shall use some literary freedom and speak of "the function $g(x) = x^3 + x^2$." Similarly, we speak of the function $y = x + 2$.

Special names are given to functions having particular forms, as the next examples indicate.

EXAMPLE 3

a. The function $f(x) = 2x + 3$ is called a **linear function.** It has the form $f(x) = ax + b$, where a and b are constants and $a \neq 0$. Its domain is all real numbers. Some functional values are

$$f(0) = 2(0) + 3 = 3,$$
$$f(t + 7) = 2(t + 7) + 3$$
$$= 2t + 14 + 3 = 2t + 17.$$

b. The equation $g(x) = 2$ defines a *constant function*. For any input the output is 2. The domain of g is all real numbers and the range is 2. For example,

$$g(.10) = 2; \qquad g(-420) = 2; \qquad g(0) = 2.$$

In general, a **constant function** g is one of the form $g(x) = c$, where c is a fixed real number.

c. $y = h(x) = -3x^2 + x - 5$ is a **quadratic function**, that is, one of the form $h(x) = ax^2 + bx + c$, where a, b, and c are constants and $a \neq 0$. The domain of h is all real numbers, and y is a function of x. Examples of functional values are

$$h(2) = -3(2)^2 + 2 - 5 = -15,$$

$$h\left(\frac{1}{t}\right) = -3\left(\frac{1}{t}\right)^2 + \left(\frac{1}{t}\right) - 5 = -\frac{3}{t^2} + \frac{1}{t} - 5,$$

$$h(r^2) = -3(r^2)^2 + r^2 - 5 = -3r^4 + r^2 - 5,$$

$$h(x + t) - h(x) = \left[-3(x + t)^2 + (x + t) - 5\right] - \left[-3x^2 + x - 5\right]$$

$$= -6xt - 3t^2 + t.$$

> **PITFALL.** *Do not be confused by notation. If $f(x) = x^2$, then $f(x + h) = (x + h)^2$. **Do not** write the function and add h:*
>
> $$f(x + h) \neq x^2 + h.$$
>
> *Also, **do not** use the distributive property on $f(x + h)$. It is not a multiplication:*
>
> $$f(x + h) \neq f(x) + f(h).$$

EXAMPLE 4

A function of the form

$$f(x) = c_n x^n + c_{n-1} x^{n-1} + \ldots + c_1 x + c_0,$$

where n is a positive integer and $c_0, c_1, \ldots, c_n$ are constants with $c_n \neq 0$, is called a **polynomial function** (in x). The term in which x has the greatest exponent is $c_n x^n$. The exponent n is called the **degree** of the function, and c_n is the **leading coefficient.** Thus $f(x) = 3x^2 - 8x + 9$ is a polynomial function of degree 2 and has a leading coefficient of 3. Likewise, $g(x) = 3x^3 - 7x^8$ has degree 8 and leading coefficient -7. Constant functions are also considered polynomial functions. A nonzero constant function, such as $f(x) = 5$, is said to have degree 0. The constant function $f(x) = 0$ has no degree assigned to it. The domain of a polynomial function is all real numbers. All the functions in Example 3 are polynomial functions. In particular, a linear function has degree one and a quadratic function has degree two. The functions in Example 5 below are not polynomial functions.

EXAMPLE 5

a. $f(t) = \dfrac{2t}{t^2 - 1}$.

Here the input number is t. The domain of f is all real numbers except ± 1 (to avoid division by 0). Some functional values are

$$f\left(\frac{1}{2}\right) = \frac{2\left(\frac{1}{2}\right)}{\left(\frac{1}{2}\right)^2 - 1} = -\frac{4}{3}; \qquad f(-t) = \frac{2(-t)}{(-t)^2 - 1} = -\frac{2t}{t^2 - 1}.$$

b. $g(x) = \sqrt{x}$.

We want input numbers to give rise only to real output numbers. For $\sqrt{x}$ to be a real number, x cannot be negative. Thus the domain of g is all $x \geqslant 0$. Examples of functional values are

$$g(0) = \sqrt{0} = 0; \qquad g(4) = \sqrt{4} = 2;$$

$$\frac{g(x + h) - g(x)}{h} = \frac{\sqrt{x + h} - \sqrt{x}}{h}.$$

c. $h(q) = \sqrt{q - 3}$.

Here we must have $q - 3 \geqslant 0$. Thus the domain of h is all $q \geqslant 3$. Some functional values are

$$h(3) = \sqrt{3 - 3} = 0; \qquad h(5) = \sqrt{5 - 3} = \sqrt{2};$$

$$h(q^2) = \sqrt{q^2 - 3}.$$

d. $f(x) = x^{2/3}$.

This function can be written $f(x) = (\sqrt[3]{x})^2$. The domain is all real numbers. Some functional values are

$$f(8) = (\sqrt[3]{8})^2 = 2^2 = 4,$$

$$f(-8) = (\sqrt[3]{-8})^2 = (-2)^2 = 4,$$

$$f(2000) = (\sqrt[3]{2000})^2 = (\sqrt[3]{1000}\,\sqrt[3]{2})^2 = (10\sqrt[3]{2})^2$$

$$= (10)^2 \sqrt[3]{2^2} = 100\sqrt[3]{4}.$$

EXAMPLE 6

The function

$$F(s) = \begin{cases} 1, & \text{if } -1 \leqslant s < 1, \\ 0, & \text{if } 1 \leqslant s \leqslant 2, \\ s - 3, & \text{if } 2 < s \leqslant 3 \end{cases}$$

is defined in three parts. Here s represents input numbers and the domain of F is all s such that $-1 \leqslant s \leqslant 3$. The value of an input number determines which part to use.

Find $F(0)$:   Since $-1 \leqslant 0 < 1$, we have $F(0) = 1$.

Find $F(2)$:   Since $1 \leqslant 2 \leqslant 2$, we have $F(2) = 0$.

Find $F(\frac{9}{4})$:   Since $2 < \frac{9}{4} \leqslant 3$, we substitute $s = \frac{9}{4}$ in $s - 3$.

$$F\left(\tfrac{9}{4}\right) = \tfrac{9}{4} - 3 = -\tfrac{3}{4}.$$

The next example refers to absolute value. Recall from Sec. 2-4 that

$$|x| = \begin{cases} x, & \text{if } x > 0, \\ 0, & \text{if } x = 0, \\ -x, & \text{if } x < 0. \end{cases}$$

EXAMPLE 7

$f(x) = |x|$ is called the *absolute value function*. Its domain is all real numbers. Some functional values are

$$f(16) = |16| = 16,$$

$$f\left(-\tfrac{4}{3}\right) = |-\tfrac{4}{3}| = -\left(-\tfrac{4}{3}\right) = \tfrac{4}{3},$$

$$f(0) = |0| = 0,$$

$$f(2x + 3) = |2x + 3|,$$

$$f(x^2 + 1) = |x^2 + 1| = x^2 + 1, \text{ since } x^2 + 1 > 0 \text{ for all } x.$$

For some equations containing two variables, either variable may be considered a function of the other. For example, if $y = 2x$, then y is a function of x and x is the independent variable. Letting f denote this function, we have $y = f(x) = 2x$. However, since $x = y/2$, then x is a function of y, where y is the independent variable. If g denotes this function, then $x = g(y) = y/2$.

We have seen that a function is essentially a *correspondence* whereby to each input number in the domain there is assigned exactly one output number in the range. The correspondence given by $f(x) = x^2$ is shown by the arrows in Fig. 3-1.

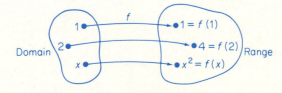

FIG. 3-1

EXAMPLE 8

The table in Fig. 3-2 is a *supply schedule*. It gives a correspondence between the price p of a certain product and the quantity q that producers will supply per week at that price. For each price there corresponds exactly one quantity.

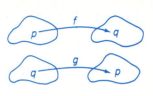

Supply schedule	
p	q
Price per unit in dollars	Quantity supplied per week
500	11
600	14
700	17
800	20

FIG. 3-2

If p is the independent variable, then q is a function of p, say $q = f(p)$, and

$$f(500) = 11, \quad f(600) = 14, \quad f(700) = 17, \quad \text{and} \quad f(800) = 20.$$

Similarly, if q is the independent variable, then p is a function of q, say $p = g(q)$, and

$$g(11) = 500, \quad g(14) = 600, \quad g(17) = 700, \quad \text{and} \quad g(20) = 800.$$

We speak of f and g as **supply functions**. Notice from the supply schedule that as price per unit increases, the producers are willing to supply more units per week.

EXERCISE 3-1

*For each function in Problems **1–26**, determine the independent variable and domain. Also find the indicated functional values.*

1. $f(x) = x - 3$; $f(0)$, $f(3)$, $f(\frac{1}{2})$, $f(q)$.

2. $F(x) = -5x$; $F(0)$, $F(10)$, $F(-\frac{1}{5})$, $F(p_1)$, $F(-x)$.

3. $g(x) = 1 - 2x$; $g(0)$, $g(-u)$, $g(7)$, $g(-2x)$, $g(x + h)$.

4. $h(x) = -\frac{9}{2}$; $h(0)$, $h[(13)^2]$, $h(t)$, $h(x + 1)$.

5. $G(t) = 8.73$; $G(5)$, $G(-107.3)$, $G(x^2)$, $G(2 + x)$.

6. $F(s) = 2(4 - s)$; $F(0)$, $F\left(\frac{1}{s}\right)$, $F(1 + \frac{1}{3})$, $F\left(\frac{s}{2}\right)$.

7. $q = h(p) = \dfrac{3(4p + 1)}{2}$; $h(1)$, $h\left(\frac{p}{2}\right)$, $h\left(\frac{1}{p}\right)$.

8. $g(x) = 3x^2$; $g(-4)$, $g(-3u)$, $g(x^3)$, $g(2/x)$.

9. $f(p) = p^2 + 2p + 1$; $f(0)$, $f(2)$, $f(x_1)$, $f(w)$, $f(p + h)$.

10. $x = H(y) = 2y^2 - 3y + 1$; $H(1)$, $H(-\frac{1}{2})$, $H(z)$, $H(z + 1)$.

11. $y = G(t) = (t + 4)^2$; $G(0)$, $G(2)$, $G(2 + h)$, $\dfrac{G(2 + h) - G(2)}{h}$.

12. $F(x) = |x - 3|$; $F(10)$, $F(3)$, $F(-3)$, $F(4t + 2)$.

13. $f(q) = |2q - 7|$; $f(6)$, $f(2)$, $f(7/2)$, $f(x^2 + 5)$.

14. $s = h(t) = \sqrt{6t}$; $h(0)$, $h(6)$, $h(\frac{2}{3})$, $h(6t)$.

15. $H(x) = \sqrt{4 + x}$; $H(0)$, $H(-4)$, $H(-3)$, $H(x + 1) - H(x)$.

16. $y = F(t) = \dfrac{t}{t - 3}$; $F(0)$, $F(4)$, $F(-1)$, $F(t + 2)$.

17. $g(s) = \dfrac{4}{s^2 - 9}$; $g(1)$, $g(-1)$, $g(-2w)$, $g(s - 1)$, $g(s) - 1$.

18. $h(z) = \left(\dfrac{z + 1}{z - 1}\right)^2$; $h(0)$, $h(1)$, $h(-\frac{1}{2})$, $h(z - 1)$.

19. $f(x) = \begin{cases} 4, & \text{if } x \geq 0 \\ 3, & \text{if } x < 0 \end{cases}$; $f(3)$, $f(-4)$, $f(\frac{17}{3})$, $f(-7.3)$.

20. $H(x) = \begin{cases} 1, & \text{if } x > 1 \\ x + 1, & \text{if } -1 \leq x \leq 1 \\ 1, & \text{if } x < -1 \end{cases}$; $H(7)$, $H(-7)$, $H(.5)$, $H(-\frac{1}{2})$.

21. $h(r) = \begin{cases} 3r - 1, & \text{if } r > 2 \\ r^2 - 4r + 7, & \text{if } r < -2 \end{cases}$; $h(3)$, $h(-3)$, $h(5)$, $h(-5)$.

22. $y = g(x) = \dfrac{1}{x - 2} + \dfrac{1}{x + 3}$; $g(-2)$, $g(3)$, $g(0)$.

23. $f(x) = x^{4/3}$; $f(0)$, $f(64)$, $f(\frac{1}{8})$, $f(-16)$.

24. $g(x) = x^{2/5}$; $g(1)$, $g(32)$, $g(-64)$, $g(t^{10})$.

25. $F(t) = \dfrac{2}{3(2t^2 - 3t - 5)}$; $F(1)$, $F(-2)$, $F(t + 1)$, $F(2t)$.

26. $y = f(x) = \dfrac{1}{\sqrt{x}}$; $f(1)$, $f(\sqrt{16}\,)$.

27. If $z = 4x^2$, can z be considered a function of x? Can x be considered a function of z?

28. If $2p = 3q - 2$, can p be considered a function of q? Can q be considered a function of p?

In Problems 29–32, find $\dfrac{f(x + h) - f(x)}{h}$.

29. $f(x) = 3x - 4$.

30. $f(x) = \dfrac{x}{2}$.

31. $f(x) = x^2 + 2x$.

32. $f(x) = 2x^2 - 3x - 5$.

33. A business with original capital of $10,000 has income and expenses each week of $2000 and $1600, respectively. If all profits are retained in the business, express the value V of the business at the end of t weeks as a function of t.

34. If a $30,000 machine depreciates 2 percent of its original value each year, find a function f which expresses its value V after t years have elapsed.

35. If q units of a certain product are sold (q is nonnegative), the profit P is given by the equation $P = 1.25q$. Is P a function of q? What is the dependent variable; the independent variable?

36. If a principal of P dollars is invested at a simple annual interest rate of r for t years, express the total accumulated amount of the principal and interest as a function of t. Is your result a linear function of t?

37. In manufacturing a component for a machine, the initial cost of a die is $850 and all other additional costs are $3 per unit produced. (a) Express the total cost C (in dollars) as a function of the number q of units produced. (b) How many units are produced if the total cost is $1600?

38. Suppose that when q units of a product can be sold, the price per unit is $f(q)$. What would be the total revenue TR obtained?

39. Table 3-1 is called a *demand schedule*. It gives a correspondence between the price p of a product and the quantity q that consumers will demand (that is, purchase) at that price. (a) If $p = f(q)$, list the numbers in the domain of f. Find $f(2900)$ and $f(3000)$. (b) If $q = g(p)$, list the numbers in the domain of g. Find $g(10)$ and $g(17)$.

TABLE 3-1   Demand Schedule

p PRICE PER UNIT IN DOLLARS	q QUANTITY DEMANDED PER WEEK
10	3000
12	2900
17	2300
20	2000

40. An insurance company examined the records of a group of individuals hospitalized for a particular illness. It was found that the total proportion who had been discharged at the end of t days of hospitalization is given by $f(t)$, where

$$f(t) = 1 - \left(\frac{300}{300 + t} \right)^3.$$

Evaluate (a) $f(0)$, (b) $f(100)$, and (c) $f(300)$. (d) At the end of how many days was .999 of the group discharged?

3-2 COMBINATIONS OF FUNCTIONS

If f and g are functions, we can combine them to create new functions. For example, suppose

$$f(x) = x^2 \quad \text{and} \quad g(x) = x + 1.$$

Adding $f(x)$ and $g(x)$ in the obvious way gives

$$f(x) + g(x) = x^2 + (x + 1).$$

This sum defines a new function—let us call it $f + g$:

$$f + g: x \to f(x) + g(x) = x^2 + (x + 1).$$

Thus,

$$(f + g)(x) = f(x) + g(x) = x^2 + x + 1.^\dagger$$

Similarly, we can create the functions $f - g$, fg, and $\dfrac{f}{g}$, where

$$(f - g)(x) = f(x) - g(x) = x^2 - (x + 1),$$
$$(fg)(x) = f(x)g(x) = x^2(x + 1),$$
$$\frac{f}{g}(x) = \frac{f(x)}{g(x)} = \frac{x^2}{x + 1}, \quad \text{if } g(x) \neq 0.$$

EXAMPLE 1

Let $f(x) = 3x - 1$ and $g(x) = x^2 + 3x + 3$.

a. $(f + g)(x) = f(x) + g(x) = (3x - 1) + (x^2 + 3x + 3)$
$$= x^2 + 6x + 2.$$

Thus, $(f + g)(2) = 2^2 + 6(2) + 2 = 18$.

b. $(f - g)(x) = f(x) - g(x) = (3x - 1) - (x^2 + 3x + 3) = -4 - x^2$.

c. $(fg)(x) = f(x)g(x) = (3x - 1)(x^2 + 3x + 3) = 3x^3 + 8x^2 + 6x - 3$.

d. $\dfrac{f}{g}(x) = \dfrac{f(x)}{g(x)} = \dfrac{3x - 1}{x^2 + 3x + 3}$.

$\dagger$We assume that x is in the domains of both f and g.

EXAMPLE 2

Let $f(x) = \sqrt{x}$ and $g(x) = 6x + 1$. Find $(f + g)(4x)$ and $(f + g)(x + h)$.

a. $(f + g)(4x) = f(4x) + g(4x)$

$$= \sqrt{4x} + 6(4x) + 1 = 2\sqrt{x} + 24x + 1.$$

b. $(f + g)(x + h) = f(x + h) + g(x + h) = \sqrt{x + h} + 6(x + h) + 1.$

It is possible to combine functions in yet another way. In Fig. 3-3 we see that x is in the domain of g. Applying g to x, we get the number $g(x)$, which we shall assume is in the *domain of f*. By applying f to $g(x)$ we get $f[g(x)]$, which is

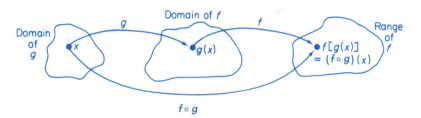

$f \circ g$

FIG. 3-3

in the range of f. This procedure of applying g and then f defines a so-called "composite" function denoted $f \circ g$. This function assigns to the input number x the output number $f[g(x)]$. Thus $(f \circ g)(x) = f[g(x)]$.

DEFINITION.   *If f and g are functions, the* **composition of f with g** *is the function f ∘ g defined by*

$$(f \circ g)(x) = f[g(x)],$$

where the domain of f ∘ g is the set of all x in the domain of g such that g(x) is in the domain of f.

EXAMPLE 3

Let $f(x) = \sqrt{x}$ and $g(x) = x + 1$. *Find the following compositions.*

a. $(f \circ g)(x)$.

$(f \circ g)(x)$ is $f[g(x)]$ and f takes the square root of an input number. But the input number for f is $g(x)$ or $x + 1$. Thus,

$$(f \circ g)(x) = f[g(x)] = f[x + 1] = \sqrt{x + 1}.$$

The domain of g is all real numbers x, and the domain of f is all nonnegative reals.

Hence, the domain of the composition is all x for which $g(x) = x + 1$ is nonnegative. That is, the domain is all $x \geqslant -1$.

b. $(g \circ f)(x)$.

$(g \circ f)(x)$ is $g[f(x)]$ and g adds 1 to an input number, which is $f(x)$ or $\sqrt{x}$. Thus g adds 1 to $\sqrt{x}$.

$$(g \circ f)(x) = g[f(x)] = g[\sqrt{x}] = \sqrt{x} + 1.$$

The domain of f is all $x \geqslant 0$ and the domain of g is all reals. Hence, the domain of the composition is all $x \geqslant 0$ for which $f(x) = \sqrt{x}$ is real, namely all $x \geqslant 0$.

From Example 3 we see that $(f \circ g)(x) \neq (g \circ f)(x)$.

PITFALL. *Do not confuse $f[g(x)]$ with the product $f(x)g(x)$. If $f(x) = \sqrt{x}$ and $g(x) = x + 1$, then*

$$f[g(x)] = \sqrt{x + 1}$$

but $\quad f(x)g(x) = \sqrt{x} \, (x + 1).$

EXAMPLE 4

If $F(p) = p^2 + 4p - 3$ and $G(p) = 2p + 1$, find $F[G(p)]$ and $G[F(p)]$.

a. $F[G(p)] = F[2p + 1] = (2p + 1)^2 + 4(2p + 1) - 3 = 4p^2 + 12p + 2.$

b. $G[F(p)] = 2(p^2 + 4p - 3) + 1 = 2p^2 + 8p - 5.$

EXAMPLE 5

If $f(x) = 5x^2 - 2$ and $g(x) = \sqrt{10 - x^2}$, find $(f \circ g)(-1)$ and $(g \circ f)(2)$

a. $(f \circ g)(-1) = f[g(-1)] = f[\sqrt{10 - (-1)^2}] = f[3] = 5(3)^2 - 2 = 43.$

b. $(g \circ f)(2) = g[f(2)] = g[5(2)^2 - 2] = g[18] = \sqrt{10 - (18)^2}$, which is not a real number. Thus you must be careful with composition; it may not be defined.

EXAMPLE 6

The function $y = (x^2 + 2x + 3)^3$ can be considered a composition. If we let

$$f(x) = x^3 \quad \text{and} \quad g(x) = x^2 + 2x + 3,$$

then

$$f[g(x)] = f[x^2 + 2x + 3] = (x^2 + 2x + 3)^3.$$

Thus $y = f[g(x)] = (f \circ g)(x)$.

EXERCISE 3-2

In each of Problems **1–6**, find

a. $(f + g)(x)$, b. $(f - g)(x)$, c. $(fg)(x)$,

d. $\dfrac{f}{g}(x)$, e. $(f \circ g)(x)$, f. $(g \circ f)(x)$.

1. $f(x) = x + 3, g(x) = x + 6$. **2.** $f(x) = -4, g(x) = 2$.

3. $f(x) = 2x + 5, g(x) = x^2 - 1$. **4.** $f(x) = x^2, g(x) = x^3 + 1$.

5. $f(x) = x^2 + 3x - 4, g(x) = 2x^2 - 7$. **6.** $f(x) = 1/x, g(x) = 4x + 5$.

7. If $f(x) = 9x - 7$ and $g(x) = 6x + 2$, find

a. $(f + g)(3)$, b. $(f + g)(2x)$, c. $(f - g)(-1)$,

d. $(f - g)(-w)$, e. $(fg)(\tfrac{1}{3})$, f. $(fg)(x + 1)$,

g. $(f/g)(0)$, h. $(f/g)(t^2)$.

8. If $f(x) = 4x$ and $g(x) = x^2 + 6x - 1$, find

a. $(f + g)(1)$, b. $(f + g)(2x + 1)$, c. $(f - g)(-3)$,

d. $(f - g)(x/2)$, e. $(fg)(-.5)$, f. $(fg)(1/x)$,

g. $(f/g)(0)$, h. $(f/g)(-z^2)$.

9. If $f(x) = 2x^2 + 3$ and $g(x) = 1 - 3x$, find

a. $(f + g)(-2)$, b. $(f + g)(x + h)$, c. $(f - g)(-2)$,

d. $(f - g)(x + h)$, e. $(fg)(0)$, f. $(fg)(\sqrt{x})$,

g. $(f/g)(1)$, h. $(f/g)(2p/3)$.

10. If $f(x) = 4x$ and $g(x) = x^2 + 6x - 1$, find $f[g(1)]$ and $g[f(1)]$.

11. If $f(x) = 2x^2 + 3$ and $g(x) = 1 - 3x$, find $f[g(2)]$ and $g[f(2)]$.

12. If $f(p) = \dfrac{4}{p}$ and $g(p) = \dfrac{p - 2}{3}$, find $(f \circ g)(p)$ and $(g \circ f)(p)$.

13. If $F(t) = t^2 + 3t + 1$ and $G(t) = \dfrac{2}{t - 1}$, find $(F \circ G)(t)$ and $(G \circ F)(t)$.

14. If $F(s) = \sqrt{s}$ and $G(t) = 3t^2 + 4t + 2$, find $(F \circ G)(t)$ and $(G \circ F)(s)$.

15. If $f(w) = \dfrac{1}{w^2 + 1}$ and $g(v) = \sqrt{v + 2}$, find $(f \circ g)(v)$ and $(g \circ f)(w)$.

16. If $f(x) = x^2 + 3$, find $(f \circ f)(x)$.

17. Let $f(x) = 3x + 8$ and $g(x) = 2$. (a) Find $(f \circ g)(x)$ and give the domain and range of this composition. (b) Find $(g \circ f)(x)$ and give the domain and range of this composition.

In Problems **18–23**, *find functions f and g such that* $h(x) = f[g(x)]$.

18. $h(x) = (x^2 + 2)^2$.

19. $h(x) = \sqrt{x - 2}$.

20. $h(x) = (x^2 - 1)^2 + 2(x^2 - 1)$.

21. $h(x) = (3x^3 - 2x)^3 - (3x^3 - 2x)^2 + 7$.

22. $h(x) = \sqrt[5]{\dfrac{x + 1}{3}}$.

23. $h(x) = \dfrac{x + 1}{(x + 1)^2 + 2}$.

24. Studies have been conducted concerning the statistical relations between a person's status, education, and income.[†] Let S denote a numerical value of status based on annual income I. For a certain population, suppose

$$S = f(I) = .45(I - 1000)^{.53}.$$

Furthermore, suppose a person's income I is a function of the number of years of education E, where

$$I = g(E) = 7202 + .29E^{3.68}.$$

Find $(f \circ g)(E)$. What does this function describe?

25. A manufacturer determines that the total number of units of output per day, q, is a function of the number of employees, m, where $q = f(m) = (40m - m^2)/4$. The total revenue, r, that is received for selling q units is given by the function g, where $r = g(q) = 40q$. Determine $g[f(m)]$. What does this composition function describe?

3-3 GRAPHS IN RECTANGULAR COORDINATES

A **rectangular** (or **Cartesian**) **coordinate system** allows us to specify and locate points in a plane. It also provides a geometric way to represent equations in two variables as well as functions.

In a plane two real number lines, called *coordinate axes*, are constructed perpendicular to each other so that their origins coincide as in Fig. 3-4. Their point of intersection is called the *origin* of the coordinate system. For now we shall call the horizontal line the *x-axis* and the vertical line the *y-axis*. The unit distance on the *x*-axis need not necessarily be the same as on the *y*-axis.

The plane on which the coordinate axes are placed is called a *rectangular coordinate plane* or, more simply, an *x,y-plane*. Every point in the *x,y*-plane can be labeled to indicate its position. To label point P in Fig. 3-5(a) we draw perpendiculars from P to the *x*-axis and *y*-axis. They meet these axes at 4 and 2, respectively. Thus P determines two numbers, 4 and 2. We say that the **rectangular coordinates** of P are given by the **ordered pair** (4, 2). The word

---

[†]R. K. Leik and B. F. Meeker. *Mathematical Sociology*. Englewood Cliffs, N.J.: Prentice-Hall, 1975.

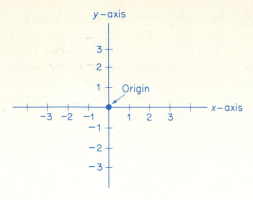

FIG. 3-4

"ordered" is important. In Fig. 3-5(b) the point corresponding to (4, 2) is not the same as that for (2, 4):

$$(4, 2) \neq (2, 4).$$

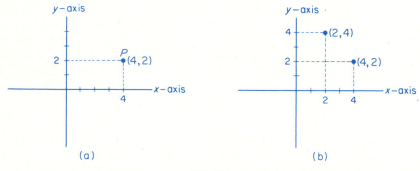

FIG. 3-5

In general, if P is any point, then its rectangular coordinates will be given by an ordered pair of the form (x, y). See Fig. 3-6. We call x the *abscissa* or *x-coordinate* of P, and y the *ordinate* or *y-coordinate* of P.

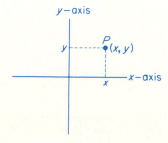

FIG. 3-6

Thus with each point in a given coordinate plane we can associate exactly one ordered pair (x, y) of real numbers. Also it should be clear that with each ordered pair (x, y) of real numbers we can associate exactly one point in that plane. Since there is a *one-to-one correspondence* between the points in the plane and all ordered pairs of real numbers, we shall refer to a point P with abscissa x and ordinate y simply as the point (x, y), or as $P(x, y)$. Moreover, we shall use the words "point" and "ordered pair" interchangeably.

In Fig. 3-7 the coordinates of various points are indicated. For example, the point $(1, -4)$ is located one unit to the right of the y-axis and four units

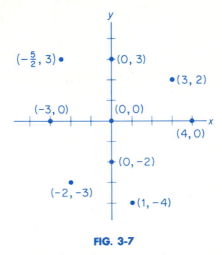

FIG. 3-7

below the x-axis. The origin is $(0, 0)$. The x-coordinate of every point on the y-axis is 0, and the y-coordinate of every point on the x-axis is 0.

The coordinate axes divide the plane into four regions called *quadrants* (Fig. 3-8). For example, quadrant I consists of all points (x_1, y_1) such that $x_1 > 0$ and $y_1 > 0$. The points on the axes do not lie in any quadrant.

Using a rectangular coordinate system, we can geometrically represent equations in two variables. For example, let us consider

$$y = x^2 + 2x - 3.$$

A solution of this equation is a value of x and a value of y that make the equation true.

$$\text{If } x = 1, \text{ then } y = 1^2 + 2(1) - 3 = 0.$$

Thus $x = 1$, $y = 0$ is a solution. Similarly,

$$\text{if } x = -2, \text{ then } y = (-2)^2 + 2(-2) - 3 = -3$$

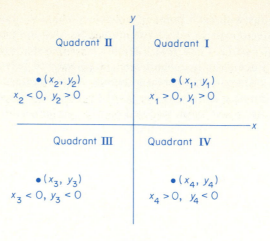

FIG. 3-8

and so $x = -2$, $y = -3$ is also a solution. By choosing other values for x we can get more solutions [see table in Fig. 3-9(a)]. It should be clear that there are infinitely many solutions.

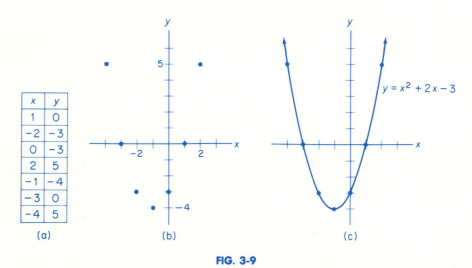

x	y
1	0
-2	-3
0	-3
2	5
-1	-4
-3	0
-4	5

(a)   (b)   (c)

$y = x^2 + 2x - 3$

FIG. 3-9

Each solution gives rise to a point (x, y). For example, to $x = 1$ and $y = 0$ corresponds $(1, 0)$. The **graph** of $y = x^2 + 2x - 3$ is the geometric representation of all its solutions. In Fig. 3-9(b) we have plotted the points corresponding to the solutions in the table.

Since the equation has infinitely many solutions, it seems impossible to determine its graph precisely. However, we are concerned only with the graph's general shape. For this reason we locate only enough points so that we may intelligently guess its general behavior. Then we join these points by a smooth curve wherever conditions permit [see Fig. 3-9(c)]. Of course, the more points we plot, the better is our graph. Here we assume that the graph extends indefinitely upward, which may be indicated by arrows.

In a later chapter you will see that calculus is a *great* aid in graphing because it helps to determine the "shape" of a graph. It provides powerful techniques for determining whether or not a curve "wiggles" between points.

EXAMPLE 1

a. *Graph* $y = 2x + 3$.

See Fig. 3-10.

x	0	$\frac{1}{2}$	$-\frac{1}{2}$	1	-1	2	-2
y	3	4	2	5	1	7	-1

$y = 2x + 3$

FIG. 3-10

b. *Graph* $s = \dfrac{100}{t}$.

Using t for the horizontal axis and s for the vertical, we get Fig. 3-11. In general, the graph of $s = k/t$, where k is a nonzero constant, is called a *hyperbola*.

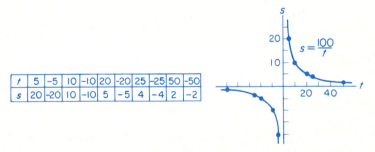

t	5	-5	10	-10	20	-20	25	-25	50	-50
s	20	-20	10	-10	5	-5	4	-4	2	-2

$s = \dfrac{100}{t}$

FIG. 3-11

c. *Graph* $x = 3$.

We can think of this as an equation in the variables x and y if we write it as $x = 3 + 0y$. Here y can be any value, but x must be 3. See Fig. 3-12.

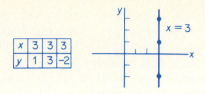

x	3	3	3
y	1	3	-2

FIG. 3-12

We can also represent functions in a coordinate plane. If f is a function and x is the independent variable, then the graph of f is all points $(x, f(x))$, where x is the domain of f.

EXAMPLE 2

a. *Graph* $f(x) = \sqrt{x}$.

The domain of f is all $x \geqslant 0$. See Fig. 3-13. We label the vertical axis as $f(x)$.

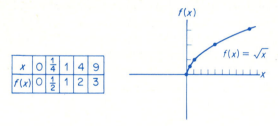

x	0	$\frac{1}{4}$	1	4	9
$f(x)$	0	$\frac{1}{2}$	1	2	3

FIG. 3-13

b. *Graph* $p = G(q) = |q|$ *(absolute value function).*

It is customary to *use the independent variable to label the horizontal axis.* See Fig. 3-14. The vertical axis can be labeled either $G(q)$ or p.

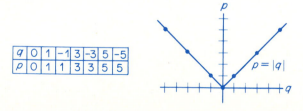

q	0	1	-1	3	-3	5	-5
p	0	1	1	3	3	5	5

FIG. 3-14

Figure 3-15 shows the graph of some function $y = f(x)$. Corresponding to the input number x on the horizontal axis is the output number $f(x)$ on the

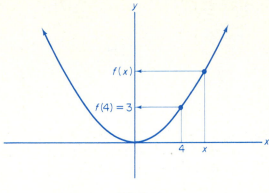

FIG. 3-15

vertical axis. For example, $f(4) = 3$. From the graph, it seems reasonable to assume that there is an output number for any value of x, and so the domain of f is all real numbers. Notice that the y-coordinates of all points on the graph are nonnegative and for any $y \geqslant 0$ there is at least one x such that $y = f(x)$. Thus the range of f is all $y \geqslant 0$. This shows that we may make an "educated" guess about the domain and range of a function by looking at its graph. In general, the domain consists of all x-values which are included in the graph, and the range is all values of y which are included in the graph. For example, from Fig. 3-13 we conclude that both the domain and range of $f(x) = \sqrt{x}$ are all nonnegative numbers. From Fig. 3-14 it is clear that the domain of $p = G(q) = |q|$ is all real numbers and the range is all $p \geqslant 0$.

EXAMPLE 3

Figure 3-16 shows the graph of a function F. To the right of 4 assume that the graph repeats itself indefinitely. Thus the domain of F is all $t \geqslant 0$. The range is $-1 \leqslant s \leqslant 1$. Some functional values are

$$F(0) = 0, \qquad F(1) = 1, \qquad F(2) = 0, \qquad F(3) = -1.$$

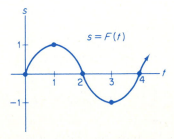

FIG. 3-16

EXAMPLE 4

Sketch the graph of

$$f(x) = \begin{cases} x, & \text{if } 0 \leqslant x < 3, \\ x - 1, & \text{if } 3 \leqslant x \leqslant 5, \\ 4, & \text{if } 5 < x \leqslant 7. \end{cases}$$

The domain of f is $0 \leqslant x \leqslant 7$. The graph is given in Fig. 3-17, where the *hollow dot* means that the point is *not* included in the graph. Observe that the range of f is all real numbers y such that $0 \leqslant y \leqslant 4$.

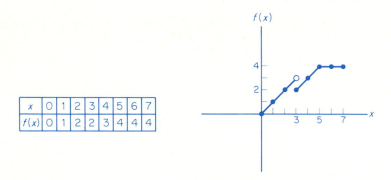

x	0	1	2	3	4	5	6	7
$f(x)$	0	1	2	2	3	4	4	4

FIG. 3-17

The leftmost diagram in Fig. 3-18 shows the graph of some equation in x and y. Notice that with the given x there are associated *two* values of y—namely y_1 and y_2. Thus the equation *does not* define y as a function of x.

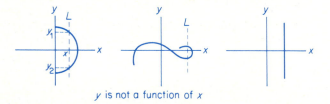

y is not a function of x

FIG. 3-18

In general, if a *vertical* line L can be drawn which meets the graph of an equation in the variables x and y in at least two points, then the equation *does not* define y as a function of x. When no such vertical line can be drawn, the graph is that of a function of x. Thus the graphs in Fig. 3-18 do not represent functions of x, but those in Fig. 3-19 do.

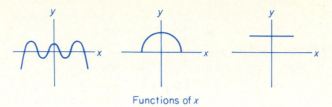

Functions of x

FIG. 3-19

EXAMPLE 5

Graph $x = 2y^2$.

Here it is easier to choose values of y and then find the corresponding values of x. The equation does *not* define a function of x. See Fig. 3-20.

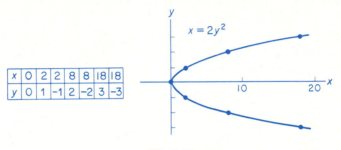

x	0	2	2	8	8	18	18
y	0	1	-1	2	-2	3	-3

FIG. 3-20

EXERCISE 3-3

In Problems **1** *and* **2**, *locate and label each of the points and give the quadrant, if possible, in which each point lies.*

1. $(2, 7), (8, -3), (-\frac{1}{2}, -2), (0, 0)$.   **2.** $(-4, 5), (3, 0), (1, 1), (0, -6)$.

3.   Figure 3-21(a) shows the graph of $y = f(x)$. (a) Estimate $f(0), f(2), f(4)$, and $f(-2)$. (b) What is the domain of f? (c) What is the range of f?

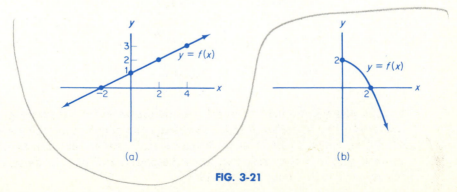

(a)   (b)

FIG. 3-21

4. Figure 3-21(b) shows the graph of $y = f(x)$. (a) Estimate $f(0)$ and $f(2)$. (b) What is the domain of f? (c) What is the range of f?

5. Figure 3-22(a) shows the graph of $y = f(x)$. (a) Estimate $f(0)$, $f(1)$, and $f(-1)$. (b) What is the domain of f? (c) What is the range of f?

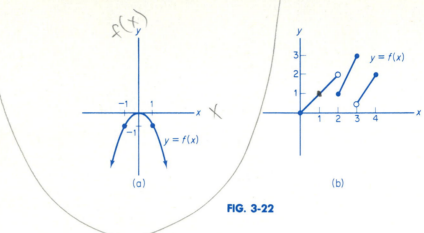

FIG. 3-22

6. Figure 3-22(b) shows the graph of $y = f(x)$. (a) Estimate $f(0)$, $f(2)$, $f(3)$, and $f(4)$. (b) What is the domain of f? (c) What is the range of f?

In Problems 7–20, sketch the graph of each equation. Based on your graph, is y a function of x and if so, what is the domain and range?

7. $y = x$.

8. $y = x + 1$.

9. $y = 3x - 5$.

10. $y = 3 - 2x$.

11. $y = x^2$.

12. $y = \dfrac{3}{x}$.

13. $x = 0$.

14. $y = x^2 - 9$.

15. $y = x^3$.

16. $x = -4$.

17. $x = -3y^2$.

18. $x^2 = y^2$.

19. $2x + y - 2 = 0$.

20. $x + y = 1$.

In Problems 21–37, sketch the graph of each function and give the domain and range.

21. $y = g(x) = 2$.

22. $v = H(u) = |u - 3|$.

23. $f(t) = -t^3$.

24. $G(s) = -8$.

25. $s = F(r) = \sqrt{r - 5}$.

26. $y = f(x) = x^2 + 2x - 8$.

27. $f(x) = |2x - 1|$.

28. $p = h(q) = q(2 - q)$.

29. $y = h(x) = x^2 - 4x + 1$.

30. $F(r) = -\dfrac{1}{r}$.

31. $F(t) = \dfrac{16}{t^2}$.

32. $y = f(x) = \dfrac{x+1}{x}$.

33. $s = f(t) = 4 - t^2$.

34. $c = g(p) = \begin{cases} p, & \text{if } 0 \leqslant p < 2, \\ 2, & \text{if } p \geqslant 2. \end{cases}$

35. $g(x) = \begin{cases} x+6, & \text{if } x \geqslant 3, \\ x^2, & \text{if } x < 3. \end{cases}$

36. $f(x) = \begin{cases} 2x+1, & \text{if } -1 \leqslant x < 2, \\ 9 - x^2, & \text{if } x \geqslant 2. \end{cases}$

37. $f(x) = \begin{cases} x+1, & \text{if } 0 < x \leqslant 3, \\ 4, & \text{if } 3 < x \leqslant 5, \\ x-1, & \text{if } x > 5. \end{cases}$

38. Given the supply schedule (see Example 8 on page 90) in Table 3-2, plot each quantity-price pair by choosing the horizontal axis for the possible quantities.

TABLE 3-2   Supply Schedule

q QUANTITY SUPPLIED PER WEEK	p PRICE PER UNIT IN DOLLARS
30	10
100	20
150	30
190	40
210	50

Approximate the points in between the data by connecting the data points with a smooth curve. Thus you get a *supply curve*. From the graph determine the relationship between price and supply. (That is, as price increases what happens to the quantity supplied?) Is price per unit a function of quantity supplied?

39. Table 3-3 is called a *demand schedule*. It indicates the quantities of Brand X that consumers will demand (that is, purchase) each week at certain prices per unit (in

TABLE 3-3   Demand Schedule

q QUANTITY DEMANDED	p PRICE PER UNIT
5	20
10	10
20	5
25	4

dollars). Plot each quantity-price pair by choosing the vertical axis for the possible prices. Connect the points with a smooth curve. In this way we approximate points in between the given data. The result is called a *demand curve*. From the graph determine the relationship between the price of Brand X and the amount that will be demanded. (That is, as price decreases what happens to the quantity demanded?) Is price per unit a function of quantity demanded?

40. Graph $y = x^2 - 4x - 5$ and then solve the equation $x^2 - 4x - 5 = 0$. What can you say about the solution of $x^2 - 4x - 5 = 0$ and the points where the graph of $y = x^2 - 4x - 5$ intersects the x-axis?

3-4 REVIEW

IMPORTANT TERMS AND SYMBOLS IN CHAPTER 3

function *(p. 84)*   domain *(p. 84)*

range *(p. 84)*   independent variable *(p. 84)*

dependent variable *(p. 84)* $f(x)$ *(p. 85)*

linear function *(p. 86)*   constant function *(p. 87)*

quadratic function *(p. 87)*   polynomial function *(p. 87)*

absolute value, $|x|$ *(p. 89)* $f(x) + g(x)$ *(p. 93)*

$f(x) - g(x)$ *(p. 93)* $f(x)g(x)$ *(p. 93)*

$f(x)/g(x)$ *(p. 93)*   composition of functions *(p. 94)*

$f[g(x)]$ *(p. 94)*   rectangular coordinate system *(p. 97)*

origin *(p. 97)*   ordered pair *(p. 97)*

(x, y) *(p. 98)*   coordinates of a point *(p. 97)*

x-coordinate, y-coordinate *(p. 98)*   abscissa, ordinate *(p. 98)*

quadrant *(p. 99)*   graph of equation *(p. 100)*

graph of function *(p. 102)*

REVIEW SECTION

1. If f is a function, the set of all input numbers is called the ___(a)___ of f. The set of all output numbers is the ___(b)___ of f.

Ans. (a) domain;   (b) range.

2. If $f(x) = -x^2 - 1$, then $f(-1) = $ ___(a)___ . If $g(x) = 3$, then $g(1) = $ ___(b)___ . If $h(x) = |6x - 15|$, then $h(2) = $ ___(c)___ .

Ans. (a) −2;   (b) 3;   (c) 3.

3. If $h(u) = 2u$, then $h(t + 1) = 2($_____$)$.

Ans. $t + 1$.

4. If $f(x) = x$, then the domain of f is _____ .

Ans. all real numbers.

5. True or false:

a.   12 is in the domain of $f(x) = 5x + 3$. _____

    b.   6 is in the domain of $g(x) = \sqrt{25 - x^2}$. _____

    c.   0 is in the domain of $h(z) = \dfrac{z}{z^2 - 9}$. _____

    d. -3 is in the domain of $F(t) = \dfrac{t}{t^2 - 9}$. _____

Ans.   (a) true;   (b) false;   (c) true;   (d) false.

6.   If $g(x) = 5$, then g is called a ___(a)___ function and all functional values are equal to ___(b)___ .

Ans.   (a) constant;   (b) 5.

7.   The domain of the function

$$f(x) = \frac{3(x - 1)(x + 6)}{(x - 4)(x + 2)}$$

consists of all real numbers except _____ and _____ .

Ans.   4 and -2.

8.   A variable representing input numbers of a function is called (a dependent) (an independent) variable.

Ans.   independent.

9.   If $h(x) = f(x) + g(x)$, where $f(x) = x + 2$ and $g(x) = 5x$, then $h(1) = $ _____ .

Ans.   8.

10.   If $f(x) = x^2$ and $g(x) = x + 1$, then $g[f(x)] = $ ___(a)___ and $f[g(x)] = $ ___(b)___ .

Ans.   (a) $x^2 + 1$;   (b) $(x + 1)^2$.

11.   The origin of a rectangular coordinate system has coordinates _____ .

Ans. $(0, 0)$.

12.   The abscissa of the point $(1, 2)$ is ___(a)___ and the ordinate is ___(b)___ .

Ans.   (a) 1;   (b) 2.

13.   A point on the horizontal axis has its (a)(abscissa)(ordinate) equal to 0. A point on the vertical axis has its (b)(abscissa)(ordinate) equal to 0.

Ans.   (a) ordinate;   (b) abscissa.

14.   The coordinate axes divide a rectangular coordinate plane into four regions called _____ .

Ans.   quadrants.

15.   The point $(3, -7)$ lies in Quadrant ___(a)___ while the point $(-3, 7)$ lies in Quadrant ___(b)___ .

Ans.   (a) IV;   (b) II.

16.   Which of the graphs in Fig. 3-23 represent functions of x? _____ .

Ans.   (a), (b), (d).

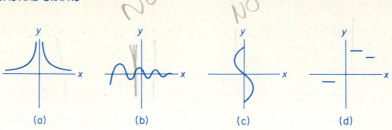

FIG. 3-23

17. The domain of the function whose graph is in Fig. 3-24 is ___(a)___ and its range is ___(b)___ .

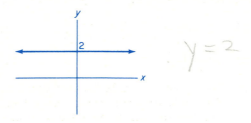

FIG. 3-24

Ans. (a) all real numbers;   (b) 2.

REVIEW PROBLEMS

*In Problems **1–8**, find the given functional expressions and the domain of the function.*

1. $f(x) = 3x^2 - 4x + 7$; $f(0)$, $f(-3)$, $f(5)$, $f(x^2)$.

2. $g(t) = \dfrac{t-3}{t+4}$; $g(3)$, $g(-1)$, $g(2)$, $g(2+h) - g(2)$.

3. $H(u) = 6u^2$; $H(\frac{1}{2})$, $H(-\sqrt{3})$, $H(\sqrt[4]{u})$, $H(u+h) - H(u)$.

4. $f(x) = \sqrt{2x-1}$; $f(\frac{1}{2})$, $f(5)$, $f(x+3)$, $f(s/2)$.

5. $g(z) = 2z^{-2/3}$; $g(1)$, $g(8)$, $g(-\frac{1}{8})$, $g(80)$.

6. $f(t) = t^{1/2} - t^{3/4}$; $f(0)$, $f(1)$, $f(16)$, $f(32)$.

7. $f(x) = \begin{cases} 4, & \text{if } x < 2 \\ 8 - x^2, & \text{if } x > 2 \end{cases}$; $f(4)$, $f(-2)$, $f(0)$, $f(10)$.

8. $h(q) = \begin{cases} q, & \text{if } -1 \leqslant q < 0 \\ 3 - q, & \text{if } 0 \leqslant q < 3 \\ 2q^2, & \text{if } 3 \leqslant q \leqslant 5 \end{cases}$; $h(0)$, $h(4)$, $h(-\frac{1}{2})$, $h(\frac{1}{2})$.

*In Problems **9** and **10**, find (a) $(f + g)(x)$, (b) $(f - g)(x)$, (c) $(fg)(x)$, (d) $(f/g)(x)$, (e) $(f \circ g)(x)$, and (f) $(g \circ f)(x)$.*

9. $f(x) = 4 - 3x;\quad g(x) = 2x - 8.$ 10. $f(x) = x^2 + 7x - 3;\quad g(x) = 2x + 1.$

11. If $f(s) = s^2 + 5s - 3$ and $g(r) = \sqrt{r + 13}$, find $f[g(10)]$ and $g[f(-2)]$.

12. If $F(x) = \dfrac{1}{2x}$ and $G(x) = \dfrac{3x + 1}{2}$, find $(F \circ G)(4x)$ and $(G \circ F)(x^2)$.

*In Problems **13** and **14**, graph the equations. If y is a function of x, determine the domain and range.*

13. $y = 9 - x^2.$ 14. $y = 3x - 7.$

*In Problems **15–18**, graph each function and give its domain and range.*

15. $y = f(x) = \begin{cases} 1 - x, & \text{if } x < 0, \\ 1, & \text{if } x > 0. \end{cases}$ 16. $y = f(x) = |x| + 1.$

17. $g(t) = \dfrac{2}{t - 4}.$ 18. $g(t) = \sqrt{4t}.$

19. The predicted annual sales S (in dollars) of a new product is given by the equation $S = 150{,}000 + 3000t$, where t is the time in years from 1983. Such an equation is called a *trend equation*. Find the predicted annual sales for 1988. Is S a function of t?

20. Graph the "post-office function"

$$c = f(x) = \begin{cases} 20, & \text{if } 0 < x \leqslant 1, \\ 37, & \text{if } 1 < x \leqslant 2, \\ 54, & \text{if } 2 < x \leqslant 3, \\ & \text{etc.,} \end{cases}$$

for $0 < x \leqslant 3$. Here c is the cost (in cents) of mailing a first-class letter of weight x (ounces) in January 1983.

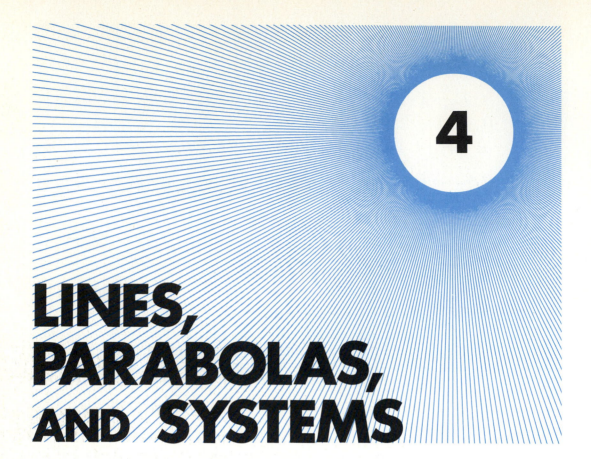

LINES, PARABOLAS, AND SYSTEMS

4-1 LINES

Many relationships in economics can be represented conveniently by straight lines. One feature of a straight line is its "steepness." For example, in Fig. 4-1 line L_1 rises faster as it goes from left to right than line L_2. In this sense it is steeper.

To measure the steepness of a line we introduce the notion of *slope*. Look at points on line L in Fig. 4-2. As the x-coordinate increases from 2 to 4, notice that the y-coordinate increases from 1 to 5. The average rate of change of y with respect to x is the ratio

$$\frac{\text{change in } y}{\text{change in } x} = \frac{5-1}{4-2} = \frac{4}{2} = 2.$$

In fact, for any two distinct points on L this ratio is 2. We say that 2 is the *slope* of L. This means that for each 1-unit increase in x, there is a 2-unit *increase* in y. Thus the line must *rise* from left to right.

112

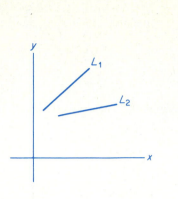

FIG. 4-1   FIG. 4-2

DEFINITION. *Suppose that (x_1, y_1) and (x_2, y_2) are two different points on a nonvertical line. The* ***slope*** *m of the line is the number*

$$m = \frac{y_2 - y_1}{x_2 - x_1} \left(= \frac{\text{vertical change}}{\text{horizontal change}} \right).$$

Slope is not defined for a vertical line since such a line has a horizontal change of 0.

PITFALL. *In the slope formula, always align the subscripts correctly.* ***Do not*** *write $m = \frac{y_2 - y_1}{x_1 - x_2}$.*

EXAMPLE 1

Suppose that the line in Fig. 4-3 shows the relationship between the price p of a widget (in dollars) and the quantity q of widgets (in thousands) that consumers will buy at that price. Find and interpret the slope.

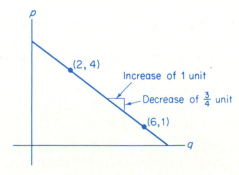

FIG. 4-3

In the slope formula we replace the y's by p's, and the x's by q's. Either point may be chosen as (q_1, p_1). Letting $(2, 4) = (q_1, p_1)$ and $(6, 1) = (q_2, p_2)$, then we have

$$m = \frac{p_2 - p_1}{q_2 - q_1} = \frac{1 - 4}{6 - 2} = \frac{-3}{4} = -\frac{3}{4}.$$

The slope is negative, $-\frac{3}{4}$. This means that for each increase in quantity of 1 (thousand widgets), there corresponds a **decrease** in price of $\frac{3}{4}$ (dollars per widget). Due to this decrease the line **falls** from left to right.

EXAMPLE 2

The slope of the *horizontal* line through $(2, 2)$ and $(3, 2)$ is (see Fig. 4-4)

$$m = \frac{y_2 - y_1}{x_2 - x_1} = \frac{2 - 2}{3 - 2} = \frac{0}{1} = 0.$$

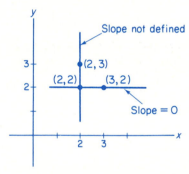

FIG. 4-4

But if we attempted to apply the slope formula to the *vertical* line through $(2, 2)$ and $(2, 3)$ [see Fig. 4-4], the result is

$$m = \frac{3 - 2}{2 - 2} = \frac{1}{0}, \text{ which is not defined.}$$

In fact,

> The slope of every horizontal line is 0. The slope of every vertical line is not defined.

PITFALL.   *Do not confuse a line with zero slope with a line that has no slope.*

In summary,

> Zero slope: horizontal line,
> No slope: vertical line,
> Positive slope: line rises from left to right,
> Negative slope: line falls from left to right.

Lines with different slopes are shown in Fig. 4-5. Notice that *the closer the slope is to* 0, *the more nearly horizontal is the line.* Also, *the greater the absolute value of the slope, the more nearly vertical is the line.* We point out that **two lines are parallel if and only if they are both vertical or both have the same slope.**

FIG. 4-5

Suppose that line L has slope m, that it passes through (x_1, y_1), and that (x, y) is *any* other point on L (see Fig. 4-6). We can find an algebraic

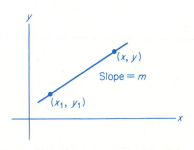

FIG. 4-6

relationship between x and y. By the slope formula,

$$\frac{y - y_1}{x - x_1} = m,$$

$$y - y_1 = m(x - x_1). \tag{1}$$

That is, every point on L satisfies Eq. (1). It is also true that any point satisfying

Eq. (1) must lie on L. Thus the graph of Eq. (1) is L and we say that

$$y - y_1 = m(x - x_1)$$

is the **point-slope form** of an equation of the line through (x_1, y_1) and having slope m.

EXAMPLE 3

Determine an equation of the line that has slope 2 and passes through $(1, -3)$ and sketch the graph.

Here $m = 2$ and $(x_1, y_1) = (1, -3)$. Using a point-slope form, we have

$$y - (-3) = 2(x - 1),$$

$$y + 3 = 2x - 2,$$

$$y = 2x - 5.$$

To sketch the line only two points need be plotted, since two points determine a straight line. See Fig. 4-7.

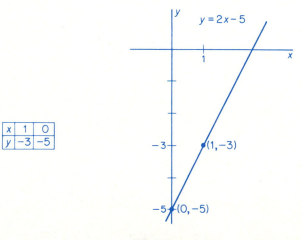

x	1	0
y	-3	-5

FIG. 4-7

An equation of the line passing through two points can be found easily. First determine the line's slope and then use a point-slope form with either point as (x_1, y_1).

EXAMPLE 4

Determine an equation of the line passing through $(-3, 8)$ *and* $(4, -2)$.

$$m = \frac{-2 - 8}{4 - (-3)} = -\frac{10}{7}.$$

Choosing $(-3, 8)$ as (x_1, y_1) gives

$$y - 8 = -\frac{10}{7}[x - (-3)],$$

$$y - 8 = -\frac{10}{7}(x + 3).$$

Choosing $(4, -2)$ as (x_1, y_1) would give an equivalent result.

A point $(0, b)$ where a graph intersects the y-axis is called a **y-intercept** (Fig. 4-8). If the slope and y-intercept of a line L are known, an equation for L

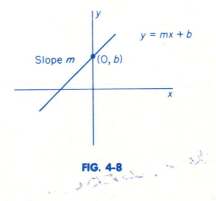

FIG. 4-8

is (using a point-slope form)

$$y - b = m(x - 0).$$

Solving for y, we have $y = mx + b$. We say that

$$y = mx + b$$

is the **slope-intercept form** of an equation of the line with slope m and y-intercept $(0, b)$.

EXAMPLE 5

a. An equation of the line with slope 3 and y-intercept $(0, -4)$ is (see Fig. 4-9)

$$y = mx + b,$$
$$y = 3x + (-4),$$
$$y = 3x - 4.$$

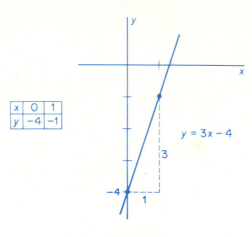

x	0	1
y	-4	-1

$y = 3x - 4$

FIG. 4-9

b. *Find the slope-intercept form of an equation of the line passing through* $(-3, 8)$ *and* $(4, -2)$.

From Example 4, a point-slope form is

$$y - 8 = -\frac{10}{7}(x + 3).$$

Solving for y gives the slope-intercept form:

$$y = -\frac{10}{7}x + \frac{26}{7} \qquad (\text{form: } y = mx + b).$$

From this we note that the y-intercept is $(0, \frac{26}{7})$.

c. The equation $y = \frac{3}{2}x - 6$ has the form $y = mx + b$, where $m = \frac{3}{2}$ and $b = -6$. Thus the graph is a line with slope $\frac{3}{2}$ and y-intercept $(0, -6)$.

In Sec. 3-1 (Example 3a) a *linear function* was described. More formally we have the following definition.

DEFINITION. *A function f is a* **linear function** *if and only if $f(x)$ can be written in the form $f(x) = ax + b$, where a and b are constants and $a \neq 0$.*

Suppose f is a linear function and we let $y = f(x)$. Then $y = ax + b$, which is an equation of a straight line with slope a and y-intercept $(0, b)$. Thus **the graph of a linear function is a straight line**. We say that the function $f(x) = ax + b$ has slope a. For example, the graph of the linear function $f(x) = 3x + 5$ is a straight line with slope 3 and y-intercept $(0, 5)$.

If a *vertical* line L passes through (a, b) [see Fig. 4-10], then any other point (x, y) lies on L if and only if $x = a$. There is no restriction on y. Hence an equation of L is $x = a$. Similarly, an equation of the *horizontal* line passing through (a, b) is $y = b$ (see Fig. 4-11). Here there is no restriction on x.

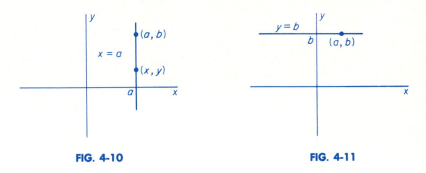

FIG. 4-10   FIG. 4-11

EXAMPLE 6

a. An equation of the vertical line through $(-2, 3)$ is $x = -2$ (see Fig. 4-12). An equation of the horizontal line through $(-2, 3)$ is $y = 3$ (see Fig. 4-13).

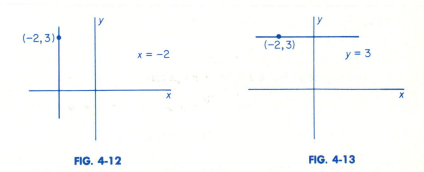

FIG. 4-12   FIG. 4-13

b. Since $(0, 0)$ lies on both the x- and y-axes, an equation of the x-axis is $y = 0$ and an equation of the y-axis is $x = 0$.

On the basis of our discussions we can show that every straight line is the graph of an equation of the form $Ax + By + C = 0$, where A, B, and C are constants and A and B are not both zero. We call this a **general linear equation** (or *an equation of the first degree*) **in the variables x and y,** and x and y are said to be **linearly related.** For example, a general linear equation for the line $y = 7x - 2$ is $(-7)x + (1)y + (2) = 0$. Conversely, the graph of a general linear equation is a straight line. Thus, since $3x + 4y + 5 = 0$ is equivalent to $y = (-\frac{3}{4})x + (-\frac{5}{4})$, its graph is a straight line with slope $-\frac{3}{4}$ and y-intercept $(0, -\frac{5}{4})$.

EXAMPLE 7

Sketch the graph of $2x - 3y + 6 = 0$.

Since this is a general linear equation, its graph is a straight line. Thus we need only determine two different points on the graph in order to sketch it. If $x = 0$, then $y = 2$. If $y = 0$, then $x = -3$. We now draw the line passing through $(0, 2)$ and $(-3, 0)$ (see Fig. 4-14). The point $(-3, 0)$ is called an **x-intercept** of the graph.

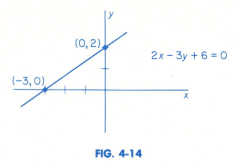

FIG. 4-14

EXAMPLE 8

Suppose a manufacturer has 100 lb of material from which he can produce two products, A and B, which require 4 lb and 2 lb of material per unit, respectively. If x and y denote the number of units produced of A and B, respectively, then all levels of production are given by the combinations of x and y that satisfy the equation

$$4x + 2y = 100 \quad \text{where} \quad x, y \geqslant 0.$$

Thus the levels of production of A and B are linearly related. Solving for y, we get the slope-intercept form

$$y = -2x + 50,$$

and hence the slope is -2. The slope reflects the rate of change of the level of production of B with respect to the level of production of A. For example, if one more unit of A is to be produced, it will require 4 more lb of material resulting in $\frac{4}{2} = 2$ *fewer* units of B. Thus as x increases by one unit, the corresponding value of y decreases two units. To sketch the graph of $y = -2x + 50$, we can use the y-intercept $(0, 50)$ and the fact that when $x = 10$, then $y = 30$ (see Fig. 4-15).

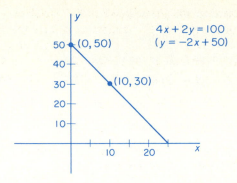

FIG. 4-15

Table 4-1 gives the various forms of equations of straight lines.

TABLE 4-1   Forms of Equations of Straight Lines

Point-slope form	$y - y_1 = m(x - x_1)$.
Slope-intercept form	$y = mx + b$.
General linear form	$Ax + By + C = 0$.
Vertical line	$x = a$.
Horizontal line	$y = b$.

EXERCISE 4-1

*In Problems **1–8**, find the slope of the straight line which passes through the given points.*

1. $(1, 2)$, $(4, 8)$.

2. $(-1, 9)$, $(1, 5)$.

3. $(6, -3)$, $(-7, 5)$.

4. $(2, -4)$, $(3, -4)$.

5. $(-2, 4)$, $(-2, 8)$.

6. $(0, -6)$, $(3, 0)$.

7. $(5, -2)$, $(4, -2)$.

8. $(1, -6)$, $(1, 0)$.

*In Problems **9–22**, determine a general linear equation $(Ax + By + C = 0)$ of the straight line that has the indicated properties and sketch each line.*

9. Passes through $(1, 2)$ and has slope 6.

10. Passes through origin and has slope -5.

11. Passes through $(-2, 5)$ and has slope $-\frac{1}{4}$.

12. Passes through $(\frac{1}{2}, 6)$ and has slope $\frac{1}{3}$.

13. Passes through $(1, 4)$ and $(8, 7)$.

14. Passes through $(7, 1)$ and $(7, -5)$.

15. Passes through $(3, -1)$ and $(-2, -9)$.

16. Passes through $(0, 0)$ and $(2, 3)$.

17. Passes through $(-2, 5)$ and $(3, 5)$.

18. Passes through $(4, 3)$ and $(2, 0)$.

19. Passes through $(2, -8)$ and is vertical.

20. Passes through $(7, 4)$ and is horizontal.

21. Passes through $(-1, 3)$ and is parallel to the line $y = 4x - 5$.

22. Passes through $(2, 1)$ and is parallel to the line $y = 3 + 2x$.

In Problems 23–36, find, if possible, the slope and y-intercept of the straight line determined by the equation and sketch the graph.

23. $y = 2x - 1$.

24. $x - 1 = 5$.

25. $3x - 8y = 8$.

26. $(x - 1) + (y - 2) = 0$.

27. $x + 2y - 3 = 0$.

28. $x + 4 = 7$.

29. $x = -5$.

30. $x - 1 = 5y + 3$.

31. $y = 3x$.

32. $y - 7 = 3(x - 4)$.

33. $y = 1$.

34. $2y - 3 = 0$.

35. $\dfrac{x}{5} - 8y = 4$.

36. $y + 7 = 0$.

In Problems 37–46, determine a general linear form and the slope-intercept form of the given equation.

37. $x = -2y + 4$.

38. $3x + 2y = 6$.

39. $4x + 9y - 5 = 0$.

40. $2(x - 3) - 4(y + 2) = 8$.

41. $\dfrac{3}{4}x = \dfrac{7}{3}y + \dfrac{1}{4}$.

42. $\dfrac{y}{-2} + \dfrac{x}{3} = 1$.

43. $\dfrac{x}{2} - \dfrac{y}{3} = -4$.

44. $y = \dfrac{1}{300}x + 8$.

45. $3x + 4y - 7 = 2x + 3y - 6$.

46. $3x - 4y = 13$.

In Problems 47–50, determine the slope and y-intercept of the given linear function and sketch the graph.

47. $f(x) = x + 1$.

48. $f(x) = x$.

49. $f(x) = -3x + 5$.

50. $f(x) = 2x - 3$.

In Problems 51–54, determine $f(x)$ if f is a linear function that has the given properties.

51. slope $= 5$, $f(3) = 1$.

52. $f(0) = 4$, $f(2) = -6$.

53. $f(2) = 3$, $f(-1) = 12$.

54. slope $= -6$, $f(\frac{1}{2}) = -2$.

55. A straight line passes through $(1, 2)$ and $(-3, 8)$. Find the point on it that has a first coordinate of 5.

56. A straight line has slope -3 and passes through $(4, -1)$. Find the point on it that has a second coordinate of -2.

57. Suppose q and p are related linearly such that $p = 12$ when $q = 40$, and $p = 18$ when $q = 25$. Find an equation that satisfies these conditions. Find p when $q = 30$. *Hint:* The given data can be represented in a q, p-coordinate plane (see Fig. 4-3) by the points $(40, 12)$ and $(25, 18)$.

58. Suppose the cost to produce 10 units of a product is \$40 and the cost of 20 units is \$70. If cost c is linearly related to output q, find a linear equation relating c and q. Find the cost to produce 35 units.

59. In production analysis, an *isocost line* is a line whose points represent all combinations of two factors of production that can be purchased for the same amount. Suppose a farmer has allocated \$20,000 for the purchase of x tons of fertilizer (costing \$200 per ton) and y acres of land (costing \$2000 per acre). Find an equation of the isocost line which describes the various combinations that can be purchased for \$20,000. Observe that neither x nor y can be negative.

60. Suppose the value of a piece of machinery decreases each year by 10 percent of its original value. If the original value is \$8000, find an equation that expresses the value v of the machinery after t years of purchase where $0 \leqslant t \leqslant 10$. Sketch the equation, choosing t as the horizontal axis and v as the vertical axis. What is the slope of the resulting line? This method of considering the value of equipment is called *straight-line depreciation*.

61. A manufacturer produces products X and Y for which the profits per unit are \$4 and \$6, respectively. If x units of X and y units of Y are sold, then the total profit P is given by $P = 4x + 6y$, where $x, y \geqslant 0$. (a) Sketch the graph of this equation for $P = 240$. The result is called an *isoprofit line* and its points represent all combinations of sales that produce a profit of \$240. (b) Determine the slope for $P = 240$. (c) If $P = 600$, determine the slope. (d) What conclusion can you draw concerning isoprofit lines for products X and Y?

4-2 PARABOLAS

In Sec. 3-1 (Example 3c) a quadratic function was described. More formally we have the following definition.

DEFINITION. *A function f is a **quadratic function** if and only if $f(x)$ can be written in the form $f(x) = ax^2 + bx + c$, where $a, b,$ and c are constants and $a \neq 0$.*

For example, $f(x) = x^2 - 3x + 2$ and $F(t) = -3t^2$ are quadratic functions. However, $g(x) = \dfrac{1}{x^2 + 1}$ is *not* a quadratic function, since it cannot be written in the form $g(x) = ax^2 + bx + c$.

The graph of the quadratic function $y = f(x) = ax^2 + bx + c$ is called a **parabola** and has a shape such as the curves in Fig. 4-16. If $a > 0$, the parabola extends upward indefinitely, and we say that the parabola *opens upward* or is *concave up* [Fig. 4-16(a)]. If $a < 0$, the parabola *opens downward* or is *concave down* [Fig. 4-16(b)].

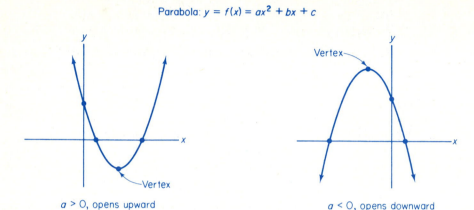

Parabola: $y = f(x) = ax^2 + bx + c$

$a > 0$, opens upward

(a)

$a < 0$, opens downward

(b)

FIG. 4-16

Figure 4-16 shows points labeled **vertex**. If $a > 0$, the vertex is the "lowest" point on the parabola. This means that at this point $f(x)$ has a minimum value. By performing algebraic manipulations on $ax^2 + bx + c$ we can determine not only this minimum value but also where it occurs.

$$f(x) = ax^2 + bx + c = (ax^2 + bx) + c.$$

Adding and subtracting $\dfrac{b^2}{4a}$ gives

$$f(x) = \left(ax^2 + bx + \frac{b^2}{4a} \right) + c - \frac{b^2}{4a}$$

$$= a\left(x^2 + \frac{b}{a}x + \frac{b^2}{4a^2} \right) + c - \frac{b^2}{4a}.$$

$$f(x) = a\left(x + \frac{b}{2a} \right)^2 + c - \frac{b^2}{4a}.$$

Since $\left(x + \dfrac{b}{2a} \right)^2 \geqslant 0$ and $a > 0$, it follows that $f(x)$ has a minimum value when $x + \dfrac{b}{2a} = 0$, that is, when $x = -\dfrac{b}{2a}$. The minimum value is $c - \dfrac{b^2}{4a}$. Thus the vertex is the point $\left(-\dfrac{b}{2a}, c - \dfrac{b^2}{4a}\right)$. Since the y-coordinate of this point is $f\left(-\dfrac{b}{2a}\right)$, we have

$$\text{vertex} = \left(-\frac{b}{2a},\ f\left(-\frac{b}{2a}\right)\right).$$

This is also the vertex of a parabola that opens downward ($a < 0$), but in this case $f\left(-\dfrac{b}{2a}\right)$ is the *maximum* value of $f(x)$ [see Fig. 4-16(b)]. In summary:

> The graph of the quadratic function $y = f(x) = ax^2 + bx + c$ is a parabola.
> 1. If $a > 0$, the parabola opens upward. If $a < 0$, it opens downward.
> 2. The vertex occurs at $\left(-\frac{b}{2a},\ f\left(-\frac{b}{2a}\right)\right)$.

We can quickly sketch the graph of a quadratic function by first locating the vertex and a few other points on the graph. Frequently it is convenient to choose these other points to be those where the parabola intersects the x- and y-axes. These are called x- and y-*intercepts*, respectively. A y-intercept $(0, y)$ is obtained by setting $x = 0$ in $y = ax^2 + bx + c$ and solving for y. The x-intercepts $(x, 0)$ are obtained by setting $y = 0$ and solving for x. Once the intercepts and vertex are found, it is then relatively easy to pass the appropriate parabola through these points. In the event that the x-intercepts are very close to the vertex, or that no x-intercepts exist, we find a point on each side of the vertex so that we can give a reasonable sketch of the parabola.

EXAMPLE 1

Graph the following quadratic functions.

a. $y = f(x) = 12 - 4x - x^2$.

Here $a = -1$, $b = -4$, and $c = 12$. Since $a < 0$ the parabola opens downward. If the vertex is (x, y), then

$$x = -\frac{b}{2a} = -\frac{-4}{2(-1)} = -2,$$

and $y = f(-2) = 12 - 4(-2) - (-2)^2 = 16$. Thus the vertex (highest point) is $(-2, 16)$.

If $x = 0$, then $y = 12 - 4(0) - 0^2 = 12$. Hence the y-intercept is $(0, 12)$. If $y = 0$, then

$$0 = 12 - 4x - x^2,$$
$$0 = (6 + x)(2 - x).$$

Thus $x = -6$ or $x = 2$, and the x-intercepts are $(-6, 0)$ and $(2, 0)$.

Now we plot the vertex and intercepts [see Fig. 4-17(a)]. Through these points we draw a parabola opening downward. See Fig. 4-17(b).

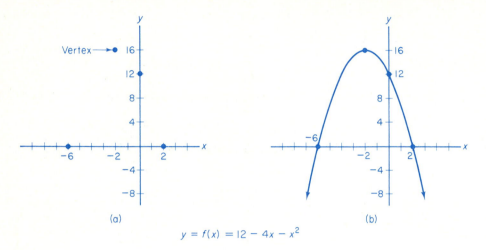

$$y = f(x) = 12 - 4x - x^2$$

FIG. 4-17

b. $p - 2q^2 = 0$.

Since $p = 2q^2 + 0q + 0$, p is a quadratic function of q where $a = 2$, $b = 0$, and $c = 0$. The parabola opens upward, since $a > 0$. If the vertex is (q, p), then

$$q = -\frac{b}{2a} = -\frac{0}{2(2)} = 0,$$

and $p = 2(0)^2 = 0$. Thus the vertex is $(0, 0)$.

A parabola opening upward with vertex at $(0, 0)$ cannot have any other intercepts. Hence, to draw a reasonable graph we plot a point on each side of the vertex and pass a parabola through the three points. See Fig. 4-18(a).

c. $g(x) = x(x - 6) + 7$.

Since $g(x) = x^2 - 6x + 7$, g is a quadratic function where $a = 1$, $b = -6$, and $c = 7$. The parabola opens upward, since $a > 0$. If the vertex is $(x, g(x))$, then

$$x = -\frac{b}{2a} = -\frac{-6}{2(1)} = 3,$$

and $g(3) = 3^2 - 6(3) + 7 = -2$. Thus the vertex is $(3, -2)$.

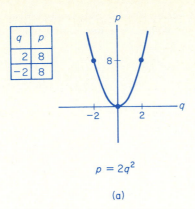

$$p = 2q^2$$

(a)

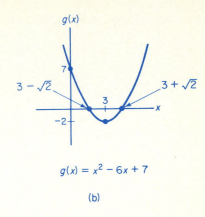

$$g(x) = x^2 - 6x + 7$$

(b)

FIG. 4-18

If $x = 0$, then $g(x) = 7$. Thus the vertical-axis intercept is $(0, 7)$. If $g(x) = 0$, then

$$0 = x^2 - 6x + 7.$$

The right side of this equation does not factor easily. Hence we shall use the quadratic formula to solve for x.

$$x = \frac{-b \pm \sqrt{b^2 - 4ac}}{2a} = \frac{-(-6) \pm \sqrt{(-6)^2 - 4(1)(7)}}{2(1)}$$

$$= \frac{6 \pm \sqrt{8}}{2} = \frac{6 \pm \sqrt{4 \cdot 2}}{2} = \frac{6 \pm 2\sqrt{2}}{2}$$

$$= \frac{6}{2} \pm \frac{2\sqrt{2}}{2} = 3 \pm \sqrt{2}.$$

Thus the x-intercepts are $(3 + \sqrt{2}, 0)$ and $(3 - \sqrt{2}, 0)$.

After plotting the vertex and intercepts, we draw a parabola opening upward. See Fig. 4-18(b).

EXAMPLE 2

Graph $y = f(x) = 2x^2 + 2x + 3$ and find the range of f.

This function is quadratic with $a = 2$, $b = 2$, and $c = 3$. Since $a > 0$, the graph is a parabola opening upward. If the vertex is (x, y), then

$$x = -\frac{b}{2a} = -\frac{2}{2(2)} = -\frac{1}{2},$$

and $y = 2(-\frac{1}{2})^2 + 2(-\frac{1}{2}) + 3 = \frac{5}{2}$. Thus the vertex is $(-\frac{1}{2}, \frac{5}{2})$

If $x = 0$, then $y = 3$, and the y-intercept is $(0, 3)$. A parabola opening upward with its vertex above the x-axis has no x-intercepts.

In Fig. 4-19 we plotted the y-intercept, the vertex, and an additional point to the left of the vertex. Passing a parabola through these points gives the desired graph. From Fig. 4-19 we see that the range of f is all $y \geqslant \frac{5}{2}$.

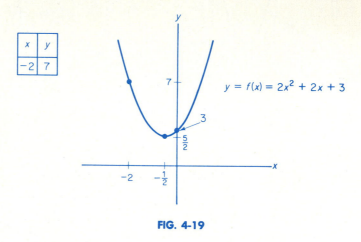

x	y
-2	7

$$y = f(x) = 2x^2 + 2x + 3$$

FIG. 4-19

EXAMPLE 3

Suppose the demand function for a manufacturer's product is $p = f(q) = 1000 - 2q$, where p is the price (in dollars) per unit when q units are demanded (per week) by consumers. Find the level of production that will maximize the manufacturer's total revenue, and determine this revenue.

Total revenue r is given by

$$\text{Total Revenue} = (\text{Price})(\text{Quantity}).$$
$$r = pq$$
$$= (1000 - 2q)q.$$
$$r = 1000q - 2q^2.$$

Note that r is a quadratic function of q, with $a = -2$, $b = 1000$, and $c = 0$. Since $a < 0$ (parabola opens downward), then r is maximum when

$$q = -\frac{b}{2a} = -\frac{1000}{2(-2)} = 250.$$

The maximum value of r is

$$r = 1000(250) - 2(250)^2$$
$$= 250{,}000 - 125{,}000 = 125{,}000.$$

Thus the maximum revenue that the manufacturer can receive is $125,000, which occurs at a production level of 250 units. Figure 4-20 shows the graph of the revenue function. Only that portion for which $q \geqslant 0$ and $r \geqslant 0$ is drawn, since quantity and revenue cannot be negative.

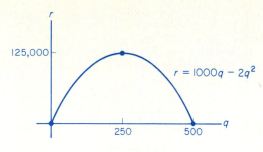

FIG. 4-20

EXERCISE 4-2

In Problems **1–8**, *state whether or not the function is quadratic.*

1. $f(x) = 26 - 3x$.

2. $g(x) = (7 - x)^2$.

3. $g(x) = 4x^2$.

4. $h(s) = 6(4s + 1)$.

5. $h(q) = \dfrac{1}{2q - 4}$.

6. $f(t) = 2t(3 - t) + 4t$.

7. $f(s) = \dfrac{s^2 - 4}{2}$.

8. $g(t) = (t^2 - 1)^2$.

In Problems **9–12**, *do not include a graph.*

9. For the parabola $y = f(x) = -4x^2 + 8x + 7$, (a) find the vertex. (b) Does the vertex correspond to the highest point, or the lowest point, on the graph?

10. Repeat Problem 9 if $y = f(x) = 8x^2 + 4x - 1$.

11. For the parabola $y = f(x) = x^2 + 2x - 8$, find (a) the y-intercept, (b) the x-intercepts, and (c) the vertex.

12. Repeat Problem 11 if $y = f(x) = 3 + x - 2x^2$.

In Problems **13–22**, *graph the functions. Give the vertex and intercepts, and state the range.*

13. $y = f(x) = x^2 - 6x + 5$.

14. $y = f(x) = -3x^2$.

15. $y = g(x) = -2x^2 - 6x$.

16. $y = f(x) = x^2 - 1$.

17. $s = h(t) = (t + 1)^2$.

18. $s = h(t) = 2t^2 + 3t - 2$.

19. $y = f(x) = 2x(4 - x) - 9$.

20. $y = H(x) = 1 - x - x^2$.

21. $t = f(s) = s^2 - 8s + 13$.

22. $t = f(s) = s^2 + 6s + 11$.

23. Determine the minimum value of $f(x) = 100x^2 - 20x + 25$.

24. Determine the maximum value of $g(t) = 4t - (50 + .1t^2)$.

25. The demand function for a manufacturer's product is $p = f(q) = 1200 - 3q$, where p is the price (in dollars) per unit when q units are demanded (per week). Find the level of production that maximizes the manufacturer's total revenue and determine this revenue.

26. A marketing firm estimates that n months after the introduction of a client's new product, $f(n)$ thousand households will use it, where

$$f(n) = \tfrac{10}{9}n(12 - n), \qquad 0 \leqslant n \leqslant 12.$$

Estimate the maximum number of households that will use the product. Sketch the graph of f.

4-3 SYSTEMS OF LINEAR EQUATIONS

In this section we shall discuss methods of solving a "set of equations." The solution will consist of values of the variables for which *all* the equations in the set are satisfied simultaneously.

To introduce this concept, suppose that a company pays its salespeople on a basis of a certain percentage of the first $100,000 in sales, plus a certain percentage of any sales beyond $100,000. If a salesperson earned $17,000 on sales of $300,000 and another earned $12,500 on sales of $225,000, what are the two rates?

Suppose we let x be the rate on the first $100,000 in sales and y be the rate on sales beyond $100,000. The salesperson who had $300,000 in sales will receive $100,000x$ (which is the pay for the first $100,000) and $200,000y$ (for the remaining sales). Thus,

$$100,000x + 200,000y = 17,000.$$

Similarly, for the other salesperson

$$100,000x + 125,000y = 12,500.$$

The problem is to find values of x and y for which *both* linear equations above are true. Let us consider this situation on a more general level. We shall return to our particular problem shortly.

The set of linear equations

$$\begin{cases} a_1x + b_1y = c_1, & \text{(1)} \\ a_2x + b_2y = c_2 & \text{(2)} \end{cases}$$

is called a **system** of two linear equations in the variables (or unknowns) x and y. Its solution consists of values of x and y which satisfy *both* equations *simultaneously*.

Geometrically, Eqs. (1) and (2) represent straight lines, say L_1 and L_2. Since the coordinates of any point on a line satisfy the equation of that line, the coordinates of any point of intersection of L_1 and L_2 will satisfy both equations. Hence a point of intersection will give a solution of the system, and vice versa. If L_1 and L_2 are sketched on the same plane, there are three possibilities as to their relative orientations:

(1) L_1 and L_2 may be parallel and have no points in common (see Fig. 4-21). Thus there is no solution.

(2) L_1 and L_2 may intersect at exactly one point, (x_0, y_0) [see Fig. 4-22]. Thus the system has the solution $x = x_0$ and $y = y_0$.

(3) L_1 and L_2 may coincide (see Fig. 4-23). Thus the coordinates of any point on L_1 are a solution of the system and so there are infinitely many solutions. In this case the given equations must be equivalent.

FIG. 4-21   FIG. 4-22   FIG. 4-23

Our main concern now is algebraic methods of solving a system of linear equations. Essentially we successively replace the system by other systems which have the same solution (that is, by *equivalent systems*), but whose equations have a progressively more desirable form for determining the solution. More precisely, we seek an equivalent system containing an equation in which one of the variables does not appear (that is, has been eliminated). We shall illustrate this procedure.

In the problem originally posed,

$$\begin{cases} 100{,}000x + 200{,}000y = 17{,}000, & (3) \\ 100{,}000x + 125{,}000y = 12{,}500, & (4) \end{cases}$$

the left and right sides of Eq. (4) are equal. Thus each side can be subtracted from the corresponding side of Eq. (3):

$$100{,}000x + 200{,}000y - (100{,}000x + 125{,}000y) = 17{,}000 - 12{,}500,$$
$$75{,}000y = 4500,$$
$$y = .06.$$

Replacing Eq. (3) with the last equation, we get the equivalent system

$$\begin{cases} y = .06, & (5) \\ 100{,}000x + 125{,}000y = 12{,}500. & (6) \end{cases}$$

Note that x does not appear in Eq. (5). Replacing y in Eq. (6) by .06, we get

$$100{,}000x + 125{,}000(.06) = 12{,}500,$$
$$100{,}000x = 5000,$$
$$x = .05.$$

Thus, the original system is equivalent to

$$\begin{cases} y = .06, \\ x = .05 \end{cases}$$

and must have the same solution, namely $x = .05$ and $y = .06$. To check our answer, substitute these values into the *original* equations.

Let us now solve the system

$$\begin{cases} 3x - 4y = 13, & (7) \\ 2x + 3y = 3 & (8) \end{cases}$$

by obtaining an equivalent system in which x does not appear in one equation. First we find an equivalent system in which the coefficients of x differ only in sign. Multiplying Eq. (7) by 2 [that is, multiplying both sides of Eq. (7) by 2] and multiplying Eq. (8) by -3 give

$$\begin{cases} 6x - 8y = 26, & (9) \\ -6x - 9y = -9. & (10) \end{cases}$$

Adding corresponding sides of Eq. (10) to Eq. (9), we can replace Eq. (9) with $-17y = 17$ or, more simply, $y = -1$:

$$\begin{cases} y = -1, & (11) \\ -6x - 9y = -9. & (12) \end{cases}$$

Notice that x does not appear in Eq. (11). Replacing y in Eq. (12) by -1 means that Eq. (12) is equivalent to $-6x - 9(-1) = -9$ or $x = 3$:

$$\begin{cases} y = -1, \\ x = 3. \end{cases}$$

The solution is $x = 3$ and $y = -1$, which should be checked by substituting these values into both Eqs. (7) and (8). Our procedure is referred to as *elimination by addition*. Although we chose to eliminate x first, we could have done the same for y by a similar procedure. Figure 4-24 shows a geometrical representation of the system.

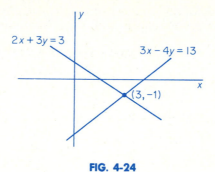

FIG. 4-24

To solve the same system

$$\begin{cases} 3x - 4y = 13, & \text{(13)} \\ 2x + 3y = 3 & \text{(14)} \end{cases}$$

by an alternate approach, we first choose an equation, for example Eq. (13), and solve it for one unknown in terms of the other, say x in terms of y. Hence Eq. (13) is equivalent to

$$x = \frac{4}{3}y + \frac{13}{3}$$

and we obtain

$$\begin{cases} x = \frac{4}{3}y + \frac{13}{3}, & \text{(15)} \\ 2x + 3y = 3. & \text{(16)} \end{cases}$$

By *substitution*, Eq. (16) is equivalent to

$$2\left(\frac{4}{3}y + \frac{13}{3}\right) + 3y = 3$$

which when solved gives

$$y = -1.$$

Replacing y in Eq. (15) by -1 gives $x = 3$, and the original system is equivalent to

$$\begin{cases} x = 3 \\ y = -1. \end{cases}$$

Our procedure in this case is called *elimination by substitution*.

EXAMPLE 1

a. *Solve the system*

$$\begin{cases} x + 2y - 8 = 0, \\ 2x + 4y + 4 = 0. \end{cases}$$

In the equivalent system

$$\begin{cases} x = -2y + 8, & \text{(17)} \\ 2x + 4y + 4 = 0 & \text{(18)} \end{cases}$$

we replace x in Eq. (18) by $-2y + 8$, obtaining

$$2(-2y + 8) + 4y + 4 = 0.$$

This simplifies to $20 = 0$.

$$\begin{cases} x = -2y + 8, & \text{(19)} \\ 20 = 0. & \text{(20)} \end{cases}$$

Since Eq. (20) is never true, there is no solution to the original system. Observe that the original equations can be written in slope-intercept form as

$$\begin{cases} y = -\dfrac{1}{2}x + 4, \\ y = -\dfrac{1}{2}x - 1. \end{cases}$$

These equations represent straight lines having slopes of $-\frac{1}{2}$ but different y-intercepts, $(0, 4)$ and $(0, -1)$. That is they determine different parallel lines (see Fig. 4-25).

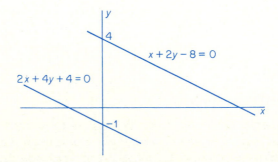

FIG. 4-25

b. *Solve*

$$\begin{cases} x + 5y = 2, & \text{(21)} \\ \dfrac{1}{2}x + \dfrac{5}{2}y = 1. & \text{(22)} \end{cases}$$

Multiplying Eq. (22) by -2, we have

$$\begin{cases} x + 5y = 2, & \text{(23)} \\ -x - 5y = -2. & \text{(24)} \end{cases}$$

Adding Eq. (23) to Eq. (24) gives

$$\begin{cases} x + 5y = 2, & \text{(25)} \\ \quad\;\; 0 = 0. & \text{(26)} \end{cases}$$

Any solution of Eq. (25) is a solution of the system, because Eq. (26) is always true. Looking at it another way, by writing Eqs. (21) and (22) in their slope-intercept forms, we get the equivalent system

$$\begin{cases} y = -\dfrac{1}{5}x + \dfrac{2}{5}, \\ y = -\dfrac{1}{5}x + \dfrac{2}{5} \end{cases}$$

in which both equations represent the same line. Hence the lines coincide (Fig. 4-26), and Eqs. (21) and (22) are equivalent. The coordinates of any point on the line $y = -\frac{1}{5}x + \frac{2}{5}$ are a solution, and so there are infinitely many solutions. For example, $x = 0$ and $y = \frac{2}{5}$ is a solution.

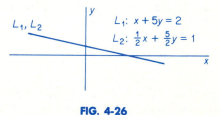

FIG. 4-26

An equation of the form $Ax + By + Cz = D$, where A, B, C, and D are constants and A, B, and C are not all zero, is called a **general linear equation in the variables x, y, and z.** Example 2 shows how to solve a system of three such equations.

EXAMPLE 2

a. *Solve*

$$
\begin{cases}
2x + y + z = 3, & (27) \\
-x + 2y + 2z = 1, & (28) \\
x - y - 3z = -6. & (29)
\end{cases}
$$

This system consists of three linear equations in three variables. From Eq. (29), $x = y + 3z - 6$. Substituting for x in Eqs. (27) and (28) and simplifying, we obtain

$$
\begin{cases}
3y + 7z = 15, & (30) \\
y - z = -5, & (31) \\
x = y + 3z - 6. & (32)
\end{cases}
$$

Note that x does not appear in Eqs. (30) and (31). Since any solution of the original system must satisfy Eqs. (30) and (31), we shall consider their solution first:

$$
\begin{cases}
3y + 7z = 15, & (30) \\
y - z = -5. & (31)
\end{cases}
$$

From Eq. (31), $y = z - 5$. This means we can replace Eq. (30) by $3(z - 5) + 7z = 15$ or $z = 3$. Since z is 3, we can replace Eq. (31) with $y = -2$. Hence the above system is equivalent to

$$
\begin{cases}
z = 3, \\
y = -2.
\end{cases}
$$

The original system becomes

$$
\begin{cases}
z = 3, \\
y = -2, \\
x = y + 3z - 6,
\end{cases}
$$

from which $x = 1$. The solution is $x = 1$, $y = -2$, and $z = 3$, which you may verify.

b. *Solve*

$$
\begin{cases}
2x + y + z = -2, & (33) \\
x - 2y = \dfrac{13}{2}, & (34) \\
3x + 2y - 2z = -\dfrac{9}{2}. & (35)
\end{cases}
$$

Since Eq. (34) can be written $x - 2y + 0z = \frac{13}{2}$, we can view Eqs. (33) to (35) as a system of three linear equations in the variables x, y, and z. From Eq. (34), $x = 2y + \frac{13}{2}$. By substituting for x in Eqs. (33) and (35) and simplifying, we obtain

$$
\begin{cases}
5y + z = -15, & (36) \\
x = 2y + \dfrac{13}{2}, & (37) \\
4y - z = -12. & (38)
\end{cases}
$$

Solving the system formed by Eqs. (36) and (38),

$$\begin{cases} 5y + z = -15, \\ 4y - z = -12, \end{cases}$$

we find that $y = -3$ and $z = 0$. Substituting these values in Eq. (37) gives $x = \frac{1}{2}$. Hence the solution of the original system is $x = \frac{1}{2}, y = -3$, and $z = 0$.

EXAMPLE 3

A chemical manufacturer wishes to fill a request for 500 gallons of a 25 percent acid solution (25 percent by volume is acid). If solutions of 30 and 18 percent are available in stock, how many gallons of each must be mixed to fill the order?

Let x and y, respectively, be the number of gallons of the 30 and 18 percent solutions which should be mixed. Then

$$x + y = 500. \tag{39}$$

In 500 gallons of a 25 percent solution, there will be .25(500) = 125 gallons of acid. This acid comes from two sources: .30x gallons of acid come from the 30 percent solution and .18y gallons of acid come from the 18 percent solution. Hence

$$.30x + .18y = 125. \tag{40}$$

Equations (39) and (40) form a system of two linear equations in two unknowns. Solving Eq. (39) for x and substituting in Eq. (40) gives

$$.30(500 - y) + .18y = 125. \tag{41}$$

Solving Eq. (41) for y, we find $y = 208\frac{1}{3}$ gallons and thus $x = 500 - 208\frac{1}{3} = 291\frac{2}{3}$ gallons.

EXERCISE 4-3

In Problems **1–16**, *solve the systems algebraically.*

1. $\begin{cases} 3x + y = 7, \\ 2x + 2y = -2. \end{cases}$

2. $\begin{cases} 2x - y = -11, \\ y + 5x = -7. \end{cases}$

3. $\begin{cases} 3x - 4y = 13, \\ 2x + 3y = 3. \end{cases}$

4. $\begin{cases} 2x - y = 1, \\ -x + 2y = 7. \end{cases}$

5. $\begin{cases} 5v + 2w = 36, \\ 8v - 3w = -54. \end{cases}$

6. $\begin{cases} p + q = 3, \\ 3p + 2q = 19. \end{cases}$

7. $\begin{cases} 4x - 3y - 2 = 3x - 7y, \\ x + 5y - 2 = y + 4. \end{cases}$

8. $\begin{cases} 5x + 7y + 2 = 9y - 4x + 6, \\ \frac{21}{2}x - \frac{4}{3}y - \frac{11}{4} = \frac{3}{2}x + \frac{2}{3}y + \frac{5}{4}. \end{cases}$

9. $\begin{cases} \frac{2}{3}x + \frac{1}{2}y = 2, \\ \frac{3}{8}x + \frac{5}{6}y = -\frac{11}{2}. \end{cases}$

10. $\begin{cases} \frac{1}{2}z - \frac{1}{4}w = \frac{1}{6}, \\ z + \frac{1}{2}w = \frac{2}{3}. \end{cases}$

11. $\begin{cases} 4p + 12q = 6, \\ 2p + 6q = 3. \end{cases}$

12. $\begin{cases} 5x - 3y = 2, \\ -10x + 6y = 4. \end{cases}$

13. $\begin{cases} 2x + y + 6z = 3, \\ x - y + 4z = 1, \\ 3x + 2y - 2z = 2. \end{cases}$

14. $\begin{cases} x + y + z = -1, \\ 3x + y + z = 1, \\ 4x - 2y + 2z = 0. \end{cases}$

15. $\begin{cases} 5x - 7y + 4z = 2, \\ 3x + 2y - 2z = 3, \\ 2x - y + 3z = 4. \end{cases}$

16. $\begin{cases} 3x - 2y + z = 0, \\ -2x + y - 3z = 5, \\ \frac{3}{2}x + \frac{4}{5}y + 4z = 10. \end{cases}$

17. A chemical manufacturer wishes to obtain 700 gallons of a 24 percent acid solution by mixing a 20 percent solution with a 30 percent solution. How many gallons of each solution should be used?

18. A company has taxable income of $312,000. The federal tax is 25 percent of that portion which is left after the state tax has been paid. The state tax is 10 percent of that portion which is left after the federal tax has been paid. Find the federal and state taxes.

19. A manufacturer of dining-room sets produces two styles, Early American and Contemporary. From past experience management has determined that 20 percent more of the Early American styles can be sold than the Contemporary styles. A profit of $250 is made on each Early American sold, while one of $350 is made on each Contemporary. If, in the forthcoming year, management desires a total profit of $130,000, how many units of each style must be sold?

20. National Surveys was awarded a contract to perform a product rating survey for Crispy Crackers. A total of 250 people were interviewed. National Surveys reported that 62.5 percent more people liked Crispy Crackers than disliked them. However, the report did not indicate that 16 percent of those interviewed had no comment. How many of those surveyed liked Crispy Crackers? How many disliked them? How many had no comment?

21. United Products Co. has plants in the cities of Exton and Whyton. Each plant is devoted to the manufacturing of calculators. At the Exton plant, fixed costs are $7000 per month, and the cost of producing each calculator is $7.50. At the Whyton plant, fixed costs are $8800 per month, and each calculator costs $6.00 to produce. Next month, United Products must manufacture 1500 calculators. Find the production order for each plant if the total cost for each plant is to be the same.

22. A coffee wholesaler blends together three types of coffee that sell for $2.20, $2.30, and $2.60 per lb so as to obtain 100 lb of coffee worth $2.40 per lb. If the wholesaler uses the same amount of the two higher-priced coffees, how much of each type must be used in the blend?

23. Company A pays its salespeople on a basis of a certain percentage of the first $100,000 in sales plus a certain percentage of any amount over $100,000 in sales. If a salesperson earned \$8500 on sales of \$175,000 and another salesperson earned \$14,800 on sales of \$280,000, find the two rates.

24. A person made two investments and the percentage return per year on each was the same. Of the total amount invested, $\frac{3}{10}$ of it plus \$600 was invested in one venture and at the end of one year the person received a return of \$384 from that venture. If the total return after one year was \$1120, find the total amount invested.

25. A company makes three types of patio furniture: chairs, rockers, and chaise lounges. Each requires wood, plastic, and aluminim, as given in Table 4-2. The company has in stock 400 units of wood, 600 units of plastic, and 1500 units of aluminum. For its end-of-the-season production run, the company wants to use up all the stock. To do this, how many chairs, rockers, and chaise lounges should it make?

TABLE 4–2

	WOOD	PLASTIC	ALUMINUM
Chair	1 unit	1 unit	2 units
Rocker	1 unit	1 unit	3 units
Chaise lounge	1 unit	2 units	5 units

26. A total of \$35,000 is invested at three interest rates: 7, 8, and 9 percent. The interest for the first year was \$2830, which was not reinvested. The second year the amount originally invested at 9 percent earned 10 percent instead, and the other rates remained the same. The total interest the second year was \$2960. How much was invested at each rate?

27. A company pays skilled workers in its assembly department \$8 per hour. Semiskilled workers in that department are paid \$4 per hour. Shipping clerks are paid \$5 per hour. Because of an increase in orders, the company needs to employ a total of 70 workers in the assembly and shipping departments. It will pay a total of \$370 per hour to these employees. Because of a union contract, twice as many semiskilled workers as skilled workers must be employed. How many semiskilled workers, skilled workers, and shipping clerks should the company hire?

4-4 NONLINEAR SYSTEMS

A system of equations in which at least one equation is not linear is called a **nonlinear system**. Solutions of such systems may often be found algebraically by substitution, as was done with linear systems. The following examples illustrate.

EXAMPLE 1

$Solve$ $\begin{cases} x^2 - 2x + y - 7 = 0, & (1) \\ 3x - y + 1 = 0. & (2) \end{cases}$

Solving Eq. (2) for y gives

$$y = 3x + 1. \qquad (3)$$

Substituting in Eq. (1) and simplifying, we have

$$x^2 - 2x + (3x + 1) - 7 = 0,$$
$$x^2 + x - 6 = 0,$$
$$(x + 3)(x - 2) = 0,$$
$$x = -3 \quad \text{or} \quad x = 2.$$

From Eq. (3), if $x = -3$ then $y = -8$; if $x = 2$ then $y = 7$. You should verify that each pair of values satisfies the given system. Hence the solutions are $x = -3$, $y = -8$ and $x = 2$, $y = 7$. These solutions can be seen geometrically in the graph of the system in Fig. 4-27. Notice that the graph of Eq. (1) is a parabola and the graph of Eq. (2) is a line. The solutions correspond to the intersection points $(-3, -8)$ and $(2, 7)$.

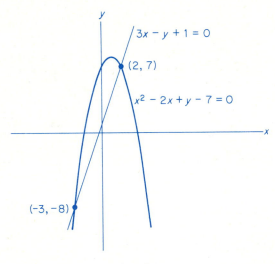

FIG. 4-27

EXAMPLE 2

Solve $\begin{cases} y = \sqrt{x + 2}, \\ x + y = 4. \end{cases}$

Solving the second equation for y gives

$$y = 4 - x. \tag{4}$$

Substituting gives

$$4 - x = \sqrt{x + 2},$$
$$16 - 8x + x^2 = x + 2, \qquad \text{[squaring both sides]}$$
$$x^2 - 9x + 14 = 0,$$
$$(x - 2)(x - 7) = 0.$$

Thus $x = 2$ or $x = 7$. From Eq. (4), if $x = 2$, then $y = 2$; if $x = 7$, then $y = -3$. Although $x = 2$ and $y = 2$ satisfy the original equations, this is not the case for $x = 7$ and $y = -3$. Thus the solution is $x = 2, y = 2$.

EXERCISE 4-4

Solve the following nonlinear systems.

1. $\begin{cases} y = 4 - x^2, \\ 3x + y = 0. \end{cases}$ 2. $\begin{cases} y = x^3, \\ x - y = 0. \end{cases}$

3. $\begin{cases} p^2 = 4 - q, \\ p = q + 2. \end{cases}$ 4. $\begin{cases} y^2 - x^2 = 28, \\ x - y = 14. \end{cases}$

5. $\begin{cases} x = y^2, \\ y = x^2. \end{cases}$ 6. $\begin{cases} p^2 - q = 0, \\ 3q - 2p - 1 = 0. \end{cases}$

7. $\begin{cases} y = 4x - x^2 + 8, \\ y = x^2 - 2x. \end{cases}$ 8. $\begin{cases} x^2 - y = 8, \\ y - x^2 = 0. \end{cases}$

9. $\begin{cases} p = \sqrt{q}, \\ p = q^2. \end{cases}$ 10. $\begin{cases} z = 4/w, \\ 3z = 2w + 2. \end{cases}$

11. $\begin{cases} x^2 = y^2 + 14, \\ y = x^2 - 16. \end{cases}$ 12. $\begin{cases} x^2 + y^2 - 2xy = 1, \\ 3x - y = 5. \end{cases}$

13. $\begin{cases} y = \dfrac{x^2}{x - 1} + 1, \\ y = \dfrac{1}{x - 1}. \end{cases}$ 14. $\begin{cases} x = y + 6, \\ y = 3\sqrt{x + 4}\,. \end{cases}$

4-5 SOME APPLICATIONS OF SYSTEMS OF EQUATIONS

For each price level of a product there is a corresponding quantity of the product that consumers will demand (that is, purchase) during some time period. Usually, the higher the price, the smaller the quantity demanded; as the price falls, the quantity demanded increases.

On the other hand, in response to various prices, there is a corresponding quantity of output of a product that producers are willing to place on the market during some time period. Usually, the higher the price per unit, the larger the quantity that producers are willing to supply; as the price falls, so will the quantity supplied.

The quantities of a product that will be demanded or supplied per unit of time at all possible alternative prices can be indicated geometrically on a coordinate plane by a *demand* or *supply curve*, as in Figs. 4-28 and 4-29. In keeping with the practice of most economists, quantity per unit of time is measured along the horizontal axis, the *q*-axis, while the vertical axis, the *p*-axis, measures price per unit. Although straight lines are not necessarily typical of these curves, their use is convenient for illustrative purposes.

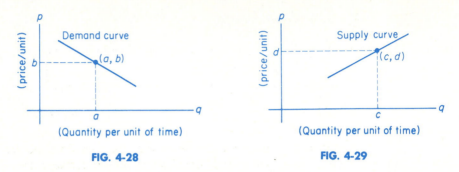

FIG. 4-28   FIG. 4-29

Throughout our discussion we shall assume that price per unit is given in dollars and that the time period is one week. The point (a, b) in Fig. 4-28 indicates that at a price of b dollars per unit, consumers will demand a units per week. Similarly, in Fig. 4-29 the point (c, d) indicates that at a price of d dollars each, producers will supply c units per week. Since negative prices or quantities are not meaningful, the coordinates of (a, b) and (c, d) must be nonnegative.

In most cases a demand curve falls from left to right (that is, has a negative slope). This reflects the relationship that consumers will buy more of a product as its price goes down. A supply curve usually rises from left to right (that is, has a positive slope). This indicates that a producer will supply more of a product at higher prices.

An equation that relates price per unit and quantity demanded (supplied) is called a *demand equation* (*supply equation*). Suppose that the linear demand equation for Product Z is

$$p = -\frac{1}{180}q + 12 \qquad (1)$$

and its linear supply equation is

$$p = \frac{1}{300}q + 8, \qquad (2)$$

where $q, p \geqslant 0$. The demand and supply curves defined by Eqs. (1) and (2) are given in Figs. 4-30 and 4-31. In analyzing Fig. 4-30, we see that consumers will purchase 540 units per week when the price is $9 per unit; 1080 units when the price is $6; etc. Figure 4-31 shows that when the price is $9 per unit, producers will place 300 units per week on the market; at $10 they will supply 600 units; etc.

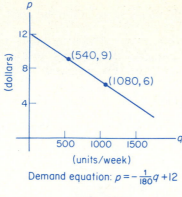

Demand equation: $p = -\frac{1}{180}q + 12$

FIG. 4-30

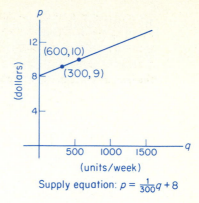

Supply equation: $p = \frac{1}{300}q + 8$

FIG. 4-31

When both the demand and supply curves of a product are represented on the same coordinate plane, the point (m, n) at which the curves intersect is called the *point of equilibrium* (see Fig. 4-32). The price n, called the *equilibrium price*, is the price at which consumers will purchase the same quantity of a

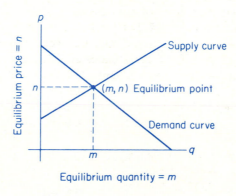

FIG. 4-32

product that producers wish to sell at that price. In short, n is the price at which stability in the producer-consumer relationship occurs. The quantity m is called the *equilibrium quantity*.

To determine precisely the equilibrium point, we solve the system formed by the supply and demand equations. Let us do this for our previous data, namely the system

$$\begin{cases} p = -\dfrac{1}{180}q + 12, & \text{(demand equation)} \\[2mm] p = \dfrac{1}{300}q + 8. & \text{(supply equation)} \end{cases}$$

By substituting $\frac{1}{300}q + 8$ for p in the demand equation we get

$$\frac{1}{300}q + 8 = -\frac{1}{180}q + 12,$$

$$\left(\frac{1}{300} + \frac{1}{180}\right)q = 4,$$

$$q = 450. \qquad \text{(equilibrium quantity)}$$

Thus,

$$p = \frac{1}{300}(450) + 8$$

$$= 9.50, \qquad \text{(equilibrium price)}$$

and the equilibrium point is (450, 9.50). Therefore, at the price of $9.50 per unit, manufacturers will produce exactly the quantity (450) of units per week that consumers will purchase at that price (see Fig. 4-33).

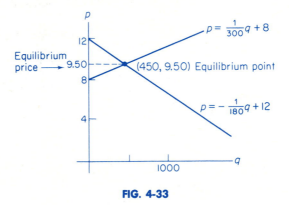

FIG. 4-33

EXAMPLE 1

Let $p = \frac{8}{100}q + 50$ be the supply equation for a certain manufacturer. Suppose the demand per week for his product is 100 units when the price is $58 per unit, and 200 units per week at $51 each.

a. Determine the demand equation, assuming that it is linear.

b. If a tax of $1.50 per unit is to be imposed on the manufacturer, how will the original equilibrium price be affected if the demand remains the same?

c. Determine the total revenue obtained by the manufacturer at the equilibrium point both before and after the tax.

a. Since the demand equation is linear, the demand curve must be a straight line. From the given data we conclude that the points (100, 58) and (200, 51) lie on this line and thus its slope is

$$m = \frac{51 - 58}{200 - 100} = -\frac{7}{100}.$$

An equation of the line is

$$p - p_1 = m(q - q_1),$$

$$p - 58 = -\frac{7}{100}(q - 100).$$

Hence the demand equation is

$$p = -\frac{7}{100}q + 65.$$

b. Before the tax, the equilibrium price is obtained by solving the system

$$\begin{cases} p = \dfrac{8}{100}q + 50, \\ p = -\dfrac{7}{100}q + 65. \end{cases}$$

By substitution,

$$-\frac{7}{100}q + 65 = \frac{8}{100}q + 50,$$

$$15 = \frac{15}{100}q,$$

$$100 = q,$$

and

$$p = \frac{8}{100}(100) + 50 = 58.$$

Thus, \$58 is the original equilibrium price. Before the tax the manufacturer supplies q units at a price of $p = \frac{8}{100}q + 50$ per unit. After the tax he will sell the same q units for an additional \$1.50 per unit. The price per unit will then be $(\frac{8}{100}q + 50) + 1.50$ and the new supply equation will be $p = \frac{8}{100}q + 51.50$. Solving the system

$$\begin{cases} p = \dfrac{8}{100}q + 51.50, \\ p = -\dfrac{7}{100}q + 65 \end{cases}$$

will give the new equilibrium price.

$$\frac{8}{100}q + 51.50 = -\frac{7}{100}q + 65,$$

$$\frac{15}{100}q = 13.50,$$

$$q = 90,$$

$$p = \frac{8}{100}(90) + 51.50 = 58.70.$$

The tax of \$1.50 per unit increases the equilibrium price by \$0.70 (Fig. 4-34). Note that there is also a decrease in the equilibrium quantity from $q = 100$ to $q = 90$ due to the

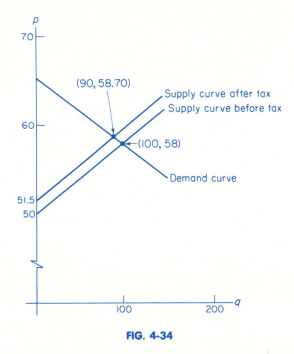

FIG. 4-34

change in the equilibrium price. (In the exercises, you are asked to find the effect of a subsidy given to the manufacturer which will reduce the price of the product.)

c. If q units of a product are sold at a price of p dollars each, then the total revenue, which we shall denote by y_{TR}, is given by

$$y_{TR} = pq.$$

Before the tax the revenue at (100, 58) is (in dollars)

$$y_{TR} = (58)(100) = 5800.$$

After the tax it is

$$y_{TR} = (58.70)(90) = 5283,$$

which is a decrease.

EXAMPLE 2

Find the equilibrium point if the supply and demand equations of a product are $p = \dfrac{q}{40} + 10$ and $p = \dfrac{8000}{q}$, respectively.

Here the demand equation is not linear. Solving the system

$$\begin{cases} p = \dfrac{q}{40} + 10, \\ p = \dfrac{8000}{q} \end{cases}$$

by substitution gives

$$\frac{8000}{q} = \frac{q}{40} + 10,$$

$$320{,}000 = q^2 + 400q,$$

$$q^2 + 400q - 320{,}000 = 0,$$

$$(q + 800)(q - 400) = 0,$$

$$q = -800 \quad \text{or} \quad q = 400.$$

We disregard $q = -800$, since q represents quantity. Choosing $q = 400$, then we have $p = (8000/400) = 20$ and the required point is $(400, 20)$ (see Fig. 4-35).

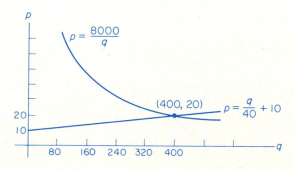

FIG. 4-35

Suppose a manufacturer produces product A and sells it at \$8.00 per unit. Then the total revenue y_{TR} he receives (in dollars) from selling q units is

$$y_{TR} = 8q. \hspace{3cm} \text{(total revenue)}$$

The difference between the total revenue received for q units and the total cost of q units is the manufacturer's profit (or loss if the difference is negative).

$$\text{Profit (or Loss)} = \text{Total Revenue} - \text{Total Cost.}$$

Total cost, y_{TC}, is the sum of total variable costs y_{VC} and total fixed costs y_{FC}.

$$y_{TC} = y_{VC} + y_{FC}.$$

Fixed costs are those costs that under normal conditions do not depend on the level of production; that is, over some period of time they remain constant at all levels of output (examples are rent, officers' salaries and normal maintenance). **Variable costs** are those costs that vary with the level of production (such as cost of materials, labor, maintenanace due to wear and tear, etc.). For q units of product A, suppose

$$y_{FC} = 5000 \hspace{3cm} \text{(fixed cost)}$$

$$\text{and} \quad y_{VC} = \frac{22}{9}q. \hspace{2.5cm} \text{(variable cost)}$$

Then

$$y_{TC} = \frac{22}{9}q + 5000. \hspace{2.5cm} \text{(total cost)}$$

The graphs of fixed cost, total cost, and total revenue appear in Fig. 4-36. The horizontal axis represents level of production q, and the vertical axis

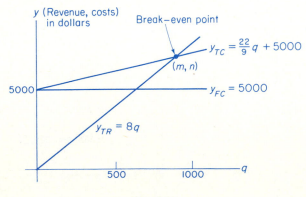

FIG. 4-36

represents the total dollar value, be it revenue or costs. The *break-even point* is the point at which Total Revenue = Total Cost ($TR = TC$). It occurs when the levels of production and sales result in neither a profit nor a loss to the manufacturer. In the diagram, called a *break-even chart*, it is the point (m, n) at which the graphs of $y_{TR} = 8q$ and $y_{TC} = \frac{22}{9}q + 5000$ intersect. We call m the *break-even quantity* and n the *break-even revenue*. When variable costs and revenue are linearly related to output, as in our case, any production level greater than m units will produce a profit, while any level less than m units will produce a loss. Thus, at an output of m units the profit is zero. In the following example we shall examine our data in more detail.

EXAMPLE 3

A manufacturer sells his product at $8 per unit, selling all that he produces. His fixed cost is $5000 and the variable cost per unit is $\frac{22}{9}$ (dollars). Find

a. *the total output and revenue at the break-even point.*

b. *the profit when 1800 units are produced.*

c. *the loss when 450 units are produced.*

d. *the output required to obtain a profit of $10,000.*

a. At an output level of q units, the variable cost is $y_{VC} = \frac{22}{9}q$ and the total revenue is $y_{TR} = 8q$. Hence,

$$y_{TR} = 8q,$$

$$y_{TC} = y_{VC} + y_{FC} = \frac{22}{9}q + 5000.$$

At the break-even point, Total Revenue = Total Cost. Thus we solve the system formed by the above equations. Since

$$y_{TR} = y_{TC},$$

we have

$$8q = \frac{22}{9}q + 5000,$$

$$\frac{50}{9}q = 5000,$$

$$q = 900.$$

Thus the desired output is 900 units, resulting in a total revenue (in dollars) of

$$y_{TR} = 8(900) = 7200.$$

b. Since Profit = Total Revenue − Total Cost, when $q = 1800$ we have

$$y_{TR} - y_{TC} = 8(1800) - \left[\frac{22}{9}(1800) + 5000\right]$$

$$= 5000.$$

The profit when 1800 units are produced and sold is $5000.

c. When $q = 450$,

$$y_{TR} - y_{TC} = 8(450) - \left[\frac{22}{9}(450) + 5000\right]$$

$$= -2500.$$

A loss of $2500 occurs when the level of production is 450 units.

d. In order to obtain a profit of $10,000, we have

$$\text{Profit} = \text{Total Revenue} - \text{Total Cost.}$$

$$10,000 = 8q - \left(\frac{22}{9}q + 5000\right),$$

$$15,000 = \frac{50}{9}q,$$

$$q = 2700.$$

Thus 2700 units must be produced.

EXAMPLE 4

Determine the break-even quantity of XYZ Manufacturing Co. given the following data: total fixed cost, $1200; variable cost per unit, $2; total revenue for selling q units, $y_{TR} = 100\sqrt{q}$.

For q units of output,

$$y_{TR} = 100\sqrt{q} ,$$

$$y_{TC} = 2q + 1200.$$

Equating total revenue to total cost gives

$$100\sqrt{q} = 2q + 1200,$$

$$50\sqrt{q} = q + 600.$$

Squaring both sides, we have

$$2500q = q^2 + 1200q + (600)^2,$$

$$0 = q^2 - 1300q + 360,000.$$

By the quadratic formula

$$q = \frac{1300 \pm \sqrt{250{,}000}}{2},$$

$$q = 400 \quad \text{or} \quad q = 900.$$

Although both $q = 400$ and $q = 900$ are break-even quantities, observe in Fig. 4-37 that there will always be a loss when $q > 900$. Thus producing more than the break-even quantity does not necessarily guarantee a profit.

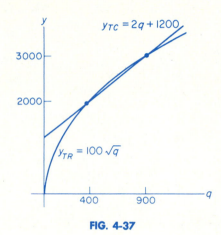

FIG. 4-37

EXERCISE 4-5

1. Suppose a manufacturer of shoes will place on the market 50 (thousand pairs) when the price is 35 (dollars per pair) and 35 when the price is 30. Find the supply equation, assuming that it is linear.

2. Suppose consumers will demand 20 (thousand) pairs of shoes when the price is 35 (dollars per pair) and 25 pairs when the price is 30. Find the demand equation, assuming that it is linear.

In Problems 3–10, the first equation is a supply equation and the second is a demand equation for a product. If p represents price per unit in dollars and q represents the number of units per unit of time, find the equilibrium point. In Problems 3 and 4 sketch the system.

3. $p = \frac{3}{100}q + 2,$

   $p = -\frac{7}{100}q + 12.$

4. $p = \frac{1}{2000}q + 3,$

   $p = -\frac{1}{2500}q + \frac{42}{5}.$

5. $35q - 2p + 250 = 0,$

   $65q + p - 537.5 = 0.$

6. $246p - 3.25q - 2460 = 0,$

   $410p + 3q - 14{,}452.5 = 0.$

7. $p = 2q + 20,$

   $p = 200 - 2q^2.$

8. $p = (q + 10)^2,$

   $p = 388 - 16q - q^2.$

9. $p = \sqrt{q + 10}$,

$p = 20 - q$.

10. $p = \frac{1}{5}q + 5$,

$p = \dfrac{3000}{q + 20}$.

In Problems 11–16, y_{TR} represents total revenue in dollars and y_{TC} represents total cost in dollars for a manufacturer. If q represents both the number of units produced and the number of units sold, find the break-even quantity. Sketch a break-even chart in Problems 11 and 12.

11. $y_{TR} = 3q$,

$y_{TC} = 2q + 4500$.

12. $y_{TR} = 14q$,

$y_{TC} = \frac{40}{3}q + 1200$.

13. $y_{TR} = .05q$,

$y_{TC} = .85q + 600$.

14. $y_{TR} = .25q$,

$y_{TC} = .16q + 360$.

15. $y_{TR} = 100 - \dfrac{1000}{q + 10}$,

$y_{TC} = q + 40$.

16. $y_{TR} = .1q^2 + 7q$,

$y_{TC} = 2q + 500$.

17. The supply and demand equations for a certain product are $3q - 200p + 1800 = 0$ and $3q + 100p - 1800 = 0$, respectively, where p represents the price per unit in dollars, and q represents the number of units per time period.

   a.   Find the equilibrium price algebraically, and derive it graphically.

   b.   Find the equilibrium price when a tax of 27 cents per unit is imposed on the supplier.

18. A manufacturer of a product sells all that he produces. His total revenue is given by $y_{TR} = 7q$ and his total cost is given by $y_{TC} = 6q + 800$, where q represents the number of units produced and sold.

   a.   Find the level of production at the break-even point and draw the break-even chart.

   b.   Find the level of production at the break-even point if the total cost increases by 5 percent.

19. A manufacturer sells his product at $8.35 per unit, selling all he produces. His fixed cost is $2116 and his variable cost is $7.20 per unit. At what level of production will he have a profit of $4600? At what level of production will he have a loss of $1150? At what level of production will he break even?

20. The market equilibrium point for a product occurs when 13,500 units are produced at a price of $4.50 per unit. The producer will supply no units at $1 and the consumers will demand no units at $20. Find the supply and demand equations if they are both linear.

21. A manufacturer of widgets will break even at a sales volume of $200,000. Fixed costs are $40,000 and each unit of output sells for $5. Determine the variable cost per unit.

22. The Footsie Sandal Co. manufactures sandals for which the material cost is $0.80 per pair and the labor cost is $0.90 per pair. Additional variable costs amount to $0.30 per pair. Fixed costs are $70,000. If each pair sells for $2.50, how many pairs must be sold for the company to break even?

23. Find the break-even point for Company Z, which sells all it produces, if the variable cost per unit is $2, fixed costs are $1050, and $y_{TR} = 50\sqrt{q}$, where q is the number of units of output.

24. A company has determined that the demand equation for its product is $p = 1000/q$, where p is the price per unit for q units in some time period. Determine the quantity demanded when the price per unit is: (a) $4; (b) $2; and (c) $0.50. For each of these prices, determine the total revenue that the company will receive. What will be the revenue regardless of the price? (*Hint:* Find the revenue when the price is p dollars.)

25. By using the data in Example 1 on page 144, determine how the original equilibrium price will be affected if the company is given a government subsidy of $1.50 per unit.

26. The Monroe Forging Company sells a corrugated steel product to the Standard Manufacturing Company and is in competition on such sales with other suppliers of the Standard Manufacturing Co. The vice president of sales of Monroe Forging Co. believes that by reducing the price of the product, a 40 percent increase in the volume of units sold to the Standard Manufacturing Co. could be secured. As the manager of the cost and analysis department, you have been asked to analyze the proposal of the vice president and submit your recommendations as to whether it is financially beneficial to the Monroe Forging Co.

You are specifically requested to determine the following.

(1) Net profit or loss based on the pricing proposal.
(2) Unit sales volume under the proposed price that is required to make the same $40,000 profit that is now earned at the current price and unit sales volume.

Table 4-3 gives data for use in your analysis.

TABLE 4-3

	CURRENT OPERATIONS	PROPOSAL OF VICE PRESIDENT OF SALES
Unit price	$2.50	$2.00
Unit sales volume	200,000 units	280,000 units
Variable cost—total	$350,000	
—per unit	$1.75	$1.75
Fixed cost	$110,000	$110,000
Profit	$40,000	???

27. Suppose products A and B have demand and supply equations that are related to each other. If q_A and q_B are quantities of A and B, respectively, and p_A and p_B are their respective prices, the demand equations are

$$q_A = 8 - p_A + p_B$$
$$\text{and} \quad q_B = 26 + p_A - p_B,$$

and the supply equations are

$$q_A = -2 + 5p_A - p_B$$
$$\text{and} \quad q_B = -4 - p_A + 3p_B.$$

Eliminate q_A and q_B to get the equilibrium prices.

4-6 REVIEW

IMPORTANT TERMS AND SYMBOLS IN CHAPTER 4

slope of a line   *(p. 113)*

x-intercept   *(p. 120)*

slope-intercept form   *(p. 117)*

linearly related   *(p. 120)*

linear function   *(p. 118)*

quadratic function   *(p. 123)*

vertex of parabola   *(p. 124)*

equivalent systems   *(p. 131)*

elimination by substitution

   (p. 134)

demand curve, demand equation

   (p. 142)

equilibrium price   *(p. 143)*

fixed cost, variable cost

   (p. 148)

y-intercept   *(p. 117)*

point-slope form   *(p. 116)*

general linear equation

  in *x* and *y*   *(p. 119)*

  in *x*, *y*, and *z*   *(p. 135)*

parabola   *(p. 124)*

system of equations   *(p. 130)*

elimination by addition   *(p. 133)*

supply curve, supply equation

   (p. 142)

point of equilibrium   *(p. 143)*

total cost   *(p. 148)*

break-even point   *(p. 149)*

REVIEW SECTION

1. A linear equation in x and y is one that can be written in the general form _____.

        Ans. $Ax + By + C = 0$; A, B not both zero.

2. The line in Fig. 4-38 has a (positive) (negative) slope.

        Ans.   negative.

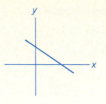

FIG. 4-38

3. The slope of a horizontal line is equal to __(a)__ and the slope of a vertical line is __(b)__ .

Ans. (a) zero; (b) not defined.

4. Two methods by which a system of two linear equations in two variables can be solved are elimination by _____ and elimination by _____ .

Ans. addition and substitution.

5. The graph of $x = 7$ is a line parallel to the _____ axis.

Ans. y.

6. If the points (7, 6) and (3, 4) lie on the graph of a straight line, then the line has a slope equal to _____ .

Ans. $\frac{1}{2}$.

7. The y-intercepts of the lines $y = x$ and $y = 2x$ are __(a)__ and __(b)__ respectively.

Ans. (a) (0, 0); (b) (0, 0).

8. If a system of two linear equations is represented geometrically by two parallel lines, what can be said about the solution of the system? _____

Ans. no solution or infinitely many solutions.

9. A point-slope form of an equation of the line through (1, −2), having slope 4, is

_____ .

Ans. $y + 2 = 4(x − 1)$.

10. The slope of the straight line $y = 4x − 3$ is _____ .

Ans. 4.

11. An equation of the vertical line passing through (2, −3) is _____ .

Ans. $x = 2$.

12. The slope of the line $2y = 3x + 2$ is _____ .

Ans. $\frac{3}{2}$.

13. A straight line whose slope is 0 has an equation of the form __(a)$(x = c)$ $(y = c)$__ and is __(b) (parallel) (perpendicular)__ to the y-axis.

Ans. (a) $y = c$; (b) perpendicular.

14. The graph of a linear function is a _____.

Ans. straight line.

15. The graph of $f(x) = 7x^2 + 3x - 5$ is called a _____.

Ans. parabola.

16. The y-intercept of the graph of $y = f(x) = 3x^2 - 5x + 2$ is the point _____.

Ans. (0, 2).

17. The parabola $g(x) = x^2 - 1$ opens _(upward) (downward)_ .

Ans. upward.

18. The vertex of the parabola $f(x) = x^2 - 6x + 1$ occurs when $x = $ _____.

Ans. 3.

19. The graph of a demand equation usually _(a) (rises) (falls)_ from left to right, while the graph of a supply equation usually _(b) (rises) (falls)_ from left to right.

Ans. (a) falls; (b) rises.

20. In the graph of a supply or demand equation, price per unit is usually measured along the _(horizontal) (vertical)_ axis.

Ans. vertical.

21. The point at which supply and demand curves intersect is called the _(a)_ . The price at which the quantity supplied is equal to the quantity demanded is called the _(b)_ price.

Ans. (a) point of equilibrium; (b) equilibrium.

22. Total cost is the sum of _____ cost and _____ cost. The point at which total revenue equals total cost is called the _____ point.

Ans. variable and fixed; break-even.

REVIEW PROBLEMS

1. The slope of the line through (2, 5) and (3, k) is 4. Find k.

2. The slope of the line through (2, 3) and (k, 3) is 0. Find k.

*In Problems **3–7**, determine the slope-intercept form and a general linear form of an equation of the straight line that has the indicated properties.*

3. Passes through (3, −2) and (−7, 8).

4. Passes through (−1, −1) and is parallel to the line $9x - 3y + 14 = 0$.

5. Passes through (10, 4) and has slope $\frac{1}{2}$.

6. Passes through (3, 5) and is vertical.

7. Passes through (−2, 4) and is horizontal.

8. Determine whether the point (0, −7) lies on the line through (1, −3) and (4, 9).

In Problems **9–12**, *write each line in the slope-intercept form and sketch. What is the slope of the line?*

9. $3x - 2y = 4.$ **10.** $x = -3y + 4.$

11. $4 - 3y = 0.$ **12.** $y = 2x.$

In Problems **13–22**, *graph each function and give its domain and range. For those that are linear, also give the slope and the vertical-axis intercept. For those that are quadratic, give all intercepts and the vertex.*

13. $y = f(x) = 4 - 2x.$ **14.** $s = g(t) = 8 - 2t - t^2.$

15. $y = f(x) = 9 - x^2.$ **16.** $y = f(x) = 3x - 7.$

17. $y = h(t) = t^2 - 4t - 5.$ **18.** $y = h(t) = 1 + 3t.$

19. $p = g(t) = 3t.$ **20.** $y = F(x) = (2x - 1)^2.$

21. $y = F(x) = -(x^2 + 2x + 3).$ **22.** $y = f(x) = \dfrac{x}{3} - 2.$

In Problems **23–32**, *solve the given system.*

23. $\begin{cases} 2x - y = 6, \\ 3x + 2y = 5. \end{cases}$ **24.** $\begin{cases} 8x - 4y = 7, \\ y = 2x - 4. \end{cases}$

25. $\begin{cases} 4x + 5y = 3, \\ 3x + 4y = 2. \end{cases}$ **26.** $\begin{cases} 3x + 6y = 9, \\ 4x + 8y = 12. \end{cases}$

27. $\begin{cases} \frac{1}{4}x - \frac{3}{2}y = -4, \\ \frac{3}{4}x + \frac{1}{2}y = 8. \end{cases}$ **28.** $\begin{cases} \frac{1}{3}x - \frac{1}{4}y = \frac{1}{12}, \\ \frac{4}{3}x + 3y = \frac{5}{3}. \end{cases}$

29. $\begin{cases} 3x - 2y + z = -2, \\ 2x + y + z = 1, \\ x + 3y - z = 3. \end{cases}$ **30.** $\begin{cases} x + \dfrac{2y + x}{6} = 14, \\ y + \dfrac{3x + y}{4} = 20. \end{cases}$

31. $\begin{cases} x^2 - y + 2x = 7, \\ x^2 + y = 5. \end{cases}$ **32.** $\begin{cases} y = \dfrac{18}{x + 4}, \\ x - y + 7 = 0. \end{cases}$

33. Suppose f is a linear function such that $f(1) = 5$ and $f(x)$ decreases by 4 units for every 3-unit increase in x. Find $f(x)$.

34. If f is a linear function such that $f(-1) = 8$ and $f(2) = 5$, find $f(x)$.

35. The demand function for a manufacturer's product is $p = f(q) = 200 - 2q$, where p is the price (in dollars) per unit when q units are demanded. Find the level of production that maximizes the manufacturer's total revenue and determine this revenue.

36. The difference in price of two items before a five-percent sales tax is imposed is $4. The difference in price after the sales tax is imposed is $4.20. Find the price of each item before the sales tax.

37. If the supply and demand equations of a certain product are $125p - q - 250 = 0$ and $100p + q - 1100 = 0$, respectively, find the equilibrium price.

38. A manufacturer of a certain product sells all that is produced. Determine the break-even point if the product is sold at \$16 per unit, fixed cost is \$10,000, and variable cost is given by $y_{VC} = 8q$, where q is the number of units produced (y_{VC} expressed in dollars).

EXPONENTIAL AND LOGARITHMIC FUNCTIONS

5-1 EXPONENTIAL AND LOGARITHMIC FUNCTIONS

There is a function which has an important role not only in mathematics but in business and economics as well. It involves a constant raised to a variable power and is called an *exponential function*. An example is $f(x) = 2^x$.

DEFINITION. *The function f defined by*

$$y = f(x) = b^x,$$

*where $b > 0$, $b \neq 1$, and the exponent x is any real number, is called an **exponential function** to the base b.*

We made the restriction $b \neq 1$ to exclude the rather simple constant function $f(x) = 1^x = 1$. Since the exponent in b^x can be any real number, the question comes up as to how we define something like $b^{\sqrt{2}}$. Stated simply, we use an approximation method. First, $b^{\sqrt{2}}$ is approximately $b^{1.4} = b^{7/5} = \sqrt[5]{b^7}$,

159

which *is* defined. Better approximations are $b^{1.41} = \sqrt[100]{b^{141}}$ and $b^{1.414}$, etc. In this way a meaning of $b^{\sqrt{2}}$ becomes clear.

In Fig. 5-1 are the graphs of the exponential functions $y = 2^x$, $y = 3^x$, and $y = (\frac{1}{2})^x = 2^{-x}$. Notice that

(1) the *domain* of an exponential function is all real numbers, and

(2) the *range* is all positive numbers.

Also, $b^0 = 1$ for every base b, as shown by the point of intersection $(0, 1)$ of the graphs.

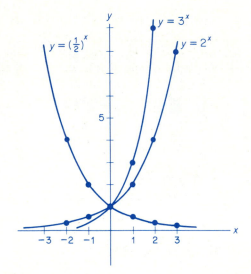

x	2^x	3^x	$(\frac{1}{2})^x$
-2	$\frac{1}{4}$	$\frac{1}{9}$	4
-1	$\frac{1}{2}$	$\frac{1}{3}$	2
0	1	1	1
1	2	3	$\frac{1}{2}$
2	4	9	$\frac{1}{4}$
3	8	27	$\frac{1}{8}$

FIG. 5-1

We also see in Fig. 5-1 that $y = b^x$ has two basic shapes, depending on whether $b > 1$ or $0 < b < 1$. If $b > 1$, then as x increases y also increases. But y can also take on values very close to 0. Now suppose $0 < b < 1$ as in $y = (\frac{1}{2})^x$. Then as x increases y *decreases*, taking on values close to zero.

One of the most **useful** numbers that is used as a base in $y = b^x$ is a certain irrational number denoted by the letter e in honor of the Swiss mathematician and physicist Leonhard Euler (1707–1783):

$$e \text{ is approximately } 2.71828 \ldots .$$

See Fig. 5-2 for the graph of $y = e^x$.

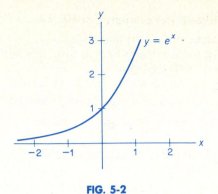

FIG. 5-2

Although e seems to be a strange number to use as a base in an exponential function, it arises quite naturally in calculus (as you will see in later chapters). It also occurs in economic analysis and problems involving natural growth (or decay), such as compound interest and population studies. A table of (approximate) values of e^x and e^{-x} is in Appendix B.

EXAMPLE 1

The predicted population $P(t)$ of a city is given by

$$P(t) = 100{,}000e^{.05t},$$

where t is the number of years after 1980. *Predict the population in the year* 2000.

Here $t = 20$ and

$$P(20) = 100{,}000e^{.05(20)} = 100{,}000e^1 = 100{,}000e.$$

Since $e \approx 2.71828$ ($\approx$ means "is approximately"), the predicted population 20 years after 1980 is approximately 271,828. Many economic forecasts are based on population studies.

EXAMPLE 2

A mail order firm finds that the proportion P of small towns in which exactly x persons respond to a magazine advertisement is given approximately by the formula

$$P = \frac{e^{-.5}(.5)^x}{1 \cdot 2 \cdot 3 \cdots x}.$$

From what proportion of small towns can the firm expect exactly two people to respond?

If we let $P = g(x)$, then we want to find $g(2)$.

$$g(2) = \frac{e^{-.5}(.5)^2}{1 \cdot 2}.$$

From the table in Appendix B, $e^{-.5} \approx .60653$ and so

$$g(2) \approx \frac{(.60653)(.25)}{2} \approx .07582.$$

Thus the firm can expect such a response from approximately 7.58 percent. The basis for the above formula is a function used in probability theory called the *Poisson probability function*.

Another important type of function is a *logarithmic function*, which is related to an exponential function. Figure 5-3 shows the graph of the exponential function $s = f(t) = 2^t$. Here f sends an input number t on the horizontal axis into a *positive* output number s:

$$f : t \to s \quad \text{where} \quad s = 2^t.$$

For example, f sends 2 into 4:

$$f : 2 \to 4.$$

Now look at the same curve in Fig. 5-4. There you can see that with each

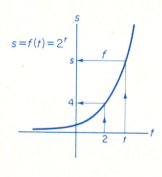

FIG. 5-3

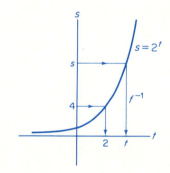

FIG. 5-4

positive number s on the vertical axis, we can associate exactly one value of t. With $s = 4$ we associate $t = 2$. Let us think here of s as an input and t as an output. Then we have a function that sends s's into t's. We shall denote this function by the symbol f^{-1} (read "f inverse").[†]

$$f^{-1} : s \to t \quad \text{where} \quad s = 2^t.$$

Thus, $f^{-1}(s) = t$.

[†] f^{-1} is a symbol for a new function. It does not mean $\frac{1}{f}$.

The functions f and f^{-1} are related. In Fig. 5-5, from the arrows you can

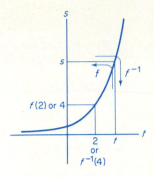

FIG. 5-5

see that f^{-1} *reverses* the action of f, and vice versa. For example,

$$f \text{ sends 2 into 4,} \quad \text{and} \quad f^{-1} \text{ sends 4 into 2.}$$

In terms of composition,

$$f^{-1}[f(2)] = f^{-1}[4] = 2.$$

More generally,

$$f^{-1}[f(t)] = t.$$

This is what is really meant by saying f^{-1} reverses the action of f. We also have $f[f^{-1}(4)] = f[2] = 4$. Generalizing this, we have

$$f[f^{-1}(s)] = s.$$

Notice that the domain of f^{-1} is the range of f (all positive numbers), and the range of f^{-1} is the domain of f (all real numbers).

We give a special name to f^{-1}. It is called the **logarithmic function base 2.** Usually we write f^{-1} as $\log_2$ (read "log base 2"). Thus $\log_2$ is merely a symbol for a special function.

In summary,

$$\text{if} \quad s = f(t) = 2^t \quad \text{then} \quad f^{-1}(s) = \log_2(s) = t. \tag{1}$$

The domain of $\log_2$ is all positive numbers and the range is all reals.

We now generalize our discussion to other bases in the following definition, where in (1) we replaced s by x and t by y. We assume $b > 0$ and $b \neq 1$.

DEFINITION. *The **logarithmic function base b**, denoted* $\log_b$, *is defined by*

$$y = \log_b x \quad \text{if and only if} \quad b^y = x.$$

The domain of $\log_b$ *is all positive numbers and its range is all real numbers.*

The logarithmic function reverses the action of the exponential function. Because of this we say that the logarithmic function is the *inverse* of the exponential function.

Always remember: when we say that the log base b of x is y, we mean that b raised to the y power is x.

$$\log_b x = y \quad \text{means} \quad b^y = x.$$

In this sense, *the logarithm of a number is an exponent*. For example,

$$\log_2 8 = 3 \quad \text{because} \quad 2^3 = 8.$$

We say that $\log_2 8 = 3$ is the **logarithmic form** of the **exponential form** $2^3 = 8$.

EXAMPLE 3

a. Since $25 = 5^2$, then $\log_5 25 = 2$.

b. Since $10^0 = 1$, then $\log_{10} 1 = 0$.

c. If $6y = e^{2r}$, then $\log_e 6y = 2r$ and so $r = [\log_e (6y)]/2$.

d. $\log_{10} 100 = 2$ means $10^2 = 100$.

e. $\log_{64} 8 = \frac{1}{2}$ means $64^{1/2} = 8$.

f. $\log_2 \frac{1}{16} = -4$ means $2^{-4} = \frac{1}{16}$.

EXAMPLE 4

Graph the function $y = \log_2 x$.

To plot points it is more convenient to use the equivalent form $2^y = x$. If $y = 0$, then $x = 1$. This gives the point $(1, 0)$. Other points are shown in Fig. 5-6. Note that the domain is all positive numbers and the range is all real numbers.

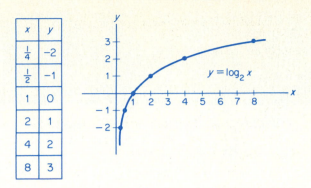

x	y
$\frac{1}{4}$	-2
$\frac{1}{2}$	-1
1	0
2	1
4	2
8	3

FIG. 5-6

Figure 5-7 shows the graph of $y = \log_e x$. Notice that it has the same shape as the graph shown in Fig. 5-6.

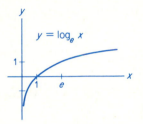

FIG. 5-7

Logarithms to the base 10, called **common logarithms**, were frequently used for computational purposes before the pocket-calculator age. The subscript 10 is generally omitted from the notation. Thus,

$$\log x \quad \text{means} \quad \log_{10} x.$$

Important in calculus are logarithms to the base e, called **natural** (or Naperian[†]) **logarithms.** We use the notation "ln" for such logarithms. Thus,

$$\ln x \quad \text{means} \quad \log_e x.$$

In Appendix C is a table of approximate values of natural logarithms. From there you can see that $\ln 2 \approx .69315$. This means that $e^{.69315} \approx 2$.

[†]After the Scottish mathematician John Napier (1550–1617), the inventor of logarithms.

EXAMPLE 5

Find each of the following.

a. log 1000.

Here the base is 10. Let log 1000 = y. Converting to exponential form, we have

$$10^y = 1000.$$

Clearly y must be 3. Thus log 1000 = 3.

b. ln 1.

Here the base is e. Let ln 1 = y. Converting to exponential form, we have

$$e^y = 1.$$

Clearly y must be 0. Thus ln 1 = 0.

c. log .1.

$$\log .1 = y,$$

$$10^y = .1 = \tfrac{1}{10} = 10^{-1}.$$

Clearly $y = -1$ and log .1 = -1.

d. ln e.

$$\ln e = y,$$

$$e^y = e.$$

Clearly $y = 1$ and ln e = 1.

EXAMPLE 6

Solve each equation for x.

a. $\log_3 x = 4$.

$$\log_3 x = 4,$$

or equivalently

$$3^4 = x,$$

$$81 = x.$$

b. $x + 1 = \log_4 16$.

$$x + 1 = \log_4 16,$$

$$4^{x+1} = 16.$$

From inspection the exponent $x + 1$ must be 2 and so $x = 1$.

c. $\log_x 49 = 2$.

$$\log_x 49 = 2,$$

$$x^2 = 49,$$

$$x = 7.$$

 We rejected $x = -7$ in solving $x^2 = 49$, since a negative number cannot be a base of a logarithmic function.

d. $12 = 5 + 3(4)^{x-1}$.

$$12 = 5 + 3(4)^{x-1},$$

$$7 = 3(4)^{x-1},$$

$$\tfrac{7}{3} = 4^{x-1}.$$

This means that the logarithm base 4 of $\frac{7}{3}$ is $x - 1$:

$$x - 1 = \log_4 \tfrac{7}{3},$$

$$x = 1 + \log_4 \tfrac{7}{3}.$$

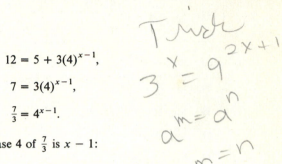

EXERCISE 5-1

*In Problems **1–6**, graph each function.*

1. $y = f(x) = 4^x$.
2. $y = f(x) = (\tfrac{1}{3})^x$.
3. $y = f(x) = 2(4^{-x})$.
4. $y = f(x) = \tfrac{1}{2}(3^{x/2})$.
5. $y = f(x) = \log_3 x$.
6. $y = f(x) = \log_{1/2} x$.

*In Problems **7–14**, use the tables in Appendices B and C to find the approximate value of each expression.*

7. $e^{1.5}$.
8. $e^{3.4}$.
9. $e^{-.4}$.
10. $e^{-3/4}$.
11. $\ln 5$.
12. $\ln 3.12$.
13. $\ln 7.39$.
14. $\ln 9.98$.

*In Problems **15–26** express each logarithmic form exponentially and each exponential form logarithmically.*

15. $25^{1/2} = 5$.
16. $2 = \log_{12} 144$.
17. $10^4 = 10{,}000$.
18. $\log_{1/2} 4 = -2$.
19. $\log_2 64 = 6$.
20. $8^{2/3} = 4$.
21. $\log_2 x = 14$.
22. $10^{.48302} = 3.041$.

23. $e^2 = 7.3891$.

24. $e^{.33647} = 1.4$.

25. $\ln 3 = 1.0986$.

26. $\log 5 = .6990$.

In Problems 27–50, find x.

27. $\log_3 x = 2$.

28. $\log_2 x = 4$.

29. $\log_5 x = 3$.

30. $\log_4 x = 0$.

31. $\log x = -1$.

32. $\ln x = 1$.

33. $\ln x = 2$.

34. $\log_x 100 = 2$.

35. $\log_x 8 = 3$.

36. $\log_x 3 = \frac{1}{2}$.

37. $\log_x \frac{1}{6} = -1$.

38. $\log_x y = 1$.

39. $\log_4 16 = x$.

40. $\log_3 1 = x$.

41. $\log 10,000 = x$.

42. $\log_2 \frac{1}{16} = x$.

43. $\log_{25} 5 = x$.

44. $\log_9 9 = x$.

45. $\log_3 x = -4$.

46. $\log_x (2x - 3) = 1$.

47. $\log_x (6 - x) = 2$.

48. $\log_8 64 = x - 1$.

49. $2 + \log_2 4 = 3x - 1$.

50. $\log_3 (x + 2) = -2$.

In Problems 51–58, find x and express your answer in terms of logarithms.

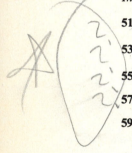

51. $2^x = 5$.

52. $4^{x+3} = 7$.

53. $e^{3x} = 2$. $\ln 2 = 3x$

54. $\dfrac{8}{3^x} = 4$.

55. $5(3^x - 6) = 10$.

56. $.1e^{.1x} = .5$.

57. $e^{2x-5} + 1 = 4$.

58. $3e^{2x} - 1 = \frac{1}{2}$.

59. The predicted population P of a city is given by $P = 125,000(1.12)^{t/20}$, where t is the number of years after 1980. Predict the population in 2000.

60. For a certain city the population P grows at the rate of 2 percent per year. The formula $P = 1,000,000(1.02)^t$ gives the population t years after 1980. Find the population in (a) 1980 and (b) 1982.

61. The probability P that a telephone operator will receive exactly x calls during a certain time interval is

$$P = \frac{e^{-3}3^x}{1 \cdot 2 \cdot 3 \cdot \ldots \cdot x}.$$

Find the probability that exactly two calls will be received. Give answer to four decimal places.

62. Repeat Problem 61 for the case of exactly one call being received.

[handwritten margin notes: R = annual rate of interest; N = # years; S = accum amt; P = principal]

63. Interest is said to be *compounded continuously*[†] if the accumulated amount S of a principal P after n years at an annual rate of r (expressed as a decimal) is given by the formula $S = Pe^{rn}$. Find the amount that $1000 will become after eight years with interest compounded continuously at an annual rate of .05.

64. Use the formula in Problem 63 to find the amount at the end of two years for $100 at an annual rate of .055, compounded continuously.

65. The formula $A = Pe^{-rn}$ gives the amount at the end of n years of a principal P which depreciates at a rate of r (expressed as a decimal) per year compounded continuously. What is the value at the end of ten years of $60,000 of machinery which depreciates at a rate of 8 percent, compounded continuously? Give your answer to the nearest dollar.

66. The demand equation for a certain product is $q = 80 - 2^p$. Sketch its graph and choose q for the horizontal axis.

67. The cost c for a firm producing q units of a product is given by the cost equation $c = (2q \ln q) + 20$. Evaluate the cost when $q = 6$. (Give your answer to two decimal places.)

68. The demand equation for a new toy is $q = 10,000(.95123)^p$. It is desired to evaluate q when $p = 10$. To convert the equation into a more desirable computational form, use Appendix B to show that $q = 10,000e^{-.05p}$. Then evaluate and give your answer to the nearest integer. *Hint:* Find a number x such that $.95123 = e^{-x}$.

69. An important function used in economic and business decisions is the *normal distribution density function*, which in standard form is

$$y = f(x) = \frac{1}{\sqrt{2\pi}} e^{-x^2/2}.$$

Evaluate $f(0)$, $f(-1)$, and $f(1)$ by using $\dfrac{1}{\sqrt{2\pi}} = .399$. Give your answers to three decimal places.

70. In a discussion of market penetration by new products, Hurter and Rubenstein[‡] refer to the function

$$F(t) = \frac{q - pe^{-(t+C)(p+q)}}{q[1 + e^{-(t+C)(p+q)}]},$$

where p, q, and C are constants. They claim that if $F(0) = 0$, then

$$C = -\frac{1}{p+q} \ln \frac{q}{p}.$$

Show that their claim is true.

[†]This concept will be developed in Sec. 7-3.

[‡]A. P. Hurter, Jr., A. H. Rubenstein, et al. "Market penetration by new innovations: the technological literature," *Technological Forecasting and Social Change*, Vol. 11 (1978), 197–221.

71. Suppose that the daily output of units, q, of a new product on the tth day of a production run is given by $q = 500(1 - e^{-.2t})$. Such an equation is called a *learning equation* and indicates that as time progresses, output per day will increase. This may be due to the gain of the workers' proficiencies at their jobs. Determine to the nearest complete unit the output on the (a) first day, and (b) tenth day after the start of a production run. (c) After how many days will a daily production run of 400 units be reached? Assume $\ln 0.2 \approx -1.6$.

5-2 PROPERTIES OF LOGARITHMS

Some basic properties of logarithms deserve mention.

PROPERTY 1. $\log_b (mn) = \log_b m + \log_b n$.

That is, the logarithm of a product is a sum of logarithms.

Proof.   Let $x = \log_b m$ and $y = \log_b n$. Then $b^x = m$ and $b^y = n$. Thus,

$$mn = b^x b^y = b^{x+y}.$$

Since $mn = b^{x+y}$, then $\log_b (mn) = x + y$. Hence,

$$\log_b (mn) = \log_b m + \log_b n.$$

PITFALL.   *The logarithm of a sum is not the sum of logarithms. That is,*

$$\log_b (m + n) \neq \log_b m + \log_b n.$$

Also, $\log_b (mn) \neq (\log_b m)(\log_b n)$.

We shall not prove the next two properties, since their proofs are similar to that of Property 1.

PROPERTY 2. $\log_b \dfrac{m}{n} = \log_b m - \log_b n$.

That is, the logarithm of a quotient is a difference of logarithms.

PITFALL.   *The logarithm of a quotient is not the quotient of logarithms. That is,*

$$\log_b \frac{m}{n} \neq \frac{\log_b m}{\log_b n}.$$

PROPERTY 3. $\log_b m^n = n \log_b m$.

That is, the logarithm of a power of a number is the exponent times a logarithm.

Table 5-1 gives values of a few common logarithms. Most entries are approximate. Notice that $\log 4 \approx .6021$, which means $10^{.6021} \approx 4$. We shall use this table in some of the examples and exercises that follow.

TABLE 5-1   Common Logarithms

x	$\log x$	x	$\log x$
2	.3010	7	.8451
3	.4771	8	.9031
4	.6021	9	.9542
5	.6990	10	1.0000
6	.7782	e	.4343

EXAMPLE 1

a. *Find* log 56.

Log 56 is not in Table 5-1. However, we can write 56 as the product $8 \cdot 7$. Thus by Property 1,

$$\log 56 = \log (8 \cdot 7) = \log 8 + \log 7 \approx .9031 + .8451 = 1.7482.$$

b. *Find* $\log \frac{9}{2}$.

By Property 2,

$$\log \frac{9}{2} = \log 9 - \log 2 \approx .9542 - .3010 = .6532.$$

c. *Find* log 64.

Since $64 = 8^2$, then by Property 3,

$$\log 64 = \log 8^2 = 2 \log 8 \approx 2(.9031) = 1.8062.$$

d. *Find* $\log \sqrt{5}$.

$$\log \sqrt{5} = \log 5^{1/2} = \frac{1}{2} \log 5 \approx \frac{1}{2}(.6990) = .3495.$$

e. *Find* $\log \frac{15}{7}$.

$$\log \frac{15}{7} = \log \frac{3 \cdot 5}{7} = \log (3 \cdot 5) - \log 7$$

$$= \log 3 + \log 5 - \log 7$$

$$\approx .4771 + .6990 - .8451 = .3310.$$

EXAMPLE 2

a. *Simplify* $\log_3 \dfrac{1}{x^2}$.

$$\log_3 \frac{1}{x^2} = \log_3 x^{-2} = -2 \log_3 x \qquad \text{[Property 3].}$$

b. *Write* $-\log \dfrac{x}{2}$ *without using a minus sign.*

$$-\log \frac{x}{2} = (-1) \log \frac{x}{2}$$

$$= \log \left(\frac{x}{2}\right)^{-1} \qquad \text{[Property 3]}$$

$$= \log \frac{2}{x}.$$

 In general, $-\log_b \dfrac{m}{n} = \log_b \dfrac{n}{m}$.

EXAMPLE 3

a. *Write* $\log_4 x - \log_4 (x + 3)$ *as a single logarithm.*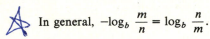

$$\log_4 x - \log_4 (x + 3) = \log_4 \frac{x}{x + 3} \qquad \text{[Property 2].}$$

b. *Write* $3 \log_2 10 + \log_2 15$ *as a single logarithm.*

$$3 \log_2 10 + \log_2 15 = \log_2 (10^3) + \log_2 15 \qquad \text{[Property 3]}$$

$$= \log_2 [(10^3)15] \qquad \text{[Property 1]}$$

$$= \log_2 15,000.$$

c. *Write* $\ln \sqrt[3]{\dfrac{x^5(x - 2)^8}{x - 3}}$ *in terms of* $\ln x, \ln (x - 2),$ *and* $\ln (x - 3)$.

$$\ln \sqrt[3]{\frac{x^5(x - 2)^8}{x - 3}} = \ln \left[\frac{x^5(x - 2)^8}{x - 3}\right]^{1/3} = \frac{1}{3} \ln \frac{x^5(x - 2)^8}{x - 3}$$

$$= \frac{1}{3} \left\{ \ln \left[x^5(x - 2)^8 \right] - \ln (x - 3) \right\}$$

$$= \frac{1}{3} \left[\ln x^5 + \ln (x - 2)^8 - \ln (x - 3) \right]$$

$$= \frac{1}{3} [5 \ln x + 8 \ln (x - 2) - \ln (x - 3)].$$

Since $b^0 = 1$ and $b^1 = b$, then by converting to logarithmic forms we have the following properties:

PROPERTY 4. $\log_b 1 = 0$.

PROPERTY 5. $\log_b b = 1$.

By Property 3, $\log_b b^n = n \log_b b$. However, by Property 5, $\log_b b = 1$. Thus we have the next property.

PROPERTY 6. $\log_b b^n = n$.

EXAMPLE 4

a. *Find* $\ln e$.

$$\ln e = \log_e e = 1 \qquad\qquad \text{[Property 5]}.$$

b. *Find* $\log 10^c$.

$$\log 10^c = \log_{10} 10^c = c \qquad\qquad \text{[Property 6]}.$$

c. *Find* $\log \frac{200}{21}$.

$$\log \frac{200}{21} = \log 200 - \log 21 = \log (2 \cdot 100) - \log (7 \cdot 3)$$

$$= \log 2 + \log 100 - (\log 7 + \log 3)$$

$$\approx .3010 + 2 - (.8451 + .4771) \qquad \text{[since } \log 100 = \log 10^2 = 2 \text{]}$$

$$= .9788.$$

d. *Find* $\log_7 \sqrt[9]{7^8}$.

$$\log_7 \sqrt[9]{7^8} = \log_7 7^{8/9} \doteq \tfrac{8}{9}.$$

e. *Find* $\log_3 \left(\frac{27}{81}\right)$.

$$\log_3 \left(\frac{27}{81}\right) = \log_3 \left(\frac{3^3}{3^4}\right) = \log_3 (3^{-1}) = -1.$$

f. *Find* $\ln e + \log \frac{1}{10}$.

$$\ln e + \log \tfrac{1}{10} = \ln e + \log 10^{-1}$$

$$= 1 + (-1) = 0.$$

For many functions f, if $f(m) = f(n)$, this does not imply that $m = n$. For example, if $f(x) = x^2$ and $m = 2$ and $n = -2$, then $f(m) = f(n)$, but $m \neq n$.

This is not the case for the logarithmic function. Notice in Fig. 5-8 that if x_1 and

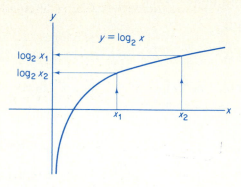

FIG. 5-8

x_2 are different, then their logarithms are different. This means that if $\log_2 m = \log_2 n$, then $m = n$. Generalizing to base b, we have the following property:

PROPERTY 7.   *If* $\log_b m = \log_b n$, *then* $m = n$.

There is a similar property for exponentials:

PROPERTY 8.   If $b^m = b^n$, then $m = n$.

EXAMPLE 5

     a. *Solve* $\log_b x + \log_b (2x) = \log_b 100$ *for x.*

$$\log_b x + \log_b (2x) = \log_b 100,$$
$$\log_b [(x)(2x)] = \log_b 100,$$
$$\log_b (2x^2) = \log_b 100.$$

By Property 7,

$$2x^2 = 100.$$
$$x^2 = 50.$$
$$x = 5\sqrt{2}.$$

(Why do we ignore $x = -5\sqrt{2}$?)

b. *Find x if $(25)^{x+2} = 5^{3x-4}$.*

Since $25 = 5^2$, we can express both sides of the equation as powers of 5.

$$(25)^{x+2} = 5^{3x-4}.$$
$$(5^2)^{x+2} = 5^{3x-4}.$$
$$5^{2x+4} = 5^{3x-4}.$$
$$2x + 4 = 3x - 4 \qquad\qquad \text{[Property 8].}$$
$$x = 8.$$

PROPERTY 9. $b^{\log_b x} = x$ and, in particular, $10^{\log x} = x$ and $e^{\ln x} = x$.

Proof.   Let $t = b^{\log_b x}$. Writing this in logarithmic form, we then have $\log_b t = \log_b x$. By Property 7, $t = x$ and so $x = b^{\log_b x}$.

EXAMPLE 6

a. *Find $2^{\log_2 6}$.*

$$2^{\log_2 6} = 6 \qquad\qquad \text{[Property 9].}$$

b. *Solve $10^{\log x^2} = 25$ for x.*

$$10^{\log x^2} = 25,$$
$$x^2 = 25,$$
$$x = \pm 5.$$

c. *Evaluate $e^{(\ln 3 + 2 \ln 4)}$.*

$$e^{(\ln 3 + 2 \ln 4)} = e^{\ln 3 + \ln 4^2} = e^{\ln(3 \cdot 4^2)} = 3 \cdot 4^2 = 48.$$

Alternately,

$$e^{(\ln 3 + 2 \ln 4)} = e^{\ln 3} e^{2 \ln 4} = e^{\ln 3} e^{\ln 4^2} = 3 \cdot 4^2 = 48.$$

EXAMPLE 7

Find $\log_5 2$.

Let $x = \log_5 2$. Then $5^x = 2$, and by taking common logarithms of both sides we get

$$\log 5^x = \log 2,$$
$$x \log 5 = \log 2,$$
$$x = \frac{\log 2}{\log 5} \approx \frac{.3010}{.6990} \approx .4306.$$

 $\dfrac{\ln 2}{\ln 5} =$ _____

If we had taken natural logarithms of both sides, the result would be $x = (\ln 2)/(\ln 5) \approx$ .69315/1.60944 ≈ .43068. This differs from our previous result due to the accuracy of the tables involved.

Generalizing the method used in Example 7, we have

$$\log_b N = \frac{\log N}{\log b}$$

and $\qquad \log_b N = \dfrac{\log_a N}{\log_a b},$ \hfill (1)

which are called **change of base formulas**. With these we can convert logarithms from one base to another.

EXAMPLE 8

Write $\log x$ *in terms of natural logarithms.*

In Eq. (1), let $b = 10$, $N = x$, and $a = e$. Then

$$\log_{10} x = \frac{\log_e x}{\log_e 10},$$

or $\qquad \log x = \dfrac{\ln x}{\ln 10}.$

EXAMPLE 9

A demand equation for a product is $p = 12^{1 - .1q}$. *Use common logarithms to express* q *in terms of* p.

Taking common logarithms of both sides of $p = 12^{1 - .1q}$ gives

$$\log p = \log (12^{1 - .1q}),$$

$$\log p = (1 - .1q) \log 12,$$

$$\frac{\log p}{\log 12} = 1 - .1q,$$

$$.1q = 1 - \frac{\log p}{\log 12},$$

$$q = 10\left(1 - \frac{\log p}{\log 3 + \log 4}\right),$$

$$q \approx 10\left(1 - \frac{\log p}{1.0792}\right).$$

EXERCISE 5-2

In Problems **1–18**, find the given values. Where necessary, use Table 5-1.

1. log 15.

2. log 16.

3. log $\frac{8}{3}$.

4. log $\frac{7}{10}$.

5. log 36.

6. log .0001.

7. log 2000.

8. log 900.

9. $\log_7 7^{48}$.

10. $\log_5 (5\sqrt{5}\,)^5$.

11. $\log_2 (2^6/2^{10})$.

12. $\log_7 4$.

13. $\log_2 3$.

14. ln e.

15. $\log_3 \sqrt[3]{3}$.

16. $\log_2 4$.

17. log 10 + ln e^3.

18. log 10^e.

In Problems **19–26**, express each of the given forms as a single logarithm.

19. log 7 + log 4.

20. $\log_3 10 - \log_3 5$.

21. $\log_2 (2x) - \log_2 (x + 1)$.

22. 2 log $x - \frac{1}{2}$ log $(x - 2)$.

23. 9 log 7 + 5 log 23.

24. 3 (log x + log y − 2 log z).

25. 2 + 10 log 1.05.

26. $\frac{1}{2}$(log 215 + 8 log 6 − 3 log 121).

In Problems **27–32**, write each expression in terms of log x, log $(x + 2)$, and log $(x - 3)$.

27. log $[x(x + 2)(x - 3)]$.

28. log $\dfrac{x^2(x + 2)}{x - 3}$.

29. log $\dfrac{\sqrt{x}}{(x + 2)(x - 3)^2}$.

30. log $[(x - 3)\sqrt{x(x + 2)}\,]$.

31. log $\sqrt{\dfrac{x^2(x - 3)^3}{x + 2}}$.

32. log $\dfrac{1}{x(x - 3)^2(x + 2)^3}$.

In Problems **33–46**, find x.

33. $e^{2x} \cdot e^{5x} = e^{14}$.

34. $(e^{5x+1})^2 = e$.

35. $(16)^{3x} = 2$.

36. $(27)^{2x+1} = \frac{1}{3}$.

37. $e^{\ln (2x)} = 5$.

38. $4^{\log_4 x + \log_4 2} = 3$.

39. $10^{\log x^2} = 4$.

40. $e^{3 \ln x} = 8$.

41. log $(2x + 1)$ = log $(x + 6)$.

42. log x + log 3 = log 5.

43. log x − log $(x - 1)$ = log 4.

44. $\log_2 x + 3 \log_2 2 = \log_2 (2/x)$.

45. $\log (x + 2)^2 = 2$, where $x > 0$.   **46.** $\ln x = \ln (3x + 1) + 1$.

In Problems 47 and 48, write each expression in terms of natural logarithms.

47. $\log (x + 8)$.   **48.** $\log_2 x$.

49. A manufacturer's supply equation is $p = \log [10 + (q/2)]$, where q is the number of units supplied at a price p per unit. At what price will the manufacturer supply (a) 1980 units; (b) 11,980 units?

50. In statistics, the sample regression equation $y = ab^x$ is reduced to a linear form by taking logarithms of both sides. Find $\log y$.

51. The demand equation for a certain product is $q = 80 - 2^p$. Solve for p and express your answer in terms of common logarithms as in Example 9. Evaluate p to two decimal places when $q = 60$.

52. After t years the number of units, q, of a product sold per year is given by $q = 1000(\frac{1}{2})^{.8^t}$. Such an equation is called a *Gompertz equation* and describes natural growth in many areas of study. Solve this equation for t in the same manner as in Example 9 and show that

$$t = \frac{\log \dfrac{3 - \log q}{\log 2}}{(3 \log 2) - 1}.$$

53. In an article, Taagepera and Hayes[†] refer to an equation of the form

$$\log T = 1.7 + .2068 \log P - .1334 \log^2 P.$$

Here T is the percentage of a country's gross national product (GNP) that corresponds to foreign trade (exports plus imports), and P is the country's population (in units of 100,000). Verify the claim that

$$T = 50 P^{(.2068 - .1334 \log P)}.$$

You may assume $\log 50 = 1.7$.

5-3 REVIEW

IMPORTANT TERMS AND SYMBOLS IN CHAPTER 5

exponential function *(p. 159)* e *(p. 160)*

logarithmic function *(p. 164)*   common logarithm *(p. 165)*

[†]R. Taagepera and J. P. Hayes. "How trade/GNP ratio decreases with country size," *Social Science Research*, Vol. 6 (1977), 108–132.

$\log x$ *(p. 165)*   natural logarithm   *(p. 165)*

$\ln x$ *(p. 165)*

REVIEW SECTION

1. The domain of the exponential function $f(x) = b^x$ is ___(a)___ and its range is
   ___(b)___ .

   > *Ans.* (a) all real numbers;   (b) all positive numbers.

2. The domain of the logarithmic function $g(x) = \log_b x$ is ___(a)___ and its range is
   ___(b)___ .

   if $b^{g(x)} = x$ $b^y = x$

   > *Ans.* (a) all positive numbers;   (b) all real numbers.

3. The graph in Fig. 5-9 is typical of a(n) (exponential)(logarithmic) function.

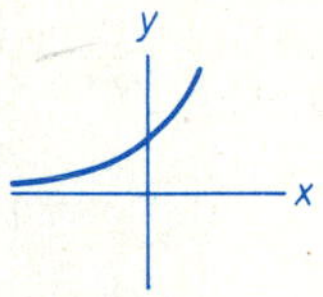

FIG. 5-9

> *Ans.* exponential.

4. $10^{\log 4} =$ _____ .

   > *Ans.* 4.

5. If $\log_2 (x + 1) = \log_2 4$, then $x =$ _____ .

   > *Ans.* 3.

6. The expression $e^{2(\ln x)}$ is equal to the square of what quantity? _____ .

   $x^2 = \#$

   > *Ans.* x.

⟨7.⟩ If $\log x = 1.2222$, then $\log \sqrt{x} =$ _____ .

   $10^{1.2222} = x$

   > *Ans.* .6111.

8. $e^{\ln x} =$ _____ .

   $\log x^{1/2}$

   > *Ans.* x.

9. $\log 10^{5x} =$ _____ .

   $2 \log x$

   > *Ans.* $5x$.

10. $\ln \dfrac{x^2 y^3}{z^4} =$ ___(a)___ $\ln x +$ ___(b)___ $\ln y -$ ___(c)___ $\ln z$.

    > *Ans.* (a) 2;   (b) 3;   (c) 4.

11. The graphs of $y = e^{x+2}$ and $y = e^2 e^x$ (are)(are not) identical.

    > *Ans.* are.

REVIEW PROBLEMS

1. Convert $3^4 = 81$ to logarithmic form.

2. Convert $\log_5 \frac{1}{5} = -1$ to exponential form.

In Problems 3–10, find x.

3. $\log_5 125 = x$.   4. $\log_x \frac{1}{8} = -3$.

5. $\log x = -2$.   6. $\ln \frac{1}{e} = x$.

7. $\log_x (2x + 3) = 2$.   8. $\log (4x + 1) = \log (x + 2)$.

9. $e^{\ln (x+4)} = 7$.   10. $\log x + \log 2 = 1$.

11. Find the value of $\log_3 4$.

12. Find the value of $\log 2500$.

13. If $\log 3 = x$ and $\log 4 = y$, express $\log (16\sqrt{3}\,)$ in terms of x and y.

14. Express

$$\log \frac{x^2 \sqrt{x + 1}}{\sqrt[3]{x^2 + 2}}$$

   in terms of $\log x$, $\log (x + 1)$, and $\log (x^2 + 2)$.

15. Simplify $e^{\ln x} + \ln e^x + \ln 1$.

16. Simplify $\log 10^2 + \log 1000 - 5$.

17. If $\ln y = x^2 + 2$, find y.

18. Sketch the graphs of $y = 3^x$ and $y = \log_3 x$.

19. Due to ineffective advertising, the Kleer-Kut Razor Company finds its annual revenues have been cut sharply. Moreover, the annual revenue R at the end of t years of business satisfies the equation $R = 200{,}000e^{-.2t}$. Find the annual revenue at the end of 2 years; at the end of 3 years.

6-1 COMPOUND INTEREST

In this chapter we shall use mathematics to model selected topics in finance that deal with the time-value of money, such as investments, loans, etc. In later chapters, when more mathematics is at our disposal, certain topics will be revisited and expanded.

Practically everyone is familiar with **compound interest**, whereby the interest earned by an invested sum of money (or **principal**) is reinvested so that it too earns interest. That is, the interest is converted (or *compounded*) into principal and hence there is "interest on interest."

For example, suppose a principal of $100 is invested for two years at the rate of 5 percent compounded annually. After one year the sum of the principal and interest is $100 + .05(100) = \$105$. This is the amount on which interest is earned for the second year, and at the end of that year the value of the investment is $105 + .05(105) = \$110.25$. The \$110.25 represents the original principal plus all accrued interest; it is called the **accumulated amount** or **compound amount**. The difference between the compound amount and the

181

original principal is called the **compound interest**. In the above case the compound interest is $110.25 - 100 = \$10.25$.

More generally, if a principal of P dollars is invested at a rate of $100r$ percent compounded annually (for example, at 5 percent, r is .05), then the compound amount after one year is $P + Pr$ or $P(1 + r)$. At the end of the second year the compound amount is

$$P(1 + r) + [P(1 + r)]r$$
$$= P(1 + r)[1 + r] \qquad \qquad [\text{factoring}]$$
$$= P(1 + r)^2.$$

Similarly, after three years the compound amount is $P(1 + r)^3$. In general, the compound amount S of a principal P at the end of n years at the rate of r compounded annually is given by

$$S = P(1 + r)^n. \qquad \qquad (1)$$

Some approximate values of $(1 + r)^n$ are given in Appendix D.

EXAMPLE 1

If \$1000 *is invested at 6 percent compounded annually,*

a. *find the compound amount after ten years.*

We use Eq. (1) with $P = 1000$, $r = .06$, and $n = 10$.

$$S = 1000(1 + .06)^{10} = 1000(1.06)^{10}.$$

In Appendix D we find that $(1.06)^{10} \approx 1.790848$. Thus,

$$S \approx 1000(1.790848) \approx \$1790.85.$$

b. *find the compound interest after ten years.*

Using the result from part (a), we have

$$\text{compound interest} = S - P$$
$$= 1790.85 - 1000$$
$$= \$790.85.$$

Suppose the principal of \$1000 in Example 1 is invested for ten years as before, but this time the compounding takes place every three months (that is, quarterly) at the rate of $1\frac{1}{2}$ percent *per quarter*. Then there are four *interest periods* or *conversion periods* per year, and in ten years there are $10(4) = 40$

conversion periods. Thus the compound amount with $r = .015$ is

$$1000(1.015)^{40} \approx 1000(1.814018)$$
$$\approx \$1814.02.$$

Usually the interest rate per conversion period is stated as an annual rate. Here we would speak of an annual rate of 6 percent compounded quarterly so that the rate per conversion period is $6\%/4 = 1\frac{1}{2}\%$. This annual rate of 6 percent is called the **nominal rate**. Unless otherwise stated, all interest rates will be assumed to be annual (nominal) rates.

PITFALL. *A nominal rate of 6 percent annually does not necessarily mean that an investment increases in value by 6 percent in a year's time.*

On the basis of our discussion, we can generalize Eq. (1). The formula

$$S = P(1 + r)^n \qquad\qquad (2)$$

gives **the compound amount S of a principal P at the end of n conversion periods at the rate of r per conversion period.**

Sometimes the phrase "money is worth" is used to express an annual interest rate. Thus, saying that money is worth 6 percent compounded quarterly refers to an annual (nominal) rate of 6 percent compounded quarterly.

EXAMPLE 2

The sum of \$3000 is placed in a savings account. If money is worth 6 percent compounded semiannually, what is the balance in the account after seven years? (Assume no other deposits and no withdrawals.)

Here $P = 3000$, the number of conversion periods in $7(2) = 14$, and the rate per conversion period is $.06/2 = .03$. By Eq. (2) we have

$$S = 3000(1.03)^{14} \approx 3000(1.512590) = \$4537.77$$

EXAMPLE 3

How long will it take for \$600 to amount to \$900 at an annual rate of 8 percent compounded quarterly?

The rate per conversion period is $.08/4 = .02$. Let n be the number of conversion periods it takes for a principal of $P = 600$ to amount to $S = 900$. Then from Eq. (2),

$$900 = 600(1.02)^n,$$

$$(1.02)^n = \frac{900}{600},$$

$$(1.02)^n = 1.5.$$

Taking the natural logarithms of both sides, we have

$$n \ln (1.02) = \ln 1.5,$$

$$n = \frac{\ln 1.5}{\ln (1.02)} \approx \frac{.40547}{.01980} \approx 20.478.$$

The number of years that corresponds to 20.478 quarterly conversion periods is $20.478/4 = 5.1195$, which is slightly more than 5 years and 1 month.

If \$1 is invested at a nominal rate of 8 percent compounded quarterly for one year, then the dollar will earn more than 8 percent that year. The compound interest is $S - P = 1(1.02)^4 - 1 \approx 1.082432 - 1 = \$.082432$, which is about 8.24 percent of the original dollar. That is, 8.24 percent is the rate of interest *compounded annually* that is actually obtained, and it is called the **effective rate**. Following this procedure, we can show that the effective rate which corresponds to a nominal rate of r compounded n times a year is given by

$$\boxed{\text{effective rate} = \left(1 + \frac{r}{n}\right)^n - 1.} \qquad (3)$$

We point out that effective rates are used to compare different interest rates, that is, which is "best."

EXAMPLE 4

What effective rate corresponds to a nominal rate of 6 percent compounded semiannually?

From (3), the effective rate is

$$\left(1 + \frac{.06}{2}\right)^2 - 1 = (1.03)^2 - 1 = .0609.$$

The effective rate is 6.09 percent.

EXAMPLE 5

To what amount will \$12,000 accumulate in 15 years if invested at an effective rate of 5 percent?

Since an effective rate is the actual rate compounded annually, we have

$$S = 12{,}000(1.05)^{15} \approx 12{,}000(2.078928)$$
$$\approx \$24{,}947.14.$$

EXAMPLE 6

How many years will it take for a principal of P to double at the effective rate of r?

Let n be the number of years it takes. When P doubles, then the compound amount S is $2P$. Thus $2P = P(1 + r)^n$ and so

$$2 = (1 + r)^n,$$
$$\ln 2 = n \ln (1 + r).$$

Hence,
$$n = \frac{\ln 2}{\ln (1 + r)} \approx \frac{.69315}{\ln (1 + r)}.$$

For example, if $r = .06$, then the number of years it takes to double a principal is approximately

$$\frac{.69315}{\ln (1.06)} \approx \frac{.69315}{.05827} \approx 11.9 \text{ years.}$$

EXAMPLE 7

Calculator Problem

Suppose that $500 amounted to $588.38 in a savings account after three years. If interest was compounded semiannually, find the nominal rate of interest, compounded semiannually, that was earned by the money.

Let r be the semiannual rate. There are six conversion periods. Thus,

$$500(1 + r)^6 = 588.38,$$

$$(1 + r)^6 = \frac{588.38}{500},$$

$$1 + r = \sqrt[6]{\frac{588.38}{500}},$$

$$r = \sqrt[6]{\frac{588.38}{500}} - 1,$$

$$r \approx 1.0275 - 1 = .0275.$$

Thus the semiannual rate was 2.75 percent, and so the nominal rate was $5\frac{1}{2}$ percent compounded semiannually.

EXERCISE 6-1

In Problems **1–8**, *find* (*a*) *the compound amount and* (*b*) *the compound interest for the given investment and annual rate.*

1. $2000 for 5 years at 6 percent compounded annually.

2. $5000 for 20 years at 5 percent compounded annually.

3. $700 for 15 years at 7 percent compounded semiannually.

4. $4000 for 12 years at 6 percent compounded semiannually.

5. $6000 for 8 years at an effective rate of 8 percent.

6. $900 for 11 years at 10 percent compounded quarterly.

7. $5000 for $2\frac{1}{2}$ years at 9 percent compounded monthly.

8. $1000 for $3\frac{3}{4}$ years at 6 percent compounded monthly.

In Problems **9** *and* **10**, *find the effective rate that corresponds to the given nominal rate.*

9. 8 percent compounded quarterly.   **10.** 12 percent compounded monthly.

In Problems **11** *and* **12**, *find how many years it would take to double a principal at the given effective rate. Give your answer to one decimal place.*

11. 8 percent.   **12.** 5 percent.

13. A $6000 certificate of deposit is purchased for $6000 and is held 7 years. If the certificate earns 8 percent compounded quarterly, what is it worth at the end of that period?

14. How many years will it take for a principal of P to triple at the effective rate of r?

15. A major credit-card company has a finance charge of $1\frac{1}{2}$ percent per month on the outstanding indebtedness. (a) What is the nominal rate compounded monthly? (b) What is the effective rate?

16. How long would it take for a principal of P to double if money is worth 12 percent compounded monthly? Give your answer to the nearest month.

17. To what sum will $2000 amount in 8 years if invested at a 6 percent effective rate for the first 4 years and 6 percent compounded semiannually thereafter?

18. How long will it take for $500 to amount to $700 if it is invested at 8 percent compounded quarterly?

19. An investor has a choice of investing a sum of money at 8 percent compounded annually, or at 7.8 percent compounded semiannually. Which is the better of the two rates?

20. *Calculator Problem.* What is the nominal rate of interest compounded quarterly that corresponds to an effective rate of 4 percent?

21. *Calculator Problem.* A bank advertises that it pays interest on savings accounts at the rate of $5\frac{1}{4}$ percent compounded daily. Find the effective rate if the bank assumes that a year consists of (a) 360 days or (b) 365 days in determining the daily rate. Assume that a year consists of 365 days and give your answer to four decimal places.

22. *Calculator Problem.* Suppose that $700 amounted to $801.06 in a savings account after 2 years. If interest was compounded quarterly, find the nominal rate of interest compounded quarterly that was earned by the money.

6-2 PRESENT VALUE

Suppose that $100 is invested for one year at a rate of 6 percent compounded annually. Then the compound amount (or *future value*) of the $100 is $106. Equivalently, the value today (or *present value*) of the $106 due in one year is $100. We can generalize this concept. If we solve the equation that gives compound amount, namely $S = P(1 + r)^n$, for P, we get $P = S/(1 + r)^n$. Thus,

$$\boxed{P = S(1 + r)^{-n}} \tag{1}$$

gives **the principal P which must be invested at the rate of r per conversion period for n conversion periods so that the compound amount is S.** We call P the **present value** of S. Approximate values of $(1 + r)^{-n}$ are given in Appendix D.

EXAMPLE 1

A trust fund for a child's education is being set up by a single payment so that at the end of 15 years there will be $24,000. If the fund earns interest at the rate of 7 percent compounded semiannually, how much money should be paid into the fund initially?

We want the present value of $24,000 due in 15 years. From Eq. (1) with $S = 24,000$, $r = .07/2 = .035$, and $n = 15(2) = 30$, we have

$$P = 24,000(1.035)^{-30} \approx 24,000(.356278)$$

$$\approx \$8550.67.$$

Suppose Mr. Smith owes Mr. Jones two sums of money: $1000 due in two years and $600 due in five years. If Mr. Smith wishes to pay off the total debt now by a single payment, how much would the payment be? Assume an interest rate of 8 percent compounded quarterly.

The single payment x due now must be such that it would grow and eventually pay off the debts when they are due. That is, it must equal the sum of the present values of the future payments. As shown in Fig. 6-1, we have

$$x = 1000(1.02)^{-8} + 600(1.02)^{-20} \qquad (2)$$
$$\approx 1000(.853490) + 600(.672971)$$
$$= 853.490 + 403.78260$$
$$\approx \$1257.27.$$

Thus, the single payment due now is $1257.27. Let us now analyze the situation

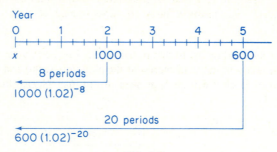

FIG. 6-1

in more detail. There are two methods of payment of the debt: a single payment
now, or two payments in the future. Notice that Eq. (2) indicates that the value
now of all payments under one method must equal the value *now* of all payments
under the other method. In general, this is true not just *now* but at *any time*. For
example, if we multiply both sides of Eq. (2) by $(1.02)^{20}$, we get

$$x(1.02)^{20} = 1000(1.02)^{12} + 600. \qquad (3)$$

The left side of Eq. (3) gives the value five years from now of the single payment
(see Fig. 6-2), while the right side gives the value five years from now of all
payments under the other method. Solving Eq. (3) for x gives the same result,
$x = \$1257.27$. Each of Eqs. (2) and (3) is called an **equation of value**. They
illustrate that when one is considering two methods of paying a debt (or other

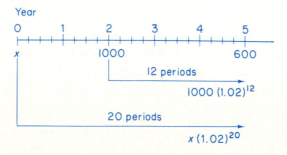

FIG. 6-2

transaction), *at any time* the value of all payments under one method must equal the value of all payments under the other method.

In certain situations one equation of value may be more convenient to use than another, as Example 2 will illustrate.

EXAMPLE 2

A debt of $3000, which is due six years from now, is instead to be paid off by three payments: $500 now, $1500 in three years, and a final payment at the end of five years. What should this payment be if an interest rate of 6 percent compounded annually is assumed?

Let x be the final payment due in five years. For computational convenience we shall set up an equation of value to represent the situation at the end of five years, for in that way the coefficient of x will be 1, as seen in Fig. 6-3. Notice that at year 5 we compute the

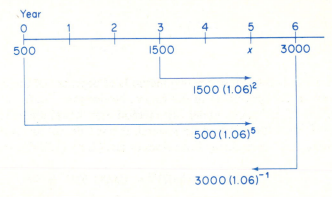

FIG. 6-3

future values of $500 and $1500, and the present value of $3000. The equation of value is

$$500(1.06)^5 + 1500(1.06)^2 + x = 3000(1.06)^{-1},$$
$$500(1.338226) + 1500(1.123600) + x \approx 3000(.943396),$$
$$x \approx \$475.68.$$

When one is considering a choice of two investments, a comparison should be made of the value of each investment at a certain time, as Example 3 shows.

EXAMPLE 3

Suppose that you had the opportunity of investing $4000 in a business such that the value of the investment after five years would be $5300. On the other hand, you could instead put the $4000 in a savings account that pays 6 percent compounded semiannually. Which investment is the better?

Let us consider the value of each investment at the end of five years. At that time the business investment would have a value of $5300, while the savings account would have a value of $4000(1.03)^{10} \approx \$5375.66$. Clearly the better choice is putting the money in the savings account.

If an initial investment will bring in payments at future times, then the payments are called **cash flows**. The **net present value**, denoted NPV, of the cash flows is defined to be the present values of the cash flows, minus the initial investment. If $NPV > 0$, then the investment is profitable; if $NPV < 0$, the investment is not profitable.

EXAMPLE 4

Suppose that you can invest $20,000 in a business that guarantees you the following cash flows at the end of the indicated years:

YEAR	CASH FLOW
2	$10,000
3	$8,000
5	$6,000

Assume an interest rate of 7 percent compounded annually and find the net present value of the cash flows.

Subtracting the initial investment from the sum of the present values of the cash flows gives

$$NPV = 10,000(1.07)^{-2} + 8000(1.07)^{-3} + 6000(1.07)^{-5} - 20,000$$

$$\approx 10,000(.873439) + 8000(.816298) + 6000(.712986) - 20,000$$

$$= 8734.39 + 6530.384 + 4277.916 - 20,000$$

$$= -\$457.31.$$

Note that since $NPV < 0$, the business venture is not profitable if one considers the time-value of money. It would be better to invest the $20,000 in a bank paying 7 percent, since the business venture is equivalent to only investing $20,000 - 457.31 = \$19,542.69$.

EXERCISE 6-2

In Problems 1–6, find the present value of the given future payment at the specified interest rate.

1. $6000 due in 20 years at 5 percent compounded annually.

2. $3500 due in eight years at 6 percent effective.

3. $4000 due in 12 years at 7 percent compounded semiannually.

4. $2500 due in 15 months at 8 percent compounded quarterly.

5. $2000 due in $2\frac{1}{2}$ years at 9 percent compounded monthly.

6. $750 due in three years at 18 percent compounded monthly.

7. A trust fund for a ten-year-old child is being set up by a single payment so that at age 21 the child will receive $27,000. Find how much the payment is if an interest rate of 6 percent compounded semiannually is assumed.

8. A debt of $550 due in four years and $550 due in five years is to be repaid by a single payment now. Find how much the payment is if an interest rate of 10 percent compounded quarterly is assumed.

9. A debt of $600 due in three years and $800 due in four years is to be repaid by a single payment two years from now. If the interest rate is 8 percent compounded semiannually, how much is the payment?

10. A debt of $5000 due in five years is to be repaid by a payment of $2000 now and a second payment at the end of six years. How much should the second payment be if the interest rate is 6 percent compounded quarterly?

11. A debt of $5000 due five years from now and $5000 due ten years from now is to be repaid by a payment of $2000 in two years, a payment of $4000 in four years, and a final payment at the end of six years. If the interest rate is 7 percent compounded annually, how much is the final payment?

12. A debt of $2000 due in three years and $3000 due in seven years is to be repaid by a single payment of $1000 now and two equal payments which are due one year from now and four years from now. If the interest rate is 6 percent compounded annually, how much are each of the equal payments?

13. An initial investment of $25,000 in a business guarantees the following cash flows.

YEAR	CASH FLOW
3	$8,000
4	$10,000
6	$14,000

Assume an interest rate of 5 percent compounded semiannually.
   (a) Find the net present value of the cash flows.
   (b) Is the investment profitable?

14. Repeat Problem 13 for the interest rate of 6 percent compounded semiannually.

15. Suppose that a person has the following choices of investing $10,000:
   (a) placing it in a savings account paying 6 percent compounded semiannually;
   (b) investing in a business such that the value of the investment after eight years is $16,000.
   Which is the better choice?

16. A owes B two sums of money: $1000 plus interest at 7 percent compounded annually, which is due in five years, and $2000 plus interest at 8 percent compounded semiannually, which is due in seven years. If both debts are to be paid off by a single payment at the end of six years, find the amount of the payment if money is worth 6 percent compounded quarterly.

17. *Calculator Problem.* Find the present value of $3000 due in two years at a bank rate of 8 percent compounded daily. Assume that the bank uses 360 days in determining the daily rate and that there are 365 days in a year.

6-3 ANNUITIES

The sequence of numbers

$$3, 6, 12, 24, 48$$

is called a (finite) *geometric sequence*. This is a sequence of numbers, called *terms*, such that each term after the first can be obtained by multiplying the preceding term by the same constant. In our case the constant is 2. If the first term of a geometric sequence is a and the constant is r, then a sequence of n terms has the form

$$a, ar, ar^2, ar^3, \ldots, ar^{n-1}.$$

Note that the ratio of every two consecutive terms is the constant r; that is,

$$\frac{ar}{a} = r, \qquad \frac{ar^2}{ar} = r, \qquad \text{etc.} \qquad (a \neq 0).$$

For this reason we call r the *common ratio*. Note also that the nth term in the sequence is ar^{n-1}.

> **DEFINITION.**   *The sequence of n numbers*
>
> $$a, ar, ar^2, \ldots, ar^{n-1}, \qquad \text{where } a \neq 0,^\dagger$$
>
> *is called a **geometric sequence** with **first term** a and **common ratio** r.*

EXAMPLE 1

a. The geometric sequence with $a = 3$, common ratio $\frac{1}{2}$, and $n = 5$ is

$$3, 3\left(\tfrac{1}{2}\right), 3\left(\tfrac{1}{2}\right)^2, 3\left(\tfrac{1}{2}\right)^3, 3\left(\tfrac{1}{2}\right)^4,$$

or $3, \frac{3}{2}, \frac{3}{4}, \frac{3}{8}, \frac{3}{16}.$

---

† If $a = 0$, the sequence is $0, 0, 0, \ldots, 0$. We shall not consider this uninteresting case.

b. The numbers

$$1, .1, .01, .001$$

form a geometric sequence with $a = 1$, $r = .1$, and $n = 4$.

If we add the terms of the geometric sequence $a, ar, ar^2, \ldots, ar^{n-1}$, the result is called a **geometric series**:

$$a + ar + ar^2 + \ldots + ar^{n-1}. \tag{1}$$

For example,

$$1 + \tfrac{1}{2} + \left(\tfrac{1}{2}\right)^2 + \ldots + \left(\tfrac{1}{2}\right)^6$$

is a geometric series with $a = 1$, common ratio $\tfrac{1}{2}$, and $n = 7$.

Let us consider the sum s of the geometric series in (1):

$$s = a + ar + ar^2 + \ldots + ar^{n-1}. \tag{2}$$

Multiplying both sides by r gives

$$rs = ar + ar^2 + ar^3 + \ldots + ar^n. \tag{3}$$

Subtracting corresponding sides of Eq. (3) from Eq. (2) gives

$$s - rs = a - ar^n,$$
$$s(1 - r) = a(1 - r^n).$$

Thus,

$$\boxed{s = \frac{a(1 - r^n)}{1 - r},} \tag{4}$$

which gives **the sum s of the first n terms of a geometric series**[†] **with first term a and common ratio r.**

EXAMPLE 2

Find the sum of the geometric series $1 + \tfrac{1}{2} + \left(\tfrac{1}{2}\right)^2 + \ldots + \left(\tfrac{1}{2}\right)^6.$

Here $a = 1$, $r = \tfrac{1}{2}$, and $n = 7$. From Eq. (4) we have

$$s = \frac{a(1 - r^n)}{1 - r} = \frac{1\left[1 - \left(\tfrac{1}{2}\right)^7\right]}{1 - \tfrac{1}{2}} = \frac{\frac{127}{128}}{\frac{1}{2}} = \frac{127}{64}.$$

[†] This formula assumes $r \neq 1$. However, if $r = 1$, then $s = a + a + \ldots + a = na$.

The notion of a geometric series is the basis of the mathematical model of an *annuity*. Basically, an **annuity** is a sequence of payments made at fixed periods of time over a given time interval. The fixed period is called the **payment period**, and the given time interval is the **term** of the annuity. An example of an annuity is the depositing of $100 in a savings account every three months for a year.

The **present value of an annuity** is the sum of the *present values* of all the payments. It represents the amount that must be invested now to purchase the payments due in the future. Unless otherwise specified, we shall assume that each payment is made at the *end* of a payment period; that is called an *ordinary annuity*. We shall also assume that interest is computed at the end of each payment period.

Let us consider an annuity of n payments of R (dollars) each, where the interest rate *per period* is r (see Fig. 6-4) and the first payment is due one period

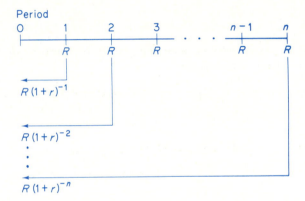

FIG. 6-4

from now. The present value A of the annuity is given by

$$A = R(1 + r)^{-1} + R(1 + r)^{-2} + \ldots + R(1 + r)^{-n}.$$

This is a geometric series of n terms with first term $R(1 + r)^{-1}$ and common ratio $(1 + r)^{-1}$. Hence from Eq. (4) we have

$$A = \frac{R(1 + r)^{-1}\left[1 - (1 + r)^{-n}\right]}{1 - (1 + r)^{-1}}$$

$$= \frac{R\left[1 - (1 + r)^{-n}\right]}{(1 + r)\left[1 - (1 + r)^{-1}\right]} = \frac{R\left[1 - (1 + r)^{-n}\right]}{(1 + r) - 1}.$$

Thus,

$$A = R\frac{1 - (1 + r)^{-n}}{r} \qquad (5)$$

gives **the present value A of an annuity of R (dollars) per payment period for n periods at the rate of r per period.** The expression $[1 - (1 + r)^{-n}]/r$ is denoted $a_{\overline{n}|r}$ and (letting $R = 1$) represents the present value of an annuity of $1 per period. The symbol $a_{\overline{n}|r}$ is read "a angle n at r." Selected values of $a_{\overline{n}|r}$ are given in Appendix D (most are approximate). Thus,

$$A = Ra_{\overline{n}|r}. \qquad (6)$$

EXAMPLE 3

Find the present value of an annuity of $100 per month for $3\frac{1}{2}$ years at an interest rate of 6 percent compounded monthly.

In Eq. (6) we let $R = 100$, $r = .06/12 = .005$, and $n = (3\frac{1}{2})(12) = 42$. Thus

$$A = 100a_{\overline{42}|.005}.$$

From Appendix D, $a_{\overline{42}|.005} \approx 37.798300$. Hence,

$$A \approx 100(37.798300) = \$3779.83.$$

EXAMPLE 4

Given an interest rate of 5 percent compounded annually, find the present value of the following annuity: $2000 due at the end of each year for three years, and $5000 due thereafter at the end of each year for four years (see Fig. 6-5).

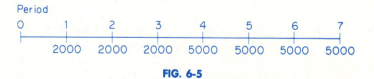

FIG. 6-5

The present value is obtained by summing the present values of all payments:

$$2000(1.05)^{-1} + 2000(1.05)^{-2} + 2000(1.05)^{-3} + 5000(1.05)^{-4} + 5000(1.05)^{-5} +$$

$$5000(1.05)^{-6} + 5000(1.05)^{-7}.$$

Rather than evaluating this expression, we can simplify our work by considering the payments to be an annuity of $5000 for seven years, minus an annuity of $3000 for three years so that the first three payments are $2000 each. Thus the present value is

$$5000a_{\overline{7}|.05} - 3000a_{\overline{3}|.05}$$

$$\approx 5000(5.786373) - 3000(2.723248)$$

$$\approx \$20{,}762.12.$$

EXAMPLE 5

If $10,000 is used to purchase an annuity consisting of equal payments at the end of each year for the next four years and the interest rate is 6 percent compounded annually, find the amount of each payment.

Here $A = \$10{,}000$ and we want to find R. Since $A = Ra_{\overline{n}|r}$, then

$$R = \frac{A}{a_{\overline{n}|r}} = \frac{10{,}000}{a_{\overline{4}|.06}} \approx \frac{10{,}000}{3.465106} \approx \$2885.91.$$

EXAMPLE 6

The premiums on an insurance policy are $50 a quarter, payable at the beginning of each quarter. If the policyholder wishes to pay one year's premiums in advance, how much should be paid provided that the interest rate is 4 percent compounded quarterly?

We want the present value of an annuity of $50 per period for four periods at a rate of 1 percent per period. However, each payment is due at the *beginning* of a payment period. Such an annuity is called an **annuity due**. The given annuity can be thought of as an initial payment of $50 followed by an ordinary annuity of $50 for three periods. Thus the present value is

$$50 + 50a_{\overline{3}|.01} \approx 50 + 50(2.940985) \approx \$197.05.$$

The **amount** (or **future value**) **of an annuity** is the value, at the end of the term, of all payments. That is, it is the sum of the compound amounts of all payments. Let us consider an ordinary annuity of n payments of R (dollars) each, where the interest rate per period is r. The compound amount of the last payment is R, since it occurs at the end of the last interest period and hence does not accrue interest. See Fig. 6-6. The $(n - 1)$th payment earns interest for one period, etc., and the first payment earns interest for $n - 1$ periods. Hence the future value of the annuity is

$$R + R(1 + r) + R(1 + r)^2 + \ldots + R(1 + r)^{n-1}.$$

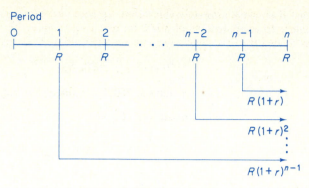

FIG. 6-6

This is a geometric series of n terms with first term R and common ratio $1 + r$. Hence its sum S is [by using Eq. (4)]

$$S = \frac{R\left[1 - (1 + r)^n\right]}{1 - (1 + r)} = R\frac{1 - (1 + r)^n}{-r}$$

$$= R\frac{(1 + r)^n - 1}{r}.$$

Thus,

$$\boxed{S = R\frac{(1 + r)^n - 1}{r}} \qquad (7)$$

gives **the amount S of an annuity of R (dollars) per payment period for n periods at the rate of r per period**. The expression $[(1 + r)^n - 1]/r$ is abbreviated $s_{\overline{n}|r}$ and approximate values of $s_{\overline{n}|r}$ are given in Appendix D. Thus,

$$\boxed{S = Rs_{\overline{n}|r}.} \qquad (8)$$

EXAMPLE 7

a. *Find the amount of an annuity consisting of payments of $50 at the end of every three months for three years at the rate of 6 percent compounded quarterly. Also find the compound interest.*

To find the amount of the annuity we use Eq. (8) with $R = 50$, $n = 4(3) = 12$, and $r = .06/4 = .015$:

$$S = 50s_{\overline{12}|.015} \approx 50(13.041211) \approx \$652.06.$$

The compound interest is the difference between the amount of the annuity and the sum of the payments, namely

$$652.06 - 12(50) = 652.06 - 600 = \$52.06.$$

b. *At the beginning of each quarter, $50 is deposited into a savings account that pays 6 percent compounded quarterly. Find the balance in the account at the end of three years.*

Since the deposits are made at the beginning of a payment period, we want the amount of an *annuity due* as defined in Example 6. See Fig. 6-7. The given annuity can

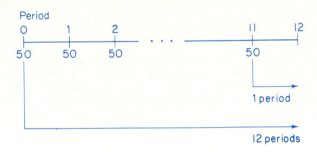

FIG. 6-7

be thought of as an ordinary annuity of $50 for 13 periods minus the final payment of $50. Thus the amount is

$$50s\,\overline{_{13|}}.015 - 50 \approx 50(14.236830) - 50 \approx \$661.84.$$

EXAMPLE 8

A **sinking fund** *is a fund into which periodic payments are made in order to satisfy a future obligation. Suppose a machine costing $7000 is to be replaced at the end of eight years, at which time it will have a salvage value of $700. In order to provide money at that time for a new machine costing the same amount, a sinking fund is set up. The amount in the fund at that time is to be the difference between the replacement cost and the salvage value. If equal payments are placed in the fund quarterly and the fund earns 8 percent compounded quarterly, what should each payment be?*

The amount needed after eight years is $7000 - 700 = \$6300$. Let R be the quarterly payment. The payments into the sinking fund form an annuity with $n = 4(8) = 32$, $r = .08/4 = .02$, and $S = 6300$. Thus, from Eq. (8) we have

$$6300 = Rs\,\overline{_{32|}}.02,$$

$$R = \frac{6300}{s\,\overline{_{32|}}.02} \approx \frac{6300}{44.227030} \approx \$142.45.$$

In general, the formula

$$R = \frac{S}{s_{\overline{n}|r}}$$

gives the periodic payment R of an annuity which is to amount to S.

EXAMPLE 9

A rental firm estimates that, if purchased, a machine will yield an annual net return of $1000 for six years, after which the machine would be worthless. How much should the firm pay for the machine if it wants to earn 7 percent on its investment and also set up a sinking fund to replace the purchase price. For the fund, assume annual payments and a rate of 5 percent compounded annually.

Let x be the purchase price. Each year the return on the investment is $.07x$. Since the machine gives a return of $1000 a year, the amount left to be placed into the fund each year is $1000 - .07x$. These payments must accumulate to x. Hence,

$$(1000 - .07x)s_{\overline{6}|.05} = x,$$

$$1000s_{\overline{6}|.05} - .07xs_{\overline{6}|.05} = x,$$

$$1000s_{\overline{6}|.05} = x(1 + .07s_{\overline{6}|.05}),$$

$$\frac{1000s_{\overline{6}|.05}}{1 + .07s_{\overline{6}|.05}} = x,$$

$$x \approx \frac{1000(6.801913)}{1 + .07(6.801913)}$$

$$\approx \$4607.92.$$

Another way to look at the problem is as follows. Each year the $1000 must account for a profit of $.07x$ and also a payment of $\dfrac{x}{s_{\overline{6}|.05}}$ into the sinking fund. Hence $1000 = .07x + \dfrac{x}{s_{\overline{6}|.05}}$, which when solved gives the same result.

EXERCISE 6-3

In Problems 1 and 2, find the sum of the given geometric series by using Eq. (4) of this section.

1. $\frac{2}{3} + (\frac{2}{3})^2 + \ldots + (\frac{2}{3})^5.$

2. $1 + \frac{1}{4} + (\frac{1}{4})^2 + \ldots + (\frac{1}{4})^5.$

In Problems 3–6, use Appendix D and find the value of the given expression.

3. $a_{\overline{35}|.04}.$

4. $a_{\overline{15}|.07}.$

5. $s_{\overline{8}|.0075}.$

6. $s_{\overline{12}|.005}.$

7. For an annuity of $200 at the end of every six months for $6\frac{1}{2}$ years, find (a) the present value and (b) the future value at an interest rate of 8 percent compounded semiannually.

8. Repeat Problem 7 for an annuity of $800 at the end of every three months for $3\frac{3}{4}$ years at an interest rate of 5 percent compounded quarterly.

9. For an interest rate of 6 percent compounded monthly, find the present value of an annuity of $50 at the end of each month for six months and $75 thereafter at the end of each month for two years.

10. Find the amount of an annuity *due* which consists of ten yearly payments of $100 provided that the interest rate is 6 percent compounded annually.

11. An annuity consisting of equal payments at the end of each quarter for three years is to be purchased for $5000. If the interest rate is 6 percent compounded quarterly, how much is each payment?

12. A person wishes to make a three-year loan and can afford payments of $50 at the end of each month. If interest is at 12 percent compounded monthly, how much can the person afford to borrow?

13. Suppose $50 is placed in a savings account at the end of each month for four years. If no further deposits are made, (a) how much is in the account after *six* years and (b) how much of this is compound interest? Assume that the savings account pays 6 percent compounded monthly.

14. A person borrows $2000 from a bank and agrees to pay it off by equal payments at the end of each month for three years. If interest is at 15 percent compounded monthly, how much is each payment?

15. The beneficiary of an insurance policy has the option of receiving a lump-sum payment of $35,000 or ten equal yearly payments, where the first payment is due at once. If interest is at four percent compounded annually, find the yearly payment.

16. A piece of machinery is purchased for $3000 down and payments of $250 at the end of every six months for six years. If interest is at 8 percent compounded semiannually, find the corresponding cash price of the machinery.

17. In ten years a $40,000 machine will have a salvage value of $4000. A new machine at that time is expected to sell for $52,000. In order to provide funds for the difference between the replacement cost and the salvage value, a sinking fund is set up into which equal payments are placed at the end of each year. If the fund earns 7 percent compounded annually, how much should each payment be?

18. A paper company is considering the purchase of a forest that is estimated to yield an annual return of $50,000 for ten years, after which the forest will have no value. The company wants to earn 8 percent on its investment and also set up a sinking fund to replace the purchase price. If money is placed in the fund at the end of each year and earns 6 percent compounded annually, find the price the manufacturer should pay for the forest. Give your answer to the nearest hundred dollars.

19. A owes B the sum of $5000 and agrees to pay B the sum of $1000 at the end of each

year for five years and a final payment at the end of the sixth year. How much should the final payment be if interest is at 8 percent compounded annually?

20. A debt of $10,000 is being repaid by ten equal semiannual payments with the first payment to be made six months from now. Interest is at the rate of 8 percent compounded semiannually. However, after two years the interest rate increases to 10 percent compounded semiannually. If the debt must be paid off on the original date agreed upon, find the new annual payment. Give your answer to the nearest dollar.

21. In order to replace a machine in the future, a company is placing equal payments into a sinking fund at the end of each year so that after ten years the amount in the fund is $25,000. The fund earns 6 percent compounded annually. After six years, the interest rate increases and the fund pays 7 percent compounded annually. Due to the higher interest rate, the company decreases the amount of the remaining payments. Find the amount of the new payment. Give your answer to the nearest dollar.

In Problems **22–25**, *use the formulas*

$$a_{\overline{n}|r} = \frac{1 - (1 + r)^{-n}}{r},$$

$$s_{\overline{n}|r} = \frac{(1 + r)^{n} - 1}{r},$$

$$R = \frac{A}{a_{\overline{n}|r}} = \frac{Ar}{1 - (1 + r)^{-n}} = \frac{Ar(1 + r)^{n}}{(1 + r)^{n} - 1},$$

$$R = \frac{S}{s_{\overline{n}|r}} = \frac{Sr}{(1 + r)^{n} - 1}.$$

22. *Calculator Problem.* Find $a_{\overline{10}|.073}$.

23. *Calculator Problem.* Find $s_{\overline{60}|.017}$.

24. *Calculator Problem.* A person borrows $2000 and will pay off the loan by equal payments at the end of each month for five years. If interest is at the rate of 16.8 percent compounded monthly, how much is each payment?

25. *Calculator Problem.* Equal payments are to be deposited in a savings account at the end of each quarter for five years so that at the end of that time there will be $3000. If interest is at $5\frac{1}{2}$ percent compounded quarterly, find the quarterly payment.

6-4 AMORTIZATION OF LOANS

Suppose a bank loans you $1500. This amount plus interest is to be repaid by equal payments of R dollars at the end of each month for three months. Furthermore, let us assume that the bank charges interest at the nominal rate of 12 percent compounded monthly. Essentially, for $1500 the bank is purchasing an annuity of three payments of R each. Using formula (6) of the last section, we find that the monthly payment R is

$$R = \frac{A}{a_{\overline{n}|r}} = \frac{1500}{a_{\overline{3}|.01}} \approx \frac{1500}{2.940985} \approx \$510.03.$$

The bank can consider each payment as consisting of two parts: (1) interest on the outstanding loan, and (2) repayment of part of the loan. This is called **amortizing**. A loan is **amortized** when part of each payment is used to pay interest and the remaining part is used to reduce the outstanding principal. Since each payment reduces the outstanding principal, the interest portion of a payment decreases as time goes on. Let us analyze the loan described above.

At the end of the first month, you pay \$510.03. The interest on the outstanding principal is .01(1500) = \$15. The balance of the payment, 510.03 − 15 = \$495.03, is then applied to reduce the principal. Hence the principal outstanding is now 1500 − 495.03 = \$1004.97. At the end of the second month, the interest is .01(1004.97) ≈ \$10.05. Thus the amount of the loan repaid is 510.03 − 10.05 = \$499.98, and the outstanding balance is 1004.97 − 499.98 = \$504.99. The interest due at the end of the third and final month is .01(504.99) ≈ \$5.05, and so the amount of the loan repaid is 510.03 − 5.05 = \$504.98. Hence the outstanding balance is 504.99 − 504.98 = \$0.01. Actually, the debt should now be paid off, and the balance of \$0.01 is due to rounding. Often, banks will change the amount of the last payment to offset this. In the above case the final payment would be \$510.04. An analysis of how each payment in the loan is handled can be given in a table called an **amortization schedule**. See Table 6-1.

TABLE 6-1   Amortization Schedule

PERIOD	PRINCIPAL OUTSTANDING AT BEGINNING OF PERIOD	INTEREST FOR PERIOD	PAYMENT AT END OF PERIOD	PRINCIPAL REPAID AT END OF PERIOD
1	1500	15	510.03	495.03
2	1004.97	10.05	510.03	499.98
3	504.99	5.05	510.03	504.98
TOTALS		30.10	1530.09	1499.99

The total interest paid is \$30.10, which is often called the *finance charge*. As mentioned before, the total of the entries in the last column would equal the original principal were it not for rounding errors.

When one is amortizing a loan, at the beginning of any period the principal outstanding is the present value of the remaining payments. Using this fact together with our previous development, we obtain the formulas listed below that describe the amortization of an interest bearing loan of A dollars, at a rate r per period, by n equal payments of R dollars each and such that a payment is made at the end of each period. Notice below that the formula for the periodic payment R involves $a_{\overline{n}|r}$, which, as you recall, is defined as $[1 - (1 + r)^{-n}]/r$.

1. Periodic payment: $R = \dfrac{A}{a_{\overline{n}|r}} = A\,\dfrac{r}{1 - (1 + r)^{-n}}$.

2. Principal outstanding at beginning of kth period:

$$Ra_{\overline{n-k+1}|r} = R\,\frac{1 - (1 + r)^{-n+k-1}}{r}.$$

3. Interest in kth payment: $Rra_{\overline{n-k+1}|r}$.

4. Principal contained in kth payment: $R[1 - ra_{\overline{n-k+1}|r}]$.

5. Total interest paid: $R(n - a_{\overline{n}|r})$ or $nR - A$.

EXAMPLE 1

Calculator Problem:

A person amortizes a loan of $30,000 for a new home by obtaining a 20-year mortgage at the rate of 9 percent compounded monthly. Find (a) the monthly payment, (b) the interest in the first payment, and (c) the principal repaid in the first payment.

The number of payment periods is $12(20) = 240$, and the interest rate per period is $r = .09/12 = .0075$. The monthly payment R is $30{,}000/a_{\overline{240}|.0075}$. Since $a_{\overline{240}|.0075}$ is not in Appendix D, we use the following equivalent formula and a calculator.

$$R = 30{,}000\left[\frac{.0075}{1 - (1.0075)^{-240}}\right]$$

$$\approx 30{,}000\left[\frac{.0075}{1 - (.166413)}\right]$$

$$\approx \$269.92.$$

The interest portion of the first payment is $30{,}000(.0075) = \$225$. Thus the principal repaid in the first payment is $269.92 - 225 = \$44.92$.

The annuity formula

$$A = R\,\frac{1 - (1 + r)^{-n}}{r}$$

can be solved for n to give the number of periods of a loan:

$$\frac{Ar}{R} = 1 - (1 + r)^{-n},$$

$$(1 + r)^{-n} = 1 - \frac{Ar}{R} = \frac{R - Ar}{R},$$

$$-n \ln (1 + r) = \ln \left(\frac{R - Ar}{R} \right),$$

$$n = - \frac{\ln \left(\dfrac{R - Ar}{R} \right)}{\ln (1 + r)},$$

$$\boxed{n = \frac{\ln \left(\dfrac{R}{R - Ar} \right)}{\ln (1 + r)}.} \qquad (1)$$

EXAMPLE 2

Mr. Smith purchases a stereo system for $1500 and agrees to pay it off by monthly payments of $75. If the store charges interest at the rate of 12 percent compounded monthly, how many months will it take to pay off the debt?

From Eq. (1),

$$n = \frac{\ln \left[\dfrac{75}{75 - 1500(.01)} \right]}{\ln (1.01)}$$

$$= \frac{\ln (1.25)}{\ln (1.01)} \approx \frac{.22314}{.00995} \approx 22.4 \text{ months.}$$

In reality there will be 23 payments; however, the final payment will be less than $75.

EXERCISE 6-4

1. A person is amortizing a 36-month car loan of $7500 with interest at the rate of 12 percent compounded monthly. Find (a) the monthly payment, (b) the interest in the first month, and (c) the principal repaid in the first payment.

2. A person is amortizing a 48-month loan of $10,000 for a house lot. If interest is at the rate of 9 percent compounded monthly, find (a) the monthly payment, (b) the interest in the first payment, and (c) the principal repaid in the first payment.

In Problems **3–6** *construct an amortization schedule for the indicated debts.*

3. $5000 repaid by four equal yearly payments with interest at 7 percent compounded annually.

4. $8000 repaid by six equal semiannual payments with interest at 8 percent compounded semiannually.

5. $900 repaid by five equal quarterly payments with interest at 10 percent compounded quarterly.

6. $10,000 repaid by five equal monthly payments with interest at 9 percent compounded monthly.

7. A loan of $1000 is being paid off by quarterly payments of $100. If interest is at the rate of 8 percent compounded quarterly, how many *full* payments will be made?

8. A loan of $2000 is being amortized over 48 months at an interest rate of 12 percent compounded monthly. Find:
   (a) the monthly payment;
   (b) the principal outstanding at the beginning of the 36th month;
   (c) the interest in the 36th payment;
   (d) the principal in the 36th payment;
   (e) the total interest paid.

9. *Calculator Problem.* A $45,000 mortgage for 25 years for a new home is obtained at the rate of 10.2 percent compounded monthly. Find (a) the monthly payment, (b) the interest in the first payment, (c) the principal repaid in the first payment, and (d) the finance charge.

10. *Calculator Problem.* An automobile loan of $8500 is to be amortized over 48 months at an interest rate of 13.2 percent compounded monthly. Find (a) the monthly payment and (b) the finance charge.

11. *Calculator Problem.* A person purchases furniture for $2000 and agrees to pay off this amount by monthly payments of $100. If interest is charged at the rate of 18 percent compounded monthly, how many *full* payments will there be?

12. *Calculator Problem.* Find the monthly payment of a five-year loan for $7000 if interest is at 12.12 percent compounded monthly.

6-5 REVIEW

IMPORTANT TERMS AND SYMBOLS IN CHAPTER 6

principal *(p. 181)*   compound amount *(p. 181)*

compound interest *(p. 182)*   conversion period *(p. 182)*

nominal rate *(p. 183)*   effective rate *(p. 184)* future value *(p. 187)*   present value *(p. 187)*

equation of value *(p. 188)*   cash flows *(p. 190)*

net present value *(p. 190)*   geometric sequence *(p. 192)*

geometric series *(p. 193)*   annuity *(p. 194)*

present value of annuity *(p. 194)* $a_{\overline{n}|r}$ *(p. 195)*

amount of annuity *(p. 196)* $s_{\overline{n}|r}$ *(p. 197)*

amortizing *(p. 202)*   amortization schedule *(p. 202)*

REVIEW SECTION

1. After three years the compound amount of an initial investment of $100 is $111. The compound interest equals _____.

    Ans. $11.

2. If interest is compounded quarterly, then in four years there are _____ conversion periods.

    Ans. 16.

3. A nominal rate of 12 percent compounded semiannually corresponds to a rate per conversion period of _____.

    Ans. 6%.

4. If compound interest is computed at the rate of $1\frac{1}{2}$ percent per quarter, then the nominal rate is _____.

    Ans. 6%.

5. If interest is compounded annually at an annual rate of 8 percent, then the effective rate is __(a)__ and the nominal rate is __(b)__ .

    Ans. (a) 8%;   (b) 8%.

6. The value today of a sum of money due in the future is called the _____ of the sum.

    Ans. present value.

7. The geometric sequence $1, 1 + r, (1 + r)^2, (1 + r)^3$ has a common ratio of _____.

    Ans. $(1 + r)$.

8. A sequence of payments made at fixed periods of time is called an _____.

    Ans. annuity.

9. A loan of $1000 is amortized over five years at the annual rate of 12 percent compounded monthly. The interest in the first monthly payment is _____.

    Ans. $10.

REVIEW PROBLEMS

1. If $2600 is invested for $6\frac{1}{2}$ years at 6 percent compounded quarterly, find (a) the compound amount and (b) the compound interest.

2. Find the effective rate that corresponds to a nominal rate of 6 percent compounded quarterly.

3. An investor has a choice of investing a sum of money at either 8.5 percent compounded annually or 8.2 percent compounded semiannually. Which is the better choice?

4. Find the net present value of the following cash flows, which can be purchased by an initial investment of $7000. Assume that interest is at 7 percent compounded semiannually.

YEAR	CASH FLOW
2	$3400
4	$3500

5. A debt of $1200 due in four years and $1000 due in six years is to be repaid by a payment of $1000 now and a second payment at the end of two years. How much should the second payment be if interest is at 8 percent compounded semiannually?

6. Find the present value of an annuity of $250 at the end of each month for four years if interest is at 6 percent compounded monthly.

7. Suppose $100 is initially placed in a savings account and $100 is deposited at the end of every six months for the next four years. If interest is at 7 percent compounded semiannually, how much is in the account at the end of four years?

8. A savings account pays interest at the rate of 5 percent compounded semiannually. What amount must be deposited now so that $250 can be withdrawn at the end of every six months for the next ten years?

9. A company borrows $5000 on which it will pay interest at the end of each year at the annual rate of 11 percent. In addition, a sinking fund is set up so that the loan can be repaid at the end of five years. Equal payments are placed in the fund at the end of each year, and the fund earns interest at the effective rate of 6 percent. Find the annual payment *in the sinking fund*.

10. A debtor is to amortize a $7000 car loan by making equal payments at the end of each month for 36 months. If interest is at 12 percent compounded monthly, find (a) the amount of each payment and (b) the finance charge.

11. A person has debts of $500 due in three years with interest at 5 percent compounded annually, and $500 due in four years with interest at 6 percent compounded semiannually. The debtor wants to pay off these debts by making two payments: the first payment now, and the second, which is double the first payment,

at the end of the third year. If money is worth 7 percent compounded annually, how much is the first payment?

12. Construct an amortization schedule for a loan of $2000 repaid by three monthly payments with interest at 12 percent compounded monthly.

13. Construct an amortization schedule for a loan of $15,000 repaid by five monthly payments with interest at 9 percent compounded monthly.

LIMITS AND CONTINUITY

7-1 LIMITS

Our study of calculus will begin in the next chapter. However, since the notion of a *limit* lies at the foundation of calculus, we must develop not only some understanding of that concept, but also insight. We shall first give you a "feeling" for limits by some examples.

Suppose we examine the function

$$f(x) = 2x - 1$$

when x is "near" 2 but not equal to 2. Some values of $f(x)$ for x less than 2 and then greater than 2 are given in Table 7-1. It is apparent that as x takes on values closer to 2, regardless of whether x approaches 2 *from the left* ($x < 2$) or *from the right* ($x > 2$), the corresponding values of $f(x)$ become closer to one number, 3. This is also clear from the graph of f in Fig. 7-1. To express our conclusion we say that 3 is the **limit** of $f(x)$ as x approaches 2. Symbolically we

TABLE 7-1

$x < 2$	$x > 2$
$f(1.7) = 2.4$	$f(2.3) = 3.6$
$f(1.8) = 2.6$	$f(2.2) = 3.4$
$f(1.9) = 2.8$	$f(2.1) = 3.2$
$f(1.99) = 2.98$	$f(2.01) = 3.02$
$f(1.999) = 2.998$	$f(2.001) = 3.002$

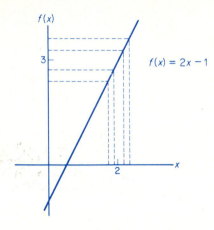

FIG. 7-1

write

$$\lim_{x \to 2} (2x - 1) = 3.$$

Actually we can make the number $f(x)$ as close to 3 as we wish by taking x sufficiently close, but not equal, to 2.

You may think that you can find the limit of a function as x approaches some number a by just evaluating the function when x is a. For the function above this is true: $f(2) = 2(2) - 1 = 3$, which was also the limit. But this method of substitution does not always work. For example, consider the function

$$g(x) = \begin{cases} 2x - 1, & \text{if } x \neq 2, \\ 1, & \text{if } x = 2. \end{cases}$$

Notice that $g(2) = 1$. Let us find the limit of $g(x)$ as x approaches 2, that is, as $x \to 2$. From the graph of g in Fig. 7-2, you can see that as x gets closer to 2 (but

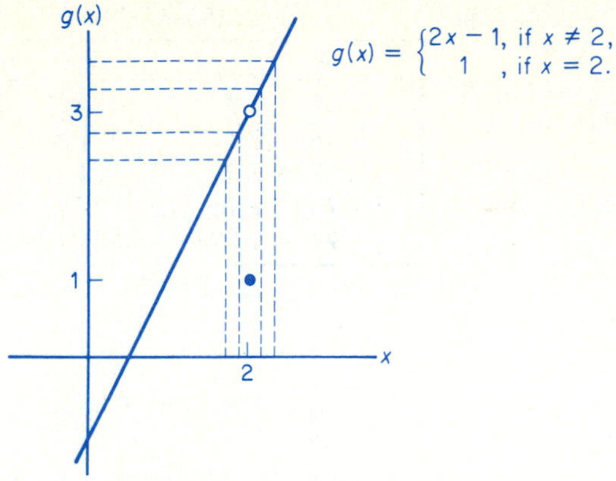

$$g(x) = \begin{cases} 2x - 1, & \text{if } x \neq 2, \\ 1, & \text{if } x = 2. \end{cases}$$

FIG. 7-2

not equal to 2), then $g(x)$ gets closer to 3. Thus,

$$\lim_{x \to 2} g(x) = 3,$$

which is *not* the same as $g(2)$.

Our results can be generalized to any function f. To say that "the limit of $f(x)$, as x approaches a, is L," written

$$\lim_{x \to a} f(x) = L,$$

means that $f(x)$ will be as close to the number L as we please for all x sufficiently close to the number a but not equal to a. Again, here we are not concerned with what happens to $f(x)$ when x equals a, but only what happens to it when x is *close to a*. We emphasize that a limit is independent of the way in which $x \to a$. The limit must be the same whether x approaches a from the left or from the right (for $x < a$ or $x > a$, respectively).

We shall now state some properties of limits which may seem reasonable to you.

I. If $f(x) = c$ is a constant function, then $\lim_{x \to a} f(x) = \lim_{x \to a} c = c$.

II. $\lim_{x \to a} x^n = a^n$, for any positive integer n.

EXAMPLE 1

a. $\lim_{x \to 2} 7 = 7; \quad \lim_{x \to -5} 7 = 7.$

b. $\lim_{x \to 6} x^2 = 6^2 = 36.$

c. $\lim_{t \to -2} t^4 = (-2)^4 = 16.$

Some other properties of limits are

If $\lim_{x \to a} f(x) = L_1$ and $\lim_{x \to a} g(x) = L_2$, where L_1 and L_2 are real numbers, then

III. $\lim_{x \to a} [f(x) \pm g(x)] = \lim_{x \to a} f(x) \pm \lim_{x \to a} g(x) = L_1 \pm L_2.$

This property can be extended to the limit of a finite number of sums and differences.

IV. $\lim_{x \to a} [f(x) \cdot g(x)] = \lim_{x \to a} f(x) \cdot \lim_{x \to a} g(x) = L_1 \cdot L_2.$

V. $\lim_{x \to a} [c\, f(x)] = c \cdot \lim_{x \to a} f(x) = cL_1,$ **where c is a constant.**

EXAMPLE 2

a. $\lim_{x \to 2} (x^2 + x) = \lim_{x \to 2} x^2 + \lim_{x \to 2} x = 2^2 + 2 = 6.$

b. $\lim_{s \to 3} (s^3 - s) = \lim_{s \to 3} s^3 - \lim_{s \to 3} s = 3^3 - 3 = 24.$

c. $\lim_{x \to -1} (x^3 - x + 1) = \lim_{x \to -1} x^3 - \lim_{x \to -1} x + \lim_{x \to -1} 1$

$= (-1)^3 - (-1) + 1 = 1.$

d. $\lim_{x \to 2} [(x + 1)(x - 3)] = \lim_{x \to 2} (x + 1) \cdot \lim_{x \to 2} (x - 3)$

$= [\lim_{x \to 2} x + \lim_{x \to 2} 1] \cdot [\lim_{x \to 2} x - \lim_{x \to 2} 3]$

$= [2 + 1] \cdot [2 - 3] = 3[-1] = -3.$

e. $\lim_{x \to -2} 3x^3 = 3 \lim_{x \to -2} x^3 = 3(-2)^3 = -24.$

EXAMPLE 3

Let $f(x) = c_n x^n + c_{n-1} x^{n-1} + \ldots + c_1 x + c_0$ define a polynomial function f. Then

$$\lim_{x \to a} f(x) = \lim_{x \to a} (c_n x^n + c_{n-1} x^{n-1} + \ldots + c_1 x + c_0)$$

$$= c_n \cdot \lim_{x \to a} x^n + c_{n-1} \cdot \lim_{x \to a} x^{n-1} + \ldots + c_1 \cdot \lim_{x \to a} x + \lim_{x \to a} c_0$$

$$= c_n a^n + c_{n-1} a^{n-1} + \ldots + c_1 a + c_0 = f(a).$$

Thus, **if f is a polynomial function, then**

$$\lim_{x \to a} f(x) = f(a).$$

The result of Example 3 allows us to find many limits by just substituting a for x. For example,

$$\lim_{x \to -3} (x^3 + 4x^2 - 7) = (-3)^3 + 4(-3)^2 - 7 = 2,$$

$$\lim_{h \to 3} \left[2(h - 1) \right] = 2(3 - 1) = 4.$$

Our final two properties will concern limits involving quotients and roots.

If $\lim_{x \to a} f(x) = L_1$ and $\lim_{x \to a} g(x) = L_2$, where L_1 and L_2 are real numbers, then

VI. $\lim_{x \to a} \dfrac{f(x)}{g(x)} = \dfrac{\lim_{x \to a} f(x)}{\lim_{x \to a} g(x)} = \dfrac{L_1}{L_2}$, if $L_2 \neq 0$.

VII. $\lim_{x \to a} \sqrt[n]{f(x)} = \sqrt[n]{\lim_{x \to a} f(x)} = \sqrt[n]{L_1}$, if $\sqrt[n]{L_1}$ is defined.†

EXAMPLE 4

a. $\displaystyle \lim_{x \to 2} \frac{x^2 + 1}{4x - 1} = \frac{\lim_{x \to 2} (x^2 + 1)}{\lim_{x \to 2} (4x - 1)} = \frac{5}{7}.$

b. $\displaystyle \lim_{x \to 1} \frac{2x^2 + x - 3}{x^3 + 4} = \frac{\lim_{x \to 1} (2x^2 + x - 3)}{\lim_{x \to 1} (x^3 + 4)} = \frac{2 + 1 - 3}{1 + 4} = \frac{0}{5} = 0.$

c. $\displaystyle \lim_{t \to 4} \sqrt{t^2 + 1} = \sqrt{\lim_{t \to 4} (t^2 + 1)} = \sqrt{17}.$

d. $\displaystyle \lim_{x \to 3} \sqrt[3]{x^2 + 7} = \sqrt[3]{\lim_{x \to 3} (x^2 + 7)} = \sqrt[3]{16} = \sqrt[3]{8 \cdot 2} = 2\sqrt[3]{2}.$

EXAMPLE 5

a. *Find* $\displaystyle \lim_{h \to 0} \frac{(2 + h)^2 - 4}{h}.$

---

†Strictly speaking, $\sqrt[n]{f(x)}$ must be defined on an open interval containing a.

As $h \to 0$, both numerator and denominator approach zero. Thus we cannot use property VI. However, since what happens when h equals zero is of no concern, we can assume $h \neq 0$ and write

$$\frac{(2 + h)^2 - 4}{h} = \frac{4 + 4h + h^2 - 4}{h} = \frac{4h + h^2}{h} = \frac{h(4 + h)}{h} = 4 + h.$$

The algebraic manipulation on the original function $\dfrac{(2 + h)^2 - 4}{h}$ yielded a new function $4 + h$, which agrees with the original function except when $h = 0$. Thus,

$$\lim_{h \to 0} \frac{(2 + h)^2 - 4}{h} = \lim_{h \to 0} (4 + h) = 4.$$

 Notice that although the original function is not defined at zero, it *does* have a limit as $h \to 0$.

b. *Find* $\displaystyle\lim_{x \to -1} \frac{x^2 - 1}{x + 1}$.

Since both numerator and denominator approach 0 as $x \to -1$, we try to express $(x^2 - 1)/(x + 1)$ in a different form for $x \neq -1$. Here we shall first factor the numerator.

$$\lim_{x \to -1} \frac{x^2 - 1}{x + 1} = \lim_{x \to -1} \frac{(x - 1)(x + 1)}{x + 1} = \lim_{x \to -1} (x - 1) = -2.$$

In Example 5 the method of finding a limit by substitution does not work. In (a), replacing h by 0 gives 0/0, which has no meaning. Similarly, in (b), replacing x by -1 gives 0/0. When the meaningless form 0/0 arises, algebraic manipulation (as in Example 5) may result in a form for which the limit can be determined. In fact, many important limits cannot be evaluated by substitution.

EXERCISE 7-1

In Problems **1–36**, *find the limits.*

1. $\displaystyle\lim_{x \to 2} 16.$

2. $\displaystyle\lim_{x \to 3} 2x.$

3. $\displaystyle\lim_{x \to 4} (x + 3).$

4. $\displaystyle\lim_{s \to 1} 2.$

5. $\displaystyle\lim_{t \to -5} (t^2 - 5).$

6. $\displaystyle\lim_{t \to 1/2} (3t - 5).$

7. $\displaystyle\lim_{x \to 0.3} (3 - 2x^2).$

8. $\displaystyle\lim_{x \to -3} (x^3 - 4).$

9. $\lim\limits_{h \to 6} (h^2 - 5h - 6)$.

10. $\lim\limits_{x \to -2} (x^2 - 2x + 1)$.

11. $\lim\limits_{x \to -1} (x^3 - 3x^2 - 2x + 1)$.

12. $\lim\limits_{r \to 9} \dfrac{4r - 3}{11}$.

13. $\lim\limits_{t \to -3} \dfrac{t - 2}{t + 5}$.

14. $\lim\limits_{x \to -6} \dfrac{x^2 + 6}{x - 6}$.

15. $\lim\limits_{h \to 0} \dfrac{h}{h^2 - 7h + 1}$.

16. $\lim\limits_{h \to 0} \dfrac{h^2 - 2h - 4}{h^3 - 1}$.

17. $\lim\limits_{p \to 4} \sqrt{p^2 + p + 5}$.

18. $\lim\limits_{y \to 9} \sqrt{y + 3}$.

19. $\lim\limits_{x \to -2} \sqrt{\dfrac{4x - 1}{x + 1}}$.

20. $\lim\limits_{x \to -1} \sqrt[3]{x^2}$.

21. $\lim\limits_{x \to 2} \dfrac{(x + 3)\sqrt{x^2 - 1}}{(x - 4)(x + 1)}$.

22. $\lim\limits_{t \to 3} \sqrt{\dfrac{2t + 3}{3t - 5}}$.

23. $\lim\limits_{t \to 1} \dfrac{t^2}{\sqrt[3]{(t^2 - 2)^2}}$.

24. $\lim\limits_{t \to 2} \dfrac{(t + 3)(t + 7)}{(t - 1)(t + 4)}$.

25. $\lim\limits_{x \to -1} \dfrac{x^2 + 2x + 1}{x + 1}$.

26. $\lim\limits_{t \to 1} \dfrac{t^2 - 1}{t - 1}$.

27. $\lim\limits_{x \to 3} \dfrac{x - 3}{x^2 - 9}$.

28. $\lim\limits_{x \to 0} \dfrac{x^2 - 2x}{x}$.

29. $\lim\limits_{x \to 4} \dfrac{x^2 - 9x + 20}{x^2 - 3x - 4}$.

30. $\lim\limits_{x \to 2} \dfrac{x^2 - 2x}{x - 2}$.

31. $\lim\limits_{x \to 1/2} \dfrac{2x^2 + 5x - 3}{4x^2 - 2x}$.

32. $\lim\limits_{x \to -4} \dfrac{x^2 + 2x - 8}{x^2 + 5x + 4}$.

33. $\lim\limits_{x \to 2} \dfrac{3x^2 - x - 10}{x^2 + 5x - 14}$.

34. $\lim\limits_{x \to 0} \dfrac{(x + 2)^2 - 4}{x}$.

35. $\lim\limits_{h \to 0} \dfrac{(2 + h)^2 - 2^2}{h}$.

36. $\lim\limits_{x \to a} \dfrac{x^4 - a^4}{x^2 - a^2}$.

37. Find $\lim\limits_{h \to 0} \dfrac{(x + h)^2 - x^2}{h}$ by treating x as a constant.

38. Find $\lim\limits_{h \to 0} \dfrac{2(x + h)^2 + 5(x + h) - 2x^2 - 5x}{h}$ by treating x as a constant.

39. If $f(x) = x + 5$, show that $\lim\limits_{h \to 0} \dfrac{f(x + h) - f(x)}{h} = 1$ by treating x as a constant.

40. If $f(x) = x^2$, show that $\lim\limits_{h \to 0} \dfrac{f(x + h) - f(x)}{h} = 2x$ by treating x as a constant.

7-2 LIMITS, CONTINUED

Figure 7-3 shows the graph of a function f. Notice that $f(x)$ is not defined when

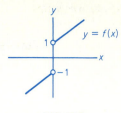

FIG. 7-3

$x = 0$. As x approaches 0 *from the right*, $f(x)$ approaches 1. We write this as

$$\lim_{x \to 0^+} f(x) = 1.$$

On the other hand, as x approaches 0 *from the left*, $f(x)$ approaches -1 and we write

$$\lim_{x \to 0^-} f(x) = -1.$$

Limits like these are called **one-sided limits**. From the last section we know that the limit of a function as $x \to a$ is independent of the way x approaches a. Thus the limit will exist if and only if both one-sided limits exist and are equal. We therefore conclude that

$$\lim_{x \to 0} f(x) \text{ does not exist.}$$

As another example of a one-sided limit, consider $f(x) = \sqrt{x - 3}$ as x approaches 3 (see Fig. 7-4). Since f is defined only when $x \geqslant 3$, we speak of the limit as x approaches 3 from the right. From the diagram it is clear that

$$\lim_{x \to 3^+} \sqrt{x - 3} = 0.$$

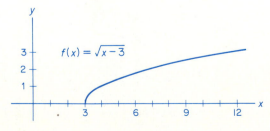

FIG. 7-4

Now let us look at $y = f(x) = 1/x^2$ near $x = 0$. Figure 7-5 shows a table of values of $f(x)$ for x near 0, together with the graph of the function. Notice that as $x \to 0$, both from the left and from the right, $f(x)$ increases without bound. Hence no limit exists at 0. We say that as $x \to 0$, $f(x)$ becomes positively infinite and symbolically we write

$$\lim_{x \to 0} \frac{1}{x^2} = \infty.$$

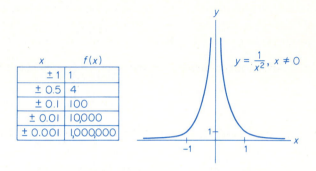

x	$f(x)$
± 1	1
± 0.5	4
± 0.1	100
± 0.01	10,000
± 0.001	1,000,000

FIG. 7-5

PITFALL. *The use of the "equals" sign in this situation does not mean that the limit exists. On the contrary, the symbolism here (∞) is a way of saying specifically that there is no limit and it indicates **why** there is no limit.*

Consider now the graph of $y = f(x) = 1/x$ for $x \neq 0$ (see Fig. 7-6). As x approaches 0 from the right, $1/x$ becomes positively infinite; as x approaches 0

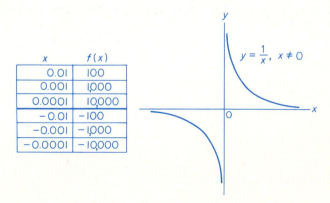

x	$f(x)$
0.01	100
0.001	1,000
0.0001	10,000
-0.01	-100
-0.001	$-1,000$
-0.0001	$-10,000$

FIG. 7-6

from the left, $1/x$ becomes negatively infinite. Symbolically we write

$$\lim_{x \to 0^+} \frac{1}{x} = \infty \quad \text{and} \quad \lim_{x \to 0^-} \frac{1}{x} = -\infty.$$

Either one of these facts implies that

$$\lim_{x \to 0} \frac{1}{x} \text{ does not exist.}$$

Now let us examine this function as x becomes infinite, first in a positive sense and then in a negative sense. From Table 7-2 you can see that as x increases without bound through positive values, the values of $f(x)$ approach 0. Likewise,

TABLE 7-2

x	$f(x)$	x	$f(x)$
1,000	.001	−1,000	−.001
10,000	.0001	−10,000	−.0001
100,000	.00001	−100,000	−.00001
1,000,000	.000001	−1,000,000	−.000001

as x decreases without bound through negative values, the values of $f(x)$ also approach 0. These observations are also apparent from the graph in Fig. 7-6. Symbolically we write

$$\lim_{x \to \infty} \frac{1}{x} = 0 \quad \text{and} \quad \lim_{x \to -\infty} \frac{1}{x} = 0.$$

EXAMPLE 1

Find the limit (if it exists).

a. $\lim\limits_{x \to -1^+} \dfrac{2}{x+1}$.

As x approaches -1 from the right, $x + 1$ approaches 0 but is always positive. Since we are dividing 2 by positive numbers approaching 0, the results, $2/(x + 1)$, are positive numbers that are becoming arbitrarily large. Thus,

$$\lim_{x \to -1^+} \frac{2}{x+1} = \infty,$$

and the limit does not exist.

b. $\lim\limits_{x \to 2} \dfrac{x+2}{x^2 - 4}$.

As $x \to 2$ the numerator approaches 4 and the denominator approaches 0. Thus we are dividing numbers near 4 by numbers near 0. The results are numbers that become arbitrarily large in magnitude. At this stage we can write

$$\lim_{x \to 2} \frac{x + 2}{x^2 - 4} \text{ does not exist.}$$

However, let us see if we can use the symbol ∞ or $-\infty$ to be more specific about "does not exist." Notice that

$$\lim_{x \to 2} \frac{x + 2}{x^2 - 4} = \lim_{x \to 2} \frac{x + 2}{(x + 2)(x - 2)} = \lim_{x \to 2} \frac{1}{x - 2}.$$

Since

$$\lim_{x \to 2^+} \frac{1}{x - 2} = \infty \quad \text{and} \quad \lim_{x \to 2^-} \frac{1}{x - 2} = -\infty,$$

then $\lim_{x \to 2} \dfrac{x + 2}{x^2 - 4}$ is neither ∞ nor $-\infty$.

c. $\lim_{t \to 2} \dfrac{t - 2}{t^2 - 4}.$

As $t \to 2$ both numerator and denominator approach 0. Thus we first simplify the fraction, as we did in Sec. 7-1.

$$\lim_{t \to 2} \frac{t - 2}{t^2 - 4} = \lim_{t \to 2} \frac{t - 2}{(t + 2)(t - 2)} = \lim_{t \to 2} \frac{1}{t + 2} = \frac{1}{4}.$$

d. $\lim_{x \to \infty} \dfrac{4}{(x - 5)^3}.$

As x becomes very large, so does $x - 5$. Since the cube of a large number is also large, $(x - 5)^3 \to \infty$. Dividing 4 by very large numbers results in numbers near 0. Thus,

$$\lim_{x \to \infty} \frac{4}{(x - 5)^3} = 0.$$

In our next discussion we shall need a certain limit, namely, $\lim_{x \to \infty} 1/x^p$ where $p > 0$. As x becomes very large, so does x^p. Dividing 1 by very large numbers results in numbers near 0. Thus,

$$\boxed{\lim_{x \to \infty} \frac{1}{x^p} = 0 \quad \text{for} \quad p > 0.}$$

Let us now turn to the limit of a quotient of two polynomials where the variable becomes infinite. For example, consider

$$\lim_{x \to \infty} \frac{8x^2 + 2x + 3}{2x^3 + 3x - 1}.$$

It is clear that as $x \to \infty$, *both* numerator and denominator become infinite. However, the form of the quotient can be changed so that we can draw a conclusion as to whether or not a limit exists. A frequently used "gimmick" in a case like this is to divide both the numerator and denominator by the largest power of x which occurs in either the numerator or denominator. In our example it is x^3. Since $x \to \infty$, we are concerned only with those values of x which are very large. Thus we can assume $x \neq 0$. This gives

$$\lim_{x \to \infty} \frac{8x^2 + 2x + 3}{2x^3 + 3x - 1} = \lim_{x \to \infty} \frac{\dfrac{8x^2 + 2x + 3}{x^3}}{\dfrac{2x^3 + 3x - 1}{x^3}}$$

$$= \lim_{x \to \infty} \frac{\dfrac{8}{x} + \dfrac{2}{x^2} + \dfrac{3}{x^3}}{2 + \dfrac{3}{x^2} - \dfrac{1}{x^3}}$$

$$= \frac{8 \cdot \lim\limits_{x \to \infty} \dfrac{1}{x} + 2 \cdot \lim\limits_{x \to \infty} \dfrac{1}{x^2} + 3 \cdot \lim\limits_{x \to \infty} \dfrac{1}{x^3}}{\lim\limits_{x \to \infty} 2 + 3 \cdot \lim\limits_{x \to \infty} \dfrac{1}{x^2} - \lim\limits_{x \to \infty} \dfrac{1}{x^3}}.$$

Since $\lim\limits_{x \to \infty} 1/x^p = 0$ for $p > 0$, then

$$\lim_{x \to \infty} \frac{8x^2 + 2x + 3}{2x^3 + 3x - 1} = \frac{8(0) + 2(0) + 3(0)}{2 + 3(0) - 0} = \frac{0}{2} = 0.$$

EXAMPLE 2

a. $\displaystyle \lim_{x \to \infty} \frac{2x + 5}{3x + 2} = \lim_{x \to \infty} \frac{\dfrac{2x + 5}{x}}{\dfrac{3x + 2}{x}} = \lim_{x \to \infty} \frac{2 + \dfrac{5}{x}}{3 + \dfrac{2}{x}} = \frac{2 + 0}{3 + 0} = \frac{2}{3}.$

b. $\displaystyle \lim_{x \to -\infty} \frac{x^2 - 5x}{x^4 + 2x^2 + 1} = \lim_{x \to -\infty} \frac{\dfrac{x^2 - 5x}{x^4}}{\dfrac{x^4 + 2x^2 + 1}{x^4}}$

$$= \lim_{x \to -\infty} \frac{\dfrac{1}{x^2} - \dfrac{5}{x^3}}{1 + \dfrac{2}{x^2} + \dfrac{1}{x^4}}$$

$$= \frac{0 - 0}{1 + 0 + 0} = \frac{0}{1} = 0.$$

c. *Find* $\displaystyle \lim_{x \to -\infty} \frac{10x^2}{x}$.

Here it is not necessary to divide both numerator and denominator by x^2. Instead, we shall first simplify the fraction.

$$\lim_{x \to -\infty} \frac{10x^2}{x} = \lim_{x \to -\infty} 10x.$$

As $x \to -\infty$, the factor 10 remains the same and the factor x becomes negatively infinite. As a result the product becomes negatively infinite. Thus,

$$\lim_{x \to -\infty} \frac{10x^2}{x} = \lim_{x \to -\infty} 10x = -\infty,$$

and no limit exists.

PITFALL.   *To find* $\displaystyle \lim_{x \to 0} \frac{2x + 5}{3x + 2}$, *we have*

$$\lim_{x \to 0} \frac{2x + 5}{3x + 2} = \frac{0 + 5}{0 + 2} = \frac{5}{2}.$$

Here we do not first divide numerator and denominator by x since x does not approach ∞ or $-\infty$.

We conclude this section with a note concerning a most important limit, namely

$$\lim_{x \to 0} (1 + x)^{1/x}.$$

Figure 7-7 shows the graph of $f(x) = (1 + x)^{1/x}$. As $x \to 0$, it is clear that the limit of $(1 + x)^{1/x}$ exists. It is approximately 2.71828 and is denoted by the letter

x	$(1+x)^{1/x}$	x	$(1+x)^{1/x}$
0.5	2.2500	−0.5	4.0000
0.1	2.5937	−0.1	2.8680
0.01	2.7048	−0.01	2.7320
0.001	2.7169	−0.001	2.7196

$f(x)$

$f(x) = (1+x)^{1/x}$

FIG. 7-7

e. This, you recall, is the base of the system of natural logarithms. The limit

$$\lim_{x \to 0} (1 + x)^{1/x} = e$$

can actually be considered the definition of *e*.

EXERCISE 7-2

1. For the function *f* given in Fig. 7-8(a), find the following limits. If the limit does not exist, so state or use the symbol ∞ or −∞ where appropriate.

(a) $\lim_{x \to 1^-} f(x)$, 2   (b) $\lim_{x \to 1^+} f(x)$, 3   (c) $\lim_{x \to 1} f(x)$, doesn't exist the som y's aren't the same

(d) $\lim_{x \to \infty} f(x)$, −∞   (e) $\lim_{x \to -2^-} f(x)$, ∞   (f) $\lim_{x \to -2^+} f(x)$, ∞   it's same coming from $x \to 1^-$ & $x \to 1^+$

find where y coord w/ere

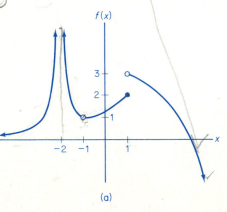

(a)

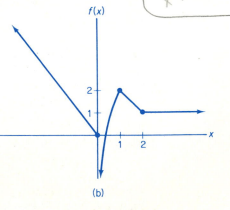

(b)

FIG. 7-8

(g) $\lim\limits_{x \to -2} f(x)$, ∞ (h) $\lim\limits_{x \to -\infty} f(x)$, $\bigcirc$ (i) $\lim\limits_{x \to -1^-} f(x)$, $|$

(j) $\lim\limits_{x \to -1^+} f(x)$, $|$ (k) $\lim\limits_{x \to -1} f(x)$. $|$

2.   For the function f given in Fig. 7-8(b), find the following limits. If the limit does not exist, so state or use the symbol ∞ or $-\infty$ where appropriate.

(a) $\lim\limits_{x \to 0^-} f(x)$,   (b) $\lim\limits_{x \to 0^+} f(x)$,   (c) $\lim\limits_{x \to 0} f(x)$,

(d) $\lim\limits_{x \to -\infty} f(x)$,   (e) $\lim\limits_{x \to 1} f(x)$,   (f) $\lim\limits_{x \to 2^-} f(x)$,

(g) $\lim\limits_{x \to 2^+} f(x)$,   (h) $\lim\limits_{x \to \infty} f(x)$.

*In each of Problems **3–46**, find the limit. If the limit does not exist, so state or use the symbol ∞ or $-\infty$ where appropriate.*

3. $\lim\limits_{x \to 3^+} (x - 2)$.

4. $\lim\limits_{x \to -1^-} (1 - x^2)$.

5. $\lim\limits_{x \to -\infty} 5x$.

6. $\lim\limits_{x \to \infty} 3$.

7. $\lim\limits_{x \to 0^-} \dfrac{6x}{x^4}$.

8. $\lim\limits_{x \to 0} \dfrac{5}{x - 1}$.

9. $\lim\limits_{x \to -\infty} x^2$.

10. $\lim\limits_{t \to \infty} (t - 1)^3$.

11. $\lim\limits_{h \to 0^+} \sqrt{h}$.

12. $\lim\limits_{h \to 5^-} \sqrt{5 - h}$.

13. $\lim\limits_{x \to 5} \dfrac{3}{x - 5}$.

14. $\lim\limits_{x \to 0^-} 2^{1/2}$.

15. $\lim\limits_{x \to 1^+} (4\sqrt{x - 1})$.

16. $\lim\limits_{x \to 2^+} (x\sqrt{x^2 - 4})$.

17. $\lim\limits_{x \to \infty} \dfrac{7}{2x + 1}$.

18. $\lim\limits_{x \to -\infty} \dfrac{1}{(4x - 1)^3}$.

19. $\lim\limits_{x \to \infty} \dfrac{x + 2}{x + 3}$.

20. $\lim\limits_{x \to \infty} \dfrac{2x - 4}{3 - 2x}$.

21. $\lim\limits_{x \to -\infty} \dfrac{x^2 - 1}{x^3 + 4x - 3}$.

22. $\lim\limits_{r \to \infty} \dfrac{r^3}{r^2 + 1}$.

23. $\lim\limits_{t \to \infty} \dfrac{5t^2 + 2t + 1}{4t + 7}$.

24. $\lim\limits_{x \to -\infty} \dfrac{2x}{3x^6 - x + 4}$.

25. $\lim\limits_{x \to \infty} \dfrac{3 - 4x - 2x^3}{5x^3 - 8x + 1}$.

26. $\lim\limits_{x \to \infty} \dfrac{7 - 2x - x^4}{9 - 3x^4 + 2x^2}$.

27. $\lim\limits_{x \to 3^-} \dfrac{x + 3}{x^2 - 9}$.

28. $\lim\limits_{x \to -2^+} \dfrac{2x}{4 - x^2}$.

29. $\lim\limits_{w \to \infty} \dfrac{2w^2 - 3w + 4}{5w^2 + 7w - 1}$.

30. $\lim\limits_{x \to \infty} \dfrac{4 - 3x^3}{x^3 - 1}$.

31. $\lim\limits_{x \to -5} \dfrac{2x^2 + 9x - 5}{x^2 + 5x}$.

32. $\lim\limits_{t \to 2} \dfrac{t^2 + 2t - 8}{2t^2 - 5t + 2}$.

33. $\lim\limits_{x \to 1} \dfrac{x^2 - 3x + 1}{x^2 + 1}$.

34. $\lim\limits_{x \to -1} \dfrac{3x^3 - x^2}{2x + 1}$.

35. $\lim\limits_{x \to 1^+} \left[1 + \dfrac{1}{x - 1} \right]$.

36. $\lim\limits_{x \to -\infty} \dfrac{x^3 + 2x^2 + 1}{x^3 - 4}$.

37. $\lim\limits_{x \to 0^+} \dfrac{2}{x + x^2}$.

38. $\lim\limits_{x \to \infty} \left(x + \dfrac{1}{x} \right)$.

39. $\lim\limits_{x \to 1} x(x - 1)^{-1}$.

40. $\lim\limits_{x \to 1/2} \dfrac{1}{2x - 1}$.

41. $\lim\limits_{x \to 0^+} \left(-\dfrac{3}{x} \right)$.

42. $\lim\limits_{x \to 0} \left(-\dfrac{3}{x} \right)$.

43. $\lim\limits_{x \to 0} |x|$.

44. $\lim\limits_{x \to 0} \left| \dfrac{1}{x} \right|$.

45. $\lim\limits_{x \to -\infty} \dfrac{x + 1}{x}$.

46. $\lim\limits_{x \to \infty} \left[\dfrac{2}{x} - \dfrac{x^2}{x^2 - 1} \right]$.

*In Problems **47–50**, sketch the graphs of the functions and find the indicated limits. If the limit does not exist, so state or use the symbol ∞ or $-\infty$ where appropriate.*

47. $f(x) = \begin{cases} 2, & \text{if } x \leqslant 2 \\ 1, & \text{if } x > 2 \end{cases}$; (a) $\lim\limits_{x \to 2^+} f(x)$, (b) $\lim\limits_{x \to 2^-} f(x)$, (c) $\lim\limits_{x \to 2} f(x)$,

(d) $\lim\limits_{x \to \infty} f(x)$, (e) $\lim\limits_{x \to -\infty} f(x)$.

48. $f(x) = \begin{cases} x, & \text{if } x \leqslant 1 \\ 2, & \text{if } x > 1 \end{cases}$; (a) $\lim\limits_{x \to 1^+} f(x)$, (b) $\lim\limits_{x \to 1^-} f(x)$, (c) $\lim\limits_{x \to 1} f(x)$,

(d) $\lim\limits_{x \to \infty} f(x)$, (e) $\lim\limits_{x \to -\infty} f(x)$.

49. $g(x) = \begin{cases} x, & \text{if } x < 0 \\ -x, & \text{if } x > 0 \end{cases}$; (a) $\lim\limits_{x \to 0^+} g(x)$, (b) $\lim\limits_{x \to 0^-} g(x)$, (c) $\lim\limits_{x \to 0} g(x)$,

(d) $\lim\limits_{x \to \infty} g(x)$, (e) $\lim\limits_{x \to -\infty} g(x)$.

50. $g(x) = \begin{cases} x^2, & \text{if } x < 0 \\ x, & \text{if } x > 0 \end{cases}$; (a) $\lim\limits_{x \to 0^+} g(x)$, (b) $\lim\limits_{x \to 0^-} g(x)$, (c) $\lim\limits_{x \to 0} g(x)$,

(d) $\lim\limits_{x \to \infty} g(x)$, (e) $\lim\limits_{x \to -\infty} g(x)$.

51. If c is the total cost in dollars to produce q units of a product, then the average cost per unit $\bar{c}$ for an output of q units is given by $\bar{c} = c/q$. Thus, if the total cost equation is $c = 5000 + 6q$, then $\bar{c} = (5000/q) + 6$. For example, the total cost of an output of 5 units is \$5030, and the average cost per unit at this level of production is \$1006. By finding $\lim\limits_{q \to \infty} \bar{c}$, show that the average cost approaches a level of stability if the producer continually increases output. What is the limiting value of the average cost? Sketch the graph of the average cost function.

52. Repeat Problem 51 given that fixed cost is $12,000 and the variable cost is given by the function $c_v = 7q$.

53. The population N of a certain small city t years from now is predicted to be

$$N = 20,000 + \frac{10,000}{(t + 2)^2}.$$

Determine the population in the long run; that is, find $\lim_{t \to \infty} N$.

In Problems **54–57,** *find* $\lim_{h \to 0} \dfrac{f(x + h) - f(x)}{h}$ *by treating x as a constant.*

54. $f(x) = 2x + 3$.   **55.** $f(x) = 4 - x$.

56. $f(x) = x^2 + x + 1$.   **57.** $f(x) = x^2 - 3$.

(Calculator Problems) In Problems **58** *and* **59,** *evaluate the given function when $x = 1, .5, .2, .1, .01, .001,$ and $.0001$. From your results draw a conclusion about* $\lim_{x \to 0^+} f(x)$.

58. $f(x) = x \ln x$.   **59.** $f(x) = x^{2x}$.

7-3 INTEREST COMPOUNDED CONTINUOUSLY

If a principal of P dollars is invested and interest is compounded k times a year at an annual rate of r, then the rate per conversion period is r/k. In t years there are kt periods. From Chapter 6 the compound amount S at the end of t years is

$$S = P\left(1 + \frac{r}{k}\right)^{kt}.$$

If $k \to \infty$, the number of conversion periods increases indefinitely and the length of each period approaches 0. In this case we say that interest is **compounded continuously**, that is, at every instant of time. The compound amount is

$$\lim_{k \to \infty} P\left(1 + \frac{r}{k}\right)^{kt},$$

which may be written

$$P\left[\lim_{k \to \infty}\left(1 + \frac{r}{k}\right)^{k/r}\right]^{rt}.$$

By letting $x = r/k$, then as $k \to \infty$ we have $x \to 0$. Thus the limit inside the brackets has the form $\lim_{x \to 0}(1 + x)^{1/x}$ which, as we saw in Sec. 7-2, is e.

Therefore,

$$S = Pe^{rt}$$

is the compound amount S of a principal of P dollars after t years at an annual interest rate r compounded continuously.

EXAMPLE 1

If $100 is invested at an annual rate of 5 percent compounded continuously, find the compound amount at the end of (a) 1 year and (b) 5 years.

a. Here $P = 100$, $r = .05$, and $t = 1$.

$$S = Pe^{rt} = 100e^{(.05)(1)} \approx 100(1.0513) = \$105.13.$$

We can compare this value with the value after one year of a $100 investment at an annual rate of 5 percent compounded semiannually—namely, $100(1.025)^2 \approx \$105.06$. The difference is not significant.

b. Here $P = 100$, $r = .05$, and $t = 5$.

$$S = 100e^{(.05)(5)} = 100e^{.25} \approx 100(1.2840) = \$128.40.$$

We can find an expression that gives the effective rate which corresponds to an annual rate of r compounded continuously. If i is the corresponding effective rate, then after one year a principal P accumulates to $P(1 + i)$. This must equal the accumulated amount under continuous interest, Pe^r. Thus $P(1 + i) = Pe^r$ or $1 + i = e^r$ and so $i = e^r - 1$. Therefore,

$$e^r - 1$$

is the **effective rate corresponding to an annual rate of r compounded continuously.**

EXAMPLE 2

Find the effective rate which corresponds to an annual rate of 5 percent compounded continuously.

The effective rate is

$$e^r - 1 = e^{.05} - 1 \approx 1.0513 - 1 = .0513 \quad \text{or} \quad 5.13\%.$$

If we solve $S = Pe^{rt}$ for P, we get $P = S/e^{rt}$ or

$$P = Se^{-rt},$$

which is the **present value of S dollars due at the end of t years at an annual rate of r compounded continuously.** That is, P is the principal that must be invested now so that it will accumulate to S after t years.

EXAMPLE 3

A trust fund is being set up by a single payment so that at the end of 20 years there will be $25,000 in the fund. If interest is compounded continuously at an annual rate of 7 percent, how much money should be paid into the fund initially?

We want the present value of $25,000 due in 20 years.

$$P = Se^{-rt} = 25{,}000e^{-(.07)(20)}$$
$$= 25{,}000e^{-1.4} \approx 25{,}000(.24660)$$
$$= 6165.$$

The present value is $6165.

EXERCISE 7-3

*In Problems **1** and **2**, find the compound amount and compound interest if $4000 is invested for six years and interest is compounded continuously at the given annual rate.*

1. $5\frac{1}{2}\%$. $=5.5$ **2.** 9%.

*In Problems **3** and **4**, find the present value of $2500 due eight years from now if interest is compounded continuously at the given annual rate.*

3. $6\frac{3}{4}\%$.   **4.** 8%.

*In Problems **5–8**, find the effective rate of interest which corresponds to the given annual rate compounded continuously.*

5. 4%.   **6.** 7%.   **7.** 10%.   **8.** 9%.

9. If $100 is deposited in a savings account that earns interest at an annual rate of $5\frac{1}{2}$ percent compounded continuously, what is the value of the account at the end of two years?

10. If $1000 is invested at an annual rate of 6 percent compounded continuously, find the compound amount at the end of eight years.

$$S = Pe^{RT}$$
$$S = 1000e^{(.06)(8)} = 1000e^{.48} = 1620$$

11. The board of directors of a corporation agrees to redeem some of its callable preferred stock in five years. At that time $1,000,000 will be required. If the corporation can invest money at an annual interest rate of 6 percent compounded continuously, how much should it presently invest so that the future value is sufficient to redeem the shares?

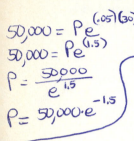

$$50,000 = Pe^{(.05)(30)}$$
$$50,000 = Pe^{(1.5)}$$
$$P = \frac{50,000}{e^{1.5}}$$
$$P = 50,000 \cdot e^{-1.5}$$

12. A trust fund is being set up by a single payment so that at the end of 30 years there will be $50,000 in the fund. If interest is compounded continuously at an annual rate of 5 percent, how much money should be paid into the fund initially?

13. What annual rate compounded continuously is equivalent to an effective rate of 5 percent?

14. What annual rate r compounded continuously is equivalent to a nominal rate of 6 percent compounded semiannually? *Hint:* First show that $r = 2 \ln 1.03$.

15. An annuity in which R dollars are paid each year by uniform payments that are payable continuously is called a *continuous annuity* or a *continuous income stream.* The present value of a continuous annuity for t years is

$$R\frac{1 - e^{-rt}}{r},$$

where r is the annual rate of interest compounded continuously. Find the present value of a continuous annuity of $100 a year for twenty years at 4 percent compounded continuously. Give your answer to the nearest dollar.

16. Suppose a business has an annual profit of $40,000 for the next five years and the profits are earned continuously throughout each year. Then the profits can be thought of as a continuous annuity (see Problem 15). If money is worth 5 percent compounded continuously, find the present value of the profits.

17. If interest is compounded continuously at an annual rate of .05, how many years would it take for a principal P to triple? Give your answer to the nearest year.

18. If interest is compounded continuously, at what annual rate will a principal of P double in ten years? Give your answer to the nearest percent.

7-4 CONTINUITY

Let us consider the functions

$$f(x) = x \quad \text{and} \quad g(x) = \begin{cases} x, & \text{if } x \neq 1, \\ 2, & \text{if } x = 1. \end{cases}$$

Their graphs appear in Fig. 7-9 and Fig. 7-10, respectively. The significant difference between the graphs is that there is a "break" in the graph of g when $x = 1$, while there is *no* "break" at all in the graph of f. Stated another way, if you were to trace both graphs with a pencil, you would have to lift the pencil on the graph of g when $x = 1$, but you would not have to lift it on the graph of f.

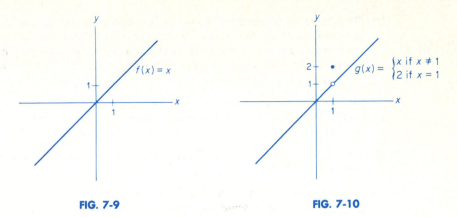

FIG. 7-9   **FIG. 7-10**

These situations can be expressed by limits. As x approaches 1,

$$\lim_{x \to 1} f(x) = 1 = f(1),$$

while $\lim_{x \to 1} g(x) = 1 \neq g(1) = 2.$

The limit of f as $x \to 1$ is the same as $f(1)$, but the limit of g as $x \to 1$ is *not* the same as $g(1)$. For these reasons we say that f is *continuous* at $x = 1$ and g is *discontinuous* at $x = 1$.

DEFINITION. *A function f is **continuous** at $x = a$ if and only if the following three conditions are met:*

(1) *$f(x)$ is defined at $x = a$, that is, a is in the domain of f,*

(2) $\lim_{x \to a} f(x)$ *exists,*

(3) $\lim_{x \to a} f(x) = f(a).$

DEFINITION. *A function is **discontinuous** at $x = a$ if and only if it is not continuous at $x = a$.*

Continuous functions have many useful properties that discontinuous functions do not have. These properties are important not only from a purely mathematical point of view, but also from the point of view of economics. We shall say more about this in Sec. 7-6.

EXAMPLE 1

a. *Show that $f(x) = 5$ is continuous at $x = 7$.*

We must verify that three conditions are met. First, f is defined at $x = 7$: $f(7) = 5$. Second,

$$\lim_{x \to 7} f(x) = \lim_{x \to 7} 5 = 5.$$

Thus f has a limit as $x \to 7$. Third,

$$\lim_{x \to 7} f(x) = 5 = f(7).$$

Therefore $f(x) = 5$ is continuous at $x = 7$.

b. *Show that $g(x) = x^2 - 3$ is continuous at $x = -4$.*

The function g is defined at $x = -4$; $g(-4) = 13$. Also,

$$\lim_{x \to -4} g(x) = \lim_{x \to -4} (x^2 - 3) = 13 = g(-4).$$

Therefore $g(x) = x^2 - 3$ is continuous at $x = -4$.

We say that a function is *continuous on an interval* if it is continuous at each point there. In such a situation, the function has a graph that is connected over the interval. For example, $f(x) = x^2$ is continuous on $[2, 5]$. In fact, in Example 3 of Sec. 7-1 we showed that for *any* polynomial function f, $\lim_{x \to a} f(x) = f(a)$. Thus, **a polynomial function is continuous at every point** and hence on every interval. We say that polynomial functions are **continuous everywhere**, or, more simply, that they are continuous.

EXAMPLE 2

The functions $f(x) = 7$ and $g(x) = x^3 - 9x + 3$ are polynomial functions. Therefore they are continuous. For example, they are continuous at $x = 3$.

If a function is not defined at a, it is automatically discontinuous there. If it *is* defined at a, then it is discontinuous at a if

(1) it has no limit as $x \to a$.

or   (2) as $x \to a$ it has a limit that is different from $f(a)$.

In Fig. 7-11 we can find points of discontinuity by inspection.

EXAMPLE 3

a. Let $f(x) = 1/x$ (see Fig. 7-12). Since f is not defined at $x = 0$, it is discontinuous there. Moreover, $\lim_{x \to 0^+} f(x) = \infty$ and $\lim_{x \to 0^-} f(x) = -\infty$. A function is said to have an **infinite discontinuity** at $x = a$ when at least one of the one-sided limits is either ∞ or $-\infty$ as $x \to a$. Hence f has an *infinite discontinuity* at $x = 0$.

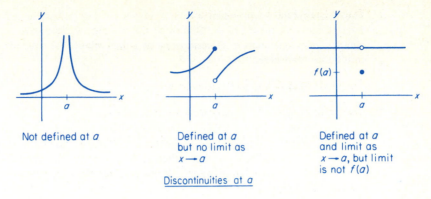

Not defined at a

Defined at a
but no limit as
$x \to a$

Defined at a
and limit as
$x \to a$, but limit
is not $f(a)$

Discontinuities at a

FIG. 7-11

b. Let $f(x) = \begin{cases} 1, & \text{if } x > 0, \\ 0, & \text{if } x = 0, \\ -1, & \text{if } x < 0. \end{cases}$ (see Fig. 7-13). Although f is defined at $x = 0$, $\lim\limits_{x \to 0} f(x)$

does not exist. Thus f is discontinuous at $x = 0$.

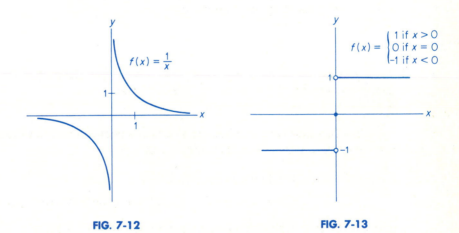

FIG. 7-12   **FIG. 7-13**

EXAMPLE 4

Find any points of discontinuity for each of the following.

a. $f(x) = \dfrac{x^4 - 3x^3 + 2x - 1}{x^2 - 4}$.

The denominator is zero when $x = \pm 2$. Hence f is not defined at ± 2 and is therefore discontinuous at these points. Otherwise the function is "well-behaved." In fact, **any quotient of polynomials is discontinuous at points where the denominator is 0, and is continuous elsewhere.**

b. $g(x) = \begin{cases} x + 6, & \text{if } x \geqslant 3, \\ x^2, & \text{if } x < 3. \end{cases}$

The only possible trouble may occur when $x = 3$ because this is the only place at which the graph of g could be disconnected. We know $g(3) = 3 + 6 = 9$. As $x \to 3^+$, then $g(x) \to 3 + 6 = 9$. As $x \to 3^-$, then $g(x) \to 3^2 = 9$. Thus the function is continuous at $x = 3$ as well as at all other x. We can reach the same conclusion by inspecting the graph of g (Fig. 7-14).

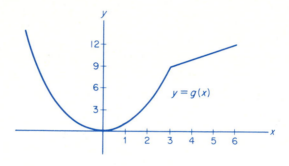

FIG. 7-14

c. $f(x) = \begin{cases} x + 2, & \text{if } x > 2, \\ x^2, & \text{if } x < 2. \end{cases}$

Since f is not defined at $x = 2$, it is discontinuous there. It is continuous for all other x.

EXAMPLE 5

Table 7-3 lists redemption values of a fifty-dollar savings bond for the first six successive periods after the date of issue. If y is the redemption value x years after the date of issue, then y is a function of x: $y = f(x)$. It is clear from the graph of this function in Fig. 7-15 that f has discontinuities when $x = \frac{1}{2}$, 1, $1\frac{1}{2}$, 2, and $2\frac{1}{2}$ and is constant for values of x between successive discontinuities. Such a function is called a *step function* because of the appearance of its graph.

There is another way to express continuity besides that given in the

TABLE 7-3

NUMBER OF YEARS AFTER ISSUE DATE		REDEMPTION VALUE
(greater than) —	(not more than)	
0 —	$\frac{1}{2}$	$37.50
$\frac{1}{2}$ —	1	38.10
1 —	$1\frac{1}{2}$	39.02
$1\frac{1}{2}$ —	2	39.90
2 —	$2\frac{1}{2}$	40.80
$2\frac{1}{2}$ —	3	41.76

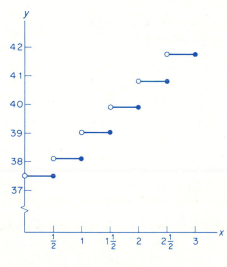

FIG. 7-15

definition. If we take the statement

$$\lim_{x \to a} f(x) = f(a)$$

and replace x by $a + h$, then as $x \to a$ we have $h \to 0$ (Fig. 7-16). Thus the statement

$$\lim_{h \to 0} f(a + h) = f(a)$$

defines continuity at a.

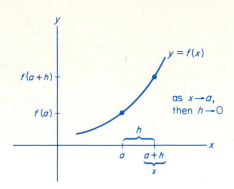

FIG. 7-16

EXERCISE 7-4

In Problems **1–6**, *use the definition of continuity to show that the given function is continuous at the indicated point.*

1. $f(x) = x^3 - 5x$, $x = 2$.

2. $f(x) = \dfrac{x - 3}{9x}$, $x = -3$.

3. $g(x) = \sqrt{2 - 3x}$, $x = 0$.

4. $f(x) = \frac{1}{8}$, $x = 2$.

5. $h(x) = \dfrac{x - 4}{x + 4}$, $x = 4$.

6. $f(x) = \sqrt[3]{x}$, $x = -1$.

In Problems **7–12** *determine whether the function is continuous at the given points.*

7. $f(x) = \dfrac{x + 4}{x - 2}$; $-2, 0$.

8. $f(x) = \dfrac{x^2 - 4x + 4}{6}$; $2, -2$.

9. $g(x) = \dfrac{x - 3}{x^2 - 9}$; $3, -3$.

10. $h(x) = \dfrac{3}{x^2 + 4}$; $2, -2$.

11. $F(x) = \begin{cases} x + 2, & \text{if } x \geqslant 2 \\ x^2, & \text{if } x < 2 \end{cases}$; $2, 0$.

12. $f(x) = \begin{cases} \dfrac{1}{x}, & \text{if } x \neq 0 \\ 0, & \text{if } x = 0 \end{cases}$; $0, -1$.

In Problems **13–16**, *state why the functions are continuous everywhere.*

13. $f(x) = 2x^2 - 3$.

14. $f(x) = \dfrac{x + 2}{5}$.

15. $f(x) = \dfrac{x - 1}{x^2 + 4}$.

16. $f(x) = x(1 - x)$.

In Problems **17–34**, *find all points of discontinuity.*

17. $f(x) = 3x^2 - 3$.

18. $h(x) = x - 2$.

19. $f(x) = \dfrac{3}{x - 4}$.

20. $f(x) = \dfrac{x^2 + 3x - 4}{x + 4}$.

21. $g(x) = \dfrac{(x^2 - 1)^2}{5}.$

22. $f(x) = \begin{cases} 5, & \text{if } x \geqslant 3, \\ 2x - 1, & \text{if } x < 3. \end{cases}$

23. $f(x) = \dfrac{x^2 + 6x + 9}{x^2 + 2x - 15}.$

24. $g(x) = \dfrac{x - 3}{x^2 + x}.$

25. $h(x) = \dfrac{x - 7}{x^3 - x}.$

26. $f(x) = \dfrac{x}{x}.$

27. $p(x) = \dfrac{x}{x^2 + 1}.$

28. $f(x) = \dfrac{x^4}{x^4 - 1}.$

29. $f(x) = \begin{cases} x^2, & \text{if } x > 2, \\ x - 1, & \text{if } x < 2. \end{cases}$

30. $f(x) = \begin{cases} \dfrac{1}{x}, & \text{if } x \neq 3, \\ 5, & \text{if } x = 3. \end{cases}$

31. $f(x) = \begin{cases} \dfrac{1}{x - 3}, & \text{if } x \geqslant 4, \\ 5 - x, & \text{if } x < 4. \end{cases}$

32. $f(x) = \begin{cases} 10x - 3, & \text{if } x \geqslant 1, \\ \dfrac{1}{x + 1}, & \text{if } x < 1. \end{cases}$

33. $f(x) = \begin{cases} \dfrac{-3}{x - 2}, & \text{if } x > 0, \\ 4 - x, & \text{if } x \leqslant 0. \end{cases}$

34. $f(x) = \dfrac{5x + 2}{3} - \dfrac{7}{x}.$

35. Suppose the long distance rate for a telephone call from Hazleton, Pa. to Washington, D.C. is $1.85 for the first three minutes and $0.30 for each additional minute or fraction thereof. If $y = f(t)$ is a function that indicates the total charge y for a call of t minutes' duration, sketch the graph of f for $0 < t \leqslant 5\frac{1}{2}$. Use your graph to determine the values of t at which discontinuities occur.

36. The *greatest integer function*, $f(x) = [x]$, is defined to be the greatest integer less than or equal to x, where x is any real number. For example, $[3] = 3$, $[1.999] = 1$, $\left[\frac{1}{4}\right] = 0$, and $[-4.5] = -5$. Sketch the graph of this function for $-3.5 \leqslant x \leqslant 3.5$. Use your sketch to determine the values of x at which discontinuities occur.

37. Sketch the graph of

$$y = f(x) = \begin{cases} -100x + 600, & \text{if } 0 \leqslant x < 5, \\ -100x + 1100, & \text{if } 5 \leqslant x < 10, \\ -100x + 1600, & \text{if } 10 \leqslant x < 15. \end{cases}$$

A function such as this might describe the inventory y of a company at time x.

38. Sketch the "post-office function"

$$c = f(x) = \begin{cases} 15, & \text{if } 0 < x \leqslant 1, \\ 28, & \text{if } 1 < x \leqslant 2, \\ 41, & \text{if } 2 < x \leqslant 3, \\ \quad \text{etc.,} \end{cases}$$

for $0 < x \leqslant 6$. Here c is the cost (in cents) of sending a parcel of weight x (ounces) in January 1980. Where do discontinuities occur?

7-5 CONTINUITY APPLIED TO INEQUALITIES

In this section you will see how continuity can be applied to solve inequalities such as $x^2 + 3x - 4 > 0$. But first we must take a moment to provide a framework on which to build our technique.

We wish to draw your attention to the relationship between the x-intercepts of the graph of a function g (that is, the points where the graph meets the x-axis) and the roots of the equation $g(x) = 0$. If the graph of g has an x-intercept $(r, 0)$, then $g(r) = 0$ and so r is a root of the equation $g(x) = 0$. Hence, from the graph of $y = g(x)$ in Fig. 7-17, we conclude that r_1, r_2, and r_3

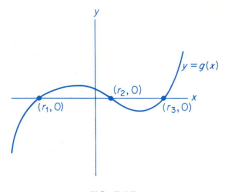

FIG. 7-17

are roots of $g(x) = 0$. On the other hand, if r is any real root of the equation $g(x) = 0$, then $g(r) = 0$ and hence $(r, 0)$ lies on the graph of g. This means that all real roots of the equation $g(x) = 0$ can be represented by the points where the graph of g meets the x-axis. Note also that in Fig. 7-17 these points determine four open intervals on the x-axis:

$$(-\infty, r_1), (r_1, r_2), (r_2, r_3), \text{ and } (r_3, \infty).$$

Now we are ready to solve inequalities. Returning to $x^2 + 3x - 4 > 0$, we shall let $f(x) = x^2 + 3x - 4 = (x + 4)(x - 1)$. Since f is a polynomial function, it is continuous everywhere. The roots of $f(x) = 0$ are -4 and 1; hence the graph of f has x-intercepts $(-4, 0)$ and $(1, 0)$ [see Fig. 7-18]. The roots, or to be more precise the intercepts, determine three intervals on the x-axis:

$$(-\infty, -4), (-4, 1), \text{ and } (1, \infty).$$

Consider the interval $(-\infty, -4)$. Since f is continuous on this interval, we claim that either $f(x) > 0$ or $f(x) < 0$ *throughout* the interval. Suppose $f(x)$ did

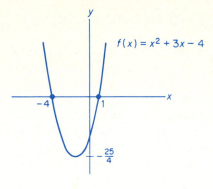

FIG. 7-18

indeed change sign there. Then by the continuity of f there would be a point where the graph intersects the x-axis, for example at $(x_0, 0)$ [refer to Fig. 7-19].

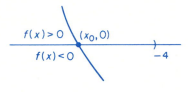

FIG. 7-19

But then x_0 would be a root of the equation $f(x) = 0$. This cannot be, since there is no root of $x^2 + 3x - 4 = 0$ that is less than -4. Hence $f(x)$ must be strictly positive or strictly negative on $(-\infty, -4)$ as well as on the other intervals.

To determine the sign of $f(x)$ on any of these intervals, it is sufficient to determine its sign at any point in the interval. For instance, -5 is in $(-\infty, -4)$ and $f(-5) = 6 > 0$. Thus $f(x) > 0$ on $(-\infty, -4)$. Since 0 is in $(-4, 1)$ and $f(0) = -4 < 0$, then $f(x) < 0$ on $(-4, 1)$. Similarly, 3 is in $(1, \infty)$ and $f(3) = 14 > 0$; thus, $f(x) > 0$ on $(1, \infty)$ [see Fig. 7-20]. Therefore, $x^2 + 3x - 4 > 0$ for $x < -4$ and for $x > 1$, and so we have solved the inequality. These results are obvious from the graph in Fig. 7-18.

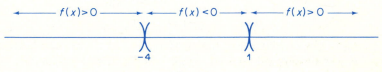

FIG. 7-20

EXAMPLE 1

Solve $x(x - 1)(x + 4) \leqslant 0$.

If $f(x) = x(x - 1)(x + 4)$, then f is continuous everywhere. The roots of $f(x) = 0$ are 0, 1, and -4, which are shown in Fig. 7-21.

FIG. 7-21

These roots determine four intervals:

$$(-\infty, -4), (-4, 0), (0, 1), \text{ and } (1, \infty).$$

Since -5 is in $(-\infty, -4)$, the sign of $f(x)$ on $(-\infty, -4)$ is the same as that of $f(-5)$. Given

$$f(x) = x(x - 1)(x + 4),$$

then

$$f(-5) = -5(-5 - 1)(-5 + 4) = (-)(-)(-) = (-),$$

so $f(x) < 0$ on $(-\infty, -4)$. For the other intervals we find that

$$f(-2) = (-)(-)(+) = (+), \quad \text{so } f(x) > 0 \text{ on } (-4, 0);$$

$$f\left(\tfrac{1}{2}\right) = (+)(-)(+) = (-), \quad \text{so } f(x) < 0 \text{ on } (0, 1);$$

$$\text{and} \quad f(2) = (+)(+)(+) = (+), \quad \text{so } f(x) > 0 \text{ on } (1, \infty).$$

A summary of our results is in the sign chart in Fig. 7-22. Thus $x(x - 1)(x + 4) \leqslant 0$

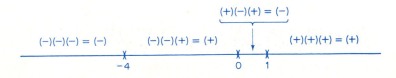

FIG. 7-22

for $x \leqslant -4$ and $0 \leqslant x \leqslant 1$. Note that -4, 0, and 1 are included in the solution because these roots satisfy the equality $(=)$ part of the inequality $(\leqslant)$.

EXAMPLE 2

Solve $\dfrac{x^2 - 6x + 5}{x} \geqslant 0.$

Let $f(x) = \dfrac{x^2 - 6x + 5}{x} = \dfrac{(x-1)(x-5)}{x}$. For a quotient we solve the inequality by considering the intervals determined by the roots of $f(x) = 0$, namely 1 and 5, and the points where f is discontinuous. The function is discontinuous at $x = 0$ and continuous otherwise. In Fig. 7-23 we have placed a hollow dot at 0 to indicate that f is not defined

FIG. 7-23

there. We thus consider the intervals

$$(-\infty, 0),\ (0, 1),\ (1, 5),\ \text{and}\ (5, \infty).$$

Determining the sign of $f(x)$ at a point in each interval, we find that

$$f(-1) = \frac{(-)(-)}{(-)} = (-), \quad \text{so } f(x) < 0 \text{ on } (-\infty, 0);$$

$$f\left(\tfrac{1}{2}\right) = \frac{(-)(-)}{(+)} = (+), \quad \text{so } f(x) > 0 \text{ on } (0, 1);$$

$$f(2) = \frac{(+)(-)}{(+)} = (-), \quad \text{so } f(x) < 0 \text{ on } (1, 5);$$

$$\text{and} \quad f(6) = \frac{(+)(+)}{(+)} = (+), \quad \text{so } f(x) > 0 \text{ on } (5, \infty).$$

The sign chart is given in Fig. 7-24. Therefore, $f(x) \geqslant 0$ for $0 < x \leqslant 1$ and $x \geqslant 5$ (see Fig. 7-25). Why are 1 and 5 included, but 0 excluded?

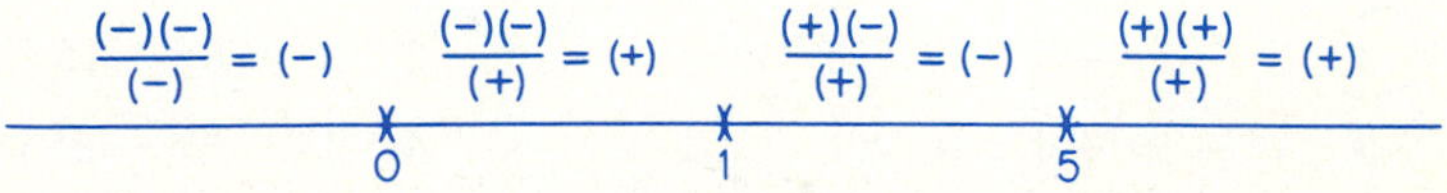

$$\frac{(-)(-)}{(-)} = (-) \qquad \frac{(-)(-)}{(+)} = (+) \qquad \frac{(+)(-)}{(+)} = (-) \qquad \frac{(+)(+)}{(+)} = (+)$$

$$\overset{\times}{\underset{0}{\rule{0pt}{0pt}}} \qquad\qquad \overset{\times}{\underset{1}{\rule{0pt}{0pt}}} \qquad\qquad \overset{\times}{\underset{5}{\rule{0pt}{0pt}}}$$

FIG. 7-24

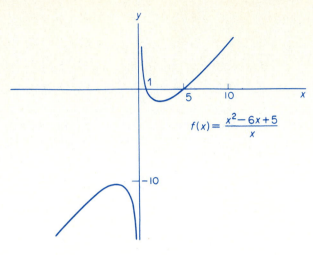

FIG. 7-25

To summarize: $f(x)$ may change sign only about points where $f(x) = 0$ or where f has a discontinuity.

→ Values that come from the denominator are discontinuous

EXAMPLE 3

Solve the following inequalities.

a. $x^2 + 1 > 0$.

The equation $x^2 + 1 = 0$ has no real roots. Thus the graph of $f(x) = x^2 + 1$ has no x-intercepts. Also, f is continuous everywhere. Thus $f(x)$ is always positive or is always negative. But x^2 is always positive or zero, so $x^2 + 1$ is always positive. Thus the solution of $x^2 + 1 > 0$ is $-\infty < x < \infty$.

b. $x^2 + 1 < 0$.

From part a, $x^2 + 1$ is always positive, so the inequality $x^2 + 1 < 0$ has no solution.

EXERCISE 7-5

By the technique discussed in this section, solve the following inequalities.

1. $x^2 - 3x - 4 > 0$.   2. $x^2 - 8x + 15 > 0$.

3. $x^2 - 5x + 6 \leqslant 0$.   4. $14 - 5x - x^2 \leqslant 0$.

5. $2x^2 + 11x + 14 < 0$.   6. $x^2 - 4 < 0$.

7. $x^2 + 4 < 0$.   8. $2x^2 - x - 2 \leqslant 0$.

9. $(x + 2)(x - 3)(x + 6) \leqslant 0.$

10. $(x - 5)(x - 2)(x + 3) \geqslant 0.$

11. $-x(x - 5)(x + 4) > 0.$

12. $(x + 2)^2 > 0.$

13. $x^3 + 4x \geqslant 0.$

14. $(x + 2)^2(x^2 - 1) < 0.$

15. $x^3 + 2x^2 - 3x > 0.$

16. $x^3 - 4x^2 + 4x > 0.$

17. $\dfrac{x}{x^2 - 1} < 0.$

18. $\dfrac{x^2 - 1}{x} < 0.$

19. $\dfrac{4}{x - 1} \geqslant 0.$

20. $\dfrac{3}{x^2 - 5x + 6} > 0.$

21. $\dfrac{x^2 - x - 6}{x^2 + 4x - 5} \geqslant 0.$

22. $\dfrac{x^2 + 2x - 8}{x^2 + 3x + 2} \geqslant 0.$

23. $\dfrac{3}{x^2 + 6x + 8} \leqslant 0.$

24. $\dfrac{2x + 1}{x^2} \leqslant 0.$

25. $x^2 + 2x \geqslant 2.$

26. $x^4 - 16 \geqslant 0.$

27. Suppose that consumers will purchase q units of a product when the price of *each* unit is $20 - .1q$ dollars. How many units must be sold in order that sales revenue will be no less than \$750?

28. A lumber company owns a forest which is of rectangular shape, 1 mi $\times$ 2 mi. The company wants to cut a uniform strip of trees along the outer edges of the forest. At most how wide can the strip be if the company wants at least $\frac{3}{4}$ sq mi of forest to remain?

29. A container manufacturer wishes to make an open box by cutting a 4-in. square from each corner of a square sheet of aluminum and then turning up the sides. The box is to contain at least 324 cu in. Find the dimensions of the smallest sheet of aluminum that can be used.

30. Imperial Educational Services (I.E.S.) is offering a workshop in data processing to key personnel at Zeta Corporation. The price per person is \$50 and Zeta Corporation guarantees that at least fifty persons will attend. Suppose I.E.S. offers to reduce the charge for *everybody* by \$0.50 for each person over the fifty who attends. How should I.E.S. limit the size of the group so that the total revenue they receive will never be less than that received for fifty persons?

7-6 WHY CONTINUOUS FUNCTIONS?

Often it is helpful to describe a situation by a continuous function. For example, the demand schedule in Table 7-4 indicates the number of units of a particular product that consumers will demand per week at various prices. This information can be given graphically as in Fig. 7-26(a) by plotting each quantity-price pair as a point. Clearly this graph does not represent a continuous function. Furthermore, it gives us no information as to the price at which, say, 35 units would be demanded. However, if we connect the points in Fig. 7-26(a) by a

TABLE 7-4   Demand Schedule

PRICE/UNIT (DOLLARS)	QUANTITY PER WEEK
p	q
20	0
10	5
5	15
4	20
2	45
1	95

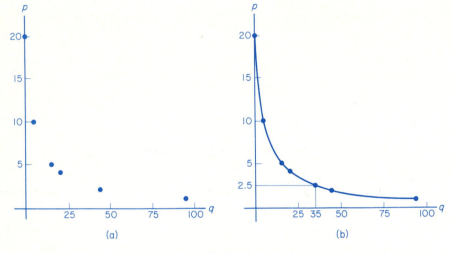

(a)   (b)

FIG. 7-26

smooth curve [see Fig. 7-26(b)], we get a so-called demand curve. From it we could guess that at about $2.50 per unit, 35 units would be demanded.

Frequently it is possible and useful to describe a graph, as in Fig. 7-26(b), by means of an explicit equation that defines a continuous function f. Such a function not only gives us a demand equation, $p = f(q)$, which allows us to anticipate corresponding prices and quantities demanded, but it also permits a convenient mathematical analysis of the nature and basic properties of demand. Of course some care must be used in working with equations such as $p = f(q)$. Mathematically, f may be defined when $q = \sqrt{37}$, but from a practical standpoint a demand of $\sqrt{37}$ units could be meaningless to our particular situation. For example, if a unit is an egg, then a demand of $\sqrt{37}$ eggs makes no sense.

In general, it will be our desire to view practical situations in terms of continuous functions whenever possible so that we may be better able to analyze their nature.

7-7 REVIEW

IMPORTANT TERMS AND SYMBOLS IN CHAPTER 7

$\lim\limits_{x \to a} f(x) = L$ *(p. 211)*   one-sided limits   *(p. 216)*

$\lim\limits_{x \to a^+} f(x) = L$ *(p. 216)* $\lim\limits_{x \to a^-} f(x) = L$ *(p. 216)*

$\lim\limits_{x \to a} f(x) = \infty$ *(p. 217)* $\lim\limits_{x \to \infty} f(x) = L$ *(p. 218)*

$\lim\limits_{x \to -\infty} f(x) = L$ *(p. 218)*   compounding continuously   *(p. 225)*

continuous   *(p. 229)*   discontinuous   *(p. 229)*

continuous everywhere   *(p. 230)*   infinite discontinuity   *(p. 230)*

REVIEW SECTION

1. $\lim\limits_{x \to a} x =$ _____ .

   Ans.   a.

2. True or false: In general, $\lim\limits_{x \to a} f(x) = f(a)$. _____

   Ans.   false.

3. True or false: For a function to have a limit at a point, the function must be defined at that point. _____

   Ans.   false.

4. If $\lim\limits_{x \to a} f(x) = \infty$, then $\lim\limits_{x \to a} \dfrac{1}{f(x)} =$ _____ .

   Ans.   0.

5. If $\lim\limits_{x \to a} f(x) = 10$, then $\lim\limits_{x \to a} [6f(x)] =$ _____ .

   Ans.   60.

6. $\lim\limits_{x \to \infty} \dfrac{1}{x} =$ __(a)__ , $\lim\limits_{x \to \infty} \dfrac{1}{x^2} =$ __(b)__ , and $\lim\limits_{x \to \infty} \dfrac{3x^2}{6x} =$ __(c)__ .

   Ans.   (a) 0,   (b) 0,   (c) ∞.

7. $\lim\limits_{h \to 0} (x + h) =$ _____ .

   Ans.   x.

8. To solve $x(x - 2) > 0$ we consider how many intervals? _____ .

   Ans.   3.

9. If $\lim\limits_{x \to a^+} f(x) = L_1$ and $\lim\limits_{x \to a^-} f(x) = L_2$ and $L_1 \neq L_2$, then f __(is) (is not)__ continuous at $x = a$. _____

   Ans.   is not.

10. The function $f(x) = (4 - 2x)/(2 + x)$ is discontinuous at $x =$ _____ .

Ans. −2.

11. True or false: A polynomial function is continuous everywhere. _____

Ans. true.

12. If f is continuous at $x = a$, then $\lim_{x \to a} f(x) =$ _____ .

Ans. $f(a)$.

REVIEW PROBLEMS

In Problems 1–20, find the limits if they exist. If the limit does not exist, so state or use the symbol ∞ or $-\infty$ where appropriate.

1. $\lim_{x \to -1} (2x^2 + 6x - 1)$.

2. $\lim_{x \to 0} \dfrac{2x^2 - 3x + 1}{2x^2 - 2}$.

3. $\lim_{x \to 3} \dfrac{x^2 - 9}{x^2 - 3x}$.

4. $\lim_{x \to -2} \dfrac{x + 1}{x^2 - 2}$.

5. $\lim_{h \to 0} (x + h)$.

6. $\lim_{x \to 2} \dfrac{x^2 - 4}{x^2 - 3x + 2}$.

7. $\lim_{x \to \infty} \dfrac{2}{x + 1}$.

8. $\lim_{x \to \infty} \dfrac{x^2 + 1}{x^2}$.

9. $\lim_{x \to \infty} \dfrac{3x - 2}{5x + 3}$.

10. $\lim_{x \to -\infty} \dfrac{1}{x^4}$.

11. $\lim_{t \to 3} \dfrac{2t - 3}{t - 3}$.

12. $\lim_{x \to -\infty} \dfrac{x^6}{x^5}$.

13. $\lim_{x \to -\infty} \dfrac{x + 3}{1 - x}$.

14. $\lim_{x \to 4} \sqrt{4}$.

15. $\lim_{y \to 5^+} \sqrt{y - 5}$.

16. $\lim_{x \to 1} f(x)$ if $f(x) = \begin{cases} x^2, & \text{if } 0 \leqslant x < 1, \\ x, & \text{if } x > 1. \end{cases}$

17. $\lim_{x \to \infty} \dfrac{x^2 - 1}{(3x + 2)^2}$.

18. $\lim_{x \to 1} \dfrac{x^2 + x - 2}{x - 1}$.

19. $\lim_{x \to 3^-} \dfrac{x + 3}{x^2 - 9}$.

20. $\lim_{x \to 2} \dfrac{2 - x}{x - 2}$.

21. For an annual interest rate of 7 percent compounded continuously, find:
   a. the compound amount of $2500 after 14 years.
   b. the present value of $2500 due in 14 years.

22. For an annual interest rate of 6 percent compounded continuously, find:
   a. the compound amount of $800 after 9 years.
   b. the present value of $800 due in 9 years.

23. Find the effective rate equivalent to an annual rate of 6 percent compounded continuously.

24. Find the effective rate equivalent to an annual rate of 1 percent compounded continuously.

25. Using the definition of continuity, show that $f(x) = x + 5$ is continuous at $x = 7$.

26. Using the definition of continuity, show that $f(x) = (x - 3)/(x^2 + 4)$ is continuous at $x = 3$.

27. State whether $f(x) = x/4$ is continuous everywhere. Give a reason for your answer.

28. State whether $f(x) = x^2 - 2$ is continuous everywhere. Give a reason for your answer.

In Problems **29–32**, *determine the points of discontinuity for each function.*

29. $f(x) = \dfrac{x^2}{x + 3}$.

30. $f(x) = \dfrac{0}{x^3}$.

31. $f(x) = \begin{cases} x + 4, & \text{if } x > -2, \\ 3x - 1, & \text{if } x \leqslant -2. \end{cases}$

32. $f(x) = \begin{cases} \dfrac{x}{x + 1}, & \text{if } x > 1, \\ \dfrac{3}{x + 4}, & \text{if } x < 1. \end{cases}$

In Problems **33–40**, *solve the given inequalities.*

33. $x^2 + 4x - 12 > 0$.

34. $2x^2 - 6x + 4 \leqslant 0$.

35. $x^3 \geqslant 2x^2$.

36. $x^3 + 8x^2 + 15x \geqslant 0$.

37. $\dfrac{x + 5}{x^2 - 1} < 0$.

38. $\dfrac{x(x + 5)(x + 8)}{3} < 0$.

39. $\dfrac{x^2 + 3x}{x^2 + 2x - 8} \geqslant 0$.

40. $\dfrac{x^2 - 4}{x^2 + 2x + 1} \geqslant 0$.

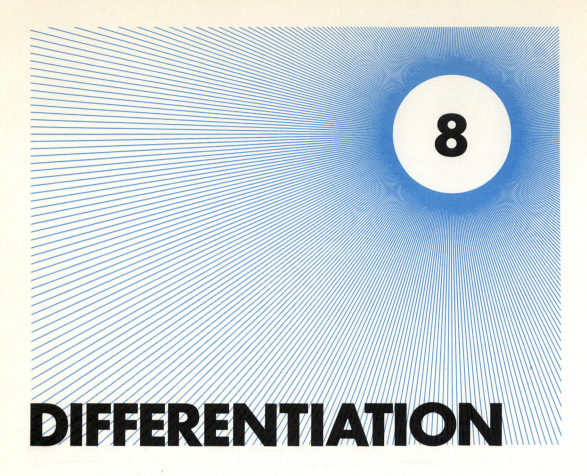

DIFFERENTIATION

Now we begin our study of calculus. The ideas involved in calculus are completely different from those of algebra and geometry. Moreover, the power and importance of these ideas and their applications will be evident to you later in the text. The objective of this chapter is not only to convey an understanding of what the so-called "derivative" of a function is, but also to teach techniques of finding derivatives by properly applying rules.

8-1 THE DERIVATIVE

One of the main problems with which calculus deals is finding the slope of the *tangent line* at a point on a curve. In geometry you probably thought of a tangent line, or *tangent*, to a circle as a line that meets the circle at exactly one point (Fig. 8-1). But this idea of a tangent is not very useful for other kinds of curves.

For example, in Fig. 8-2(a) the lines L_1 and L_2 intersect the curve at exactly one point. Intuitively we would not think of L_2 as the tangent at this point, but it seems natural that L_1 is. Also, in Fig. 8-2(b) we would consider L_3

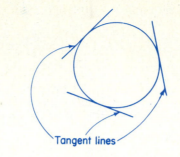

FIG. 8-1

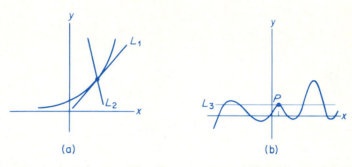

(a)   (b)

FIG. 8-2

to be the tangent at point P even though it intersects the curve at other points. From these examples you can see that we must drop the idea that a tangent must intersect a curve at only one point. To develop a suitable definition of tangent line we use the limit concept.

Look at the graph of the function $y = f(x)$ in Fig. 8-3. Here $P(x_1, y_1)$ and $Q(x_2, y_2)$ are two different points on the curve. The line PQ passing through

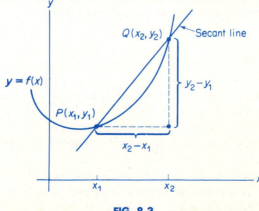

FIG. 8-3

them is called a *secant line*. By the slope formula, the slope of PQ is

$$m_{PQ} = \frac{y_2 - y_1}{x_2 - x_1}.$$

If Q moves along the curve and approaches P, the secant line has a limiting position as shown in Fig. 8-4. As Q approaches P from the right, the positions of the secant lines are PQ', PQ'', etc. As Q approaches P from the left, they are

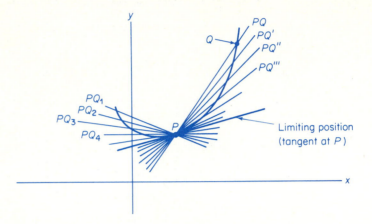

FIG. 8-4

PQ_1, PQ_2, etc. *In both cases, the **same** limiting position is obtained.* This common limiting position of the secant lines is called the **tangent line** to the curve at P. This definition seems reasonable and avoids the difficulties mentioned at the beginning of this section.

Not every curve has a tangent at each of its points. For example, the curve $y = |x|$ does not have a tangent at $(0, 0)$. In Fig. 8-5, at $(0, 0)$ a secant line from the right must always be the line $y = x$, and from the left it is the line $y = -x$. Since there is no common limiting position, there is no tangent.

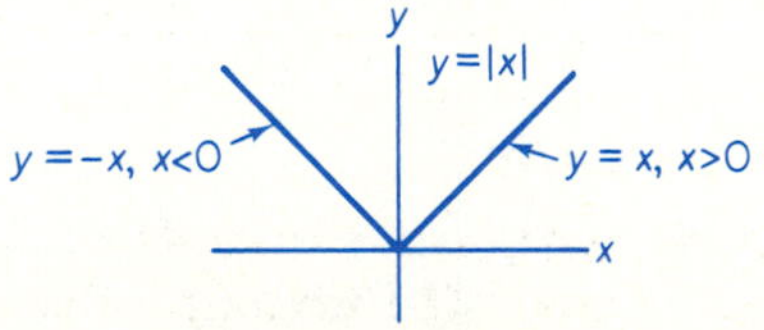

$y = -x, x < 0$ $y = |x|$ $y = x, x > 0$

FIG. 8-5

Now that we have a suitable definition of a tangent to a curve at a point, we can define the *slope of a curve* at a point.

DEFINITION. *The **slope of a curve** at a point P is the slope of the tangent line at P.*

Since the tangent is a limiting position of secant lines, the slope of the tangent is the limiting value of the slopes of the secant lines PQ as Q approaches P. We shall now find an expression for the slope of the curve $y = f(x)$ at the point $(x_1, f(x_1))$. In Fig. 8-6 the slope of the secant line PQ is

$$m_{PQ} = \frac{f(x_2) - f(x_1)}{x_2 - x_1}.$$

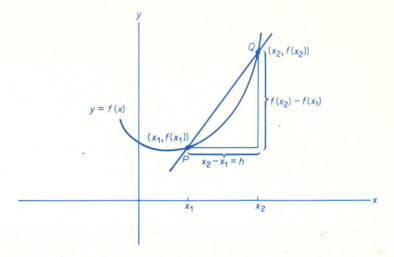

FIG. 8-6

If the difference $x_2 - x_1$ is h, then we can write x_2 as $x_1 + h$. Here $h \neq 0$, for if $h = 0$ then $x_2 = x_1$ and no secant line would exist. Thus,

$$m_{PQ} = \frac{f(x_1 + h) - f(x_1)}{(x_1 + h) - x_1} = \frac{f(x_1 + h) - f(x_1)}{h}.$$

As Q moves along the curve towards P, then $x_2 \to x_1$. This means that h is getting closer to zero. The limiting value of the slopes of the secant lines —which is the slope of the tangent line at $(x_1, f(x_1))$—is the limit:

$$\lim_{h \to 0} \frac{f(x_1 + h) - f(x_1)}{h}. \tag{1}$$

EXAMPLE 1

The slope of the curve $y = f(x) = x^2$ at the point $(1, 1)$ is

$$\lim_{h \to 0} \frac{f(1 + h) - f(1)}{h} = \lim_{h \to 0} \frac{(1 + h)^2 - (1)^2}{h}$$

$$= \lim_{h \to 0} \frac{1 + 2h + h^2 - 1}{h} = \lim_{h \to 0} \frac{2h + h^2}{h}$$

$$= \lim_{h \to 0} \frac{h(2 + h)}{h} = \lim_{h \to 0} (2 + h) = 2.$$

Thus the tangent line to $y = x^2$ at $(1, 1)$ has a slope of 2 (Fig. 8-7).

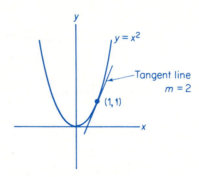

FIG. 8-7

We can generalize the result in (1) to any point $(x, f(x))$ on the curve by replacing x_1 by x. We thus have the following definition, which forms the basis of differential calculus.

DEFINITION. *If $y = f(x)$ defines a function f, the limit*

$$\lim_{h \to 0} \frac{f(x + h) - f(x)}{h},$$

*if it exists, is called the **derivative** of f at x and is denoted $f'(x)$, which is read "f prime of x." The process of finding the derivative is called **differentiation**.*

EXAMPLE 2

If $f(x) = x^2$, find the derivative of f.

$$f'(x) = \lim_{h \to 0} \frac{f(x + h) - f(x)}{h}$$

$$= \lim_{h \to 0} \frac{(x + h)^2 - x^2}{h} = \lim_{h \to 0} \frac{x^2 + 2xh + h^2 - x^2}{h}$$

$$= \lim_{h \to 0} \frac{2xh + h^2}{h} = \lim_{h \to 0} \frac{h(2x + h)}{h} = \lim_{h \to 0} (2x + h) = 2x.$$

*Note that in taking the limit we treated x as a constant because it was h, not x, that was changing. Also note that $f'(x) = 2x$ defines a function of x. In all cases, **the derivative of a function f is also a function, f'.***

Besides writing $f'(x)$, other ways of denoting the derivative of $y = f(x)$ at x are

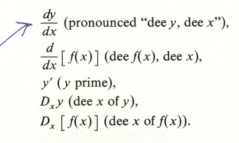

$$\frac{dy}{dx} \text{ (pronounced "dee } y, \text{ dee } x\text{"),}$$

$$\frac{d}{dx}\left[f(x) \right] \text{ (dee } f(x), \text{ dee } x\text{),}$$

y' (y prime),

$D_x y$ (dee x of y),

$D_x\left[f(x) \right]$ (dee x of f(x)).

PITFALL. $\dfrac{dy}{dx}$ *is not a fraction, but is a symbol for a derivative. We have not yet attached any meaning to individual symbols such as dy and dx.*

The derivative of $y = f(x)$ at x is also referred to as the derivative *with respect to x.*

If the derivative of f can be evaluated at $x = x_1$, the resulting *number* is called the derivative of f at x_1, denoted $f'(x_1)$. In this case we say that f is *differentiable* at x_1. Note that

> $f'(x_1)$ is the slope of the tangent to $y = f(x)$ at $(x_1, f(x_1))$.

In addition to the notation $f'(x_1)$, we can also write

$$\left.\frac{dy}{dx}\right|_{x=x_1} \quad \text{and} \quad y'(x_1).$$

EXAMPLE 3

If $f(x) = 2x^2 + 2x + 3$, find $f'(1)$.

We shall find the derivative $f'(x)$ and then evaluate it at $x = 1$.

(handwritten: Put this into)

$$f'(x) = \lim_{h \to 0} \frac{f(x + h) - f(x)}{h}$$

$$= \lim_{h \to 0} \frac{\left[2(x + h)^2 + 2(x + h) + 3\right] - (2x^2 + 2x + 3)}{h}$$

$$= \lim_{h \to 0} \frac{2x^2 + 4xh + 2h^2 + 2x + 2h + 3 - 2x^2 - 2x - 3}{h}$$

$$= \lim_{h \to 0} \frac{4xh + 2h^2 + 2h}{h} = \lim_{h \to 0} (4x + 2h + 2).$$

$$f'(x) = 4x + 2.$$
$$f'(1) = 4(1) + 2 = 6.$$

EXAMPLE 4

a. *Find the slope of the curve $y = x^2$ at the point $(3, 9)$. Then find an equation of the tangent line at $(3, 9)$.*

From Example 2, $y' = 2x$. Thus, $y'(3) = 2(3) = 6$. That is, the tangent line to the curve $y = x^2$ at $(3, 9)$ has a slope of 6. A point-slope form of the tangent line is $y - 9 = 6(x - 3)$, from which $y = 6x - 9$ (see Fig. 8-8).

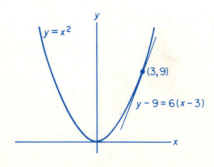

FIG. 8-8

PITFALL.   *It is **not** correct to say that since the derivative of $y = x^2$ is $2x$, the tangent at (3, 9) is $y - 9 = 2x(x - 3)$. The derivative must be **evaluated** at the point of tangency to determine the slope of the tangent line.*

b. *Find the slope of the curve $y = 2x + 3$ at the point where $x = 6$.*

Letting $y = f(x) = 2x + 3$, we have

$$y' = \lim_{h \to 0} \frac{f(x + h) - f(x)}{h} = \lim_{h \to 0} \frac{[2(x + h) + 3] - (2x + 3)}{h}$$

$$= \lim_{h \to 0} \frac{2h}{h} = \lim_{h \to 0} 2 = 2.$$

Since $y' = 2$, the slope when $x = 6$, or at any point, is 2. Note that the curve is a straight line and thus has the same slope at each point.

EXAMPLE 5

Find $D_x (\sqrt{x})$.

If $f(x) = \sqrt{x}$, then

$$D_x (\sqrt{x}) = \lim_{h \to 0} \frac{f(x + h) - f(x)}{h} = \lim_{h \to 0} \frac{\sqrt{x + h} - \sqrt{x}}{h}.$$

As $h \to 0$, both the numerator and denominator approach zero. This can be avoided by rationalizing the numerator.

$$\frac{\sqrt{x + h} - \sqrt{x}}{h} = \frac{\sqrt{x + h} - \sqrt{x}}{h} \cdot \frac{\sqrt{x + h} + \sqrt{x}}{\sqrt{x + h} + \sqrt{x}} = \frac{(x + h) - x}{h(\sqrt{x + h} + \sqrt{x})}$$

$$= \frac{h}{h(\sqrt{x + h} + \sqrt{x})} = \frac{1}{\sqrt{x + h} + \sqrt{x}}.$$

Thus,

$$D_x (\sqrt{x}) = \lim_{h \to 0} \frac{1}{\sqrt{x + h} + \sqrt{x}} = \frac{1}{\sqrt{x} + \sqrt{x}} = \frac{1}{2\sqrt{x}}.$$

Note that the original function, $\sqrt{x}$, is defined for $x \geqslant 0$. But the derivative, $1/(2\sqrt{x})$, is defined only when $x > 0$. From the graph of $y = \sqrt{x}$ in Fig. 8-9, it is clear that when $x = 0$, the tangent is a vertical line and hence does not have a slope.

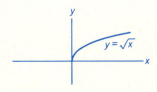

FIG. 8-9

If a variable, say q, is a function of some variable, say p, then we would speak of the derivative of q with respect to p and could write dq/dp.

EXAMPLE 6

must get common denom.

If $q = f(p) = \dfrac{1}{2p}$, find dq/dp.

$$\frac{dq}{dp} = \lim_{h \to 0} \frac{f(p+h) - f(p)}{h}$$

$$= \lim_{h \to 0} \frac{\dfrac{1}{2(p+h)} - \dfrac{1}{2p}}{h} = \lim_{h \to 0} \frac{p - (p+h)}{h[2p(p+h)]}$$

$$= \lim_{h \to 0} \frac{-h}{h[2p(p+h)]} = \lim_{h \to 0} \frac{-1}{2p(p+h)} = -\frac{1}{2p^2}.$$

Note that when $p = 0$, neither the function nor its derivative exists.

As a final note we point out that the derivative of $y = f(x)$ at x, namely $f'(x)$, is nothing more than the limit

$$\lim_{h \to 0} \frac{f(x+h) - f(x)}{h}.$$

However, we can interpret f' as a function that gives the slope of the tangent line to the curve $y = f(x)$ at the point $(x, f(x))$. This interpretation is simply a geometric convenience that assists our understanding. The above limit may exist aside from any geometric consideration at all. As you will see later, there are other useful interpretations.

EXERCISE 8-1

In Problems **1–16**, use the definition of the derivative to find each of the following.

1. $\dfrac{d}{dx}[f(x)]$ if $f(x) = x$.

2. $\dfrac{d}{dx}[f(x)]$ if $f(x) = 4x - 1$.

3. $\dfrac{dy}{dx}$ if $y = 2x + 4$.

4. $\dfrac{dy}{dx}$ if $y = -3x$.

5. $\dfrac{d}{dx}(3 - 2x)$.

6. $\dfrac{d}{dx}\left(4 - \dfrac{x}{2}\right)$.

7. $f'(x)$ if $f(x) = 3$.

8. $f'(x)$ if $f(x) = 7.01$.

9. $D_x\,(x^2 + 4x - 8)$.   10. $D_x\,y$ if $y = x^2 + 5$.

11. $\dfrac{dq}{dp}$ if $q = 2p^2 + 5p - 1$.   12. $D_x\,(x^2 - x - 3)$.

13. $D_x\,y$ if $y = \dfrac{1}{x}$.   14. $\dfrac{dC}{dq}$ if $C = 7 + 2q - 3q^2$.

15. $f'(x)$ if $f(x) = \sqrt{x + 2}$.   16. $g'(x)$ if $g(x) = \dfrac{2}{x - 3}$.

17. Find the slope of the curve $y = x^2 + 4$ at the point $(-2, 8)$.

18. Find the slope of the curve $y = 2 - 3x^2$ at the point $(1, -1)$.

19. Find the slope of the curve $y = 4x^2 - 5$ when $x = 0$.

20. Find the slope of the curve $y = \sqrt{x}$ when $x = 1$.

*In Problems **21–26**, find an equation of the tangent line to the curve at the given point.*

21. $y = x + 4$; $(3, 7)$.   22. $y = 2x^2 - 5$; $(-2, 3)$.

23. $y = 3x^2 + 3x - 4$; $(-1, -4)$.   24. $y = (x - 1)^2$; $(0, 1)$.

25. $y = \dfrac{3}{x + 1}$; $(2, 1)$.   26. $y = \dfrac{5}{1 - 3x}$; $(2, -1)$.

8-2 RULES FOR DIFFERENTIATION

By now you would probably agree that differentiating a function by direct use of the definition of a derivative can be tedious. Fortunately, there are rules that give us completely mechanical and efficient procedures for differentiation. They also avoid direct use of limits. We shall look at some rules in this section.

To begin with, recall that the graph of the constant function $f(x) = c$ is a horizontal line (Fig. 8-10), which has a slope of zero everywhere. This means that $f'(x) = 0$, which is our first rule. We shall give a formal proof.

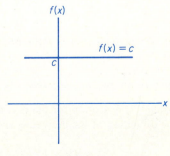

FIG. 8-10

RULE 1. *If $f(x) = c$, where c is a constant, then $f'(x) = 0$.*
That is, the derivative of a constant function is zero.

Proof. If $f(x) = c$, applying the definition of the derivative gives

$$f'(x) = \lim_{h \to 0} \frac{f(x + h) - f(x)}{h} = \lim_{h \to 0} \frac{c - c}{h}$$

$$= \lim_{h \to 0} \frac{0}{h} = \lim_{h \to 0} 0 = 0.$$

Therefore,

> If $f(x) = c$, then $f'(x) = 0$.

EXAMPLE 1

a. If $f(x) = 3$, then $f'(x) = 0$.

b. If $g(x) = \sqrt{5}$, then $g'(x) = 0$. For example, the derivative of g when $x = 4$ is $g'(4) = 0$.

c. If $s(t) = (1{,}938{,}623)^{807.4}$, then $ds/dt = 0$ since s is a constant function.

The proof of the next rule requires that we expand a binomial. Recall that

$$(x + h)^2 = x^2 + 2xh + h^2,$$
$$\text{and} \quad (x + h)^3 = x^3 + 3x^2h + 3xh^2 + h^3.$$

In both expansions, notice in reading from left to right that the exponents of x decrease while those of h increase. This is true for the general case $(x + h)^n$, for n a positive integer. That is,

$$(x + h)^n = x^n + nx^{n-1}h + (\ )x^{n-2}h^2 + \ldots + (\ )xh^{n-1} + h^n,$$

where the missing numbers inside the parentheses are constants whose values are not needed in the proof of the next rule, which involves the derivative of x raised to a constant power.

RULE 2. *If $f(x) = x^n$, where n is any real number, then $f'(x) = nx^{n-1}$.*

Proof. We shall give a proof for the case that n is a positive integer. If $f(x) = x^n$, applying the definition of the derivative gives

$$f'(x) = \lim_{h \to 0} \frac{f(x + h) - f(x)}{h} = \lim_{h \to 0} \frac{(x + h)^n - x^n}{h}.$$

By our discussion above,

$$(x + h)^n = x^n + nx^{n-1}h + (\quad)x^{n-2}h^2 + \ldots + h^n.$$

Thus,

$$f'(x) = \lim_{h \to 0} \frac{x^n + nx^{n-1}h + (\quad)x^{n-2}h^2 + \ldots + h^n - x^n}{h}.$$

In the numerator the sum of the first and last terms is 0. Dividing each of the remaining terms by h gives

$$f'(x) = \lim_{h \to 0} [nx^{n-1} + (\quad)x^{n-2}h + \ldots + h^{n-1}].$$

Each term after the first has h as a factor and must approach 0 as $h \to 0$. Hence

$$f'(x) = nx^{n-1}.$$

Therefore,

$$\boxed{\text{If } f(x) = x^n, \text{ then } f'(x) = nx^{n-1}.}$$

Rule 2 indicates that *the derivative of a constant power of x is the exponent times x raised to a power one less than the given power.*

EXAMPLE 2

a. If $f(x) = x^2$, then by Rule 2,

$$f'(x) = 2x^{2-1} = 2x.$$

b. If $g(w) = w^{9/4}$, then by Rule 2,

$$g'(w) = \tfrac{9}{4}w^{(9/4)-1} = \tfrac{9}{4}w^{5/4}.$$

c. If $F(x) = x = x^1$, then

$$\frac{d}{dx}[F(x)] = \frac{d}{dx}(x) = 1 \cdot x^{1-1} = 1 \cdot x^0 = 1.$$

Thus the derivative of x with respect to x is 1.

d. Suppose $y = x\sqrt{x}$. To find $D_x y$ we first write $x\sqrt{x}$ as $x^{3/2}$. Thus $y = x^{3/2}$ and

$$D_x y = \tfrac{3}{2}x^{(3/2)-1} = \tfrac{3}{2}x^{1/2} = \tfrac{3}{2}\sqrt{x}.$$

e. Suppose $h(x) = \dfrac{1}{x^{3/2}}$. To apply Rule 2 we *must* write $h(x)$ as $h(x) = x^{-3/2}$.

$$D_x \left(\frac{1}{x^{3/2}}\right) = D_x (x^{-3/2}) = -\frac{3}{2}x^{(-3/2)-1} = -\frac{3}{2}x^{-5/2}.$$

PITFALL. $D_x \left(\dfrac{1}{x^{3/2}} \right) \neq \dfrac{1}{\frac{3}{2}x^{1/2}}$. *Do not merely differentiate the denominator.*

RULE 3.   *If $g(x) = c\, f(x)$ and $f'(x)$ exists, then $g'(x) = c\, f'(x)$.*

That is, the derivative of a constant times a function is the constant times the derivative of the function.

Proof.   If $g(x) = c\, f(x)$, applying the definition of the derivative of g gives

$$g'(x) = \lim_{h \to 0} \frac{g(x+h) - g(x)}{h} = \lim_{h \to 0} \frac{c\, f(x+h) - c\, f(x)}{h}$$

$$= \lim_{h \to 0} \left[c \cdot \frac{f(x+h) - f(x)}{h} \right] = c \cdot \lim_{h \to 0} \frac{f(x+h) - f(x)}{h}.$$

But $\displaystyle \lim_{h \to 0} \frac{f(x+h) - f(x)}{h} = f'(x)$, and thus $g'(x) = c\, f'(x)$. Therefore,

$$\boxed{\text{If } g(x) = c\, f(x), \text{ then } g'(x) = c\, f'(x).}$$

EXAMPLE 3

Find the derivative of each of the following functions.

a. $g(x) = 5x^3$.

If we let $f(x) = x^3$, then $g(x) = 5f(x)$ and

$$g'(x) = 5f'(x) \hspace{3cm} \text{(Rule 3)}$$
$$= 5\, D_x\, (x^3) = 5(3x^{3-1}) \hspace{1.5cm} \text{(Rule 2)}$$
$$= 15x^2.$$

b. $g(p) = \frac{13}{2}p$.

If we let $f(p) = p$, then $g(p) = \frac{13}{2}f(p)$ and by Rule 3,

$$g'(p) = \tfrac{13}{2}f'(p) = \tfrac{13}{2}\, D_p\, (p) = \tfrac{13}{2}(1) = \tfrac{13}{2}.$$

c. $y = \dfrac{.702}{x^2\sqrt{x}} = .702x^{-5/2}$.

Note that y can be considered a constant times a function.

$$D_x y = .702\, D_x\, (x^{-5/2}) \hspace{2.5cm} \text{(Rule 3)}$$
$$= .702\left(-\tfrac{5}{2}x^{-7/2} \right) = -1.755x^{-7/2} \hspace{1cm} \text{(Rule 2).}$$

PITFALL. *If $f(x) = (4x)^3$, you may be tempted to write $f'(x) = 3(4x)^2$. This is **incorrect**! The reason is that Rule 2 applies to a power of the variable x, **not** a power of a function such as $4x$. To apply our rules we must get a suitable form for $f(x)$. We can write $(4x)^3$ as $4^3 x^3$ or $64x^3$. Thus,*

$$f'(x) = 64 \, D_x \, (x^3) = 64(3x^2) = 192x^2.$$

The next two rules involve derivatives of sums and differences of functions.

RULE 4.   *If $F(x) = f(x) + g(x)$ and $f'(x)$ and $g'(x)$ exist, then $F'(x) = f'(x) + g'(x)$.*
   That is, the derivative of a sum of two functions is the sum of the derivatives of the functions.

   Proof.   If $F(x) = f(x) + g(x)$, applying the definition of the derivative of F gives

$$F'(x) = \lim_{h \to 0} \frac{F(x + h) - F(x)}{h}$$

$$= \lim_{h \to 0} \frac{[f(x + h) + g(x + h)] - [f(x) + g(x)]}{h}$$

$$= \lim_{h \to 0} \frac{[f(x + h) - f(x)] + [g(x + h) - g(x)]}{h} \qquad \text{(regrouping)}$$

$$= \lim_{h \to 0} \left[\frac{f(x + h) - f(x)}{h} + \frac{g(x + h) - g(x)}{h} \right].$$

Since the limit of a sum is the sum of the limits,

$$F'(x) = \lim_{h \to 0} \frac{f(x + h) - f(x)}{h} + \lim_{h \to 0} \frac{g(x + h) - g(x)}{h}.$$

But these two limits are $f'(x)$ and $g'(x)$. Thus,

$$F'(x) = f'(x) + g'(x).$$

Therefore,

$$\boxed{\begin{array}{c} \text{If } F(x) = f(x) + g(x), \text{ then} \\[2mm] F'(x) = f'(x) + g'(x). \end{array}}$$

The proof of the next rule is similar to that of Rule 4.

RULE 5.   *If $F(x) = f(x) - g(x)$ and $f'(x)$ and $g'(x)$ exist, then $F'(x) = f'(x) - g'(x)$.*
   That is, the derivative of a difference of two functions is the difference of the derivatives of the functions.

$$\boxed{\begin{array}{c} \text{If } F(x) = f(x) - g(x), \text{ then} \\[2mm] F'(x) = f'(x) - g'(x). \end{array}}$$

Rules 4 and 5 can be extended to the derivative of any finite number of sums and differences of functions. For example, if $F(x) = f(x) - g(x) + h(x) + k(x)$, then $F'(x) = f'(x) - g'(x) + h'(x) + k'(x)$.

EXAMPLE 4

Differentiate each of the following functions.

a. $F(x) = 3x^5 + \sqrt{x}$.

Let $f(x) = 3x^5$ and $g(x) = \sqrt{x} = x^{1/2}$. Then $F(x) = f(x) + g(x)$, a sum of two functions. Thus,

$$F'(x) = f'(x) + g'(x) = D_x (3x^5) + D_x (x^{1/2}) \qquad \text{(Rule 4)}$$

$$= 3 D_x (x^5) + D_x (x^{1/2}) \qquad \text{(Rule 3)}$$

$$= 3(5x^4) + \frac{1}{2}x^{-1/2} = 15x^4 + \frac{1}{2\sqrt{x}} \qquad \text{(Rule 2).}$$

b. $f(x) = x^5 - \sqrt[3]{x^2}$.

Since f is the difference of two functions,

$$f'(x) = D_x (x^5) - D_x \left(\sqrt[3]{x^2}\right) = D_x (x^5) - D_x (x^{2/3}) \qquad \text{(Rule 5)}$$

$$= 5x^4 - \frac{2}{3}x^{-1/3} = 5x^4 - \frac{2}{3\sqrt[3]{x}} \qquad \text{(Rule 2).}$$

c. $f(z) = \frac{z^4}{4} - \frac{5}{z^{1/3}}$.

Note that we can write $f(z) = \frac{1}{4}z^4 - 5z^{-1/3}$.

$$\frac{d}{dz}[f(z)] = D_z \left(\frac{1}{4}z^4\right) - D_z (5z^{-1/3}) \qquad \text{(Rule 5)}$$

$$= \frac{1}{4} D_z (z^4) - 5 D_z (z^{-1/3}) \qquad \text{(Rule 3)}$$

$$= \frac{1}{4}(4z^3) - 5\left(-\frac{1}{3}z^{-4/3}\right) \qquad \text{(Rule 2)}$$

$$= z^3 + \frac{5}{3}z^{-4/3}.$$

d. $y = 6x^3 - 2x^2 + 7x - 8$.

$$D_x y = D_x (6x^3) - D_x (2x^2) + D_x (7x) - D_x (8)$$

$$= 6 D_x (x^3) - 2 D_x (x^2) + 7 D_x (x) - D_x (8)$$

$$= 6(3x^2) - 2(2x) + 7(1) - 0$$

$$= 18x^2 - 4x + 7.$$

EXAMPLE 5

Find the derivative of $f(x) = 2x(x^2 - 5x + 2)$ when $x = 2$.

By the distributive property,

$$f(x) = 2x^3 - 10x^2 + 4x.$$

Thus,

$$f'(x) = 2(3x^2) - 10(2x) + 4(1)$$
$$= 6x^2 - 20x + 4$$

and

$$f'(2) = 6(2)^2 - 20(2) + 4 = -12.$$

EXAMPLE 6

Find an equation of the tangent line to the curve $y = \dfrac{3x^2 - 2}{x}$ when $x = 1$.

By writing y as a difference of two functions, we have $y = \dfrac{3x^2}{x} - \dfrac{2}{x} = 3x - 2x^{-1}$. Thus,

$$\frac{dy}{dx} = 3(1) - 2[(-1)x^{-2}] = 3 + \frac{2}{x^2}.$$

The slope of the tangent line to the curve when $x = 1$ is

$$\frac{dy}{dx}\Big|_{x=1} = 3 + \frac{2}{1^2} = 5.$$

To find the y-coordinate of the point on the curve where $x = 1$, we substitute this value of x into the *original equation* of the curve. When $x = 1$, then $y = [3(1)^2 - 2]/1 = 1$. Hence the point $(1, 1)$ lies on both the curve and the tangent line. Therefore an equation of the tangent line is

$$y - 1 = 5(x - 1),$$
$$y = 5x - 4.$$

EXERCISE 8-2

*In Problems **1–58**, differentiate the functions.*

1. $f(x) = 7$.

2. $f(x) = \left(\frac{933}{465}\right)^{2/3}$.

3. $f(x) = x^8$.

4. $f(x) = .7x$.

5. $f(x) = 9x^5$.

6. $f(x) = \sqrt{2}\, x^{83/4}$.

7. $g(w) = w^{-7}$.

8. $f(t) = 3t^{-2}$.

9. $f(x) = 4x^{-14/5}$.

10. $v(x) = x^e$.

11. $f(x) = 3x - 2$.

12. $f(w) = 5w - 7 \ln \frac{4}{5}$.

13. $f(p) = \frac{13p}{5} + \frac{7}{3}$.

14. $q(x) = \frac{5x + 2}{8}$.

15. $g(x) = 3x^2 - 5x - 2$.

16. $f(q) = 7q^2 - 5q + 3$.

17. $f(t) = -13t^2 + 14t + 2$.

18. $p(x) = 97x^2 - 383x + 205$.

19. $f(x) = 14x^3 - 6x^2 + 7x - e^3$.

20. $f(r) = -8r^3 + 6$.

21. $f(q) = -3q^3 + \frac{9}{2}q^2 + 9q + 9$.

22. $f(x) = 100x^{-3} - 50x^{-1/2} + 10x - 1$.

23. $f(x) = x^8 - 7x^6 + 3x^2 + 9^{2/3}$.

24. $g(u) = 3u^{12} - 8u^8 + 3u^{-3} + 2u^2 - 9$.

25. $f(x) = 2x^{501} - 125x^{100} + .2x^{3.4}$.

26. $f(x) = 17 + 8x^{1/7} - 10x^{12} - 3x^{-15}$.

27. $f(x) = 2(13 - x^4)$.

28. $f(s) = 5(s^4 - 3)$.

29. $g(x) = \frac{13 - x^4}{3}$.

30. $f(x) = \frac{5(x^4 - 3)}{2}$.

31. $f(x) = x^{-4} - 9x^{1/3} + 5x^{-2/5}$.

32. $f(z) = 3z^{1/4} - 12^2 - 8z^{-3/4}$.

33. $h(x) = -2(27x - 14x^5)$.

34. $f(x) = \frac{-(1 + x - x^2 + x^3 + x^4 - x^5)}{2}$.

35. $f(x) = -2x^2 + \frac{3}{2}x + \frac{x^4}{4} + 2$.

36. $p(x) = \frac{x^7}{7} + \frac{x}{2}$.

37. $f(x) = \frac{1}{x}$.

38. $f(x) = \frac{7}{x^3}$.

39. $f(s) = \frac{1}{4s^5}$.

40. $g(w) = \frac{2}{3w^3}$.

41. $f(t) = 4\sqrt{t}$.

42. $f(x) = \frac{4}{\sqrt{x}}$.

43. $q(x) = \frac{1}{\sqrt[5]{x}}$.

44. $f(x) = \frac{3}{\sqrt[4]{x^3}}$.

45. $f(x) = x(3x^2 - 7x + 7)$.

46. $f(x) = x^3(3x^6 - 5x^2 + 4)$.

47. $g(t) = \frac{t^2}{2} - \frac{2}{t^2}$.

48. $f(x) = x\sqrt{x}$.

49. $f(x) = x^3(3x)^2$.

50. $f(x) = \sqrt{x}\,(5 - 6x + 3\sqrt[4]{x})$.

51. $v(x) = x^{-2/3}(x + 5)$.

52. $f(x) = x^{3/5}(x^2 + 7x + 1)$.

53. $f(q) = \frac{4q^3 + 7q - 4}{q}$.

54. $f(w) = \frac{w - 5}{w^5}$.

55. $f(x) = (x + 1)(x + 3)$.

56. $f(x) = x^2(x - 2)(x + 4)$.

57. $w(x) = \dfrac{x^2 + x^3}{x^2}.$ **58.** $f(x) = \dfrac{7x^3 + x}{2\sqrt{x}}.$

For each curve in Problems **59–62,** *find the slope at the indicated points.*

59. $y = 3x^2 + 4x - 8;$ $(0, -8), (2, 12), (-3, 7).$

60. $y = 5 - 6x - 2x^3;$ $(0, 5), (\frac{3}{2}, -\frac{43}{4}), (-3, 77).$

61. $y = 4;$ when $x = -4, x = 7, x = 22.$

62. $y = 2x - 3\sqrt{x};$ when $x = 1, x = 16, x = 25.$

In Problems **63** *and* **64,** *find an equation of the tangent line to the curve at the indicated point.*

63. $y = 4x^2 + 5x + 2;$ $(1, 11).$ **64.** $y = (1 - x^2)/5;$ $(4, -3).$

65. Find an equation of the tangent line to the curve $y = 3 + x - 5x^2 + x^4$ when $x = 0.$

66. Repeat Problem 65 for the curve $y = \dfrac{\sqrt{x}\,(2 - x^2)}{x}$ when $x = 4.$

67. Find all points on the curve $y = \frac{1}{3}x^3 - x^2$ where the tangent line is horizontal.

68. Find all points on the curve $y = x^2 - 5x + 3$ where the slope is 1.

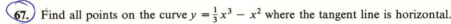

8-3 THE DERIVATIVE AS A RATE OF CHANGE

To denote the change in a variable such as x, the symbol Δx (read "delta x") is commonly used. For example, if x changes from $x = 1$ to $x = 3$, then the change is $\Delta x = 3 - 1 = 2$. The new value of x $(= 3)$ can be written $1 + \Delta x$. Similarly, if q increases by Δq, the new value is $q + \Delta q$. We shall use Δ-notation in the following discussion.

     A manufacturer's **total cost function,** $c = f(q)$, gives the total cost c of producing and marketing q units of a product. Figure 8-11 shows the graph of f, which is called a *total cost curve.* Suppose a manufacturer produces q units at a total cost of $f(q)$. If the production level is increased by Δq units to $q + \Delta q$, then the total cost is $f(q + \Delta q)$. The average cost per unit for these Δq additional units is

$$\frac{\text{change in total cost } c}{\text{change in output } q} = \frac{f(q + \Delta q) - f(q)}{\Delta q}.$$

Denoting the change in total cost by Δc, we have

$$\frac{\Delta c}{\Delta q} = \frac{f(q + \Delta q) - f(q)}{\Delta q}, \qquad\qquad (1)$$

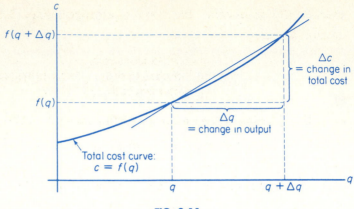

FIG. 8-11

which is called the **average rate of change of cost c with respect to output q** over the interval $[q, q + \Delta q]$. Notice in Fig. 8-11 that $\Delta c/\Delta q$ is the slope of a secant line.

To be more specific, suppose $c = f(q) = .1q^2 + 3$, where c is in dollars and q is in pounds. The cost of 4 lb is $f(4) = 4.6$. If output increases by 2 lb ($\Delta q = 2$), the new level of production is $4 + \Delta q = 6$, and the total cost is $f(4 + \Delta q) = f(6) = 6.6$. Thus the average cost per lb of the additional output on the interval $[4, 6]$ is

$$\frac{\Delta c}{\Delta q} = \frac{f(6) - f(4)}{2} = \frac{6.6 - 4.6}{2} = \$1 \text{ per lb.}$$

Similarly, for changes in output of 1, .1, and .01, we obtain the results given in Table 8-1. Notice that as $\Delta q \to 0$ it appears that $\Delta c/\Delta q \to .80$. This would

TABLE 8-1

CHANGE IN OUTPUT Δq	INTERVAL $[4, 4 + \Delta q]$	AVERAGE COST PER LB OF ADDITIONAL OUTPUT $\Delta c/\Delta q$
2	[4, 6]	1
1	[4, 5]	.90
.1	[4, 4.1]	.81
.01	[4, 4.01]	.801

indicate that for a small increase in output above 4 lb, the cost per lb of that additional output is approximately $0.80. We say that $\lim\limits_{\Delta q \to 0} \Delta c/\Delta q$ is the *instantaneous* rate of change of cost c with respect to output q when $q = 4$.

For any cost function $c = f(q)$, we say that $\lim\limits_{\Delta q \to 0} \Delta c/\Delta q$ is the **instantaneous rate of change of c with respect to q.** More simply, this limit is called **marginal cost.** Recall that $\Delta c/\Delta q$ is the slope of a secant line. Thus if $\Delta q \to 0$, then $\Delta c / \Delta q$ approaches the slope of a tangent line; that is, $\lim\limits_{\Delta q \to 0} \Delta c / \Delta q = dc / dq$.[†] Hence,

$$\frac{dc}{dq} = \left\{ \begin{array}{c} \text{instantaneous rate} \\ \text{of change of } c \\ \text{with respect to } q \end{array} \right\} = \text{marginal cost.}$$

For the case of the previous cost function $c = f(q) = .1q^2 + 3$,

$$\frac{dc}{dq} = .2q.$$

To find marginal cost when 4 lb are produced, we evaluate dc / dq when $q = 4$:

$$\left. \frac{dc}{dq} \right|_{q=4} = .2(4) = \$0.80.$$

From Table 8-1 the cost of producing one more pound beyond 4 lb is .90. The marginal cost, .80, is close to this number because the slope of the tangent line to the cost curve at $q = 4$ is a good approximation to the slope of the secant line through $(4, f(4))$ and $(5, f(5))$. See Fig. 8-12. For this reason *we interpret marginal cost as the approximate change in cost resulting from one additional unit of output.*

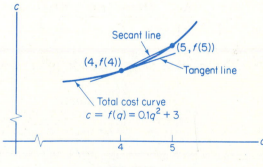

FIG. 8-12

Our discussion concerning rate of change applies not only to cost functions, but also to *any* function $y = f(x)$. In general we can interpret dy / dx as

$$\frac{dy}{dx} = \left\{ \begin{array}{c} \text{instantaneous rate of} \\ \text{change of } y \text{ with respect to } x. \end{array} \right.$$

[†]This can also be seen by replacing Δq in Eq. (1) by h:

$$\lim_{\Delta q \to 0} \frac{\Delta c}{\Delta q} = \lim_{\Delta q \to 0} \frac{f(q + \Delta q) - f(q)}{q} = \lim_{h \to 0} \frac{f(q + h) - f(q)}{h} = \frac{dc}{dq}.$$

Thus the instantaneous rate of change of $y = f(x)$ at a point is the slope of the tangent line to the graph of $y = f(x)$ at that point. For convenience we usually refer to instantaneous rate of change simply as *rate of change*.

EXAMPLE 1

Find the (instantaneous) rate of change of $y = x^4$ with respect to x. Evaluate when $x = 2$ and when $x = -1$.

The rate of change is given by dy/dx:

$$\frac{dy}{dx} = 4x^3.$$

When $x = 2$, then $dy/dx = 4(2)^3 = 32$. This means that if x increases by a small amount, then y increases approximately 32 times as much. More simply, we say that y is increasing 32 times as fast as x does. When $x = -1$, then $dy/dx = 4(-1)^3 = -4$. The significance of the minus sign on -4 is that y is *decreasing* 4 times as fast as x increases.

EXAMPLE 2

Let $p = 100 - q^2$ be the demand function for a manufacturer's product. Find the rate of change of price p per unit with respect to quantity q. How fast is the price changing with respect to q when $q = 5$? Assume that p is in dollars.

The rate of change of p with respect to q is dp/dq.

$$\frac{dp}{dq} = \frac{d}{dq}(100 - q^2) = -2q.$$

Thus,

$$\left.\frac{dp}{dq}\right|_{q=5} = -2(5) = -10.$$

This means that when 5 units are demanded, an *increase* of one extra unit demanded will *decrease* the price per unit that consumers are willing to pay by approximately $10.

If c is the total cost of producing q units of a product, then the *average cost per unit*, $\bar{c}$, for producing q units is

$$\bar{c} = \frac{c}{q}. \tag{2}$$

For example, if the total cost of 20 units is $100, then the average cost per unit is $\bar{c} = 100/20 = \$5$. By multiplying both sides of Eq. (2) by q, we have

$$c = q\bar{c}.$$

That is, total cost is the product of the number of units produced and the average cost per unit.

EXAMPLE 3

If

$$\bar{c} = .0001q^2 - .02q + 5 + \frac{5000}{q}$$

is a manufacturer's average cost equation, find the marginal cost function. What is the marginal cost when 50 units are produced?

We first find total cost c. Since $c = q\bar{c}$, then

$$c = q\bar{c}$$

$$= q\left[.0001q^2 - .02q + 5 + \frac{5000}{q}\right].$$

$$c = .0001q^3 - .02q^2 + 5q + 5000.$$

Differentiating c, we have the marginal cost function:

$$\frac{dc}{dq} = .0001(3q^2) - .02(2q) + 5(1) + 0$$

$$= .0003q^2 - .04q + 5.$$

The marginal cost when 50 units are produced is

$$\left.\frac{dc}{dq}\right|_{q=50} = .0003(50)^2 - .04(50) + 5 = 3.75.$$

Let us interpret this result. If c is in dollars and production is increased by one unit from $q = 50$ to $q = 51$, then the cost of the additional unit is approximately \$3.75. If production is increased by $\frac{1}{3}$ unit from $q = 50$, then the cost of the additional output is approximately $(\frac{1}{3})(3.75) = \$1.25$.

Suppose $r = f(q)$ is the total revenue function of a manufacturer. The equation $r = f(q)$ states that the total dollar value received for selling q units of the product is r. The **marginal revenue** is defined as the rate of change of the total dollar value received with respect to the total number of units sold. Hence, marginal revenue is merely the derivative of r with respect to q.

$$\textbf{marginal revenue} = \frac{dr}{dq}.$$

Marginal revenue indicates the rate at which revenue changes with respect to units sold. We interpret it as the approximate change in revenue that results from selling one additional unit of output.

EXAMPLE 4

Suppose a manufacturer sells a product at \$5 per unit. If q units are sold, the total revenue is given by

$$r = 5q.$$

Thus the marginal revenue function is

$$\frac{dr}{dq} = \frac{d}{dq}(5q) = 5.$$

The marginal revenue when $q = 10$ is

$$\left.\frac{dr}{dq}\right|_{q=10} = 5.$$

Suppose $r = f(q) = 2q$ gives the total revenue r (in dollars) that a manufacturer receives for selling q units of his product. The rate of change of revenue with respect to number of units sold is

$$\underbrace{\frac{\text{Total Rev.}}{\text{units sold}}}\ \frac{dr}{dq} = 2.$$

This means that revenue is changing at the rate of \$2 per unit, regardless of the number of units sold. Although this is valuable information, it may be more significant when compared to r itself. For example, if $q = 50$ then $r = 2(50) =$ \$100. Thus the rate of change of revenue is $2/100 = .02$ **of r**. On the other hand, if $q = 5000$ then $r = 2(5000) =$ \$10,000, so the rate of change of r is $2/10,000 = .0002$ **of r**. Although r changes at the same rate at each level, when compared to r itself this rate is relatively smaller when $r = 10,000$ than when $r = 100$. By considering the ratio

$$\frac{dr/dq}{r},$$

we have a means of comparing the rate of change of r with r itself. This ratio is called the *relative rate of change* of the revenue function $r = f(q)$. We have shown above that the relative rate of change when $q = 50$ is

$$\frac{dr/dq}{r} = \frac{2}{100} = .02,$$

and when $q = 5000$, it is

$$\frac{dr/dq}{r} = \frac{2}{10,000} = .0002.$$

By multiplying these relative rates by 100, we obtain so-called *percentage rates of change*. Thus the percentage rate of change when $q = 50$ is $(.02)(100) = 2$ percent; when $q = 5000$ it is $(.0002)(100) = .02$ percent.

In general, for any function f we have the following definition.

DEFINITION. *The **relative rate of change** of* $f(x)$ *is*

$$\frac{f'(x)}{f(x)}.$$

*The **percentage rate of change** of* $f(x)$ *is*

$$\frac{f'(x)}{f(x)} \cdot 100.$$

EXAMPLE 5

Find the relative and percentage rates of change of $y = f(x) = 3x^2 - 5x + 25$ *when* $x = 5$.

Here

$$f'(x) = 3(2x) - 5(1) + 0 = 6x - 5.$$

Since $f'(5) = 6(5) - 5 = 25$ and $f(5) = 3(5)^2 - 5(5) + 25 = 75$, the relative rate of change of y when $x = 5$ is

$$\frac{f'(5)}{f(5)} = \frac{25}{75} \approx .333.$$

Multiplying .333 by 100 gives the percentage rate of change: $(.333)(100) = 33.3$ percent.

EXERCISE 8-3

In Problems **1–6**, find (a) the rate of change of y with respect to x, and (b) the relative rate of change of y. At the given value of x, find (c) the rate of change of y, (d) the relative rate of change of y, and (e) the percentage rate of change of y.

1. $y = f(x) = x + 4; \quad x = 5.$ **2.** $y = f(x) = 4 - 2x; \quad x = 3.$

3. $y = 3x^2 + 6; \quad x = 2.$ **4.** $y = 2 - x^2; \quad x = 0.$

5. $y = 8 - x^3; \quad x = 1.$ **6.** $y = x^2 + 3x - 4; \quad x = -1.$

In Problems **7–12**, *cost functions are given where* c *is the cost of producing* q *units of a product. In each case find the marginal cost function. What is the marginal cost at the given value(s) of* q?

7. $c = 500 + 10q; \quad q = 100.$ **8.** $c = 5000 + 6q; \quad q = 36.$

9. $c = .3q^2 + 2q + 850; \quad q = 3.$ **10.** $c = .1q^2 + 3q + 2; \quad q = 3.$

11. $c = q^2 + 50q + 1000$; $q = 15, q = 16, q = 17$.

12. $c = .03q^3 - .6q^2 + 4.5q + 7700$; $q = 10, q = 20, q = 100$.

*In Problems **13–16**, $\bar{c}$ represents average cost, which is a function of the number q of units produced. Find the marginal cost function and the marginal cost for the indicated values of q.*

13. $\bar{c} = .01q + 5 + \dfrac{500}{q}$; $q = 50, q = 100$.

14. $\bar{c} = 2 + \dfrac{1000}{q}$; $q = 25, q = 235$.

15. $\bar{c} = .00002q^2 - .01q + 6 + \dfrac{20,000}{q}$; $q = 100, q = 500$.

16. $\bar{c} = .001q^2 - .3q + 40 + \dfrac{7000}{q}$; $q = 10, q = 20$.

*In Problems **17–20**, r represents total revenue and is a function of the number q of units sold. Find the marginal revenue function and the marginal revenue for the indicated values of q.*

17. $r = .7q$; $q = 8, q = 100, q = 200$.

18. $r = q(15 - \frac{1}{30}q)$; $q = 5, q = 15, q = 150$.

19. $r = 250q + 45q^2 - q^3$; $q = 5, q = 10, q = 25$.

20. $r = 2q(30 - .1q)$; $q = 10, q = 20$.

21. For the cost function $c = .2q^2 + 1.2q + 4$, how fast does c change with respect to q when $q = 5$? Determine the percentage rate of change of c with respect to q when $q = 5$.

22. For the cost function $c = .4q^2 + 4q + 5$, find the rate of change of c with respect to q when $q = 2$. Also, what is $\Delta c/\Delta q$ over the interval $[2, 3]$?

23. The total cost function for a hosiery mill is estimated by Dean:[†]

$$c = -10,484.69 + 6.750q - .000328q^2,$$

where q is output in dozens of pairs and c is total cost in dollars. Find the marginal cost function and evaluate it when $q = 5000$.

24. The total cost function for an electric light and power plant is estimated by Nordin:[‡]

$$c = 32.07 - .79q + .02142q^2 - .0001q^3, \qquad 20 \leqslant q \leqslant 90$$

[†]J. Dean, "Statistical cost functions of a hosiery mill," *Studies in Business Administration,* XI, no. 4 (Chicago: University of Chicago Press, 1941).

[‡]J. A. Nordin, "Note on a light plant's cost curves," *Econometrica,* vol. 15 (1947), 231–235.

where q is eight-hour total output (as percentage of capacity) and c is total fuel cost in dollars. Find the marginal cost function and evaluate it when $q = 70$.

25. For a certain manufacturer, the revenue r obtained from the sale of q units of his product is given by $r = 30q - .3q^2$. (a) How fast does r change with respect to q? When $q = 10$, (b) find the relative rate of change of r, and (c) to the nearest percent, find the percentage rate of change of r.

26. Repeat Problem 25 for the revenue function $r = 20q - .1q^2$ and $q = 100$.

27. Suppose the 100 largest cities in the U.S. in 1920 are ranked according to magnitude (areas of cities). From Lotka[†] the following relation approximately holds:

$$PR^{.93} = 5,000,000,$$

where P is the population of the city having respective rank R. This relation is called the *law of urban concentration* for 1920. Solve for P in terms of R and then find how fast population is changing with respect to rank.

28. Using the straight-line method of depreciation, the value v of a certain machine after t years have elapsed is given by $v = 50,000 - 5000t$ where $0 \leqslant t \leqslant 10$. How fast is v changing with respect to t when $t = 2$? $t = 3$? at any time t?

8-4 DIFFERENTIABILITY AND CONTINUITY

In the next section we shall make use of an important relationship between differentiability and continuity, namely

> If f is differentiable at $x = a$,
> then f is continuous at $x = a$.

Intuitively we can see why this is true in Fig. 8-13. If the derivative exists at $x = a$, then as $h \to 0$ the limiting position of the secant line is the tangent line. For this position to be approached, $f(a + h)$ must approach $f(a)$. That is,

$$\lim_{h \to 0} f(a + h) = f(a),$$

which is another way of saying that f is continuous at $x = a$.

If a function is not continuous at a point, then it cannot have a derivative there. For example, the function in Fig. 8-14 is discontinuous at $x = a$. The curve has no tangent at that point, so the function is not differentiable there.

[†]A. J. Lotka, *Elements of Mathematical Biology* (New York: Dover, 1956).

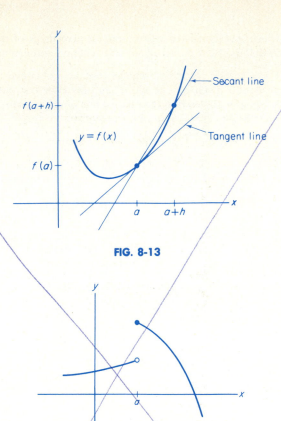

FIG. 8-13

FIG. 8-14

EXAMPLE 1

a. Let $f(x) = x^2$. Since $f'(x) = 2x$ is defined for all values of x, then $f(x) = x^2$ must be continuous for all values of x.

b. The function $f(p) = \dfrac{1}{2p}$ is not continuous at $p = 0$ because f is not defined there. Thus the derivative does not exist at $p = 0$.

The converse of the statement that differentiability implies continuity is *false*. In Example 2 you will see a function that is continuous at a point but not differentiable there.

EXAMPLE 2

The function $y = f(x) = |x|$ is continuous at $x = 0$. See Fig. 8-15. As we mentioned in Sec. 8-1, there is no tangent line at $x = 0$. Thus the derivative does not exist there. This shows that continuity does *not* imply differentiability.

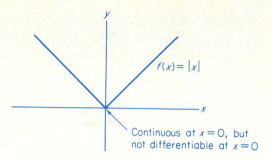

$f(x) = |x|$

Continuous at $x = 0$, but
not differentiable at $x = 0$

FIG. 8-15

8-5 PRODUCT AND QUOTIENT RULES

The equation $F(x) = (x^2 + 3x)(4x + 5)$ expresses $F(x)$ as a product of two functions: $x^2 + 3x$ and $4x + 5$. To find $F'(x)$ by using only Rules 1–5, we first multiply the functions, which gives $F(x) = 4x^3 + 17x^2 + 15x$. Then we differentiate term by term:

$$F'(x) = 4(3x^2) + 17(2x) + 15(1) = 12x^2 + 34x + 15. \qquad (1)$$

In many problems that involve differentiating a product of functions, the multiplication is not so simple as it was here. Often it is not even practical to attempt it. Fortunately there is a rule for differentiating a product, and the rule avoids such multiplications. Since the derivative of a sum of functions is the sum of their derivatives, you might think that the derivative of a product of two functions is the product of their derivatives. This is **not** the case, as Rule 6 shows.

RULE 6. (*Product Rule.*) *Let* $F(x) = f(x)g(x)$. *If* $f'(x)$ *and* $g'(x)$ *exist, then*

$$F'(x) = f(x)g'(x) + g(x)f'(x).$$

That is, the derivative of the product of two functions is the first function times the derivative of the second, plus the second function times the derivative of the first.

Proof. By the definition of the derivative of F, we have

$$F'(x) = \lim_{h \to 0} \frac{F(x + h) - F(x)}{h}$$

$$= \lim_{h \to 0} \frac{f(x + h)g(x + h) - f(x)g(x)}{h}.$$

Now we use a "trick." Adding and subtracting $f(x + h)g(x)$, we have

$$F'(x) = \lim_{h \to 0} \frac{f(x + h)g(x + h) - f(x)g(x) + [f(x + h)g(x) - f(x + h)g(x)]}{h}$$

Regrouping gives

$$F'(x) = \lim_{h \to 0} \frac{[f(x + h)g(x + h) - f(x + h)g(x)] + [f(x + h)g(x) - f(x)g(x)]}{h}$$

$$= \lim_{h \to 0} \frac{f(x + h)[g(x + h) - g(x)] + g(x)[f(x + h) - f(x)]}{h}$$

$$= \lim_{h \to 0} \frac{f(x + h)[g(x + h) - g(x)]}{h} + \lim_{h \to 0} \frac{g(x)[f(x + h) - f(x)]}{h}$$

$$= \lim_{h \to 0} f(x + h) \cdot \lim_{h \to 0} \frac{g(x + h) - g(x)}{h} + \lim_{h \to 0} g(x) \cdot \lim_{h \to 0} \frac{f(x + h) - f(x)}{h}.$$

Since we assumed that f and g are differentiable, then

$$\lim_{h \to 0} \frac{f(x + h) - f(x)}{h} = f'(x)$$

and

$$\lim_{h \to 0} \frac{g(x + h) - g(x)}{h} = g'(x).$$

Moreover, the differentiability of f implies that f is continuous, and from Sec. 8-4,

$$\lim_{h \to 0} f(x + h) = f(x).$$

Thus,

$$F'(x) = f(x)g'(x) + g(x)f'(x).$$

(See how important continuity is!) Therefore, we have the **product rule:**

$$\boxed{\begin{array}{l} \text{If } F(x) = f(x)g(x), \text{ then} \\ F'(x) = f(x)g'(x) + g(x)f'(x). \end{array}}$$

EXAMPLE 1

If $F(x) = (x^2 + 3x)(4x + 5)$, find $F'(x)$.

Here F can be considered a product of two functions: $f(x) = x^2 + 3x$ and $g(x) = 4x + 5$. By Rule 6, the product rule,

$$F'(x) = f(x)g'(x) + g(x)f'(x)$$

$$= (x^2 + 3x) \, D_x(4x + 5) + (4x + 5) \, D_x(x^2 + 3x)$$

$$= (x^2 + 3x)(4) + (4x + 5)(2x + 3)$$

$$= 12x^2 + 34x + 15.$$

This agrees with our previous result [see Eq. (1)].

> **PITFALL.** *We repeat: the derivative of a product of two functions **is not** the product of the derivatives. For example, $D_x(x^2 + 3x) = 2x + 3$ and $D_x(4x + 5) = 4$, but*
>
> $$D_x[(x^2 + 3x)(4x + 5)] = 12x^2 + 34x + 15 \neq (2x + 3)4.$$

EXAMPLE 2

a. *Find the slope of the graph of $F(x) = (7x^3 - 5x + 2)(2x^4 + x + 7)$ when $x = 1$.*

Let $f(x) = 7x^3 - 5x + 2$ and $g(x) = 2x^4 + x + 7$. Then $F(x) = f(x) \cdot g(x)$, a product of two functions. By the product rule,

$$F'(x) = (7x^3 - 5x + 2)\, D_x(2x^4 + x + 7) + (2x^4 + x + 7)\, D_x(7x^3 - 5x + 2)$$
$$= (7x^3 - 5x + 2)(8x^3 + 1) + (2x^4 + x + 7)(21x^2 - 5).$$

Evaluating $F'(x)$ at $x = 1$ gives the slope of the graph at that point:

$$F'(1) = 4(9) + 10(16) = 196.$$

Note: **we do not have to simplify $F'(x)$ before evaluating it.**

b. *If $y = (x^{2/3} + 3)(x^{-1/3} + 5x)$, find $D_x y$.*

$$D_x y = (x^{2/3} + 3)\, D_x(x^{-1/3} + 5x) + (x^{-1/3} + 5x)\, D_x(x^{2/3} + 3)$$
$$= (x^{2/3} + 3)\left(-\tfrac{1}{3}x^{-4/3} + 5\right) + (x^{-1/3} + 5x)\left(\tfrac{2}{3}x^{-1/3}\right)$$
$$= \tfrac{25}{3}x^{2/3} + \tfrac{1}{3}x^{-2/3} - x^{-4/3} + 15.$$

c. *If $y = (x + 2)(x + 3)(x + 4)$, find y'.*

By grouping, we can consider y to be a product of two functions:

$$y = [(x + 2)(x + 3)](x + 4).$$

The product rule gives

$$y' = [(x + 2)(x + 3)]\, D_x(x + 4) + (x + 4)\, D_x[(x + 2)(x + 3)]$$
$$= [(x + 2)(x + 3)](1) + (x + 4)\, D_x[(x + 2)(x + 3)].$$

Applying the product rule again, we have

$$y' = [(x + 2)(x + 3)](1) + (x + 4)[(x + 2)\, D_x(x + 3) + (x + 3)\, D_x(x + 2)]$$
$$= [(x + 2)(x + 3)](1) + (x + 4)[(x + 2)(1) + (x + 3)(1)].$$

After simplifying, we obtain

$$y' = 3x^2 + 18x + 26.$$

Usually we do not use the product rule when simpler ways are obvious. For example, if $f(x) = 2x(x + 3)$, then it is quicker to write $f(x) = 2x^2 + 6x$, from which $f'(x) = 4x + 6$. Similarly, we do not usually use the product rule to differentiate $y = 4(x^2 - 3)$. Since the 4 is a constant multiplier, by Rule 3 we have $y' = 4(2x) = 8x$.

The next rule is used for differentiating a *quotient* of two functions.

RULE 7.   (*Quotient Rule.*) *Let* $F(x) = \dfrac{f(x)}{g(x)}$ *such that* $g(x) \neq 0$. *If* $f'(x)$ *and* $g'(x)$ *exist, then*

$$F'(x) = \frac{g(x)f'(x) - f(x)g'(x)}{[g(x)]^2}.$$

Proof.   Since $F(x) = \dfrac{f(x)}{g(x)}$,

$$F(x)g(x) = f(x).$$

By the product rule,

$$F(x)g'(x) + g(x)F'(x) = f'(x).$$

Solving for $F'(x)$, we have

$$F'(x) = \frac{f'(x) - F(x)g'(x)}{g(x)}.$$

But $F(x) = f(x)/g(x)$. Thus,

$$F'(x) = \frac{f'(x) - \dfrac{f(x)g'(x)}{g(x)}}{g(x)}$$

$$= \frac{g(x)f'(x) - f(x)g'(x)}{[g(x)]^2}.$$

Therefore[†], we have the **quotient rule**:

$$\boxed{\begin{array}{l} \text{If } F(x) = f(x)/g(x), \text{ then} \\[2mm] F'(x) = \dfrac{g(x)f'(x) - f(x)g'(x)}{[g(x)]^2}. \end{array}}$$

PITFALL.   *The derivative of a quotient of two functions **is not** the quotient of the derivatives of the functions.*

[†]You may have observed that this proof assumes the existence of $F'(x)$. However, the quotient rule can be proven without this assumption.

EXAMPLE 3

a. *If* $F(x) = \dfrac{4x^2 - 2x + 3}{2x - 1}$, *find* $F'(x)$.

Let $f(x) = 4x^2 - 2x + 3$ and $g(x) = 2x - 1$. Then $F(x) = f(x)/g(x)$ and by Rule 7, the quotient rule,

$$F'(x) = \frac{g(x)f'(x) - f(x)g'(x)}{[g(x)]^2}$$

$$= \frac{(2x - 1)\, D_x(4x^2 - 2x + 3) - (4x^2 - 2x + 3)\, D_x(2x - 1)}{(2x - 1)^2}$$

$$= \frac{(2x - 1)(8x - 2) - (4x^2 - 2x + 3)(2)}{(2x - 1)^2}$$

$$= \frac{8x^2 - 8x - 4}{(2x - 1)^2} = \frac{4(2x^2 - 2x - 1)}{(2x - 1)^2}.$$

b. *If* $y = \dfrac{1}{x^2}$, *find* y'.

Here y can be considered a quotient, and by the quotient rule,

$$y' = \frac{(x^2)\, D_x(1) - (1)\, D_x(x^2)}{(x^2)^2}$$

$$= \frac{x^2(0) - 1(2x)}{x^4}$$

$$= \frac{-2x}{x^4} = -\frac{2}{x^3}.$$

A simpler and more direct method of differentiating y is by writing $y = x^{-2}$ and using Rule 2: $y' = -2x^{-3} = -2/x^3$.

PITFALL. *Note that* $D_x \dfrac{1}{x^2} \neq \dfrac{1}{2x}$.

c. *Find an equation of the tangent line to the curve* $y = \dfrac{(x + 1)(x^2 + 2x + 5)}{1 - x}$ *at* $(0, 5)$.

By the quotient rule,

$$y' = \frac{(1 - x)\, D_x[(x + 1)(x^2 + 2x + 5)] - [(x + 1)(x^2 + 2x + 5)]\, D_x(1 - x)}{(1 - x)^2}.$$

Using the product rule to find $D_x[(x + 1)(x^2 + 2x + 5)]$, we have

$$y' = \frac{(1 - x)[(x + 1)(2x + 2) + (x^2 + 2x + 5)(1)] - [(x + 1)(x^2 + 2x + 5)](-1)}{(1 - x)^2}.$$

The slope of the curve at $(0, 5)$ is $y'(0) = 12$. An equation of the tangent line is

$$y - 5 = 12(x - 0),$$
$$y = 12x + 5.$$

EXAMPLE 4

If the demand equation for a manufacturer's product is $p = 1000/(q + 5)$, find the marginal revenue function and evaluate it when $q = 45$.

The revenue r received for selling q units is

revenue = (price)(quantity),
$$r = pq.$$

Thus the revenue function is

$$r = \left(\frac{1000}{q + 5} \right) q$$

or

$$r = \frac{1000q}{q + 5}.$$

To find the marginal revenue function, all we must determine is dr/dq.

$$\frac{dr}{dq} = \frac{(q + 5) D_q(1000q) - (1000q) D_q(q + 5)}{(q + 5)^2}$$

$$= \frac{(q + 5)1000 - (1000q)(1)}{(q + 5)^2} = \frac{5000}{(q + 5)^2}.$$

When $q = 45$,

$$\frac{dr}{dq} = \frac{5000}{(45 + 5)^2} = \frac{5000}{2500} = 2.$$

This means that selling one additional unit beyond 45 results in approximately $2 more in revenue.

A function that plays an important role in economic analysis is the **consumption function**. The consumption function $C = f(I)$ expresses a relationship between the total national income I and the total national consumption C. Usually, both I and C are expressed in billions of dollars and I is restricted to some interval. The *marginal propensity to consume* is defined as the rate of change of consumption with respect to income. It is merely the derivative of C with respect to I.

$$\textbf{marginal propensity to consume} = \frac{dC}{dI}.$$

If we assume that the difference between income I and consumption C is savings S, then

$$S = I - C.$$

Differentiating both sides with respect to I gives

$$\frac{dS}{dI} = \frac{d}{dI}(I) - \frac{d}{dI}(C) = 1 - \frac{dC}{dI}.$$

We define dS/dI as the **marginal propensity to save**. Thus, the marginal propensity to save indicates how fast savings change with respect to income.

EXAMPLE 5

If the consumption function is given by

$$C = \frac{5(2\sqrt{I^3} + 3)}{I + 10},$$

determine the marginal propensity to consume and the marginal propensity to save when $I = 100$.

$$\frac{dC}{dI} = \frac{(I + 10)\, D_I[5(2I^{3/2} + 3)] - 5(2\sqrt{I^3} + 3)\, D_I[I + 10]}{(I + 10)^2}$$

$$= \frac{(I + 10)[5(3I^{1/2})] - 5(2\sqrt{I^3} + 3)[1]}{(I + 10)^2}.$$

When $I = 100$ the marginal propensity to consume is

$$\left.\frac{dC}{dI}\right|_{I=100} = \frac{6485}{12{,}100} \approx .536.$$

The marginal propensity to save when $I = 100$ is $1 - .536 = .464$. This means that if a current income of \$100 billion increases by \$1 billion, then the nation will consume approximately 53.6 percent (536/1000) and save 46.4 percent (464/1000) of that increase.

EXERCISE 8-5

In Problems **1–50,** *differentiate the functions.*

1. $f(x) = (4x + 1)(6x + 3)$.

2. $f(x) = (3x - 1)(7x + 2)$.

3. $s(t) = (8 - 7t)(t^2 - 2)$.

4. $Q(x) = (5 - 2x)(x^2 + 1)$.

5. $f(r) = (3r^2 - 4)(r^2 - 5r + 1)$.

6. $C(I) = (2I^2 - 3)(3I^2 - 4I + 1)$.

7. $y = (x^2 + 3x - 2)(2x^2 - x - 3)$.

8. $y = (2 - 3x + 4x^2)(1 + 2x - 3x^2)$.

9. $f(w) = (8w^2 + 2w - 3)(5w^3 + 2).$ **10.** $f(x) = (3x - x^2)(3 - x - x^2).$

11. $g(x) = 3(x^3 - 2x^2 + 5x - 4)(x^4 - 2x^3 + 7x + 1).$

12. $y = -\frac{3}{2}(2x^4 - 3x + 1)(3x^3 - 6x^2 + 2x - 4).$

13. $y = (x^2 - 1)(3x^3 - 6x + 5) - (x + 4)(4x^2 + 2x + 1).$

14. $h(x) = 4(x^5 - 3)(2x^3 + 4) + 3(8x^2 - 5)(3x + 2).$

15. $f(p) = \frac{3}{2}(\sqrt{p} - 4)(4p - 5).$ **16.** $g(x) = (\sqrt{x} - 3x + 1)(\sqrt[4]{x} - 2\sqrt{x}).$

17. $y = (2x^{.45} - 3)(x^{1.3} - 7x).$ **18.** $y = (x - 1)(x - 2)(x - 3).$

19. $y = (2x - 1)(3x + 4)(x + 7).$ **20.** $y = \dfrac{x}{x - 3}.$

21. $y = 7 \cdot \frac{2}{3}.$ **22.** $y = \dfrac{2x - 3}{4x + 1}.$

23. $f(x) = \dfrac{x}{x - 1}.$ **24.** $f(x) = \dfrac{-2x}{1 - x}.$

25. $y = \dfrac{x + 2}{x - 1}.$ **26.** $h(w) = \dfrac{3w^2 + 5w - 1}{w - 3}.$

27. $h(z) = \dfrac{5 - 2z}{z^2 - 4}.$ **28.** $y = \dfrac{x^2 - 4x + 2}{x^2 + x + 1}.$

29. $y = \dfrac{8x^2 - 2x + 1}{x^2 - 5x}.$ **30.** $f(x) = \dfrac{x^3 - x^2 + 1}{x^2 + 1}.$

31. $y = \dfrac{x^2 - 4x + 3}{2x^2 - 3x + 2}.$ **32.** $F(z) = \dfrac{z^4 + 4}{3z}.$

33. $g(x) = \dfrac{1}{x^{100} + 1}.$ **34.** $y = \dfrac{3}{7x^3}.$

35. $u(v) = \dfrac{v^5 - 8}{v}.$ **36.** $y = \dfrac{x - 5}{2\sqrt{x}}.$

37. $y = \dfrac{3x^2 - x - 1}{\sqrt[3]{x}}.$ **38.** $y = \dfrac{x^3 - 2}{2x^{2.1} + 1}.$

39. $y = 7 - \dfrac{4}{x - 8} + \dfrac{2x}{3x + 1}.$ **40.** $q(x) = 13x^2 + \dfrac{x - 1}{2x + 3} - \dfrac{4}{x}.$

41. $H(s) = \dfrac{(s + 2)(s - 4)}{s - 5}.$ **42.** $y = \dfrac{(2x - 1)(3x + 2)}{4 - 5x}.$

43. $y = \dfrac{x - 5}{(x + 2)(x - 4)}.$ **44.** $y = \dfrac{4 - 5x}{(2x - 1)(3x + 2)}.$

45. $s(t) = \dfrac{t^2 + 3t}{(t^2 - 1)(t^3 + 7)}.$ **46.** $y = \dfrac{(2x - 3)(x^2 - 4x + 1)}{3x^3 + 1}.$

47. $y = \dfrac{(x - 1)(x - 2)}{(x - 3)(x - 4)}.$ **48.** $f(s) = \dfrac{17}{s(5s^2 - 10s + 4)}.$

49. $y = 3x - \dfrac{\dfrac{2}{x} - \dfrac{3}{x-1}}{x-2}$.

50. $y = 7 - 10x^2 + \dfrac{1 - \dfrac{7}{x^2+3}}{x+2}$.

51. Find the slope of the curve $y = (4x^2 + 2x - 5)(x^3 + 7x + 4)$ at $(-1, 12)$.

52. Find the slope of the curve $y = \dfrac{x^3}{x^4 + 1}$ at $(1, \frac{1}{2})$.

In Problems 53–56, find an equation of the tangent line to the curve at the given point.

53. $y = 6/(x - 1)$; $(3, 3)$.

54. $y = \dfrac{4x + 5}{x^2}$; $(-1, 1)$.

55. $y = (2x + 3)[2(x^4 - 5x^2 + 4)]$; $(0, 24)$.

56. $y = \dfrac{x + 1}{x^2(x - 4)}$; $(2, -\frac{3}{8})$.

In Problems 57 and 58, determine the relative rate of change of y with respect to x for the given value of x.

57. $y = \dfrac{x}{2x - 6}$; $x = 1$.

58. $y = \dfrac{1 - x}{1 + x}$; $x = 5$.

59. For the U.S. (1922–42), the consumption function is estimated by[†]

$$C = .672I + 113.1.$$

Find the marginal propensity to consume.

60. Repeat Problem 59 if $C = .712I + 95.05$ for the U.S. for 1929–41.[†]

In Problems 61–64, each equation represents a consumption function. Find the marginal propensity to consume and the marginal propensity to save for the given value of I.

61. $C = 2 + 2\sqrt{I}$; $I = 9$.

62. $C = 6 + \dfrac{3I}{4} - \dfrac{\sqrt{I}}{3}$; $I = 25$.

63. $C = \dfrac{16\sqrt{I} + .8\sqrt{I^3} - .2I}{\sqrt{I} + 4}$; $I = 36$.

64. $C = \dfrac{20\sqrt{I} + .5\sqrt{I^3} - .4I}{\sqrt{I} + 5}$; $I = 100$.

[†]T. Haavelmo, "Methods of measuring the marginal propensity to consume," *Journal of the American Statistical Association*, vol. XLII (1947), 105–122.

*In Problems **65–68**, each equation represents a demand function for a certain product where p denotes price per unit for q units. Find the marginal revenue function in each case. Recall that revenue = pq.*

65. $p = 25 - .02q$.

66. $p = 500/q$.

67. $p = \dfrac{108}{q + 2} - 3$.

68. $p = \dfrac{q + 750}{q + 50}$.

69. If the total cost function for a manufacturer is given by

$$c = \frac{5q^2}{q + 3} + 5000,$$

find the marginal cost function.

8-6 THE CHAIN RULE AND POWER RULE

Our next rule, the chain rule, is one of the most important rules for finding derivatives.

To begin, suppose

$$y = f(u) = 2u^2 - 3u - 2 \tag{1}$$

and

$$u = g(x) = x^2 + x. \tag{2}$$

Here y is a function of u and u is a function of x. If we substitute $x^2 + x$ for u in Eq. (1), we can consider y to be a function of x:

$$y = 2(x^2 + x)^2 - 3(x^2 + x) - 2.$$

After expanding we can find dy/dx in the usual way.

$$y = 2x^4 + 4x^3 + 2x^2 - 3x^2 - 3x - 2$$

$$= 2x^4 + 4x^3 - x^2 - 3x - 2.$$

$$\frac{dy}{dx} = 8x^3 + 12x^2 - 2x - 3.$$

From this example you can see that finding dy/dx by first performing a substitution could be quite involved, especially if $2u^2$ in Eq. (1) were $2u^{200}$. The chain rule will allow us to handle such situations with ease.

RULE 8. (*Chain Rule.*) *If $y = f(u)$ is a differentiable function of u and u is a differentiable function of x, then y is a differentiable function of x and*

$$\frac{dy}{dx} = \frac{dy}{du} \cdot \frac{du}{dx}.$$

Let us see why the chain rule is reasonable. Suppose $y = 8u + 5$ and $u = 2x - 3$. Let x change by one unit. How does u change? Answer: $du/dx = 2$. But for *each* one-unit change in u there is a change in y of $dy/du = 8$. Therefore, what is the change in y if x changes by one unit, that is, what is dy/dx? Answer: $8 \cdot 2$, which is $\frac{dy}{du} \cdot \frac{du}{dx}$. Thus $\frac{dy}{dx} = \frac{dy}{du} \cdot \frac{du}{dx}$.

EXAMPLE 1

a. *If $y = 2u^2 - 3u - 2$ and $u = x^2 + x$, find dy/dx.*

By Rule 8, the chain rule,

$$\frac{dy}{dx} = \frac{dy}{du} \cdot \frac{du}{dx} = \frac{d}{du}(2u^2 - 3u - 2) \cdot \frac{d}{dx}(x^2 + x)$$

$$= (4u - 3)(2x + 1).$$

We can write our answer exclusively in terms of x by replacing u by $x^2 + x$.

$$\frac{dy}{dx} = [4(x^2 + x) - 3](2x + 1)$$

$$= [4x^2 + 4x - 3](2x + 1)$$

$$= 8x^3 + 12x^2 - 2x - 3 \qquad \text{(as we saw before)}.$$

b. *If $y = \sqrt{u}$ and $u = 7 - x^3$, find dy/dx.*

By the chain rule,

$$\frac{dy}{dx} = \frac{dy}{du} \cdot \frac{du}{dx}$$

$$= \frac{d}{du}(\sqrt{u}) \cdot \frac{d}{dx}(7 - x^3)$$

$$= \frac{1}{2\sqrt{u}} \cdot (-3x^2)$$

$$= -\frac{3x^2}{2\sqrt{u}} = -\frac{3x^2}{2\sqrt{7 - x^3}}.$$

c. *If $y = u^{10}$ and $u = 8 - t^2 + t^5$, find dy/dt.*

Here y is a function of u, and u is a function of t. Hence we can consider y as a function of t. By the chain rule,

$$\frac{dy}{dt} = \frac{dy}{du} \cdot \frac{du}{dt} = \frac{d}{du}(u^{10}) \cdot \frac{d}{dt}(8 - t^2 + t^5)$$

$$= (10u^9)(-2t + 5t^4)$$

$$= 10(8 - t^2 + t^5)^9(-2t + 5t^4).$$

d. *If $y = 4u^3 + 10u^2 - 3u - 7$ and $u = 4/(3x - 5)$, find dy/dx when $x = 1$.*

By the chain rule,

$$\frac{dy}{dx} = \frac{dy}{du} \cdot \frac{du}{dx} = \frac{d}{du}(4u^3 + 10u^2 - 3u - 7) \cdot \frac{d}{dx}\left(\frac{4}{3x - 5}\right)$$

$$= (12u^2 + 20u - 3) \cdot \frac{(3x - 5)\,D_x(4) - 4\,D_x(3x - 5)}{(3x - 5)^2}$$

$$= (12u^2 + 20u - 3) \cdot \frac{-12}{(3x - 5)^2}.$$

Even though dy/dx is in terms of x's and u's, we can evaluate it when $x = 1$ if we determine the corresponding value of u. When $x = 1$, then $u = \dfrac{4}{3(1) - 5} = -2$. Thus,

$$\left.\frac{dy}{dx}\right|_{x=1} = \left[12(-2)^2 + 20(-2) - 3\right] \cdot \frac{-12}{[3(1) - 5]^2}$$

$$= 5 \cdot (-3) = -15.$$

Suppose we wanted to differentiate $y = (x^3 - x^2 + 6)^{100}$. We can think of the right side as u^{100} where $x^3 - x^2 + 6$ plays the role of u. This suggests a substitution. Let $x^3 - x^2 + 6$ be u. Then

$$y = u^{100} \quad \text{where} \quad u = x^3 - x^2 + 6.$$

By the chain rule,

$$\frac{dy}{dx} = \frac{dy}{du} \cdot \frac{du}{dx}$$

$$= (100u^{99})(3x^2 - 2x)$$

$$= 100(x^3 - x^2 + 6)^{99}(3x^2 - 2x).$$

EXAMPLE 2

a. *If* $y = \sqrt[5]{8x^2 - 7x}$, *find* y'.

Let $u = 8x^2 - 7x$. Then $y = \sqrt[5]{u} = u^{1/5}$.

$$y' = \frac{dy}{dx} = \frac{dy}{du} \cdot \frac{du}{dx}$$

$$= \left(\tfrac{1}{5}u^{-4/5}\right)(16x - 7)$$

$$= \frac{(16x - 7)(8x^2 - 7x)^{-4/5}}{5}.$$

b. *If* $y = \dfrac{1}{(x^2 - 2)^4}$, *find* $\dfrac{dy}{dx}$.

Let $u = x^2 - 2$. Then $y = 1/u^4 = u^{-4}$. Thus,

$$\frac{dy}{dx} = \frac{dy}{du} \cdot \frac{du}{dx} = (-4u^{-5})(2x)$$

$$= -\frac{8x}{u^5} = -\frac{8x}{(x^2 - 2)^5}.$$

In Example 2 we used the chain rule to differentiate a power of a *function* of x. In (a) we differentiated $y = (8x^2 - 7x)^{1/5}$ and in (b) we differentiated $y = (x^2 - 2)^{-4}$. The following rule generalizes our results and is known as the *power rule*.

RULE 9. (*Power Rule.*) *If* $y = u^n$, *where* n *is any real number and* u *is a differentiable function of* x, *then*

$$\frac{dy}{dx} = nu^{n-1}\frac{du}{dx}.$$

Proof. By the chain rule,

$$\frac{dy}{dx} = \frac{dy}{du} \cdot \frac{du}{dx}.$$

But by Rule 2, $$\frac{dy}{du} = \frac{d}{du}(u^n) = nu^{n-1},$$

so $$\frac{dy}{dx} = nu^{n-1}\frac{du}{dx}.$$

Therefore we have the *power rule:*

> If $y = u^n$ where u is a function of x, then
>
> $$\frac{dy}{dx} = nu^{n-1}\frac{du}{dx}.$$

Another way of writing the power rule is

> $$\frac{d}{dx}([u(x)]^n) = n[u(x)]^{n-1}u'(x).$$

EXAMPLE 3

a. *If $y = (x^3 - 1)^7$, find y'.*

Since y is a power of a *function* of x, the power rule applies. Letting $u(x) = x^3 - 1$ and $n = 7$, we have

$$y' = n[u(x)]^{n-1}u'(x)$$

$$= 7(x^3 - 1)^{7-1}\frac{d}{dx}(x^3 - 1)$$

$$= 7(x^3 - 1)^6(3x^2) = 21x^2(x^3 - 1)^6.$$

b. *If $y = \sqrt{4x^2 + 3x - 1}$, find dy/dx when $x = -2$.*

Since $y = (4x^2 + 3x - 1)^{1/2}$, we use the power rule with $u(x) = 4x^2 + 3x - 1$ and $n = \frac{1}{2}$.

$$\frac{dy}{dx} = \frac{1}{2}(4x^2 + 3x - 1)^{(1/2)-1}\frac{d}{dx}(4x^2 + 3x - 1)$$

$$= \frac{8x + 3}{2\sqrt{4x^2 + 3x - 1}}.$$

$$\left.\frac{dy}{dx}\right|_{x=-2} = \frac{-13}{2\sqrt{9}} = -\frac{13}{6}.$$

c. *If $z = \left(\dfrac{2s + 5}{s^2 + 1}\right)^4$, find $\dfrac{dz}{ds}$.*

Since z is a power of a function, we first use the power rule.

$$\frac{dz}{ds} = 4\left(\frac{2s + 5}{s^2 + 1}\right)^{4-1}\frac{d}{ds}\left(\frac{2s + 5}{s^2 + 1}\right).$$

By the quotient rule,

$$\frac{dz}{ds} = 4\left(\frac{2s+5}{s^2+1}\right)^3 \frac{(s^2+1)(2) - (2s+5)(2s)}{(s^2+1)^2}.$$

Simplifying, we have

$$\frac{dz}{ds} = 4 \cdot \frac{(2s+5)^3}{(s^2+1)^3} \cdot \frac{(-2s^2-10s+2)}{(s^2+1)^2}$$

$$= \frac{-8(s^2+5s-1)(2s+5)^3}{(s^2+1)^5}.$$

d. *If $y = (x^2-4)^5(3x+5)^4$, find y'.*

Since y is a product, we first apply the product rule.

$$y' = (x^2-4)^5 \, D_x\left[(3x+5)^4\right] + (3x+5)^4 \, D_x\left[(x^2-4)^5\right].$$

Now we can use the power rule.

$$y' = (x^2-4)^5\left[4(3x+5)^3(3)\right] + (3x+5)^4\left[5(x^2-4)^4(2x)\right].$$

Simplifying, we have

$$y' = 12(x^2-4)^5(3x+5)^3 + 10x(3x+5)^4(x^2-4)^4$$

$$= 2(x^2-4)^4(3x+5)^3[6(x^2-4) + 5x(3x+5)] \qquad \text{[factoring]}$$

$$= 2(x^2-4)^4(3x+5)^3(21x^2+25x-24).$$

Usually, the power rule should be used to differentiate $y = [u(x)]^n$. Although a function such as $y = (x^2+2)^2$ may be written $y = x^4 + 4x^2 + 4$ and differentiated easily, this method is impractical for a function such as $y = (x^2+2)^{1000}$. Since $y = (x^2+2)^{1000}$ is of the form $y = [u(x)]^n$, we have

$$y' = 1000(x^2+2)^{999}(2x).$$

Let us now use our knowledge of calculus to develop analytically a concept relevant to economic studies. Suppose a manufacturer hires m employees who produce a total of q units of a product per day. We can think of q as a function of m. If r is the total revenue the manufacturer receives for selling the q units produced by the m employees, then r can be considered a function of m. Thus we can look at dr/dm, the rate of change of revenue with respect to the number of employees. We know that total revenue is given by

$$r = pq,$$

where p is the price per unit. Here p is a function of q and is determined by the product's demand equation. By the product rule,

$$\frac{dr}{dm} = p\frac{d}{dm}(q) + q\frac{d}{dm}(p) = p\frac{dq}{dm} + q\frac{dp}{dm}.$$

But by the chain rule,

$$\frac{dp}{dm} = \frac{dp}{dq} \cdot \frac{dq}{dm}.$$

Therefore,

$$\frac{dr}{dm} = p\frac{dq}{dm} + q\frac{dp}{dq} \cdot \frac{dq}{dm}$$

or

$$\frac{dr}{dm} = \frac{dq}{dm}\left(p + q\frac{dp}{dq}\right). \tag{3}$$

The derivative dr/dm is called the **marginal revenue product**. It is approximately the change in revenue that results when a manufacturer hires an extra employee.

EXAMPLE 4

A manufacturer determines that m employees will produce a total of q units of a product per day where $q = 10m^2/\sqrt{m^2 + 19}$. If the demand equation for the product is $p = 900/(q + 9)$, determine the marginal revenue product when $m = 9$.

First we find dq/dm and dp/dq. Using the quotient and power rules, we obtain

$$\frac{dq}{dm} = \frac{(m^2 + 19)^{1/2} D_m(10m^2) - (10m^2) D_m\left[(m^2 + 19)^{1/2}\right]}{\left[(m^2 + 19)^{1/2}\right]^2}$$

$$= \frac{(m^2 + 19)^{1/2}(20m) - 10m^2\left[(1/2)(m^2 + 19)^{-1/2}(2m)\right]}{m^2 + 19}$$

$$= \frac{10m(m^2 + 19)^{-1/2}[(m^2 + 19)(2) - m(m)]}{m^2 + 19}$$

$$= \frac{10m(m^2 + 38)}{(m^2 + 19)^{3/2}}.$$

Since $p = 900(q + 9)^{-1}$, then by the power rule,

$$\frac{dp}{dq} = 900\left[(-1)(q + 9)^{-2}(1)\right] = -\frac{900}{(q + 9)^2}.$$

Substituting into Eq. (3), we have the marginal revenue product:

$$\frac{dr}{dm} = \frac{10m(m^2 + 38)}{(m^2 + 19)^{3/2}}\left[p + q\left(-\frac{900}{[q + 9]^2}\right)\right].$$

When $m = 9$, then $q = 81$ and $p = 10$. Thus,

$$\left.\frac{dr}{dm}\right|_{m=9} = 10.71.$$

EXERCISE 8-6

In Problems 1–8, use the chain rule.

1. If $y = u^2 - 2u$ and $u = x^2 - x$, find dy/dx.

2. If $y = 2u^3 - 8u$ and $u = 7x - x^3$, find dy/dx.

3. If $y = \dfrac{1}{w^2}$ and $w = 2 - x$, find dy/dx.

4. If $y = \sqrt[3]{z}$ and $z = x^6 - x^2 + 1$, find dy/dx.

5. If $w = u^2$ and $u = \dfrac{t + 1}{t - 1}$, find dw/dt when $t = 3$.

6. If $z = u^2 + \sqrt{u} + 9$ and $u = 2s^2 - 1$, find dz/ds when $s = -1$.

7. If $y = 3w^2 - 8w + 4$ and $w = 3x^2 + 1$, find dy/dx when $x = 0$.

8. If $y = 3u^3 - u^2 + 7u - 2$ and $u = 3x - 2$, find dy/dx when $x = 1$.

In Problems 9–44, find y'.

9. $y = (7x + 4)^8$.

10. $y = (4 - 3x)^{25}$.

11. $y = (3 - 2p^2)^{14}$.

12. $y = (x^2 - 8x)^{40}$.

13. $y = \dfrac{(4x^3 - 8x + 2)^{10}}{3}$.

14. $y = \dfrac{(7 - q^2 + q)^{12}}{9}$.

15. $y = (4r^2 - 10r + 3)^{-15}$.

16. $y = (t^2 - 5)^{-4}$.

17. $y = \dfrac{7}{(x^3 - x^2 + 2)^7}$.

18. $y = \dfrac{6}{(2 - x^2 + x)^4}$.

19. $y = 15(4z^3 - z^2 + 2)^{1/5}$.

20. $y = 2(8x - 2)^{2/3}$.

21. $y = \sqrt{2x^2 - x + 3}$.

22. $y = \sqrt[3]{8s^2 - 1}$.

23. $y = \sqrt[5]{(x^2 + 1)^3}$.

24. $y = \dfrac{1}{(3x^2 - x)^{2/3}}$.

25. $y = \left(\dfrac{x - 7}{x + 4}\right)^{10}$.

26. $y = \left(\dfrac{2w}{w + 2}\right)^4$.

[handwritten note: Division doesn't take precedence because the entire function is raised to the 9th power]

27. $y = 2\left(\dfrac{q^3 - 2q + 4}{5q^2 + 1}\right)^5$.

28. $y = 3\left(\dfrac{x^2 + 2x - 2}{x^3 + x}\right)^8$.

29. $y = \sqrt{\dfrac{x - 2}{x + 3}}$.

30. $y = \sqrt[3]{\dfrac{8x^2 - 3}{x^2 + 2}}$.

31. $y = (x^2 + 2x - 1)^3(5x + 7)$.

32. $y = (8x^3 - 1)^3(2x^2 + 1)^2$.

33. $y = [(4x + 3)(6x^2 + x + 8)]^8$.

34. $y = \dfrac{2t - 5}{(t^2 + 4)^3}$.

35. $y = \dfrac{(2w + 3)^3}{w^2 + 4}$.

36. $y = \sqrt{(x - 1)(x + 2)^3}$.

37. $y = 6(5x^2 + 2)\sqrt{x^4 + 5}$.

38. $y = \sqrt[3]{\dfrac{8x - 7}{5x^2 + 6}}$.

39. $y = (4 - 3x^2)^2(2 - 3x)^3$.

40. $y = 6 + 3x - 4x(7x + 1)^2$.

41. $y = 8t + \dfrac{t - 1}{t + 4} - \left(\dfrac{8t - 7}{4}\right)^2$.

42. $y = 4[(3p - 8)(3p^2 - 2p + 1)^3]^4$.

43. $y = \dfrac{(8x - 1)^5}{(3x - 1)^3}$.

44. $y = \dfrac{(4x^2 - 2)(8x - 1)}{(3x - 1)^2}$.

In Problems 45 and 46, use the quotient rule and power rule to find y'. Do not simplify your answer.

45. $y = \dfrac{(2x + 1)(3x - 5)^2}{(x^2 - 7)^4}$.

46. $y = \dfrac{\sqrt{x + 2}\,(4x^2 - 1)^2}{9x - 3}$.

47. If $y = (5u + 6)^3$ and $u = (x^2 + 1)^4$, find dy/dx when $x = 0$.

48. If $z = 2y^2 - 4y + 5$, $y = 6x - 5$, and $x = 2t$, find dz/dt when $t = 1$.

49. Find the slope of the curve $y = (x^2 - 7x - 8)^3$ at the point $(8, 0)$.

50. Find the slope of the curve $y = \sqrt{x + 1}$ at the point $(8, 3)$.

In Problems 51–54, find an equation of the tangent line to the curve at the given point.

51. $y = \sqrt[3]{(x^2 - 8)^2}$; $(3, 1)$.

52. $y = (2x + 3)^2$; $(-2, 1)$.

53. $y = \dfrac{\sqrt{7x + 2}}{x + 1}$; $(1, \tfrac{3}{2})$.

54. $y = \dfrac{-3}{(3x^2 + 1)^3}$; $(0, -3)$.

In Problems 55 and 56, determine the percentage rate of change of y with respect to x for the given value of x.

55. $y = (x^2 + 9)^3$; $x = 4$.

56. $y = \dfrac{1}{(x^2 + 1)^2}$; $x = -3$.

In Problems 57–60, q is the total number of units produced per day by m employees of a manufacturer, and p is the price per unit at which the q units are sold. In each case find the marginal revenue product for the given value of m.

57. $q = 2m, p = -.5q + 20$; $m = 5$.

58. $q = (200m - m^2)/20, p = -.1q + 70$; $m = 40$.

59. $q = 10m^2/\sqrt{m^2 + 9}, p = 525/(q + 3)$; $m = 4$.

60. $q = 100m/\sqrt{m^2 + 19}, p = 4500/(q + 10)$; $m = 9$.

61. Suppose $p = 100 - \sqrt{q^2 + 20}$ is a demand equation for a manufacturer's product. (a) Find the rate of change of p with respect to q. (b) Find the relative rate of change of p with respect to q. (c) Find the marginal revenue function.

62. If $p = c/q$, where c is a constant, is the demand equation for a manufacturer's product, and $q = f(m)$ defines a function that gives the total number of units produced per day by m employees, show that the marginal revenue product is always zero.

63. Suppose the cost c of producing q units of a product is given by $c = 4000 + 10q + .1q^2$. If the price per unit p is given by the equation $q = 800 - 2.5p$, use the chain rule to find the rate of change of cost with respect to price per unit when $p = 80$.

64. Suppose that for a certain group of 20,000 births, the number l_x of people surviving to age x years is

$$l_x = 2000\sqrt{100 - x}, \qquad 0 \leqslant x \leqslant 100.$$

(a) Find the rate of change of l_x with respect to x and evaluate your answer for $x = 36$. (b) Find the relative rate of change of l_x when $x = 36$.

In Problems 65 and 66, each equation represents a consumption function. Find the marginal propensity to consume and the marginal propensity to save for the given value of I.

65. $C = \dfrac{20\sqrt{I} + .5\sqrt{I^3} - .4I}{\sqrt{I} + 100}$; $I = 100$.   **66.** $C = \dfrac{5(2I + \sqrt{I + 9})}{\sqrt{I + 9}}$; $I = 135$.

67. If the total cost function for a manufacturer is given by

$$c = \frac{5q^2}{\sqrt{q^2 + 3}} + 5000,$$

find the marginal cost function.

8-7 DERIVATIVES OF LOGARITHMIC FUNCTIONS

In this section the derivatives of logarithmic functions will be found. We begin with the derivative of $\ln x$. Let

$$y = f(x) = \ln x,$$

where x is positive. By the definition of the derivative,

$$\frac{d}{dx}(\ln x) = \lim_{h \to 0} \frac{f(x + h) - f(x)}{h} = \lim_{h \to 0} \frac{\ln(x + h) - \ln x}{h}.$$

Since the difference of logarithms is a logarithm of a quotient, we can write

$$\frac{d}{dx}(\ln x) = \lim_{h \to 0} \frac{\ln\left(\dfrac{x + h}{x}\right)}{h}$$

$$= \lim_{h \to 0}\left[\frac{1}{h}\ln\left(\frac{x + h}{x}\right)\right] = \lim_{h \to 0}\left[\frac{1}{h}\ln\left(1 + \frac{h}{x}\right)\right].$$

Writing $\dfrac{1}{h}$ as $\dfrac{1}{x}\cdot\dfrac{x}{h}$, we have

$$\frac{d}{dx}(\ln x) = \lim_{h \to 0}\left[\frac{1}{x}\cdot\frac{x}{h}\ln\left(1 + \frac{h}{x}\right)\right]$$

$$= \lim_{h \to 0}\left[\frac{1}{x}\ln\left(1 + \frac{h}{x}\right)^{x/h}\right] \quad (\text{since } n \ln r = \ln r^n)$$

$$= \frac{1}{x}\cdot\lim_{h \to 0}\left[\ln\left(1 + \frac{h}{x}\right)^{x/h}\right].$$

It can be shown that we can write this as

$$\frac{d}{dx}(\ln x) = \frac{1}{x}\ln\left[\lim_{h \to 0}\left(1 + \frac{h}{x}\right)^{x/h}\right]. \tag{1}$$

To evaluate $\displaystyle\lim_{h \to 0}\left(1 + \frac{h}{x}\right)^{x/h}$, first note that as $h \to 0$, then $\dfrac{h}{x} \to 0$. Thus, if we replace $\dfrac{h}{x}$ by k, the limit has the form

$$\lim_{k \to 0}(1 + k)^{1/k}.$$

As mentioned in Sec. 7-2, this limit is e, the base of natural logarithms. Thus Eq. (1) becomes

$$\frac{d}{dx}(\ln x) = \frac{1}{x}\ln e = \frac{1}{x}(1) = \frac{1}{x}.$$

Hence,

$$\frac{d}{dx}(\ln x) = \frac{1}{x}.$$

(2)

EXAMPLE 1

If $y = x \ln x$, then by the product rule and Eq. (2),

$$y' = x\frac{d}{dx}(\ln x) + (\ln x)\frac{d}{dx}(x) = x\left(\frac{1}{x}\right) + (\ln x)(1) = 1 + \ln x.$$

We now extend Eq. (2) to cover a broader class of functions. Let

$$y = \ln u \quad \text{where} \quad u = f(x)$$

and u is positive and differentiable. By the chain rule,

$$\frac{d}{dx}(\ln u) = \frac{dy}{du} \cdot \frac{du}{dx} = \frac{d}{du}(\ln u) \cdot \frac{du}{dx} = \frac{1}{u} \cdot \frac{du}{dx}.$$

Thus,

$$\frac{d}{dx}(\ln u) = \frac{1}{u} \cdot \frac{du}{dx}.$$

(3)

EXAMPLE 2

Differentiate each of the following.

a. $y = \ln (x^2 + 1)$.

This function has the form $\ln u$ with $u = x^2 + 1$. Using Eq. (3), we have

$$\frac{dy}{dx} = \frac{1}{x^2 + 1}\frac{d}{dx}(x^2 + 1) = \frac{1}{x^2 + 1}(2x) = \frac{2x}{x^2 + 1}.$$

b. $y = x^2 \ln (4x + 2)$.

Using the product rule and then Eq. (3) with $u = 4x + 2$, we obtain

$$D_x y = x^2 D_x[\ln (4x + 2)] + [\ln (4x + 2)] D_x(x^2)$$

$$= x^2\left(\frac{1}{4x + 2}\right)(4) + [\ln (4x + 2)](2x)$$

$$= \frac{4x^2}{4x + 2} + 2x \ln (4x + 2).$$

c. $y = \ln (\ln x)$.

This has the form $y = \ln u$ where $u = \ln x$. Using Eqs. (3) and (2), we obtain

$$y' = \frac{1}{\ln x} \frac{d}{dx} (\ln x) = \frac{1}{\ln x} \left(\frac{1}{x} \right) = \frac{1}{x \ln x}.$$

To differentiate some functions involving logarithms, like $y = \ln (2x + 5)^3$, it may be easier to simplify the function *before* the differentiation by using properties of logarithms. Example 3 will illustrate this.

EXAMPLE 3

Differentiate each of the following.

a. $y = \ln (2x + 5)^3$.

First we simplify the right side by using properties of logarithms.

$$y = \ln (2x + 5)^3 = 3 \ln (2x + 5).$$

$$\frac{dy}{dx} = 3 \left(\frac{1}{2x + 5} \right) (2) = \frac{6}{2x + 5}.$$

If the simplification were not performed first,

$$\frac{dy}{dx} = \frac{1}{(2x + 5)^3} D_x \left[(2x + 5)^3 \right]$$

$$= \frac{1}{(2x + 5)^3} (3)(2x + 5)^2 (2) = \frac{6}{2x + 5}.$$

b. $f(p) = \ln [(p + 1)^2 (p + 2)^3 (p + 3)^4]$.

We simplify the right side and then differentiate.

$$f(p) = 2 \ln (p + 1) + 3 \ln (p + 2) + 4 \ln (p + 3).$$

$$f'(p) = 2 \left(\frac{1}{p + 1} \right) (1) + 3 \left(\frac{1}{p + 2} \right) (1) + 4 \left(\frac{1}{p + 3} \right) (1)$$

$$= \frac{2}{p + 1} + \frac{3}{p + 2} + \frac{4}{p + 3}.$$

c. $f(w) = \ln \sqrt{\dfrac{1 + w^2}{w^2 - 1}}$.

Again, using properties of logarithms will simplify our work.

$$f(w) = \frac{1}{2}[\ln(1 + w^2) - \ln(w^2 - 1)].$$

$$f'(w) = \frac{1}{2}\left[\frac{1}{1 + w^2}(2w) - \frac{1}{w^2 - 1}(2w)\right]$$

$$= \frac{w}{1 + w^2} - \frac{w}{w^2 - 1} = \frac{2w}{1 - w^4}.$$

find common denom - (handwritten annotation)

d. $f(x) = \ln^3[(2x + 1)^4]$.

The exponent 3 refers to the cubing of $\ln(2x + 1)^4$. That is,

$$\ln^3\left[(2x + 1)^4\right] \quad \text{means} \quad \left[\ln(2x + 1)^4\right]^3.$$

Thus,

$$f(x) = \ln^3\left[(2x + 1)^4\right] = \left[\ln(2x + 1)^4\right]^3 = [4\ln(2x + 1)]^3.$$

By the power rule,

$$f'(x) = 3[4\ln(2x + 1)]^2 D_x[4\ln(2x + 1)]$$

$$= 3[4\ln(2x + 1)]^2\left[4\left(\frac{1}{2x + 1}\right)(2)\right]$$

$$= \frac{384}{2x + 1}[\ln(2x + 1)]^2$$

$$= \frac{384}{2x + 1}\ln^2(2x + 1).$$

We can generalize Eq. (3) to any base b. Since $\ln u = \dfrac{\log_b u}{\log_b e}$ (from Sec. 5-2), then $\log_b u = (\log_b e)\ln u$. Hence,

$$\frac{d}{dx}(\log_b u) = \frac{d}{dx}\left[(\log_b e)\ln u\right]$$

$$= (\log_b e)\frac{d}{dx}(\ln u)$$

$$= (\log_b e)\left(\frac{1}{u}\frac{du}{dx}\right).$$

Thus,

$$\frac{d}{dx}(\log_b u) = \frac{1}{u}(\log_b e)\frac{du}{dx}. \qquad (4)$$

Since the use of natural logarithms (that is, $b = e$) gives a value of 1 to the factor $\log_b e$ in Eq. (4), natural logarithms are used extensively in calculus.

EXAMPLE 4

If $y = \log(2x + 1)$, *find the rate of change of y with respect to x.*

We want to find dy/dx. By Eq. (4) with $u = 2x + 1$ and $b = 10$,

$$\frac{dy}{dx} = \frac{1}{2x+1}(\log e)D_x(2x+1) = \frac{1}{2x+1}(\log e)(2) = \frac{2\log e}{2x+1}.$$

EXERCISE 8-7

In Problems **1–34**, *differentiate the functions.*

1. $y = \ln(3x - 4)$.

2. $y = \ln(5x - 6)$.

3. $y = \ln x^2$.

4. $y = \ln(ax^2 + b)$.

5. $y = \ln(1 - x^2)$.

6. $y = \ln(-x^2 + 6x)$.

7. $f(p) = \ln(2p^3 + 3p)$.

8. $f(r) = \ln(2r^4 - 3r^2 + 2r + 1)$.

9. $y = \ln^4(ax)$.

10. $y = \ln^2(2x + 3)$.

11. $y = \ln(x^2 + 4x + 5)$.

12. $y = \ln x^{100}$.

13. $f(t) = t \ln t$.

14. $y = x^2 \ln x$.

15. $y = \log_3(2x - 1)$.

16. $f(w) = \log(w^2 + w)$.

17. $y = (x^2 + 1)\ln(2x + 1)$.

18. $y = (ax + b)\ln(ax)$.

19. $y = \ln[(x^2 + 2)^2(x^3 + x - 1)]$.

20. $y = \ln[(5x + 2)^4(8x - 3)^6]$.

21. $f(l) = \ln\left(\frac{1+l}{1-l}\right)$.

22. $y = \ln\left(\frac{2x+3}{3x-4}\right)$.

23. $y = \ln\sqrt{1 + x^2}$.

24. $f(s) = \ln\left(\frac{s^2}{1+s^2}\right)$.

25. $y = \ln[(x + 1)^5 + (x + 2)^4 + x^8]$.

26. $y = \ln x^3 + \ln^3 x$.

27. $y = \ln\sqrt[4]{\frac{1+x^2}{1-x^2}}$.

28. $y = \ln\sqrt{\frac{x^4-1}{x^4+1}}$.

29. $f(z) = \dfrac{\ln z}{z}$.

30. $y = \dfrac{x^2 - 1}{\ln x}$.

31. $y = x \ln \sqrt{x - 1}$.

32. $y = \ln (x^2 \sqrt{3x - 2})$.

33. $y = \sqrt{4 + \ln x}$.

34. $y = \ln (x + \sqrt{1 + x^2})$.

35. Find an equation of the tangent line to the curve $y = \ln (x^2 - 2x - 2)$ when $x = 3$.

36. If $y = \ln^2 x$, find the relative rate of change of y.

37. Find the marginal revenue function if the demand function is $p = 25/\ln (q + 2)$.

38. A total cost function is given by $c = 25 \ln (q + 1) + 12$. Find the marginal cost when $q = 6$.

39. Show that the relative rate of change of $y = f(x)$ with respect to x is equal to the derivative of $y = \ln f(x)$.

8-8 DERIVATIVES OF EXPONENTIAL FUNCTIONS

We now turn to the derivative of the exponential function $y = e^u$ where u is a differentiable function of x. Since $y = e^u$, then in logarithmic form we have

$$u = \ln y. \tag{1}$$

Now we take derivatives with respect to x of both sides of Eq. (1):

$$\frac{d}{dx}(u) = \frac{d}{dx}(\ln y).$$

Using Eq. (3) of the last section where y is a function of x, we obtain

$$\frac{du}{dx} = \frac{1}{y}\frac{dy}{dx}.$$

Solving for dy/dx and replacing y by e^u gives

$$\frac{dy}{dx} = y\frac{du}{dx} = e^u\frac{du}{dx}.$$

Thus,

$$\boxed{\frac{d}{dx}(e^u) = e^u\frac{du}{dx}.} \tag{2}$$

In particular, if $u = x$, then $du/dx = 1$ and

$$\frac{d}{dx}(e^x) = e^x. \qquad (3)$$

PITFALL. *Do **not** use the power rule to find $D_x(e^x)$. That is, $D_x(e^x) \neq xe^{x-1}$.*

EXAMPLE 1

a. *Find* $\frac{d}{dx}(e^{x^3+3x})$.

The function has the form e^u with $u = x^3 + 3x$. Using Eq. (2), we obtain

$$\frac{d}{dx}(e^{x^3+3x}) = e^{x^3+3x} D_x(x^3 + 3x) = e^{x^3+3x}(3x^2 + 3)$$

$$= 3(x^2 + 1)e^{x^3+3x}.$$

b. *If* $y = \dfrac{x}{e^x}$, *find* y'.

We *first* use the quotient rule and then use Eq. (3).

$$\frac{dy}{dx} = \frac{e^x D_x(x) - x D_x(e^x)}{(e^x)^2} = \frac{e^x(1) - x(e^x)}{(e^x)^2} = \frac{e^x(1-x)}{e^{2x}} = \frac{1-x}{e^x}.$$

c. *If* $f(w) = w^4 e^{2w}$, *find* $f'(w)$.

We *first* use the product rule and then use Eq. (2) with $u = 2w$.

$$f'(w) = w^4 D_w(e^{2w}) + e^{2w} D_w(w^4)$$

$$= w^4(e^{2w})(2) + e^{2w}(4w^3) = 2e^{2w}w^3(w + 2).$$

d. *Find* $D_x[e^{x+1} \ln (x^2 + 1)]$.

By the product rule,

$$D_x[e^{x+1} \ln (x^2 + 1)] = e^{x+1} D_x[\ln (x^2 + 1)] + [\ln (x^2 + 1)] D_x(e^{x+1})$$

$$= e^{x+1}\left(\frac{1}{x^2 + 1}\right)(2x) + [\ln (x^2 + 1)]e^{x+1}(1)$$

$$= e^{x+1}\left[\frac{2x}{x^2 + 1} + \ln (x^2 + 1)\right].$$

e. *If $y = e^2 + e^x + \ln 3$, find y'.*

Since e^2 and $\ln 3$ are constants,

$$y' = 0 + e^x + 0 = e^x.$$

We can generalize Eq. (2) by considering the derivative of a^u where $a > 0$, $a \neq 1$, and u is a differentiable function of x. First we shall write a^u as an exponential function with base e. By Property 9 of Sec. 5-2, we have $a = e^{\ln a}$. Thus,

$$D_x(a^u) = D_x\left[(e^{\ln a})^u\right] = D_x(e^{u \ln a})$$

$$= e^{u \ln a} D_x(u \ln a) \qquad\qquad [\text{by Eq. (2)}]$$

$$= e^{u \ln a}\left(\frac{du}{dx}\right) \ln a \qquad\qquad [\ln a \text{ is constant}]$$

$$= a^u(\ln a)\frac{du}{dx} \qquad\qquad [\text{since } e^{u \ln a} = a^u].$$

Thus,

$$\boxed{\frac{d}{dx}(a^u) = a^u(\ln a)\frac{du}{dx}.} \qquad\qquad (4)$$

EXAMPLE 2

a. *Find dy/dx if $y = 4^{2x^3 + 5x}$.*

Using Eq. (4) with $a = 4$ and $u = 2x^3 + 5x$, we obtain

$$\frac{dy}{dx} = 4^{2x^3 + 5x}(\ln 4)\frac{d}{dx}(2x^3 + 5x)$$

$$= 4^{2x^3 + 5x}(\ln 4)(6x^2 + 5) = (\ln 4)(6x^2 + 5)4^{2x^3 + 5x}.$$

b. *If $y = x^{100} + 100^x$, find $D_x y$.*

Note that this function involves a variable to a constant power and a constant raised to a variable power. Do not confuse these forms!

$$D_x y = 100x^{99} + 100^x(\ln 100) D_x(x) = 100x^{99} + 100^x \ln 100.$$

EXAMPLE 3

*An important function used in economic and business decisions is the **normal distribution density function***

$$y = f(x) = \frac{1}{\sigma\sqrt{2\pi}} e^{-(1/2)[(x-\mu)/\sigma]^2}$$

where σ (a Greek letter read "sigma") and μ (a Greek letter read "mu") are constants. Its graph, called the normal curve, is "bell-shaped" (see Fig. 8-16). Determine the rate of

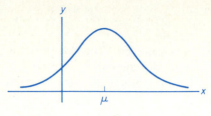

FIG. 8-16

change of y with respect to x when x = μ.

$$\frac{dy}{dx} = \frac{1}{\sigma\sqrt{2\pi}} \left[e^{-(1/2)[(x-\mu)/\sigma]^2} \right] \left[-\frac{1}{2}(2)\left(\frac{x-\mu}{\sigma}\right)\left(\frac{1}{\sigma}\right) \right].$$

Evaluating dy/dx when $x = \mu$, we obtain

$$\left.\frac{dy}{dx}\right|_{x=\mu} = 0.$$

For your convenience, Table 8-2 gives the important differentiation formulas of this chapter. We emphasize that you should not only be totally familiar with these formulas and the mechanics involved in applying them, but you should also know both the definition and the interpretations of a derivative.

EXERCISE 8-8

In Problems 1–30, differentiate the functions.

1. $y = e^{x^2+1}$.

2. $y = e^{2x^2+5}$.

3. $y = e^{3-5x}$.

4. $f(q) = e^{-q^3+6q-1}$.

5. $f(r) = e^{3r^2+4r+4}$.

6. $y = e^{9x^2+5x^3-6}$.

7. $y = xe^x$.

8. $y = x^2e^{-x}$.

9. $y = x^2e^{-x^2}$.

10. $y = xe^{2x}$.

11. $y = \dfrac{e^x + e^{-x}}{2}$.

12. $y = \dfrac{e^x - e^{-x}}{2}$.

13. $y = 4^{3x^2}$.

14. $y = 4^{3x+1}$.

15. $f(w) = \dfrac{e^{2w}}{w^2}$.

16. $y = 2^x x^2$.

17. $y = e^{1+\sqrt{x}}$.

18. $y = e^{x-\sqrt{x}}$.

19. $y = x^3 - 3^x$.

20. $y = (e^{3x} + 1)^4$.

TABLE 8-2　Differentiation Formulas

$$\frac{d}{dx}(c) = 0, \text{ where } c \text{ is any constant.}$$

$$\frac{d}{dx}(x^n) = nx^{n-1}, \text{ where } n \text{ is any real number.}$$

$$\frac{d}{dx}[cf(x)] = cf'(x).$$

$$\frac{d}{dx}[f(x) + g(x)] = f'(x) + g'(x).$$

$$\frac{d}{dx}[f(x) - g(x)] = f'(x) - g'(x).$$

$$\frac{d}{dx}[f(x)\,g(x)] = f(x)g'(x) + g(x)f'(x).$$

$$\frac{d}{dx}\left[\frac{f(x)}{g(x)}\right] = \frac{g(x)f'(x) - f(x)g'(x)}{[g(x)]^2}.$$

$$\frac{dy}{dx} = \frac{dy}{du} \cdot \frac{du}{dx}, \text{ where } y \text{ is a function of } u \text{ and } u \text{ is a function of } x.$$

$$\frac{d}{dx}(u^n) = nu^{n-1}\frac{du}{dx}.$$

$$\frac{d}{dx}(\log_b u) = \frac{1}{u}(\log_b e)\frac{du}{dx}.$$

$$\frac{d}{dx}(\ln u) = \frac{1}{u}\frac{du}{dx}.$$

$$\frac{d}{dx}(a^u) = a^u(\ln a)\frac{du}{dx}.$$

$$\frac{d}{dx}(e^u) = e^u\frac{du}{dx}.$$

[Handwritten annotations: "power rule", "log rule", "exp rule", and: $y = f^n(x),\ y' = nf^{n-1}(x)\,f'(x)$; $y = \ln f(x),\ y' = \dfrac{f'(x)}{f(x)}$; $\dfrac{1}{f(x)}(f'(x))$; $y = e^{fx},\ y' = e^{f(x)}f'(x)$]

21. $y = \dfrac{e^x - 1}{e^x + 1}.$ 　　　　　　　**22.** $f(z) = e^{1/z}.$

23. $y = e^{e^x}.$ 　　　　　　　　　　**24.** $y = e^{2x}(x + 1).$

25. $y = e^{\ln x}.$ 　　　　　　　　　**26.** $y = e^{\ln(x^2 + 1)}.$

27. $y = e^{x \ln x}.$ 　　　　　　　　**28.** $y = e^{-x} \ln x.$

29. $y = (\log 2)^x.$ 　　　　　　　　**30.** $y = \ln e^{4x + 1}.$

31. Find an equation of the tangent line to the graph of $y = e^x$ when $x = 2.$

32. Find the slope of the tangent line to the graph of $y = 2e^{-4x^2}$ when $x = 0.$

For each of the demand equations in Problems 33 and 34, find the rate of change of price p with respect to quantity q. What is the rate of change for the indicated value of q?

33. $p = 15e^{-.001q};\quad q = 500.$ 　　　　**34.** $p = 8e^{-3q/800};\quad q = 400.$

35. The population P of a city t years from now is given by $P = 20{,}000e^{.03t}$. Show that $dP/dt = kP$ where k is a constant. This means that the rate of change of population at any time is proportional to the population at that time.

In Problems 36 and 37, $\bar{c}$ is the average cost of producing q units of a product. Find the marginal cost function and the marginal cost for the given values of q.

36. $\bar{c} = (7000e^{q/700})/q$; $q = 350, q = 700$.

37. $\bar{c} = \dfrac{850}{q} + 4000\dfrac{e^{(2q+6)/800}}{q}$; $q = 97, q = 197$.

38. For a firm the daily output q on the t-th day of a production run is given by $q = 500(1 - e^{-2t})$. Find the instantaneous rate of change of output q with respect to t on the tenth day.

39. For the normal density function

$$f(x) = \frac{1}{\sqrt{2\pi}}\, e^{-x^2/2},$$

find $f'(0)$.

40. After t years, the value S of a principal of P dollars which is invested at the annual rate of r compounded continuously is given by $S = Pe^{rt}$. Show that the relative rate of change of S with respect to t is r.

41. In a discussion of diffusion of a new process into a market, Hurter and Rubenstein[†] refer to an equation of the form

$$Y = k\alpha^{\beta^t},$$

where Y is the cumulative level of diffusion of the new process at time t, and k, α, and β are positive constants. Verify their claim that

$$\frac{dY}{dt} = k\alpha^{\beta^t}(\beta^t \ln \alpha) \ln \beta.$$

8-9 IMPLICIT DIFFERENTIATION

To introduce implicit differentiation, we shall find the slope of a tangent line to a circle. Let us take the circle of radius 2 whose center is at the origin (Fig. 8-17). Its equation is

$$x^2 + y^2 = 4,$$
$$x^2 + y^2 - 4 = 0. \tag{1}$$

[†]A. P. Hurter, Jr., A. H. Rubenstein, et al. "Market penetration by new innovations: the technological literature," *Technological Forecasting and Social Change*, vol. 11 (1978), 197–221.

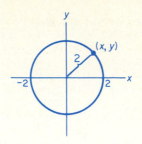

FIG. 8-17

The point $(\sqrt{2}, \sqrt{2})$ lies on the circle. To find the slope at this point, we need to find dy/dx there. Until now we have always had y given explicitly (directly) in terms of x before determining y'; that is, in the form $y = f(x)$. In Eq. (1) this is not so. We say that Eq. (1) has the form $F(x, y) = 0$ where $F(x, y)$ denotes a function of two variables. The obvious thing to do is solve Eq. (1) for y in terms of x:

$$x^2 + y^2 - 4 = 0,$$

$$y^2 = 4 - x^2,$$

$$y = \pm\sqrt{4 - x^2}. \tag{2}$$

A problem now occurs—Eq. (2) may give two values of y for a value of x. It does not define y explicitly as a function of x. We can, however, "consider" Eq. (1) as defining y as one of two different functions of x:

$$y = +\sqrt{4 - x^2} \qquad \text{and} \qquad y = -\sqrt{4 - x^2},$$

whose graphs are given in Fig. 8-18. Since $(\sqrt{2}, \sqrt{2})$ lies on the graph of

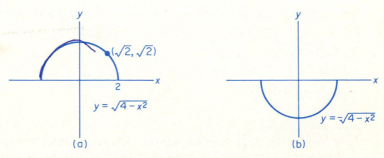

(a)   (b)

FIG. 8-18

$y = \sqrt{4 - x^2}$, we should differentiate that function:

$$y = \sqrt{4 - x^2},$$

$$\frac{dy}{dx} = \frac{1}{2}(4 - x^2)^{-1/2}(-2x)$$

$$= -\frac{x}{\sqrt{4 - x^2}}.$$

$$\left.\frac{dy}{dx}\right|_{x=\sqrt{2}} = -\frac{\sqrt{2}}{\sqrt{4-2}} = -1.$$

Thus the slope of the circle $x^2 + y^2 - 4 = 0$ at the point $(\sqrt{2}, \sqrt{2})$ is -1.

Let us summarize the difficulties we had. First, y was not originally given explicitly in terms of x. Second, after we tried to find such a relation, we ended up with more than one function of x. In fact, depending on the equation given, it may be very complicated or even impossible to find an explicit expression for y. For example, it would be difficult to solve $ye^x + \ln(x + y) = 0$ for y. We shall now consider a method which avoids such difficulties.

An equation of the form $F(x, y) = 0$, such as we had originally, is said to express y *implicitly* as a function of x. The word "implicitly" is used since y is not given explicitly as a function of x. However, it is assumed or *implied* that the equation defines y as at least one differentiable function of x. Thus we assume that the former equation $x^2 + y^2 - 4 = 0$ defines at least one function of x, say $y = f(x)$. Hence, to find dy/dx we treat y as a function of x and differentiate both sides of the equation with respect to x.

$$\frac{d}{dx}(x^2 + y^2 - 4) = \frac{d}{dx}(0),$$

$$\frac{d}{dx}(x^2) + \frac{d}{dx}(y^2) - \frac{d}{dx}(4) = \frac{d}{dx}(0).$$

We know that $\frac{d}{dx}(x^2) = 2x$ and that both $\frac{d}{dx}(4)$ and $\frac{d}{dx}(0)$ are 0. But $\frac{d}{dx}(y^2)$ is **not** $2y$ because we are differentiating with respect to x, not y. That is, y is not the independent variable. Since y is assumed to be a function of x, the y^2-term has the form u^n, where y plays the role of u. Just as the power rule says that $\frac{d}{dx}(u^2) = 2u\frac{du}{dx}$, we have $\frac{d}{dx}(y^2) = 2y\frac{dy}{dx} = 2yy'$. Hence the above equation becomes

$$2x + 2yy' = 0.$$

Solving for y', we obtain

$$2yy' = -2x,$$

$$y' = -\frac{x}{y}. \tag{3}$$

Notice that the expression for y' involves the variable y as well as x. This means that to find y' at a point, both coordinates of the point must be substituted into y'. Thus,

$$y'\Big|_{(\sqrt{2}, \sqrt{2})} = -\frac{\sqrt{2}}{\sqrt{2}} = -1 \text{ as before.}$$

This method of finding dy/dx is called **implicit differentiation**. We note that Eq. (3) is not defined when $y = 0$. Geometrically this is clear, since the tangent line to the circle at either $(2, 0)$ or $(-2, 0)$ is vertical and the slope is not defined.

EXAMPLE 1

For each of the following, find y' by implicit differentiation.

a. $y + y^3 = x$.

We treat y as a function of x and differentiate both sides with respect to x.

$$D_x(y) + D_x(y^3) = D_x(x).$$

Note that $D_x(x) = 1$, and $D_x(y)$ can be written y'. By the power rule, $D_x(y^3) = 3y^2 \frac{dy}{dx} = 3y^2 y'$. Thus,

$$y' + 3y^2 y' = 1.$$

Solving for y', we get

$$y'(1 + 3y^2) = 1,$$

$$y' = \frac{1}{1 + 3y^2}.$$

b. $x^3 + 4xy^2 - y^4 - 27 = 0$.

We assume that y is a function of x and differentiate both sides with respect to x.

$$D_x(x^3) + 4D_x(xy^2) - D_x(y^4) - D_x(27) = D_x(0).$$

To find $D_x(xy^2)$ we use the product rule.

$$[3x^2] + 4[x D_x(y^2) + y^2 D_x(x)] - [4y^3 y'] - 0 = 0,$$
$$[3x^2] + 4[x(2yy') + y^2(1)] - [4y^3 y'] = 0,$$
$$3x^2 + 8xyy' + 4y^2 - 4y^3 y' = 0.$$

Solving for y' gives

$$y'(8xy - 4y^3) = -3x^2 - 4y^2,$$
$$y' = \frac{-3x^2 - 4y^2}{8xy - 4y^3}$$
$$= \frac{3x^2 + 4y^2}{4y^3 - 8xy}.$$

EXAMPLE 2

For each of the following, find y' by implicit differentiation.

a. $e^{xy} = x + y$.

$$D_x(e^{xy}) = D_x(x) + D_x(y),$$
$$e^{xy} D_x(xy) = 1 + D_x(y),$$
$$e^{xy}(x D_x y + y D_x x) = 1 + D_x(y),$$
$$e^{xy}(xy' + y) = 1 + y',$$
$$y'(xe^{xy} - 1) = 1 - ye^{xy},$$
$$y' = \frac{1 - ye^{xy}}{xe^{xy} - 1}.$$

b. $x^3 = (y - x^2)^2$.

$$D_x(x^3) = D_x[(y - x^2)^2],$$
$$3x^2 = 2(y - x^2)(y' - 2x),$$
$$3x^2 = 2(yy' - 2xy - x^2y' + 2x^3),$$
$$3x^2 + 4xy - 4x^3 = 2y'(y - x^2),$$
$$y' = \frac{3x^2 + 4xy - 4x^3}{2(y - x^2)}.$$

If we want to find the slope of the curve $x^3 = (y - x^2)^2$ at the point $(1, 2)$, we have

$$y'|_{(1, 2)} = \frac{3(1)^2 + 4(1)(2) - 4(1)^3}{2[2 - (1)^2]} = \frac{7}{2}.$$

EXAMPLE 3

If $q - p = \ln q + \ln p$, find dq/dp, the rate of change of q with respect to p.

$$D_p(q) - D_p(p) = D_p(\ln q) + D_p(\ln p),$$

$$D_p(q) - 1 = \frac{1}{q} D_p(q) + \frac{1}{p},$$

$$\frac{dq}{dp} - 1 = \frac{1}{q} \frac{dq}{dp} + \frac{1}{p},$$

$$\frac{dq}{dp} = \left(\frac{1}{p} + 1\right)\frac{q}{q - 1}.$$

EXERCISE 8-9

In Problems **1–22** find dy/dx by implicit differentiation.

1. $x^2 + 4y^2 = 4$.

2. $3x^2 + 6y^2 = 1$.

3. $xy = 4$.

4. $x + xy - 2 = 0$.

5. $xy - y - 4x = 5$.

6. $x^2 + y^2 = 2xy + 3$.

7. $x^3 + y^3 - 12xy = 0$.

8. $2x^2 - 3y^2 = 4$.

9. $x^{3/4} + y^{3/4} = 7$.

10. $y^3 = 4x$.

11. $3y^4 - 5x = 0$.

12. $x^{1/5} + y^{1/5} = 4$.

13. $\sqrt{x} + \sqrt{y} = 3$.

14. $2x^3 + 3xy + y^3 = 0$.

15. $x = \sqrt{y} + \sqrt[3]{y}$.

16. $x^3y^3 + x = 9$.

17. $3x^2y^3 - x + y = 25$.

18. $y^2 + y = \ln x$.

19. $y \ln x = xe^y$.

20. $\ln (xy) + x = 4$.

21. $xe^y + y = 4$.

22. $ax^2 - by^2 = c$.

23. If $x + xy + y^2 = 7$, find y' at $(1, 2)$.

24. Find the slope of the curve $4x^2 + 9y^2 = 1$ at the point $(0, \frac{1}{3})$; at the point (x_0, y_0).

25. Find an equation of the tangent line to the curve $x^3 + y^2 = 3$ at the point $(-1, 2)$.

26. Repeat Problem 25 for the curve $y^2 + xy - x^2 = 5$ at the point $(4, 3)$.

For the demand equations in Problems **27–30**, find the rate of change of q with respect to p.

27. $p = 100 - q^2$.

28. $p = 400 - \sqrt{q}$.

29. $p = 20/(q + 5)^2$.

30. $p = 20/(q^2 + 5)$.

31. New products or technologies often tend to replace old ones. For example, today most commercial airlines use jet engines rather then prop engines. In discussing the

forecasting of technological substitution, Hurter and Rubenstein[†] refer to the equation

$$\ln \frac{f(t)}{1 - f(t)} + \sigma \frac{1}{1 - f(t)} = C_1 + C_2 t,$$

where $f(t)$ is the market share of the substitute over time t, and C_1, C_2, and σ (a Greek letter read "sigma") are constants. Verify their claim that the rate of substitution is

$$f'(t) = \frac{C_2 f(t)[1 - f(t)]^2}{\sigma f(t) + [1 - f(t)]}.$$

8-10 LOGARITHMIC DIFFERENTIATION

There is a technique that may be used to simplify the differentiation of $y = f(x)$ when $f(x)$ involves products, quotients, or powers. We first take the natural logarithm of both sides of $y = f(x)$. After simplifying by using properties of logarithms, we then differentiate both sides with respect to x. The next example illustrates this method of **logarithmic differentiation**.

EXAMPLE 1

Find y' if $y = \sqrt[4]{\dfrac{(x - 1)(x^2 + 2)^2}{(x + 3)(x - 4)^3}}$.

This function has the form $y = u^{1/4}$, so the power rule may be used. However, finding du/dx would involve considerable work, since we would get quite involved with the quotient and product rules. Logarithmic differentiation makes the work less tedious. First we take the natural logarithm of both sides of the original equation:

$$\ln y = \ln \sqrt[4]{\frac{(x - 1)(x^2 + 2)^2}{(x + 3)(x - 4)^3}} .$$

Using properties of logarithms to simplify the right side, we have

$$\ln y = \frac{1}{4} \ln \frac{(x - 1)(x^2 + 2)^2}{(x + 3)(x - 4)^3} ,$$

$$\ln y = \frac{1}{4}[\ln (x - 1) + 2 \ln (x^2 + 2) - \ln (x + 3) - 3 \ln (x - 4)].$$

[†]A. P. Hurter, Jr., A. H. Rubenstein, et al. "Market penetration by new innovations: the technological literature," *Technological Forecasting and Social Change*, vol. 11 (1978), 197–221.

Differentiating both sides with respect to x gives

$$\left(\frac{1}{y}\right)y' = \frac{1}{4}\left[\frac{1}{x-1} + (2)\cdot\frac{1}{x^2+2}(2x) - \frac{1}{x+3} - (3)\cdot\frac{1}{x-4}\right],$$

$$\frac{y'}{y} = \frac{1}{4}\left[\frac{1}{x-1} + \frac{4x}{x^2+2} - \frac{1}{x+3} - \frac{3}{x-4}\right].$$

Solving for y' yields

$$y' = \frac{y}{4}\left[\frac{1}{x-1} + \frac{4x}{x^2+2} - \frac{1}{x+3} - \frac{3}{x-4}\right],$$

where y is given in the original equation.

Logarithmic differentiation can also be used to differentiate a function of the form $y = u^v$, where both u and v are differentiable functions of x.

EXAMPLE 2

Find y' for each of the following.

a. $y = x^x$.

This has the form $y = u^v$, where u and v are functions of x. Taking the natural logarithm of each side, we obtain

$$\ln y = \ln x^x.$$

Using properties of logarithms, we have

$$\ln y = x \ln x.$$

Differentiating both sides with respect to x gives

$$\left(\frac{1}{y}\right)y' = x\left(\frac{1}{x}\right) + (\ln x)(1),$$

$$\frac{y'}{y} = 1 + \ln x.$$

Solving for y', we obtain

$$y' = y(1 + \ln x).$$

Since $y = x^x$ (as originally given), by substitution we can write the answer in terms of x only:

$$y' = x^x(1 + \ln x).$$

b. $y = x^{e^{-x^2}}$.

This has the form $y = u^v$ where $v = e^{-x^2}$. Using logarithmic differentiation, we have

$$\ln y = \ln x^{e^{-x^2}} = e^{-x^2} \ln x,$$

$$\left(\frac{1}{y}\right)y' = e^{-x^2}\left(\frac{1}{x}\right) + (\ln x)\left[(e^{-x^2})(-2x)\right],$$

$$\frac{y'}{y} = \frac{e^{-x^2}}{x} - 2xe^{-x^2}\ln x = e^{-x^2}\left(\frac{1}{x} - 2x \ln x\right),$$

$$y' = ye^{-x^2}\left(\frac{1}{x} - 2x \ln x\right)$$

$$= x^{e^{-x^2}}e^{-x^2}\left(\frac{1}{x} - 2x \ln x\right) \qquad \text{[by substitution].}$$

PITFALL. *When using properties of logarithms, you must at all times be able to justify your steps. Thus, if $y = \ln (x + y)$, then $\ln y \neq \ln (x + y)$. Similarly, if $y = x^x + x^5$, then $\ln y \neq \ln x^x + \ln x^5$.*

Be sure you know how to differentiate each of the following forms:

$$y = \begin{cases} [f(x)]^n, & \text{(a)} \\ a^{f(x)}, & \text{(b)} \\ [f(x)]^{g(x)}. & \text{(c)} \end{cases}$$

For type (a), you may use use the power rule; for type (b), use the differentiation formula for exponential functions; for type (c), use logarithmic differentiation. It would be nonsense to write $D_x(x^x) = x \cdot x^{x-1}$. Be sure that all steps are justified by the basic concepts that have been developed.

EXERCISE 8-10

In Problems 1–20, find y' by using logarithmic differentiation.

1. $y = (x + 1)^2(x - 1)(x^2 + 3)$.

2. $y = (3x + 4)(8x - 1)^2(3x^2 + 1)^4$.

3. $y = (3x^3 - 1)^2(2x + 5)^3$.

4. $y = (3x + 1)\sqrt{8x - 1}$.

5. $y = \sqrt{x + 1}\ \sqrt{x^2 - 2}\ \sqrt{x + 4}$.

6. $y = (x + 2)\sqrt{x^2 + 9}\ \sqrt[3]{2x + 1}$.

7. $y = \dfrac{(2x^2 + 2)^2}{(x + 1)^2(3x + 2)}$.

8. $y = \sqrt{\dfrac{(x - 1)(x + 1)}{3x - 4}}$.

9. $y = \dfrac{(8x + 3)^{1/2}(x^2 + 2)^{1/3}}{(1 + 2x)^{1/4}}$.

10. $y = \dfrac{x(1 + x^2)^2}{\sqrt{2 + x^2}}$.

11. $y = \dfrac{\sqrt{1 - x^2}}{1 - 2x}$.

12. $y = \sqrt{\dfrac{x^2 + 5}{x + 9}}$.

13. $y = x^{2x+1}$.

14. $y = x^{\sqrt{x}}$.

15. $y = x^{1/x}$.

16. $y = \left(\dfrac{2}{x}\right)^x$.

17. $y = x^{x^2}$.

18. $y = x^{e^x}$.

19. $y = e^x x^{3x}$.

20. $y = (\ln x)^{e^x}$.

21. Find an equation of the tangent line to $y = (x + 1)(x + 2)^2(x + 3)^2$ at the point where $x = 0$.

22. If $y = x^{2x}$, find the relative rate of change of y with respect to x when $x = 2$.

23. Without using logarithmic differentiation, find the derivative of $y = x^x$. *Hint:* First show that $y = x^x = e^{x \ln x}$.

8-11　HIGHER-ORDER DERIVATIVES

Since the derivative of a function is itself a function, it too may be differentiated. When this is done, the result (being a function itself) may also be differentiated. Continuing in this way, we obtain *higher-order derivatives*.

If $y = f(x)$, then $f'(x)$ is called the **first derivative** of f with respect to x. The derivative of $f'(x)$, denoted $f''(x)$, is called the **second derivative** of f with respect to x, etc. Some of the various ways in which higher-order derivatives may be written are given in Table 8-3.

TABLE 8-3

first derivative:	y',	$f'(x)$,	$\dfrac{dy}{dx}$,	$\dfrac{d}{dx}[f(x)]$,	$D_x\, y$.
second derivative:	y'',	$f''(x)$,	$\dfrac{d^2y}{dx^2}$,	$\dfrac{d^2}{dx^2}[f(x)]$,	$D_x^2 y$.
third derivative:	y''',	$f'''(x)$,	$\dfrac{d^3y}{dx^3}$,	$\dfrac{d^3}{dx^3}[f(x)]$,	$D_x^3\, y$.
fourth derivative:	$y^{(4)}$,	$f^{(4)}(x)$,	$\dfrac{d^4y}{dx^4}$,	$\dfrac{d^4}{dx^4}[f(x)]$,	$D_x^4\, y$.

PITFALL. *The symbol $D_x^2 y$ represents the second derivative of y. It is not the same as $[D_x\, y]^2$, the square of the first derivative of y.*

$$D_x^2 y \neq [D_x\, y]^2.$$

EXAMPLE 1

a. *If* $y = 2x^4 + 6x^3 - 12x^2 + 6x - 2$, *find* y'''.

Differentiating y, we obtain

$$y' = 8x^3 + 18x^2 - 24x + 6.$$

Differentiating y', we obtain

$$y'' = 24x^2 + 36x - 24.$$

Differentiating y'', we obtain

$$y''' = 48x + 36.$$

b. *If* $f(x) = 7$, *find* $f''(x)$.

$$f'(x) = 0.$$
$$f''(x) = 0.$$

c. *If* $y = e^{x^2}$, *find* $\dfrac{d^2y}{dx^2}$.

$$\frac{dy}{dx} = e^{x^2}(2x) = 2xe^{x^2}.$$

$$\frac{d^2y}{dx^2} = 2\big[x(e^{x^2})(2x) + e^{x^2}(1)\big]$$

$$= 2e^{x^2}(2x^2 + 1).$$

d. *If* $y = f(x) = \dfrac{x^2}{x+4}$, *find* $\dfrac{d^2y}{dx^2}$ *and evaluate it when* $x = 4$.

$$\frac{dy}{dx} = \frac{(x+4)(2x) - (x^2)(1)}{(x+4)^2} = \frac{x^2 + 8x}{(x+4)^2}.$$

$$\frac{d^2y}{dx^2} = \frac{(x+4)^2(2x+8) - (x^2+8x)(2)(x+4)}{(x+4)^4}$$

$$= \frac{(x+4)[(x+4)(2x+8) - (x^2+8x)(2)]}{(x+4)^4}$$

$$= \frac{32}{(x+4)^3}.$$

$$\frac{d^2y}{dx^2}\bigg|_{x=4} = \frac{1}{16}.$$

The second derivative evaluated at $x = 4$ can also be denoted $f''(4)$ or $y''(4)$.

e. *If $f(x) = x \ln x$, find the rate of change of $f''(x)$.*

To find the rate of change of any function, we must find its derivative. Thus we want $D_x[f''(x)]$ which is $f'''(x)$.

$$f'(x) = x\left(\frac{1}{x}\right) + (\ln x)(1) = 1 + \ln x.$$

$$f''(x) = 0 + \frac{1}{x} = \frac{1}{x}.$$

$$f'''(x) = -\frac{1}{x^2}.$$

We shall now find a higher-order derivative by means of implicit differentiation. Keep in mind that we shall assume y to be a function of x.

EXAMPLE 2

a. *Find y'' if $x^2 + 4y^2 = 4$.*

$$x^2 + 4y^2 = 4.$$

Differentiating both sides with respect to x, we obtain

$$2x + 8yy' = 0,$$

$$y' = \frac{-x}{4y}. \tag{1}$$

$$y'' = \frac{4y\, D_x(-x) - (-x)\, D_x(4y)}{(4y)^2}$$

$$= \frac{4y(-1) - (-x)(4y')}{16y^2}$$

$$= \frac{-4y + 4xy'}{16y^2}.$$

$$y'' = \frac{-y + xy'}{4y^2}. \tag{2}$$

Since $y' = \frac{-x}{4y}$ from Eq. (1), by substituting into Eq. (2) we have

$$y'' = \frac{-y + x\left(\frac{-x}{4y}\right)}{4y^2} = \frac{-4y^2 - x^2}{16y^3}$$

$$= -\frac{4y^2 + x^2}{16y^3}.$$

Since $x^2 + 4y^2 = 4$,

$$y'' = -\frac{4}{16y^3} = -\frac{1}{4y^3}.$$

b. *Find y'' if $y^2 = e^{x+y}$.*

$$y^2 = e^{x+y},$$

$$2yy' = e^{x+y}(1 + y').$$

Solving for y', we obtain

$$y' = \frac{e^{x+y}}{2y - e^{x+y}}.$$

Since $y^2 = e^{x+y}$.

$$y' = \frac{y^2}{2y - y^2} = \frac{y}{2 - y}.$$

$$y'' = \frac{(2 - y)(y') - y(-y')}{(2 - y)^2}$$

$$= \frac{2y'}{(2 - y)^2}.$$

Since $y' = \dfrac{y}{2 - y}$,

$$y'' = \frac{2y}{(2 - y)^3}.$$

EXERCISE 8-11

In Problems 1–20, find the indicated derivatives.

1. $y = 4x^3 - 12x^2 + 6x + 2, y'''$.

2. $y = 2x^4 - 6x^2 + 7x - 2, y'''$.

3. $y = 7 - x, \dfrac{d^2y}{dx^2}$.

4. $y = -x - x^2, \dfrac{d^2y}{dx^2}$.

5. $y = x^3 + e^x, y^{(4)}$.

6. $f(q) = \ln q, f'''(q)$.

7. $f(x) = x^2 \ln x, D_x^2[f(x)]$.

8. $y = \dfrac{1}{x}, y'''$.

9. $f(p) = \dfrac{1}{6p^3}, f'''(p)$.

10. $f(x) = \sqrt{x}, D_x^2[f(x)]$.

11. $f(r) = \sqrt{1 - r}, f''(r)$.

12. $y = e^{-4x^2}, y''$.

13. $y = \dfrac{1}{5x - 6}, \dfrac{d^2y}{dx^2}$.

14. $y = (2x + 1)^4, y''$.

15. $y = \dfrac{x+1}{x-1}, y''.$ **16.** $y = 2x^{1/2} + (2x)^{1/2}, y''.$

17. $y = \ln[x(x+1)], y''.$ **18.** $y = \ln\dfrac{(2x-3)(4x-5)}{x+3}, y''.$

19. $f(z) = z^2 e^z, f''(z).$ **20.** $y = \dfrac{x}{e^x}, \dfrac{d^2y}{dx^2}.$

In Problems **21–30**, *find* y''.

21. $x^2 + 4y^2 - 16 = 0.$ **22.** $x^2 - y^2 = 16.$

23. $y^2 = 4x.$ **24.** $4x^2 + 3y^2 = 4.$

25. $\sqrt{x} + 4\sqrt{y} = 4.$ **26.** $y^2 - 6xy = 4.$

27. $xy + y - x = 4.$ **28.** $xy + y^2 = 1.$

29. $y^2 = e^{x+y}.$ **30.** $e^x - e^y = x^2 + y^2.$

31. Find the rate of change of $f'(x)$ if $f(x) = (5x - 3)^4.$

32. Find the rate of change of $f''(x)$ if $f(x) = 6\sqrt{x} + \dfrac{1}{6\sqrt{x}}.$

33. If $c = .3q^2 + 2q + 850$ is a cost function, how fast is marginal cost changing when $q = 100$?

34. If $p = 1000 - 45q - q^2$ is a demand equation, how fast is marginal revenue changing when $q = 10$?

35. If $f(x) = x^4 - 6x^2 + 5x - 6$, determine the values of x for which $f''(x) = 0.$

8-12 REVIEW

IMPORTANT TERMS AND SYMBOLS IN CHAPTER 8

tangent line   *(p. 248)*

derivative   *(p. 250)*

$\lim\limits_{h \to 0} \dfrac{f(x+h) - f(x)}{h}$ *(p. 250)*

Δx *(p. 263)*

rate of change   *(p. 266)*

marginal revenue   *(p. 267)*

percentage rate of change   *(p. 269)*

quotient rule   *(p. 276)*

marginal propensity to consume

   (p. 278)

slope of a curve   *(p. 249)*

$y', f'(x), D_x y$ *(p. 251)*

$\dfrac{dy}{dx}, D_x^2 y, \dfrac{d^3y}{dx^3}, \dfrac{d^4}{dx^4}[f(x)]$ *(p. 311)*

marginal cost   *(p. 264)*

average cost   *(p. 266)*

relative rate of change   *(p. 269)*

product rule   *(p. 273)*

consumption function   *(p. 278)*

marginal propensity to save

   (p. 279)

chain rule (p. 283)   power rule (p. 285)

marginal revenue product (p. 288)   implicit differentiation (p. 305)

logarithmic differentiation (p. 308)   higher-order derivatives (p. 311)

REVIEW SECTION

1. In terms of a limit, the definition of the derivative of $y = f(x)$ with respect to x is

$$\lim_{h \to 0} \frac{f(x+h) - f(x)}{h}$$

   Ans. $\displaystyle\lim_{h \to 0} \frac{f(x+h) - f(x)}{h}$.

2. Geometrically, $f'(x)$ evaluated at $x = a$ is the _____ of the tangent line to the graph of $y = f(x)$ at $x = a$.

   Ans. slope.

3. If $f'(x) = 2x^3 + 7$, then $f''(x) = $ _____.

   Ans. $6x^2$.

4. The slope of the tangent line to $y = x^2$ at $(1, 1)$ is _____.

   Ans. 2.

5. If $f(x) = \frac{1}{3}$, then $f'''(4)$ _____.

   Ans. 0.

6. If $y = f(x)$, then dy/dx denotes an _(average) (instantaneous)_ rate of change of y with respect to x.

   Ans. instantaneous.

7. If $y = e^{8x^2+3}$, then $y' = \left(e^{8x^2+3} \right)(16x)$ _____.

   Ans. $16xe^{8x^2+3}$.

8. If $y = \ln(8x^2 + 3)$, then $dy/dx = \dfrac{1}{8x^2+3}(16x) = \dfrac{16x}{8x^2+3}$ _____.

   Ans. $16x/(8x^2 + 3)$.

9. Does the derivative of $y = |x|$ exist at $x = 0$? _____

   Ans. No.

10. True or false: If a function is continuous at a point, then it is differentiable there. __(a)__ . If a function is differentiable at a point, then it is continuous there. __(b)__ .

   Ans. (a) false; (b) true.

11. The expression $f'(x)/f(x)$ is called the _____ rate of change of $f(x)$.

   Ans. relative.

12. The slope of the curve $y = (x + 2)^2$ at $x = 1$ is ___6___ .

   Ans. 6.

$$2(x+2)^1 \cdot 1 =$$
$$2x+4$$
$$2+4=6$$

13. If $y = e^x$, then $d^4y/dx^4 = \underline{\hspace{1cm}}$.

Ans. e^x.

14. The slope of the curve $y = x + e^x$ at $x = 0$ is $\underline{\hspace{1cm}}$.

$1 + e^x(1) = 1 + e^0 = 1 + 1 = 2$ *Ans.* 2.

15. If $y = \frac{x}{7}$, then $D_x^2 y = \dfrac{(7)(1) - (0)(x)}{49} = \dfrac{7}{49} = \dfrac{1}{7}$

Ans. 0.

16. If u is a function of x, then the power rule asserts that $D_x(u^n) = \underline{\hspace{1cm}}$.

Ans. $nu^{n-1}\dfrac{du}{dx}$.

17. If $f'(x) = 7$, then $D_x(8f(x)) = \underline{\hspace{1cm}}$.

Ans. 56.

18. If 27 percent is the percentage rate of change of $f(x)$, then the relative rate of change is $\underline{.27}$.

Ans. .27.

19. The derivative of a function (is) (is not) a function.

Ans. is.

20. True or false: If $y = f(x) \cdot g(x)$, then $y' = f'(x) \cdot g'(x)$. $\underline{\hspace{0.8cm}}$ (a) . If $y = f(x)/g(x)$, then $y' = f'(x)/g'(x)$. (b) false

Ans. (a) false; (b) false.

21. If $f(x) = 7x + 1$, then $f'(1) = \underline{7}$.

Ans. 7.

22. If the marginal propensity to consume is .6, then the marginal propensity to save is $\underline{\hspace{1cm}}$.

Ans. .4.

23. True or false: $D_x(x^2 + 7)^3 = 3(x^2 + 7)^2$. false

$3(x^2 + 7)^2 \cdot (2x)$ *Ans.* false.

24. The rate of change of total revenue with respect to the number of units sold is called $\underline{\hspace{0.8cm}}$ (a) , and the rate of change of total cost with respect to the number of units produced is called $\underline{\hspace{0.8cm}}$ (b) .

Ans. (a) marginal revenue; (b) marginal cost.

REVIEW PROBLEMS

In Problems 1–54, differentiate.

1. $y = 6^3$.

2. $y = x$.

3. $y = 7x^4 - 6x^3 + 5x^2 + 1$.

4. $y = \sqrt{x + 3}$.

5. $y = 2e^x + e^2 + e^{x^2}$.

6. $y = 1/x^3$.

7. $y = \dfrac{x^2 + 3}{5}$.

8. $y = \dfrac{1}{2x + 1}$.

9. $f(r) = \ln (r^2 + 5r)$.

10. $y = e^{\ln x}$.

11. $y = (x^2 + 6x)(x^3 - 6x^2 + 4)$.

12. $y = (x^2 + 1)^{100}(x - 6)$.

13. $f(x) = (2x^2 + 4x)^{100}$.

14. $y = 2^{7x^2}$.

15. $y = (8 + 2x)(x^2 + 1)^4$.

16. $f(t) = \log_6 \sqrt{t^2 + 1}$.

17. $y = \sqrt[3]{4x - 1}$.

18. $y = \sqrt[3]{(1 - 3x^2)^2}$.

19. $y = e^x(x^2 + 2)$.

20. $f(w) = we^w + w^2$.

21. $f(z) = \dfrac{z^2 - 1}{z^2 + 1}$.

22. $y = \dfrac{x - 5}{(x + 2)^2}$.

23. $y = \dfrac{\ln x}{e^x}$.

24. $y = \dfrac{e^x + e^{-x}}{x^2}$.

25. $y = e^{x^2 + 4x + 5}$.

26. $y = (2x)^{3/5} + e$.

27. $y = \dfrac{x(x + 1)}{2x^2 + 3}$. $x^2 + x$

28. $g(z) = \dfrac{-7z}{(z - 1)^{-1}}$.

Implicit Diff 29. $2xy + y^2 = 6$ (find y').

30. $y = (x - 6)^4(x + 4)^3(6 - x)^2$.

log diff 31. $y = \sqrt{(x - 6)(x + 5)(9 - x)}$.

32. $4x^2 - 9y^2 = 4$ (find y').

33. $f(q) = \ln [(q + 1)^2(q + 2)^3]$.

34. $y = x^{x^3}$.

35. $y = \dfrac{1}{\sqrt{1 - x}}$.

36. $y = \sqrt{\dfrac{(x - 2)(x + 3)}{\sqrt{x - 1}}}$.

2. 37. $y = \log_2 (8x + 5)^2$.

38. $y + xy + y^2 = 1$ (find y').

log diff 39. $y = (x + 1)^{x+1}$.

40. $y = (x + 2)^{\ln x}$.

41. $y = \dfrac{x^2 + 6}{\sqrt{x^2 + 5}}$.

42. $y = \dfrac{(x + 3)^5}{x}$.

Implicit Diff 43. $x^2y^2 = 1$ (find y').

44. $f(x) = 5x\sqrt{1 - 2x}$.

45. $y = 2x^{-3/8} + (2x)^{-3/8}$.

46. $f(t) = e^{\sqrt{t}}$.

47. $f(l) = \ln (1 + l + l^2 + l^3)$.

48. $y = \sqrt{\dfrac{x}{2}} + \sqrt{\dfrac{2}{x}}$.

49. $y = (x^3 + 6x^2 + 9)^{3/5}$.

50. $y = (e + e^2)^0$.

51. $f(u) = \ln (u^2\sqrt{1 - u})$.

52. $y = \dfrac{1 + e^x}{1 - e^x}$.

log diff 53. $y = \dfrac{(x^2 + 2)^{3/2}(x^2 + 9)^{4/9}}{(x^3 + 6x)^{4/11}}$.

54. $y = \dfrac{\ln x}{\sqrt{x}}$.

In Problems 55–62, find the indicated derivative at the given point. It is not necessary to simplify the derivative before substituting the coordinates.

55. $y = x^4 - 2x^3 + 6x$, y''', $(1, 5)$.

56. $y = x^2 e^x$, y''', $(1, e)$.

57. $y = \dfrac{x}{\sqrt{x-1}}$, y'', $\left(5, \dfrac{5}{2}\right)$.

58. $y = \dfrac{2}{1-x}$, y'', $\left(-2, \dfrac{2}{3}\right)$.

59. $x + xy + y = 5$, y'', $(2, 1)$.

60. $xy + y^2 = 2$, y'', $(1, 1)$.

61. $y = \dfrac{4x}{x^2 + 4}$, y'', $(2, 1)$.

62. $y = (x + 1)^3(x - 1)$, y'', $(-1, 0)$.

In Problems 63–68, find an equation of the tangent line to the curve at the point corresponding to the given value of x.

63. $y = x^2 - 6x + 4$, $x = 1$.

64. $y = -2x^3 + 6x + 1$, $x = 2$.

65. $y = e^x$, $x = \ln 2$.

66. $y = \dfrac{x}{1-x}$, $x = 3$.

67. $x^2 - y^2 = 9$, $x = 7$, $y > 0$.

68. $xy = 6$, $x = 1$.

69. If $f(x) = 4x^2 + 2x + 8$, find the relative and percentage rates of change of $f(x)$ when $x = 1$.

70. If $f(x) = x/(x + 4)$, find the relative and percentage rates of change of $f(x)$ when $x = 1$.

71. If $r = q(20 - .1q)$ is a total revenue function, find the marginal revenue function.

72. If $c = .0001q^3 - .02q^2 + 3q + 6000$ is a total cost function, find the marginal cost when $q = 100$.

73. If $C = 7 + .6I - .25\sqrt{I}$ is a consumption function, find the marginal propensity to consume and the marginal propensity to save when $I = 16$.

74. If $p = (q + 14)/(q + 4)$ is a demand equation, find the rate of change of price p with respect to quantity q.

75. If $p = -.5q + 450$ is a demand equation, find the marginal revenue function.

76. If $\bar{c} = (500/q)e^{q/300}$ is an average cost function, find the marginal cost function.

77. The total cost function for an electric light and power plant is estimated by[†]

$$c = 16.68 + .125q + .00439q^2, \qquad 20 \leqslant q \leqslant 90$$

where q is eight-hour total output (as percentage of capacity) and c is total fuel cost in dollars. Find the marginal cost function and evaluate it when $q = 70$.

78. A manufacturer has determined that m employees will produce a total of q units per day where $q = m(50 - m)$. If his demand function is given by $p = -.01q + 9$, find the marginal revenue product when $m = 10$.

[†]J. A. Nordin, "Note on a light plant's cost curves," *Econometrica*, vol. 15 (1947), 231–235.

APPLICATIONS OF DIFFERENTIATION

9-1 INTERCEPTS AND SYMMETRY

Examining the graphical behavior of equations is a basic part of mathematics and has applications to business and economic analysis. In this section we shall examine equations to determine whether their graphs have certain features. Specifically, we shall consider *intercepts* and *symmetry*.

A point where a graph intersects the x-axis is called an x-*intercept* of the graph and has the form $(x, 0)$. A y-*intercept* is a point $(0, y)$ where the graph intersects the y-axis.

EXAMPLE 1

Find the x- and y-intercepts of the graphs of the following equations.

a. $x^2 + y^2 = 25$.

If $(x, 0)$ is an x-intercept, its coordinates must satisfy $x^2 + y^2 = 25$. Setting $y = 0$ and solving for x gives

$$x^2 + 0^2 = 25.$$
$$x = \pm 5.$$

The x-intercepts are thus $(5, 0)$ and $(-5, 0)$. Similarly, to determine the y-intercepts $(0, y)$, we set $x = 0$ in $x^2 + y^2 = 25$ and solve for y.

$$0^2 + y^2 = 25.$$
$$y = \pm 5.$$

Thus the y-intercepts are $(0, 5)$ and $(0, -5)$. See Fig. 9-1.

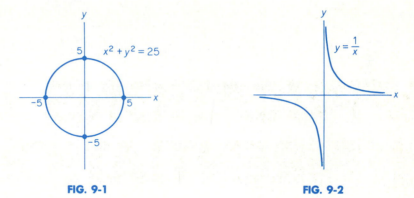

FIG. 9-1   **FIG. 9-2**

b. $y = \dfrac{1}{x}$.

Since x cannot be 0, the graph has no y-intercept. If $y = 0$, then $0 = 1/x$ and this equation has no solution. Thus no x-intercepts exist either (see Fig. 9-2).

At times it may be quite difficult or even impossible to find intercepts. For example, the x-intercepts of the graph of $y - \sqrt{2}\,x^5 - 4x^2 - 7 = 0$ would be difficult to find, although the y-intercept is easily found to be $(0, 7)$. In cases such as this, we settle for those intercepts that we can find conveniently.

Some graphs may have *symmetry*. For example, consider the graph of $y = x^2$ in Fig. 9-3. The portion for which $x \leqslant 0$ is the reflection (or mirror image) through the y-axis of that portion for which $x \geqslant 0$, and vice versa. More precisely, if (x_0, y_0) is any point on this graph, then the point $(-x_0, y_0)$ must also lie on the graph. We say that this graph is *symmetric with respect to the y-axis*.

DEFINITION.   *A graph is **symmetric with respect to the y-axis** if and only if $(-x_0, y_0)$ lies on the graph when (x_0, y_0) does.*

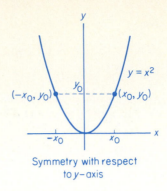

Symmetry with respect
to y-axis

FIG. 9-3

EXAMPLE 2

Use the definition above to show that the graph of $y = x^2$ is symmetric with respect to the y-axis.

Suppose (x_0, y_0) is *any* point on the graph of $y = x^2$. Then

$$y_0 = x_0^2.$$

We must show that the coordinates of $(-x_0, y_0)$ satisfy $y = x^2$:

$$y_0 = (-x_0)^2?$$

$$y_0 = x_0^2?$$

But from above we know that $y_0 = x_0^2$. Thus the graph *is* symmetric with respect to the y-axis.

When one is testing for symmetry in Example 2, (x_0, y_0) can be any point on the graph. In the future, for convenience we shall omit the subscripts. This means that a graph is symmetric with respect to the y-axis if replacing x by $-x$ in its equation results in an equivalent equation.

The graph of $x = y^2$ appears in Fig. 9-4. Here the portion for which $y \leqslant 0$ is the reflection through the x-axis of that portion for which $y \geqslant 0$, and vice versa. If (x, y) lies on the graph, then $(x, -y)$ also lies on it. We say that this graph is *symmetric with respect to the x-axis.*

DEFINITION. *A graph is **symmetric with respect to the x-axis** if and only if $(x, -y)$ lies on the graph when (x, y) does.*

A third type of symmetry, *symmetry with respect to the origin*, is illustrated by the graph of $y = x^3$ (Fig. 9-5). Whenever (x, y) lies on the graph, then $(-x, -y)$ also lies on it. Note that the line segment joining (x, y) and $(-x, -y)$ is bisected by the origin.

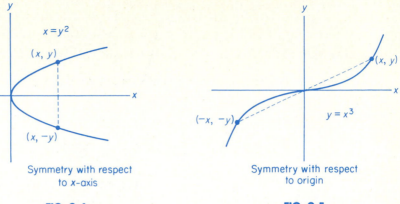

Symmetry with respect
to x-axis

FIG. 9-4

Symmetry with respect
to origin

FIG. 9-5

DEFINITION. *A graph is **symmetric with respect to the origin** if and only if $(-x, -y)$ lies on the graph
when (x, y) does.*

A summary of tests for symmetry is given in Table 9-1.

TABLE 9-1   Tests for Symmetry

Symmetry with respect to x-axis.	**Replace y by −y in given equation. Symmetric if equivalent equation is obtained.**
Symmetry with respect to y-axis.	**Replace x by −x in given equation. Symmetric if equivalent equation is obtained.**
Symmetry with respect to origin.	**Replace x by −x and y by −y in given equation. Symmetric if equivalent equation is obtained.**

EXAMPLE 3

*Determine whether or not the graph of $y = f(x) = 1 - x^4$ is symmetric with respect to the
x-axis, the y-axis, or the origin. Then find the intercepts and sketch the graph.*

Symmetry.

x-axis: Replacing y by $-y$ in $y = 1 - x^4$ gives

$$-y = 1 - x^4,$$

$$y = -1 + x^4,$$

which is not equivalent to the given equation. Thus the graph is *not* symmetric with
respect to the x-axis.

y-axis: Replacing x by $-x$ in $y = 1 - x^4$ gives

$$y = 1 - (-x)^4,$$
$$y = 1 - x^4,$$

which is equivalent to the given equation. Thus the graph *is* symmetric with respect to the *y*-axis.

origin: Replacing x by $-x$ and y by $-y$ in $y = 1 - x^4$ gives

$$-y = 1 - (-x)^4,$$
$$-y = 1 - x^4,$$
$$y = -1 + x^4,$$

which is not equivalent to the given equation. Thus the graph is *not* symmetric with respect to the origin.

Intercepts.   Testing for *x*-intercepts, we set $y = 0$ in $y = 1 - x^4$. Then

$$1 - x^4 = 0,$$
$$(1 - x^2)(1 + x^2) = 0,$$
$$(1 - x)(1 + x)(1 + x^2) = 0,$$
$$x = 1 \text{ or } x = -1.$$

The *x*-intercepts are thus $(1, 0)$ and $(-1, 0)$. Testing for *y*-intercepts, we set $x = 0$. Then $y = 1$ and so $(0, 1)$ is the only *y*-intercept.

Discussion.   If the intercepts and some points (x, y) where $x > 0$ are plotted, we can sketch the entire graph by using the property of symmetry with respect to the *y*-axis (Fig. 9-6).

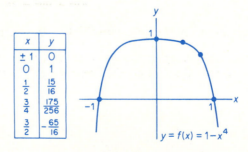

x	y
± 1	0
0	1
$\frac{1}{2}$	$\frac{15}{16}$
$\frac{3}{4}$	$\frac{175}{256}$
$\frac{3}{2}$	$-\frac{65}{16}$

$y = f(x) = 1 - x^4$

FIG. 9-6

In Example 3 you saw that the graph of $y = f(x) = 1 - x^4$ does not have *x*-axis symmetry. With the exception of the constant function $f(x) = 0$, *the graph of any function $y = f(x)$ cannot be symmetric with respect to the x-axis.* To see why, let (x_0, y_0) be any point on the graph. Then $y_0 = f(x_0)$. If the graph has *x*-axis symmetry, then $(x_0, -y_0)$ also lies on it. Thus $-y_0 = f(x_0)$. Hence, with the input number x_0 the function f associates two output numbers, y_0 and $-y_0$. But a function can associate only one output with a given input. Thus $y_0 = -y_0$, which implies $y_0 = 0$. Since (x_0, y_0) is *any* point on the graph, then $y = f(x) = 0$.

EXAMPLE 4

Test the graph of $4x^2 + 9y^2 = 36$ for intercepts and symmetry. Sketch the graph.

Intercepts. If $y = 0$, then $4x^2 = 36$, and so $x = \pm 3$. Thus the x-intercepts are $(3, 0)$ and $(-3, 0)$. If $x = 0$, then $9y^2 = 36$, and so $y = \pm 2$. Thus the y-intercepts are $(0, 2)$ and $(0, -2)$.

Symmetry. Testing for x-axis symmetry, we replace y by $-y$:

$$4x^2 + 9(-y)^2 = 36,$$
$$4x^2 + 9y^2 = 36.$$

Since we obtain the original equation, the graph is symmetric with respect to the x-axis.

Testing for y-axis symmetry, we replace x by $-x$:

$$4(-x)^2 + 9y^2 = 36,$$
$$4x^2 + 9y^2 = 36.$$

Again we have the original equation, and so the graph is also symmetric with respect to the y-axis.

Testing for symmetry with respect to the origin, we replace x by $-x$ and y by $-y$:

$$4(-x)^2 + 9(-y)^2 = 36,$$
$$4x^2 + 9y^2 = 36.$$

Since this is the original equation, the graph is also symmetric with respect to the origin.

Discussion. In Fig. 9-7 the intercepts and some points in the first quadrant are plotted. The points in that quadrant are then connected by a smooth curve. By symmetry with

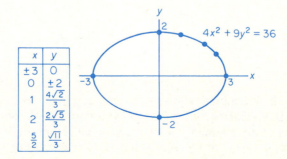

x	y
± 3	0
0	± 2
1	$\dfrac{4\sqrt{2}}{3}$
2	$\dfrac{2\sqrt{5}}{3}$
$\dfrac{5}{2}$	$\dfrac{\sqrt{11}}{3}$

FIG. 9-7

respect to the x-axis, the points in the fourth quadrant are obtained. Then by symmetry with respect to the y-axis the complete graph is found. There are other ways of graphing the equation by using symmetry. For example, after plotting the intercepts and some points in the first quadrant, then by symmetry with respect to the origin we can obtain the points in the third quadrant. By symmetry with respect to the x-axis (or y-axis) we can then obtain the entire graph.

In Example 4 the graph of $4x^2 + 9y^2 = 36$ was symmetric with respect to the x-axis, the y-axis, and the origin. **For any graph, if any two of the three types of symmetry that we have discussed exist, then the remaining type must also exist.**

EXERCISE 9-1

In Problems 1–16, find the x- and y-intercepts of the graphs of the equations. Also determine whether or not the graphs are symmetric with respect to the x-axis, the y-axis, or the origin. Do not sketch the graphs.

1. $y = 5x$.

2. $y = f(x) = x^2 - 4$.

3. $2x^2 + y^2x^4 = 8 - y$.

4. $x = y^3$.

5. $4x^2 - 9y^2 = 36$.

6. $y = 7$.

7. $x = -2$.

8. $y = |2x| - 2$.

9. $x = -y^{-4}$.

10. $y = \sqrt{x^2 - 4}$.

11. $x - 4y - y^2 + 21 = 0$.

12. $x^3 - xy + y^2 = 0$.

13. $y = f(x) = x^3/(x^2 + 5)$.

14. $x^2 + xy + y^2 = 0$.

15. $e^{x^2+y^2} - 5 = 0$.

16. $y = e^{x^2}$.

In Problems 17–24, find the x- and y-intercepts of the graphs of the equations. Also determine whether or not the graphs are symmetric with respect to the x-axis, the y-axis, or the origin. Then sketch the graphs.

17. $|x| - |y| = 0$.

18. $x = y^4$.

19. $2x + y^2 = 4$.

20. $y = x - x^3$.

21. $y = f(x) = x^3 - 4x$.

22. $x^2 + y^2 = 16$.

23. $4x^2 + y^2 = 16$.

24. $x^2 - y^2 = 1$.

9-2 ASYMPTOTES

In Example 1(b) of the last section, we showed that the graph of the function $y = 1/x$ has no intercepts. Although this graph is symmetric with respect to the origin (as you may verify), it has other distinguishing features (see Fig. 9-8). As

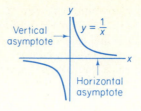

FIG. 9-8

x approaches zero from the right, $1/x$ becomes positively infinite; as x approaches zero from the left, $1/x$ becomes negatively infinite. In terms of limits,

$$\lim_{x \to 0^+} \frac{1}{x} = \infty \quad \text{and} \quad \lim_{x \to 0^-} \frac{1}{x} = -\infty.$$

We say that the line $x = 0$ (the y-axis) is a *vertical asymptote* of the graph of $y = 1/x$. This means it is a vertical line near which the graph "blows up" (that is, the graph rises or falls without bound).

On the other hand, as x approaches ∞, as well as $-\infty$, $1/x$ approaches 0. That is,

$$\lim_{x \to \infty} \frac{1}{x} = 0 \quad \text{and} \quad \lim_{x \to -\infty} \frac{1}{x} = 0.$$

We say that the line $y = 0$ (the x-axis) is a *horizontal asymptote* of the graph of $y = 1/x$. This means that it is a horizontal line near which the graph "settles down" as $x \to \infty$ or $x \to -\infty$.

DEFINITION. *The line $x = a$ is a **vertical asymptote** of the graph of the function f if and only if*

$$\lim_{x \to a^+} f(x) = \infty \ (\text{or} -\infty)$$

or $$\lim_{x \to a^-} f(x) = \infty \ (\text{or} -\infty).$$

*The line $y = b$ is a **horizontal asymptote** of the graph of f if and only if*

$$\lim_{x \to \infty} f(x) = b \quad or \quad \lim_{x \to -\infty} f(x) = b.$$

Note that if $x = a$ is a vertical asymptote, the function cannot be continuous at a—in fact, it has an infinite discontinuity at a.

EXAMPLE 1

Determine the horizontal and vertical asymptotes for the graphs of the following functions.

a. $y = \dfrac{1}{x - 2} + 3$.

To test for horizontal asymptotes, we find the limits of y as $x \to \infty$ and as $x \to -\infty$.

$$\lim_{x \to \infty} \left[\frac{1}{x - 2} + 3 \right] = \lim_{x \to \infty} \frac{1}{x - 2} + \lim_{x \to \infty} 3.$$

As x increases without bound, so does $x - 2$. Hence $1/(x - 2)$ approaches zero. Thus,

$$\lim_{x \to \infty} \left[\frac{1}{x - 2} + 3 \right] = 3,$$

and so the line $y = 3$ is a horizontal asymptote. Also,

$$\lim_{x \to -\infty} \left[\frac{1}{x - 2} + 3 \right] = \lim_{x \to -\infty} \frac{1}{x - 2} + \lim_{x \to -\infty} 3$$

$$= 0 + 3 = 3.$$

Hence the graph will settle down near the line $y = 3$ both as $x \to \infty$ and $x \to -\infty$.

To determine vertical asymptotes, we must find where $\dfrac{1}{x - 2} + 3$ blows up. Note that the denominator of $1/(x - 2)$ is 0 when $x = 2$. If x is slightly larger than 2, then $x - 2$ is both close to 0 and positive. Thus $1/(x - 2)$ is very large and so

$$\lim_{x \to 2^+} \left[\frac{1}{x - 2} + 3 \right] = \infty.$$

Hence, the line $x = 2$ is a vertical asymptote. If x is slightly less than 2, then $x - 2$ is very close to 0 but negative. Thus $1/(x - 2)$ is "very negative" and so

$$\lim_{x \to 2^-} \left[\frac{1}{x - 2} + 3 \right] = -\infty.$$

We conclude that the function increases without bound as $x \to 2^+$ and decreases without bound as $x \to 2^-$. The graph appears in Fig. 9-9. The dotted lines indicate the asymptotes and are not part of the graph. However, they can be useful guides in sketching it.

Usually, good places to look for vertical asymptotes of a function $y = f(x)$ are at those values of x for which a denominator of $f(x)$ is 0.

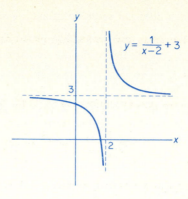

FIG. 9-9

b. $y = f(x) = x^3 + 2x$.

Testing for horizontal asymptotes, we have

$$\lim_{x \to \infty} (x^3 + 2x) = \infty \quad \text{and} \quad \lim_{x \to -\infty} (x^3 + 2x) = -\infty.$$

Thus the graph does not settle down as $x \to \infty$ or $x \to -\infty$. Hence, there are no horizontal asymptotes.

We previously said that if the line $x = a$ is a vertical asymptote of the graph of a function f, then f must have an infinite discontinuity at a. But $y = f(x) = x^3 + 2x$ is a polynomial function and is continuous everywhere. Thus its graph has no vertical asymptotes. See Fig. 9-10.

In general, **any polynomial function that is not a constant has neither horizontal nor vertical asymptotes.**

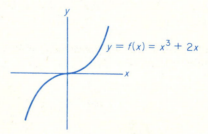

FIG. 9-10

c. $y = e^x - 1$.

Testing for horizontal asymptotes, we let $x \to \infty$. Then e^x increases without bound and so

$$\lim_{x \to \infty} (e^x - 1) = \infty.$$

Thus the graph does not settle down as $x \to \infty$. However, as $x \to -\infty$, then $e^x \to 0$ and so

$$\lim_{x \to -\infty} (e^x - 1) = \lim_{x \to -\infty} e^x - \lim_{x \to -\infty} 1 = 0 - 1 = -1.$$

Therefore the line $y = -1$ is a horizontal asymptote. The graph has no vertical asymptotes, since $e^x - 1$ neither increases nor decreases without bound around any fixed value of x (see Fig. 9-11).

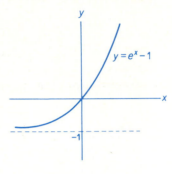

FIG. 9-11

EXAMPLE 2

Sketch the graph of $y = \dfrac{x^2}{x^2 - 1}$ *with the aid of intercepts, symmetry, and asymptotes.*

Intercepts.   If $x = 0$, then $y = 0$; if $y = 0$, then $x = 0$. Therefore the x-intercept, as well as the y-intercept, is (0, 0).

Symmetry.   Since y is a function of x and it is not the zero function, then the graph is *not* symmetric with respect to the x-axis.

Testing for y-axis symmetry, we replace x by $-x$:

$$y = \frac{(-x)^2}{(-x)^2 - 1},$$

$$y = \frac{x^2}{x^2 - 1}.$$

The graph *is* symmetric with respect to the y-axis. *Symmetry with respect to exactly one axis implies that the graph cannot be symmetric with respect to the origin.*

Asymptotes.

Horizontal:

$$\lim_{x \to \infty} \frac{x^2}{x^2 - 1} = \lim_{x \to \infty} \frac{1}{1 - \dfrac{1}{x^2}}$$

(dividing both numerator and denominator by x^2)

$$= \frac{1}{1 - 0} = 1.$$

Therefore, as $x \to \infty$, the graph approaches the line $y = 1$, a horizontal asymptote. By symmetry, as $x \to -\infty$, the graph again approaches the line $y = 1$.

Vertical: Since

$$\frac{x^2}{x^2 - 1} = \frac{x^2}{(x - 1)(x + 1)}.$$

it is easy to see from the denominator that the graph will blow up when x is close to 1 or -1. In fact,

$$\lim_{x \to 1^-} \frac{x^2}{(x - 1)(x + 1)} = -\infty \quad \text{and} \quad \lim_{x \to 1^+} \frac{x^2}{(x - 1)(x + 1)} = \infty.$$

By symmetry,

$$\lim_{x \to -1^+} \frac{x^2}{(x - 1)(x + 1)} = -\infty \quad \text{and} \quad \lim_{x \to -1^-} \frac{x^2}{(x - 1)(x + 1)} = \infty.$$

Thus the lines $x = 1$ and $x = -1$ are vertical asymptotes.

Discussion.   By plotting the intercept and other points on the graph when $x > 0$, and by using properties of symmetry and asymptotes, it is relatively easy to sketch the graph (see Fig. 9-12).

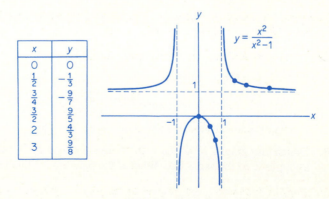

x	y
0	0
$\frac{1}{2}$	$-\frac{1}{3}$
$\frac{3}{4}$	$-\frac{9}{7}$
$\frac{3}{2}$	$\frac{9}{5}$
2	$\frac{4}{3}$
3	$\frac{9}{8}$

$$y = \frac{x^2}{x^2 - 1}$$

FIG. 9-12

EXERCISE 9-2

In Problems 1–14, determine the horizontal and vertical asymptotes of the graphs of the functions. Do not sketch the graphs.

1. $y = \dfrac{4}{x}$.

2. $y = -\dfrac{4}{x^2}$.

3. $y = \dfrac{4}{x - 6} + 4$.

4. $y = \dfrac{2x + 1}{2x - 1}$.

5. $y = x^3 - 5x + 8$.

6. $y = \dfrac{x^3}{x^2 - 9}$.

7. $f(x) = \dfrac{x - 1}{2x + 3}$.

8. $f(x) = \dfrac{x^2}{5}$.

9. $f(x) = \dfrac{2x^2}{x^2 + x - 6}$.

10. $f(x) = \dfrac{5}{2x^2 - 9x + 4}$.

11. $f(x) = \sqrt[3]{x^2}$.

12. $y = \dfrac{x^2(x^2 - 9)}{x^2}$.

13. $y = 2e^{x+2} + 4$.

14. $f(x) = e^{x^3}$.

In Problems 15–26, find the x- and y-intercepts of the graphs of the functions; determine whether the graphs are symmetric with respect to the x-axis, y-axis, or origin; determine horizontal and vertical asymptotes; sketch the graphs.

15. $y = \dfrac{3}{x - 1}$.

16. $y = \dfrac{x}{4 - x}$.

17. $f(x) = \dfrac{8}{x^3}$.

18. $f(x) = \dfrac{1}{x^4}$.

19. $f(x) = \dfrac{1}{x^2 - 1}$.

20. $f(x) = \dfrac{x^2}{x^2 - 4}$.

21. $y = \dfrac{x^2 - 1}{x^2 - 4}$.

22. $y = \dfrac{x^3 - x}{x}$.

23. $y = \dfrac{x^2(x^2 - 9)}{x^2}$.

24. $f(x) = \begin{cases} 1/x, & \text{if } x > 0, \\ (x + 1)/x, & \text{if } x < 0. \end{cases}$

25. $y = 3 - e^{2x}$.

26. $y = e^{-x} - 1$.

27. In discussing the time pattern of purchasing, Mantell and Sing[†] use the curve

$$y = \dfrac{x}{a + bx}$$

as a mathematical model. They claim that $y = 1/b$ is an asymptote. Verify this.

28. Sketch the graphs of $y = 6 - 3e^{-x}$ and $y = 6 + 3e^{-x}$. Show that they are both asymptotic to the same line. What is the equation of this line?

[†] L. H. Mantell and F. P. Sing, *Economics for Business Decisions* (New York: McGraw-Hill Book Co., 1972), p. 107.

29. For a new product the yearly number of thousand packages sold, y, after t years from its introduction is predicted to be

$$y = f(t) = 150 - 76e^{-t}.$$

Show that $y = 150$ is a horizontal asymptote of the graph. This shows that after the product is established with consumers, the market tends to be constant.

9-3 RELATIVE MAXIMA AND MINIMA

In curve sketching, plotting points at random usually is not good enough to properly determine a curve's shape. For example, the points $(-1, 0)$, $(0, -1)$, and $(1, 0)$ satisfy the equation $y = (x + 1)^3(x - 1)$. You might hastily conclude that its graph should appear as in Fig. 9-13, but in fact the actual shape is given

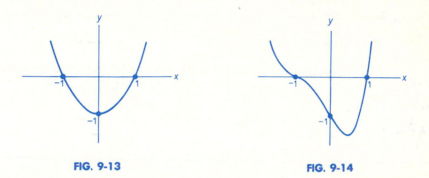

FIG. 9-13   **FIG. 9-14**

in Fig. 9-14. In this section, as well as in the following one, we shall explore the powerful role that differentiation plays in analyzing a function so that we may determine the true shape and behavior of its graph.

Suppose Fig. 9-15 gives the graph of $y = f(x)$. As x increases (goes from left to right) on the interval I_1 determined by a and b, the corresponding values

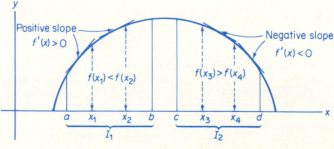

FIG. 9-15

of y increase and the curve is rising. Symbolically, if x_1 and x_2 are any two points in I_1 such that $x_1 < x_2$, then $f(x_1) < f(x_2)$ and f is said to be an *increasing function* on I_1. Moreover, for this portion of the curve, any tangent line will have a positive slope, and thus $f'(x)$ must be positive for all x in I_1. On the other hand, as x increases on the interval I_2 determined by c and d, the curve is falling. Here $x_3 < x_4$ implies $f(x_3) > f(x_4)$ and f is said to be a *decreasing function* on I_2. In this case any tangent line has a negative slope, and thus $f'(x) < 0$ for all x in I_2.

DEFINITION. *A function f is an **increasing** [**decreasing**] **function** on the interval I if and only if for any two points x_1, x_2 in I such that $x_1 < x_2$, then $f(x_1) < f(x_2)$ [$f(x_1) > f(x_2)$].*

RULE 1. *If $f'(x) > 0$ on an interval I, then f is an increasing function on I. If $f'(x) < 0$ on I, then f is a decreasing function on I.*

A function f is said to be increasing (decreasing) at a *point* x_0 if there is an open interval containing x_0 on which f is increasing (decreasing). Thus in Fig. 9-15, f is increasing at x_1 and is decreasing at x_3.

To illustrate these notions we shall use Rule 1 to find the intervals on which $y = 18x - \dfrac{2x^3}{3}$ is increasing or decreasing. Letting $y = f(x)$, we must determine when $f'(x)$ is positive and when $f'(x)$ is negative.

$$f'(x) = 18 - 2x^2 = 2(9 - x^2) = 2(3 + x)(3 - x).$$

[handwritten margin note: open interval means does not include the endpts.]

Using the technique of Sec. 7-5, we can find the sign of $f'(x)$ by considering the intervals determined by the roots of $2(3 + x)(3 - x) = 0$, namely 3 and -3 (see Fig. 9-16). In each interval the sign of $f'(x)$ is determined by the signs of its

FIG. 9-16

factors:

if $x < -3$, then $f'(x) = 2(-)(+) = (-)$ and f is decreasing;
if $-3 < x < 3$, then $f'(x) = 2(+)(+) = (+)$ and f is increasing;
if $x > 3$, then $f'(x) = 2(+)(-) = (-)$ and f is decreasing [see Fig. 9-17(a)].

Thus f is decreasing on $(-\infty, -3)$ and $(3, \infty)$, and is increasing on $(-3, 3)$ as seen in Fig. 9-17(b). These results could be sharpened. Actually, by definition, f is decreasing on $(-\infty, -3]$ and $[3, \infty)$, and increasing on $[-3, 3]$. However, for our purposes open intervals are sufficient. *It will be our practice to determine* **open** *intervals on which a function is increasing or decreasing.*

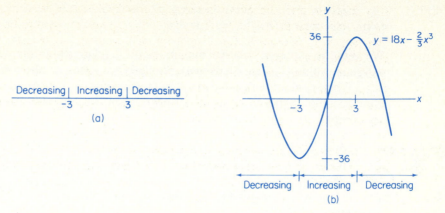

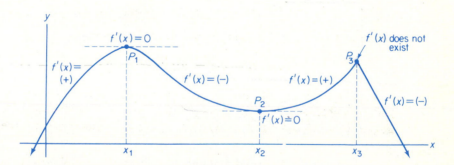

FIG. 9-17

Look now at the graph of $y = f(x)$ in Fig. 9-18. Three observations can be made.

FIG. 9-18

horizontal line has zero slope
vertical line has slope that doesn't exist

First, there is something special about the points P_1, P_2, and P_3. Notice that P_1 is *higher* than any other "nearby" points on the curve—likewise for P_3. The point P_2 is *lower* than any other "nearby" points on the curve. Since P_1, P_2, and P_3 may not necessarily be the highest or lowest points on the *entire* curve, we simply say that f has a *relative maximum* when $x = x_1$ and when $x = x_3$, and a *relative minimum* when $x = x_2$. Actually, there is an *absolute maximum* (highest point on the entire curve) when $x = x_1$, but there is no *absolute minimum* (lowest point on the entire curve), since the curve is assumed to extend downward indefinitely. We define these new terms as follows:

DEFINITION. *A function f has a **relative maximum** [**relative minimum**] when $x = x_0$ if there is an open interval containing x_0 on which $f(x_0) > f(x)$ [$f(x_0) < f(x)$] for all x in the interval.*

> **DEFINITION.** *A function f has an **absolute maximum** [**absolute minimum**] when $x = x_0$ if $f(x_0) \geqslant f(x)$ [$f(x_0) \leqslant f(x)$] for all x in the domain of f.*

We refer to either a relative maximum or a relative minimum as a **relative extremum** (plural: *relative extrema*). Similarly we speak of **absolute extrema**.

When dealing with relative extrema, we compare the functional value at a point to those of nearby points; however, when dealing with absolute extrema, we compare the functional value at a point to all others determined by the domain. Thus, relative extrema are "local" in nature, while absolute extrema are "global" in nature.

Our second observation is that at a relative extremum the derivative may not be defined (as when $x = x_3$). But whenever it is defined, it is 0 (as when $x = x_1$ and $x = x_2$), and hence the tangent line is horizontal as shown in Fig. 9-18. We may state:

RULE 2. *If f has a relative extremum when $x = x_0$, then $f'(x_0) = 0$ or $f'(x_0)$ is not defined.*

Third, each relative extremum occurs at a point around which the sign of $f'(x)$ is changing, regardless of whether or not the derivative is defined at the point. For the relative maximum when $x = x_1$, $f'(x)$ goes from $(+)$ for $x < x_1$ to $(-)$ for $x > x_1$, *as long as x is near x_1.* At the relative minimum when $x = x_2$, $f'(x)$ goes from $(-)$ to $(+)$, and at the relative maximum when $x = x_3$, it again goes from $(+)$ to $(-)$. Thus, *around relative maxima, f is increasing and then decreasing, and the reverse holds for relative minima.*

RULE 3. *If x_0 is in the domain of f and $f'(x)$ changes from positive to negative as x increases through x_0, then f has a relative maximum when $x = x_0$. If $f'(x)$ changes from negative to positive as x increases through x_0, then f has a relative minimum when $x = x_0$.*

From Rules 1, 2, and 3, it should be clear that relative extrema may occur at values of x for which $f'(x) = 0$ or is not defined, for it is there that $f'(x)$ may change sign. These values of x are called *critical values*. To find relative extrema, you should examine the signs of $f'(x)$ over the intervals determined by the critical values.

> **DEFINITION.** *If $f'(x_0) = 0$ or $f'(x_0)$ is not defined, x_0 is called a **critical value** of f. If x_0 is a critical value and is in the domain of f, then $(x_0, f(x_0))$ is called a **critical point**.*

> **PITFALL.** *Not every critical value corresponds to a relative extremum.*

For example, if $y = f(x) = x^3$, then $f'(x) = 3x^2$. Since $f'(0) = 0$, then 0 is a critical value. Now if $x < 0$, then $3x^2 > 0$. If $x > 0$, then $3x^2 > 0$. Since $f'(x)$ does not change sign, no relative maximum or minimum exists. Indeed, since $f'(x) \geqslant 0$ for all x, the graph of f never falls and f is said to be *nondecreasing* (see Fig. 9-19). On the other hand, if $y = f(x) = 1/x^2$, then $y' = -2/x^3$. Since y' is not defined when $x = 0$, then 0 is a critical value. If $x < 0$, then $y' > 0$. If $x > 0$, then $y' < 0$. Although a change in sign of

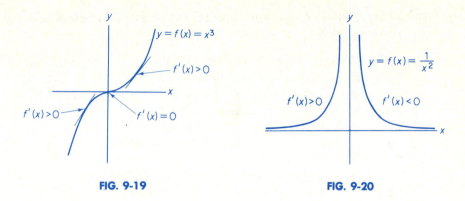

FIG. 9-19   **FIG. 9-20**

y' occurs around $x = 0$, no relative maximum exists there since 0 is not in the domain of f. Nevertheless, this critical value is important in determining the intervals over which f is increasing or decreasing. Here f is increasing on $(-\infty, 0)$ and decreasing on $(0, \infty)$ [see Fig. 9-20].

From our discussions and the "Pitfall" above, we can conclude that a critical value is only a "candidate" for a relative extremum. It may correspond to a relative maximum, a relative minimum, or neither.

Summarizing the results of this section, we have the *first-derivative test* for the relative extrema of $y = f(x)$:

FIRST-DERIVATIVE TEST.

1. **Find $f'(x)$.**
2. **Determine critical values.** (open intervals)
3. **On the intervals suggested by the critical values, determine whether f is increasing ($f'(x) > 0$) or decreasing ($f'(x) < 0$).**
4. **For each critical value x_0 in the domain of f, determine whether $f'(x)$ changes sign as x increases through x_0. There is a relative maximum when $x = x_0$ if $f'(x)$ changes from $(+)$ to $(-)$, and a relative minimum if $f'(x)$ changes from $(-)$ to $(+)$. If $f'(x)$ does not change sign, there is no relative maximum or minimum when $x = x_0$.**

EXAMPLE 1

If $y = f(x) = x + \dfrac{4}{x + 1}$, use the first-derivative test to determine intervals on which f is increasing or decreasing and locate all relative extrema.

1. $f'(x) = 1 - \dfrac{4}{(x + 1)^2} = \dfrac{(x + 1)^2 - 4}{(x + 1)^2} = \dfrac{x^2 + 2x - 3}{(x + 1)^2} = \dfrac{(x + 3)(x - 1)}{(x + 1)^2}.$

2. Setting $f'(x) = 0$ gives the critical values $x = -3, 1$. Since $f'(-1)$ does not exist, $x = -1$ is also a critical value.

3. There are four intervals to consider (Fig. 9-21):

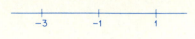

FIG. 9-21

if $x < -3$, then $f'(x) = \dfrac{(-)(-)}{(+)} = (+)$ and f is increasing;

if $-3 < x < -1$, then $f'(x) = \dfrac{(+)(-)}{(+)} = (-)$ and f is decreasing;

if $-1 < x < 1$, then $f'(x) = \dfrac{(+)(-)}{(+)} = (-)$ and f is decreasing;

if $x > 1$, then $f'(x) = \dfrac{(+)(+)}{(+)} = (+)$ and f is increasing (Fig. 9-22).

Thus f is increasing on $(-\infty, -3)$ and $(1, \infty)$ and is decreasing on $(-3, -1)$ and $(-1, 1)$.

FIG. 9-22

4. When $x = -3$, there is a relative maximum since $f'(x)$ changes from $(+)$ to $(-)$. When $x = 1$, there is a relative minimum since $f'(x)$ changes from $(-)$ to $(+)$. We ignore $x = -1$ since -1 is not in the domain of f (Fig. 9-23).

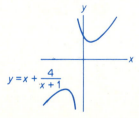

$$y = x + \frac{4}{x+1}$$

FIG. 9-23

EXAMPLE 2

Test $y = f(x) = x^{2/3}$ for relative extrema.

We have $f'(x) = \frac{2}{3}x^{-1/3} = 2/(3\sqrt[3]{x})$. When $x = 0$, then $f'(x)$ is not defined and thus $x = 0$ is a critical value. If $x < 0$, then $f'(x) < 0$. If $x > 0$, then $f'(x) > 0$. Since 0 is also in the domain of f, there is a relative (as well as an absolute) minimum when $x = 0$ (see Fig. 9-24).

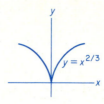

FIG. 9-24

EXAMPLE 3

Test $y = f(x) = x^2e^x$ for relative extrema.

By the product rule,

$$f'(x) = x^2e^x + e^x(2x) = xe^x(x + 2).$$

Since e^x is always positive, the critical values are $x = 0, -2$. From the signs of $f'(x)$ given in Fig. 9-25, we conclude that there is a relative maximum when $x = -2$ and a relative minimum when $x = 0$.

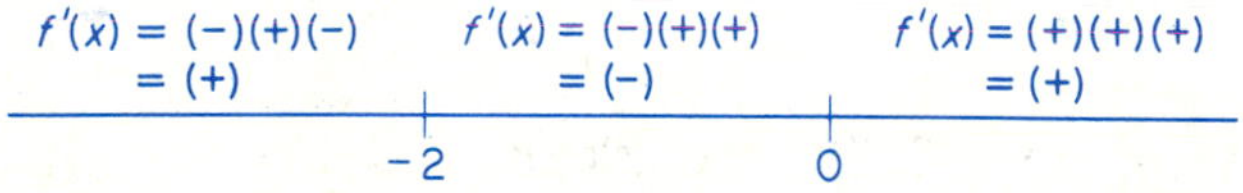

$$
\begin{array}{ccc}
f'(x) = (-)(+)(-) & f'(x) = (-)(+)(+) & f'(x) = (+)(+)(+) \\
= (+) & = (-) & = (+)
\end{array}
$$

$$
\underset{-2}{\rule{3cm}{0.4pt}}\quad\underset{0}{\rule{3cm}{0.4pt}}
$$

FIG. 9-25

EXAMPLE 4

Sketch the graph of $y = f(x) = 2x^2 - x^4$.

Intercepts.   If $x = 0$, then $y = 0$. If $y = 0$, then

$$0 = 2x^2 - x^4 = x^2(2 - x^2) = x^2(\sqrt{2} + x)(\sqrt{2} - x),$$

and thus $x = 0, \pm\sqrt{2}$. The intercepts are $(0, 0)$ $(\sqrt{2}, 0)$, and $(-\sqrt{2}, 0)$.

Symmetry.   Testing for y-axis symmetry, we have

$$y = 2(-x)^2 - (-x)^4,$$
$$y = 2x^2 - x^4.$$

Since this is the original equation, there is y-axis symmetry. It can be shown that there is no x-axis symmetry and hence no symmetry with respect to the origin.

Asymptotes. No horizontal or vertical asymptotes exist, since f is a nonconstant polynomial function.

First-Derivative Test.

1. $y' = 4x - 4x^3 = 4x(1 - x^2) = 4x(1 + x)(1 - x)$.

2. Setting $y' = 0$ gives the critical values $x = 0, \pm 1$. The critical points are $(-1, 1)$, $(0, 0)$, and $(1, 1)$. The y-coordinates of these points were found by substituting $x = 0, \pm 1$ into the *original* equation, $y = 2x^2 - x^4$.

3. There are four intervals to consider in Fig. 9-26:

FIG. 9-26

if $x < -1$, then $y' = 4(-)(-)(+) = (+)$ and f is increasing;

if $-1 < x < 0$, then $y' = 4(-)(+)(+) = (-)$ and f is decreasing;

if $0 < x < 1$, then $y' = 4(+)(+)(+) = (+)$ and f is increasing;

if $x > 1$, then $y' = 4(+)(+)(-) = (-)$ and f is decreasing (Fig. 9-27).

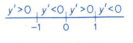

FIG. 9-27

4. Relative maxima occur at $(-1, 1)$ and $(1, 1)$; a relative minimum occurs at $(0, 0)$.

Discussion. In Fig. 9-28(a) we have plotted the intercepts and the horizontal tangents at the relative maximum and minimum points. We know the curve rises from the left, has a

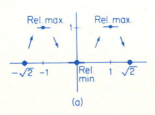

(a)

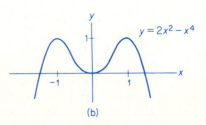

(b)

FIG. 9-28

relative maximum, then falls, has a relative minimum, then rises to a relative maximum, and falls thereafter. A sketch is shown in Fig. 9-28(b).

In Example 4 relative maxima, as well as absolute maxima, occur at $x = \pm 1$ [see Fig. 9-28(b)]. Although there is a relative minimum, there is no absolute minimum.

If the domain of a function is an interval that contains an endpoint, to determine absolute extrema we must not only examine the function for relative extrema, but we must also take into consideration the values of $f(x)$ at the endpoints. Although endpoints are not considered when we look for relative maxima or minima, they may yield *absolute* maxima or minima. Example 5 will illustrate.

EXAMPLE 5

Find all extrema (relative and absolute) for $y = f(x) = x^2 - 4x + 5$ on the closed interval [1, 4].

1. $f'(x) = 2x - 4 = 2(x - 2)$.

2. Setting $f'(x) = 0$ gives the critical value $x = 2$, which is in the domain of f.

3. The intervals to consider are when $x < 2$ and when $x > 2$.

4. If $x < 2$, then $f'(x) < 0$ and f is decreasing; if $x > 2$, then $f'(x) > 0$ and f is increasing. Thus there is a relative minimum when $x = 2$. It occurs on the graph at the point (2, 1) [see Fig. 9-29].

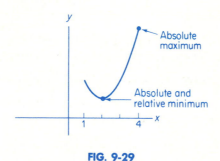

FIG. 9-29

5. Since f is decreasing for $x < 2$, then an absolute maximum may *possibly* occur at the left-hand endpoint of the domain of f, that is, when $x = 1$. Similarly, since f is increasing for $x > 2$, an absolute maximum may *possibly* occur at the right-hand endpoint, that is, when $x = 4$. Testing the endpoints, we have $f(1) = 2$ and $f(4) = 5$. Noting that $f(4) > f(1)$, we conclude that an absolute maximum occurs when $x = 4$. When $x = 2$ there is an absolute, as well as a relative, minimum.

EXERCISE 9-3

In Problems **1–28** determine when the function is increasing or decreasing and locate all relative maxima and minima. Do not sketch the graph.

1. $y = x^2 + 2$.

2. $y = x^2 + 4x + 3$.

3. $y = x - x^2 + 2$.

4. $y = 4x - x^2$.

5. $y = -\dfrac{x^3}{3} - 2x^2 + 5x - 2$.

6. $y = 4x^3 - 3x^4$.

7. $y = x^4 - 2x^2$.

8. $y = -2 + 12x - x^3$.

9. $y = x^3 - 6x^2 + 9x$.

10. $y = x^3 - 6x^2 + 12x - 6$.

11. $y = 3x^5 - 5x^3$.

12. $y = 5x - x^5$.

13. $y = -x^5 - 5x^4 + 200$.

14. $y = 3x^4 - 4x^3 + 1$.

15. $y = \dfrac{1}{x - 1}$.

16. $y = \dfrac{3}{x}$.

17. $y = \dfrac{10}{\sqrt{x}}$.

18. $y = \dfrac{x}{x + 1}$.

19. $y = \dfrac{x^2}{1 - x}$.

20. $y = x + \dfrac{4}{x}$.

21. $y = (x + 2)^3(x - 5)^2$.

22. $y = x^2(x + 3)^4$.

23. $y = e^{-2x}$.

24. $y = x \ln x$.

25. $y = x^2 - 2 \ln x$.

26. $y = xe^x$.

27. $y = e^x + e^{-x}$.

28. $y = e^{-x^2}$.

In Problems **29–40**, determine: intervals on which the functions are increasing or decreasing; relative maxima and minima; symmetry; horizontal and vertical asymptotes; those intercepts that can be obtained conveniently. Then sketch the graphs.

29. $y = x^2 - 6x - 7$.

30. $y = 2x^2 - 5x - 12$.

31. $y = 3x - x^3$.

32. $y = x^4 - 16$.

33. $y = 2x^3 - 9x^2 + 12x$.

34. $y = x^3 - 9x^2 + 24x - 19$.

35. $y = x^4 + 4x^3 + 4x^2$.

36. $y = x^5 - \frac{5}{4}x^4$.

37. $y = \dfrac{x + 1}{x - 1}$.

38. $y = \dfrac{x^2}{x^2 + 1}$.

39. $y = \dfrac{x^2}{x + 3}$.

40. $y = x + \dfrac{1}{x}$.

In Problems **41–46**, find when absolute maxima and minima occur for the given function on the given interval.

41. $f(x) = x^2 - 2x + 3$, $[-1, 2]$.   42. $f(x) = -2x^2 - 6x + 5$, $[-2, 3]$.

43. $f(x) = \frac{1}{3}x^3 - x^2 - 3x + 1$, $[0, 2]$.   44. $f(x) = \frac{1}{4}x^4 - \frac{3}{2}x^2$, $[0, 1]$.

45. $f(x) = 4x^3 + 3x^2 - 18x + 3$, $[\frac{1}{2}, 3]$.   46. $f(x) = x^{4/3}$, $[-8, 8]$.

47. If $c_f = 25{,}000$ is a fixed cost function, show that the average fixed cost function $\bar{c}_f = c_f/q$ is a decreasing function for $q > 0$. Thus, as output q increases, each unit's portion of fixed cost declines.

48. If $c = 4q - q^2 + 2q^3$ is a cost function, when is marginal cost increasing?

49. Given the demand function $p = 400 - 2q$, find when marginal revenue is increasing.

50. For the cost function $c = \sqrt{q}$, show that marginal and average costs are always decreasing for $q > 0$.

51. For a manufacturer's product, the revenue function is given by $r = 240q + 57q^2 - q^3$. Determine the output for maximum revenue.

52. In his model for storage and shipping costs of materials for a manufacturing process, Lancaster[†] derives the following cost function:

$$C(k) = 100\left[100 + 9k + \frac{144}{k}\right], \qquad 1 \leqslant k \leqslant 100,$$

where $C(k)$ is the total cost (in dollars) of storage and transportation for 100 days of operation if a load of k tons of material is moved every k days. (a) Find $C(1)$. (b) For what value of k does $C(k)$ have a minimum? (c) What is the minimum value?

9-4 CONCAVITY

We have seen how the first derivative is used to determine when a function is increasing or decreasing and to locate relative maxima and minima. However, for us to know with assurance the actual shape of a curve, we must have additional information. For example, consider the curve $y = f(x) = x^2$. Since $f'(x) = 2x$ and $f'(0) = 0$, then $x = 0$ is a critical value. If $x < 0$, then $f'(x) < 0$ and f is decreasing; if $x > 0$, then $f'(x) > 0$ and f is increasing. Thus there is a relative minimum when $x = 0$. Figures 9-30 (a) and (b) both satisfy the preceding conditions. But which one truly describes the curve? This question will easily be settled by using the second derivative and the notion of *concavity*.

[†] P. Lancaster, *Mathematics: Models of the Real World* (Englewood Cliffs, N.J.: Prentice-Hall, Inc. 1976).

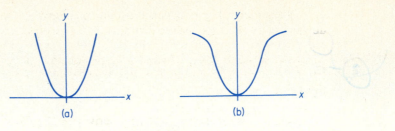

FIG. 9-30

In Fig. 9-3l, note that in each case the curve $y = f(x)$ "bends" (or opens) upward. Moreover, for each curve, if tangent lines are drawn their slopes increase in value as x increases. In (a) the slopes go from small positive values to

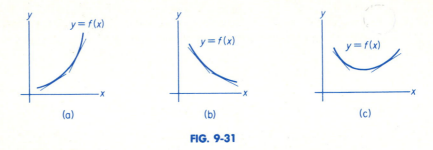

FIG. 9-31

larger values; in (b) they are negative and approaching zero (thus increasing); in (c) they pass from negative values to positive values. Since $f'(x)$ gives the slope at a point, f' is an increasing function here. In each case we say that the curve (or function f) is *concave up*. By a similar analysis, in Fig. 9-32 it can be seen in each case that as x increases, the slopes of the tangent lines are decreasing and

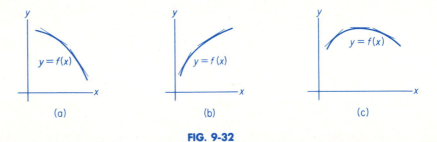

FIG. 9-32

the curves are bending downward. Thus f' is a decreasing function here, and we say f is *concave down*.

DEFINITION. *A function f is said to be **concave up** [**concave down**] on an interval I if f' is an increasing [decreasing] function on I.*

Remember: If f is concave up on an interval I, then its graph is bending upward there. If f is concave down, then its graph is bending downward.

PITFALL.   *Concavity relates to whether f', not f, is increasing or decreasing. Thus in Fig. 9-31(b), note that f is concave up and decreasing, but in Fig. 9-32(a) note that f is concave down and decreasing.*

Since f' is increasing when its derivative $D_x[f'(x)] = f''(x)$ is positive and f' is decreasing when $f''(x)$ is negative, we can state the following rule:

RULE 4.   *If $f''(x) > 0$ on an interval I, then f is concave up on I. If $f''(x) < 0$ on I, then f is concave down on I.*

A function f is also said to be concave up at a *point* x_0 if there exists an interval around x_0 on which f is concave up. In fact, for the functions that we shall consider, if $f''(x_0) > 0$, then f is concave up at x_0.[†] Similarly, f is concave down at x_0 if $f''(x_0) < 0$.

EXAMPLE 1

a. *Test $y = f(x) = (x - 1)^3 + 1$ for concavity.*

We must find y''. Since $y' = 3(x - 1)^2$, then $y'' = 6(x - 1)$. Thus f is concave up when $6(x - 1) > 0$; that is, when $x > 1$. Also, f is concave down when $6(x - 1) < 0$; that is, when $x < 1$. See Fig. 9-33.

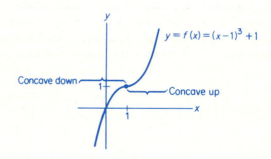

FIG. 9-33

b. *Test $y = x^2$ for concavity.*

Since $y' = 2x$, then $y'' = 2 > 0$. Thus the graph of $y = x^2$ must always be concave up, as in Fig. 9-30(a). The graph of $y = x^2$ cannot appear as in Fig. 9-30(b), for in that situation there are intervals on which the curve is concave down.

[†] This is guaranteed for functions f such that f'' is continuous.

A point on a graph, such as (1, 1) in Fig. 9-33, where concavity changes from downward to upward, or vice versa, is called a **point of inflection**. For this to occur, the sign of $f''(x)$ must go from $(-)$ to $(+)$ or from $(+)$ to $(-)$. *Possible inflection points are points where $f''(x) = 0$ or is not defined.* In fact, such points determine intervals on which $f''(x)$ should be examined when one is testing for concavity. We use the same method as that for determining when a function is increasing or decreasing.

EXAMPLE 2

Test $y = 6x^4 - 8x^3 + 1$ for concavity and points of inflection.

$$y' = 24x^3 - 24x^2.$$

$$y'' = 72x^2 - 48x = 24x(3x - 2) = 72x\left(x - \tfrac{2}{3}\right).$$

Setting $y'' = 0$ gives $x = 0, \tfrac{2}{3}$ as *possible* points of inflection. There are three intervals to consider (Fig. 9-34):

FIG. 9-34

if $x < 0$, then $y'' = 72(-)(-) = (+)$ and the curve is concave up;

if $0 < x < \tfrac{2}{3}$, then $y'' = 72(+)(-) = (-)$ and the curve is concave down;

if $x > \tfrac{2}{3}$, then $y'' = 72(+)(+) = (+)$ and the curve is concave up (see Fig. 9-35).

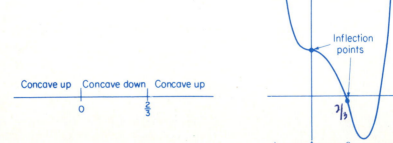

FIG. 9-35   **FIG. 9-36**

Since concavity changes when $x = 0$ and $x = \frac{2}{3}$, inflection points occur for these values of x. See Fig. 9-36. In summary, the curve is concave up on $(-\infty, 0)$ and $(\frac{2}{3}, \infty)$ and is concave down on $(0, \frac{2}{3})$. Inflection points occur when $x = 0$ or $x = \frac{2}{3}$.

PITFALL. *If $f''(x_0) = 0$, this does not prove that the graph of f has an inflection point when $x = x_0$. For if $f(x) = x^4$, then $f''(x) = 12x^2$ and $f''(0) = 0$. But $x < 0$ implies $f''(x) > 0$, and $x > 0$ implies $f''(x) > 0$. Thus concavity does not change and there are no inflection points. See Fig. 9-37.*

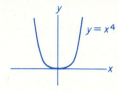

FIG. 9-37

EXAMPLE 3

Sketch the graph of $y = 2x^3 - 9x^2 + 12x$.

Intercepts.   When $x = 0$, then $y = 0$. Setting $y = 0$ gives $0 = x(2x^2 - 9x + 12)$. Clearly $x = 0$, and using the quadratic formula on $2x^2 - 9x + 12 = 0$ gives no real roots. Thus the only intercept is $(0, 0)$.

Symmetry.   None.

Asymptotes.   None.

Letting $y = f(x)$, we have

$$f'(x) = 6x^2 - 18x + 12 = 6(x^2 - 3x + 2) = 6(x - 1)(x - 2).$$

$$f''(x) = 12x - 18 = 12\left(x - \frac{3}{2}\right).$$

Maxima and Minima.   From $f'(x)$ the critical values are $x = 1, 2$. See Fig. 9-38.

FIG. 9-38

If $x < 1$, then $f'(x) = 6(-)(-) = (+)$ and f is increasing;
if $1 < x < 2$, then $f'(x) = 6(+)(-) = (-)$ and f is decreasing;
if $x > 2$, then $f'(x) = 6(+)(+) = (+)$ and f is increasing (see Fig. 9-39).

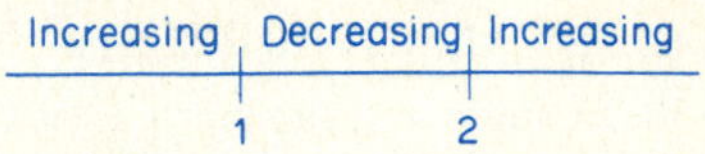

Increasing | Decreasing | Increasing

1   2

FIG. 9-39

There is a relative maximum when $x = 1$ and a relative minimum when $x = 2$.

Concavity.   Setting $f''(x) = 0$ gives a possible inflection point at $x = \frac{3}{2}$. When $x < \frac{3}{2}$, then $f''(x) < 0$ and f is concave down. When $x > \frac{3}{2}$, then $f''(x) > 0$ and f is concave up. See Fig. 9-40.

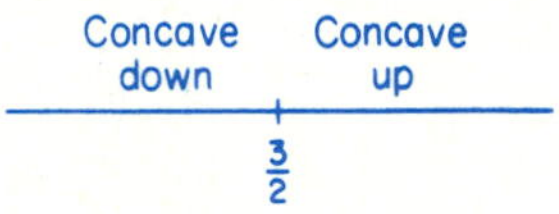

Concave down | Concave up

$\frac{3}{2}$

FIG. 9-40

Since concavity changes, there is a point of inflection when $x = \frac{3}{2}$.

Discussion.   We now find the coordinates of the important points on the graph (and

x	0	1	$\frac{3}{2}$	2
y	0	5	$\frac{9}{2}$	4

any other points if there is doubt as to the behavior of the curve). As x increases, the function is first concave down and increases to a relative maximum at $(1, 5)$; it then decreases to $(\frac{3}{2}, \frac{9}{2})$; it then becomes concave up but continues to decrease until it reaches a relative minimum at $(2, 4)$; thereafter it increases and is still concave up. See Fig. 9-41.

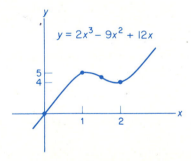

$y = 2x^3 - 9x^2 + 12x$

FIG. 9-41

EXAMPLE 4

Sketch the graph of $y = \dfrac{4x}{x^2 + 1}$.

Intercepts.   When $x = 0$, then $y = 0$; when $y = 0$, then $x = 0$. Thus, $(0, 0)$ is the only intercept.

Symmetry.   There is symmetry only with respect to the origin: replacing x by $-x$ and y by $-y$ gives

$$-y = \frac{4(-x)}{(-x)^2 + 1},$$

which is equivalent to

$$y = \frac{4x}{x^2 + 1}.$$

Asymptotes.   Testing for horizontal asymptotes, we have

$$\lim_{x \to \infty} \frac{4x}{x^2 + 1} = \lim_{x \to \infty} \frac{\dfrac{4}{x}}{1 + \dfrac{1}{x^2}} = \frac{0}{1} = 0,$$

and $\lim\limits_{x \to -\infty} \dfrac{4x}{x^2 + 1} = 0.$

The x-axis ($y = 0$) is a horizontal asymptote. There are no vertical asymptotes because the given function is never discontinuous.

Letting $y = f(x)$, we have

$$f'(x) = \frac{(x^2 + 1)(4) - 4x(2x)}{(x^2 + 1)^2} = \frac{4 - 4x^2}{(x^2 + 1)^2} = \frac{4(1 + x)(1 - x)}{(x^2 + 1)^2}.$$

$$f''(x) = \frac{(x^2 + 1)^2(-8x) - (4 - 4x^2)(2)(x^2 + 1)(2x)}{(x^2 + 1)^4}$$

$$= \frac{8x(x^2 + 1)(x^2 - 3)}{(x^2 + 1)^4} = \frac{8x(x + \sqrt{3})(x - \sqrt{3})}{(x^2 + 1)^3}.$$

Maxima and Minima.   From $f'(x)$, the critical values are $x = \pm 1$.

If $x < -1$, then $f'(x) = \dfrac{4(-)(+)}{(+)} = (-)$ and f is decreasing;

if $-1 < x < 1$, then $f'(x) = \dfrac{4(+)(+)}{(+)} = (+)$ and f is increasing;

if $x > 1$, then $f'(x) = \dfrac{4(+)(-)}{(+)} = (-)$ and f is decreasing (see Fig. 9-42).

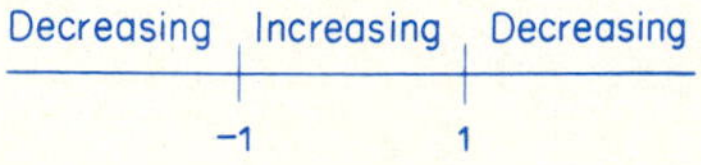

Decreasing | Increasing | Decreasing
— | — | —
-1 | | 1

FIG. 9-42

There is a relative minimum when $x = -1$ and a relative maximum when $x = 1$.

Concavity. Setting $f''(x) = 0$, we see that the possible points of inflection are when $x = \pm \sqrt{3}, 0$.

If $x < -\sqrt{3}$, then $f''(x) = \dfrac{8(-)(-)(-)}{(+)} = (-)$ and f is concave down;

if $-\sqrt{3} < x < 0$, then $f''(x) = \dfrac{8(-)(+)(-)}{(+)} = (+)$ and f is concave up;

if $0 < x < \sqrt{3}$, then $f''(x) = \dfrac{8(+)(+)(-)}{(+)} = (-)$ and f is concave down;

if $x > \sqrt{3}$, then $f''(x) = \dfrac{8(+)(+)(+)}{(+)} = (+)$ and f is concave up (see Fig. 9-43).

FIG. 9-43

Inflection points occur when $x = 0, \pm \sqrt{3}$.

Discussion. After consideration of all of the above information, the graph of $y = 4x/(x^2 + 1)$ is given in Fig. 9-44 together with a table of coordinates of important points.

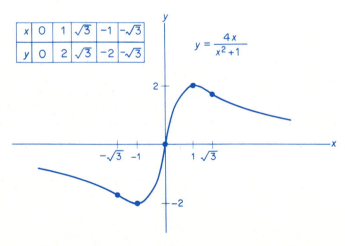

FIG. 9-44

EXAMPLE 5

Sketch the graph of $y = \dfrac{1}{4 - x^2}$.

Intercepts. $(0, \frac{1}{4})$.

Symmetry.   Symmetric to the y-axis.

Asymptotes.   As $x \to \infty$, then $y \to 0$; as $x \to -\infty$, then $y \to 0$. Thus $y = 0$ (the x-axis) is a horizontal asymptote. Since $1/(4 - x^2)$ blows up near $x = \pm 2$, the vertical asymptotes are the lines $x = 2$ and $x = -2$.

Maxima and Minima.   Since $y = (4 - x^2)^{-1}$,

$$y' = -1(4 - x^2)^{-2}(-2x) = \frac{2x}{(4 - x^2)^2}.$$

The critical values are $x = 0, \pm 2$. If $x < -2$, then $y' < 0$; if $-2 < x < 0$, then $y' < 0$; if $0 < x < 2$, then $y' > 0$; if $x > 2$, then $y' > 0$. The function is decreasing on $(-\infty, -2)$ and $(-2, 0)$ and increasing on $(0, 2)$ and $(2, \infty)$. See Fig. 9-45. There is a relative minimum when $x = 0$.

FIG. 9-45

Concavity.

$$y'' = \frac{(4 - x^2)^2(2) - (2x)2(4 - x^2)(-2x)}{(4 - x^2)^4} = \frac{2(4 - x^2)(4 + 3x^2)}{(4 - x^2)^4} = \frac{8 + 6x^2}{(4 - x^2)^3}.$$

Setting $y'' = 0$, we get no real roots. However, y'' is undefined when $x = \pm 2$. Thus concavity may change around these values. If $x < -2$, then $y'' < 0$; if $-2 < x < 2$, then $y'' > 0$; if $x > 2$, then $y'' < 0$. The graph is concave up on $(-2, 2)$ and concave down on $(-\infty, -2)$ and $(2, \infty)$. See Fig. 9-46. Although concavity changes around $x = \pm 2$, these values of x do not give points of inflection since y *itself* is not defined at $x = \pm 2$.

FIG. 9-46

Discussion.   Plotting the points in the table in Fig. 9-47, some arbitrarily chosen, and using the above information, we get the indicated graph. Due to symmetry our table has only $x > 0$.

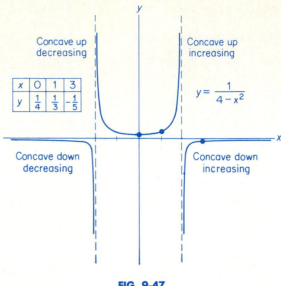

FIG. 9-47

EXERCISE 9-4

In Problems **1–14**, determine concavity and where points of inflection occur. Do not sketch the graphs.

1. $y = -2x^2 + 4x$.

2. $y = 3x^2 - 6x + 5$.

3. $y = 4x^3 + 12x^2 - 12x$.

4. $y = x^3 - 6x^2 + 9x + 1$.

5. $y = x^4 - 6x^2 + 5x - 6$.

6. $y = -\dfrac{x^4}{4} + \dfrac{9x^2}{2} + 2x$.

7. $y = \dfrac{x + 1}{x - 1}$.

8. $y = x + \dfrac{1}{x}$.

9. $y = \dfrac{x^2}{x^2 + 1}$.

10. $y = \dfrac{x^2}{x + 3}$.

11. $y = e^x$.

12. $y = e^x - e^{-x}$.

13. $y = xe^x$.

14. $y = xe^{-x}$.

In Problems **15–44** sketch each curve. Determine: intervals on which the function is increasing, decreasing, concave up, concave down; relative maxima and minima; inflection points; symmetry; horizontal and vertical asymptotes; those intercepts which can be obtained conveniently.

15. $y = x^2 + 4x + 3$.

16. $y = x^2 + 2$.

17. $y = 4x - x^2$.

18. $y = x - x^2 + 2$.

19. $y = 2x^2 - 5x - 12$.

20. $y = x^2 - 6x - 7$.

21. $y = x^3 - 9x^2 + 24x - 19.$

22. $y = 3x - x^3.$

23. $y = \dfrac{x^3}{3} - 3x.$

24. $y = x^3 - 6x^2 + 9x.$

25. $y = x^3 - 3x^2 + 3x - 3.$

26. $y = 2x^3 - 9x^2 + 12x.$

27. $y = 4x^2 - x^4.$

28. $y = -\dfrac{x^3}{3} - 2x^2 + 5x - 2.$

29. $y = 4x^3 - 3x^4.$

30. $y = x^4 - 2x^2.$

31. $y = -2 + 12x - x^3.$

32. $y = (3 + 2x)^3.$

33. $y = x^3 - 6x^2 + 12x - 6.$

34. $y = 3x^5 - 5x^3.$

35. $y = 5x - x^5.$

36. $y = \dfrac{x^5}{100} - \dfrac{x^4}{20}.$

37. $y = 3x^4 - 4x^3 + 1.$

38. $y = x(1 - x)^3.$

39. $y = \dfrac{3}{x}.$

40. $y = \dfrac{1}{x - 1}.$

41. $y = \dfrac{x}{x + 1}.$

42. $y = \dfrac{10}{\sqrt{x}}.$

43. $y = x^2 + \dfrac{1}{x^2}.$

44. $y = \dfrac{x^2}{1 - x}.$

45. Show that the graph of the demand equation $p = 100/(q + 2)$ is decreasing and concave up for $q > 0.$

46. For the cost function $c = 3q^2 + 5q + 6$, show that the graph of the average cost function $\bar{c}$ is always concave up for $q > 0.$

9-5 THE SECOND-DERIVATIVE TEST

The second derivative may be used to test certain critical values for relative extrema. Observe in Fig. 9-48 that when $x = x_0$ there is a horizontal tangent; that is, $f'(x_0) = 0$. This suggests a relative maximum or minimum. However, we see also that the curve is bending upward there (that is, $f''(x_0) > 0$). This leads

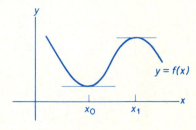

FIG. 9-48

us to conclude that there is a relative minimum at x_0. On the other hand, $f'(x_1) = 0$ but the curve is bending downward at x_1 (that is, $f''(x_1) < 0$). From this we conclude that a relative maximum exists there. This technique of examining the second derivative at points where $f'(x) = 0$ is called the *second-derivative test* for relative maxima and minima.

SECOND-DERIVATIVE TEST.

Suppose $f'(x_0) = 0$.

If $f''(x_0) < 0$, then f has a relative maximum at x_0;

if $f''(x_0) > 0$, then f has a relative minimum at x_0;

if $f''(x_0) = 0$, the test gives no information, that is, at x_0 there may be a relative maximum, a relative minimum, or neither (use the first-derivative test).

PITFALL. *In the case where $f'(x_0) = f''(x_0) = 0$, some students conclude that there is no relative maximum or minimum at x_0. This conclusion may be **false** as Example 1(c) will show.*

EXAMPLE 1

Use the second-derivative test to examine the following for relative maxima and minima.

a. $y = 18x - \frac{2}{3}x^3$.

$$y' = 18 - 2x^2 = 2(9 - x^2) = 2(3 + x)(3 - x).$$
$$y'' = -4x.$$

Solving $y' = 0$ gives the critical values $x = \pm 3$. If $x = 3$, then $y'' = -4(3) = -12 < 0$, so there is a relative maximum when $x = 3$. If $x = -3$, then $y'' > 0$, so there is a relative minimum when $x = -3$. See Fig. 9-17.

b. $y = 6x^4 - 8x^3 + 1$.

$$y' = 24x^3 - 24x^2 = 24x^2(x - 1).$$
$$y'' = 72x^2 - 48x.$$

Solving $y' = 0$ gives the critical values $x = 0, 1$. Since $y'' > 0$ if $x = 1$, there is a relative minimum when $x = 1$. For $x = 0$, then $y'' = 0$ and the second-derivative test gives no information. We now turn to the first-derivative test. If $x < 0$, then $y' < 0$; if $0 < x < 1$, then $y' < 0$. Thus no relative maximum or minimum exists when $x = 0$. See Fig. 9-36.

c. $y = x^4$.

$$y' = 4x^3.$$
$$y'' = 12x^2.$$

Solving $y' = 0$ gives the critical value $x = 0$. But if $x = 0$, then $y'' = 0$ and the second-derivative test fails. Since $y' < 0$ for $x < 0$, and $y' > 0$ for $x > 0$, by the first-derivative test there is a relative minimum when $x = 0$. See Fig. 9-37.

If a continuous function has *exactly one* relative extremum on an interval, it can be shown that the relative extremum must also be an *absolute* extremum on the interval. To illustrate, in Example 1(c), $y = x^4$ has a relative minimum when $x = 0$ and there are no other relative extrema. Since $y = x^4$ is continuous, this relative minimum is also an absolute minimum for the function.

EXERCISE 9-5

In Problems **1–10**, *test for relative maxima and minima by using the second-derivative test. In Problems* **1–4**, *state whether the relative extrema are also absolute extrema.*

1. $y = x^2 - 5x + 6$.

2. $y = -2x^2 + 6x + 12$.

3. $y = -4x^2 + 2x - 8$.

4. $y = 3x^2 - 5x + 6$.

5. $y = x^3 - 27x + 1$.

6. $y = x^3 - 12x + 1$.

7. $y = -x^3 + 3x^2 + 1$.

8. $y = x^4 - 2x^2 + 4$.

9. $y = 2x^4 + 2$.

10. $y = -x^7$.

9-6 APPLIED MAXIMA AND MINIMA

By using techniques of the previous sections, we can examine situations that require determining the value of a variable that will maximize or minimize a function. For example, we might want to maximize profit or minimize cost. The crucial part is setting up the function to be investigated. Then we find its derivative and test the resulting critical values. For this the first-derivative test or the second-derivative test may be used, although it is often obvious from the nature of the problem whether or not a critical value represents an appropriate answer. For the problems that we shall consider, our interest is in *absolute* maxima and minima. In some cases they occur at endpoints.

Read each of the examples carefully so that you may gain insight into setting up the function to be analyzed.

EXAMPLE 1

For the total cost function $c = \dfrac{q^2}{4} + 3q + 400$, where q is the number of units produced, at what level of output will average cost per unit be a minimum? What is this minimum?

The quantity to be minimized is average cost $\bar{c}$. The average cost function is given by

$$\bar{c} = \frac{c}{q} = \frac{\frac{q^2}{4} + 3q + 400}{q} = \frac{q}{4} + 3 + \frac{400}{q}. \qquad (1)$$

$$D_q\bar{c} = \frac{1}{4} - \frac{400}{q^2} = \frac{q^2 - 1600}{4q^2}.$$

Setting $D_q\bar{c} = 0$, we get

$$q^2 - 1600 = 0,$$

$$(q - 40)(q + 40) = 0,$$

$$q = 40 \quad \text{(since we assume } q > 0\text{)}.$$

To determine if this level of output gives a relative minimum, we shall use the second-derivative test.

$$D_q^2\bar{c} = \frac{800}{q^3},$$

which is positive for $q = 40$. Thus $\bar{c}$ has a relative minimum when $q = 40$. We note that $\bar{c}$ is continuous for $q > 0$. Since $q = 40$ is the only relative extrema for $q > 0$, we conclude that this relative minimum is indeed an absolute minimum. Substituting $q = 40$ in (1) gives the minimum average cost $\bar{c} = 23$.

EXAMPLE 2

The demand equation for a manufacturer's product is $p = (80 - q)/4$ where q is the number of units and p is price per unit. At what value of q will there be maximum revenue?

Let r be total revenue. Then revenue = (price) (quantity). Thus,

$$\text{Revenue Formula} \quad r = pq = \frac{80 - q}{4} \cdot q = \frac{80q - q^2}{4}.$$

Setting $dr/dq = 0$:

$$\frac{dr}{dq} = \frac{80 - 2q}{4} = 0,$$

$$q = 40.$$

Since $d^2r/dq^2 = -\frac{1}{2}$, the second derivative is always negative. Thus $q = 40$ gives a relative maximum. Also, since r is a continuous function of q, we conclude that $q = 40$ gives the absolute maximum revenue. This revenue is $[80(40) - (40)^2]/4 = 400$.

Calculus can be applied to inventory decisions, as the following example shows.

EXAMPLE 3

A manufacturer annually produces and sells 10,000 *units of a product. Sales are uniformly distributed throughout the year. He wishes to determine the number of units to be manufactured in each production run in order to minimize annual set-up costs and carrying costs. The size of such production runs is referred to as the* **economic lot size** *or* **economic order quantity**. *The production cost of each unit is* $20 *and carrying costs (insurance, interest, storage, etc.) are* 10 *percent of the value of the average inventory. Set-up costs per production run are* $40. *Find the economic lot size.*

Let q be the number of units in a production run. Since sales are distributed at a uniform rate, we shall assume that inventory varies uniformly from q to 0 between production runs. Thus we take the average inventory to be $q/2$ units. The production costs are $20 per unit, and so the value of the average inventory is $20(q/2)$. Carrying costs are 10 percent of this value:

$$.10(20)\left(\frac{q}{2}\right).$$

The number of production runs per year is $10{,}000/q$. Thus the total set-up costs are

$$40\left(\frac{10{,}000}{q}\right).$$

Hence the total annual carrying costs and set-up costs C are

$$C = .10(20)\left(\frac{q}{2}\right) + 40\left(\frac{10{,}000}{q}\right)$$

$$= q + \frac{400{,}000}{q}.$$

$$\frac{dC}{dq} = 1 - \frac{400{,}000}{q^2} = \frac{q^2 - 400{,}000}{q^2}.$$

Setting $dC/dq = 0$, we get

$$q^2 = 400{,}000.$$

Since $q > 0$, we choose

$$q = \sqrt{400{,}000} = 200\sqrt{10} \approx 632.5.$$

To determine if this value of q gives minimum cost, we shall examine the first derivative. If $0 < q < \sqrt{400{,}000}$, then $dC/dq < 0$. If $q > \sqrt{400{,}000}$, then $dC/dq > 0$. Since C always decreases between $q = 0$ and $q = \sqrt{400{,}000}$, and always increases to the right of $q = \sqrt{400{,}000}$, we conclude that there is an *absolute* minimum at $q = 632.5$. The number of production runs is $10{,}000/632.5 \approx 15.8$. For practical purposes, there would be 16 lots, each having an economic lot size of 625 units.

EXAMPLE 4

The Vista TV Cable Co. currently has 2000 subscribers who are paying a monthly rate of $5. A survey reveals that there will be 50 more subscribers for each $.10 decrease in the rate. At what rate will maximum revenue be obtained and how many subscribers will there be at this rate?

Let x be the rate. Then the total decrease in the rate is $5 - x$ and the number of $.10 decreases is $\dfrac{5 - x}{.10}$. For *each* of these decreases there will be 50 more subscribers. Thus the total of *new* subscribers is $50\left(\dfrac{5 - x}{.10}\right)$ and the total of all subscribers is

$$2000 + 50\left(\frac{5 - x}{.10}\right). \tag{2}$$

The revenue r is given by $r =$ (rate) (number of subscribers):

$$r = x\left[2000 + 50\left(\frac{5 - x}{.10}\right)\right]$$

$$= 4500x - 500x^2.$$

Setting $r' = 0$, we have

$$r' = 4500 - 1000x = 0.$$

$$x = 4.50.$$

Since $r'' = -1000 < 0$, we have a relative maximum when $x = 4.50$. Since r is a continuous function of x, we conclude that when $x = \$4.50$ there is an absolute maximum. Substituting $x = 4.50$ in (2) gives 2250 subscribers.

EXAMPLE 5

A health economist determined that if a particular health-care program for the elderly were initiated, then t years after its start, n thousand elderly people would receive direct benefits, where

$$n = \frac{t^3}{3} - 6t^2 + 32t, \qquad 0 \le t \le 12.$$

For what value of t does the maximum number receive benefits?

Setting $dn/dt = 0$, we have

$$\frac{dn}{dt} = t^2 - 12t + 32 = 0,$$

$$(t - 4)(t - 8) = 0,$$

$$t = 4 \quad \text{or} \quad t = 8.$$

Now, $d^2n/dt^2 = 2t - 12$, which is negative for $t = 4$ and positive for $t = 8$. Thus there is a relative maximum when $t = 4$. This gives $n = 53\frac{1}{3}$. To determine whether this is an absolute maximum, we must find n at the endpoints of the domain. If $t = 0$, then $n = 0$. If $t = 12$, then $n = 96$. Thus an absolute maximum occurs when $t = 12$. A graph of the function is given in Fig. 9-49.

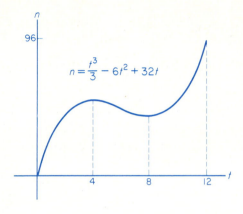

FIG. 9-49

Suppose $p = f(q)$ is the demand function for a firm's product, where p is price per unit and q is the number of units produced and sold. Then the total revenue $r = qp = q f(q)$ is a function of q. Let the total cost c of producing q units be given by the cost function $c = g(q)$. Thus, the total profit P, which is total revenue − total cost, is also a function of q:

$$P = r - c = q f(q) - g(q).$$

Let us consider the most profitable output for the firm. Ignoring special cases, we know that profit is maximized when $dP/dq = 0$ and $d^2P/dq^2 < 0$.

$$\frac{dP}{dq} = \frac{d}{dq}(r - c) = \frac{dr}{dq} - \frac{dc}{dq} = 0.$$

Thus,

$$\frac{dr}{dq} = \frac{dc}{dq}.$$

That is, at the level of maximum profit the slope of the tangent to the total revenue curve must equal the slope of the tangent to the total cost curve (Fig. 9-50). But dr/dq is marginal revenue MR, and dc/dq is marginal cost MC.

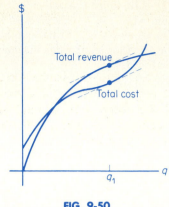

FIG. 9-50

Thus, under typical conditions, *to maximize profit it is necessary that*

$$MR = MC.$$

For this to indeed correspond to a maximum, it is necessary that $d^2P/dq^2 < 0$.

$$\frac{d^2P}{dq^2} = \frac{d^2r}{dq^2} - \frac{d^2c}{dq^2} < 0, \quad \text{or} \quad \frac{d^2r}{dq^2} < \frac{d^2c}{dq^2}.$$

That is, when $MR = MC$, in order to insure maximum profit the slope of the marginal revenue curve must be less than the slope of the marginal cost curve.

The condition that $d^2P/dq^2 < 0$ when $dP/dq = 0$ can be viewed another way. Equivalently, to have $MR = MC$ correspond to a maximum it is necessary that dP/dq go from $(+)$ to $(-)$; that is, from $dr/dq - dc/dq > 0$ to $dr/dq - dc/dq < 0$. Hence, as output increases, we must have $MR > MC$ and then $MR < MC$. This means that at the point q_1 of maximum profit *the marginal cost curve must cut the marginal revenue curve from below* (Fig. 9-51). For production

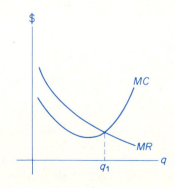

FIG. 9-51

up to q_1, the revenue from additional output would be greater than the cost of such output and total profit would increase. For output beyond q_1, $MC > MR$ and each unit of output would add more to total costs than to total revenue. Hence, total profits would decline.

In the next example we use the word *monopolist*. Under a situation of monopoly, there is only one seller of a product for which there are no similar substitutes, and the seller, that is, the monopolist, controls the market. By considering the demand equation for the product, the monopolist may set the price (or volume of output) so that maximum profit will be obtained.

EXAMPLE 6

Suppose that the demand equation for a monopolist's product is $p = 400 - 2q$ and the average cost function is $\bar{c} = .2q + 4 + (400/q)$, where q is number of units, and both p and $\bar{c}$ are expressed in dollars.

a. *Determine the level of output at which profit is maximized.*

b. *Determine the price at which maximum profit occurs.*

c. *Determine the maximum profit.*

d. *If, as a regulatory device, the government imposes a tax of $22 per unit on the monopolist, what is the new price for profit maximization?*

Since revenue $r = pq = 400q - 2q^2$ and total cost $c = q\bar{c} = .2q^2 + 4q + 400$, profit P is

$$P = r - c = 400q - 2q^2 - (.2q^2 + 4q + 400).$$
$$P = 396q - 2.2q^2 - 400. \tag{3}$$

a. Setting $dP/dq = 0$, we have

$$\frac{dP}{dq} = 396 - 4.4q = 0.$$

$$q = 90.$$

Since $d^2P/dq^2 = -4.4 < 0$, we conclude that $q = 90$ gives a maximum.

b. From the demand equation, $p = 400 - 2(90) = 220$.

c. Substituting $q = 90$ in (3) gives $P = 17,420$.

d. The tax of $22 per unit means that for q units the total cost increases by $22q$. The new cost function is $c_1 = .2q^2 + 4q + 400 + 22q$, and the profit P_1 is given by

$$P_1 = 400q - 2q^2 - (.2q^2 + 4q + 400 + 22q).$$
$$P_1 = 374q - 2.2q^2 - 400.$$

Setting $dP_1/dq = 0$ gives

$$\frac{dP_1}{dq} = 374 - 4.4q = 0.$$

$$q = 85.$$

Thus to maximize profit, the monopolist restricts output to 85 units at a higher price of $p_1 = 400 - 2(85) = 230$. Since this is only $10 more than before, only part of the tax has been shifted to the consumer, and the monopolist must bear the cost of the balance. The profit now is $15,495, which is lower than the former profit.

EXAMPLE 7

For insurance purposes a manufacturer plans to fence in a 10,800-sq-ft rectangular storage area adjacent to a building by using the building as one side of the enclosed area (see Fig. 9-52). The fencing parallel to the building faces a highway and will cost $3 per ft installed, while the fencing for the other two sides costs $2 per ft installed. Find the amount of each type of fence so that the total cost of the fence will be a minimum. What is the minimum cost?

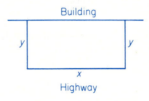

FIG. 9-52

In Fig. 9-52 we have labeled the length of the side parallel to the building as x and the lengths of the other two sides as y, where x and y are in feet. The cost (in dollars) of the fencing along the highway is $3x$, and along each of the other sides it is $2y$. Thus the total cost C of the fencing is

$$C = 3x + 2y + 2y = 3x + 4y.$$

We wish to minimize C. In order to differentiate, we first express C in terms of one variable only. To do this we find a relationship between x and y. Since the storage area xy must be 10,800,

$$xy = 10,800$$

$$\text{or} \quad y = \frac{10,800}{x}.$$

By substitution we have

$$C = 3x + 4\left(\frac{10,800}{x}\right) = 3x + \frac{43,200}{x}.$$

To minimize C we set $dC/dx = 0$ and solve for x:

$$\frac{dC}{dx} = 3 - \frac{43{,}200}{x^2} = 0$$

from which

$$x^2 = \frac{43{,}200}{3} = 14{,}400.$$

$$x = 120 \quad (\text{since } x > 0).$$

Now, $d^2C/dx^2 = 86{,}400/x^3 > 0$ for $x = 120$, and we conclude that $x = 120$ indeed gives the minimum value of C. When $x = 120$, then $y = 10{,}800/120 = 90$. Thus 120 ft of the \$3 fencing and 180 ft of the \$2 fencing are needed. This gives a cost of \$720.

EXERCISE 9-6 16

In each of the following, p is price per unit (in dollars) and q is output per unit of time.

1. A manufacturer finds that the total cost c of producing his product is given by the cost function $c = .05q^2 + 5q + 500$. At what level of output will average cost per unit be at a minimum?

2. The cost per hour C (in dollars) of operating an automobile is given by

$$C = .12s - .0012s^2 + .08, \qquad 0 < s < 60,$$

   where s is the speed in miles per hour. At what speed is the cost per hour a minimum?

3. The demand equation for a monopolist's product is $p = -5q + 30$. At what price will revenue be maximized?

4. For a monopolist's product, the demand function is $q = 10{,}000e^{-.02p}$. Find the value of p for which maximum revenue is obtained.

5. For a monopolist's product, the demand equation is $p = 72 - .04q$ and the cost function is $c = 500 + 30q$. At what level of output will profit be maximized? At what price does this occur and what is the profit?

6. For a monopolist, the cost per unit of producing a product is \$3 and the demand equation is $p = 10/\sqrt{q}$. What price will give the greatest profit?

7. For a monopolist, the demand equation is $p = 42 - 4q$ and the average cost function is $\bar{c} = 2 + (80/q)$. Find the profit-maximizing price.

8. For a monopolist's product, the demand function is $p = 50/\sqrt{q}$ and the average cost function is $\bar{c} = .50 + (1000/q)$. Find the profit-maximizing price and output. At this level, show marginal revenue is equal to marginal cost.

9. For XYZ Manufacturing Co., total fixed costs are \$1200, material and labor costs combined are \$2 per unit, and the demand equation is $p = 100/\sqrt{q}$. What level of

output will maximize profit? Show that this occurs when marginal revenue is equal to marginal cost. What is the price at profit maximization?

10. A real estate firm owns 70 garden-type apartments. At $125 per month each apartment can be rented. However, for each $5 per month increase, there will be two vacancies with no possibility of filling them. What rent per apartment will maximize monthly revenue?

11. A manufacturer finds that for the first 500 units of his product that are produced and sold, the profit is $50 per unit. The profit on each of the units beyond 500 is decreased by $.10 times the number of additional units produced. For example, the total profit when 502 units are produced and sold is 500(50) + 2(49.80). What level of output will maximize profit?

12. A TV cable company has 1000 subscribers who are paying $5 per month. It can get 100 more subscribers for each $.10 decrease in the monthly fee. What rate will yield maximum revenue and what will this revenue be?

13. Find two numbers whose sum is 40 and whose product is a maximum.

14. Find two nonnegative numbers whose sum is 20 and such that the product of twice one number and the square of the other number will be a maximum.

15. A company has set aside $3000 to fence in a rectangular portion of land adjacent to a stream by using the stream for one side of the enclosed area. The cost of the fencing parallel to the stream is $5 per foot installed, and the fencing for the remaining two sides is $3 per foot installed. Find the dimensions of the maximum enclosed area.

16. The owner of the Laurel Nursery Garden Center wants to fence in 1000 square feet of land in a rectangular plot to be used for different types of shrubs. The plot is to be divided into four equal plots with three fences parallel to the same pair of sides as shown in Fig. 9-53. What is the least number of feet of fence needed?

FIG. 9-53

17. A container manufacturer is designing a rectangular box, open at the top and with a square base, that is to have a volume of 32 cu ft. If the box is to require the least amount of material, what must be the dimensions of the box? See Fig. 9-54.

FIG. 9-54

18. An open-top box with a square base is to be constructed from 192 sq ft of material. What should be the dimensions of the box if the volume is to be a maximum? What is the maximum volume? See Fig 9-54.

19. A rectangular cardboard poster is to have 150 sq in. for printed matter. It is to have a 3-in. margin at the top and bottom and a 2-in. margin on each side. Find the dimensions of the poster so that the amount of cardboard used is minimized. See Fig. 9-55.

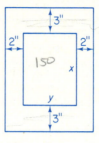

FIG. 9-55

20. An open box is to be made by cutting equal squares from each corner of a 12-in. square piece of cardboard and then folding up the sides. Find the length of the side of the square that must be cut out if the volume of the box is to be maximized. What is the maximum volume? See Fig. 9-56.

FIG. 9-56

21. A cylindrical can, open at the top, is to have a fixed volume of K. Show that if the least amount of material is to be used, then both the radius and height are equal to $\sqrt[3]{K/\pi}$. See Fig. 9-57.

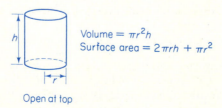

Volume $= \pi r^2 h$
Surface area $= 2\pi r h + \pi r^2$

Open at top

FIG. 9-57

22. A cylindrical can, open at the top, is to be made from a fixed amount of material, K. If the volume is to be a maximum, show that both the radius and height are equal to $\sqrt{K/(3\pi)}$. See Fig. 9-57.

23. The demand equation for a monopolist's product is $p = 600 - 2q$ and the total cost function is $c = .2q^2 + 28q + 200$. Find the profit-maximizing output and price, and determine the corresponding profits. If the government were to impose a tax of $22 per unit on the manufacturer, what would be the new profit-maximizing output and price? What is the profit now?

24. Use the *original* data in Problem 23 and assume that the government imposes a license fee of $100 on the manufacturer. This is a lump-sum amount without regard to output. Show that marginal revenue and marginal cost do not change and, hence, the profit maximizing price and output remain the same. Show, however, that there will be less profit.

25. A manufacturer has to produce annually 1000 units of a product that is sold at a uniform rate during the year. The production cost of each unit is $10 and carrying costs (insurance, interest, storage, etc.) are estimated to be 12.8 percent of the value of average inventory. Set-up costs per production run are $40. Find the economic lot size.

26. For a monopolist's product, the cost function is $c = .004q^3 + 20q + 5000$ and the demand function is $p = 450 - 4q$. Find the profit-maximizing output. At this level, show that marginal cost = marginal revenue.

27. Imperial Educational Services (I.E.S.) is considering offering a workshop in resource allocation to key personnel at Acme Corp. To make the offering economically feasible, I.E.S. feels that at least thirty persons must attend at a cost of $50 each. Moreover, I.E.S. will agree to reduce the charge for *everybody* by $1.25 for each person over the thirty who attends. How many people should be in the group for I.E.S. to maximize revenue? Assume that the maximum allowable number in the group is forty.

28. The Kiddie Toy Company plans to lease an electric motor to be used 90,000 horsepower-hours per year in manufacturing. One horsepower-hour is the work done in one hour by a one-horsepower motor. The annual cost to lease a suitable motor is $150 plus $.60 per horsepower. The cost per horsepower-hour of operating the motor is $.006/N$ where N is the horsepower. What size motor, in horsepower, should be leased in order to minimize cost?

29. For a manufacturer, the cost of making a part is $3 per unit for labor and $1 per unit for materials; overhead is fixed at $2000 per week. If more than 5000 units are made each week, labor is $4.50 per unit for those units in excess of 5000. At what level of production will average cost per unit be at a minimum?

30. The cost of operating a truck on a throughway (excluding the salary of the driver) is $.11 + (s/600)$ dollars per mile, where s is the (steady) speed of the truck in miles per hour. The truck driver's salary is $6 per hour. At what speed should the truck driver operate the truck to make a 700-mile trip most economical?

31. A company produces daily x tons of chemical A ($x \leqslant 4$) and y tons of chemical B where $y = (24 - 6x)/(5 - x)$. The profit on chemical A is \$2000 per ton and on B it is \$1000 per ton. How much of chemical A should be produced per day to maximize profit? Answer the same question if the profit on A is P per ton and that on B is $P/2$ per ton.

32. To erect an office building, fixed costs are \$250,000 and include land, architect's fee, basement, foundation, etc. If x floors are to be constructed, the cost (excluding fixed costs) is $c = (x/2)[100{,}000 + 5000(x - 1)]$. The revenue per month is \$5000 per floor. Find the number of floors that will yield a maximum rate of return on investment (rate of return = total revenue/total cost).

9-7 DIFFERENTIALS

We shall now give a reason for using the symbol dy/dx to denote the derivative of y with respect to x. To do this we introduce the notion of the *differential* of a function.

> **DEFINITION.** *Suppose that* $y = f(x)$ *is a differentiable function of* x. *Then the **differential of** y, denoted* dy *or* $d[f(x)]$, *is*
>
> $$dy = f'(x) \cdot h,$$
>
> *where the variable* h *can be any real number. Note that* dy *is a function of two variables,* x *and* h.

EXAMPLE 1

Find the differential of $y = x^3 - 2x^2 + 3x - 4$ *and evaluate it when* $x = 1$ *and* $h = .04$.

$$dy = D_x(x^3 - 2x^2 + 3x - 4) \cdot h,$$

$$dy = (3x^2 - 4x + 3)h.$$

If $x = 1$ and $h = .04$, then

$$dy = \left[3(1)^2 - 4(1) + 3\right](.04) = .08.$$

If $f(x) = x$, then $d[f(x)] = d[x] = D_x[x]h = 1h = h$. Hence the differential of x is h. We write $d[x]$ as dx. Thus $dx = h$. From now on it will be our practice to write dx for h. For example,

$$d(x^2 + 5) = D_x(x^2 + 5) \cdot h = D_x(x^2 + 5) \cdot dx = 2x\, dx.$$

Summarizing, if $y = f(x)$ defines a differentiable function of x, we have

$$\boxed{dy = f'(x)\, dx,}$$

where dx is any real number.

EXAMPLE 2

a. If $f(x) = \sqrt{x}$, then

$$d(\sqrt{x}) = D_x(\sqrt{x}) \cdot dx$$

$$= \frac{1}{2} x^{-1/2}\, dx = \frac{1}{2\sqrt{x}}\, dx.$$

b. If $u = (x^2 + 3)^5$, then

$$du = 5(x^2 + 3)^4(2x)\, dx = 10x(x^2 + 3)^4\, dx.$$

c. If $u = x$, then $du = 1\, dx = dx$.

If $y = f(x)$, then $dy = f'(x)\, dx$. Provided that $dx \neq 0$, we can divide both sides by dx:

$$\frac{dy}{dx} = f'(x).$$

That is, dy/dx can be viewed either as the quotient of two differentials, namely dy divided by dx, or as the derivative of f at x. It is for this reason that we introduced the symbol dy/dx to denote the derivative.

The differential can be interpreted geometrically. In Fig. 9-58 the point $P(x, f(x))$ is on the curve $y = f(x)$. Suppose x changes by dx, a real number, to $x + dx$. Then the new functional value is $f(x + dx)$ and the corresponding point on the curve is $Q(x + dx, f(x + dx))$. Passing through P and Q are horizontal and vertical lines, respectively, that intersect at S. A line L tangent to the curve at P intersects QS at R, forming the right triangle PRS. The slope of L is $f'(x)$ or, equivalently, it is $\overline{SR}/\overline{PS}$:

$$f'(x) = \frac{\overline{SR}}{\overline{PS}}.$$

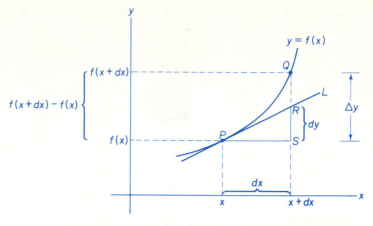

FIG. 9-58

Since $dy = f'(x)\, dx$, and $dx = \overline{PS}$,

$$dy = f'(x)\, dx$$

$$= \frac{\overline{SR}}{\overline{PS}} \cdot \overline{PS}$$

$$= \overline{SR}.$$

Thus, if dx is a change in x at P, dy is the corresponding vertical change along the **tangent line** at P. Note that for the same dx, the vertical change along the **curve** is $\Delta y = \overline{SQ} = f(x + dx) - f(x)$. Do not confuse Δy with dy. However, it is apparent from Fig. 9-58 that

when dx is close to 0, dy is an approximation to Δy.

Symbolically we write $\Delta y \approx dy$ to indicate that Δy is approximately equal to dy. Observe that the graph of f near P is approximated by the tangent line at P.

EXAMPLE 3

Suppose the total profit P (in dollars) of producing q units of a product is

$$P = P(q) = \frac{q^3}{1000} - \frac{1}{4}q^2 + 30q.$$

Use differentials to find the approximate change in profit if the level of production changes from $q = 200$ to $q = 205$.

We want to approximate the change in P, that is, approximate ΔP, when q goes from 200 to 205. We approximate ΔP by dP when $q = 200$ and $dq = 5$.

$$\Delta P \approx dP = P'(q)\, dq = \left(\frac{3q^2}{1000} - \frac{q}{2} + 30 \right) dq$$

$$= \left[\frac{3(200)^2}{1000} - \frac{200}{2} + 30 \right] 5$$

$$= 250.$$

The actual change ΔP is $P(205) - P(200) = 258.875$.

We said that if $y = f(x)$, then $\Delta y \approx dy$ if dx is small. Thus

$$\Delta y = f(x + dx) - f(x) \approx dy$$

or

$$\boxed{f(x + dx) \approx f(x) + dy.}$$ (1)

This gives us a way of approximating a functional value. For example, suppose we approximate ln (1.06). If we let $y = f(x) = \ln x$, then we want to approximate $f(1.06)$. By (1) and the fact that $d(\ln x) = (1/x)\, dx$, we have

$$f(x + dx) \approx f(x) + dy,$$

$$\ln (x + dx) \approx \ln x + \frac{1}{x}\, dx.$$

Note that $1.06 = 1 + .06$. Since .06 is small and we know the exact value of ln $1 = 0$, we shall let $x = 1$ and $dx = .06$:

$$\ln (1.06) = \ln (1 + .06) \approx \ln (1) + \frac{1}{1}(.06) = .06.$$

The actual value of ln (1.06) to five decimal places is .05827.

EXAMPLE 4

The demand function for a product is given by $p = f(q) = 20 - \sqrt{q}$, where p is the price per unit in dollars for q units. By using differentials, approximate the price when 99 units are demanded.

We want to approximate $f(99)$. By (1),

$$f(q + dq) \approx f(q) + dp,$$

where
$$dp = -\frac{1}{2\sqrt{q}}\, dq.$$

We choose $q = 100$ because 100 is near 99 and it is then easy to compute $f(100) = 20 - \sqrt{100} = 10$. Choosing $dq = -1$ (note the negative sign), we have

$$f(99) = f[100 + (-1)] \approx f(100) - \frac{1}{2\sqrt{100}}(-1),$$

$$f(99) \approx 10 + .05 = 10.05.$$

Thus the price per unit when 99 units are demanded is approximately $10.05.

The equation $y = x^3 + 4x + 5$ defines y as a function of x. However, it also defines x implicitly as a function of y. Thus we can look at the derivative of x with respect to y, dx/dy. Since dx/dy can be considered a quotient of differentials, we are motivated to write (and it is indeed true) that

$$\boxed{\frac{dx}{dy} = \frac{1}{\dfrac{dy}{dx}}, \quad dy/dx \neq 0.}$$

But dy/dx is the derivative of y with respect to x and equals $3x^2 + 4$. Thus,

$$\frac{dx}{dy} = \frac{1}{3x^2 + 4}.$$

This is the *reciprocal* of dy/dx.

EXAMPLE 5

Find dp/dq if $q = \sqrt{2500 - p^2}$.

Since $q = (2500 - p^2)^{1/2}$, then

$$\frac{dq}{dp} = \frac{1}{2}(2500 - p^2)^{-1/2}(-2p) = -\frac{p}{\sqrt{2500 - p^2}}.$$

Hence,
$$\frac{dp}{dq} = \frac{1}{\dfrac{dq}{dp}} = -\frac{\sqrt{2500 - p^2}}{p}.$$

In Problems **1–10**, *find the differentials of the functions in terms of x and dx.*

1. $y = 3x - 4$.

2. $y = 2$.

3. $f(x) = \sqrt{x^4 + 2}$.

4. $f(x) = (4x^2 - 5x + 2)^3$.

5. $u = \dfrac{1}{x^2}$.

6. $u = \dfrac{1}{\sqrt{x}}$.

7. $p = \ln(x^2 + 7)$.

8. $p = e^{x^3 + 5}$.

9. $y = (4x + 3)e^{2x^2 + 3}$.

10. $y = \ln\sqrt{x^4 + 1}$.

In Problems **11–14**, *evaluate* $d[f(x)]$ *for the indicated values of x and dx.*

11. $f(x) = 4 - 7x;\quad x = 3, dx = .02$.

12. $f(x) = 4x^2 - 3x + 10;\quad x = -1, dx = .25$.

13. $f(x) = \sqrt{25 - x^2}$;\quad $x = 4, dx = -.1$.

14. $f(x) = e^{x^2};\quad x = 0, dx = -.01$.

In Problems **15–22**, *approximate each expression by using differentials.*

15. $\sqrt{101}$.

16. $\sqrt{120}$.

17. $\sqrt[3]{63}$.

18. $\sqrt[4]{17}$.

19. $\ln .97$.

20. $\ln 1.01$.

21. $e^{.01}$.

22. $e^{-.01}$.

In Problems **23–28**, *find* dx/dy *or* dp/dq.

23. $y = 2x - 1$.

24. $y = 5x^2 + 3x + 2$.

25. $q = (p^2 + 5)^3$.

26. $q = \sqrt{p + 5}$.

27. $q = \dfrac{1}{p}$.

28. $q = e^{5-p}$.

In Problems **29** *and* **30**, *find the rate of change of q with respect to p for the indicated value of q.*

29. $p = \dfrac{500}{q + 2}; q = 18$.

30. $p = 50 - \sqrt{q}$; $q = 100$.

31. Suppose the profit P (in dollars) of producing q units of a product is

$$P = 396q - 2.2q^2 - 400.$$

Using differentials, find the approximate change in profit if the level of production changes from $q = 80$ to $q = 81$. Find the actual change.

32. Given the revenue function

$$r = 250q + 45q^2 - q^3,$$

use differentials to find the approximate change in revenue if the number of units increases from $q = 40$ to $q = 41$. Find the actual change.

33. The demand equation for a monopolist is $p = \dfrac{10}{\sqrt{q}}$. Using differentials, approximate the price when 24 units are demanded.

34. Answer the same question in Problem 33 if 101 units are demanded.

35. If $y = f(x)$, then the *proportional change* in y is defined to be $\Delta y / y$, which can be approximated with differentials by dy / y. Use this last form to approximate the proportional change in the cost function $c = f(q) = (q^4/2) + 3q + 400$ when $q = 10$ and $dq = 2$. Give your answer correct to one decimal place.

9-8 ELASTICITY OF DEMAND

Elasticity of demand is a means by which economists measure how a change in price of a product will affect quantity demanded. That is, it refers to consumer response to price changes. Loosely speaking, elasticity of demand is the ratio of the resulting percentage change in quantity demanded to a given percentage change in price:

$$\frac{\text{percentage change in quantity}}{\text{percentage change in price}}.$$

For example, if for a price increase of 5 percent, quantity demanded were to decrease by 2 percent, we would *loosely say* that elasticity of demand is $-.02/.05 = -\frac{2}{5}$.

Suppose $p = f(q)$ is the demand function for a product. Consumers will demand q units at a price of $f(q)$ per unit, and will demand $q + h$ units at a price of $f(q + h)$ per unit (Fig. 9-59). The *percentage* change in quantity demanded from q to $q + h$ is $\dfrac{(q + h) - q}{q} \cdot 100 = \dfrac{h}{q} \cdot 100$. The percentage

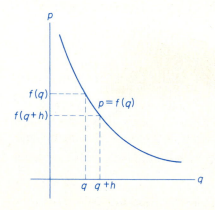

FIG. 9-59

change in price per unit from $f(q)$ to $f(q + h)$ is $\dfrac{f(q + h) - f(q)}{f(q)} \cdot 100$. The ratio of these percentage changes is

$$\frac{\dfrac{h}{q} \cdot 100}{\dfrac{f(q + h) - f(q)}{f(q)} \cdot 100} = \frac{h}{q} \cdot \frac{f(q)}{f(q + h) - f(q)}$$

$$= \frac{f(q)}{q} \cdot \frac{h}{f(q + h) - f(q)}$$

$$= \frac{\dfrac{f(q)}{q}}{\dfrac{f(q + h) - f(q)}{h}} . \qquad (1)$$

If f is differentiable, then as $h \to 0$ the limit of $[f(q + h) - f(q)]/h$ is dp/dq. Thus (1) approaches

$$\frac{\dfrac{f(q)}{q}}{\dfrac{dp}{dq}} = \frac{\dfrac{p}{q}}{\dfrac{dp}{dq}} .$$

DEFINITION. *If $p = f(q)$ is a differentiable demand function, the **point elasticity of demand**, denoted by the Greek letter η (eta), at (q, p) is*

$$\eta = \frac{\dfrac{p}{q}}{\dfrac{dp}{dq}} .$$

Let us determine the point elasticity of demand for the function $p = 1200 - q^2$.

$$\eta = \frac{\dfrac{p}{q}}{\dfrac{dp}{dq}} = \frac{\dfrac{1200 - q^2}{q}}{-2q} = -\frac{1200 - q^2}{2q^2} = -\left[\frac{600}{q^2} - \frac{1}{2}\right]. \qquad (2)$$

For example, if $q = 10$, then $\eta = -[(600/10^2) - \frac{1}{2}] = -5\frac{1}{2}$. Thus, if price were increased by 1 percent when $q = 10$, the quantity demanded would decrease by approximately $5\frac{1}{2}$ percent. Increasing price by $\frac{1}{2}$ percent results in a decrease of approximately 2.75 percent in demand.

Note that when elasticity is evaluated, no units are attached to it—it is nothing more than a real number. For normal behavior of demand, a price increase (decrease) corresponds to a quantity decrease (increase). Thus, dp/dq will always be negative or 0, and η will always be negative or 0. Some economists disregard the minus sign; in the above situation they would consider the elasticity to be $5\frac{1}{2}$. We shall not adopt this practice.

There are three categories of elasticity:

(1) When $|\eta| > 1$, demand is *elastic*.

(2) When $|\eta| = 1$, demand has *unit elasticity*.

(3) When $|\eta| < 1$, demand is *inelastic*.

In Eq. (2), since $|\eta| = 5\frac{1}{2}$ when $q = 10$, demand is elastic. If $q = 20$, then $|\eta| = |-[(600/20^2) - \frac{1}{2}]| = 1$ and so demand has unit elasticity. If $q = 25$, then $|\eta| = |-\frac{23}{50}|$ and demand is inelastic.

Loosely speaking, for a given percentage change in price, there will be a greater percentage change in quantity demanded if demand is elastic, a smaller percentage change if demand is inelastic, and an equal percentage change if demand has unit elasticity.

For example, suppose that there is a price increase of 5 percent. If this results in a decrease in demand of 10 percent, then demand is elastic. Here the price change caused a proportionately larger change in demand. On the other hand, if the demand decreases 2 percent, then demand is inelastic. Here the price change caused a proportionately smaller change in demand. Finally, if the price increase causes the same proportionate change in demand, then demand has unit elasticity.

EXAMPLE 1

Determine the point elasticity of the following demand equations for $q > 0$.

a. $p = \dfrac{k}{q}$ where $k > 0$.

$$\eta = \frac{\dfrac{p}{q}}{\dfrac{dp}{dq}} = \frac{\dfrac{k}{q^2}}{\dfrac{-k}{q^2}} = -1.$$

Thus the demand has unit elasticity for all $q > 0$. The graph of $p = k/q$ is called an *equilateral hyperbola* and is often found in economics texts in discussions of elasticity. See Section 3-3, Example 1(b) for a graph of such a curve.

b. $q = p^2 - 40p + 400$.

This equation defines p implicitly as a function of q. From Sec. 9-7,

$$\frac{dp}{dq} = \frac{1}{\dfrac{dq}{dp}}.$$

Therefore, $dp/dq = 1/(2p - 40)$ and

$$\eta = \frac{\dfrac{p}{q}}{\dfrac{dp}{dq}} = \frac{\dfrac{p}{q}}{\dfrac{1}{2p - 40}} = \frac{p(2p - 40)}{q}.$$

For example, if $p = 15$, then $q = 25$; hence $\eta = [15(-10)]/25 = -6$ and demand is elastic.

Point elasticity of a *linear* demand equation is quite interesting. Suppose the equation has the form

$$p = mq + b \quad \text{where} \quad m < 0 \text{ and } b > 0.$$

See Fig. 9-60. We shall assume $q > 0$; thus $p < b$. The point elasticity of demand is

$$\eta = \frac{\dfrac{p}{q}}{\dfrac{dp}{dq}} = \frac{\dfrac{p}{q}}{m} = \frac{p}{mq} = \frac{p}{p - b}.$$

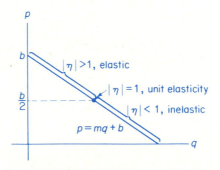

FIG. 9-60

By considering $d\eta/dp$, we shall show that η is a decreasing function of p.

$$\frac{d\eta}{dp} = \frac{(p - b) - p}{(p - b)^2} = -\frac{b}{(p - b)^2}.$$

Since $b > 0$, then $d\eta/dp < 0$ and thus η is a decreasing function of p—as p increases, η must decrease. However p ranges between 0 and b, and at the midpoint of this range, $b/2$,

$$\eta = \frac{\dfrac{b}{2}}{\dfrac{b}{2} - b} = -1.$$

Therefore, if $p < b/2$, then $\eta > -1$; if $p > b/2$, then $\eta < -1$. Stating this another way, when $p < b/2$, $|\eta| < 1$ and demand is inelastic; when $p = b/2$, $|\eta| = 1$ and demand has unit elasticity; when $p > b/2$, $|\eta| > 1$ and demand is elastic. This shows that the slope of a demand curve is not a measure of elasticity. The slope of the above line is m everywhere, but elasticity varies with the point on the line.

It is possible to relate how elasticity of demand affects changes in revenue (marginal revenue). If $p = f(q)$ is a manufacturer's demand function, his total revenue is

$$r = pq.$$

To find marginal revenue dr/dq we differentiate r by using the product rule.

$$\frac{dr}{dq} = p + q\frac{dp}{dq}. \tag{3}$$

Factoring the right side of Eq. (3), we have

$$\frac{dr}{dq} = p\left(1 + \frac{q}{p}\frac{dp}{dq}\right).$$

But

$$\frac{q}{p}\frac{dp}{dq} = \frac{\dfrac{dp}{dq}}{\dfrac{p}{q}} = \frac{1}{\eta}.$$

Thus

$$\frac{dr}{dq} = p\left(1 + \frac{1}{\eta}\right). \tag{4}$$

If demand is elastic, then $\eta < -1$ and $1 + \dfrac{1}{\eta} > 0$. If demand is inelastic, then $\eta > -1$ and $1 + \dfrac{1}{\eta} < 0$. Let us assume that $p > 0$. From Eq. (4) we can conclude that $dr/dq > 0$ on intervals for which demand is elastic; hence, total revenue r is increasing there. On the other hand, marginal revenue is negative on intervals for which demand is inelastic; hence, total revenue is decreasing there.

Thus we conclude from the above argument that as more units are sold, a manufacturer's total revenue increases if demand is elastic, but decreases if demand is inelastic. That is, if demand is elastic, a lower price will increase revenue. This means that a lower price will cause a large enough increase in demand to actually increase revenue. If inelastic, a lower price will decrease revenue. For unit elasticity, a lower price leaves total revenue unchanged.

EXERCISE 9-8

*In Problems **1–14**, find the point elasticity of the demand equations for the indicated values of q or p and determine whether demand is elastic, inelastic, or has unit elasticity.*

1. $p = 40 - 2q$; $q = 5$.

2. $p = 12 - .03q$; $q = 300$.

3. $q = 600 - 100p$; $p = 3$.

4. $q = 100 - p$; $p = 50$.

5. $p = \dfrac{1000}{q}$; $q = 288$.

6. $p = \dfrac{1000}{q^2}$; $q = 156$.

7. $p = \dfrac{500}{q + 2}$; $q = 100$.

8. $p = \dfrac{800}{2q + 1}$; $q = 25$.

9. $q = \sqrt{2500 - p}$; $p = 900$.

10. $q = \sqrt{2500 - p^2}$; $p = 20$.

11. $q = \dfrac{(p - 100)^2}{2}$; $p = 20$.

12. $q = p^2 - 60p + 898$; $p = 10$.

13. $p = 150 - e^{q/100}$; $q = 100$.

14. $p = 100e^{-q/200}$ $q = 200$.

15. For the linear demand equation $p = 13 - .05q$, verify that demand is elastic when $p = 10$, is inelastic when $p = 3$, and demand has unit elasticity when $p = 6.50$.

16. For what value (or values) of q do the following demand equations have unit elasticity?

   a. $p = 26 - .10q$,   b. $p = 1200 - q^2$.

17. The demand equation for a product is

$$q = 500 - 40p + p^2,$$

where p is the price per unit (in dollars) and q is the quantity of units demanded (in thousands). Find the point elasticity of demand when $p = 15$. If this price of 15 is increased by $\frac{1}{2}$ percent, what is the approximate change in demand?

18. The demand equation of a product is

$$q = \sqrt{2500 - p^2} .$$

Find the point elasticity of demand when $p = 30$. If the price of 30 decreases $2/3$ percent, what is the approximate change in demand?

19. For the demand equation $p = 500 - 2q$ verify that demand is elastic and total revenue is increasing for $0 < q < 125$. Verify that demand is inelastic and total revenue is decreasing for $125 < q < 250$.

20. Verify that $\dfrac{dr}{dq} = p\left(1 + \dfrac{1}{\eta}\right)$ if $p = 40 - 2q$.

21. Repeat Problem 20 for $p = \dfrac{1000}{q^2}$.

22. Let $p = mq + b$ be a linear demand equation where $m \neq 0$ and $b > 0$.
    a. Show that $\lim_{p \to b^-} \eta = -\infty$.
    b. Show that $\eta = 0$ when $p = 0$.

23. For the demand equation $p = 1000 - q^2$, if $5 \leqslant q \leqslant 30$, for what value of q is $|\eta|$ a maximum? For what value is it a minimum?

24. Repeat Problem 23 for $p = 200/(q + 5)$ and $5 \leqslant q \leqslant 95$.

9-9 REVIEW

IMPORTANT TERMS AND SYMBOLS IN CHAPTER 9

x-intercept (p. 320)	y-intercept (p. 320)
x-axis symmetry (p. 322)	y-axis symmetry (p. 321)
symmetry about origin (p. 323)	vertical asymptote (p. 327)
horizontal asymptote (p. 327)	increasing function (p. 334)
decreasing function (p. 334)	relative maximum (p. 335)
relative minimum (p. 335)	relative extrema (p. 336)
absolute extrema (p. 336)	critical value (p. 336)
first-derivative test (p. 337)	concave up (p. 344)
concave down (p. 344)	inflection point (p. 346)
second-derivative test (p. 354)	differential (p. 367)
dy, dx (p. 367)	point elasticity of demand (p. 374)
η (p. 374)	elastic (p. 375)
inelastic (p. 375)	unit elasticity (p. 375)

REVIEW SECTION

1. The graph of $y = x^2 + 1$ has no (a) (x) (y) -intercept but has one (b) (x) (y) -intercept.

    Ans. (a) x; (b) y.

2. The y-intercept of the graph of $y = 8x^3 - 7x + 4$ is _____ .

Ans. (0, 4).

3. The graph of $y = x^4 - x^2 + 3$ is symmetric with respect to the __(a)__ -axis but not the __(b)__ -axis.

Ans. (a) y; (b) x.

4. If a graph is symmetric with respect to both the x-axis and y-axis, then it __(is) (is not)__ symmetric with respect to the origin.

Ans. is.

5. In the graph of $y = \dfrac{6}{x + 2} + 3$ the line $x =$ __(a)__ is a vertical asymptote and the line $y =$ __(b)__ is a horizontal asymptote.

Ans. (a) -2; (b) 3.

6. The graph of $y = e^x$ has a __(a) (horizontal) (vertical)__ asymptote but no __(b) (horizontal) (vertical)__ asymptote.

Ans. (a) horizontal; (b) vertical.

7. Suppose f is a function defined on an interval I. If $f'(x) < 0$ on I, then f is __(a) (increasing) (decreasing)__ on I and the graph of f is __(b) (rising) (falling)__ . If $f'(x) > 0$ on I, then f is __(c) (increasing) (decreasing)__ on I. If $f'(x_1) = 0$ or is not defined, then $x = x_1$ is called a __(d)__ value. At such a point, f may have a relative __(e)__ or __(f)__ or neither.

Ans. (a) decreasing; (b) falling; (c) increasing; (d) critical; (e) maximum; (f) minimum.

8. If $f''(x) > 0$ on I, then f is concave __(up) (down)__ on I.

Ans. up.

9. If $f''(x_1) = 0$ or is not defined, then there may be a point of __(a)__ when $x = x_1$. If $f'(x_1) = 0$ and $f''(x_1) > 0$, then f has a relative __(b)__ when $x = x_1$.

Ans. (a) inflection; (b) minimum.

10. If $f(x_1) \geqslant f(x)$ for all x in the domain of f, then when $x = x_1$ there occurs a(n) _____ maximum.

Ans. absolute

11. Under typical conditions, profit is maximized at the level of output for which marginal revenue = _____ .

Ans. marginal cost.

12. If f has a relative maximum when $x = x_1$, then as x increases through x_1, $f'(x)$ changes from __(a) (+) (−)__ to __(b) (+) (−)__ . At a relative minimum, $f'(x)$ changes from __(c) (+) (−)__ to __(d) (+) (−)__ .

Ans. (a) + ; (b) − ; (c) − ; (d) + .

13. True or false: If $f'(x_1) = 0$, then f has a relative maximum or minimum when $x = x_1$, __(a)__ . If $f''(x_2) = 0$, then f has a point of inflection when $x = x_2$, __(b)__ .

Ans. (a) false; (b) false.

14. In the graph of $y = f(x)$ in Fig. 9-61, an absolute maximum occurs at point(s) __(a)__ , an absolute minimum at __(b)__ , a relative maximum at __(c)__ , a relative minimum at __(d)__ , and inflection point(s) at __(e)__ .

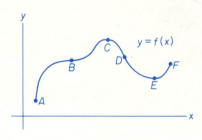

FIG. 9-61

Ans. (a) C; (b) A; (c) C; (d) E; (e) B, D.

15. The differential of $y = x^2 + 3$ is _____ .

Ans. $2x\,dx$.

16. If $\dfrac{dy}{dx} = \dfrac{2}{3}$, then $\dfrac{dx}{dy} =$ _____ .

Ans. $\dfrac{3}{2}$.

17. The differential of $y = x$ when $x = 6$ and $dx = .001$ is _____ .

Ans. $.001$.

18. If the elasticity of a demand function at a point is -1.301, then the demand is said to be (elastic) (inelastic) there.

Ans. elastic.

REVIEW PROBLEMS

In Problems 1–10, sketch the graphs of the functions. Indicate intervals on which the function is increasing, decreasing, concave up, concave down; indicate relative maximum points, relative minimum points, points of inflection, horizontal asymptotes, vertical asymptotes, symmetry, and those intercepts that can be obtained conveniently.

1. $y = x^2 - 2x - 24$.

2. $y = x^3 - 27x$.

3. $y = x^3 - 12x + 20$.

4. $y = x^4 - 4x^3 - 20x^2 + 150$.

5. $y = x^3 + x.$

6. $y = \dfrac{x + 2}{x - 3}.$

7. $f(x) = \dfrac{100(x + 5)}{x^2}.$

8. $y = \dfrac{x^2 - 4}{x^2 - 1}.$

9. $f(x) = \dfrac{e^x + e^{-x}}{2}.$

10. $f(x) = 1 + \ln(x^2).$

11. A manufacturer determines that m employees on a certain production line will produce q units per month where $q = 80m^2 - .1m^4$. To obtain maximum monthly production, how many employees should be assigned to the production line?

12. The demand function for a monopolist's product is $p = \sqrt{600 - q}$. If he wants to produce at least 100 units but not more than 300 units, how many units should he produce to maximize total revenue?

13. The demand function for a monopolist's product is $p = 400 - 2q$; p is given in dollars. If the average cost per unit for producing q units is $\bar{c} = q + 160 + (2000/q)$, find the maximum profit that the monopolist can achieve.

14. If $c = .01q^2 + 5q + 100$ is a cost equation, find the average cost equation. At what level of production q is there a minimum average cost? Show algebraically that the graph of the marginal cost function intersects the graph of the average cost function at this point.

In Problems 15 and 16, determine the differentials of the functions in terms of x and dx.

15. $f(x) = x^2 \ln(x + 5).$

16. $f(x) = (x^2 + 5)/(x - 7).$

Approximate the expressions in Problems 17 and 18 by use of differentials.

17. $e^{-.01}.$

18. $\sqrt{25.5}.$

For the demand equations in Problems 19–21, determine whether demand is elastic, inelastic, or has unit elasticity for the indicated value of q.

19. $p = 18 - .02q; \quad q = 600.$

20. $p = \dfrac{500}{q}; \quad q = 200.$

21. $p = 900 - q^2; \quad q = 10.$

22. In a model for the market penetration of a new product, sales S of the product at time t is given by[†]

$$S = g(t) = \frac{m(p + q)^2}{p} \left[\frac{e^{-(p + q)t}}{\left(\dfrac{q}{p} e^{-(p + q)t} + 1 \right)^2} \right],$$

where p, q, and m are nonzero constants.

[†] A. P. Hurter, Jr., A. H. Rubenstein, et al. "Market penetration by new innovations: the technological literature," *Technological Forecasting and Social Change*, vol. 11 (1978), 197–221.

(a) Show that

$$\frac{dS}{dt} = \frac{\frac{m}{p}(p + q)^3 e^{-(p+q)t}\left[\frac{q}{p}e^{-(p+q)t} - 1\right]}{\left(\frac{q}{p}e^{-(p+q)t} + 1\right)^3}.$$

(b) Determine the value of t for which maximum sales occur. You may assume that S attains a maximum when $dS/dt = 0$.

23. A rectangular box is to be made by cutting out equal squares from each corner of a piece of cardboard 10 in. by 16 in. and then folding up the sides. What must be the length of the side of the square cut out if the volume of the box is to be a maximum?

24. A rectangular field is to be enclosed by a fence and divided equally into three parts by two fences parallel to one pair of the sides. If a total of 800 ft of fencing is to be used, find the dimensions of the field if its area is to be maximized.

25. A rectangular poster having an area of 500 square inches is to have a 4-in. margin at each side and at the bottom and a 6-in. margin at the top. The remainder of the poster is for printed matter. Find the dimensions of the poster so that the area for the printed matter is maximized.

INTEGRATION

Chapters 8 and 9 dealt with differential calculus. We differentiated a function and obtained another function called its derivative. *Integral calculus* is concerned with the opposite process. We are given the derivative of a function and must find the original function. The need for doing this arises in a natural way. For example, we may have a marginal revenue function and want to find the revenue function from it. Furthermore, you will see that there is a limit concept involved with integral calculus, and it is related to that used in differential calculus. This limit concept allows us to take the limit of a special kind of sum as the number of terms in the sum becomes infinite. This is the real power of integral calculus! With this notion we may find such things as areas of regions that cannot be found by any other convenient method.

10-1 THE INDEFINITE INTEGRAL

If F is a function such that

$$F'(x) = f(x), \tag{1}$$

then F is called an *antiderivative* of f. Thus an antiderivative of f is merely a function F which when differentiated gives f. Multiplying both sides of Eq. (1) by the differential dx gives $F'(x)\,dx = f(x)\,dx$. However, $F'(x)\,dx$ is the differential dF:

$$dF = f(x)\,dx.$$

Thus we can think of an antiderivative of f as a function whose differential is $f(x)\,dx$.

DEFINITION. *An **antiderivative** of a function f is a function F such that*

$$F'(x) = f(x)$$

or equivalently, in differential notation,

$$dF = f(x)\,dx.$$

For example, since

$$D_x(x^2) = 2x,$$

x^2 is an antiderivative of $2x$. Is this the only antiderivative of $2x$? The answer is no! Since

$$D_x(x^2 + 1) = 2x$$
$$\text{and} \quad D_x(x^2 - 5) = 2x,$$

then $x^2 + 1$ and $x^2 - 5$ are also antiderivatives of $2x$. In fact, it can be shown that any antiderivative of $2x$ must have the form $x^2 + C$, where C is a constant. Thus *any two antiderivatives of $2x$ differ only by a constant.*

The symbol we shall use for an arbitrary antiderivative of $2x$ is $\int 2x\,dx$, which is read "the *indefinite integral* of $2x$ with respect to x." Since all antiderivatives of $2x$ have the form $x^2 + C$, we write

$$\int 2x\,dx = x^2 + C.$$

The symbol $\int$ is called the **integral sign**, $2x$ is the **integrand**, and C is the **constant of integration**. The dx is part of the integral notation and indicates the variable involved. We say here that x is the **variable of integration.**

More generally, the **indefinite integral** with respect to x of any function f is written $\int f(x)\,dx$ and denotes an arbitrary antiderivative of f. It can be shown that all antiderivatives of f differ at most by a constant. Thus if F is an antiderivative of f, then

$$\int f(x)\,dx = F(x) + C, \qquad \text{where C is a constant.}$$

To *integrate f* means to find $\int f(x)\, dx$.

$$\int f(x)\, dx = F(x) + C \text{ if and only if } F'(x) = f(x).$$

EXAMPLE 1

Find $\int 5\, dx$.

First we must find (perhaps better words are "guess at") a function whose derivative is 5. Since $D_x(5x) = 5$, then $5x$ is an antiderivative of 5. Thus

$$\int 5\, dx = 5x + C.$$

PITFALL. *It is incorrect to write* $\int 5\, dx = 5x$. *Do not forget the constant of integration.*

Using differentiation formulas from Chapter 8, we have compiled a list of basic integration formulas in Table 10-1. These formulas are easily verified. For

TABLE 10-1

BASIC INTEGRATION FORMULAS

1. $\int k\, dx = kx + C$, k a constant.

2. $\int x^n\, dx = \dfrac{x^{n+1}}{n+1} + C$, $n \neq -1$. ✳ because denom w/be 0

3. $\int e^x\, dx = e^x + C$.

4. $\int k\, f(x)\, dx = k \int f(x)\, dx$, k a constant.

5. $\int [f(x) \pm g(x)]\, dx = \int f(x)\, dx \pm \int g(x)\, dx$.

[handwritten: $\int x^{-1}\, dx = \int \frac{1}{x}\, dx = \ln x$ use this for rule 2 when $n = -1$]

example, formula (2) is true since the derivative of $x^{n+1}/(n+1)$ is x^n for $n \neq -1$. To verify formula (4) we must show that the derivative of $k\int f(x)\, dx$ is $kf(x)$. Now,

$$D_x\left[k \int f(x)\, dx \right] = k\, D_x\left[\int f(x)\, dx \right].$$

But $\int f(x)\, dx$ is a function whose derivative is $f(x)$. Thus, $D_x[\int f(x)\, dx] = f(x)$ and so

$$D_x\left[k \int f(x)\, dx \right] = kf(x).$$

You should verify the other formulas. Formula (5) can be extended to any finite number of sums and/or differences.

We point out that formula (2) states that the indefinite integral of most powers of x is obtained by increasing the exponent of x by one, dividing by the new exponent, and adding on the constant of integration. The case when $n = -1$ results in division by zero; it will be discussed in the next section.

EXAMPLE 2

Find the following indefinite integrals.

a. $\int 1 \, dx$.

By formula (1) in Table 10-1 with $k = 1$,

$$\int 1 \, dx = 1x + C = x + C.$$

Usually we write $\int 1 \, dx$ as $\int dx$. Thus, $\int dx = x + C$.

b. $\int x^5 \, dx$.

By formula (2) with $n = 5$,

$$\int x^5 \, dx = \frac{x^{5+1}}{5+1} + C = \frac{x^6}{6} + C.$$

c. $\int 7x \, dx$.

By formula (4) with $k = 7$ and $f(x) = x$,

$$\int 7x \, dx = 7 \int x \, dx.$$

Since $\int x \, dx = \int x^1 \, dx$, by formula (2) with $n = 1$ we have

$$\int x^1 \, dx = \frac{x^{1+1}}{1+1} + C_1 = \frac{x^2}{2} + C_1,$$

where C_1 is the constant of integration. Therefore,

$$\int 7x \, dx = 7 \int x \, dx = 7\left[\frac{x^2}{2} + C_1\right] = \frac{7}{2}x^2 + 7C_1.$$

Replacing the constant $7C_1$ by C, we have

$$\int 7x \, dx = \frac{7}{2}x^2 + C.$$

It is not necessary to write all intermediate steps when integrating. More simply we write

$$\int 7x \, dx = 7 \int x \, dx = (7)\frac{x^2}{2} + C = \frac{7}{2}x^2 + C.$$

d. $\int -\frac{3}{5}e^x \, dx.$

$$\int -\frac{3}{5}e^x \, dx = -\frac{3}{5}\int e^x \, dx \qquad\qquad \text{(formula 4)}$$

$$= -\frac{3}{5}e^x + C \qquad\qquad \text{(formula 3).}$$

PITFALL. *Only a constant factor can "jump" in front of an integral sign. It is correct to write $\int 7x \, dx = 7 \int x \, dx = (7/2)x^2 + C$, but it is **incorrect** to write $\int 7x \, dx = 7x \int \, dx = (7x)(x + C) = 7x^2 + 7Cx.$*

EXAMPLE 3

Find the following indefinite integrals.

a. $\int \frac{1}{\sqrt{t}} \, dt.$

Here t is the *variable of integration*. We rewrite the integrand so that one of the basic forms can be used.

$$\int \frac{1}{\sqrt{t}} \, dt = \int t^{-1/2} \, dt.$$

By formula (2) in Table 10-1 with $n = -\frac{1}{2}$,

$$\int \frac{1}{\sqrt{t}} \, dt = \int t^{-1/2} \, dt = \frac{t^{(-1/2)+1}}{-\frac{1}{2}+1} + C = \frac{t^{1/2}}{\frac{1}{2}} + C = 2\sqrt{t} + C.$$

b. $\int \frac{1}{6x^3} \, dx.$

$$\int \frac{1}{6x^3} \, dx = \frac{1}{6}\int x^{-3} \, dx = \left(\frac{1}{6}\right)\frac{x^{-3+1}}{-3+1} + C$$

$$= -\frac{x^{-2}}{12} + C = -\frac{1}{12x^2} + C.$$

EXAMPLE 4

Find the following indefinite integrals.

a. $\int (x^2 + 2x - 4)\, dx.$

By formula (5) in Table 10-1,

$$\int (x^2 + 2x - 4)\, dx = \int x^2\, dx + \int 2x\, dx - \int 4\, dx.$$

Now,

$$\int x^2\, dx = \frac{x^{2+1}}{2+1} + C_1 = \frac{x^3}{3} + C_1,$$

$$\int 2x\, dx = 2\int x\, dx = (2)\frac{x^{1+1}}{1+1} + C_2 = x^2 + C_2,$$

and

$$\int 4\, dx = 4x + C_3.$$

Thus,

$$\int (x^2 + 2x - 4)\, dx = \frac{x^3}{3} + x^2 - 4x + C_1 + C_2 - C_3.$$

Letting $C = C_1 + C_2 - C_3$, we have

$$\int (x^2 + 2x - 4)\, dx = \frac{x^3}{3} + x^2 - 4x + C.$$

Omitting intermediate steps, we simply write

$$\int (x^2 + 2x - 4)\, dx = \int x^2\, dx + 2\int x\, dx - \int 4\, dx$$

$$= \frac{x^3}{3} + (2)\frac{x^2}{2} - 4x + C$$

$$= \frac{x^3}{3} + x^2 - 4x + C.$$

b. $\int (2\sqrt[5]{x^4} - 7x^3 + 10e^x - 1)\, dx.$

$$\int (2\sqrt[5]{x^4} - 7x^3 + 10e^x - 1)\, dx = 2\int x^{4/5}\, dx - 7\int x^3\, dx + 10\int e^x\, dx - \int 1\, dx$$

$$= (2)\frac{x^{9/5}}{\frac{9}{5}} - (7)\frac{x^4}{4} + 10e^x - x + C$$

$$= \frac{10}{9}x^{9/5} - \frac{7}{4}x^4 + 10e^x - x + C.$$

Sometimes, in order to apply the basic integration formulas it is necessary to first perform algebraic manipulations on an integrand, as Example 5 shows.

EXAMPLE 5

Find the following indefinite integrals.

a. $\int y^2 \left(y + \frac{2}{3} \right) dy.$

By multiplying the integrand we get

$$\int y^2 \left(y + \frac{2}{3} \right) dy = \int \left(y^3 + \frac{2}{3} y^2 \right) dy$$

$$= \frac{y^4}{4} + \left(\frac{2}{3} \right) \frac{y^3}{3} + C = \frac{y^4}{4} + \frac{2y^3}{9} + C.$$

b. $\int \frac{(3x + \sqrt{x})(\sqrt[3]{x} - 2)}{6} dx.$

By factoring out $\frac{1}{6}$ and multiplying the binomials we get

$$\int \frac{(3x + x^{1/2})(x^{1/3} - 2)}{6} dx$$

$$= \frac{1}{6} \int (3x^{4/3} + x^{5/6} - 6x - 2x^{1/2}) dx$$

$$= \frac{1}{6} \left[(3)\frac{x^{7/3}}{\frac{7}{3}} + \frac{x^{11/6}}{\frac{11}{6}} - (6)\frac{x^2}{2} - (2)\frac{x^{3/2}}{\frac{3}{2}} \right] + C$$

$$= \frac{3x^{7/3}}{14} + \frac{x^{11/6}}{11} - \frac{x^2}{2} - \frac{2x^{3/2}}{9} + C.$$

PITFALL. *In Example 5(a), we first multiplied the factors in the integrand. We point out that*

$$\int y^2 \left(y + \frac{2}{3} \right) dy \neq \left[\int y^2 \, dy \right] \left[\int \left(y + \frac{2}{3} \right) dy \right].$$

More generally,

$$\int f(x)g(x) \, dx \neq \int f(x) \, dx \cdot \int g(x) \, dx.$$

EXAMPLE 6

For a particular urban group, economists studied the current average yearly income y (in dollars) that a person can expect to receive with x years of education before seeking regular employment. They estimated that the rate at which income changes with respect to education is given by

$$\frac{dy}{dx} = 10x^{3/2}, \qquad 4 \leqslant x \leqslant 16,$$

where y = 5872 when x = 9. Find y.

Since $dy/dx = 10x^{3/2}$, y is an antiderivative of $10x^{3/2}$. Thus

$$y = \int 10x^{3/2} \, dx = 10 \int x^{3/2} \, dx$$

$$= (10) \frac{x^{5/2}}{\frac{5}{2}} + C = 4x^{5/2} + C.$$

Thus, *any* function of the form

$$y = 4x^{5/2} + C \tag{2}$$

satisfies $dy/dx = 10x^{3/2}$. To determine the *specific* one for which $y = 5872$ when $x = 9$, we must find the right C. Substituting $y = 5872$ and $x = 9$ into Eq. (2), we obtain

$$5872 = 4(9)^{5/2} + C$$
$$= 4(243) + C.$$
$$5872 = 972 + C.$$

Therefore $C = 4900$ and

$$y = 4x^{5/2} + 4900.$$

EXAMPLE 7

Given that $y'' = x^2 - 6$, $y'(0) = 2$, and $y(1) = -1$, find y.

Since $y'' = \frac{d}{dx}(y') = x^2 - 6$, then y' is an antiderivative of $x^2 - 6$. Thus

$$y' = \int (x^2 - 6) \, dx = \frac{x^3}{3} - 6x + C_1. \tag{3}$$

Since $y'(0) = 2$ means that $y' = 2$ when $x = 0$, from Eq. (3) we have

$$2 = \frac{0^3}{2} - 6(0) + C_1.$$

Hence $C_1 = 2$ and

$$y' = \frac{x^3}{3} - 6x + 2.$$

By integration we can find y:

$$y = \int \left(\frac{x^3}{3} - 6x + 2 \right) dx$$

$$= \left(\frac{1}{3} \right) \frac{x^4}{4} - (6) \frac{x^2}{2} + 2x + C_2.$$

$$y = \frac{x^4}{12} - 3x^2 + 2x + C_2. \tag{4}$$

Since $y = -1$ when $x = 1$, from Eq. (4) we have

$$-1 = \frac{1^4}{12} - 3(1)^2 + 2(1) + C_2.$$

Therefore $C_2 = -\frac{1}{12}$ and

$$y = \frac{x^4}{12} - 3x^2 + 2x - \frac{1}{12}.$$

EXAMPLE 8

Given the marginal revenue function

$$dr/dq = 2000 - 20q - 3q^2,$$

determine the demand function.

Since dr/dq is the derivative of total revenue r,

$$r = \int (2000 - 20q - 3q^2) \, dq$$

$$= 2000q - (20) \frac{q^2}{2} - (3) \frac{q^3}{3} + C.$$

$$r = 2000q - 10q^2 - q^3 + C. \tag{5}$$

When no units are sold, total revenue is 0; that is, $r = 0$ when $q = 0$. From Eq. (5),

$$0 = 2000(0) - 10(0)^2 - 0^3 + C.$$

Hence $C = 0$ and

$$r = 2000q - 10q^2 - q^3.$$

But $r = pq$ where p is the price per unit. Solving for p gives the demand function:

$$p = \frac{r}{q} = \frac{2000q - 10q^2 - q^3}{q}.$$

$$p = 2000 - 10q - q^2.$$

EXAMPLE 9

The chief executive officer of a corporation asked the vice president of production to determine the total cost of producing 10,000 pounds of product in one week. The vice president found that the marginal cost function dc/dq is

$$\frac{dc}{dq} = \frac{1}{1,000,000}\left[\frac{1}{500}q^2 - 25q\right] + .2,$$

where q is pounds of product per week, and fixed costs per week are $4000. What reply should the vice president give?

Since dc/dq is the derivative of total cost c,

$$c = \int \left\{\frac{1}{1,000,000}\left[\frac{1}{500}q^2 - 25q\right] + .2\right\} dq$$

$$= \frac{1}{1,000,000}\int\left[\frac{1}{500}q^2 - 25q\right] dq + \int .2 \, dq.$$

$$c = \frac{1}{1,000,000}\left[\frac{q^3}{1500} - \frac{25q^2}{2}\right] + .2q + C.$$

Fixed costs are constant regardless of output. Therefore, setting $q = 0$ and $c = 4000$ gives $C = 4000$. Thus

$$c = \frac{1}{1,000,000}\left[\frac{q^3}{1500} - \frac{25q^2}{2}\right] + .2q + 4000. \qquad (6)$$

From Eq. (6), when $q = 10,000$ then $c = 5416\frac{2}{3}$. The vice president should reply that the total cost for producing 10,000 pounds of product in one week is $5416.67.

EXERCISE 10-1

In Problems 1–64, find the indefinite integrals.

1. $\int 5 \, dx.$

2. $\int \frac{1}{2} \, dx.$

3. $\int x^8 \, dx.$

4. $\int 2x^{25} \, dx.$

5. $\int t^{13/2} \, dt.$

6. $\int \frac{1}{2}x^{5/3} \, dx.$

7. $\int x^{-7} \, dx.$

8. $\int \frac{z^{-3}}{3} \, dz.$

9. $\int \frac{1}{x^{10}} \, dx.$

10. $\int \frac{7}{x^4} \, dx.$

11. $\int \frac{1}{y^{11/5}} \, dy.$

12. $\int \frac{7}{2x^{9/4}} \, dx.$

13. $\int \sqrt[5]{x^6} \, dx.$

14. $\int -\frac{3}{2}\sqrt{x} \, dx.$

15. $\int \frac{1}{\sqrt[8]{x^7}} \, dx.$

16. $\int \frac{3}{4\sqrt[8]{x}} \, dx.$

17. $\int (7 + e) \, dx.$

18. $\int (5 - 6) \, dx.$

19. $\int x^{\sqrt{2}} \, dx.$

20. $\int x^{1.2} \, dx.$

21. $\int 3x^7 \, dx.$

22. $\int 5x^4 \, dx.$

23. $\int (8 + u) \, du.$

24. $\int (r^3 + 2r) \, dr.$

25. $\int (y^5 + 5y) \, dy.$

26. $\int (7 - 3w - 2w^2) \, dw.$

27. $\int (3t^2 - 4t - 5) \, dt.$

28. $\int (1 + u + u^2 + u^3) \, du.$

29. $\int \left(\frac{x}{7} - \frac{3}{4}x^4 \right) dx.$

30. $\int \left(\frac{2x^2}{7} - \frac{8}{3}x^4 \right) dx.$

31. $\int \frac{2\sqrt{x}}{3} \, dx.$

32. $\int dw.$

33. $\int 3e^x \, dx.$

34. $\int \left(\frac{e^x}{3} + 2x \right) dx.$

35. $\int (6 - \frac{5}{4}z^2 + 2e^z) \, dz.$

36. $\int \frac{1}{12}(\frac{1}{3}x^5) \, dx.$

37. $\int \left(\frac{e^u}{4} + 1 \right) du.$

38. $\int \left(3y^3 - 2y^2 + \frac{e^y}{6} \right) dy.$

39. $\int (x^{-2} - 5x^{-3} + 2x^{-4}) \, dx.$

40. $\int (-3x^{-2} - 2x^{-3}) \, dx.$

41. $\int (x^{8.3} - 9x^6 + 3x^{-4} + x^{-3}) \, dx.$

42. $\int (.3y^4 - 8y^{-3} + 2) \, dy.$

43. $\int \left(\frac{x^3}{3} - \frac{3}{x^3} \right) dx.$

44. $\int \left(\frac{1}{2x^3} - \frac{1}{x^4} \right) dx.$

45. $\int \left(\frac{3w^2}{2} - \frac{2}{3w^2} \right) dw.$

46. $\int \frac{2}{e^{-s}} \, ds.$

47. $\int (\sqrt[3]{x} - \sqrt[4]{x} + \sqrt[5]{x}) \, dx.$

48. $\int \left(\frac{\sqrt[5]{x}}{2} - \frac{2}{3}\sqrt[3]{x} \right) dx.$

49. $\int (2\sqrt{x} - 3\sqrt[4]{x}) \, dx.$

50. $\int 0 \, dx.$

51. $\int 2x^{-6/5} \, dx.$

52. $\int (x^{-4/5} + 2) \, dx.$

53. $\int \left(-\dfrac{\sqrt[3]{x^2}}{5} - \dfrac{7}{2\sqrt{x}} + 6x \right) dx.$ **54.** $\int \left(\sqrt[3]{x} - \dfrac{1}{\sqrt[3]{x}} \right) dx.$

55. $\int (x^2 + 5)(x - 3)\, dx.$ **56.** $\int x^4(x^3 + 3x^2 + 7)\, dx.$

57. $\int \sqrt{x}\,(x + 3)\, dx.$ **58.** $\int (z + 2)^2\, dz.$

59. $\int (2u + 1)^2\, du.$ **60.** $\int \left(\dfrac{1}{\sqrt[3]{x}} + 1 \right)^2 dx.$

61. $\int v^{-2}(2v^4 + 3v^2 - 2v^{-3})\, dv.$ **62.** $\int [6e^u - u^3(\sqrt{u} + 1)]\, du.$

63. $\int \dfrac{e^6 + e^x}{2}\, dx.$ **64.** $\int \dfrac{\sqrt{x}\,(x^5 - \sqrt[3]{x} + 2)}{3}\, dx.$

In Problems 65–70, find y subject to the given conditions.

65. $dy/dx = 3x - 4; \quad y(-1) = \frac{13}{2}.$ **66.** $dy/dx = x^2 - x; \quad y(3) = 4.$

67. $y'' = -x^2 - 2x; \quad y'(1) = 0, y(1) = 1.$

68. $y'' = x + 1; \quad y'(0) = 0, y(0) = 5.$

69. $y''' = 2x; \quad y''(-1) = 3, y'(3) = 10, y(0) = 2.$

70. $y''' = e^x + 1; \quad y''(0) = 1, y'(0) = 2, y(0) = 3.$

In Problems 71–74, dc/dq is a marginal cost function and fixed costs are indicated in braces. For Problems 71 and 72 find the total cost function. For Problems 73 and 74, find the total cost for the indicated value of q.

71. $dc/dq = 1.35; \quad \{200\}.$ **72.** $dc/dq = 2q + 50; \quad \{1000\}.$

73. $dc/dq = .09q^2 - 1.2q + 4.5; \quad \{7700\}; \quad q = 10.$

74. $dc/dq = .000102q^2 - .034q + 5; \quad \{10{,}000\}; \quad q = 100.$

In Problems 75–77, dr/dq is a marginal revenue function. Find the demand function.

75. $dr/dq = .7.$ **76.** $dr/dq = 15 - \frac{1}{15}q.$

77. $dr/dq = 275 - q - .3q^2.$

78. The sole producer of a product has determined that the marginal revenue function is $dr/dq = 100 - 3q^2$. Determine the point elasticity of demand for the product when $q = 5$. *Hint:* First find the demand function.

79. A manufacturer has determined that the marginal cost function is $dc/dq = .003q^2 - .4q + 40$, where q is the number of units produced. If marginal cost is $27.50 when $q = 50$, and fixed costs are $5000, what is the *average* cost of producing 100 units?

10-2 MORE INTEGRATION FORMULAS

The formula

$$\int x^n \, dx = \frac{x^{n+1}}{n+1} + C, \qquad \text{if } n \neq -1,$$

which applies to a power of x, can be generalized to handle a power of a *function* of x. Let u be a differentiable function of x. By the power rule, if $n \neq -1$ then

$$\frac{d}{dx}\left(\frac{[u(x)]^{n+1}}{n+1}\right) = \frac{(n+1)[u(x)]^n \cdot u'(x)}{n+1} = [u(x)]^n \cdot u'(x).$$

Thus,

$$\int [u(x)]^n \cdot u'(x) \, dx = \frac{[u(x)]^{n+1}}{n+1} + C, \qquad n \neq -1. \qquad (1)$$

We call this the *power rule for integration*. Since $u'(x) \, dx$ is the differential of u, namely du, using mathematical shorthand we can write this rule as

$$\boxed{\int u^n \, du = \frac{u^{n+1}}{n+1} + C, \qquad \text{if } n \neq -1.}$$

EXAMPLE 1

Use the power rule for integration to find the following indefinite integrals.

a. $\int (x+1)^{20} \, dx$.

Since we have a power of $x + 1$, we shall set $u = x + 1$. Then $du = dx$ and $\int (x+1)^{20} \, dx$ has the form $\int u^{20} \, du$. Hence,

$$\int (x+1)^{20} \, dx = \int u^{20} \, du = \frac{u^{21}}{21} + C = \frac{(x+1)^{21}}{21} + C.$$

Note that we give our answer not in terms of u but explicitly in terms of x.

b. $\int 3x^2(x^3 + 7)^3 \, dx$.

Let $u = x^3 + 7$. Then $du = 3x^2 \, dx$. Fortunately, $3x^2$ appears in the integrand and can be used as part of du.

$$\int 3x^2(x^3 + 7)^3 \, dx = \int (x^3 + 7)^3 [3x^2 \, dx] = \int u^3 \, du$$

$$= \frac{u^4}{4} + C = \frac{(x^3 + 7)^4}{4} + C.$$

EXAMPLE 2

Find $\int x\sqrt{x^2 + 5}\ dx$.

We can write this integral as

$$\int x(x^2 + 5)^{1/2}\ dx.$$

If $u = x^2 + 5$, then $du = 2x\ dx$. Since the *constant* factor 2 in du does *not* appear in the integrand, this integral does not have the form $\int u^n\ du$. However, we can put it in this form by first multiplying and dividing the integrand by 2. This does not change its value. Thus,

$$\int x(x^2 + 5)^{1/2}\ dx = \int \frac{2}{2} x(x^2 + 5)^{1/2}\ dx = \int \frac{1}{2}(x^2 + 5)^{1/2}[2x\ dx].$$

Since the integrand now has $\frac{1}{2}$ as a *constant* factor, we have

$$\int x(x^2 + 5)^{1/2}\ dx = \frac{1}{2}\int (x^2 + 5)^{1/2}[2x\ dx]$$

$$= \frac{1}{2}\int u^{1/2}\ du = \frac{1}{2}\left[\frac{u^{3/2}}{\frac{3}{2}}\right] + C.$$

Going back to x, we have

$$\int x\sqrt{x^2 + 5}\ dx = \frac{(x^2 + 5)^{3/2}}{3} + C.$$

In Example 2 it was necessary to adjust for a *constant* factor in the integrand. In fact, by formula (4) of Table 10-1, if c is a nonzero constant, then

$$\int f(x)\ dx = \frac{c}{c}\int f(x)\ dx = \frac{1}{c}\int cf(x)\ dx.$$

That is, we can multiply the integrand by a nonzero constant c as long as we compensate for this by multiplying the entire integral by $1/c$. Such a manipulation cannot be performed with variable factors.

EXAMPLE 3

Find the following indefinite integrals.

a. $\int \sqrt[3]{6y}\ dy$.

If we set $u = 6y$, then $du = 6\ dy$. Since the factor 6 does not appear in the integrand, we insert a factor of 6 and adjust for it with a factor of $\frac{1}{6}$.

$$\int \sqrt[3]{6y}\, dy = \int (6y)^{1/3}\, dy = \frac{1}{6} \int (6y)^{1/3}[6\, dy] = \frac{1}{6} \int u^{1/3}\, du$$

$$= \left(\frac{1}{6}\right) \frac{u^{4/3}}{\frac{4}{3}} + C = \frac{(6y)^{4/3}}{8} + C.$$

b. $\displaystyle\int \frac{2x^3 + 3x}{(x^4 + 3x^2 + 7)^4}\, dx.$

We can write this as $\int (x^4 + 3x^2 + 7)^{-4}(2x^3 + 3x)\, dx$. If $u = x^4 + 3x^2 + 7$, then $du = (4x^3 + 6x)\, dx$, which is two times the quantity $(2x^3 + 3x)\, dx$ in the integral. Thus we insert a factor of 2 and adjust for it with a factor of $\frac{1}{2}$.

$$\int (x^4 + 3x^2 + 7)^{-4}(2x^3 + 3x)\, dx$$

$$= \frac{1}{2} \int (x^4 + 3x^2 + 7)^{-4}[2(2x^3 + 3x)\, dx]$$

$$= \frac{1}{2} \int (x^4 + 3x^2 + 7)^{-4}[(4x^3 + 6x)\, dx]$$

$$= \frac{1}{2} \int u^{-4}\, du = \frac{1}{2} \cdot \frac{u^{-3}}{-3} + C = -\frac{1}{6u^3} + C$$

$$= -\frac{1}{6(x^4 + 3x^2 + 7)^3} + C.$$

When using the power rule for integration, take care when making your choice for u. In Example 3(b) you would *not* be able to proceed very far if, for instance, you let $u = 2x^3 + 3x$. At times you may find it necessary to try many different choices. **Skill at integration comes only after many hours of practice and conscientious study.**

EXAMPLE 4

Find $\displaystyle\int 4x^2(x^4 + 1)^2\, dx.$

If we set $u = x^4 + 1$, then $du = 4x^3\, dx$. Here we are not able to get the form for du in the integral because we need an additional factor of the *variable* x. Remember: you can insert only a **constant** factor in an integral and can adjust for it in front of the integral sign. In our situation we can find the integral by first expanding the integrand.

$$\int 4x^2(x^4 + 1)^2 \, dx = 4 \int x^2(x^8 + 2x^4 + 1) \, dx$$

$$= 4 \int (x^{10} + 2x^6 + x^2) \, dx$$

$$= 4\left(\frac{x^{11}}{11} + \frac{2x^7}{7} + \frac{x^3}{3} \right) + C.$$

We now turn our attention to integrating exponential functions. If $u = u(x)$ is a differentiable function, then $d(e^u) = e^u \, du$. Thus,

$$\boxed{\int e^u \, du = e^u + C.}$$

(2)

EXAMPLE 5

Find the following integrals.

a. $\int 2xe^{x^2} \, dx.$

Let $u = x^2$. Then $du = 2x \, dx$ and by Eq. (2),

$$\int 2xe^{x^2} \, dx = \int e^{x^2}[2x \, dx] = \int e^u \, du$$

$$= e^u + C = e^{x^2} + C.$$

b. $\int (x^2 + 1)e^{x^3 + 3x} \, dx.$

If $u = x^3 + 3x$, then $du = (3x^2 + 3) \, dx = 3(x^2 + 1) \, dx$. If the integrand contained a factor of 3, the integral would have the form $\int e^u \, du$. Thus we write

$$\int (x^2 + 1)e^{x^3 + 3x} \, dx = \frac{1}{3} \int e^{x^3 + 3x}[3(x^2 + 1) \, dx]$$

$$= \frac{1}{3} \int e^u \, du = \frac{1}{3} e^u + C$$

$$= \frac{1}{3} e^{x^3 + 3x} + C.$$

PITFALL. *Do not apply the formula for $\int u^n \, du$ to $\int e^u \, du$. For example,*

$$\int e^x \, dx \neq \frac{e^{x+1}}{x + 1} + C.$$

As you know, the formula $\int u^n \, du = u^{n+1}/(n+1) + C$ assumes that $n \neq -1$. To find $\int u^{-1} \, du = \int \dfrac{1}{u} \, du$, we first observe that

$$d(\ln u) = \frac{d}{du}(\ln u) \, du = \frac{1}{u} \, du.$$

It would seem that $\int \dfrac{1}{u} \, du = \ln u + C$. However, the logarithm of a number u is defined if and only if u is positive. If $u < 0$, then $\ln u$ is not defined. Thus, $\int \dfrac{1}{u} \, du = \ln u + C$ as long as $u > 0$. On the other hand, if $u < 0$, then $-u > 0$ and $\ln(-u)$ *is* defined. Moreover,

$$d[\ln(-u)] = \frac{d}{du}[\ln(-u)] \, du = \frac{1}{-u}(-1) \, du = \frac{1}{u} \, du.$$

In this case ($u < 0$), $\int \dfrac{1}{u} \, du = \ln(-u) + C$. In summary, if $u > 0$, then $\int \dfrac{1}{u} \, du = \ln u + C$; if $u < 0$, then $\int \dfrac{1}{u} \, du = \ln(-u) + C$. Combining these cases, we write

$$\boxed{\int \frac{1}{u} \, du = \ln |u| + C, \qquad \text{if } u \neq 0.} \tag{3}$$

EXAMPLE 6

Find the following integrals.

a. $\int \dfrac{7}{x} \, dx.$

$$\int \frac{7}{x} \, dx = 7 \int \frac{1}{x} \, dx.$$

If $u = x$, then $du = dx$ and by Eq. (3) we have

$$\int \frac{7}{x} \, dx = 7 \ln |x| + C.$$

Using properties of logarithms, we can write this answer another way:

$$\int \frac{7}{x} \, dx = \ln |x^7| + C.$$

b. $\int \dfrac{2x}{x^2 + 5}\, dx.$

Let $u = x^2 + 5$. Then $du = 2x\, dx$. Thus

$$\int \frac{2x}{x^2 + 5}\, dx = \int \frac{1}{x^2 + 5}[2x\, dx] = \int \frac{1}{u}\, du$$

$$= \ln |u| + C = \ln |x^2 + 5| + C.$$

Since $x^2 + 5$ is always positive we can omit the absolute-value bars and write

$$\int \frac{2x}{x^2 + 5}\, dx = \ln (x^2 + 5) + C.$$

c. $\int \dfrac{(2x^3 + 3x)\, dx}{x^4 + 3x^2 + 7}.$

If $u = x^4 + 3x^2 + 7$, then $du = (4x^3 + 6x)\, dx$, which is two times the given numerator. Thus we insert a factor of 2 and adjust for it with a factor of $\frac{1}{2}$.

$$\int \frac{2x^3 + 3x}{x^4 + 3x^2 + 7}\, dx = \frac{1}{2} \int \frac{2(2x^3 + 3x)}{x^4 + 3x^2 + 7}\, dx$$

$$= \frac{1}{2} \int \frac{1}{x^4 + 3x^2 + 7}[(4x^3 + 6x)\, dx]$$

$$= \frac{1}{2} \int \frac{1}{u}\, du = \frac{1}{2} \ln |u| + C$$

$$= \frac{1}{2} \ln |x^4 + 3x^2 + 7| + C$$

$$= \ln \sqrt{x^4 + 3x^2 + 7} + C.$$

d. $\int \left[\dfrac{1}{(1 - w)^2} + \dfrac{1}{w - 1} \right] dw.$

$$\int \left[\frac{1}{(1 - w)^2} + \frac{1}{w - 1} \right] dw = \int (1 - w)^{-2}\, dw + \int \frac{1}{w - 1}\, dw$$

$$= -1 \int (1 - w)^{-2}[-dw] + \int \frac{1}{w - 1}\, dw.$$

On the last line the first integral has the form $\int u^{-2}\, du$, and the second has the form

$\int \dfrac{1}{v}\, dv$. Thus,

$$\int \left[\frac{1}{(1-w)^2} + \frac{1}{w-1} \right] dw = -\frac{(1-w)^{-1}}{-1} + \ln|w-1| + C$$

$$= \frac{1}{1-w} + \ln|w-1| + C.$$

For your convenience we list in Table 10-2 the basic integration formulas so far discussed. We assume that u is a function of x.

TABLE 10-2

BASIC INTEGRATION FORMULAS
1. $\int k\,du = ku + C, \quad k$ a constant.
2. $\int u^n\,du = \dfrac{u^{n+1}}{n+1} + C, \quad n \neq -1.$
3. $\int e^u\,du = e^u + C.$
4. $\int \dfrac{1}{u}\,du = \ln|u| + C, \quad u \neq 0.$
5. $\int kf(x)\,dx = k\int f(x)\,dx.$
6. $\int [f(x) \pm g(x)]\,dx = \int f(x)\,dx \pm \int g(x)\,dx.$

EXERCISE 10-2

In Problems **1–76**, find the indefinite integrals.

1. $\int (x+4)^8\, dx.$

2. $\int 2(x+3)^3\, dx.$

3. $\int 2x(x^2+16)^3\, dx.$

4. $\int (3x^2 + 14x)(x^3 + 7x^2 + 1)\, dx.$

5. $\int (3y^2 + 6y)(y^3 + 3y^2 + 1)^{2/3}\, dy.$

6. $\int (-12z^2 - 12z + 1)(-4z^3 - 6z^2 + z)^{18}\, dz.$

7. $\int \dfrac{3}{(3x-1)^3}\, dx.$

8. $\int \dfrac{4x}{(2x^2-7)^{10}}\, dx.$

9. $\int 3e^{3x}\, dx.$

10. $\int 2e^{2t+5}\, dt.$

11. $\int (2t + 1)e^{t^2+t}\, dt.$

12. $\int -3w^2 e^{-w^3}\, dw.$

13. $\int \sqrt{x + 10}\, dx.$

14. $\int \dfrac{1}{\sqrt{x - 2}}\, dx.$

15. $\int (7x - 6)^4\, dx.$

16. $\int x^2(3x^3 + 7)^3\, dx.$

17. $\int x(x^2 + 3)^{12}\, dx.$

18. $\int x\sqrt{1 + 2x^2}\, dx.$

19. $\int x^4(27 + x^5)^{1/3}\, dx.$

20. $\int x^3 e^{4x^4}\, dx.$

21. $\int xe^{5x^2}\, dx.$

22. $\int (3 - 2x)^{10}\, dx.$

23. $\int 6e^{-2x}\, dx.$

24. $\int x^4 e^{-6x^5}\, dx.$

25. $\int \dfrac{1}{x + 5}\, dx.$

26. $\int \dfrac{2x + 1}{x + x^2}\, dx.$

27. $\int \dfrac{3x^2 + 4x^3}{x^3 + x^4}\, dx.$

28. $\int \dfrac{3x^2 - 2x}{1 - x^2 + x^3}\, dx.$

29. $\int \dfrac{6z}{(z^2 - 6)^5}\, dz.$

30. $\int \dfrac{1}{(8y - 3)^3}\, dy.$

31. $\int \dfrac{4}{x}\, dx.$

32. $\int \dfrac{3}{1 + 2y}\, dy.$

33. $\int \dfrac{s^2}{s^3 + 5}\, ds.$

34. $\int \dfrac{2x^2}{3 - 4x^3}\, dx.$

35. $\int \dfrac{7}{5 - 3x}\, dx.$

36. $\int \dfrac{7t}{5t^2 - 6}\, dt.$

37. $\int \sqrt{5x}\, dx.$

38. $\int \dfrac{1}{(4x)^7}\, dx.$

39. $\int \dfrac{x}{\sqrt{x^2 - 4}}\, dx.$

40. $\int \dfrac{7}{3 - 2x}\, dx.$

41. $\int 2y^3 e^{y^4+1}\, dy.$

42. $\int \sqrt{4x - 3}\, dx.$

43. $\int v^2 e^{-2v^3+1}\, dv.$

44. $\int \dfrac{x^2}{\sqrt[3]{2x^3 + 9}}\, dx.$

45. $\int (e^{-5x} + 2e^x)\, dx.$

46. $\int 4\sqrt[3]{y + 1}\, dy.$

47. $\int (x + 1)(3 - 3x^2 - 6x)^3\, dx.$

48. $\int 2ye^{3y^2}\, dy.$

49. $\int \dfrac{x^2 + 2}{x^3 + 6x}\, dx.$

50. $\int (e^x - e^{-x} + e^{2x})\, dx.$

51. $\int \dfrac{16s - 4}{3 - 2s + 4s^2}\, ds.$

52. $\int (t^2 + 4t)(t^3 + 6t^2)^6\, dt.$

53. $\int x(2x^2 + 1)^{-1}\, dx.$

54. $\int (w^3 - 8w^7 + 1)(w^4 - 4w^8 + 4w)^{-6}\, dw.$

55. $\int -(x^2 - 2x^5)(x^3 - x^6)^{-10}\, dx.$

56. $\int \tfrac{3}{7}(v - 2)e^{2 - 4v + v^2}\, dv.$

57. $\int (2x^3 + x)(x^4 + x^2)\, dx,$

58. $\int (e^{3.1})^2\, dx.$

59. $\int \dfrac{18 + 12x}{(4 - 9x - 3x^2)^5}\, dx.$

60. $\int (e^x - e^{-x})^2\, dx.$

61. $\int x(2x + 1)e^{4x^3 + 3x^2 - 4}\, dx.$

62. $\int (u^2 + 3 - ue^{7 - u^2})\, du.$

63. $\int x\sqrt{(7 - 5x^2)^3}\, dx.$

64. $\int e^{-x/4}\, dx.$

65. $\int \dfrac{dx}{\sqrt{2x}}.$

66. $\int \dfrac{x^3}{e^{x^4}}\, dx.$

67. $\int (x^2 + 1)^2\, dx.$

68. $\int \left[x(x^2 - 16)^2 - \dfrac{1}{2x + 5} \right] dx.$

69. $\int \left[\dfrac{x}{x^2 + 1} + \dfrac{x^5}{(x^6 + 1)^2} \right] dx.$

70. $\int \left[\dfrac{1}{x - 1} + \dfrac{1}{(x - 1)^2} \right] dx.$

71. $\int \left[\dfrac{1}{3x - 5} - (x^2 - 2x^5)(x^3 - x^6)^{-10} \right] dx.$

72. $\int (r^3 + 5)^2\, dr.$

73. $\int \left[\sqrt{2x + 3} - \dfrac{x}{x^2 + 3} \right] dx.$

74. $\int \left[\dfrac{2x}{x^2 + 3} - \dfrac{x^3}{(x^4 + 2)^2} \right] dx.$

75. $\int \dfrac{e^{\sqrt{x}}}{\sqrt{x}}\, dx.$

76. $\int (e^4 - 2^e)\, dx.$ ← *constants*

In Problems **77–80**, find y subject to the given conditions.

77. $D_x y = (3 - 2x)^2; \quad y(0) = 1.$

78. $D_x y = x/(x^2 + 4); \quad y(1) = 0.$

79. $y'' = 1/x^2; \quad y'(-1) = 1, y(1) = 0.$

80. $y'' = \sqrt{x + 2}\,; \quad y'(2) = \tfrac{1}{3}, y(2) = -\tfrac{7}{15}.$

10-3 TECHNIQUES OF INTEGRATION

Now that you have had some practice in determining indefinite integrals, suppose we consider some problems of a greater degree of difficulty.

When one is integrating fractions, sometimes a preliminary division is needed to get familiar integration forms, as the next example shows.

EXAMPLE 1

a. $\int \dfrac{x^3 + x - 1}{x^2}\, dx.$

A familiar integration form is not apparent. However, we can write the integrand as the sum and difference of three fractions by dividing each term in the numerator by the denominator.

$$\int \frac{x^3 + x - 1}{x^2}\, dx = \int \left[\frac{x^3}{x^2} + \frac{x}{x^2} - \frac{1}{x^2} \right] dx = \int \left[x + \frac{1}{x} - \frac{1}{x^2} \right] dx$$

$$= \frac{x^2}{2} + \ln |x| - \int x^{-2}\, dx = \frac{x^2}{2} + \ln |x| + \frac{1}{x} + C.$$

b. $\int \dfrac{2x^3 + 3x^2 + x + 1}{2x + 1}\, dx.$

Here the integrand is a quotient of polynomials in which the degree of the numerator is greater than or equal to that of the denominator, and the denominator has more than one term. In such a situation, in order to integrate we first use long division until the degree of the remainder is less than that of the divisor.

$$\int \frac{2x^3 + 3x^2 + x + 1}{2x + 1}\, dx = \int \left(x^2 + x + \frac{1}{2x + 1} \right) dx$$

$$= \frac{x^3}{3} + \frac{x^2}{2} + \int \frac{1}{2x + 1}\, dx$$

$$= \frac{x^3}{3} + \frac{x^2}{2} + \frac{1}{2} \int \frac{1}{2x + 1} [2\, dx]$$

$$= \frac{x^3}{3} + \frac{x^2}{2} + \frac{1}{2} \ln |2x + 1| + C.$$

EXAMPLE 2

Find the following indefinite integrals.

a. $\int \dfrac{1}{\sqrt{x}\,(\sqrt{x} - 2)^3}\, dx.$

We can write this integral as

$$\int \frac{(\sqrt{x} - 2)^{-3}}{\sqrt{x}} dx.$$

Let $u = \sqrt{x} - 2$. Then $du = \dfrac{1}{2\sqrt{x}} dx$ and

$$\int \frac{(\sqrt{x} - 2)^{-3}}{\sqrt{x}} dx = 2 \int (\sqrt{x} - 2)^{-3} \left[\frac{1}{2\sqrt{x}} dx \right]$$

$$= 2 \int u^{-3} du = 2 \left(\frac{u^{-2}}{-2} \right) + C$$

$$= -u^{-2} + C = -(\sqrt{x} - 2)^{-2} + C.$$

b. $\int \dfrac{1}{x \ln x} dx.$

If $u = \ln x$, then $du = \dfrac{1}{x} dx$ and

$$\int \frac{1}{x \ln x} dx = \int \frac{1}{\ln x} \left(\frac{1}{x} dx \right) = \int \frac{1}{u} du$$

$$= \ln |u| + C = \ln |\ln x| + C.$$

c. $\int \dfrac{5}{w(\ln w)^{3/2}} dw.$

If $u = \ln w$, then $du = \dfrac{1}{w} dw$ and

$$\int \frac{5}{w(\ln w)^{3/2}} dw = 5 \int (\ln w)^{-3/2} \left(\frac{1}{w} dw \right)$$

$$= 5 \int u^{-3/2} du = 5 \cdot \frac{u^{-1/2}}{-\frac{1}{2}} + C$$

$$= \frac{-10}{u^{1/2}} + C = \frac{-10}{(\ln w)^{1/2}} + C.$$

Integrals of the form $\int a^u \, du$ can be handled by writing a as $e^{\ln a}$ (see Property 9 in Sec. 5-2), as the next example shows.

EXAMPLE 3

Determine $\int 2^{3-x} \, dx.$

Since $2 = e^{\ln 2}$,

$$\int 2^{3-x} \, dx = \int (e^{\ln 2})^{3-x} \, dx = \int e^{(\ln 2)(3-x)} \, dx.$$

If we let $u = (\ln 2)(3 - x)$, then $du = (-\ln 2) \, dx$.

$$\int e^{(\ln 2)(3-x)} \, dx = -\frac{1}{\ln 2} \int e^{(\ln 2)(3-x)}[(-\ln 2) \, dx]$$

$$= -\frac{1}{\ln 2} \int e^u \, du = -\frac{1}{\ln 2} e^u + C$$

$$= -\frac{1}{\ln 2} e^{(\ln 2)(3-x)} + C.$$

Since $e^{(\ln 2)(3-x)} = 2^{3-x}$, we have

$$\int 2^{3-x} \, dx = -\frac{2^{3-x}}{\ln 2} + C.$$

EXAMPLE 4

The Minister of Economic Affairs of a country determined that the marginal propensity to consume for his country is given by

$$\frac{dC}{dI} = \frac{3}{4} - \frac{1}{2\sqrt{3I}},$$

where consumption C is a function of national income I, written $C = C(I)$. Here I is expressed in billions of slugs (50 slugs = \$.01). Determine the consumption function for the country if it is known that consumption is 10 billion slugs ($C = 10$) when $I = 12$.

Since the marginal propensity to consume is the derivative of the consumption function C, we have

$$C(I) = \int \left(\frac{3}{4} - \frac{1}{2\sqrt{3I}} \right) dI = \int \frac{3}{4} \, dI - \frac{1}{2} \int (3I)^{-1/2} \, dI$$

$$= \frac{3}{4} I - \frac{1}{2} \int (3I)^{-1/2} \, dI.$$

If we let $u = 3I$, then $du = 3 \, dI$ and

$$C(I) = \frac{3}{4} I - \left(\frac{1}{2} \right) \frac{1}{3} \int (3I)^{-1/2}[3 \, dI]$$

$$= \frac{3}{4} I - \frac{1}{6} \frac{(3I)^{1/2}}{\frac{1}{2}} + C_1.$$

$$C(I) = \frac{3}{4} I - \frac{\sqrt{3I}}{3} + C_1.$$

When $I = 12$, then $C(I) = 10$ and thus

$$10 = \frac{3}{4}(12) - \frac{\sqrt{3(12)}}{3} + C_1.$$

$$10 = 9 - 2 + C_1.$$

$$C_1 = 3.$$

The consumption function is

$$C(I) = \frac{3}{4}I - \frac{\sqrt{3I}}{3} + 3.$$

EXERCISE 10-3

In Problems 1–42, determine the indefinite integrals.

1. $\displaystyle\int \frac{3x^3 + x^2 - x}{x^2}\, dx.$

2. $\displaystyle\int \frac{3x^2 - 7x}{4x}\, dx.$

3. $\displaystyle\int (3x^2 + 2)\sqrt{2x^3 + 4x + 1}\, dx.$

4. $\displaystyle\int \frac{x}{\sqrt[3]{x^2 + 5}}\, dx.$

5. $\displaystyle\int \frac{4}{\sqrt{2 - 3x}}\, dx.$

6. $\displaystyle\int \frac{xe^{x^2}\, dx}{e^{x^2} - 2}.$

7. $\displaystyle\int 4^{7x}\, dx.$

8. $\displaystyle\int 3^x\, dx.$

9. $\displaystyle\int 2x(7 - e^{x^2/4})\, dx.$

10. $\displaystyle\int \left(e^x + x^e + ex + \frac{e}{x}\right) dx.$

11. $\displaystyle\int \frac{3e^{2x}}{e^{2x} + 1}\, dx.$

12. $\displaystyle\int (e^{4 - 3x})^2\, dx.$

13. $\displaystyle\int \frac{e^{7/x}}{x^2}\, dx.$

14. $\displaystyle\int \frac{2x^4 - 6x^3 + x - 2}{x - 2}\, dx.$

15. $\displaystyle\int \frac{(\sqrt{x} + 2)^2}{3\sqrt{x}}\, dx.$

16. $\displaystyle\int \frac{3e^s}{6 + 5e^s}\, ds.$

17. $\displaystyle\int \frac{\ln x}{x}\, dx.$

18. $\displaystyle\int \sqrt{t}\,(5 - t\sqrt{t}\,)^4\, dt.$

19. $\displaystyle\int \frac{\ln^2 (r + 1)}{r + 1}\, dr.$

20. $\displaystyle\int \frac{8x^3 - 6x^2 - ex^4}{3x^3}\, dx.$

21. $\displaystyle\int x\sqrt{e^{x^2 + 3}}\, dx.$

22. $\displaystyle\int \frac{x + 3}{x + 6}\, dx.$

23. $\displaystyle\int \frac{1}{(x + 3)\ln (x + 3)}\, dx.$

24. $\displaystyle\int (x^{e^2} + 2x)\, dx.$

25. $\int \left(\dfrac{x^3}{\sqrt{x^4 - 1}} - \ln 4 \right) dx.$

26. $\int \dfrac{x - x^{-2}}{x^2 + 2x^{-1}} \, dx.$

27. $\int \dfrac{2x^4 - 8x^3 - 6x^2 + 4}{x^3} \, dx.$

28. $\int \dfrac{e^x + e^{-x}}{e^x - e^{-x}} \, dx.$

29. $\int \dfrac{6x^2 - 11x + 5}{3x - 1} \, dx.$

30. $\int \dfrac{(2x - 1)(x + 3)}{x - 5} \, dx.$

31. $\int \dfrac{x}{x - 1} \, dx.$

32. $\int \dfrac{x}{(x^2 + 1) \ln (x^2 + 1)} \, dx.$

33. $\int \dfrac{xe^{x^2}}{\sqrt{e^{x^2} + 2}} \, dx.$

34. $\int \dfrac{7}{(2x + 1)[1 + \ln (2x + 1)]^2} \, dx.$

35. $\int \dfrac{(e^{-x} + 6)^2}{e^x} \, dx.$

36. $\int \left[\dfrac{1}{8x + 1} - \dfrac{1}{e^x(8 + e^{-x})^2} \right] dx.$

37. $\int \sqrt{x} \, \sqrt{(8x)^{3/2} + 3} \, dx.$

38. $\int \dfrac{3}{x(\ln x)^{1/2}} \, dx.$

39. $\int \dfrac{\sqrt{s}}{e^{\sqrt{s^3}}} \, ds.$

40. $\int \dfrac{\ln^3 x}{3x} \, dx.$

41. $\int e^{\ln (x + 2)} \, dx.$

42. $\int dx.$

In Problems 43 and 44, dr/dq is a marginal revenue function. Find the demand function.

43. $\dfrac{dr}{dq} = \dfrac{200}{(q + 2)^2}.$

44. $\dfrac{dr}{dq} = \dfrac{900}{(2q + 3)^3}.$

In Problems 45 and 46, $\dfrac{dc}{dq}$ is a marginal cost function. Find the total cost function if fixed costs in each case are 2000.

45. $\dfrac{dc}{dq} = \sqrt{\dfrac{20}{q + 5}}.$

46. $\dfrac{dc}{dq} = 2e^{.001q}.$

In Problems 47–49, dC/dI represents marginal propensity to consume. Find the consumption function subject to the given condition.

47. $\dfrac{dC}{dI} = \dfrac{1}{\sqrt{I}}; \quad C(9) = 8.$

48. $\dfrac{dC}{dI} = \dfrac{3}{4} - \dfrac{1}{2\sqrt{3I}}; \quad C(3) = \dfrac{11}{4}.$

49. $\dfrac{dC}{dI} = \dfrac{3}{4} - \dfrac{1}{6\sqrt{I}}; \quad C(25) = 23.$

10-4 SUMMATION

To prepare you for further applications of integration, we need to discuss certain sums.

Consider finding the sum S of the first n positive integers:

$$S = 1 + 2 + \ldots + (n - 1) + n. \qquad (1)$$

Writing the right side of Eq. (1) in reverse order, we have

$$S = n + (n - 1) + \ldots + 2 + 1. \qquad (2)$$

Adding the corresponding sides of Eqs. (1) and (2) gives

$$
\begin{aligned}
S &= 1 &&+ 2 &&+ \ldots + (n - 1) + && n \\
S &= n &&+ (n - 1) &&+ \ldots + 2 &&+ 1 \\
\hline
2S &= (n + 1) &&+ (n + 1) &&+ \ldots + (n + 1) &&+ (n + 1).
\end{aligned}
$$

On the right side of the last equation the term $(n + 1)$ occurs n times. Thus $2S = n(n + 1)$, and so

$$S = \frac{n(n + 1)}{2} \qquad [\text{the sum of the first } n \text{ positive integers}]. \quad (3)$$

For example, the sum of the first 100 positive integers corresponds to $n = 100$ and is $100(100 + 1)/2$ or 5050.

For convenience, to indicate a sum we shall introduce *sigma notation*, so named because the Greek letter Σ (sigma) is used. For example,

$$\sum_{k=1}^{3} (2k + 5)$$

denotes the sum of those numbers obtained from the expression $2k + 5$ by first replacing k by 1, then by 2, and finally by 3. Thus

$$\sum_{k=1}^{3} (2k + 5) = [2(1) + 5] + [2(2) + 5] + [2(3) + 5]$$

$$= 7 + 9 + 11 = 27.$$

The letter k is called the *index of summation*; the numbers 1 and 3 are the *limits of summation* (1 is the *lower limit* and 3 is the *upper limit*). The symbol used for the index is a "dummy" symbol in the sense that it does not affect the sum of the terms. Any other letter can be used. For example,

$$\sum_{j=1}^{3} (2j + 5) = 7 + 9 + 11 = \sum_{k=1}^{3} (2k + 5).$$

EXAMPLE 1

Evaluate each of the following.

a. $\displaystyle\sum_{k=4}^{7} \frac{k^2 + 3}{2}$.

Here the sum begins with $k = 4$.

$$\sum_{k=4}^{7} \frac{k^2 + 3}{2} = \frac{4^2 + 3}{2} + \frac{5^2 + 3}{2} + \frac{6^2 + 3}{2} + \frac{7^2 + 3}{2}$$

$$= \frac{19}{2} + \frac{28}{2} + \frac{39}{2} + \frac{52}{2} = 69.$$

b. $\displaystyle\sum_{j=0}^{2} (-1)^{j+1}(j - 1)^2$.

$$\sum_{j=0}^{2} (-1)^{j+1}(j - 1)^2$$

$$= (-1)^{0+1}(0 - 1)^2 + (-1)^{1+1}(1 - 1)^2 + (-1)^{2+1}(2 - 1)^2$$

$$= (-1) + 0 + (-1) = -2.$$

To express the sum of the first n positive integers in sigma notation, we can write

$$\sum_{k=1}^{n} k = 1 + 2 + \ldots + n.$$

By Eq. (3),

$$\boxed{\sum_{k=1}^{n} k = \frac{n(n + 1)}{2}.} \tag{4}$$

Note in (4) that $\displaystyle\sum_{k=1}^{n} k$ is a function of n alone, not of k.

EXAMPLE 2

Evaluate each of the following.

a. $\displaystyle\sum_{k=1}^{60} k$.

Here we must find the sum of the first sixty positive integers. By Eq. (4) with $n = 60$,

$$\sum_{k=1}^{60} k = \frac{60(60 + 1)}{2} = 1830.$$

b. $\displaystyle\sum_{k=1}^{n-1} k.$

Here we must add the first $n - 1$ positive integers. Replacing n by $n - 1$ in Eq. (4), we obtain

$$\sum_{k=1}^{n-1} k = \frac{(n - 1)[(n - 1) + 1]}{2} = \frac{(n - 1)n}{2}.$$

Another useful formula is that for the sum of the *squares* of the first n positive integers. We shall use it in the next section.

$$\boxed{\sum_{k=1}^{n} k^2 = \frac{n(n + 1)(2n + 1)}{6}.} \tag{5}$$

EXAMPLE 3

Evaluate $1 + 4 + 9 + 16 + 25 + 36.$

This sum can be written as $\displaystyle\sum_{k=1}^{6} k^2$. By Eq. (5) with $n = 6$,

$$\sum_{k=1}^{6} k^2 = \frac{6(6 + 1)[2(6) + 1]}{6} = 91.$$

We conclude with a property of sigma. If $x_1, x_2, \ldots, x_n$ are real numbers and c is a constant, then

$$\sum_{i=1}^{n} cx_i = cx_1 + cx_2 + \ldots + cx_n$$

$$= c(x_1 + x_2 + \ldots + x_n) = c \sum_{i=1}^{n} x_i.$$

Thus,

$$\boxed{\sum_{i=1}^{n} cx_i = c \sum_{i=1}^{n} x_i.}$$

This means that a constant factor can "jump" before a sigma. For example,

$$\sum_{i=1}^{5} 3i^2 = 3\sum_{i=1}^{5} i^2.$$

By Eq. (5) we have

$$\sum_{i=1}^{5} 3i^2 = 3\sum_{i=1}^{5} i^2 = 3\left[\frac{5(6)(11)}{6}\right] = 165.$$

PITFALL. *Although constant factors can "jump" before sigma, nothing else can.*

EXERCISE 10-4

In Problems **1–10**, *evaluate the given sum.*

1. $\displaystyle\sum_{k=1}^{5} (k + 4)$.

2. $\displaystyle\sum_{k=12}^{15} (5 - 2k)$.

3. $\displaystyle\sum_{j=1}^{10} (-1)^j$.

4. $\displaystyle\sum_{j=0}^{5} 2^j$.

5. $\displaystyle\sum_{n=2}^{3} (3n^2 - 7)$.

6. $\displaystyle\sum_{n=2}^{4} \frac{n + 1}{n - 1}$.

7. $\displaystyle\sum_{k=3}^{4} \frac{(-1)^k(k + 1)}{2^k}$.

8. $\displaystyle\sum_{n=1}^{5} 1$.

9. $\displaystyle\sum_{k=1}^{3} \frac{(-1)^{k-1}(1 - k^2)}{k}$.

10. $\displaystyle\sum_{n=1}^{4} (n^2 + n)$.

In Problems **11–16**, *express the given sums in sigma notation.*

11. $1 + 2 + 3 + \ldots + 15$.

12. $7 + 8 + 9 + 10$.

13. $1 + 3 + 5 + 7$.

14. $2 + 4 + 6 + 8$.

15. $1^2 + 2^2 + 3^2 + \ldots + 12^2$.

16. $3 + 6 + 9 + 12$.

In Problems **17–22**, *by using Eqs. (4) and (5) evaluate the sums.*

17. $\displaystyle\sum_{k=1}^{450} k$.

18. $\displaystyle\sum_{k=1}^{10} k^2$.

19. $\displaystyle\sum_{j=1}^{6} 4j$.

20. $\displaystyle\sum_{i=1}^{40} \frac{i}{2}$.

21. $\displaystyle\sum_{i=1}^{6} 3i^2$.

22. $\displaystyle\sum_{j=1}^{8} \left(\frac{j}{2}\right)^2$.

23. A company has an asset whose original value is $3200 and which has no salvage value. The maintenance cost each year is $100 and increases by $100 each year. Show that the average annual total cost C over a period of n years is

$$C = \frac{3200}{n} + 50(n + 1).$$

Find the value of n that minimizes C. What is the average annual cost at this value of n?

10-5 THE DEFINITE INTEGRAL

Figure 10-1 shows the region bounded by the lines $y = f(x) = 2x$, $y = 0$ (the x-axis), and $x = 1$. It is simply a right triangle. If b and h are the lengths of the base and the height, respectively, then from geometry the area A of the triangle is $A = \frac{1}{2}bh = \frac{1}{2}(1)(2) = 1$ square unit. We shall now find this area by another method which, as you will see later, applies to more complex regions. This method involves summation of areas of rectangles.

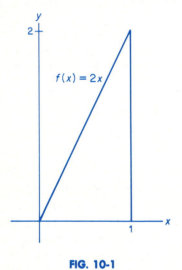

FIG. 10-1

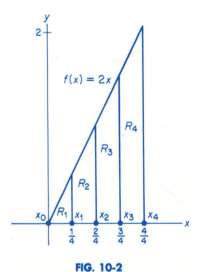

FIG. 10-2

Let us divide the interval $[0, 1]$ on the x-axis into four subintervals of equal length by means of the equally spaced points $x_0 = 0$, $x_1 = \frac{1}{4}$, $x_2 = \frac{2}{4}$, $x_3 = \frac{3}{4}$, and $x_4 = \frac{4}{4} = 1$ (see Fig. 10-2). Each subinterval has length $\Delta x = \frac{1}{4}$. These subintervals determine four subregions: R_1, R_2, R_3, and R_4, as indicated.

With each subregion we can associate a *circumscribed* rectangle (Fig. 10-3); that is, a rectangle whose base is the corresponding subinterval and whose height is the *maximum* value of $f(x)$ on that subinterval. Since f is an increasing

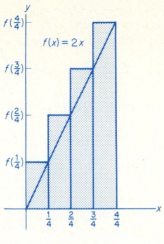

FIG. 10-3

function, the maximum value of $f(x)$ on each subinterval occurs when x is the right-hand endpoint. Thus the areas of the circumscribed rectangles associated with regions R_1, R_2, R_3, and R_4 are $\frac{1}{4}f(\frac{1}{4})$, $\frac{1}{4}f(\frac{2}{4})$, $\frac{1}{4}f(\frac{3}{4})$, and $\frac{1}{4}f(\frac{4}{4})$, respectively. The area of each rectangle is an approximation to the area of its corresponding subregion. Thus the sum of the areas of these rectangles, denoted by $\overline{S}_4$, (read "S sub 4 upper bar" or "the fourth upper sum"), approximates the area A of the triangle.

$$\overline{S}_4 = \tfrac{1}{4}f\left(\tfrac{1}{4}\right) + \tfrac{1}{4}f\left(\tfrac{2}{4}\right) + \tfrac{1}{4}f\left(\tfrac{3}{4}\right) + \tfrac{1}{4}f\left(\tfrac{4}{4}\right)$$

$$= \tfrac{1}{4}\left[2\left(\tfrac{1}{4}\right) + 2\left(\tfrac{2}{4}\right) + 2\left(\tfrac{3}{4}\right) + 2\left(\tfrac{4}{4}\right)\right] = \tfrac{5}{4}.$$

You may verify that we can write $\overline{S}_4$ as $\overline{S}_4 = \sum_{i=1}^{4} f(x_i)\,\Delta x$. The fact that $\overline{S}_4$ is greater than the actual area of the triangle might have been expected, since $\overline{S}_4$ includes areas of shaded regions that are not in the triangle (see Fig. 10-3).

On the other hand, with each subregion we can also associate an *inscribed* rectangle (see Fig. 10-4); that is, a rectangle whose base is the corresponding subinterval but whose height is the *minimum* value of $f(x)$ on that subinterval. Since f is an increasing function, the minimum value of $f(x)$ on each subinterval will occur when x is the left-hand endpoint. Thus the areas of the four inscribed rectangles associated with R_1, R_2, R_3, and R_4 are $\frac{1}{4}f(0)$, $\frac{1}{4}f(\frac{1}{4})$, $\frac{1}{4}f(\frac{2}{4})$, and $\frac{1}{4}f(\frac{3}{4})$, respectively. Their sum, denoted $\underline{S}_4$ (read "S sub 4 lower bar" or "the fourth lower sum"), is also an approximation to the area A of the triangle.

$$\underline{S}_4 = \tfrac{1}{4}f(0) + \tfrac{1}{4}f\left(\tfrac{1}{4}\right) + \tfrac{1}{4}f\left(\tfrac{2}{4}\right) + \tfrac{1}{4}f\left(\tfrac{3}{4}\right)$$

$$= \tfrac{1}{4}\left[2(0) + 2\left(\tfrac{1}{4}\right) + 2\left(\tfrac{2}{4}\right) + 2\left(\tfrac{3}{4}\right)\right] = \tfrac{3}{4}.$$

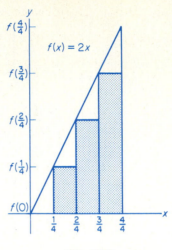

FIG. 10-4

Using sigma notation, we can write $\underline{S}_4$ as $\underline{S}_4 = \sum\limits_{i=0}^{3} f(x_i)\,\Delta x$. Note that $\underline{S}_4$ is less than the area of the triangle because the rectangles do not account for that portion of the triangle which is not shaded in Fig. 10-4.

Since $\frac{3}{4} = \underline{S}_4 \leqslant A \leqslant \overline{S}_4 = \frac{5}{4}$, we say that $\underline{S}_4$ is an approximation to A from *below* and $\overline{S}_4$ is an approximation to A from *above*.

If $[0, 1]$ is divided into more subintervals, we expect that better approximations to A will occur. To test this out, let us use six subintervals of equal length $\Delta x = \frac{1}{6}$. Then $\overline{S}_6$, the total area of six circumscribed rectangles (see Fig. 10-5),

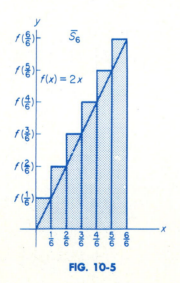

FIG. 10-5   **FIG. 10-6**

and $\underline{S}_6$, the total area of six inscribed rectangles (see Fig. 10-6) are

$$\overline{S}_6 = \tfrac{1}{6}f\left(\tfrac{1}{6}\right) + \tfrac{1}{6}f\left(\tfrac{2}{6}\right) + \tfrac{1}{6}f\left(\tfrac{3}{6}\right) + \tfrac{1}{6}f\left(\tfrac{4}{6}\right) + \tfrac{1}{6}f\left(\tfrac{5}{6}\right) + \tfrac{1}{6}f\left(\tfrac{6}{6}\right)$$

$$= \tfrac{1}{6}\left[2\left(\tfrac{1}{6}\right) + 2\left(\tfrac{2}{6}\right) + 2\left(\tfrac{3}{6}\right) + 2\left(\tfrac{4}{6}\right) + 2\left(\tfrac{5}{6}\right) + 2\left(\tfrac{6}{6}\right)\right] = \tfrac{7}{6}$$

and

$$\underline{S}_6 = \tfrac{1}{6}f(0) + \tfrac{1}{6}f\left(\tfrac{1}{6}\right) + \tfrac{1}{6}f\left(\tfrac{2}{6}\right) + \tfrac{1}{6}f\left(\tfrac{3}{6}\right) + \tfrac{1}{6}f\left(\tfrac{4}{6}\right) + \tfrac{1}{6}f\left(\tfrac{5}{6}\right)$$

$$= \tfrac{1}{6}\left[2(0) + 2\left(\tfrac{1}{6}\right) + 2\left(\tfrac{2}{6}\right) + 2\left(\tfrac{3}{6}\right) + 2\left(\tfrac{4}{6}\right) + 2\left(\tfrac{5}{6}\right)\right] = \tfrac{5}{6}.$$

Note that $\underline{S}_6 \leqslant A \leqslant \overline{S}_6$ and, with appropriate labelling, both $\overline{S}_6$ and $\underline{S}_6$ will be of the *form* $\Sigma f(x)\,\Delta x$. Using six subintervals gave better approximations to the area than did four subintervals, as expected.

More generally, if we divide $[0, 1]$ into n subintervals of equal length Δx, then $\Delta x = 1/n$ and the endpoints of the subintervals are $x = 0, 1/n, 2/n, \ldots, (n-1)/n$, and $n/n = 1$. See Fig. 10-7. The total area of n

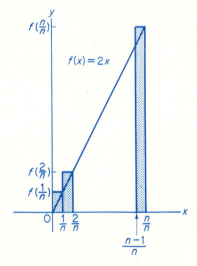

FIG. 10-7

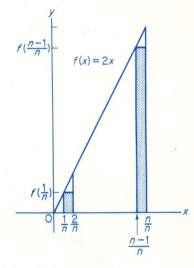

FIG. 10-8

circumscribed rectangles is

$$\overline{S}_n = \frac{1}{n}f\left(\frac{1}{n}\right) + \frac{1}{n}f\left(\frac{2}{n}\right) + \ldots + \frac{1}{n}f\left(\frac{n}{n}\right) \tag{1}$$

$$= \frac{1}{n}\left[2\left(\frac{1}{n}\right) + 2\left(\frac{2}{n}\right) + \ldots + 2\left(\frac{n}{n}\right)\right]$$

$$= \frac{2}{n^2}[1 + 2 + \ldots + n] \qquad \left(\text{by factoring } \frac{2}{n} \text{ from each term}\right).$$

From Sec. 10-4, the sum of the first n positive integers is $\dfrac{n(n+1)}{2}$. Thus

$$\bar{S}_n = \left(\frac{2}{n^2}\right)\frac{n(n+1)}{2} = \frac{n+1}{n}.$$

For n *inscribed* rectangles, the total area determined by the subintervals (see Fig. 10-8) is

$$\underline{S}_n = \frac{1}{n}f(0) + \frac{1}{n}f\left(\frac{1}{n}\right) + \ldots + \frac{1}{n}f\left(\frac{n-1}{n}\right) \qquad (2)$$

$$= \frac{1}{n}\left[2(0) + 2\left(\frac{1}{n}\right) + \ldots + 2\left(\frac{n-1}{n}\right)\right]$$

$$= \frac{2}{n^2}[1 + \ldots + (n-1)].$$

Summing the first $n-1$ positive integers as we did in Example 2(b) of Sec. 10-4, we obtain

$$\underline{S}_n = \left(\frac{2}{n^2}\right)\frac{(n-1)n}{2} = \frac{n-1}{n}.$$

From Equations (1) and (2) we again see that both $\bar{S}_n$ and $\underline{S}_n$ are sums of the *form* $\Sigma f(x)\,\Delta x$, namely $\bar{S}_n = \displaystyle\sum_{k=1}^{n} f\left(\frac{k}{n}\right)\Delta x$ and $\underline{S}_n = \displaystyle\sum_{k=0}^{n-1} f\left(\frac{k}{n}\right)\Delta x$.

From the nature of $\bar{S}_n$ and $\underline{S}_n$, it seems reasonable and it is indeed true that

$$\underline{S}_n \leqslant A \leqslant \bar{S}_n.$$

As n becomes larger, $\underline{S}_n$ and $\bar{S}_n$ become better approximations to A. In fact, let us take the limit of $\underline{S}_n$ and $\bar{S}_n$ as n approaches ∞ through positive integral values.

$$\lim_{n\to\infty} \underline{S}_n = \lim_{n\to\infty} \frac{n-1}{n} = \lim_{n\to\infty}\left(1 - \frac{1}{n}\right) = 1.$$

$$\lim_{n\to\infty} \bar{S}_n = \lim_{n\to\infty} \frac{n+1}{n} = \lim_{n\to\infty}\left(1 + \frac{1}{n}\right) = 1.$$

Since $\bar{S}_n$ and $\underline{S}_n$ have the same common limit, namely

$$\lim_{n\to\infty} \bar{S}_n = \lim_{n\to\infty} \underline{S}_n = 1, \qquad (3)$$

and since

$$\underline{S}_n \leqslant A \leqslant \bar{S}_n,$$

we shall take this limit to be the area of the triangle. Thus $A = 1$ square unit which agrees with our prior finding.

We define the common limit of $\bar{S}_n$ and $\underline{S}_n$, namely 1, to be the **definite integral** of $f(x) = 2x$ on the interval from $x = 0$ to $x = 1$, and we denote this by writing

$$\int_0^1 2x \, dx = 1. \tag{4}$$

The reason for using the term "definite integral" and the symbolism in Eq. (4) will become apparent in the next section. The numbers 0 and 1 appearing with the integral sign $\int$ in Eq. (4) are called the *limits of integration;* 0 is the *lower limit* and 1 is the *upper limit.*

In general, the definite integral of a function f over the interval from $x = a$ to $x = b$, where $a \leqslant b$, is the common limit of $\bar{S}_n$ and $\underline{S}_n$, if it exists, and is written

$$\int_a^b f(x) \, dx.$$

The symbol x is the *variable of integration* and $f(x)$ is the *integrand.* In terms of a limiting process we have

$$\sum f(x) \, \Delta x \to \int_a^b f(x) \, dx.$$

Two points must be made about the definite integral. First, the definite integral is a limit of a sum of the form $\sum f(x) \, \Delta x$. In fact, one can think of the integral sign as an elongated "S", the first letter of "Summation." Second, for an arbitrary function f defined on an interval, we may be able to calculate sums of the form $\sum f(x) \, \Delta x$ and determine their common limit if it exists. However, some terms in the sums may be negative if $f(x)$ is negative at points in the interval. These terms are not areas of rectangles (an area is never negative), and so the common limit may not represent area. Thus, the definite integral is nothing more than a real number; it may or may not represent area.

As you saw in (3), $\lim_{n \to \infty} \underline{S}_n$ is equal to $\lim_{n \to \infty} \bar{S}_n$. For an arbitrary function this is not always true. However, for the functions that we shall consider, these limits will be equal and the definite integral will always exist. To save time we shall just use the **right-hand endpoint** of each subinterval in computing a sum. For the functions in this section, this sum will be denoted S_n and will correspond to either $\underline{S}_n$ or $\bar{S}_n$.

EXAMPLE 1

a. *Find the area of the region in the first quadrant bounded by* $y = f(x) = 4 - x^2$ *and the lines* $x = 0$ *and* $y = 0$.

A sketch of the region appears in Fig. 10-9. The interval over which x varies in this region is seen to be $[0, 2]$, which we divide into n subintervals of equal length Δx. Since the length of $[0, 2]$ is 2, we take $\Delta x = 2/n$. The endpoints of the subintervals are $x = 0, 2/n, 2(2/n), \ldots, (n - 1)(2/n)$, and $n(2/n) = 2$ (see Fig. 10-10).

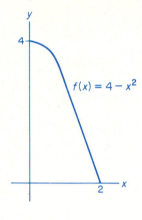

FIG. 10-9   **FIG. 10-10**

Using right-hand endpoints, we get

$$S_n = \frac{2}{n} f\left(\frac{2}{n}\right) + \frac{2}{n} f\left[2\left(\frac{2}{n}\right)\right] + \ldots + \frac{2}{n} f\left[n\left(\frac{2}{n}\right)\right]$$

$$= \frac{2}{n}\left[f\left(\frac{2}{n}\right) + f\left[2\left(\frac{2}{n}\right)\right] + \ldots + f\left[n\left(\frac{2}{n}\right)\right]\right]$$

$$= \frac{2}{n}\left[\left\{4 - \left[\frac{2}{n}\right]^2\right\} + \left\{4 - \left[2\left(\frac{2}{n}\right)\right]^2\right\} + \ldots + \left\{4 - \left[n\left(\frac{2}{n}\right)\right]^2\right\}\right].$$

Since the number 4 occurs n times in the sum, we can simplify S_n.

$$S_n = \frac{2}{n}\left[4n - \left(\frac{2}{n}\right)^2 - 2^2\left(\frac{2}{n}\right)^2 - \ldots - n^2\left(\frac{2}{n}\right)^2\right]$$

$$= \frac{2}{n}\left[4n - \left(\frac{2}{n}\right)^2\{1^2 + 2^2 + \ldots + n^2\}\right].$$

From Sec. 10-4, $\sum\limits_{k=1}^{n} k^2 = \dfrac{n(n+1)(2n+1)}{6}$ and thus

$$S_n = \frac{2}{n}\left[4n - \left(\frac{2}{n}\right)^2 \frac{n(n+1)(2n+1)}{6}\right]$$

$$= 8 - \frac{4(n+1)(2n+1)}{3n^2}$$

$$= 8 - \frac{4}{3}\left(\frac{2n^2+3n+1}{n^2}\right).$$

Finally we take the limit of S_n as $n \to \infty$.

$$\lim_{n\to\infty} S_n = \lim_{n\to\infty}\left[8 - \frac{4}{3}\left(\frac{2n^2+3n+1}{n^2}\right)\right]$$

$$= \lim_{n\to\infty}\left[8 - \frac{4}{3}\left(2 + \frac{3}{n} + \frac{1}{n^2}\right)\right]$$

$$= 8 - \frac{8}{3} = \frac{16}{3}.$$

Hence the area of the region is $\frac{16}{3}$ square units.

b. *Evaluate* $\int_0^2 (4 - x^2)\,dx.$

Since $\int_0^2 (4 - x^2)\,dx = \lim\limits_{n\to\infty} S_n$, from part (a) we conclude that

$$\int_0^2 (4 - x^2)\,dx = \frac{16}{3}.$$

EXAMPLE 2

Integrate $f(x) = x - 5$ *from* $x = 0$ *to* $x = 3$; *that is, evaluate* $\int_0^3 (x - 5)\,dx.$

A sketch of $f(x) = x - 5$ over $[0, 3]$ appears in Fig. 10-11. We divide $[0, 3]$ into n subintervals of equal length $\Delta x = 3/n$. The endpoints are $x = 0, 3/n, 2(3/n), \ldots,$ $(n-1)(3/n)$, and $n(3/n) = 3$. Note that $f(x)$ is negative at each endpoint. We form the sum

$$S_n = \frac{3}{n}f\left(\frac{3}{n}\right) + \frac{3}{n}f\left[2\left(\frac{3}{n}\right)\right] + \cdots + \frac{3}{n}f\left[n\left(\frac{3}{n}\right)\right].$$

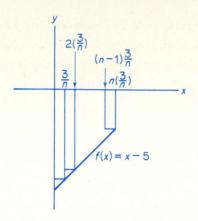

FIG. 10-11

Since all terms are negative, they do *not* represent areas of rectangles; in fact, they are the negatives of areas of rectangles. Simplifying, we have

$$S_n = \frac{3}{n}\left[\left\{\frac{3}{n} - 5\right\} + \left\{2\left(\frac{3}{n}\right) - 5\right\} + \ldots + \left\{n\left(\frac{3}{n}\right) - 5\right\}\right]$$

$$= \frac{3}{n}\left[-5n + \frac{3}{n}\{1 + 2 + \ldots + n\}\right]$$

$$= \frac{3}{n}\left[-5n + \left(\frac{3}{n}\right)\frac{n(n+1)}{2}\right]$$

$$= -15 + \frac{9}{2}\cdot\frac{n+1}{n}$$

$$= -15 + \frac{9}{2}\left(1 + \frac{1}{n}\right).$$

Taking the limit, we obtain

$$\lim_{n\to\infty} S_n = \lim_{n\to\infty}\left[-15 + \frac{9}{2}\left(1 + \frac{1}{n}\right)\right] = -15 + \frac{9}{2} = -\frac{21}{2}.$$

Thus

$$\int_0^3 (x - 5)\, dx = -\frac{21}{2}.$$

This definite integral is **not** the area of the region bounded by $f(x) = x - 5$, $y = 0$, $x = 0$, and $x = 3$. It represents the negative of that area.

In Example 2 it was shown that *the definite integral does not have to represent area.* In fact, there the definite integral was negative. However, if f is continuous and $f(x) \geqslant 0$ on $[a, b]$, then $S_n \geqslant 0$ for all n. Hence $\lim_{n\to\infty} S_n \geqslant 0$

and so $\int_a^b f(x)\, dx \geq 0$. Furthermore, this definite integral gives the area of the region bounded by $y = f(x)$, $y = 0$, $x = a$ and $x = b$ (see Fig. 10-12).

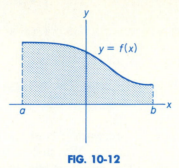

FIG. 10-12

Although the approach that we took to discuss the definite integral is sufficient for our purposes, it is by no means rigorous. **The important thing to remember about the definite integral is that it is the limit of a sum.**

EXERCISE 10-5

In Problems 1–4, sketch the region in the first quadrant that is bounded by the given curves. Approximate the area of the region by the indicated sum. Use the right-hand endpoint of each subinterval.

1. $f(x) = x$, $y = 0$, $x = 1$; S_3.

2. $f(x) = 3x$, $y = 0$, $x = 1$; S_5.

3. $f(x) = x^2$, $y = 0$, $x = 1$; S_3.

4. $f(x) = x^2 + 1$, $y = 0$, $x = 0$, $x = 1$; S_2.

In Problems 5–10, sketch the region in the first quadrant that is bounded by the given curves. Determine the exact area of the region by considering the limit of S_n as $n \to \infty$. Use the right-hand endpoint of each subinterval.

5. Region as described in Problem 1.

6. Region as described in Problem 2.

7. Region as described in Problem 3.

8. Region as described in Problem 4.

9. $f(x) = 2x^2$, $y = 0$, $x = 2$.

10. $f(x) = 9 - x^2$, $y = 0$, $x = 0$.

For each of the following problems, evaluate the given definite integral by taking the limit of

S_n. *Use the right-hand endpoint of each subinterval. Sketch the graph, over the given interval, of the function to be integrated.*

11. $\int_0^2 3x \, dx.$ **12.** $\int_0^4 9 \, dx.$ **13.** $\int_0^3 -4x \, dx.$

14. $\int_0^3 (2x - 9) \, dx.$ **15.** $\int_0^1 (x^2 + x) \, dx.$

10-6 THE FUNDAMENTAL THEOREM OF INTEGRAL CALCULUS

Until now the limiting processes of both the derivative and definite integral have been considered separately. We shall now bring these fundamental ideas together and develop the important relationship that exists between them. As a result, we may evaluate definite integrals more efficiently.

The graph of a function f is given in Fig. 10-13. Assume that f is continuous on the interval $[a, b]$ and its graph does not fall below the x-axis.

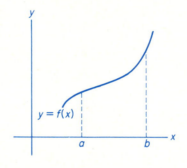

FIG. 10-13

That is, $f(x) \geqslant 0$. From the last section we know that the area of the region below the graph and above the x-axis from $x = a$ to $x = b$ is given by $\int_a^b f(x) \, dx$. We shall now consider another way to determine this area.

Suppose that there is a function $A = A(x)$, which we shall refer to as an "area" function, that gives the area of the region below the graph of f and above the x-axis from a to x, where $a \leqslant x \leqslant b$. This region is shaded in Fig. 10-14. Do not confuse $A(x)$, which is an area, with $f(x)$, which is the height of the graph at x.

From its definition we can state two properties of A immediately:

(1) $A(a) = 0$ since there is no area from a to a;

(2) $A(b)$ is the area from a to b; that is,

$$A(b) = \int_a^b f(x) \, dx.$$

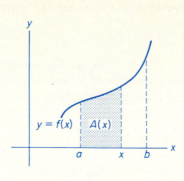

FIG. 10-14   **FIG. 10-15**

If x is increased by h units, then $A(x + h)$ is the area of the shaded region in Fig. 10-15. Hence $A(x + h) - A(x)$ is the difference of the areas in Figs. 10-15 and 10-14: namely, the area of the shaded region in Fig. 10-16. For h sufficiently close to zero, the area of this region is the same as the area of a

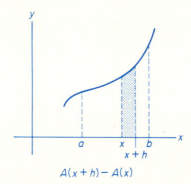

FIG. 10-16   **FIG. 10-17**

rectangle (Fig. 10-17) whose base is h and whose height is some value $\bar{y}$ between $f(x)$ and $f(x + h)$. Here $\bar{y}$ is a function of h. Thus the area of the rectangle is, on the one hand, $A(x + h) - A(x)$, and on the other hand it is $h\bar{y}$:

$$A(x + h) - A(x) = h\bar{y}$$

or

$$\frac{A(x + h) - A(x)}{h} = \bar{y} \qquad\qquad \left[\text{dividing by } h\right].$$

As $h \to 0$, then $\bar{y}$ approaches the number $f(x)$, so

$$\lim_{h \to 0} \frac{A(x + h) - A(x)}{h} = f(x). \tag{1}$$

But the left side is merely the derivative of A. Thus Eq. (1) becomes

$$A'(x) = f(x).$$

We conclude that the area function A has the additional property that its derivative A' is f. That is, A is an antiderivative of f. Now, suppose that F is *any* antiderivative of f. Since both A and F are antiderivatives of the same function, they must differ at most by a constant C:

$$A(x) = F(x) + C. \tag{2}$$

Recall that $A(a) = 0$. Evaluating both sides of Eq. (2) when $x = a$ gives

$$0 = F(a) + C$$

or

$$C = -F(a).$$

Thus Eq. (2) becomes

$$A(x) = F(x) - F(a). \tag{3}$$

If $x = b$, then from Eq. (3)

$$A(b) = F(b) - F(a). \tag{4}$$

But recall that

$$A(b) = \int_a^b f(x)\, dx. \tag{5}$$

From Eqs. (4) and (5) we get

$$\int_a^b f(x)\, dx = F(b) - F(a).$$

Thus a relationship between a definite integral and antidifferentiation has become clear. To find $\int_a^b f(x)\, dx$ it suffices to find an antiderivative of f, say F, and subtract the value of F at the lower limit a from its value at the upper limit b. We assumed here that f was continuous and $f(x) \geqslant 0$ so that we could appeal to the "area" concept. However, our result is true for any continuous function[†] and is known as the *Fundamental Theorem of Integral Calculus*.

[†] If f is continuous on $[a, b]$, it can be shown that $\int_a^b f(x)\, dx$ does indeed exist.

> **FUNDAMENTAL THEOREM OF INTEGRAL CALCULUS.**
>
> If f is continuous on the interval $[a, b]$ and F is any antiderivative of f there, then
>
> $$\int_a^b f(x)\, dx = F(b) - F(a).$$

It is crucial that you understand the difference between a definite integral and an indefinite integral. The **definite integral** $\int_a^b f(x)\, dx$ is a **number** defined to be the limit of a sum. The Fundamental Theorem states that the **indefinite integral** $\int f(x)\, dx$ (an antiderivative of f), which is a **function** of x and is related to the differentiation process, can be used to determine this limit.

Suppose we apply the Fundamental Theorem to evaluate $\int_0^2 (4 - x^2)\, dx$. Here $f(x) = 4 - x^2$, $a = 0$, and $b = 2$. Since an antiderivative of $4 - x^2$ is $F(x) = 4x - (x^3/3)$, then

$$\int_0^2 (4 - x^2)\, dx = F(2) - F(0) = \left(8 - \frac{8}{3}\right) - (0) = \frac{16}{3}.$$

This confirms our result in Example 1(b) of Sec. 10-5. If we had chosen $F(x)$ to be $4x - (x^3/3) + C$, then $F(2) - F(0) = [(8 - \frac{8}{3}) + C] - [0 + C] = \frac{16}{3}$ as before. Since the choice of the value of C is immaterial, for convenience we shall always choose it to be 0, as originally done. Usually, $F(b) - F(a)$ is abbreviated by writing

$$F(x)\Big|_a^b.$$

Hence we have

$$\int_0^2 (4 - x^2)\, dx = \left(4x - \frac{x^3}{3}\right)\Big|_0^2 = \left(8 - \frac{8}{3}\right) - 0 = \frac{16}{3}.$$

For a definite integral, we have the following convention:

$$\int_b^a f(x)\, dx = -\int_a^b f(x)\, dx.$$

That is, interchanging the limits of integration changes the integral's sign. For example,

$$\int_2^0 (4 - x^2)\, dx = -\int_0^2 (4 - x^2)\, dx.$$

Some properties of the definite integral deserve mention:

(1) $\int_a^b kf(x)\,dx = k\int_a^b f(x)\,dx$, where k is a constant.

(2) $\int_a^b [f(x) \pm g(x)]\,dx = \int_a^b f(x)\,dx \pm \int_a^b g(x)\,dx$.

(3) $\int_a^b f(x)\,dx = \int_a^b f(t)\,dt$. The variable of integration x used in $\int_a^b f(x)\,dx$ is a "dummy variable" in the sense that any other variable would produce the same result, that is, the same number. You may verify, for example, that $\int_0^2 x^2\,dx = \int_0^2 t^2\,dt$.

(4) If f is continuous on an interval I and a, b, and c are in I, then

$$\int_a^c f(x)\,dx = \int_a^b f(x)\,dx + \int_b^c f(x)\,dx.$$

This means that you may subdivide the interval over which a definite integral is to be evaluated. Thus,

$$\int_0^2 (4 - x^2)\,dx = \int_0^1 (4 - x^2)\,dx + \int_1^2 (4 - x^2)\,dx.$$

We shall look at some examples of definite integration now and compute some areas in the next section.

EXAMPLE 1

Evaluate each of the following definite integrals.

a. $\int_{-1}^3 (3x^2 - x + 6)\,dx$.

$$\int_{-1}^3 (3x^2 - x + 6)\,dx = \left(x^3 - \frac{x^2}{2} + 6x \right)\Big|_{-1}^3$$

$$= \left[3^3 - \frac{3^2}{2} + 6(3) \right] - \left[(-1)^3 - \frac{(-1)^2}{2} + 6(-1) \right]$$

$$= \left(\frac{81}{2} \right) - \left(-\frac{15}{2} \right) = 48.$$

b. $\int_0^1 \dfrac{x^3}{\sqrt{1+x^4}}\, dx.$

$$\int_0^1 \frac{x^3}{\sqrt{1+x^4}}\, dx = \int_0^1 x^3 (1+x^4)^{-1/2}\, dx$$

$$= \frac{1}{4}\int_0^1 (1+x^4)^{-1/2}[4x^3\, dx] = \left(\frac{1}{4}\right)\frac{(1+x^4)^{1/2}}{\frac{1}{2}}\Big|_0^1$$

$$= \frac{1}{2}(1+x^4)^{1/2}\Big|_0^1 = \frac{1}{2}(2)^{1/2} - \frac{1}{2}(1)^{1/2}$$

$$= \frac{1}{2}(\sqrt{2}-1).$$

PITFALL. *In part (b), the value of the antiderivative $\frac{1}{2}(1+x^4)^{1/2}$ at the lower limit 0 is $\frac{1}{2}(1)^{1/2}$. Do not assume that an evaluation at the limit zero will yield 0.*

c. $\int_1^2 [4t^{1/3} + t(t^2+1)^3]\, dt.$

$$\int_1^2 \left[4t^{1/3} + t(t^2+1)^3\right] dt = 4\int_1^2 t^{1/3}\, dt + \frac{1}{2}\int_1^2 (t^2+1)^3[2t\, dt]$$

$$= (4)\frac{t^{4/3}}{\frac{4}{3}}\Big|_1^2 + \left(\frac{1}{2}\right)\frac{(t^2+1)^4}{4}\Big|_1^2$$

$$= 3(2^{4/3}-1) + \frac{1}{8}(5^4 - 2^4)$$

$$= 3\cdot 2^{4/3} - 3 + \frac{609}{8}$$

$$= 6\sqrt[3]{2} + \frac{585}{8}.$$

d. $\int_0^1 e^{3t}\, dt.$

$$\int_0^1 e^{3t}\, dt = \frac{1}{3}\int_0^1 e^{3t}[3\, dt]$$

$$= \left(\frac{1}{3}\right)e^{3t}\Big|_0^1 = \frac{1}{3}(e^3 - e^0) = \frac{1}{3}(e^3 - 1).$$

EXAMPLE 2

Evaluate $\int_{-2}^{1} x^3 \, dx.$

$$\int_{-2}^{1} x^3 \, dx = \frac{x^4}{4}\Big|_{-2}^{1} = \frac{1^4}{4} - \frac{(-2)^4}{4} = \frac{1}{4} - \frac{16}{4} = -\frac{15}{4}.$$

The reason the result is negative is clear from the graph of $y = x^3$ on the interval $[-2, 1]$ (see Fig. 10-18). For $-2 < x < 0$, $f(x)$ is negative. Since a definite integral is a limit of a sum of the form $\Sigma f(x) \, \Delta x$, then $\int_{-2}^{0} x^3 \, dx$ is not only a negative number, but it is also the

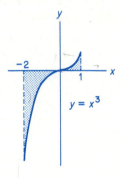

FIG. 10-18

negative of the area of the shaded region in the third quadrant. On the other hand, $\int_{0}^{1} x^3 \, dx$ is the area of the shaded region in the first quadrant. However, the definite integral over the entire interval $[-2, 1]$ is the *algebraic* sum of these numbers since

$$\int_{-2}^{1} x^3 \, dx = \int_{-2}^{0} x^3 \, dx + \int_{0}^{1} x^3 \, dx.$$

Thus $\int_{-2}^{1} x^3 \, dx$ does not represent the area between the curve and the x-axis. However, the area can be given in the form

$$\left| \int_{-2}^{0} x^3 \, dx \right| + \int_{0}^{1} x^3 \, dx.$$

PITFALL. *Remember that $\int_{a}^{b} f(x) \, dx$ is a limit of a sum. In some cases this limit represents area. In others it does not. Do not attach units of area to every definite integral.*

Since f is an antiderivative of f', by the Fundamental Theorem we have

$$\int_{a}^{b} f'(x) \, dx = f(b) - f(a).$$

But $f'(x)$ is the rate of change of f with respect to x. Thus, if we know the rate of change of f and want to find the difference in functional values $f(b) - f(a)$, it suffices to evaluate $\int_a^b f'(x)\, dx$.

EXAMPLE 3

A manufacturer's marginal cost function is

$$\frac{dc}{dq} = .6q + 2.$$

If production is presently set at $q = 80$ units per week, how much more would it cost to increase production to 100 units per week?

Since $c = c(q)$ is the total cost function, we want to find the difference $c(100) - c(80)$. However, the rate of change of c is dc/dq, and thus

$$c(100) - c(80) = \int_{80}^{100} \frac{dc}{dq}\, dq = \int_{80}^{100} (.6q + 2)\, dq$$

$$= \left[\frac{.6q^2}{2} + 2q \right]\Bigg|_{80}^{100} = [.3q^2 + 2q]\Bigg|_{80}^{100}$$

$$= \left[.3(100)^2 + 2(100) \right] - \left[.3(80)^2 + 2(80) \right]$$

$$= 3200 - 2080 = 1120.$$

If c is in dollars, then the cost of increasing production from 80 units to 100 units is $1120.

EXERCISE 10-6

In Problems **1–42**, evaluate the definite integral.

1. $\int_0^3 4\, dx.$

2. $\int_1^3 (2 + e)\, dx.$

3. $\int_1^2 3x\, dx.$

4. $\int_0^2 -5x\, dx.$

 5. $\int_{-1}^3 \frac{5}{3} x^3\, dx.$

6. $\int_0^{10} .04x^3\, dx.$

7. $\int_{-2}^1 (4x - 6)\, dx.$

8. $\int_{-1}^1 (5y + 2)\, dy.$

9. $\int_0^2 (t^2 + t)\, dt.$

10. $\int_1^3 (2w^2 + 1)\, dw.$

11. $\int_2^3 (y^2 - 2y + 1)\, dy.$

12. $\int_3^2 (2t - t^2)\, dt.$

13. $\int_{-2}^{-1} (3w^2 - w - 1)\, dw.$

14. $\int_8^9 dt.$

15. $\int_1^2 -4t^{-4}\, dt.$

16. $\int_1^2 \dfrac{x^{-2}}{2}\, dx.$

17. $\int_{-1}^1 \sqrt[3]{x^5}\, dx.$

18. $\int_{1/2}^{3/2} (x^2 + x + 1)\, dx.$

19. $\int_{1/2}^3 \dfrac{1}{x^2}\, dx.$

20. $\int_4^9 \left(\dfrac{1}{\sqrt{x}} - 2 \right) dx.$

21. $\int_{-1}^1 (z + 1)^5\, dz.$

22. $\int_1^8 (x^{1/3} - x^{-1/3})\, dx.$

23. $\int_0^1 2x^2(x^3 - 1)^3\, dx.$

24. $\int_1^3 (x + 3)^3\, dx.$

25. $\int_1^8 \dfrac{4}{y}\, dy.$

26. $\int_0^{e-1} \dfrac{1}{x + 1}\, dx.$

27. $\int_0^2 x^2\, e^{x^3}\, dx.$

28. $\int_0^1 (3x^2 + 4x)(x^3 + 2x^2)^4\, dx.$

29. $\int_4^5 \dfrac{2}{(x - 3)^3}\, dx.$

30. $\int_0^6 \sqrt{2x + 4}\, dx.$

31. $\int_{1/3}^2 \sqrt{10 - 3p}\, dp.$

32. $\int_{-1}^1 q\sqrt{q^2 + 3}\, dq.$

33. $\int_0^1 x^2 \sqrt[3]{7x^3 + 1}\, dx.$

34. $\int_0^{\sqrt{7}} \left[2x - \dfrac{x}{(x^2 + 1)^{5/3}} \right] dx.$

35. $\int_0^1 \dfrac{2x^3 + x}{x^2 + x^4 + 1}\, dx.$

36. $\int_a^b (m + ny)\, dy.$

37. $\int_0^1 (e^x - e^{-2x})\, dx.$

38. $\int_{-2}^1 |x|\, dx.$

39. $\int_1^e (x^{-1} + x^{-2} - x^{-3})\, dx.$

40. $\int_1^2 \left(6\sqrt{x} - \dfrac{1}{\sqrt{2x}} \right) dx.$

41. $\int_1^3 (x + 1)e^{x^2 + 2x}\, dx.$

42. $\int_3^4 \dfrac{e^{\ln x}}{x}\, dx.$

43. In statistics, the mean μ (a Greek letter read "mu") of the continuous probability density function f defined on the interval $[a, b]$ is

$$\mu = \int_a^b [\, x \cdot f(x)]\, dx,$$

and the variance σ^2 (σ is a Greek letter read "sigma") is

$$\sigma^2 = \int_a^b (x - \mu)^2 f(x)\, dx.$$

Compute μ and then σ^2 if $a = 0$, $b = 1$, and $f(x) = 1$.

44. In statistics, the cumulative probability function F is obtained from the continuous

probability density function f by the formula

$$F(t) = \int_0^t f(x)\,dx.$$

If $f(x) = 6x - 6x^2$ on the interval $[0, 1]$, find $F(t)$.

45. The economist Pareto[†] has stated an empirical law of distribution of higher incomes that gives the number N of persons receiving x or more dollars. If $dN/dx = -Ax^{-B}$, where A and B are constants, set up a definite integral that gives the total number of persons having incomes between a and b, where $a < b$.

46. If c_0 is the yearly consumption of a mineral at time $t = 0$, then under continuous consumption the total amount of the mineral used in the interval $[0, t_1]$ is

$$\int_0^{t_1} c_0\, e^{kt}\, dt,$$

where k is the rate of consumption. For a rare-earth mineral it has been determined that $c_0 = 3000$ units and $k = .05$. Evaluate the above integral for these data.

47. A manufacturer's marginal cost function is $dc/dq = .2q + 3$. If c is in dollars, determine the cost involved to increase production from 60 to 70 units.

48. Repeat Problem 47 if $dc/dq = .003q^2 - .6q + 40$ and production increases from 100 to 200 units.

49. A manufacturer's marginal revenue function is $dr/dq = 1000/\sqrt{100q}$. If r is in dollars, find the change in the manufacturer's total revenue if production is increased from 400 to 900 units.

50. Repeat Problem 49 if $dr/dq = 250 + 90q - 3q^2$ and production is increased from 10 to 20 units.

51. The present value (in dollars) of a continuous flow of income of \$2000 a year for five years at 6 percent compounded continuously is given by

$$\int_0^5 2000 e^{-.06t}\, dt.$$

Evaluate the present value to the nearest dollar.

52. The total expenditures (in dollars) of a business over the next five years is given by

$$\int_0^5 4000 e^{.05t}\, dt.$$

Evaluate the expenditures.

53. For a certain population, suppose l is a function such that $l(x)$ is the number of persons who reach the age of x in any year of time. This function is called a *life*

[†]G. Tintner, *Methodology of Mathematical Economics and Econometrics*, University of Chicago Press, Chicago, 1967, p. 16.

table function. Under appropriate conditions, the integral

$$\int_x^{x+n} l(t)\, dt$$

gives the expected number of people in the population between the exact ages of x and $x + n$, inclusive. If $l(x) = 10{,}000\sqrt{100 - x}$, determine the number of people between the exact ages of 36 and 64 inclusive. Give your answer to the nearest integer, since fractional answers make no sense.

54. Taagepera[†] considers a "one-dimensional" country of length $2R$ (see Fig. 10-19).

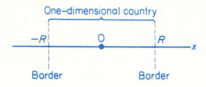

FIG. 10-19

Suppose the production of goods for this country is continuously distributed from border to border. If the amount produced each year per unit of distance is $f(x)$, then the country's total yearly production is given by

$$G = \int_{-R}^{R} f(x)\, dx.$$

Evaluate G if $f(x) = i$, where i is constant.

55. For the "one-dimensional" country of Problem 54, under certain conditions the amount E of the country's exports is given by

$$E = \int_{-R}^{R} \frac{i}{2}\left[e^{-k(R-x)} + e^{-k(R+x)}\right] dx,$$

where i and k are constants ($k \neq 0$). Evaluate E.

56. In a discussion of inventory, Barbosa and Friedman[‡] refer to the function

$$g(x) = \frac{1}{k}\int_1^{1/x} ku^r\, du,$$

where k and r are constants, $k > 0$ and $r > -2$, and $x > 0$. Verify the claim that

$$g'(x) = -\frac{1}{x^{r+2}}.$$

(*Hint.* Consider two cases: when $r \neq -1$, and when $r = -1$.)

[†]R. Taagepera, "Why the trade/GNP ratio decreases with country size," *Social Science Research*, vol. 5 (1976), 385–404.

[‡]L. C. Barbosa and M. Friedman. "Deterministic inventory lot size models—a general root law," *Management Science*, vol. 24, no. 8 (1978), 819–826.

10-7 AREA

In Sec. 10-5 we found the area of a region by evaluating the limit of a sum of the form $\Sigma f(x)\,\Delta x$. Since this limit also is a definite integral, we can use the Fundamental Theorem to evaluate the limit.

When using the definite integral to determine area, you should make a rough sketch of the region involved. Let us consider the area of the region bounded by $y = f(x)$ and the x-axis from $x = a$ to $x = b$, as shown in Fig. 10-20. Since we are summing areas of rectangles, a sample rectangle should be included in the sketch. This will help you understand the integration process. Such a rectangle (see Fig. 10-20) is called a **vertical element of area** (or a **vertical strip**). In the diagram the width of the vertical element is Δx. The length is the

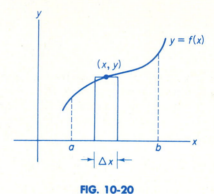

FIG. 10-20

y-value of the curve. Hence the rectangle has area $y\,\Delta x$ or $f(x)\,\Delta x$. We want to add the areas of all such elements between $x = a$ and $x = b$ and find the limit of this sum by means of definite integration:

$$\Sigma f(x)\,\Delta x \to \int_a^b f(x)\,dx.$$

For example, let us find the area of the region in the first quadrant that is bounded by the curve $y = x^2 - 1$, the x-axis, and the line $x = 2$ (see Fig. 10-21). The width of the sample element is Δx and its length is y. Thus the area of the element is $y\,\Delta x$. The limit of the sum of all such areas of elements between $x = 1$ and $x = 2$ is found by the definite integral. The limits of integration are $x = 1$ and $x = 2$.

$$\Sigma y\,\Delta x \to \int_1^2 y\,dx.$$

To evaluate this integral we must express the integrand in terms of the variable

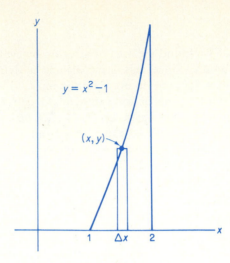

$$y = x^2 - 1$$

(x, y)

FIG. 10-21

of integration x. Since $y = x^2 - 1$,

$$\text{area} = \int_1^2 (x^2 - 1)\, dx = \left(\frac{x^3}{3} - x \right)\Bigg|_1^2$$

$$= \left(\frac{8}{3} - 2 \right) - \left(\frac{1}{3} - 1 \right)$$

$$= \frac{4}{3} \text{ sq units.}$$

EXAMPLE 1

Find the area of the region bounded by the curve $y = 6 - x - x^2$ and the x-axis.

First we must sketch the curve so that we can visualize the region. Since $y = -(x^2 + x - 6) = -(x - 2)(x + 3)$, the x-intercepts are $(2, 0)$ and $(-3, 0)$. Using techniques of graphing that were previously discussed, we obtain the graph shown in Fig. 10-22. Note that for this region, it is crucial that the x-intercepts of the curve be found in order to determine the values of x over which the areas of the vertical elements must be summed. For the vertical element shown, the width is Δx and the length is y. Hence the area of the element is $(6 - x - x^2)\, \Delta x$. Summing these from $x = -3$ to $x = 2$ and taking the limit gives

$$\text{area} = \int_{-3}^2 (6 - x - x^2)\, dx = \left(6x - \frac{x^2}{2} - \frac{x^3}{3} \right)\Bigg|_{-3}^2$$

$$= \left(12 - \frac{4}{2} - \frac{8}{3} \right) - \left(-18 - \frac{9}{2} - \frac{-27}{3} \right)$$

$$= \frac{125}{6} \text{ sq units.}$$

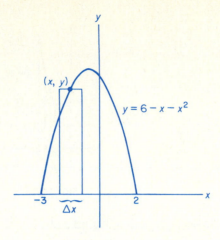

FIG. 10-22

EXAMPLE 2

Find the area of the region bounded by $y = x^2 + 2x + 2$, the x-axis, and the lines $x = -2$ and $x = 1$.

A sketch of the region is given in Fig. 10-23.

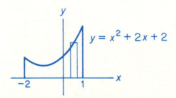

FIG. 10-23

$$\text{area} = \int_{-2}^{1} y \, dx = \int_{-2}^{1} (x^2 + 2x + 2) \, dx$$

$$= \left(\frac{x^3}{3} + x^2 + 2x \right)\Big|_{-2}^{1} = \left(\frac{1}{3} + 1 + 2 \right) - \left(-\frac{8}{3} + 4 - 4 \right)$$

$$= 6 \text{ sq units.}$$

EXAMPLE 3

Find the area between $y = e^x$ and the x-axis from $x = 1$ to $x = 2$.

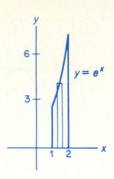

FIG. 10-24

A sketch of the region is given in Fig. 10-24.

$$\text{area} = \int_1^2 y\,dx = \int_1^2 e^x\,dx = e^x\Big|_1^2 = e^2 - e = e(e-1) \text{ sq units.}$$

EXAMPLE 4

Find the area of the region bounded by the curves $y = x^2 - x - 2$ and $y = 0$ (the x-axis) from $x = -2$ to $x = 2$.

A sketch of the region is given in Fig. 10-25. Notice that the x-intercepts are $(-1, 0)$ and $(2, 0)$.

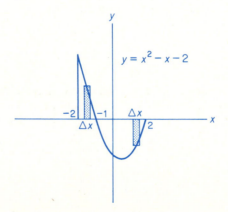

FIG. 10-25

PITFALL. *It is wrong to write hastily that the area is $\int_{-2}^{2}(x^2 - x - 2)\,dx$ for the following reason. For the left rectangle the length is y. However, for the rectangle on the right, y is negative and so the rectangle has length $-y$. Remember that an area is never negative. This points out the importance of sketching the region.*

On the interval $[-2, -1]$, the area of the element is

$$y \, \Delta x = (x^2 - x - 2) \, \Delta x.$$

On $[-1, 2]$ it is

$$-y \, \Delta x = -(x^2 - x - 2) \, \Delta x.$$

Thus

$$\text{area} = \int_{-2}^{-1} (x^2 - x - 2) \, dx + \int_{-1}^{2} -(x^2 - x - 2) \, dx$$

$$= \left(\frac{x^3}{3} - \frac{x^2}{2} - 2x \right) \Big|_{-2}^{-1} - \left(\frac{x^3}{3} - \frac{x^2}{2} - 2x \right) \Big|_{-1}^{2}$$

$$= \left[\left(-\frac{1}{3} - \frac{1}{2} + 2 \right) - \left(-\frac{8}{3} - \frac{4}{2} + 4 \right) \right] -$$

$$\left[\left(\frac{8}{3} - \frac{4}{2} - 4 \right) - \left(-\frac{1}{3} - \frac{1}{2} + 2 \right) \right] = \frac{19}{3} \text{ sq units.}$$

The next example shows the use of area as a probability in statistics.

EXAMPLE 5

*In statistics, a (probability) **density function** f of a variable x, where x assumes all values in the interval $[a, b]$, has the following properties:*

1. $f(x) \geqslant 0$.

2. $\int_{a}^{b} f(x) \, dx = 1$.

3. *The probability that x assumes a value between c and d, written $P(c \leqslant x \leqslant d)$, where $a \leqslant c \leqslant d \leqslant b$, is represented by the area of the region bounded by the graph of f and the x-axis between $x = c$ and $x = d$. Hence (see Fig. 10-26)*

$$P(c \leqslant x \leqslant d) = \int_{c}^{d} f(x) \, dx.$$

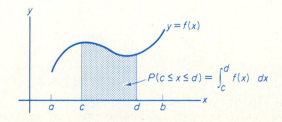

FIG. 10-26

For the density function $f(x) = 6(x - x^2)$, *where* $0 \leqslant x \leqslant 1$, *find* (a) $P(0 \leqslant x \leqslant \frac{1}{4})$ *and* (b)
$P(x \geqslant \frac{1}{2})$.

a. Here $[a, b]$ is $[0, 1]$, c is 0 and d is $\frac{1}{4}$. By property 3, we have

$$P\left(0 \leqslant x \leqslant \tfrac{1}{4}\right) = \int_0^{1/4} 6(x - x^2)\, dx = 6\int_0^{1/4}(x - x^2)\, dx$$

$$= 6\left(\frac{x^2}{2} - \frac{x^3}{3}\right)\Big|_0^{1/4}$$

$$= (3x^2 - 2x^3)\big|_0^{1/4}$$

$$= \left[3\left(\tfrac{1}{4}\right)^2 - 2\left(\tfrac{1}{4}\right)^3\right] - 0$$

$$= 3\left(\tfrac{1}{16}\right) - 2\left(\tfrac{1}{64}\right)$$

$$= \tfrac{3}{16} - \tfrac{1}{32}$$

$$= \tfrac{5}{32}.$$

b. Since f is defined for $0 \leqslant x \leqslant 1$, to say that $x \geqslant \frac{1}{2}$ means that $\frac{1}{2} \leqslant x \leqslant 1$. Thus,

$$P\left(x \geqslant \tfrac{1}{2}\right) = \int_{1/2}^1 6(x - x^2)\, dx$$

$$= 6\int_{1/2}^1 (x - x^2)\, dx$$

$$= 6\left(\frac{x^2}{2} - \frac{x^3}{3}\right)\Big|_{1/2}^1$$

$$= (3x^2 - 2x^3)\big|_{1/2}^1$$

$$= \left[3(1)^2 - 2(1)^3\right] - \left[3\left(\tfrac{1}{2}\right)^2 - 2\left(\tfrac{1}{2}\right)^3\right]$$

$$= (3 - 2) - \left(\tfrac{3}{4} - \tfrac{1}{4}\right)$$

$$= \tfrac{1}{2}.$$

EXERCISE 10-7

In Problems 1–34, find the area of the region bounded by the given curves, the x-axis, and the given lines. In each case first sketch the region.

1. $y = 4x$, $x = 2$.   2. $y = 3x + 1$, $x = 0$, $x = 4$.

3. $y = 3x + 2$, $x = 2$, $x = 3$.   4. $y = x + 5$, $x = 2$, $x = 4$.

5. $y = x - 1$, $x = 5$.   6. $y = 2x^2$, $x = 1$, $x = 2$.

7. $y = x^2$, $x = 2$, $x = 3$.   8. $y = 2x^2 - x$, $x = -2$, $x = -1$.

9. $y = x^2 + 2$, $x = -1$, $x = 2$.   10. $y = 2x + x^3$, $x = 1$.

11. $y = x^2 - 2x$, $x = -3$, $x = -1$.   12. $y = 3x^2 - 4x$, $x = -2$, $x = -1$.

13. $y = 9 - x^2$.   14. $y = \dfrac{4}{x}$, $x = 1$, $x = 2$.

15. $y = 1 - x - x^3$, $x = -2$, $x = 0$.   16. $y = e^x$, $x = 1$, $x = 3$.

17. $y = 3 + 2x - x^2$.   18. $y = \dfrac{1}{x^2}$, $x = 2$, $x = 3$.

19. $y = \dfrac{1}{x}$, $x = 1$, $x = e$.   20. $y = \dfrac{1}{x}$, $x = 1$, $x = e^2$.

21. $y = \sqrt{x + 9}$, $x = -9$, $x = 0$.   22. $y = x^2 - 2x$, $x = 1$, $x = 3$.

23. $y = \sqrt{2x - 1}$, $x = 1$, $x = 5$.   24. $y = x^3 + 3x^2$, $x = -2$, $x = 2$.

25. $y = \sqrt[3]{x}$, $x = 2$.   26. $y = x^2 - 4$, $x = -2$, $x = 2$.

27. $y = e^x$, $x = 0$, $x = 2$.   28. $y = |x|$, $x = -2$, $x = 2$.

29. $y = x + \dfrac{2}{x}$, $x = 1$, $x = 2$.   30. $y = 6 - x - x^2$.

31. $y = x^3$, $x = -2$, $x = 4$.   32. $y = \sqrt{x - 2}$, $x = 2$, $x = 6$.

33. $y = 2x - x^2$, $x = 1$, $x = 3$.   34. $y = x^2 - x + 1$, $x = 0$, $x = 1$.

35. Given

$$f(x) = \begin{cases} 3x^2, & \text{if } 0 \leqslant x \leqslant 2, \\ 16 - 2x, & \text{if } x > 2, \end{cases}$$

find the area of the region bounded by the graph of $y = f(x)$, the x-axis, and the line $x = 3$. Include a sketch of the region.

36. Under conditions of a continuous uniform distribution, a topic in statistics, the proportion of persons with incomes between a and t, where $a \leqslant t \leqslant b$, is the area of the region between the curve $y = 1/(b - a)$ and the x-axis from $x = a$ to $x = t$. Sketch the graph of the curve and determine the area of the given region.

37. Suppose $f(x) = \frac{1}{8}x$, where $0 \leqslant x \leqslant 4$. If f is a density function (see Example 5), find (a) $P(0 \leqslant x \leqslant 1)$, (b) $P(2 \leqslant x \leqslant 4)$, and (c) $P(x \geqslant 3)$.

38. Suppose $f(x) = 3(1 - x)^2$, where $0 \leqslant x \leqslant 1$. If f is a density function (see Example 5), find (a) $P(\frac{1}{2} \leqslant x \leqslant 1)$, (b) $P(\frac{1}{3} \leqslant x \leqslant \frac{1}{2})$, and (c) $P(x \leqslant \frac{1}{3})$. (d) Use your result from part c to determine $P(x \geqslant \frac{1}{3})$.

39. Suppose $f(x) = 1/x$, where $e \leqslant x \leqslant e^2$. If f is a density function (see Example 5), find (a) $P(3 \leqslant x \leqslant 5)$, (b) $P(x \leqslant 4)$, and (c) $P(x \geqslant 3)$. (d) Verify that $P(e \leqslant x \leqslant e^2) = 1$.

10-8 AREA BETWEEN CURVES

We shall now consider finding the area of a region enclosed by several curves. As before, our procedure will be to draw a sample element of area and use the definite integral to "add together" the areas of all such elements.

EXAMPLE 1

Find the area of the region bounded by the curves $y = \sqrt{x}$ and $y = x$.

A sketch of the region appears in Fig. 10-27. To determine where the curves intersect, we solve the system formed by the equations $y = \sqrt{x}$ and $y = x$. Eliminating y by substitu-

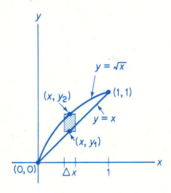

FIG. 10-27

tion, we obtain

$$\sqrt{x} = x.$$
$$x = x^2 \qquad \text{[squaring both sides]}.$$
$$0 = x^2 - x = x(x - 1).$$
$$x = 0 \qquad \text{or} \qquad x = 1.$$

If $x = 0$, then $y = 0$; if $x = 1$, then $y = 1$. Thus the curves intersect at $(0, 0)$ and $(1, 1)$. The width of the indicated element of area is Δx. The length is the y-value on the upper curve minus the y-value on the lower curve. If we distinguish between the curves by writing $y_1 = x$ and $y_2 = \sqrt{x}$, then the length of the element is

$$y_{\text{upper}} - y_{\text{lower}} = y_2 - y_1 = \sqrt{x} - x.$$

Thus the area of the element is $(\sqrt{x} - x)\,\Delta x$. Summing all such areas from $x = 0$ to $x = 1$ by the definite integral, we get the area of the entire region.

$$\Sigma (\sqrt{x} - x)\,\Delta x \rightarrow \int_0^1 (\sqrt{x} - x)\,dx.$$

$$\text{area} = \int_0^1 (x^{1/2} - x)\,dx = \left(\frac{x^{3/2}}{\frac{3}{2}} - \frac{x^2}{2} \right)\Bigg|_0^1$$

$$= \left(\frac{2}{3} - \frac{1}{2} \right) - (0 - 0) = \frac{1}{6} \text{ sq unit.}$$

It should be obvious to you that the points of intersection are important in determining the limits of integration.

Sometimes area can more easily be determined by summing areas of horizontal elements rather than vertical elements. In the following example an area will be found by both methods. In each case the element of area determines the form of the integral.

EXAMPLE 2

Find the area of the region bounded by the curve $y^2 = 4x$ and the lines $y = 3$ and $x = 0$ (the y-axis).

The region is sketched in Fig. 10-28. When the curves $y = 3$ and $y^2 = 4x$ intersect, then $9 = 4x$ and so $x = \frac{9}{4}$. Thus the point of intersection is $(\frac{9}{4}, 3)$. Since the width of the vertical strip is Δx, we integrate with respect to the variable x. Thus y_{upper} and y_{lower} must

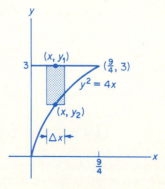

FIG. 10-28

be expressed as functions of x. For the curve $y^2 = 4x$, we have $y = \pm 2\sqrt{x}$. But for the portion of this curve that bounds the region we must have $y \geqslant 0$, so we use $y = 2\sqrt{x}$. Thus the length of the strip is $y_{\text{upper}} - y_{\text{lower}} = 3 - 2\sqrt{x}$. Hence the strip has an area of $(3 - 2\sqrt{x})\,\Delta x$ and we wish to sum up all such areas from $x = 0$ to $x = \frac{9}{4}$.

$$\text{area} = \int_0^{9/4} (3 - 2\sqrt{x})\, dx = \left(3x - \frac{4x^{3/2}}{3}\right)\Bigg|_0^{9/4}$$

$$= \left[3\left(\frac{9}{4}\right) - \frac{4}{3}\left(\frac{9}{4}\right)^{3/2}\right] - [0 - 0]$$

$$= \frac{27}{4} - \frac{4}{3}\left[\left(\frac{9}{4}\right)^{1/2}\right]^3 = \frac{27}{4} - \frac{4}{3}\left(\frac{3}{2}\right)^3 = \frac{9}{4} \text{ sq units.}$$

Let us now approach this problem from the point of view of a **horizontal element of area** (or **horizontal strip**) as shown in Fig. 10-29. The width of the element is Δy. The length of

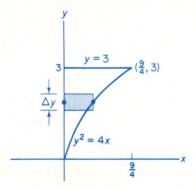

FIG. 10-29

the element is the *x-value on the right curve minus the x-value on the left curve*. Thus the area of the element is $(x_{\text{right}} - x_{\text{left}})\,\Delta y$. We wish to sum all such areas from $y = 0$ to $y = 3$:

$$\Sigma (x_{\text{right}} - x_{\text{left}})\, \Delta y \to \int_0^3 (x_{\text{right}} - x_{\text{left}})\, dy.$$

Since the variable of integration is y, we must express x_{right} and x_{left} as functions of y. The right curve is $y^2 = 4x$ or, equivalently, $x = y^2/4$. The left curve is $x = 0$. Thus

$$\text{area} = \int_0^3 (x_{\text{right}} - x_{\text{left}})\, dy$$

$$= \int_0^3 \left(\frac{y^2}{4} - 0\right) dy = \frac{y^3}{12}\bigg|_0^3 = \frac{9}{4} \text{ sq units.}$$

Note that for this region, horizontal strips make the definite integral easier to evaluate (and set up) than vertical strips. In any case, remember that **the limits of integration are those limits for the variable of integration.**

EXAMPLE 3

Find the area of the region bounded by the curves $y = 4x - x^2 + 8$ and $y = x^2 - 2x$.

A sketch of the region appears in Fig. 10-30. To find when the curves intersect we solve

$$4x - x^2 + 8 = x^2 - 2x.$$
$$-2x^2 + 6x + 8 = 0.$$
$$x^2 - 3x - 4 = 0.$$
$$(x + 1)(x - 4) = 0.$$
$$x = -1 \quad \text{or} \quad x = 4.$$

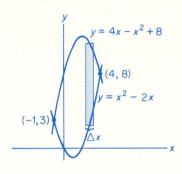

FIG. 10-30

When $x = -1$, then $y = 3$; when $x = 4$, then $y = 8$. Thus the curves intersect at $(-1, 3)$ and $(4, 8)$. We shall use vertical strips since they appear to present no difficulty. The area of the element is

$$(y_{\text{upper}} - y_{\text{lower}}) \, \Delta x = [(4x - x^2 + 8) - (x^2 - 2x)] \, \Delta x$$
$$= (-2x^2 + 6x + 8) \, \Delta x.$$

Summing all such areas from $x = -1$ to $x = 4$, we have

$$\text{area} = \int_{-1}^{4} (-2x^2 + 6x + 8) \, dx = 41\tfrac{2}{3} \text{ sq units.}$$

EXAMPLE 4

Find the area of the region bounded by $y^2 = x$ and $x - y = 2$.

A sketch of the region appears in Fig. 10-31. The curves intersect when $y^2 = y + 2$. Thus $y^2 - y - 2 = 0$, or equivalently $(y + 1)(y - 2) = 0$, from which $y = -1$ or $y = 2$. The points of intersection are $(1, -1)$ and $(4, 2)$. Let us consider vertical elements of area [see Fig. 10-31(a)]. Solving $y^2 = x$ for y gives $y = \pm \sqrt{x}$. As seen in Fig. 10-31(a), to the *left* of $x = 1$ the upper end of the element lies on $y = \sqrt{x}$ and the lower end lies on $y = -\sqrt{x}$. To the *right* of $x = 1$, the upper curve is $y = \sqrt{x}$ and the lower curve is

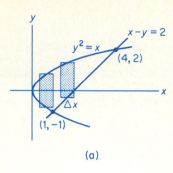

FIG. 10-31

$x - y = 2$ (or $y = x - 2$). Thus, with vertical strips *two* integrals are needed to evaluate the area.

$$\text{area} = \int_0^1 \left[\sqrt{x} - (-\sqrt{x}\,) \right] dx + \int_1^4 \left[\sqrt{x} - (x - 2) \right] dx.$$

Let us consider horizontal strips to see if we can simplify our work. In Fig. 10-31(b), the width of the strip is Δy. The rightmost curve is *always* $x - y = 2$ (or $x = y + 2$) and the leftmost curve is *always* $y^2 = x$ (or $x = y^2$). Thus the area of the horizontal strip is $[(y + 2) - y^2] \Delta y$ and the total area is

$$\text{area} = \int_{-1}^2 (y + 2 - y^2) \, dy = \frac{9}{2} \text{ sq units}.$$

Clearly the use of horizontal strips is the more desirable approach for the problem.

EXERCISE 10-8

In Problems 1–22, find the area of the region bounded by the graphs of the given equations.

1. $y = x^2, y = 2x$.　　　　　　　　　　2. $y = x, y = -x + 3, y = 0$.

3. $y = x^2, x = 0, y = 4\ (x \geqslant 0)$.　　4. $y = x^2, y = x$.

5. $y = x^2 + 3, y = 9$.　　　　　　　　6. $y^2 = x, x = 2$.

7. $x = 8 + 2y, x = 0, y = -1, y = 3$.　8. $y = x - 4, y^2 = 2x$.

9. $y = 4 - x^2, y = -3x$.　　　　　　10. $x = y^2 + 2, x = 6$.

11. $y^2 = x, 3x - 2y = 1$.　　　　　　12. $y = x^2, y = x + 2$.

13. $2y = 4x - x^2, 2y = x - 4$.　　　　14. $y = \sqrt{x}, y = x^2$.

15. $y^2 = x, y = x - 2$.　　　　　　　16. $y = 2 - x^2, y = x$.

17. $y = 8 - x^2, y = x^2, x = -1, x = 1$.　18. $y^2 = 4 - x, y = x + 2$.

19. $y = x^2, y = 2, y = 5$.　　　　　　20. $y = x^3 - x, x\text{-axis}$.

21. $y = x^3, y = x.$ **22.** $y = x^3, y = \sqrt{x}$.

23. A *Lorentz curve* is used in studying income distributions. If x is the cumulative percentage of income recipients, ranked from poorest to richest, and y is the cumulative percentage of income, then equality of income distribution is given by the line $y = x$ in Fig. 10-32, where x and y are expressed as decimals. For example, 10 percent of the people receive 10 percent of total income, 20 percent of the people

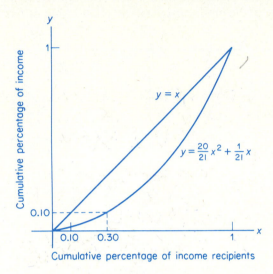

FIG. 10-32

receive 20 percent of the income, etc. Suppose the actual distribution is given by the Lorentz curve defined by $y = \frac{20}{21}x^2 + \frac{1}{21}x$. Note, for example, that 30 percent of the people receive only 10 percent of total income. The degree of deviation from equality is measured by the *coefficient of inequality*[†] for a Lorentz curve. This coefficient is defined to be the area between the curve and the diagonal, divided by the area under the diagonal:

$$\frac{\text{area between curve and diagonal}}{\text{area under diagonal}} .$$

For example, when all incomes are equal, the coefficient of inequality is zero. Find the coefficient of inequality for the Lorentz curve defined above.

24. Find the coefficient of inequality as in Problem 23 for the Lorentz curve defined by $y = \frac{11}{12}x^2 + \frac{1}{12}x.$

[†]G. Stigler, *The Theory of Price*, 3rd ed., The Macmillan Co., New York, 1966, p. 293–294.

10-9 CONSUMERS' AND PRODUCERS' SURPLUS

Determining the area of a region in the plane has applications in economics. Figure 10-33 shows the supply and demand curves for a product. The point (q_0, p_0) where these curves intersect is called the *point of equilibrium*. Here p_0 is the price per unit at which consumers will purchase the same quantity q_0 of a product that producers wish to sell at that price. In short, p_0 is the price at which stability in the producer-consumer relationship occurs.

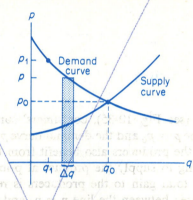

FIG. 10-33

Let us assume that the market is indeed at equilibrium and, consequently, the price per unit of the product is p_0. According to the demand curve, there are consumers who would be willing to pay *more* than p_0 for the product. For example, at the price per unit of p_1 there are consumers who would be willing to buy a total of q_1 units. Thus all consumers who are willing to pay more than p_0 are benefiting from the lower equilibrium price.

The vertical strip in Fig. 10-33 has area $p \, \Delta q$. This expression can also be thought of as the total amount of money that consumers would spend by buying Δq units of the product if the price per unit were p. Since the price is actually p_0, these consumers spend only $p_0 \, \Delta q$ for these Δq units and thus benefit by the amount $p \, \Delta q - p_0 \, \Delta q$. But $p \, \Delta q - p_0 \, \Delta q = (p - p_0) \, \Delta q$ is the area of a rectangle of width Δq and length $p - p_0$ (see Fig. 10-34). Summing the areas of all such rectangles from $q = 0$ to $q = q_0$ by integration, we have $\int_0^{q_0} (p - p_0) \, dq$. This integral, under certain conditions, represents the total gain to consumers who are willing to pay more than the equilibrium price. This total gain is called **consumers' surplus**, abbreviated CS. If the demand function is given by $p = f(q)$, then

$$CS = \int_0^{q_0} [f(q) - p_0] \, dq.$$

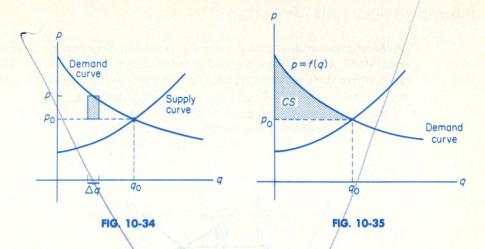

FIG. 10-34   FIG. 10-35

Geometrically (see Fig. 10-35), consumers' surplus is represented by the area between the line $p = p_0$ and the demand curve $p = f(q)$ from $q = 0$ to $q = q_0$.

Some of the producers also benefit from the equilibrium price, since they would be willing to supply the product at prices *lower* than p_0. Under certain conditions the total gain to the producers is represented geometrically in Fig. 10-36 by the area between the line $p = p_0$ and the supply curve $p = g(q)$ from

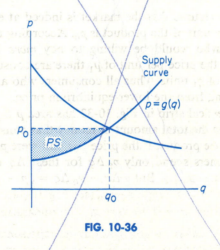

FIG. 10-36

$q = 0$ to $q = q_0$. This gain, called **producers' surplus** and abbreviated *PS*, is given by

$$PS = \int_0^{q_0} \left[p_0 - g(q) \right] dq.$$

EXAMPLE 1

The demand function for a product is $p = f(q) = 100 - .05q$, where p is the price per unit (in dollars) for q units. The supply function is $p = g(q) = 10 + .1q$. Determine consumers' surplus and producers' surplus when market equilibrium has been established.

First we must find the equilibrium point by solving the system formed by $p = 100 - .05q$ and $p = 10 + .1q$. Eliminating p, we have

$$10 + .1q = 100 - .05q,$$

$$.15q = 90,$$

$$q = 600.$$

When $q = 600$, then $p = 10 + .1(600) = 70$. Thus, $q_0 = 600$ and $p_0 = 70$. Consumers' surplus is

$$CS = \int_0^{q_0} [f(q) - p_0] \, dq = \int_0^{600} (100 - .05q - 70) \, dq$$

$$= \left(30q - .05 \frac{q^2}{2} \right) \Big|_0^{600} = 18,000 - 9000 = 9000.$$

Producers' surplus is

$$PS = \int_0^{q_0} [p_0 - g(q)] \, dq = \int_0^{600} [70 - (10 + .1q)] \, dq$$

$$= \left(60q - .1 \frac{q^2}{2} \right) \Big|_0^{600} = 36,000 - 18,000 = 18,000.$$

Therefore, consumers' surplus is $9000 and producers' surplus is $18,000.

EXAMPLE 2

The demand equation for a product is $q = f(p) = (90/p) - 2$ and the supply equation is $q = g(p) = p - 1$. Determine the consumers' surplus and producers' surplus when market equilibrium has been established.

Determining the equilibrium point, we have

$$p - 1 = \frac{90}{p} - 2,$$

$$p^2 + p - 90 = 0,$$

$$(p + 10)(p - 9) = 0.$$

Thus $p_0 = 9$ and $q_0 = 9 - 1 = 8$ (see Fig. 10-37). Note that the demand equation expresses q as a function of p. Since consumers' surplus can be considered as an area, this

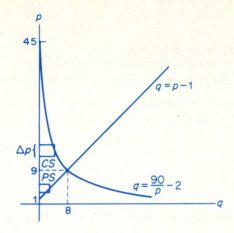

FIG. 10-37

area can be determined by means of horizontal strips of width Δp and length $q = f(p)$. These strips are summed from $p = 9$ to $p = 45$.

$$CS = \int_{9}^{45} \left(\frac{90}{p} - 2 \right) dp = (90 \ln |p| - 2p) \Big|_{9}^{45}$$

$$= 90 \ln 5 - 72 \approx 72.85.$$

Using horizontal strips for producers' surplus, we have

$$PS = \int_{1}^{9} (p - 1) \, dp = \frac{(p - 1)^2}{2} \Big|_{1}^{9} = 32.$$

EXERCISE 10-9

In Problems 1–6, the first equation is a demand equation and the second is a supply equation of a product. In each case determine the consumers' surplus and producers' surplus when market equilibrium has been established.

1. $p = 20 - .8q$,
    $p = 4 + 1.2q$.

2. $p = 900 - q^2$,
    $p = 100 + q^2$.

3. $p = \dfrac{50}{q + 5}$,
    $p = \dfrac{q}{10} + 4.5$.

4. $p = 400 - q^2$,
    $p = 20q + 100$.

5. $q = 100(10 - p)$,
    $q = 80(p - 1)$.

6. $q = \sqrt{100 - p}$,
    $q = \dfrac{p}{2} - 10$.

10-10 REVIEW

IMPORTANT TERMS AND SYMBOLS IN CHAPTER 10

antiderivative *(p. 385)*   indefinite integral *(p. 385)*

integrand *(p. 385)*   constant of integration *(p. 385)*

Σ *(p. 410)*   index of summation *(p. 410)*

limits of summation *(p. 410)*   definite integral *(p. 419)*

limits of integration *(p. 419)*   Fundamental Theorem of Integral
                                                   Calculus *(p. 427)*

$\int f(x)\, dx$, $\int_a^b f(x)\, dx$ *(p. 427)*   element of area *(p. 435)*

consumers' surplus *(p. 448)*   producers' surplus *(p. 449)*

REVIEW SECTION

1. If F is an antiderivative of f, then $F'(x) = $ _____ .

> *Ans.* $f(x)$.

2. $\int 5x^4\, dx = $ _____ .

> *Ans.* $x^5 + C$.

3. In $\int f(x)\, dx$, $f(x)$ is called the _____ .

> *Ans.* integrand.

4. $\int_{10}^{20} 2\, dx = $ _____ .

> *Ans.* 20.

5. True or false: $\int x^{-1}\, dx = \dfrac{x^{-1+1}}{-1+1} + C,$ __(a)__ ; $\int e^x\, dx = \dfrac{e^{x+1}}{x+1} + C,$ __(b)__ .

> *Ans.* (a) false; (b) false.

6. $\int (2x - 5)^9\, dx = $ _____ .

> *Ans.* $\frac{1}{20}(2x - 5)^{10} + C$.

7. $\int e^{2x+1}\, dx = $ _____ .

> *Ans.* $\frac{1}{2}e^{2x+1} + C$.

8. If $\int f(x)\, dx = F(x) + C$, then C is called the _____ .

> *Ans.* constant of integration.

9. $\int \dfrac{2}{x+5}\,dx =$ _____ .

Ans. $2\ln|x+5| + C$.

10. A definite integral of a function is a (function) (number) .

Ans.   number.

11. $\displaystyle\sum_{k=0}^{2}(k+1) =$ _____ .

Ans.   6.

12. True or false: $\displaystyle\sum_{k=1}^{15}(k^2 + 3k + 5) = \sum_{j=1}^{15}(j^2 + 3j + 5)$. _____

Ans.   true.

13. If $D_x y = 1$ and $y(0) = 1$, then $y =$ _____ .

Ans. $x + 1$.

14. All antiderivatives of a given function differ at most by a _____ .

Ans.   constant.

15. If $F'(x) = f(x)$, then $\displaystyle\int_2^3 f(x)\,dx =$ _____ .

Ans. $F(3) - F(2)$.

16. If $f(x) = 0$, then $\displaystyle\int f(x)\,dx =$ _____ .

Ans. C, a constant.

17. True or false: $\displaystyle\int x(x^2)\,dx = \dfrac{x^2}{2}\cdot\dfrac{x^3}{3} + C.$ _____

Ans.   false.

18. True or false: $\displaystyle\int \sqrt{x}\,dx = \tfrac{3}{2}x^{3/2} + C.$ _____

Ans.   false.

19. $\displaystyle\int \dfrac{x+1}{x}\,dx =$ _____ .

Ans. $x + \ln|x| + C$.

REVIEW PROBLEMS

Determine the integrals in Problems 1–32.

1. $\displaystyle\int (x^3 + 2x - 7)\,dx.$

2. $\displaystyle\int dx.$

3. $\displaystyle\int_0^9 (\sqrt{x} + x)\,dx.$

4. $\displaystyle\int \dfrac{2}{5 - 3x}\,dx.$

5. $\displaystyle\int \dfrac{2}{(x+5)^3}\,dx.$

6. $\displaystyle\int_4^{12}(y - 8)^{501}\,dy.$

7. $\int \dfrac{6x^2 - 12}{x^3 - 6x + 1} \, dx.$

8. $\int_0^2 x e^{4 - x^2} \, dx.$

9. $\int_0^1 \sqrt[3]{3t + 8} \, dt.$

10. $\int \dfrac{5 - 3x}{2} \, dx.$

11. $\int y(y + 1)^2 \, dy.$

12. $\int_0^1 10^{-8} \, dx.$

13. $\int \dfrac{\sqrt[4]{z} - \sqrt[3]{z}}{\sqrt{z}} \, dz.$

14. $\int \dfrac{(.5x - .1)^4}{.4} \, dx.$

15. $\int_1^2 \dfrac{t^2}{2 + t^3} \, dt.$

16. $\int \dfrac{4x^2 - x}{x} \, dx.$

17. $\int x^2 \sqrt{3x^3 + 2} \, dx.$

18. $\int (2x^3 + x)(x^4 + x^2)^{3/4} \, dx.$

19. $\int (e^{2y} - e^{-2y}) \, dy.$

20. $\int \dfrac{8x}{3\sqrt[3]{7 - 2x^2}} \, dx.$

21. $\int \left(\dfrac{1}{x} + \dfrac{2}{x^2} \right) dx.$

22. $\int_0^1 \dfrac{e^{2x}}{1 + e^{2x}} \, dx.$

23. $\int_{-2}^1 (y^4 - y + 1) \, dy.$

24. $\int_7^{70} dx.$

25. $\int_{\sqrt{3}}^2 7x \sqrt{4 - x^2} \, dx.$

26. $\int_0^1 (2x + 1)(x^2 + x)^4 \, dx.$

27. $\int_0^1 \left[2x - \dfrac{1}{(x + 1)^{2/3}} \right] dx.$

28. $\int_2^8 (\sqrt{2x} - x + 4) \, dx.$

29. $\int \dfrac{\sqrt{t} - 3}{t^2} \, dt.$

30. $\int \dfrac{z^2}{z - 1} \, dz.$

31. $\int_{-1}^0 \dfrac{x^2 + 4x - 1}{x + 2} \, dx.$

32. $\int \dfrac{(x^2 + 4)^2}{x^2} \, dx.$

In Problems **33–40**, determine the area of the region bounded by the given curves, the x-axis, and the given lines.

33. $y = x^2 - 1$, $x = 2$ $(y \geqslant 0)$.

34. $y = 4e^{2x}$, $x = 0$, $x = 3$.

35. $y = \sqrt{x + 4}$, $x = 0$.

36. $y = x^2 - x - 2$, $x = -2$, $x = 2$.

37. $y = 5x - x^2$.

38. $y = \sqrt[4]{x}$, $x = 1$, $x = 16$.

39. $y = \dfrac{1}{x} + 3$, $x = 1$, $x = 3$.

40. $y = x^3 - 1$, $x = -1$.

In Problems **41–46**, find the area of the region bounded by the given curves.

41. $y^2 = 4x$, $x = 0$, $y = 2$.

42. $y = 2x^2$, $x = 0$, $y = 2$ $(x \geqslant 0)$.

43. $y = x^2 + 4x - 5$, $y = 0$.

44. $y = 2x^2$, $y = x^2 + 9$.

45. $y = x^2 - 2x, y = 12 - x^2$.   **46.** $y = \sqrt{x}$, $x = 0, y = 3$.

47. If marginal revenue is given by $dr/dq = 100 - (3/2)\sqrt{2q}$, determine the corresponding demand equation.

48. If marginal cost is given by $dc/dq = q^2 + 7q + 6$, and fixed costs are 2500, determine the total cost for producing 6 units.

49. A manufacturer's marginal revenue function is $dr/dq = 275 - q - .3q^2$. If r is in dollars, find the increase in the manufacturer's total revenue if production is increased from 10 to 20 units.

50. A manufacturer's marginal cost function is $dc/dq = 500/\sqrt{2q + 25}$. If c is in dollars, determine the cost involved to increase production from 100 to 300 units.

51. For a product the demand equation is $p = .01q^2 - 1.1q + 30$ and its supply equation is $p = .01q^2 + 8$. Determine consumers' surplus and producers' surplus when market equilibrium has been established.

52. Suppose $f(x) = k + (x/10)$, where k is a constant and $0 < x < 3$. If f is a density function (see Example 5 in Sec. 10-7), find (a) the value of k, (b) $P(1 < x < 2)$, and (c) the distribution function F.

METHODS AND APPLICATIONS OF INTEGRATION

11-1 INTEGRATION BY PARTS[†]

Many integrals cannot be found by our previous methods. However, there are ways of changing certain integrals to forms that are easier to integrate. Of these methods, we shall discuss two: *integration by parts*, and (in the next section) *integration using partial fractions*.

If u and v are differentiable functions of x, by the product rule we have

$$(uv)' = uv' + vu'.$$

By rearranging, we have

$$uv' = (uv)' - vu'.$$

[†] May be omitted without loss of continuity.

Integrating both sides with respect to x, we get

$$\int uv' \, dx = \int (uv)' \, dx - \int vu' \, dx. \tag{1}$$

For $\int (uv)' \, dx$, we must find a function whose derivative with respect to x is $(uv)'$. Clearly uv is such a function. Hence $\int (uv)' \, dx = uv + C_1$ and Eq. (1) becomes

$$\int uv' \, dx = uv + C_1 - \int vu' \, dx.$$

Incorporating C_1 into the constant of integration for $\int vu' \, dx$ and replacing $v' \, dx$ by dv and $u' \, dx$ by du, we have the *integration by parts formula*:

> **INTEGRATION BY PARTS FORMULA**
>
> $$\int u \, dv = uv - \int v \, du.$$

(2)

This formula expresses a given integral, $\int u \, dv$, in terms of another integral, $\int v \, du$, which may be easier to find.

To apply the formula to $\int f(x) \, dx$, we must write $f(x) \, dx$ as the product of two factors (or *parts*) by choosing a function u and a differential dv such that $f(x) \, dx = u \, dv$. For the formula to be useful, we must be able to integrate the part chosen for dv. To illustrate, consider

$$\int xe^x \, dx.$$

This integral cannot be determined by previous integration formulas. We can write $xe^x \, dx$ in the form $u \, dv$ by letting

$$u = x \quad \text{and} \quad dv = e^x \, dx.$$

To apply the integration by parts formula, we must find du and v:

$$du = dx \quad \text{and} \quad v = \int e^x \, dx = e^x + C_1.$$

Thus,

$$\int \underset{u}{\underline{x}} \; \underset{dv}{\underline{e^x \, dx}} = uv - \int v \, du$$

$$= x(e^x + C_1) - \int (e^x + C_1) \, dx$$

$$= xe^x + C_1 x - e^x - C_1 x + C$$

$$= xe^x - e^x + C = e^x(x - 1) + C.$$

The first constant C_1 does not appear in the final answer. This is a characteristic of integration by parts and from now on this constant will not be written when finding v from dv.

When you are using the integration by parts formula, sometimes the "best choice" for u and dv may not be obvious. In some cases one choice may be as good as another; in other cases only one choice may be suitable. Insight into making a good choice (if any exists) will come only with practice and, of course, trial and error.

EXAMPLE 1

Find $\int \dfrac{\ln x}{\sqrt{x}}\, dx$ by integration by parts.

We try

$$u = \ln x \quad \text{and} \quad dv = \frac{1}{\sqrt{x}}\, dx.$$

Then

$$du = \frac{1}{x}\, dx \quad \text{and} \quad v = \int x^{-1/2}\, dx = 2x^{1/2}.$$

Thus,

$$\int \underbrace{\ln x}_{u}\ \underbrace{\left(\frac{1}{\sqrt{x}}\, dx\right)}_{dv} = uv - \int v\, du$$

$$= (\ln x)(2\sqrt{x}\,) - \int (2x^{1/2})\left(\frac{1}{x}\, dx\right)$$

$$= 2\sqrt{x}\, \ln x - 2\int x^{-1/2}\, dx$$

$$= 2\sqrt{x}\, \ln x - 2(2\sqrt{x}\,) + C$$

$$= 2\sqrt{x}\, (\ln x - 2) + C.$$

EXAMPLE 2

Evaluate $\displaystyle\int_{1}^{2} x \ln x\, dx.$

Let $u = x$ and $dv = \ln x\, dx$. Then $du = dx$, but $v = \int \ln x\, dx$ is not apparent by inspection. We shall make a different choice for u and dv. Let

$$u = \ln x \quad \text{and} \quad dv = x\, dx.$$

Then

$$du = \frac{1}{x} dx \quad \text{and} \quad v = \int x \, dx = \frac{x^2}{2}.$$

Thus,

$$\int_1^2 x \ln x \, dx = (\ln x)\left(\frac{x^2}{2}\right)\Big|_1^2 - \int_1^2 \left(\frac{x^2}{2}\right)\frac{1}{x} dx$$

$$= (\ln x)\left(\frac{x^2}{2}\right)\Big|_1^2 - \frac{1}{2}\int_1^2 x \, dx$$

$$= \frac{x^2 \ln x}{2}\Big|_1^2 - \frac{1}{2}\left(\frac{x^2}{2}\right)\Big|_1^2$$

$$= (2 \ln 2 - 0) - \left(1 - \frac{1}{4}\right) = 2 \ln 2 - \frac{3}{4}.$$

EXAMPLE 3

Determine $\int \ln y \, dy$.

Let $u = \ln y$ and $dv = dy$. Then $du = (1/y) \, dy$ and $v = y$.

$$\int \ln y \, dy = (\ln y)(y) - \int y\left(\frac{1}{y} dy\right)$$

$$= y \ln y - \int dy = y \ln y - y + C$$

$$= y(\ln y - 1) + C.$$

EXAMPLE 4

Determine $\int x e^{x^2} \, dx$.

PITFALL. *Do not forget about basic integration forms. Integration by parts is not needed here!*

$$\int x e^{x^2} \, dx = \frac{1}{2} \int e^{x^2}(2x \, dx)$$

$$= \frac{1}{2} \int e^u \, du \quad \text{(where } u = x^2\text{)}$$

$$= \frac{1}{2} e^u + C = \frac{1}{2} e^{x^2} + C.$$

Sometimes integration by parts must be used more than once, as shown in the following example.

EXAMPLE 5

Determine $\int x^2 e^{2x+1}\, dx$.

Let $u = x^2$ and $dv = e^{2x+1}\, dx$. Then $du = 2x\, dx$ and $v = e^{2x+1}/2$.

$$\int x^2 e^{2x+1}\, dx = \frac{x^2 e^{2x+1}}{2} - \int \frac{e^{2x+1}}{2}(2x\, dx)$$

$$= \frac{x^2 e^{2x+1}}{2} - \int x e^{2x+1}\, dx.$$

To find $\int x e^{2x+1}\, dx$, we shall again use integration by parts. Here, let $u = x$ and $dv = e^{2x+1}\, dx$. Then $du = dx$ and $v = e^{2x+1}/2$.

$$\int x e^{2x+1}\, dx = \frac{x e^{2x+1}}{2} - \int \frac{e^{2x+1}}{2}\, dx$$

$$= \frac{x e^{2x+1}}{2} - \frac{e^{2x+1}}{4} + C_1.$$

Thus,

$$\int x^2 e^{2x+1}\, dx = \frac{x^2 e^{2x+1}}{2} - \frac{x e^{2x+1}}{2} + \frac{e^{2x+1}}{4} + C \qquad (\text{where } C = -C_1)$$

$$= \frac{e^{2x+1}}{2}\left(x^2 - x + \frac{1}{2}\right) + C.$$

EXERCISE 11-1

In Problems 1–20, find the integrals.

1. $\int x e^{-x}\, dx$.

2. $\int x e^{2x}\, dx$.

3. $\int y^3 \ln y\, dy$.

4. $\int x^2 \ln x\, dx$.

5. $\int \ln (4x)\, dx$.

6. $\int \frac{t}{e^t}\, dt$.

7. $\int x\sqrt{x + 1}\, dx$.

8. $\int \frac{x}{\sqrt{1 + 4x}}\, dx$.

9. $\int \sqrt{x} \ln x\, dx$.

10. $\int \frac{\ln (x + 1)}{2(x + 1)}\, dx$.

11. $\displaystyle\int_1^2 xe^{2x}\,dx.$

12. $\displaystyle\int_0^1 xe^{-x}\,dx.$

13. $\displaystyle\int_0^1 xe^{-x^2}\,dx.$

14. $\displaystyle\int \frac{x^3}{\sqrt{4-x^2}}\,dx.$

15. $\displaystyle\int_1^2 \frac{x}{\sqrt{4-x}}\,dx.$

16. $\displaystyle\int (\ln x)^2\,dx.$

17. $\displaystyle\int x^2 e^x\,dx.$

18. $\displaystyle\int x^2 e^{-2x}\,dx.$

19. $\displaystyle\int (x - e^{-x})^2\,dx.$

20. $\displaystyle\int x^3 e^{x^2}\,dx.$

21. Find the area of the region bounded by the x-axis, the curve $y = \ln x$ and the line $x = e^3$.

22. Find the area of the region bounded by the x-axis and the curve $y = xe^{-x}$ between $x = 0$ and $x = 4$.

11-2 INTEGRATION BY PARTIAL FRACTIONS†

We now consider the integral of a quotient of two polynomials—that is, the integral of a *rational function*. Without loss of generality, we may assume that the numerator $N(x)$ and denominator $D(x)$ have no common polynomial factor and that the degree of $N(x)$ is less than the degree of $D(x)$ [that is, $N(x)/D(x)$ defines a *proper rational function*]. For if the numerator were not of lower degree, we could use long division to divide $N(x)$ by $D(x)$:

$$D(x)\,\overline{\smash{\big)}\,N(x)}\quad;\quad \text{thus}\quad \frac{N(x)}{D(x)} = P(x) + \frac{R(x)}{D(x)}.$$

$P(x)$ would be a polynomial (easily integrable) and $R(x)$ would be a polynomial of lower degree than $D(x)$. Thus $R(x)/D(x)$ would define a proper rational function. For example,

$$\int \frac{2x^4 - 3x^3 - 4x^2 - 17x - 6}{x^3 - 2x^2 - 3x}\,dx = \int \left(2x + 1 + \frac{4x^2 - 14x - 6}{x^3 - 2x^2 - 3x}\right)dx$$

$$= x^2 + x + \int \frac{4x^2 - 14x - 6}{x^3 - 2x^2 - 3x}\,dx.$$

---

† May be omitted without loss of continuity.

Therefore, we shall consider

$$\int \frac{4x^2 - 14x - 6}{x^3 - 2x^2 - 3x} \, dx = \int \frac{4x^2 - 14x - 6}{x(x + 1)(x - 3)} \, dx.$$

Observe that the denominator of the integrand consists only of **distinct linear factors,** each factor occurring exactly once. It can be shown that to each such factor $x - a$ there corresponds a *partial fraction* of the form

$$\frac{A}{x - a} \qquad\qquad (A \text{ a constant})$$

such that the integrand is the sum of the partial fractions. If there are n such *distinct* linear factors, there will be n such partial fractions, each of which is easily integrated. Applying these facts, we can write

$$\frac{4x^2 - 14x - 6}{x(x + 1)(x - 3)} = \frac{A}{x} + \frac{B}{x + 1} + \frac{C}{x - 3}. \qquad (1)$$

To determine the constants A, B, and C, we first combine the terms on the right side:

$$\frac{4x^2 - 14x - 6}{x(x + 1)(x - 3)} = \frac{A(x + 1)(x - 3) + Bx(x - 3) + Cx(x + 1)}{x(x + 1)(x - 3)}.$$

Since the denominators of both sides are equal, we may equate their numerators:

$$4x^2 - 14x - 6 = A(x + 1)(x - 3) + Bx(x - 3) + Cx(x + 1). \quad (2)$$

Although Eq. (1) is not defined for $x = 0$, $x = -1$, and $x = 3$, we want to find values for A, B, and C that will make Eq. (2) true for all values of x. That is, it will be an identity. By successively setting x in Eq. (2) equal to any three different numbers, we can obtain a system of equations which can be solved for A, B, and C. In particular, the work can be simplified by letting x be the roots of $D(x) = 0$, in our case $x = 0$, $x = -1$, and $x = 3$. Using Eq. (2), if $x = 0$ we have

$$-6 = A(1)(-3) + B(0) + C(0) = -3A \quad \text{and} \quad A = 2;$$

if $x = -1$,

$$12 = A(0) + B(-1)(-4) + C(0) = 4B \quad \text{and} \quad B = 3;$$

if $x = 3$,

$$-12 = A(0) + B(0) + C(3)(4) = 12C \quad \text{and} \quad C = -1.$$

Thus Eq. (1) becomes

$$\frac{4x^2 - 14x - 6}{x(x + 1)(x - 3)} = \frac{2}{x} + \frac{3}{x + 1} - \frac{1}{x - 3}.$$

Hence

$$\int \frac{4x^2 - 14x - 6}{x(x + 1)(x - 3)} \, dx = \int \left(\frac{2}{x} + \frac{3}{x + 1} - \frac{1}{x - 3} \right) dx$$

$$= 2 \int \frac{dx}{x} + 3 \int \frac{dx}{x + 1} - \int \frac{dx}{x - 3}$$

$$= 2 \ln |x| + 3 \ln |x + 1| - \ln |x - 3| + C$$

$$= \ln \left| \frac{x^2(x + 1)^3}{x - 3} \right| + C \quad \text{(using properties of logarithms)}.$$

For the *original* integral we can now state

$$\int \frac{2x^4 - 3x^3 - 4x^2 - 17x - 6}{x^3 - 2x^2 - 3x} \, dx = x^2 + x + \ln \left| \frac{x^2(x + 1)^3}{x - 3} \right| + C.$$

There is an alternative method of determining A, B, and C. It involves expanding the right side of Eq. (2) and combining similar terms:

$$4x^2 - 14x - 6 = A(x^2 - 2x - 3) + B(x^2 - 3x) + C(x^2 + x)$$

$$= Ax^2 - 2Ax - 3A + Bx^2 - 3Bx + Cx^2 + Cx.$$

$$4x^2 - 14x - 6 = (A + B + C)x^2 + (-2A - 3B + C)x + (-3A).$$

For this identity, coefficients of corresponding powers of x on the left and right sides of the equation must be equal:

$$\begin{cases} 4 = A + B + C, \\ -14 = -2A - 3B + C, \\ -6 = -3A. \end{cases}$$

Solving gives $A = 2$, $B = 3$, and $C = -1$ as before.

EXAMPLE 1

Determine $\int \dfrac{2x + 1}{3x^2 - 27} \, dx.$

Since the degree of $N(x)$ is less than the degree of $D(x)$, no long division is necessary.

The integral can be written as

$$\frac{1}{3} \int \frac{2x + 1}{x^2 - 9} \, dx.$$

Expressing $(2x + 1)/(x^2 - 9)$ as a sum of partial fractions, we have

$$\frac{2x + 1}{x^2 - 9} = \frac{2x + 1}{(x + 3)(x - 3)} = \frac{A}{x + 3} + \frac{B}{x - 3}.$$

Combining terms and equating numerators gives

$$2x + 1 = A(x - 3) + B(x + 3).$$

If $x = 3$,

$$7 = 6B \quad \text{and} \quad B = \frac{7}{6};$$

if $x = -3$,

$$-5 = -6A \quad \text{and} \quad A = \frac{5}{6}.$$

Thus,

$$\int \frac{2x + 1}{3x^2 - 27} \, dx = \frac{1}{3} \left[\int \frac{\frac{5}{6} \, dx}{x + 3} + \int \frac{\frac{7}{6} \, dx}{x - 3} \right]$$

$$= \frac{1}{3} \left[\frac{5}{6} \ln |x + 3| + \frac{7}{6} \ln |x - 3| \right] + C$$

$$= \frac{1}{18} \ln \left| (x + 3)^5 (x - 3)^7 \right| + C.$$

If the denominator of $N(x)/D(x)$ contains only linear factors, some of which are repeated, then for each factor $(x - a)^k$, where k is the maximum number of times $x - a$ occurs as a factor, there will correspond the sum of k partial fractions:

$$\frac{A}{x - a} + \frac{B}{(x - a)^2} + \ldots + \frac{K}{(x - a)^k}.$$

EXAMPLE 2

Determine $\int \dfrac{6x^2 + 13x + 6}{(x + 2)(x + 1)^2} \, dx.$

Since the degree of $N(x)$ is less than that of $D(x)$, no long division is necessary. In $D(x)$

the factor $x + 2$ occurs once and the factor $x + 1$ occurs twice. There will be three partial fractions and three constants to determine.

$$\frac{6x^2 + 13x + 6}{(x + 2)(x + 1)^2} = \frac{A}{x + 2} + \frac{B}{x + 1} + \frac{C}{(x + 1)^2}.$$

$$6x^2 + 13x + 6 = A(x + 1)^2 + B(x + 2)(x + 1) + C(x + 2).$$

Let us choose $x = -2$, $x = -1$, and for convenience $x = 0$. Then if $x = -2$,

$$4 = A;$$

if $x = -1$,

$$-1 = C;$$

if $x = 0$,

$$6 = A + 2B + 2C = 4 + 2B - 2 = 2 + 2B,$$
$$2 = B.$$

Thus,

$$\int \frac{6x^2 + 13x + 6}{(x + 2)(x + 1)^2} \, dx = 4\int \frac{dx}{x + 2} + 2\int \frac{dx}{x + 1} - \int \frac{dx}{(x + 1)^2}$$

$$= 4 \ln |x + 2| + 2 \ln |x + 1| + \frac{1}{x + 1} + C$$

$$= \ln \left[(x + 2)^4 (x + 1)^2 \right] + \frac{1}{x + 1} + C.$$

Suppose a quadratic factor $x^2 + bx + c$ occurs in $D(x)$ and $x^2 + bx + c$ cannot be expressed as a product of two linear factors with real coefficients. Such a factor is called *irreducible over the real numbers*. To each irreducible quadratic factor that occurs exactly once in $D(x)$ there will correspond a partial fraction of the form

$$\frac{Ax + B}{x^2 + bx + c}.$$

EXAMPLE 3

Determine $\int \dfrac{-2x - 4}{x^3 + x^2 + x} \, dx.$

Since $x^3 + x^2 + x = x(x^2 + x + 1)$, we have the linear factor x and the quadratic factor $x^2 + x + 1$, which does not seem factorable on inspection. If it were factorable into

$(x - r_1) (x - r_2)$, where r_1 and r_2 are real, then r_1 and r_2 would be roots of the equation $x^2 + x + 1 = 0$. By the quadratic formula, the roots are

$$x = \frac{-1 \pm \sqrt{1 - 4}}{2}.$$

Since there are no real roots, we conclude that $x^2 + x + 1$ is irreducible. Thus there will be two partial fractions and *three* constants to determine:

$$\frac{-2x - 4}{x(x^2 + x + 1)} = \frac{A}{x} + \frac{Bx + C}{x^2 + x + 1}.$$

$$-2x - 4 = A(x^2 + x + 1) + (Bx + C)x$$

$$= Ax^2 + Ax + A + Bx^2 + Cx.$$

$$0x^2 - 2x - 4 = (A + B)x^2 + (A + C)x + A.$$

Equating coefficients of like powers of x, we obtain

$$\begin{cases} 0 = A + B, \\ -2 = A + C, \\ -4 = A. \end{cases}$$

Solving gives $A = -4$, $B = 4$, and $C = 2$. Thus,

$$\int \frac{-2x - 4}{x(x^2 + x + 1)}\, dx = \int \left(\frac{-4}{x} + \frac{4x + 2}{x^2 + x + 1} \right) dx$$

$$= -4 \int \frac{dx}{x} + 2 \int \frac{2x + 1}{x^2 + x + 1}\, dx.$$

Both integrals have the form $\int \frac{du}{u}$, and so

$$\int \frac{-2x - 4}{x(x^2 + x + 1)}\, dx = -4 \ln |x| + 2 \ln |x^2 + x + 1| + C$$

$$= \ln \left[\frac{(x^2 + x + 1)^2}{x^4} \right] + C.$$

Suppose $D(x)$ contains factors of the form $(x^2 + bx + c)^k$, where k is the maximum number of times the irreducible factor $x^2 + bx + c$ occurs. Then to each such factor there will correspond a sum of k partial fractions of the form

$$\frac{A + Bx}{x^2 + bx + c} + \frac{C + Dx}{(x^2 + bx + c)^2} + \ldots + \frac{M + Nx}{(x^2 + bx + c)^k}.$$

EXAMPLE 4

Determine $\int \dfrac{x^5}{(x^2 + 4)^2}\, dx$.

Since $N(x)$ has degree 5 and $D(x)$ has degree 4, we first divide $N(x)$ by $D(x)$.

$$\frac{x^5}{x^4 + 8x^2 + 16} = x - \frac{8x^3 + 16x}{(x^2 + 4)^2}.$$

The quadratic factor $x^2 + 4$ in the denominator of $(8x^3 + 16x)/(x^2 + 4)^2$ is irreducible and occurs as a factor twice. Thus to $(x^2 + 4)^2$ there correspond two partial fractions and *four* coefficients to be determined.

$$\frac{8x^3 + 16x}{(x^2 + 4)^2} = \frac{Ax + B}{x^2 + 4} + \frac{Cx + D}{(x^2 + 4)^2}.$$

$$8x^3 + 16x = (Ax + B)(x^2 + 4) + Cx + D.$$

$$8x^3 + 0x^2 + 16x + 0 = Ax^3 + Bx^2 + (4A + C)x + 4B + D.$$

Equating like powers of x, we obtain

$$\begin{cases} 8 = A, \\ 0 = B, \\ 16 = 4A + C, \\ 0 = 4B + D. \end{cases}$$

Solving gives $A = 8$, $B = 0$, $C = -16$, and $D = 0$. Thus,

$$\int \frac{x^5}{(x^2 + 4)^2}\, dx = \int \left(x - \left[\frac{8x}{x^2 + 4} - \frac{16x}{(x^2 + 4)^2} \right] \right) dx$$

$$= \int x\, dx - 4 \int \frac{2x}{x^2 + 4}\, dx + 8 \int \frac{2x}{(x^2 + 4)^2}\, dx.$$

The second integral on the last line has the form $\int \dfrac{du}{u}$ and the third integral has the form $\int \dfrac{du}{u^2}$.

$$\int \frac{x^5}{(x^2 + 4)^2}\, dx = \frac{x^2}{2} - 4 \ln (x^2 + 4) - \frac{8}{x^2 + 4} + C.$$

From our examples you may have deduced that the number of constants needed to express $N(x)/D(x)$ by partial fractions is equal to the degree of $D(x)$, if it is assumed that $N(x)/D(x)$ defines a proper rational function. This is indeed the case. It should be added that the representation of a proper rational function by partial fractions is unique; that is, there is only one choice of constants that can be made. Furthermore, regardless of the complexity of the polynomial $D(x)$, it can always (theoretically) be expressed as a product of linear and irreducible quadratic factors with real coefficients.

EXAMPLE 5

Find $\displaystyle\int \frac{2x + 3}{x^2 + 3x + 1}\,dx$.

PITFALL. *Do not forget about basic integration forms.*

$$\int \frac{2x + 3}{x^2 + 3x + 1}\,dx = \ln|x^2 + 3x + 1| + C.$$

EXERCISE 11-2

In Problems **1-22**, determine the integrals.

1. $\displaystyle\int \frac{5x - 2}{x^2 - x}\,dx$.

2. $\displaystyle\int \frac{3x + 8}{x^2 + 2x}\,dx$.

3. $\displaystyle\int \frac{x + 10}{x^2 - x - 2}\,dx$.

4. $\displaystyle\int \frac{dx}{x^2 - 5x + 6}$.

5. $\displaystyle\int \frac{3x^3 - 3x + 4}{4x^2 - 4}\,dx$.

6. $\displaystyle\int \frac{4 - x^2}{(x - 4)(x - 2)(x + 3)}\,dx$.

7. $\displaystyle\int \frac{17x - 12}{x^3 - x^2 - 12x}\,dx$.

8. $\displaystyle\int \frac{4 - x}{x^4 - x^2}\,dx$.

9. $\displaystyle\int \frac{3x^5 + 4x^3 - x}{x^6 + 2x^4 - x^2 - 2}\,dx$.

10. $\displaystyle\int \frac{x^4 - 3x^3 - 5x^2 + 8x - 1}{x^3 - 2x^2 - 8x}\,dx$.

11. $\displaystyle\int \frac{2x^2 - 5x - 2}{(x - 2)^2(x - 1)}\,dx$.

12. $\displaystyle\int \frac{-3x^3 + 2x - 3}{x^2(x^2 - 1)}\,dx$.

13. $\displaystyle\int \frac{x^2 + 8}{x^3 + 4x}\,dx$.

14. $\displaystyle\int \frac{2x^3 - 6x^2 - 10x - 6}{x^4 - 1}\,dx$.

15. $\displaystyle\int \frac{-x^3 + 8x^2 - 9x + 2}{(x^2 + 1)(x - 3)^2}\,dx$.

16. $\displaystyle\int \frac{2x^4 + 9x^2 + 8}{x(x^2 + 2)^2}\,dx$.

17. $\displaystyle\int \frac{14x^3 + 24x}{(x^2 + 1)(x^2 + 2)}\,dx$.

18. $\displaystyle\int \frac{12x^3 + 20x^2 + 28x + 4}{(x^2 + 2x + 3)(x^2 + 1)}\,dx$.

19. $\displaystyle\int \frac{3x^3 + x}{(x^2 + 1)^2}\,dx$.

20. $\displaystyle\int \frac{3x^2 - 8x + 4}{x^3 - 4x^2 + 4x - 6}\,dx$.

21. $\int_0^1 \dfrac{2 - 2x}{x^2 + 7x + 12} \, dx.$ **22.** $\int_1^2 \dfrac{2x^2 + 1}{(x + 3)(x + 2)} \, dx.$

23. Find the area bounded by $y = (x^2 + 1)/(x + 2)^2$ and the x-axis from $x = 0$ to $x = 1$.

11-3 INTEGRATION USING TABLES

Certain forms of integrals that occur frequently may be found in standard tables of integration formulas.[†] A short table appears in Appendix E and its use will be illustrated in this section.

No table of integrals is exhaustive. We may still have to appeal to prior techniques of integration when a suitable form is not listed in the tables. Moreover, a given integral may have to be replaced by an equivalent form before it will fit a formula in the table. The equivalent form must match the formula *exactly*. Consequently, the steps that you perform should *not* be done mentally. *Write them down!* Failure to do this can easily lead to incorrect results. Before proceeding with the exercises, be sure you understand the illustrative examples *thoroughly*.

In the following examples the formula numbers refer to the Table of Selected Integrals given in Appendix E.

EXAMPLE 1

Find $\int \dfrac{x \, dx}{(2 + 3x)^2}.$

Scanning the tables, we identify the integrand with formula 7:

$$\int \frac{u \, du}{(a + bu)^2} = \frac{1}{b^2}\left(\ln |a + bu| + \frac{a}{a + bu}\right) + C.$$

For the given integrand, let $u = x$, $a = 2$, and $b = 3$. Then $du = dx$ and we have

$$\int \frac{x \, dx}{(2 + 3x)^2} = \int \frac{u \, du}{(a + bu)^2}.$$

By the formula,

$$\int \frac{x \, dx}{(2 + 3x)^2} = \int \frac{u \, du}{(a + bu)^2} = \frac{1}{b^2}\left(\ln |a + bu| + \frac{a}{a + bu}\right) + C.$$

[†] See, for example, S. M. Selby, *Standard Mathematical Tables*, 22nd ed. Cleveland: Chemical Rubber Co., 1974.

Returning to the variable x and replacing a by 2 and b by 3, we obtain

$$\int \frac{x \, dx}{(2 + 3x)^2} = \frac{1}{9}\left(\ln |2 + 3x| + \frac{2}{2 + 3x}\right) + C.$$

EXAMPLE 2

Find $\int x^2 \sqrt{x^2 - 1} \, dx$.

This integral is identified with formula 24:

$$\int u^2 \sqrt{u^2 \pm a^2} \, du = \frac{u}{8}(2u^2 \pm a^2)\sqrt{u^2 \pm a^2} - \frac{a^4}{8} \ln \left| u + \sqrt{u^2 \pm a^2} \right| + C.$$

In applying this formula, if the bottommost sign in the dual symbol "$\pm$" on the left side is used, the bottommost sign in the dual symbols on the right must also be used. Letting $u = x$ and $a = 1$, we have $du = dx$. Thus,

$$\int x^2 \sqrt{x^2 - 1} \, dx = \int u^2 \sqrt{u^2 - a^2} \, du$$

$$= \frac{u}{8}(2u^2 - a^2)\sqrt{u^2 - a^2} - \frac{a^4}{8} \ln \left| u + \sqrt{u^2 - a^2} \right| + C.$$

Since $u = x$ and $a = 1$,

$$\int x^2 \sqrt{x^2 - 1} \, dx = \frac{x}{8}(2x^2 - 1)\sqrt{x^2 - 1} - \frac{1}{8} \ln \left| x + \sqrt{x^2 - 1} \right| + C.$$

EXAMPLE 3

Find $\int \frac{dx}{x\sqrt{16x^2 + 3}}$.

The integrand can be identified with formula 28:

$$\int \frac{du}{u\sqrt{u^2 + a^2}} = \frac{1}{a} \ln \left| \frac{\sqrt{u^2 + a^2} - a}{u} \right| + C.$$

If we let $u = 4x$ and $a = \sqrt{3}$, then $du = 4 \, dx$. Thus (watch closely),

$$\int \frac{dx}{x\sqrt{16x^2 + 3}} = \int \frac{(4 \, dx)}{(4x)\sqrt{(4x)^2 + (\sqrt{3})^2}} = \int \frac{du}{u\sqrt{u^2 + a^2}}.$$

By formula 28, the last integral is

$$\frac{1}{a} \ln \left| \frac{\sqrt{u^2 + a^2} - a}{u} \right| + C.$$

Hence, replacing u by $4x$ and a by $\sqrt{3}$, we have

$$\int \frac{dx}{x\sqrt{16x^2 + 3}} = \frac{1}{\sqrt{3}} \ln \left| \frac{\sqrt{16x^2 + 3} - \sqrt{3}}{4x} \right| + C.$$

EXAMPLE 4

Find $\displaystyle\int \frac{dx}{x^2(2 - 3x^2)^{1/2}}.$

The integrand is identified with formula 21:

$$\int \frac{du}{u^2\sqrt{a^2 - u^2}} = -\frac{\sqrt{a^2 - u^2}}{a^2 u} + C.$$

Letting $u = \sqrt{3}\,x$ and $a^2 = 2$, we have $du = \sqrt{3}\,dx$. Hence

$$\int \frac{dx}{x^2(2 - 3x^2)^{1/2}} = \sqrt{3} \int \frac{(\sqrt{3}\,dx)}{(\sqrt{3}x)^2(2 - 3x^2)^{1/2}} = \sqrt{3} \int \frac{du}{u^2(a^2 - u^2)^{1/2}}$$

$$= \sqrt{3} \left[-\frac{\sqrt{a^2 - u^2}}{a^2 u} \right] + C = \sqrt{3} \left[-\frac{\sqrt{2 - 3x^2}}{2(\sqrt{3}\,x)} \right] + C$$

$$= -\frac{\sqrt{2 - 3x^2}}{2x} + C.$$

EXAMPLE 5

Find $\displaystyle\int 7x^2 \ln(4x)\,dx.$

This is similar to formula 42, where $n = 2$:

$$\int u^n \ln u\,du = \frac{u^{n+1}\ln u}{n + 1} - \frac{u^{n+1}}{(n + 1)^2} + C.$$

If we let $u = 4x$, then $du = 4\ dx$. Hence

$$\int 7x^2 \ln (4x)\ dx = \frac{7}{64} \int (4x)^2 \ln (4x)(4\ dx)$$

$$= \frac{7}{64} \int u^2 \ln u\ du = \frac{7}{64} \left(\frac{u^3 \ln u}{3} - \frac{u^3}{9} \right) + C$$

$$= \frac{7}{64} \left[\frac{(4x)^3 \ln (4x)}{3} - \frac{(4x)^3}{9} \right] + C$$

$$= 7x^3 \left[\frac{\ln (4x)}{3} - \frac{1}{9} \right] + C.$$

EXAMPLE 6

Find $\int \dfrac{e^{2x}\ dx}{7 + e^{2x}}$.

At first glance we do not identify the integrand with any form in the table. Perhaps rewriting the integral will help. Let $u = 7 + e^{2x}$; then $du = 2e^{2x}\ dx$.

$$\int \frac{e^{2x}\ dx}{7 + e^{2x}} = \frac{1}{2} \int \frac{(2e^{2x}\ dx)}{7 + e^{2x}} = \frac{1}{2} \int \frac{du}{u} = \frac{1}{2} \ln |u| + C$$

$$= \frac{1}{2} \ln |7 + e^{2x}| + C = \frac{1}{2} \ln (7 + e^{2x}) + C.$$

Thus we had only to use our knowledge of basic integration forms. Actually, this form appears as formula 2 in the tables.

EXAMPLE 7

Evaluate $\int_1^4 \dfrac{dx}{(4x^2 + 2)^{3/2}}$.

We shall use formula 32 to first get the indefinite integral:

$$\int \frac{du}{(u^2 \pm a^2)^{3/2}} = \frac{\pm u}{a^2 \sqrt{u^2 \pm a^2}} + C.$$

Letting $u = 2x$ and $a^2 = 2$, then we have $du = 2\ dx$. Thus,

$$\int \frac{dx}{(4x^2 + 2)^{3/2}} = \frac{1}{2} \int \frac{(2\ dx)}{[(2x)^2 + 2]^{3/2}} = \frac{1}{2} \int \frac{du}{(u^2 + 2)^{3/2}}$$

$$= \frac{1}{2} \left[\frac{u}{2\sqrt{u^2 + 2}} \right] + C.$$

Instead of substituting back to x and evaluating from $x = 1$ to $x = 4$, we can determine the corresponding limits of integration with respect to u. Since $u = 2x$, then when $x = 1$ we have $u = 2$; when $x = 4$ we have $u = 8$. Thus,

$$\int_1^4 \frac{dx}{(4x^2 + 2)^{3/2}} = \frac{1}{2} \int_2^8 \frac{du}{(u^2 + 2)^{3/2}}$$

$$= \frac{1}{2} \left(\frac{u}{2\sqrt{u^2 + 2}} \right) \Big|_2^8 = \frac{2}{\sqrt{66}} - \frac{1}{2\sqrt{6}}.$$

PITFALL. *In Example 7, when changing the variable of integration x to the variable of integration u, be certain to change the limits of integration so that they agree with u. That is,*

$$\int_1^4 \frac{dx}{(4x^2 + 2)^{3/2}} \neq \frac{1}{2} \int_1^4 \frac{du}{(u^2 + 2)^{3/2}}.$$

Suppose that you must pay out \$100 at the end of each year for the next two years. Recall that a series of payments over a period of time, such as this, is called an *annuity*. If you were to pay off the debt now instead, you would pay the present value of the \$100 that is due at the end of the first year, plus the present value of the \$100 that is due at the end of the second year. (Present value of an annuity is discussed in Sec. 6-3.) The sum of these present values is the present value of the annuity. We shall now consider the present value of payments made continuously over the time interval from $t = 0$ to $t = T$, t in years, when interest is compounded continuously at an annual rate of r.

Suppose a payment is made at time t such that on an annual basis this payment is $f(t)$. Then over a small time interval $[t_i, t_{i+1}]$ of length Δt, the total amount of all payments is approximately $f(t_i)\,\Delta t$. [For example, if $f(t) = 2000$ and Δt were one day, then the total amount of the payments would be $2000(\frac{1}{365})$.] The present value of these payments is approximately $e^{-rt_i}f(t_i)\,\Delta t$ (see Sec. 7-3). Over the interval $[0, T]$, the total of all such present values is

$$\sum e^{-rt_i}f(t_i)\,\Delta t.$$

This sum approximates the present value A of the annuity. The smaller Δt is, the better is the approximation. That is, as $\Delta t \to 0$ the limit of the sum *is* the present value. However, this limit is also a definite integral. That is,

$$\boxed{A = \int_0^T f(t)e^{-rt}\,dt.} \tag{1}$$

This gives the **present value of a continuous annuity** at an annual rate r (compounded continuously) for T years if a payment at time t is at the rate of

$f(t)$ per year. Sometimes we speak of this integral as the **present value of a continuous income stream.**

    We can also look at the *future* value of an annuity rather than its present value. If a payment is made at time t, then it has a certain value at the *end* of the period of the annuity, that is, $T - t$ years later. This value is

$$\left(\begin{array}{c} \text{Amount of} \\ \text{payment} \end{array} \right) + \left(\begin{array}{c} \text{Interest on this} \\ \text{payment for } T - t \text{ years} \end{array} \right).$$

If S is the total of such values for all payments, then S is called the **accumulated amount of a continuous annuity** and is given by

$$\boxed{S = \int_0^T f(t) e^{r(T-t)}\, dt.}$$

EXAMPLE 8

Find the present value (to the nearest dollar) of a continuous annuity at an annual rate of 4 percent for 10 years if the payment at time t is at the rate of t^2 dollars per year.

The present value is given by

$$A = \int_0^T f(t) e^{-rt}\, dt = \int_0^{10} t^2 e^{-.04t}\, dt.$$

We shall use formula 39,

$$\int u^n e^{au}\, du = \frac{u^n e^{au}}{a} - \frac{n}{a} \int u^{n-1} e^{au}\, du,$$

called a *reduction formula* since it reduces an integral into an expression that involves an integral which is easier to determine. If $u = t$, $n = 2$, and $a = -.04$, then $du = dt$ and we have

$$A = \frac{t^2 e^{-.04t}}{-.04} \bigg|_0^{10} - \frac{2}{-.04} \int_0^{10} t e^{-.04t}\, dt.$$

In the new integral the exponent of t has been reduced to 1. Now, applying formula 38,

$$\int u e^{au}\, du = \frac{e^{au}}{a^2} (au - 1) + C,$$

to this integral where $u = t$, $du = dt$, and $a = -.04$, we have

$$A = \int_0^{10} t^2 e^{-.04t}\, dt = \frac{t^2 e^{-.04t}}{-.04}\Big|_0^{10} - \frac{2}{-.04}\left[\frac{e^{-.04t}}{(-.04)^2}(-.04t - 1)\right]\Big|_0^{10}$$

$$= \frac{100e^{-.4}}{-.04} + \frac{2}{.04}\left[\frac{e^{-.4}}{(-.04)^2}(-.4 - 1) - \frac{1}{(-.04)^2}(-1)\right]$$

$$= -2500e^{-.4} + 50[-875e^{-.4} + 625]$$

$$= -46{,}250e^{-.4} + 31{,}250 \approx -46{,}250(.67032) + 31{,}250$$

$$\approx 248.$$

The present value is $248.

Equation (1) may be used to find the present value of future profits of a business. In this situation, $f(t)$ would be the annual rate of profit at time t.

EXERCISE 11-3

In Problems 1-34, find the integrals by using the table in Appendix E.

1. $\int \dfrac{dx}{x(6 + 7x)}$.

2. $\int \dfrac{x^2\, dx}{(1 + 2x)^2}$.

3. $\int \dfrac{dx}{x\sqrt{x^2 + 9}}$.

4. $\int \dfrac{dx}{(x^2 + 7)^{3/2}}$.

5. $\int \dfrac{x\, dx}{(2 + 3x)(4 + 5x)}$.

6. $\int 2^{5x}\, dx$.

7. $\int \dfrac{dx}{4 + 3e^{2x}}$.

8. $\int x^2\sqrt{1 + x}\, dx$.

9. $\int \dfrac{2\, dx}{x(1 + x)^2}$.

10. $\int \dfrac{dx}{x\sqrt{5 - 11x^2}}$.

11. $\int_0^1 \dfrac{x\, dx}{2 + x}$.

12. $\int \dfrac{x^2\, dx}{2 + 5x}$.

13. $\int \sqrt{x^2 - 3}\, dx$.

14. $\int \dfrac{dx}{(4 + 3x)(4x + 3)}$.

15. $\int_0^{1/12} xe^{12x}\, dx$.

16. $\int \sqrt{\dfrac{2 + 3x}{5 + 3x}}\, dx$.

17. $\int x^2 e^x \, dx.$

18. $\int_1^2 \dfrac{dx}{x^2(1+x)}.$

19. $\int \dfrac{\sqrt{4x^2+1}}{x^2} \, dx.$

20. $\int \dfrac{dx}{x\sqrt{2-x}}.$

21. $\int \dfrac{x \, dx}{(1+3x)^2}.$

22. $\int \dfrac{dx}{\sqrt{(1+2x)(3+2x)}}.$

23. $\int \dfrac{dx}{7-5x^2}.$

24. $\int x^2 \sqrt{2x^2-9} \, dx.$

25. $\int x^5 \ln(3x) \, dx.$

26. $\int \dfrac{dx}{x^2(1+x)^2}.$

27. $\int 2x\sqrt{1+3x} \, dx.$

28. $\int x^2 \ln x \, dx.$

29. $\int \dfrac{dx}{\sqrt{4x^2-13}}.$

30. $\int \dfrac{dx}{x \ln(2x)}.$

31. $\int x \ln(2x) \, dx.$

32. $\int \dfrac{\sqrt{2-3x^2}}{x} \, dx.$

33. $\int \dfrac{dx}{x^2\sqrt{9-4x^2}}.$

34. $\int_0^1 \dfrac{x^3 \, dx}{1+x^4}.$

In Problems 35–52, find the integrals by any method.

35. $\int \dfrac{x \, dx}{x^2+1}.$

36. $\int \sqrt{x} \, e^{x^{3/2}} \, dx.$

37. $\int x\sqrt{2x^2+1} \, dx.$

38. $\int \dfrac{4x^2-\sqrt{x}}{x} \, dx.$

39. $\int \dfrac{dx}{x^2-5x+6}.$

40. $\int \dfrac{e^{2x}}{\sqrt{e^{2x}+3}} \, dx.$

41. $\int x^3 \ln x \, dx.$

42. $\int_0^3 xe^{-x} \, dx.$

43. $\int xe^{2x} \, dx.$

44. $\int_1^2 x^2\sqrt{3+2x} \, dx.$

45. $\int \ln^2 x \, dx.$

46. $\int_1^e \ln x \, dx.$

47. $\int_1^2 \dfrac{x \, dx}{\sqrt{4-x}}.$

48. $\int_1^2 x\sqrt{1+2x} \, dx.$

49. $\int_0^1 \dfrac{2x \, dx}{\sqrt{8-x^2}}.$

50. $\int_0^{\ln 2} x^3 e^{2x} \, dx.$

51. $\int_1^2 x \ln(2x) \, dx.$

52. $\int_1^2 dx.$

53. Find the present value, to the nearest dollar, of a continuous annuity at an annual rate of r for T years if the payment at time t is at the annual rate of $f(t)$ dollars given that

(a) $r = .06$, $T = 10$, $f(t) = 5000$,

(b) $r = .05$, $T = 8$, $f(t) = 200t$.

54. If $f(t) = k$, where k is a positive constant, show that the value of the integral in Eq. (1) of this section is

$$k\left(\frac{1 - e^{-rT}}{r}\right).$$

55. Find the accumulated amount, to the nearest dollar, of a continuous annuity at an annual rate of r for T years if the payment at time t is at an annual rate of $f(t)$ dollars given that

(a) $r = .06$, $T = 10$, $f(t) = 400$,

(b) $r = .04$, $T = 5$, $f(t) = 40t$.

56. Over the next five years the profits of a business at time t are estimated to be $20,000t$ dollars per year. The business is to be sold at a price equal to the present value of these future profits. If interest is compounded continuously at the annual rate of 6 percent, to the nearest ten dollars at what price should the business be sold?

11-4 AVERAGE VALUE OF A FUNCTION

If we are given the three numbers 1, 2, and 9, then their average value, or *mean*, is their sum divided by 3. Calling this average $\bar{y}$, we have

$$\bar{y} = \frac{1 + 2 + 9}{3} = 4.$$

Similarly, suppose we are given a function f defined on the interval $[a, b]$ and the points $x_1, x_2, \ldots, x_n$ are in the interval. Then the average value of the n corresponding functional values $f(x_1), f(x_2), \ldots, f(x_n)$ is

$$\bar{y} = \frac{f(x_1) + f(x_2) + \ldots + f(x_n)}{n} = \frac{\sum\limits_{k=1}^{n} f(x_k)}{n}. \tag{1}$$

We can go a step further. Let us divide the interval $[a, b]$ into n subintervals of equal length. We shall choose x_1 to be in the first subinterval, x_2 to be in the second, etc. Since $[a, b]$ has length $b - a$, each subinterval has length $\dfrac{b - a}{n}$, which we shall call Δx. Thus, (1) can be written

$$\bar{y} = \frac{\displaystyle\sum_{k=1}^{n} f(x_k)\left(\frac{\Delta x}{\Delta x}\right)}{n} = \frac{1}{\Delta x}\left[\frac{\displaystyle\sum_{k=1}^{n} f(x_k)\,\Delta x}{n}\right] = \frac{1}{n\,\Delta x}\sum_{k=1}^{n} f(x_k)\,\Delta x. \quad (2)$$

Since $\Delta x = \dfrac{b-a}{n}$, the expression $n\,\Delta x$ in (2) can be replaced by $b-a$. Moreover, as $n \to \infty$ the number of functional values used in computing $\bar{y}$ increases, and we get the so-called *average value of the function f*, denoted $\bar{f}$:

$$\bar{f} = \lim_{n\to\infty} \frac{1}{b-a}\sum_{k=1}^{n} f(x_k)\,\Delta x = \frac{1}{b-a}\lim_{n\to\infty}\sum_{k=1}^{n} f(x_k)\,\Delta x.$$

But the limit on the right is just the definite integral $\displaystyle\int_a^b f(x)\,dx$.

DEFINITION. *The **average** (or **mean**) **value of a function** $y = f(x)$ over the interval $[a, b]$ is denoted $\bar{f}$ (or $\bar{y}$) and is given by*

$$\bar{f} = \frac{1}{b-a}\int_a^b f(x)\,dx.$$

EXAMPLE 1

Find the average value of the function $f(x) = x^2$ over the interval $[1, 2]$.

$$\bar{f} = \frac{1}{b-a}\int_a^b f(x)\,dx$$

$$= \frac{1}{2-1}\int_1^2 x^2\,dx = \frac{x^3}{3}\bigg|_1^2 = \frac{7}{3}.$$

average height

In Example 1 we found that the average value of $y = f(x) = x^2$ over $[1, 2]$ was $\frac{7}{3}$. We can interpret this value geometrically. Since

$$\frac{1}{2-1}\int_1^2 x^2\,dx = \frac{7}{3},$$

then

$$\int_1^2 x^2\,dx = \frac{7}{3}(2-1).$$

However, this integral gives the area of the region bounded by $f(x) = x^2$ and the x-axis from $x = 1$ to $x = 2$. See Fig. 11-1. This area also equals $(\frac{7}{3})(2-1)$, which is the area of a rectangle with height $\bar{f} = \frac{7}{3}$ and width $b - a = 2 - 1 = 1$.

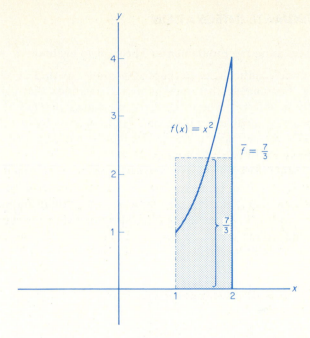

FIG. 11-1

EXERCISE 11-4

In Problems **1–8**, find the average value of the function over the given interval.

1. $f(x) = x^2$;   [0, 4].
                                       2. $f(x) = 3x - 1$;   [1, 2].

3. $f(x) = 2 - 3x^2$;   [−1, 2].
             4. $f(x) = x^2 + x + 1$;   [1, 3].

5. $f(t) = 4t^3$;   [−2, 2].
                               6. $f(i) = i\sqrt{i^2 + 9}$;   [0, 4].

7. $f(x) = \sqrt{x}$;   [1, 9].
                               8. $f(x) = \dfrac{1}{x}$;   [2, 4].

9. The profit P (in dollars) of a business is given by

$$P = P(q) = 396q - 2.1q^2 - 400,$$

where q is the number of units of the product sold. Find the average profit on the interval from $q = 0$ to $q = 100$.

10. Suppose the cost c (in dollars) of producing q units of a product is given by

$$c = 4000 + 10q + .1q^2.$$

Find the average cost on the interval from $q = 100$ to $q = 500$.

11. The value S (in dollars) of an investment of \$3000 at 5 percent compounded continuously for t years is given by $S = 3000e^{.05t}$. Find the average value of a two-year investment.

11-5 APPROXIMATE INTEGRATION

When using the Fundamental Theorem to evaluate $\int_a^b f(x)\, dx$, you may find it extremely difficult, or perhaps impossible, to find an elementary antiderivative of f. Fortunately there are numerical methods that can be used to estimate a definite integral. These methods use values of $f(x)$ at various points and are especially suitable for electronic computers or hand calculators. We shall consider two methods: the *trapezoidal rule* and *Simpson's rule*. In both cases we assume that f is continuous on $[a, b]$.

In developing the trapezoidal rule, for convenience we shall also assume that $f(x) \geqslant 0$ on $[a, b]$ so that we can think in terms of area. Basically, this rule involves approximating the graph of f by straight line segments.

In Fig. 11-2, the interval $[a, b]$ is divided into n subintervals of equal length by the points $a = x_0, x_1, x_2, \ldots,$ and $x_n = b$. Since the length of $[a, b]$ is

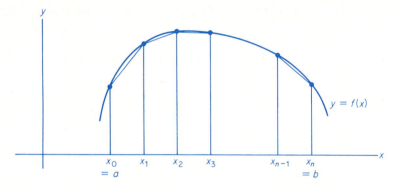

FIG. 11-2

$b - a$, then the length of each subinterval is $(b - a)/n$, which we shall call h. Clearly, $x_1 = a + h$, $x_2 = a + 2h, \ldots, x_n = a + nh = b$. With each subinterval we can associate a trapezoid (a four-sided figure with two parallel sides). The area of the region bounded by the curve, the x-axis, and the lines $x = a$ and $x = b$ is approximated by the sum of the areas of the trapezoids determined by the subintervals.

Consider the first trapezoid, which is redrawn in Fig. 11-3. Since the area of a trapezoid is equal to one-half the base times the sum of the parallel sides, this trapezoid has area

$$\tfrac{1}{2} h [\, f(a) + f(a + h)\,].$$

Similarly, the second trapezoid has area

$$\tfrac{1}{2} h [\, f(a + h) + f(a + 2h)\,].$$

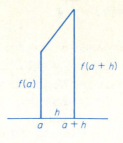

FIG. 11-3

With n trapezoids the area A under the curve is approximated by

$$A \approx \tfrac{1}{2}h\big[f(a) + f(a + h)\big] + \tfrac{1}{2}h\big[f(a + h) + f(a + 2h)\big] +$$

$$\tfrac{1}{2}h\big[f(a + 2h) + f(a + 3h)\big] + \ldots + \tfrac{1}{2}h\big[f(a + (n - 1)h) + f(b)\big].$$

Since $A = \int_a^b f(x)\,dx$, by simplifying the above we have

> **THE TRAPEZOIDAL RULE**
>
> $$\int_a^b f(x)\,dx \approx \frac{h}{2}\{f(a) + 2f(a + h) + 2f(a + 2h) +$$
>
> $$\ldots + 2f[a + (n - 1)h] + f(b)\}.$$

Usually, the more subintervals, the better is the approximation.

EXAMPLE 1

Use the trapezoidal rule to estimate the value of

$$\int_0^1 \frac{1}{1 + x^2}\,dx$$

by using $n = 5$. Compute each term to four decimal places and round off the answer to three decimal places.

Here $f(x) = 1/(1 + x^2)$, $n = 5$, $a = 0$, and $b = 1$. Thus,

$$h = \frac{b - a}{n} = \frac{1 - 0}{5} = \frac{1}{5} = .2.$$

The terms to be added are

$$
\begin{aligned}
f(a) = \ f(0) \ &= 1.0000 \\
2f(a + h) = 2f(.2) \ &= 1.9231 \\
2f(a + 2h) = 2f(.4) \ &= 1.7241 \\
2f(a + 3h) = 2f(.6) \ &= 1.4706 \\
2f(a + 4h) = 2f(.8) \ &= 1.2195 \\
f(b) = \ f(1) \quad &= \underline{\ .5000} \\
&\ \ 7.8373 = \text{Sum.}
\end{aligned}
$$

Thus our estimate for the integral is

$$
\int_0^1 \frac{1}{1 + x^2}\, dx \approx \frac{.2}{2}(7.8373) \approx .784.
$$

The actual value of the integral is approximately .785.

Another method for estimating $\int_a^b f(x)\, dx$ is given by Simpson's rule, which involves approximating the graph of f by parabolic segments. We shall omit the derivation.

SIMPSON'S RULE (n even)

$$
\int_a^b f(x)\, dx \approx \frac{h}{3}\{f(a) + 4f(a + h) + 2f(a + 2h) +
$$

$$
\dots + 4f[a + (n - 1)h] + f(b)\}.
$$

The pattern of coefficients inside the braces is 1, 4, 2, 4, 2, . . . , 2, 4, 1, and this requires that **n be even**. Let us use this rule for the integral in Example 1.

EXAMPLE 2

Use Simpson's rule to estimate the value of $\int_0^1 \frac{1}{1 + x^2}\, dx$ by using $n = 4$. Compute each term to four decimal places and round off the answer to three decimal places.

Here $f(x) = 1/(1 + x^2)$, $n = 4$, $a = 0$, and $b = 1$. Thus $h = (b - a)/n = 1/4 = .25$.

The terms to be added are

$$f(a) = \quad f(0) \quad = 1.0000$$
$$4f(a + h) = 4f(.25) = 3.7647$$
$$2f(a + 2h) = 2f(.5) \quad = 1.6000$$
$$4f(a + 3h) = 4f(.75) = 2.5600$$
$$f(b) = \quad f(1) \quad = \underline{\ .5000}$$

$$9.4247 = \text{Sum.}$$

Thus, by Simpson's rule,

$$\int_0^1 \frac{1}{1 + x^2}\, dx \approx \frac{.25}{3}(9.4247) \approx .785.$$

This is a better approximation than we obtained by using the trapezoidal rule.

Both Simpson's rule and the trapezoidal rule may be used if we know only $f(a), f(a + h)$, etc.; we need not know f itself. Example 3 will illustrate.

EXAMPLE 3

*A function often used in demography (the study of births, marriages, mortality, etc. in a population) is the **life table function**, denoted l. In a population having 100,000 births in any year of time, $l(x)$ represents the number of persons who reach the age of x in any year of time. If $l(20) = 95,961$, then the number of persons who attain age 20 in any year of time is 95,961. Suppose that the function l applies to all people born over an extended period of time. It can be shown that at any time, the expected number of persons in the population between the exact ages of x and $x + n$ inclusive is given by*

$$\int_x^{x+n} l(t)\, dt.$$

In Table 11-1[†] are values of $l(x)$ for males and females in a certain population. Approximate the number of women in the 20–35 age group by using the trapezoidal rule with $n = 3$.

[†] For United States, 1967. Adapted from *Population: Facts and Methods of Demography* by Nathan Keyfitz and William Flieger. W. H. Freeman and Company. Copyright © 1971.

TABLE 11-1

Life Table		
	Males	Females
Age, x	l(x)	l(x)
0	100,000	100,000
5	97,158	97,791
10	96,921	97,618
15	96,672	97,473
20	95,961	97,188
25	95,000	96,839
30	94,097	96,429
35	93,067	95,844
40	91,628	94,961
45	89,489	93,667
50	86,195	91,726
55	81,154	88,935
60	73,830	84,971
65	64,108	79,445
70	52,007	71,196
75	38,044	59,946
80	24,900	45,662

We want to evaluate

$$\int_{20}^{35} l(t)\ dt.$$

We have $h = \dfrac{b - a}{n} = \dfrac{35 - 20}{3} = 5$. The terms to be added are

$$l(20) = 97,188$$
$$2l(25) = 2(96,839) = 193,678$$
$$2l(30) = 2(96,429) = 192,858$$
$$l(35) = \underline{95,844}$$
$$579,568 = \text{Sum.}$$

Thus,

$$\int_{20}^{35} l(t)\ dt \approx \tfrac{5}{2}(579,568) = 1,448,920.$$

There are formulas that are used to determine the accuracy of answers obtained by using the trapezoidal or Simpson's rule. They may be found in standard texts on numerical analysis.

EXERCISE 11-5

In each problem, compute each term to four decimal places and round off the answer to three decimal places.

*Problems **1–6**, use the trapezoidal rule or Simpson's rule (as indicated) and the given value of n to estimate the integral. In Problems **1–4**, also find the answer by antidifferentiation (the Fundamental Theorem of Integral Calculus).*

1. $\int_0^1 x^2 \, dx$;   trapezoidal rule, $n = 5$.

2. $\int_0^1 x^2 \, dx$;   Simpson's rule, $n = 4$.

3. $\int_1^4 \dfrac{dx}{x}$;   Simpson's rule, $n = 6$.

4. $\int_1^4 \dfrac{dx}{x}$;   trapezoidal rule, $n = 6$.

5. $\int_0^2 \dfrac{x \, dx}{x + 1}$;   trapezoidal rule, $n = 4$.

6. $\int_2^4 \dfrac{dx}{x + x^2}$;   Simpson's rule, $n = 4$.

*In Problems **7** and **8**, use the life table (Table 11-1) in Example 3 to estimate the given integrals by the trapezoidal rule.*

7. $\int_{15}^{40} l(t) \, dt$, males, $n = 5$.   8. $\int_{35}^{55} l(t) \, dt$, females, $n = 4$.

*In Problems **9** and **10**, suppose the graph of a continuous function f, where $f(x) \geqslant 0$, contains the given points. Use Simpson's rule and all of the points to approximate the area between the graph and the x-axis on the given interval.*

9. (1, .4), (2, .6), (3, 1.2), (4, .8), (5, .5);   [1, 5].

10. (2, 0), (2.5, 3.6), (3, 10), (3.5, 19.9), (4, 34);   [2, 4].

*Problems **11** and **12** are designed for students with calculators having the square-root function. Estimate the given integrals.*

11. $\int_0^1 \sqrt{1 - x^2} \, dx$;   Simpson's rule, $n = 4$.

12. $\int_4^6 \dfrac{1}{\sqrt{1 + x}} \, dx$;   Simpson's rule, $n = 4$. Also find the answer by antidifferentiation.

11-6 IMPROPER INTEGRALS†

A. INFINITE LIMITS

Any integral of the form

$$\int_a^\infty f(x)\, dx, \tag{1}$$

$$\int_{-\infty}^b f(x)\, dx, \tag{2}$$

or $$\int_{-\infty}^\infty f(x)\, dx \tag{3}$$

is called an **improper integral**. In each case the interval over which the integral is evaluated has infinite length.

We define (1) as follows:

$$\int_a^\infty f(x)\, dx = \lim_{r \to \infty} \int_a^r f(x)\, dx.$$

When this limit exists, $\int_a^\infty f(x)\, dx$ is said to be *convergent* or to *converge to that limit*. When the limit does not exist, the integral is said to be *divergent*.

We can give a geometric interpretation of this improper integral for the case where f is nonnegative for $a \leqslant x < \infty$. See Fig. 11-4. The integral $\int_a^r f(x)\, dx$ is the area under the curve and above the x-axis from $x = a$ to

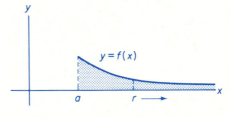

FIG. 11-4

$x = r$. As $r \to \infty$, we may think of $\int_a^r f(x)\, dx$ as the area of the unbounded region that is shaded in Fig. 11-4. If $\int_a^\infty f(x)\, dx$ converges, then the unbounded

----

† May be omitted without loss of continuity.

region is considered to have a finite area, and this area is represented by $\int_a^\infty f(x)\,dx$. If the improper integral is divergent, then the region does not have a finite area.

The improper integral in (2) is defined as

$$\int_{-\infty}^b f(x)\,dx = \lim_{r \to -\infty} \int_r^b f(x)\,dx.$$

If this limit exists, $\int_{-\infty}^b f(x)\,dx$ is said to be convergent. Otherwise it is divergent.

EXAMPLE 1

Determine whether the following improper integrals are convergent or divergent. If convergent, determine the value of the integral.

a. $\int_1^\infty \dfrac{1}{x^3}\,dx.$

$$\int_1^\infty \frac{1}{x^3}\,dx = \lim_{r \to \infty} \int_1^r x^{-3}\,dx = \lim_{r \to \infty} -\frac{x^{-2}}{2}\Big|_1^r$$

$$= \lim_{r \to \infty}\left[-\frac{1}{2r^2} + \frac{1}{2}\right] = -0 + \frac{1}{2} = \frac{1}{2}.$$

Therefore $\int_1^\infty \dfrac{1}{x^3}\,dx$ converges to $\dfrac{1}{2}$.

b. $\int_{-\infty}^0 e^x\,dx.$

$$\int_{-\infty}^0 e^x\,dx = \lim_{r \to -\infty} \int_r^0 e^x\,dx = \lim_{r \to -\infty} e^x\Big|_r^0$$

$$= \lim_{r \to -\infty} (1 - e^r) = 1 - 0 = 1.$$

Therefore $\int_{-\infty}^0 e^x\,dx$ converges to 1.

c. $\int_1^\infty \dfrac{1}{\sqrt{x}}\,dx.$

$$\int_1^\infty \frac{1}{\sqrt{x}}\,dx = \lim_{r \to \infty} \int_1^r x^{-1/2}\,dx = \lim_{r \to \infty} 2x^{1/2}\Big|_1^r$$

$$= \lim_{r \to \infty} 2(\sqrt{r} - 1) = \infty.$$

Therefore, the improper integral diverges.

The improper integral $\int_{-\infty}^{\infty} f(x)\, dx$ is defined in terms of improper integrals of the forms (1) and (2):

$$\int_{-\infty}^{\infty} f(x)\, dx = \int_{-\infty}^{0} f(x)\, dx + \int_{0}^{\infty} f(x)\, dx. \tag{4}$$

If *both* integrals on the right side of (4) are convergent, then $\int_{-\infty}^{\infty} f(x)\, dx$ is said to be convergent; otherwise, it is divergent.

EXAMPLE 2

Determine whether $\int_{-\infty}^{\infty} e^{x}\, dx$ *is convergent or divergent.*

$$\int_{-\infty}^{\infty} e^{x}\, dx = \int_{-\infty}^{0} e^{x}\, dx + \int_{0}^{\infty} e^{x}\, dx.$$

By Example 1(b) $\int_{-\infty}^{0} e^{x}\, dx = 1.$ On the other hand,

$$\int_{0}^{\infty} e^{x}\, dx = \lim_{r \to \infty} \left[\int_{0}^{r} e^{x}\, dx \right] = \lim_{r \to \infty} e^{x} \Big|_{0}^{r}$$

$$= \lim_{r \to \infty} (e^{r} - 1) = \infty.$$

Since $\int_{0}^{\infty} e^{x}\, dx$ is divergent, $\int_{-\infty}^{\infty} e^{x}\, dx$ is also divergent.

EXAMPLE 3

In statistics, a function f is called a density function if $f(x) \geqslant 0$ *and*

$$\int_{-\infty}^{\infty} f(x)\, dx = 1.$$

Suppose

$$f(x) = \begin{cases} ke^{-x}, & \text{for } x > 0, \\ 0, & \text{elsewhere} \end{cases}$$

is a density function. Find k.

We can write $\int_{-\infty}^{\infty} f(x)\, dx = 1$ as

$$\int_{-\infty}^{\infty} f(x)\, dx = \int_{-\infty}^{0} f(x)\, dx + \int_{0}^{\infty} f(x)\, dx = 1.$$

Since $f(x) = 0$ for $x < 0$, $\int_{-\infty}^{0} f(x)\, dx = 0$. Thus,

$$\int_{0}^{\infty} ke^{-x}\, dx = 1,$$

$$\lim_{r \to \infty} \int_{0}^{r} ke^{-x}\, dx = 1,$$

$$\lim_{r \to \infty} -ke^{-x}\Big|_{0}^{r} = 1,$$

$$\lim_{r \to \infty} (-ke^{-r} + k) = 1,$$

$$0 + k = 1,$$

$$k = 1.$$

B. DISCONTINUOUS INTEGRAND

Another type of improper integral has the form $\int_{a}^{b} f(x)\, dx$, where f is either

    i. continuous on $(a, b]$ with an infinite discontinuity as $x \to a^+$, or
    ii. continuous on $[a, b)$ with an infinite discontinuity as $x \to b^-$.

When f meets the conditions of i, we define

$$\int_{a}^{b} f(x)\, dx = \lim_{r \to a^+} \int_{r}^{b} f(x)\, dx.$$

When f meets the conditions of ii, we define

$$\int_{a}^{b} f(x)\, dx = \lim_{r \to b^-} \int_{a}^{r} f(x)\, dx.$$

As usual, these integrals are said to be convergent when the limits exist and divergent otherwise.

EXAMPLE 4

Evaluate each of the following integrals if possible.

a. $\int_{0}^{1} \dfrac{1}{x}\, dx.$

The function $f(x) = 1/x$ has an infinite discontinuity as $x \to 0^+$ and the integral is improper (see Fig. 11-5).

$$\int_0^1 \frac{1}{x}\,dx = \lim_{r \to 0^+} \int_r^1 \frac{1}{x}\,dx = \lim_{r \to 0^+} [\ln|x|]\Big|_r^1$$

$$= \lim_{r \to 0^+} (\ln 1 - \ln r) = \infty.$$

Here we used the fact that $\lim_{r \to 0^+} (\ln r) = -\infty$, which is clear from Fig. 5-7. Therefore, $\int_0^1 \frac{1}{x}\,dx$ does not exist and is divergent.

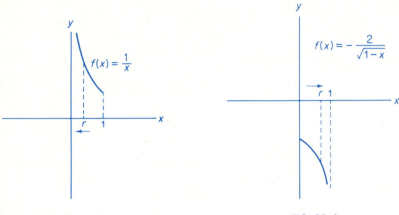

FIG. 11-5   **FIG. 11-6**

b. $\int_0^1 \dfrac{-2}{\sqrt{1-x}}\,dx.$

The integrand has an infinite discontinuity as $x \to 1^-$, and so the integral is improper. See Fig. 11-6.

$$\int_0^1 \frac{-2}{\sqrt{1-x}}\,dx = \lim_{r \to 1^-} \int_0^r -2(1-x)^{-1/2}\,dx = \lim_{r \to 1^-} 4(1-x)^{1/2}\Big|_0^r$$

$$= \lim_{r \to 1^-} 4(\sqrt{1-r} - 1) = -4.$$

Therefore, the improper integral converges to -4.

Another type of improper integral has the form $\int_a^b f(x)\,dx$ where f is continuous on $[a, b]$ *except* at c, where $a < c < b$, and such that f has an infinite discontinuity at $x = c$. We define

$$\int_a^b f(x)\, dx = \int_a^c f(x)\, dx + \int_c^b f(x)\, dx.$$

If *both* of the improper integrals on the right are convergent, then $\int_a^b f(x)\, dx$ is convergent; otherwise it is divergent.

EXAMPLE 5

Determine whether $\int_1^5 \dfrac{1}{(x-3)^2}\, dx$ *is convergent or divergent.*

The function $f(x) = 1/(x-3)^2$ has an infinite discontinuity when $x = 3$ and so the integral is improper (see Fig. 11-7).

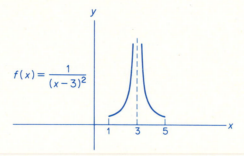

FIG. 11-7

$$\int_1^5 \frac{1}{(x-3)^2}\, dx = \int_1^3 \frac{1}{(x-3)^2}\, dx + \int_3^5 \frac{1}{(x-3)^2}\, dx.$$

However,

$$\int_1^3 \frac{1}{(x-3)^2}\, dx = \lim_{r \to 3^-} \int_1^r \frac{1}{(x-3)^2}\, dx$$

$$= \lim_{r \to 3^-} \left. -\frac{1}{x-3} \right|_1^r$$

$$= \lim_{r \to 3^-} \left[-\frac{1}{r-3} - \frac{1}{2} \right] = \infty.$$

Since $\int_1^3 (x-3)^{-2}\, dx$ is divergent, $\int_1^5 (x-3)^{-2}\, dx$ must also be divergent. In this case it is immaterial whether $\int_3^5 (x-3)^{-2}\, dx$ is convergent or divergent (although it actually is divergent).

EXERCISE 11-6

In Problems **1–12**, *determine the integrals, if they exist. Indicate those that are divergent.*

1. $\int_{3}^{\infty} \frac{1}{x^2}\, dx.$

2. $\int_{2}^{\infty} \frac{1}{(2x-1)^3}\, dx.$

3. $\int_{1}^{\infty} \frac{1}{x}\, dx.$

4. $\int_{1}^{\infty} \frac{1}{\sqrt[3]{x+1}}\, dx.$

5. $\int_{1}^{\infty} e^{-x}\, dx.$

6. $\int_{0}^{\infty} (5 + e^{-x})\, dx.$

7. $\int_{1}^{\infty} \frac{1}{\sqrt{x}}\, dx.$

8. $\int_{4}^{\infty} \frac{x\, dx}{\sqrt{(x^2+9)^3}}.$

9. $\int_{-\infty}^{-2} \frac{1}{(x+1)^3}\, dx.$

10. $\int_{-\infty}^{3} \frac{1}{\sqrt{7-x}}\, dx.$

11. $\int_{-\infty}^{\infty} xe^{-x^2}\, dx.$

12. $\int_{-\infty}^{\infty} (5 - 3x)\, dx.$

In Problems **13–24**, *determine the integrals, if they exist. Indicate those that are divergent.*

13. $\int_{0}^{9} \frac{1}{\sqrt{x}}\, dx.$

14. $\int_{1}^{2} \frac{1}{x-1}\, dx.$

15. $\int_{0}^{3} \frac{x}{\sqrt{9-x^2}}\, dx.$

16. $\int_{0}^{1} \frac{1}{\sqrt{1-x}}\, dx.$

17. $\int_{0}^{3} \frac{x}{9-x^2}\, dx.$

18. $\int_{0}^{1} \frac{1}{x^3}\, dx.$

19. $\int_{0}^{1} \frac{1}{x+7}\, dx.$

20. $\int_{-1}^{1} \frac{dx}{\sqrt[3]{x}}.$

21. $\int_{-1}^{1} \frac{1}{x^2}\, dx.$

22. $\int_{-2}^{2} \frac{1}{x-1}\, dx.$

23. $\int_{-1}^{1} \frac{1}{x^{2/3}}\, dx.$

24. $\int_{-100}^{0} \frac{1}{\sqrt{-x}}\, dx.$

In Problems **25** *and* **26**, *determine the integrals, if they exist. Indicate those that are divergent.*

25. $\int_{1}^{\infty} xe^{x}\, dx.$

26. $\int_{1}^{\infty} x \ln x\, dx.$

27. The density function of the life in hours x of an electronic component in a calculator is given by

$$f(x) = \begin{cases} \dfrac{k}{x^2}, & \text{for } x \geqslant 800, \\ 0, & \text{for } x < 800. \end{cases}$$

(a) If k satisfies the condition that $\int_{800}^{\infty} f(x)\, dx = 1$, find k. (b) The probability that the component will last at least 1200 hours is given by $\int_{1200}^{\infty} f(x)\, dx$. Evaluate this integral.

28. Given the density function

$$f(x) = \begin{cases} ke^{-4x}, & \text{for } x \geqslant 0, \\ 0, & \text{elsewhere,} \end{cases}$$

find k.

29. For a business the present value of all future profits at an annual interest rate r compounded continuously is given by

$$\int_0^\infty p(t)e^{-rt}\, dt,$$

where $p(t)$ is the profit per year in dollars at time t. If $p(t) = 240,000$ and $r = .06$, evaluate the above integral.

30. In discussing entrance of a firm into an industry, Stigler[†] uses the equation

$$V = \pi_0 \int_0^\infty e^{\theta t} e^{-\rho t}\, dt,$$

where π_0, θ (a Greek letter read "theta"), and ρ (a Greek letter read "rho") are constants. Show that $V = \pi_0/(\rho - \theta)$ if $\theta < \rho$.

31. Find the area of the region in the first quadrant bounded by the curve $y = e^{-2x}$ and the x-axis.

32. The predicted rate of growth per year of a population of a certain small city is given by $10,000/(t + 2)^2$, where t is the number of years from now. In the long run (that is, as $t \to \infty$), what is the expected change in population from today's level?

11-7 DIFFERENTIAL EQUATIONS

Occasionally you may have to solve an equation that involves the derivative of an unknown function. For example,

$$y' = xy^2 \tag{1}$$

is such an equation. It is called a **differential equation**. More precisely, it is a *first order differential equation* since it involves a derivative of the first order and none of higher order. A solution of Eq. (1) is any function $y = f(x)$ that is defined on an interval and that satisfies Eq. (1) for all x in the interval.

To solve $y' = xy^2$ or, equivalently,

$$\frac{dy}{dx} = xy^2, \tag{2}$$

we consider dy/dx to be a quotient of differentials and algebraically "separate variables" by rewriting the equation so that each side contains only one variable and a differential is not in a denominator:

$$\frac{dy}{y^2} = x\, dx.$$

[†] G. Stigler, *The Theory of Price*, 3rd ed., The Macmillan Co., New York, 1966, p. 344.

Integrating both sides and combining the constants of integration, we obtain

$$\int \frac{1}{y^2}\, dy = \int x \, dx.$$

$$-\frac{1}{y} = \frac{x^2}{2} + C_1,$$

$$-\frac{1}{y} = \frac{x^2 + 2C_1}{2}. \tag{3}$$

Letting $2C_1 = C$ and solving Eq. (3) for y, we have

$$y = \frac{-2}{x^2 + C}. \tag{4}$$

On any interval on which y is defined, we can verify by substitution that y is a solution to differential equation (2):

$$\frac{dy}{dx} = xy^2 \; ?$$

$$\frac{4x}{(x^2 + C)^2} = x\left[\frac{-2}{x^2 + C}\right]^2 \; ?$$

$$\frac{4x}{(x^2 + C)^2} = \frac{4x}{(x^2 + C)^2}.$$

Note in Eq. (4) that for *each* value of C, a different solution is obtained. We call Eq. (4) the **general solution** of the differential equation. The method that we used to get it is called **separation of variables**.

In the example above, suppose we are given the condition that $y = -\frac{2}{3}$ when $x = 1$; that is $y(1) = -\frac{2}{3}$. Then the *particular* function that satisfies Eq. (2) can be found by substituting the values $x = 1$ and $y = -\frac{2}{3}$ into Eq. (4) and solving for C:

$$-\frac{2}{3} = -\frac{2}{1^2 + C},$$

$$C = 2.$$

Therefore, the solution of $dy/dx = xy^2$ such that $y(1) = -\frac{2}{3}$ is

$$y = -\frac{2}{x^2 + 2}. \tag{5}$$

We call Eq. (5) a **particular solution** to the differential equation.

EXAMPLE 1

Solve $y' = -\dfrac{y}{x}$ *if* $x, y > 0$.

Writing y' as dy/dx, separating variables, and integrating, we have

$$\frac{dy}{y} = -\frac{dx}{x},$$

$$\int \frac{1}{y}\,dy = -\int \frac{1}{x}\,dx,$$

$$\ln |y| = C_1 - \ln |x|.$$

Since $x, y > 0$,

$$\ln y = C_1 - \ln x. \tag{6}$$

The constant of integration, C_1, can be any real number. Since the range of the logarithmic function is all real numbers, C_1 can be replaced by $\ln C$, where $C > 0$. Therefore, Eq. (6) becomes

$$\ln y = \ln C - \ln x,$$

$$\ln y = \ln \frac{C}{x}.$$

This can only happen if

$$y = \frac{C}{x}, \qquad C, x > 0.$$

In Sec. 7-3 interest compounded continuously was developed. Let us now take a different approach to this topic. Suppose P dollars are invested at an annual rate r compounded n times a year. Let the function $S = S(t)$ give the compound amount S after t years from the date of the initial investment. Then the initial principal is $S(0) = P$. Furthermore, since there are n interest periods per year, each period has length $1/n$ years, which we shall denote by Δt. At the end of the first period, the accrued interest for that period is added to the principal, and the sum acts as the principal for the second period, etc. Hence, if the beginning of an interest period occurs at time t, then the increase in the amount present (that is, the interest earned) at the end of a period of Δt is $S(t + \Delta t) - S(t) = \Delta S$. Equivalently, the interest earned is principal times rate times time:

$$\Delta S = S \cdot r \cdot \Delta t.$$

Dividing both sides by Δt, we obtain

$$\frac{\Delta S}{\Delta t} = rS. \tag{7}$$

As $\Delta t \to 0$, then $n = \dfrac{1}{\Delta t} \to \infty$ and consequently interest is being *compounded continuously*; that is, the principal is subject to continuous growth at every instant. However as $\Delta t \to 0$, then $\Delta S / \Delta t \to dS / dt$ and Eq. (7) takes the form

$$\frac{dS}{dt} = rS. \tag{8}$$

This differential equation means that *when interest is compounded continuously, the rate of change of the amount of money present at time t is proportional to the amount present at time t.*

To determine the actual function S, we solve differential equation (8) by the method of separation of variables.

$$\frac{dS}{dt} = rS,$$

$$\frac{dS}{S} = r \, dt,$$

$$\int \frac{1}{S} \, dS = \int r \, dt,$$

$$\ln |S| = rt + C_1.$$

Since it can be assumed that $S > 0$, then $\ln |S| = \ln S$.

$$\ln S = rt + C_1.$$

Solving for S, we have

$$S = e^{rt + C_1} = e^{C_1} e^{rt}.$$

Replacing e^{C_1} by C, we obtain

$$S = Ce^{rt}.$$

Since $S(0) = P$,

$$P = Ce^{r(0)} = C(1).$$

Hence $C = P$ and

$$S = Pe^{rt}. \tag{9}$$

Equation (9) gives the total value after t years of an initial investment of P dollars, compounded continuously at an annual rate r.

In our compound interest discussion we saw from Eq. (8) that the rate of change in the amount present was proportional to the amount present. There are many natural quantities, such as population, whose rate of growth or decay at any time is considered proportional to the amount of that quantity present.

If N denotes the amount of such a quantity at time t, then the above rate of growth means that

$$\frac{dN}{dt} = kN.$$

If we separate variables and solve for N as we did for Eq. (8), we get

$$N = N_0 e^{kt}, \tag{10}$$

where N_0 and k are constants. Due to the form of Eq. (10), we say that the quantity follows an **exponential law of growth** if k is positive and **exponential decay** if k is negative.

EXAMPLE 2

In a certain city the rate at which the population grows at any time is proportional to the size of the population. If the population was 125,000 in 1950 and 140,000 in 1970, what is the expected population in 1990?

Let N be the size of the population at time t. Since the exponential law of growth applies,

$$N = N_0 e^{kt}.$$

We must first find the constants N_0 and k. The year 1950 will correspond to $t = 0$. Thus $t = 20$ is 1970 and $t = 40$ is 1990. Now, if $t = 0$, then $N = 125,000$. Thus,

$$N = N_0 e^{kt},$$
$$125,000 = N_0 e^0 = N_0.$$

Hence, $N_0 = 125,000$ and

$$N = 125,000 e^{kt}.$$

But if $t = 20$, then $N = 140,000$. This means

$$140,000 = 125,000 e^{20k}.$$

Thus,

$$e^{20k} = \frac{140,000}{125,000} = 1.12,$$

$$20k = \ln (1.12) \qquad [\text{logarithmic form}],$$

$$k = \tfrac{1}{20} \ln (1.12).$$

Therefore,

$$N = 125,000e^{(t/20)\ln 1.12} \qquad\qquad (11)$$

$$= 125,000[e^{\ln 1.12}]^{t/20}.$$

$$N = 125,000(1.12)^{t/20}. \qquad\qquad (12)$$

If $t = 40$,

$$N = 125,000(1.12)^2 = 156,800.$$

Note that we can write Eq. (11) in a form different from Eq. (12). Since $\ln 1.12 \approx .11333$, then $k \approx .11333/20 \approx .0057$. Thus,

$$N \approx 125,000e^{.0057t}.$$

Suppose that the number N of individuals in a population at time t follows an exponential law of growth. From Eq. (10), $N = N_0e^{kt}$, where $k > 0$ and N_0 is the population when $t = 0$. This law assumes that at time t the rate of growth, dN/dt, of the population is proportional to the number of individuals in the population. That is, $dN/dt = kN$.

Under exponential growth, a population would get infinitely large as time goes on. In reality, however, when the population gets large enough, there are environmental factors that slow down the rate of growth. Examples are food supply, overcrowding, etc. Since these factors cause dN/dt to eventually decrease, it is reasonable to assume that population size is limited to some maximum number M, where $0 < N < M$, and as $N \to M$, then $dN/dt \to 0$ and the population size tends to be stable.

In summary, we want a population model that has exponential growth initially but which also includes the effects of environmental resistance to large population growth. Such a model is obtained by multiplying the right side of $dN/dt = kN$ by the factor $(M - N)/M$:

$$\frac{dN}{dt} = kN\left(\frac{M - N}{M}\right).$$

Notice that if N is small, then $(M - N)/M$ is close to 1 and we have growth that is approximately exponential. As $N \to M$, then $M - N \to 0$ and $dN/dt \to 0$

as we wanted in our model. Replacing k/M by K, we have

$$\frac{dN}{dt} = KN(M - N). \tag{13}$$

This states that the rate of growth is proportional to the product of the population size and the difference between the maximum size and the population size. We can solve for N in Eq. (13) by the method of separation of variables.

$$\frac{dN}{N(M - N)} = K\,dt,$$

$$\int \frac{1}{N(M - N)}\,dN = \int K\,dt. \tag{14}$$

The integral on the left side can be found by using formula (5) in the table of integrals. Thus Eq. (14) becomes

$$\frac{1}{M} \ln \left| \frac{N}{M - N} \right| = Kt + C,$$

$$\ln \left| \frac{N}{M - N} \right| = MKt + MC.$$

Since $N > 0$ and $M - N > 0$, we can write

$$\ln \frac{N}{M - N} = MKt + MC.$$

In exponential form we have

$$\frac{N}{M - N} = e^{MKt + MC} = e^{MKt}e^{MC}.$$

Replacing the positive constant e^{MC} by A gives

$$\frac{N}{M - N} = Ae^{MKt},$$

$$N = (M - N)Ae^{MKt},$$

$$N = MAe^{MKt} - NAe^{MKt},$$

$$N(Ae^{MKt} + 1) = MAe^{MKt},$$

$$N = \frac{MAe^{MKt}}{Ae^{MKt} + 1}.$$

Dividing numerator and denominator by Ae^{MKt}, we have

$$N = \frac{M}{1 + \dfrac{1}{Ae^{MKt}}} = \frac{M}{1 + \dfrac{1}{A}e^{-MKt}}.$$

Replacing $1/A$ by b and MK by c gives

$$N = \frac{M}{1 + be^{-ct}}. \tag{15}$$

Equation (15) is called the **logistic function** or the **Verhulst-Pearl logistic function**. Its graph, called a *logistic curve*, is S-shaped and appears in Fig. 11-8.

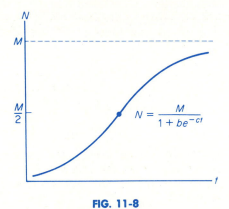

FIG. 11-8

Notice in the graph that $N = M$ is a horizontal asymptote; that is,

$$\lim_{t\to\infty} \frac{M}{1 + be^{-ct}} = \frac{M}{1 + b(0)} = M.$$

Moreover, from Eq. (13) the rate of growth is

$$KN(M - N).$$

To find when the maximum rate of growth occurs, we solve

$$\frac{d}{dN}\big[KN(M - N)\big] = \frac{d}{dN}\big[K(MN - N^2)\big]$$
$$= K[M - 2N] = 0.$$

Thus $N = M/2$. The rate of growth increases until the population size is $M/2$ and decreases thereafter. The maximum rate of growth occurs when $N = M/2$ and corresponds to a point of inflection in the graph of N. To find the value of t for which this occurs, we substitute $M/2$ for N in Eq. (15) and solve for t.

$$\frac{M}{2} = \frac{M}{1 + be^{-ct}},$$

$$1 + be^{-ct} = 2,$$

$$e^{-ct} = \frac{1}{b},$$

$$e^{ct} = b,$$

$$ct = \ln b,$$

$$t = \frac{\ln b}{c}.$$

Thus the maximum rate of growth occurs at the point $([\ln b]/c, M/2)$. We point out that in Eq. (15), we may replace e^c by C and then the logistic function has the form

$$N = \frac{M}{1 + bC^{-t}}.$$

EXAMPLE 3

Suppose the membership in a new country club is to be a maximum of 800 persons due to limitations of the physical plant. One year ago the initial membership was 50 persons and now there are 200. Provided that enrollment follows a logistic function, how many members will there be three years from now?

Let N be the number of members enrolled t years after the formation of the club. Then

$$N = \frac{M}{1 + be^{-ct}}.$$

Here $M = 800$, and when $t = 0$ we have $N = 50$.

$$50 = \frac{800}{1 + b},$$

$$1 + b = \frac{800}{50} = 16,$$

$$b = 15.$$

Thus,

$$N = \frac{800}{1 + 15e^{-ct}}. \tag{16}$$

When $t = 1$, then $N = 200$.

$$200 = \frac{800}{1 + 15e^{-c}},$$

$$1 + 15e^{-c} = \frac{800}{200} = 4,$$

$$e^{-c} = \frac{3}{15} = \frac{1}{5}.$$

Hence $c = -\ln\left(\frac{1}{5}\right) = \ln 5$. However, it is more convenient to substitute the value of e^{-c} in Eq. (16).

$$N = \frac{800}{1 + 15\left(\frac{1}{5}\right)^{t}}.$$

Three years from now, $t = 4$. Thus,

$$N = \frac{800}{1 + 15\left(\frac{1}{5}\right)^{4}} \approx 781.$$

EXERCISE 11-7

In Problems **1–8**, *solve the differential equations.*

1. $y' = 2xy^2$.

2. $y' = x^3y^3$.

3. $\dfrac{dy}{dx} = y, \quad y > 0$.

4. $\dfrac{dy}{dx} = \dfrac{x}{y}$.

5. $y' = \dfrac{y}{x}, \quad x, y > 0$.

6. $y' = e^x y^2$.

7. $\dfrac{dy}{dx} - x\sqrt{x^2 + 1} = 0$.

8. $\dfrac{dy}{dx} + xe^x = 0$.

In Problems **9–14**, *solve each of the differential equations subject to the given conditions.*

9. $y' = \dfrac{1}{y}; \quad y > 0, y(2) = 2$.

10. $y' = e^{x-y}; \quad y(0) = 0$. *Hint:* $e^{x-y} = e^x/e^y$.

11. $e^y y' - x^2 = 0; \quad y = 0$ when $x = 0$.

12. $x^2 y' + \dfrac{1}{y^2} = 0; \quad y(1) = 2$.

13. $(4x^2 + 3)^2 y' - 4xy^2 = 0; \quad y(0) = \dfrac{3}{2}$.

14. $y' + x^2 y = 0; \quad y > 0, y = 1$ when $x = 0$.

15. In a certain town the population at any time changes at a rate proportional to the population. If the population in 1970 was 20,000 and in 1980 it was 24,000, find an

equation for the population at time t, where t is the number of years past 1970. Write your answer in two forms, one involving e. You may assume $\ln 1.2 = .18$. What is the expected population in 1990?

16. The population of a town increases by natural growth at a rate which is proportional to the number N of persons present. If the population at time $t = 0$ is 10,000, find two expressions for the population N, t years later, if the population doubles in 50 years. Assume $\ln 2 = .69$. Also, find N for $t = 100$.

17. Suppose that the population of the world in 1930 was 2 billion and in 1960 it was 3 billion. If the exponential law of growth is assumed, what is the expected population in 2000? Give your answer in terms of e.

18. If exponential growth is assumed, in approximately how many years will a population triple if it doubles in 50 years? *Hint*: Let the population at $t = 0$ be N_0.

19. Suppose that a population follows exponential growth given by $dN/dt = kN$ for $t > t_0$, and $N = N_0$ when $t = t_0$. Find N, the population size at time t.

20. The logistic curve for the United States population from 1790 to 1910 is estimated[†] to be

$$N = \frac{197.30}{1 + 35.60e^{-.031186t}},$$

where N is the population in millions and t is in years counted from 1800. If this logistic function were valid for years after 1910, for what year would the point of inflection occur? Give your answer to one decimal place. Assume that $\ln 35.6 = 3.5723$.

21. A small town decides to conduct a fund-raising drive for a new fire engine whose cost is $70,000. The initial amount in the fund is $10,000. On the basis of past drives, it is determined that t months after the beginning of the drive, the rate dx/dt at which people contribute to such a fund is proportional to the difference between the desired goal of $70,000 and the total amount x in the fund at that time. After one month a total of $40,000 is in the fund. How much will be in the fund after three months?

22. In a discussion of unexpected properties of mathematical models of population, Bailey[‡] considers the case in which the birth rate per *individual* is proportional to the population size N at time t. Since the growth rate per individual is $\dfrac{1}{N}\dfrac{dN}{dt}$, this means that

$$\frac{1}{N}\frac{dN}{dt} = kN$$

or $$\frac{dN}{dt} = kN^2, \qquad \text{(subject to } N = N_0 \text{ at } t = 0\text{),}$$

[†]N. Keyfitz, *Introduction to the Mathematics of Population*. Reading, Mass.: Addison-Wesley, 1968.

[‡]N. T. J. Bailey, *The Mathematical Approach to Biology and Medicine*. New York: John Wiley, 1967.

where $k > 0$. Show that

$$N = \frac{N_0}{1 - kN_0 t}.$$

Use this result to show that

$$\lim N = \infty \quad \text{as} \quad t \to \left(\frac{1}{kN_0}\right)^-.$$

This means that over a finite interval of time there is an infinite amount of growth. Such a model might be useful only for rapid growth over a short interval of time.

23. Suppose that the rate of growth of a population is proportional to the difference between some maximum size M and the number N of individuals in the population at time t. Suppose that when $t = 0$ the population size is N_0. Find a formula for N.

[†]**24.** Suppose that $A(t)$ is the amount of a product that is consumed at time t, and A follows an exponential law of growth. If $t_1 < t_2$ and at time t_2 the amount consumed, $A(t_2)$, is double the amount consumed at time t_1, $A(t_1)$, then $t_2 - t_1$ is called a doubling period. In a discussion of exponential growth, Shonle[‡] states that under exponential growth, " ... the amount of a product consumed during one doubling period is equal to the total used for all time up to the beginning of the doubling period in question." To justify this statement, reproduce his argument as follows. The amount of the product used up to time t_1 is given by

$$\int_{-\infty}^{t_1} A_0 e^{kt}\, dt, \qquad k > 0,$$

where A_0 is the amount when $t = 0$. Show that this is equal to $(A_0/k)e^{kt_1}$. Next, the amount used during the time interval from t_1 to t_2 is

$$\int_{t_1}^{t_2} A_0 e^{kt}\, dt.$$

Show that this is equal to

$$\frac{A_0}{k} e^{kt_1}[e^{k(t_2 - t_1)} - 1]. \qquad (17)$$

If the interval $[t_1, t_2]$ is a doubling period, then

$$A_0 e^{kt_2} = 2A_0 e^{kt_1}.$$

Show that this implies $e^{k(t_2 - t_1)} = 2$. Substitute this into (17); your result should be the same as the total used during all time up to t_1, namely $(A_0/k)e^{kt_1}$.

[†] Refers to Sec. 11-6.

[‡] J. I. Shonle, *Environmental Applications of General Physics*. Reading, Mass.: Addison-Wesley, 1975.

11-8 REVIEW

IMPORTANT TERMS AND SYMBOLS IN CHAPTER 11

integration by parts   *(p. 457)*

continuous annuity

present value   *(p. 473)*

future value   *(p. 474)*

trapezoidal rule   *(p. 481)*

improper integral   *(p. 486)*

first-order differential equation   *(p. 493)*

exponential growth   *(p. 497)*

partial fractions   *(p. 462)*

Simpson's rule   *(p. 482)*

$\int_{a}^{\infty} f(x)\, dx, \int_{-\infty}^{b} f(x)\, dx,$

$\int_{-\infty}^{\infty} f(x)\, dx$ *(p. 486)*

average value of function   *(p. 478)*

separation of variables   *(p. 494)*

logistic function   *(p. 500)*

REVIEW SECTION

†**1.** $\int u\, dv = uv - \int v\, du$ is called the _____ formula.

Ans.   integration by parts.

†**2.**   To express $5/[x^3(x^2 + 9)]$ as a sum of partial fractions, the number of constants that you must determine is _____ .

Ans.   5.

†**3.**   If $\lim_{r \to \infty} \int_{a}^{r} f(x)\, dx$ exists, then $\int_{a}^{\infty} f(x)\, dx$ is said to be (convergent)(divergent) .

Ans.   convergent.

†**4.** $\int_{0}^{\infty} e^x\, dx$ is (convergent)(divergent) .

Ans.   divergent

5.   The equation $x^3 y' + y^2 = 0$ is called a _____ equation.

Ans.   first-order differential.

6.   Exponential growth of a quantity means that the quantity at any time changes at a rate proportional to _____ .

Ans.   the amount of the quantity.

7.   The average value of $f(x) = x^3$ over the interval $[a, b]$ is given by the integral

_____ .

Ans. $\dfrac{1}{b - a} \int_{a}^{b} x^3\, dx.$

---

† Refers to Sections 11-1, 11-2, or 11-6.

8. Two methods of estimating definite integrals are by Simpson's rule and the ___(a)___ rule. Of these, the rule that requires an even number of subintervals is ___(b)___ .

Ans. (a) trapezoidal,   (b) Simpson's rule.

9. If the rate of growth of a quantity follows the law $N = N_0 e^{kt}$ where $k > 0$, we say that the quantity has exponential (growth)(decay) .

Ans. growth.

10. The logistic function $N = \dfrac{M}{1 + be^{-ct}}$ has the horizontal asymptote $N = $ _____ .

Ans. M.

REVIEW PROBLEMS

In Problems 1–18, determine the integrals.

1. $\displaystyle\int x \ln x \, dx.$

2. $\displaystyle\int \frac{1}{\sqrt{4x^2 + 1}} \, dx.$

3. $\displaystyle\int_0^2 \sqrt{4x^2 + 9} \, dx.$

4. $\displaystyle\int \frac{2x}{3 - 4x} \, dx.$

5. $\displaystyle\int \frac{x \, dx}{(2 + 3x)(3 + x)}.$

6. $\displaystyle\int_e^{e^2} \frac{1}{x \ln x} \, dx.$

7. $\displaystyle\int \frac{dx}{x(x + 2)^2}.$

8. $\displaystyle\int \frac{dx}{x^2 - 1}.$

9. $\displaystyle\int \frac{dx}{x^2 \sqrt{9 - 16x^2}}.$

10. $\displaystyle\int x^2 \ln 4x \, dx.$

11. $\displaystyle\int \frac{9 \, dx}{x^2 - 9}.$

12. $\displaystyle\int \frac{3x}{\sqrt{1 + 3x}} \, dx.$

13. $\displaystyle\int xe^{7x} \, dx.$

14. $\displaystyle\int \frac{dx}{2 + 3e^{4x}}.$

15. $\displaystyle\int \frac{dx}{2x \ln 2x}.$

16. $\displaystyle\int \frac{dx}{x(2 + x)}.$

17. $\displaystyle\int \frac{2x}{3 + 2x} \, dx.$

18. $\displaystyle\int \frac{dx}{\sqrt{4x^2 - 9}}.$

19. Find the average value of $f(x) = 3x^2 + 2x$ over the interval $[2, 4]$.

20. Find the average value of $f(t) = te^{t^2}$ over the interval $[2, 5]$.

In Problems 21 and 22, use (a) the trapezoidal rule and (b) Simpson's rule to estimate the integral. Use the given value of n. Give your answer to three decimal places.

21. $\displaystyle\int_0^3 \frac{1}{x + 1} \, dx, \; n = 6.$

22. $\displaystyle\int_0^1 \frac{1}{2 - x^2} \, dx, \; n = 4.$

In Problems **23–28**, *determine the improper integrals, if they exist[†]. Indicate those that are divergent.*

23. $\displaystyle\int_{3}^{\infty} \frac{1}{x^3}\, dx.$

24. $\displaystyle\int_{-\infty}^{0} e^{3x}\, dx.$

25. $\displaystyle\int_{0}^{2} \frac{1}{x-2}\, dx.$

26. $\displaystyle\int_{0}^{1} \frac{x}{\sqrt{1-x^2}}\, dx.$

27. $\displaystyle\int_{1}^{\infty} \frac{1}{2x}\, dx.$

28. $\displaystyle\int_{-\infty}^{\infty} xe^{1-x^2}\, dx.$

In Problems **29** *and* **30**, *solve the differential equations.*

29. $y' = 3x^2y + 2xy, \quad y > 0.$

30. $y' - 2xe^{x^2 - y + 3} = 0, \quad y(0) = 3.$

31. The population of a city in 1960 was 100,000 and in 1975 it was 120,000. Assuming exponential growth, project the population in 1990.

32. The population of Sigmatown doubles every 10 years due to exponential growth. At a certain time the population is 10,000. Find an expression for the number of people N at time t years later. Assume $\ln 2 = .69$.

33. Find the present value, to the nearest dollar, of a continuous annuity at an annual rate of 5 percent for 10 years if the payment at time t is at the annual rate of $f(t) = 40t$ dollars.

[†] Refers to Sec. 11-6.

MULTIVARIABLE CALCULUS

12-1 FUNCTIONS OF SEVERAL VARIABLES

Suppose a manufacturer produces two products, X and Y. His total cost is dependent on the levels of production of *both* X and Y. Table 12-1 is a schedule that indicates his total cost at various levels of production. For example, when 5 units of X and 6 units of Y are produced, the total cost is 17. Corresponding to this situation, it seems natural to associate the number 17 with the *ordered pair* (5, 6):

$$(5, 6) \to 17.$$

TABLE 12-1

NO. OF UNITS OF X PRODUCED (x)	NO. OF UNITS OF Y PRODUCED (y)	TOTAL COST OF PRODUCTION (c)
5	6	17
5	7	19
6	6	18
6	7	20

The first element of the ordered pair, 5, represents the number of units of X produced, while the second element, 6, represents the number of units of Y produced. Corresponding to the other production situations, we have

$$(5, 7) \rightarrow 19,$$

$$(6, 6) \rightarrow 18,$$

and $\quad (6, 7) \rightarrow 20.$

This correspondence can be considered an input-output relation where the inputs are ordered pairs. Note that with each input we associate exactly one output. Thus the correspondence defines a function f such that

the domain of f consists of (5, 6), (5, 7), (6, 6), (6, 7),

and

the range of f consists of 17, 19, 18, 20.

In function notation,

$$f(5, 6) = 17, \quad f(6, 6) = 18,$$
$$f(5, 7) = 19, \quad f(6, 7) = 20.$$

We say that the total cost schedule of this manufacturer can be described by $c = f(x, y)$, a function of the two independent variables x and y. The letter c is the dependent variable.

Turning to another function of two variables, we see that the equation

$$z = \frac{2}{x^2 + y^2}$$

defines z as a function of x and y:

$$z = f(x, y) = \frac{2}{x^2 + y^2}.$$

The domain of f is all ordered pairs of real numbers (x, y) for which the equation has meaning when the first and second elements of (x, y) are substituted for x and y, respectively, in the equation. Thus the domain of f is all ordered pairs except (0, 0). To find $f(2, 3)$, for example, we substitute $x = 2$ and $y = 3$ into $2/(x^2 + y^2)$. Hence $f(2, 3) = 2/(2^2 + 3^2) = \frac{2}{13}$.

EXAMPLE 1

a. $f(x, y) = \dfrac{x + 3}{y - 2}$ is a function of two variables. Since $(x + 3)/(y - 2)$ is not defined when $y = 2$, the domain of f is all (x, y) such that $y \neq 2$. Some functional values are

$$f(0, 3) = \frac{0 + 3}{3 - 2} = 3,$$

$$f(3, 0) = \frac{3 + 3}{0 - 2} = -3.$$

Note that $f(0, 3) \neq f(3, 0)$.

b. $g(r, s) = 2r - 3s + 5$ is a function of two variables, r and s. The domain of g is all ordered pairs (r, s). Some functional values are

$$g(4, 7) = 2(4) - 3(7) + 5 = -8,$$

$$g(r + h, s) = 2(r + h) - 3s + 5.$$

c. $h(x, y) = 4x$ defines h as a function of x and y. The domain is all ordered pairs of real numbers. Some functional values are

$$h(2, 5) = 4(2) = 8,$$

$$h(2, 6) = 4(2) = 8.$$

d. If $z^2 = x^2 + y^2$ and $x = 3$ and $y = 4$, then $z^2 = 3^2 + 4^2 = 25$. Consequently $z = \pm 5$. Thus, with the ordered pair $(3, 4)$ we *cannot* associate exactly one output number. Hence z is *not* a function of x and y.

If $y = f(x)$ is a function of one variable, the domain of f can be geometrically represented by points on the real number line. The function itself can be represented by its graph in a coordinate plane, sometimes called a two-dimensional coordinate system. However, for a function of two variables, $z = f(x, y)$, its domain (consisting of ordered pairs of real numbers) can be geometrically represented by a *region* in the plane. The function itself can be geometrically represented in a *three*-dimensional coordinate system. Such a system is formed when three mutually perpendicular real number lines in space intersect at the origin of each line (Fig. 12-1). The three number lines are commonly called the x-, y-, and z-axes, and their point of intersection is called the origin of the system.

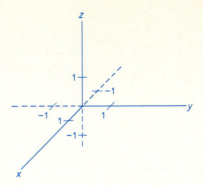

FIG. 12-1

With each point P in space we can associate a unique ordered triple of numbers. To do this [see Fig. 12-2(a)], a perpendicular line is constructed from P to the x, y-plane, that is, the plane determined by the x- and y-axes. Letting Q be the point where the line intersects this plane, we construct perpendiculars to

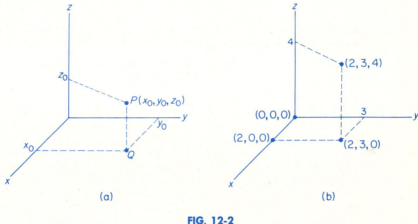

(a)   (b)

FIG. 12-2

the x- and y-axes from Q. These lines intersect the x-axis and the y-axis at points corresponding to the numbers x_0 and y_0, respectively. From P a perpendicular to the z-axis is constructed which intersects the z-axis at a point corresponding to the number z_0. Thus with the point P we associate the ordered triple (x_0, y_0, z_0). It should also be evident that with each ordered triple of numbers we can associate a unique point in space. Due to this one-to-one correspondence between points in space and ordered triples, an ordered triple may be called a point. In Fig. 12-2(b) the points $(2, 0, 0)$, $(2, 3, 0)$, and $(2, 3, 4)$ are shown. Note that the origin corresponds to $(0, 0, 0)$.

Suppose we wish to represent geometrically a function of two variables, $z = f(x, y)$. Then to each ordered pair (x, y) in the domain of f we assign the point $(x, y, f(x, y))$. The set of all such points is called the *graph* of f. Such a graph appears in Fig. 12-3. You can consider $z = f(x, y)$ as representing a surface in space.[†]

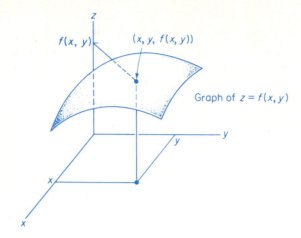

FIG. 12-3

In Chapter 7, continuity of a function of one variable was discussed. If $y = f(x)$ is continuous at $x = x_0$, then points near x_0 will have their functional values near $f(x_0)$. Extending this concept to a function of two variables, we say that the function $z = f(x, y)$ is continuous at (x_0, y_0) when points near (x_0, y_0) have their functional values near $f(x_0, y_0)$. Loosely interpreting this and without delving into the concept in great depth, we can say that a function of two variables will be continuous on its domain (that is, continuous at each point in its domain) if its graph is a "connected surface." In the following sections of this chapter we shall see that when a function is continuous, we can make important mathematical generalizations.

Until now, we have considered only functions of either one or two variables. In general, a function of n variables is one whose domain consists of ordered n-tuples $(x_1, x_2, \ldots, x_n)$. For example $f(x, y, z) = 2x + 3y + 4z$ defines a function of three variables with a domain consisting of all ordered triples. The function $g(x_1, x_2, x_3, x_4) = x_1 x_2 x_3 x_4$ is a function of four variables with a domain consisting of all ordered 4-tuples. Although functions of several variables are extremely important and useful, we cannot geometrically represent functions of more than two variables.

[†]We shall freely use the term "surface" in the intuitive sense.

We now give a brief discussion of sketching surfaces in space. We begin with planes that are parallel to a coordinate plane. By a "coordinate plane" we mean a plane containing two coordinate axes. For example, the plane determined by the x- and y-axes is the x,y-plane. Similarly we speak of the x,z-plane and the y,z-plane. The coordinate planes divide space into eight parts, called *octants*. In particular, the part containing all points (x, y, z) where x, y, and $z > 0$ is called the **first octant**.

Suppose S is a plane that is parallel to the x,y-plane and that also passes through the point $(0, 0, 5)$ [see Fig. 12-4(a)]. Then the point (x, y, z) will lie on S if and only if $z = 5$; that is, x and y can be any real numbers, but z must equal

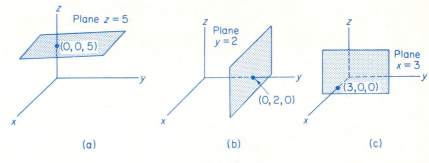

(a)   (b)   (c)

FIG. 12-4

5. For this reason we say $z = 5$ is an equation of S. Similarly, an equation of the plane parallel to the x,z-plane and passing through the point $(0, 2, 0)$ is $y = 2$ [Fig. 12-4(b)]. The equation $x = 3$ is an equation of the plane passing through $(3, 0, 0)$ and parallel to the y,z-plane [Fig. 12-4(c)]. Now let us look at planes in general.

In space the graph of an equation of the form

$$Ax + By + Cz + D = 0,$$

where D is a constant and A, B and C are constants that are not all zero, is a plane. Since three distinct points determine a plane, a convenient way to sketch a plane is to first determine the points, if any, where the plane intersects the x-, y-, or z-axes. These points are called *intercepts*.

EXAMPLE 2

To find all intercepts fill zeros in for 2 variables at a time, then

Sketch the plane $2x + 3y + z = 6$. *solve for the variable that's left*

The plane intersects the x-axis when $y = 0$ and $z = 0$. Thus $2x = 6$, which gives $x = 3$. Similarly, if $x = z = 0$, then $y = 2$; if $x = y = 0$, then $z = 6$. Thus the intercepts are $(3, 0, 0)$, $(0, 2, 0)$, and $(0, 0, 6)$. The portion of the plane in the first octant is shown in Fig. 12-5(a).

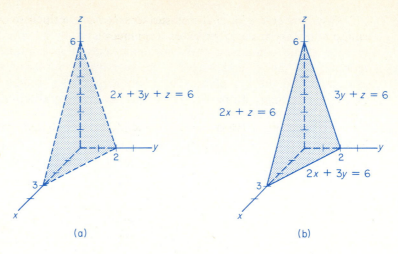

FIG. 12-5

A surface can also be sketched with the aid of its **traces**. These are the intersections of the surface with the coordinate planes. For the plane $2x + 3y + z = 6$ in Example 2, the trace in the x, y-plane is obtained by setting $z = 0$. This gives $2x + 3y = 6$, which is an equation of a *line* in the x, y-plane. Similarly, setting $x = 0$ gives the trace in the y, z-plane: the line $3y + z = 6$. The x, z-trace is the line $2x + z = 6$. See Fig. 12-5(b).

EXAMPLE 3

Sketch the surface $2x + z = 4$.

This equation has the form of a plane. The x- and z-intercepts are $(2, 0, 0)$ and $(0, 0, 4)$, and there is no y-intercept, since x and z cannot both be zero. Setting $y = 0$ gives the x, z-trace $2x + z = 4$, which is a line. In fact, the intersection of the surface with *any* plane $y = k$ is also $2x + z = 4$. Hence the plane appears as in Fig. 12-6.

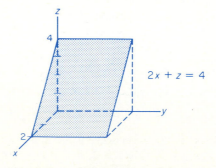

FIG. 12-6

Our final examples deal with surfaces that are not planes but whose graphs can be easily obtained by previous techniques.

EXAMPLE 4

Sketch the surface $z = x^2$.

The x, z-trace is the parabola $z = x^2$. In fact, for *any* fixed value of y we get $z = x^2$. Thus the graph appears as in Fig. 12-7.

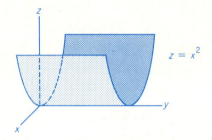

FIG. 12-7

EXAMPLE 5

Sketch the surface $x^2 + y^2 + z^2 = 25$.

Setting $z = 0$ gives the x, y-trace $x^2 + y^2 = 25$, which is a circle of radius 5. Similarly, the y, z- and x, z-traces are the circles $y^2 + z^2 = 25$ and $x^2 + z^2 = 25$, respectively. Note also that since $x^2 + y^2 = 25 - z^2$, the intersection of the surface with the plane $z = k$, where $-5 \leqslant k \leqslant 5$, is a circle. For example, if $z = 3$ the intersection is the circle $x^2 + y^2 = 16$. If $z = 4$, the intersection is $x^2 + y^2 = 9$. That is, cross sections of the surface that are parallel to the x, y-plane are circles. A portion of the surface appears in Fig. 12-8. The entire surface is a sphere.

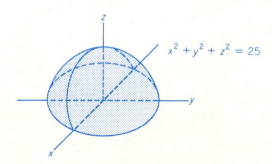

FIG. 12-8

EXERCISE 12-1

*In Problems **1–12**, determine the given functional values for the indicated functions.*

1. $f(x, y) = 3x + y - 1$; $f(0, 4)$.

2. $f(x, y) = xy^2 + 2$; $f(1, -4)$.

3. $g(x, y, z) = ze^{x+y}$; $g(-2, 2, 6)$.

4. $g(x, y, z) = xy + xz + yz$; $g(1, 2, -3)$.

5. $h(r, s, t, u) = \dfrac{r + s^2}{t - u}$; $h(-3, 3, 5, 4)$.

6. $h(r, s, t, u) = \ln (ru)$; $h(1, 5, 3, 1)$.

7. $g(p_A, p_B) = 2p_A(p_A^2 - 5)$; $g(4, 8)$.

8. $g(p_A, p_B) = p_A \sqrt{p_B} + 10$; $g(8, 4)$.

9. $F(x, y, z) = 3$; $F(2, 0, -1)$.

10. $F(x, y, z) = \dfrac{x}{yz}$; $F(0, 0, 3)$.

11. $f(x, y) = 2x - 5y + 4$; $f(x_0 + h, y_0)$.

12. $f(x, y) = x^2y - 3y^3$; $f(r + t, r)$.

*In Problems **13–16**, find equations of the planes that satisfy the given conditions.*

13. Parallel to the x, z-plane and passes through the point $(0, -4, 0)$.

14. Parallel to the y, z-plane and passes through the point $(8, 0, 0)$.

15. Parallel to the x, y-plane and passes through the point $(2, 7, 6)$.

16. Parallel to the y, z-plane and passes through the point $(-4, -2, 7)$.

*In Problems **17–26**, sketch the given surfaces.*

17. $x + y + z = 1$.   **18.** $2x + y + 2z = 6$.

19. $3x + 6y + 2z = 12$.   **20.** $x + 2y + 3z = 4$.

21. $x + 2y = 2$.   **22.** $y + z = 1$.

23. $z = 4 - x^2$.   **24.** $y = x^2$.

25. $x^2 + y^2 + z^2 = 1$.   **26.** $x^2 + y^2 = 1$.

12-2 PARTIAL DERIVATIVES

Figure 12-9 shows the graph of a function $z = f(x, y)$ and a plane that is parallel to the x, z-plane and passes through the point $(x_0, y_0, f(x_0, y_0))$ on the graph. An equation of this plane is $y = y_0$. Hence any point on the curve cut

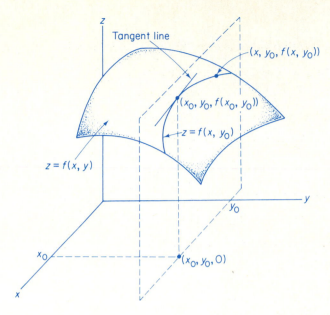

FIG. 12-9

from the surface by the plane must have the form $(x, y_0, f(x, y_0))$. Thus the curve can be described by $z = f(x, y_0)$. Since y_0 is constant, $z = f(x, y_0)$ can be considered a function of one variable, x. When the derivative of this function of x is evaluated at x_0, it gives the slope of the tangent line to this curve at $(x_0, y_0, f(x_0, y_0))$. See Fig. 12-9. This slope is called the *partial derivative of f with respect to x* at (x_0, y_0) and is denoted $f_x(x_0, y_0)$. In terms of limits,

$$f_x(x_0, y_0) = \lim_{h \to 0} \frac{f(x_0 + h, y_0) - f(x_0, y_0)}{h}. \qquad (1)$$

On the other hand, in Fig. 12-10 the plane $x = x_0$ is parallel to the y, z-plane and cuts the surface $z = f(x, y)$ in a curve given by $z = f(x_0, y)$, a function of y. When the derivative of this function of y is evaluated at y_0, it gives the slope of the tangent line to this curve at the point $(x_0, y_0, f(x_0, y_0))$. This slope is called the *partial derivative of f with respect to y* at (x_0, y_0) and is denoted $f_y(x_0, y_0)$. In terms of limits,

$$f_y(x_0, y_0) = \lim_{h \to 0} \frac{f(x_0, y_0 + h) - f(x_0, y_0)}{h}. \qquad (2)$$

Sometimes $f_x(x_0, y_0)$ is said to be the slope at $(x_0, y_0, f(x_0, y_0))$ of the tangent line to the graph of f *in the x-direction*; similarly $f_y(x_0, y_0)$ is the slope of the tangent line *in the y-direction*.

For generality, by replacing x_0 and y_0 in Eqs. (1) and (2) by x and y, respectively, we get the following definition.

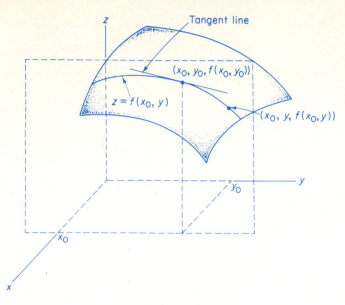

FIG. 12-10

DEFINITION. *If $z = f(x, y)$, the **partial derivative of f with respect to** x, denoted f_x, is*

$$f_x(x, y) = \lim_{h \to 0} \frac{f(x + h, y) - f(x, y)}{h}$$

provided this limit exists.
*The **partial derivative of f with respect to** y, denoted f_y, is*

$$f_y(x, y) = \lim_{h \to 0} \frac{f(x, y + h) - f(x, y)}{h}$$

provided this limit exists.

EXAMPLE 1

If $f(x, y) = xy^2 + x^2y$, find $f_x(x, y)$ and $f_y(x, y)$. Also find $f_x(3, 4)$ and $f_y(3, 4)$.

By the definition of $f_x(x, y)$.

$$
\begin{aligned}
f_x(x, y) &= \lim_{h \to 0} \frac{f(x + h, y) - f(x, y)}{h} \\
&= \lim_{h \to 0} \frac{\left[(x + h)y^2 + (x + h)^2 y\right] - [xy^2 + x^2 y]}{h} \\
&= \lim_{h \to 0} \frac{xy^2 + hy^2 + x^2 y + 2xhy + h^2 y - xy^2 - x^2 y}{h} \\
&= \lim_{h \to 0} (y^2 + 2xy + hy) = y^2 + 2xy.
\end{aligned}
$$

Therefore,

$$f_x(x, y) = y^2 + 2xy.$$

To find $f_x(3, 4)$ we evaluate $f_x(x, y)$ when $x = 3$ and $y = 4$.

$$f_x(3, 4) = 4^2 + 2(3)(4) = 40.$$

By the definition of $f_y(x, y)$.

$$
\begin{aligned}
f_y(x, y) &= \lim_{h \to 0} \frac{f(x, y + h) - f(x, y)}{h} \\
&= \lim_{h \to 0} \frac{\left[x(y + h)^2 + x^2(y + h) \right] - [xy^2 + x^2y]}{h} \\
&= \lim_{h \to 0} \frac{xy^2 + 2xyh + xh^2 + x^2y + x^2h - xy^2 - x^2y}{h} \\
&= \lim_{h \to 0} (2xy + xh + x^2) = 2xy + x^2.
\end{aligned}
$$

Therefore,

$$f_y(x, y) = 2xy + x^2.$$

Evaluating when $x = 3$ and $y = 4$, we have

$$f_y(3, 4) = 2(3)(4) + 3^2 = 33.$$

Note that $f_x(x, y)$ and $f_y(x, y)$ are each functions of the two variables x and y.

From its definition we see that to find f_x we treat y as a constant and differentiate f with respect to x in the usual way. For example, if $f(x, y) = xy^2 + x^2y$, then by treating y as a constant and differentiating with respect to x we have

$$f_x(x, y) = (1)y^2 + (2x)y = y^2 + 2xy,$$

as was shown in Example 1.

Similarly, to find f_y we treat x as a constant and differentiate f with respect to y in the usual way. Thus for $f(x, y) = xy^2 + x^2y$,

$$f_y(x, y) = x(2y) + x^2(1) = 2xy + x^2,$$

as was shown in Example 1.

Notations for partial derivatives of $z = f(x, y)$ are in Table 12-2.

TABLE 12-2

PARTIAL DERIVATIVE OF f (OR z) WITH RESPECT TO x	PARTIAL DERIVATIVE OF f (OR z) WITH RESPECT TO y
$f_x(x, y)$	$f_y(x, y)$
$\dfrac{\partial}{\partial x}[f(x, y)]$	$\dfrac{\partial}{\partial y}[f(x, y)]$
$\dfrac{\partial z}{\partial x}$	$\dfrac{\partial z}{\partial y}$

Table 12-3 gives notations for partial derivatives evaluated at (x_0, y_0).

TABLE 12-3

PARTIAL DERIVATIVE OF f (OR z) WITH RESPECT TO x EVALUATED AT (x_0, y_0)	PARTIAL DERIVATIVE OF f (OR z) WITH RESPECT TO y EVALUATED AT (x_0, y_0)
$f_x(x_0, y_0)$	$f_y(x_0, y_0)$
$\dfrac{\partial z}{\partial x}\bigg|_{(x_0,\, y_0)}$	$\dfrac{\partial z}{\partial y}\bigg|_{(x_0,\, y_0)}$
$\dfrac{\partial z}{\partial x}\bigg|_{\substack{x=x_0 \\ y=y_0}}$	$\dfrac{\partial z}{\partial y}\bigg|_{\substack{x=x_0 \\ y=y_0}}$

EXAMPLE 2

a. *If $z = 3x^3y^3 - 9x^2y + xy^2 + 4y$, find* $\dfrac{\partial z}{\partial x}$, $\dfrac{\partial z}{\partial y}$, $\dfrac{\partial z}{\partial x}\bigg|_{(1,\, 0)}$ *and* $\dfrac{\partial z}{\partial y}\bigg|_{(1,\, 0)}$.

To find $\partial z/\partial x$ we differentiate z with respect to x while treating y as a constant:

$$\frac{\partial z}{\partial x} = 3(3x^2)y^3 - 9(2x)y + (1)y^2 + 0$$

$$= 9x^2y^3 - 18xy + y^2.$$

Evaluating at $(1, 0)$, we obtain

$$\frac{\partial z}{\partial x}\bigg|_{(1,\, 0)} = 9(1)^2(0)^3 - 18(1)(0) + 0^2 = 0.$$

To find $\partial z / \partial y$ we differentiate z with respect to y while treating x as a constant.

$$\frac{\partial z}{\partial y} = 3x^3(3y^2) - 9x^2(1) + x(2y) + 4(1)$$

$$= 9x^3y^2 - 9x^2 + 2xy + 4.$$

Thus,

$$\left.\frac{\partial z}{\partial y}\right|_{(1, 0)} = 9(1)^3(0)^2 - 9(1)^2 + 2(1)(0) + 4 = -5.$$

b. *If $w = x^2e^{2x+3y}$, find $\partial w / \partial x$ and $\partial w / \partial y$.*

To find $\partial w / \partial x$, we treat y as a constant and differentiate with respect to x. Since x^2e^{2x+3y} is a product of two functions, each involving x, we use the product rule.

$$\frac{\partial w}{\partial x} = x^2 \frac{\partial}{\partial x}(e^{2x+3y}) + e^{2x+3y} \frac{\partial}{\partial x}(x^2)$$

$$= x^2(2e^{2x+3y}) + e^{2x+3y}(2x)$$

$$= 2x(x + 1)e^{2x+3y}.$$

To find $\partial w / \partial y$, we treat x as a constant and differentiate with respect to y.

$$\frac{\partial w}{\partial y} = x^2 \frac{\partial}{\partial y}(e^{2x+3y}) = 3x^2e^{2x+3y}.$$

We have seen that for a function of two variables, two partial derivatives can be considered. Actually the concept of partial derivatives can be extended to functions of more than two variables. For example, with $w = f(x, y, z)$ we have three partial derivatives:

the partial with respect to x, denoted $f_x(x, y, z)$, $\partial w / \partial x$, etc.;

the partial with respect to y, denoted $f_y(x, y, z)$, $\partial w / \partial y$, etc.;

and the partial with respect to z, denoted $f_z(x, y, z)$, $\partial w / \partial z$, etc.

 To determine $\partial w / \partial x$, treat y and z as constants and differentiate with respect to x. For $\partial w / \partial y$, treat x and z as constants and differentiate with respect to y. For $\partial w / \partial z$, treat x and y as constants and differentiate with respect to z. With a function of n variables we have n partial derivatives, which are determined in the obvious way.

EXAMPLE 3

a. *If $f(x, y, z) = x^2 + y^2z + z^3$, find $f_x(x, y, z)$, $f_y(x, y, z)$, and $f_z(x, y, z)$.*

Treating y and z as constants and differentiating with respect to x, we have

$$f_x(x, y, z) = 2x.$$

Treating x and z as constants and differentiating with respect to y, we have

$$f_y(x, y, z) = 2yz.$$

Treating x and y as constants and differentiating with respect to z, we have

$$f_z(x, y, z) = y^2 + 3z^2.$$

b. *If $p = g(r, s, t, u) = \dfrac{rsu}{rt^2 + s^2t}$, determine $\dfrac{\partial p}{\partial s}$, $\dfrac{\partial p}{\partial t}$, and $\dfrac{\partial p}{\partial t}\Big|_{(0,\,1,\,1,\,1)}$.*

To find $\partial p / \partial s$, first note that p is a quotient of two functions, each involving the variable s. Thus we use the quotient rule and treat r, t, and u as constants.

$$\frac{\partial p}{\partial s} = \frac{(rt^2 + s^2t)\dfrac{\partial}{\partial s}(rsu) - rsu\dfrac{\partial}{\partial s}(rt^2 + s^2t)}{(rt^2 + s^2t)^2}$$

$$= \frac{(rt^2 + s^2t)(ru) - (rsu)(2st)}{(rt^2 + s^2t)^2}.$$

Simplifying gives

$$\frac{\partial p}{\partial s} = \frac{ru(rt - s^2)}{t(rt + s^2)^2}.$$

To find $\partial p / \partial t$ we can first write g as

$$g(r, s, t, u) = rsu(rt^2 + s^2t)^{-1}.$$

Next we use the power rule and treat r, s, and u as constants.

$$\frac{\partial p}{\partial t} = rsu(-1)(rt^2 + s^2t)^{-2}\frac{\partial}{\partial t}(rt^2 + s^2t)$$

$$= -rsu(rt^2 + s^2t)^{-2}(2rt + s^2).$$

$$\frac{\partial p}{\partial t} = -\frac{rsu(2rt + s^2)}{(rt^2 + s^2t)^2}.$$

Letting $r = 0$, $s = 1$, $t = 1$ and $u = 1$ gives

$$\left.\frac{\partial p}{\partial t}\right|_{(0,\,1,\,1,\,1)} = -\frac{0(1)(1)\left[2(0)(1) + (1)^2\right]}{\left[0(1)^2 + (1)^2(1)\right]^2} = 0.$$

EXERCISE 12-2

*In each of Problems **1–26**, find all partial derivatives.*

1. $f(x, y) = x - 5y + 3$.

2. $f(x, y) = 4 - 5x^2 + 6y^3$.

3. $f(x, y) = 3x - 4$.

4. $f(x, y) = \sqrt{7}$.

5. $g(x, y) = x^5y^4 - 3x^4y^3 + 7x^3 + 2y^2 - 3xy + 4$.

6. $g(x, y) = x^8 - 2x^6y^5 + 3x^5y^3 + x^3y^3 + 3x - 4$.

7. $g(p, q) = \sqrt{pq}$.

8. $g(w, z) = \sqrt[3]{w^2 + z^2}$.

9. $h(s, t) = \dfrac{s^2 + 4}{t - 3}$.

10. $h(u, v) = \dfrac{4uv^2}{u^2 + v^2}$.

11. $u(q_1, q_2) = \frac{3}{4}\ln q_1 + \frac{1}{4}\ln q_2$.

12. $Q(l, k) = 3l^{.41}k^{.59}$.

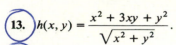

13. $h(x, y) = \dfrac{x^2 + 3xy + y^2}{\sqrt{x^2 + y^2}}$.

14. $h(x, y) = \dfrac{\sqrt{x + 4}}{x^2y + y^2x}$.

15. $z = e^{5xy}$.

16. $z = (x^2 + y)e^{3x + 4y}$.

17. $z = 5x \ln (x^2 + y)$.

18. $z = \ln (3x^2 + 4y^4)$.

19. $f(r, s) = \sqrt{r + 2s}\,(r^3 - 2rs + s^2)$.

20. $f(r, s) = \sqrt{rs}\,e^{2+r}$.

21. $f(r, s) = e^{3-r}\ln (7 - s)$.

22. $f(r, s) = (5r^2 + 3s^3)(2r - 5s)$.

23. $g(x, y, z) = 3x^2y + 2xy^2z + 3z^3$.

24. $g(x, y, z) = x^2y^3z^5 - 3x^2y^4z^3 + 5xz$.

25. $g(r, s, t) = e^{s+t}(r^2 + 7s^3)$.

26. $g(r, s, t, u) = rs \ln (2t + 5u)$.

*In Problems **27–32**, evaluate the given partial derivatives.*

27. $f(x, y) = x^3y + 7x^2y^2$; $f_x(1, -2)$.

28. $z = \sqrt{5x^2 + 3xy + 2y}$; $\left.\dfrac{\partial z}{\partial x}\right|_{\substack{x=0 \\ y=2}}$.

29. $g(x, y, z) = e^x\sqrt{y + 2z}$; $g_z(0, 1, 4)$.

30. $g(x, y, z) = \dfrac{3x^2 + 2y}{xy + xz}$; $g_y(1, 1, 1)$.

31. $h(r, s, t, u) = (s^2 + tu) \ln (2r + 7st)$; $h_s(1, 0, 0, 1)$.

32. $h(r, s, t, u) = \dfrac{7r + 3s^2u^2}{s}$; $h_t(4, 3, 2, 1)$.

12-3 APPLICATIONS OF PARTIAL DERIVATIVES

Suppose a manufacturer produces x units of product X and y units of product Y. Then the total cost c of these units is a function of x and y and is called a *joint-cost function*. If such a function is $c = f(x, y)$, then $\partial c / \partial x$ is called the (*partial*) *marginal cost with respect to* x. It is the rate of change of c with respect to x when y is held fixed. On the other hand, $\partial c / \partial y$ is the (*partial*) *marginal cost with respect to* y. It is the rate of change of c with respect to y when x is held fixed.

For example, if c is expressed in dollars and $\partial c / \partial y = 2$, then the cost of producing an extra unit of Y when the level of production of X is fixed is approximately two dollars.

If a manufacturer produces n products, his joint-cost function is a function of n variables and there are n (partial) marginal cost functions.

EXAMPLE 1

A company manufactures two types of sleds, the Lightning and the Alaskan models. Suppose the joint-cost function for producing x sleds of the Lightning model and y sleds of the Alaskan model is $c = f(x, y) = .06x^2 + 7x + 15y + 1000$ where c is expressed in dollars. Determine the marginal costs $\partial c / \partial x$ and $\partial c / \partial y$ when $x = 100$ and $y = 50$ and interpret the results.

Treating y as a constant and differentiating c with respect to x, we obtain

$$\frac{\partial c}{\partial x} = .12x + 7.$$

Treating x as a constant and differentiating c with respect to y, we have

$$\frac{\partial c}{\partial y} = 15.$$

Thus,

$$\left. \frac{\partial c}{\partial x} \right|_{(100, 50)} = .12(100) + 7 = 19,$$

and

$$\left. \frac{\partial c}{\partial y} \right|_{(100, 50)} = 15.$$

This means that increasing the output of the Lightning model from 100 to 101, while maintaining production of the Alaskan model at 50, increases costs by approximately $19. On the other hand, increasing output of the Alaskan model from 50 to 51 and holding production of the Lightning model at 100 will increase costs by approximately $15. In fact, since $\partial c / \partial y$ is a constant function, the marginal cost with respect to y is $15 at all levels of production.

Output of a product depends on many factors of production. Among these may be labor, capital, land, machinery, etc. If the function $P = f(l, k)$ gives the output P produced when a producer uses l units of labor and k units of capital, then this function is called a *production function*. We define the *marginal productivity with respect to l* to be $\partial P / \partial l$. This is the rate of change of P with respect to l when k is held fixed. Likewise, the *marginal productivity with respect to k* is $\partial P / \partial k$. It is the rate of change of P with respect to k when l is held fixed.

EXAMPLE 2

A manufacturer of a popular toy has determined that its production function is $P = \sqrt{lk}$, where l is the number of man-hours per week and k is the capital (expressed in hundreds of dollars per week) required for a weekly production of P gross of the toy. Determine the marginal productivity functions and evaluate them when $l = 400$ and $k = 16$. Interpret the results.

Since $P = (lk)^{1/2}$,

$$\frac{\partial P}{\partial l} = \frac{1}{2}(lk)^{-1/2}k = \frac{k}{2\sqrt{lk}}$$

and

$$\frac{\partial P}{\partial k} = \frac{1}{2}(lk)^{-1/2}l = \frac{l}{2\sqrt{lk}}.$$

Evaluating when $l = 400$ and $k = 16$, we obtain

$$\frac{\partial P}{\partial l}\bigg|_{\substack{l=400 \\ k=16}} = \frac{16}{2\sqrt{400(16)}} = \frac{1}{10},$$

$$\frac{\partial P}{\partial k}\bigg|_{\substack{l=400 \\ k=16}} = \frac{400}{2\sqrt{400(16)}} = \frac{5}{2}.$$

Thus if $l = 400$ and $k = 16$, increasing l to 401 and holding k at 16 will increase output by approximately $\frac{1}{10}$ gross. But if k is increased to 17 while l is held at 400, the output increases by approximately $\frac{5}{2}$ gross.

Sometimes two products may be related so that changes in the price of one of them can affect the demand for the other. A typical example is that of butter and margarine. If such a relationship exists between products A and B, then the demand for each product is dependent on the prices of both. Suppose q_A and q_B are the quantities demanded for A and B respectively, and p_A and p_B are their respective prices. Then both q_A and q_B are functions of p_A and p_B:

$$q_A = f(p_A, p_B), \quad \text{demand function for } A$$
$$q_B = g(p_A, p_B), \quad \text{demand function for } B.$$

We can find four partial derivatives:

$\dfrac{\partial q_A}{\partial p_A}$,   *the marginal demand for A with respect to p_A;*

$\dfrac{\partial q_A}{\partial p_B}$,   *the marginal demand for A with respect to p_B;*

$\dfrac{\partial q_B}{\partial p_A}$,   *the marginal demand for B with respect to p_A;*

$\dfrac{\partial q_B}{\partial p_B}$,   *the marginal demand for B with respect to p_B.*

Under typical conditions, if the price of B is fixed and the price of A increases, then the quantity of A demanded will decrease. Thus $\partial q_A / \partial p_A < 0$. Similarly, $\partial q_B / \partial p_B < 0$. However, $\partial q_A / \partial p_B$ and $\partial q_B / \partial p_A$ may be either positive or negative. If

$$\frac{\partial q_A}{\partial p_B} > 0 \quad \text{and} \quad \frac{\partial q_B}{\partial p_A} > 0,$$

then A and B are said to be **competitive products** or **substitutes**. In this situation an increase in the price of B causes an increase in the demand for A, if it is assumed that the price of A does not change. Likewise, an increase in the price of A causes an increase in the demand for B when the price of B is held fixed. Butter and margarine are examples of substitutes.

Proceeding to a different situation, we say that if

$$\frac{\partial q_A}{\partial p_B} < 0 \quad \text{and} \quad \frac{\partial q_B}{\partial p_A} < 0,$$

then A and B are **complementary products**. In this case an increase in the price of B causes a decrease in the demand for A if the price of A does not change. Similarly, an increase in the price of A causes a decrease in the demand for B when the price of B is held fixed. For example, cameras and film are complementary products. An increase in the price of film will make picture-taking more expensive. Hence the demand for cameras will decrease.

EXAMPLE 3

The demand functions for products A and B are each a function of the prices of A and B and are given by

$$q_A = \frac{50\sqrt[3]{p_B}}{\sqrt{p_A}} \quad and \quad q_B = \frac{75p_A}{\sqrt[3]{p_B^2}},$$

respectively. Find the four marginal demand functions and also determine whether A and B are competitive products, complementary products, or neither.

Writing $q_A = 50p_A^{-1/2}p_B^{1/3}$ and $q_B = 75p_Ap_B^{-2/3}$, we have

$$\frac{\partial q_A}{\partial p_A} = 50\left(-\frac{1}{2}\right)p_A^{-3/2}p_B^{1/3} = -25p_A^{-3/2}p_B^{1/3},$$

$$\frac{\partial q_A}{\partial p_B} = 50p_A^{-1/2}\left(\frac{1}{3}\right)p_B^{-2/3} = \frac{50}{3}p_A^{-1/2}p_B^{-2/3},$$

$$\frac{\partial q_B}{\partial p_A} = 75(1)p_B^{-2/3} = 75p_B^{-2/3},$$

$$\frac{\partial q_B}{\partial p_B} = 75p_A\left(-\frac{2}{3}\right)p_B^{-5/3} = -50p_Ap_B^{-5/3}.$$

Since p_A and p_B represent prices, they are both positive. Hence $\partial q_A/\partial p_B > 0$ and $\partial q_B/\partial p_A > 0$. We conclude that A and B are competitive products.

EXERCISE 12-3

For the joint-cost functions in Problems 1–3, find the indicated marginal cost at the given production level.

1. $c = 4x + .3y^2 + 2y + 500;$ $\frac{\partial c}{\partial y}, x = 20, y = 30.$

2. $c = x\sqrt{x + y} + 1000;$ $\frac{\partial c}{\partial x}, x = 40, y = 60.$

3. $c = .03(x + y)^3 - .6(x + y)^2 + 4.5(x + y) + 7700;$ $\frac{\partial c}{\partial x}, x = 50, y = 50.$

For the production functions in Problems 4 and 5, find the marginal production functions $\partial P/\partial k$ and $\partial P/\partial l$.

4. $P = 20lk - 2l^2 - 4k^2 + 800.$

5. $P = 1.582l^{.192}k^{.764}.$

6. A Cobb-Douglas production function is a production function of the form $P = Al^\alpha k^\beta$ where A, α, and β are constants and $\alpha + \beta = 1$. For such a function, show that

   a. $\partial P/\partial l = \alpha P/l.$

   b. $\partial P/\partial k = \beta P/k.$

c. $l\dfrac{\partial P}{\partial l} + k\dfrac{\partial P}{\partial k} = P$. This means that summing the marginal productivities of each factor times the amount of each factor results in the total product P.

In Problems 7–9, q_A and q_B are demand functions for products A and B, respectively. In each case find $\partial q_A/\partial p_A$, $\partial q_A/\partial p_B$, $\partial q_B/\partial p_A$, $\partial q_B/\partial p_B$ and determine whether A and B are competitive, complementary, or neither.

7. $q_A = 1000 - 50p_A + 2p_B; \quad q_B = 500 + 4p_A - 20p_B.$

8. $q_A = 20 - p_A - 2p_B; \quad q_B = 50 - 2p_A - 3p_B.$

9. $q_A = \dfrac{100}{p_A\sqrt{p_B}}; \quad q_B = \dfrac{500}{p_B\sqrt[3]{p_A}}.$

10. The production function for the Canadian manufacturing industries for 1927 is estimated by[†] $P = 33.0l^{.46}k^{.52}$, where P is product, l is labor, and k is capital. Find the marginal productivities for labor and capital and evaluate when $l = 1$ and $k = 1$.

11. An estimate of the production function for dairy farming in Iowa (1939) is given by[‡]

$$P = A^{.27}B^{.01}C^{.01}D^{.23}E^{.09}F^{.27},$$

where P is product, A is land, B is labor, C is improvements, D is liquid assets, E is working assets, and F is cash operating expenses. Find the marginal productivities for labor and improvements.

12. In a study of success among master of business administration (MBA) graduates, it was estimated that for staff managers (which includes accountants, analysts, etc.) current annual compensation z (in dollars) was given by

$$z = 10{,}990 + 1120x + 873y,$$

where x and y are the number of years of work experience before and after receiving the MBA degree, respectively.[§] Find $\partial z/\partial x$ and interpret your result.

[†]P. Daly and P. Douglas. "The production function for Canadian manufactures," *Journal of the American Statistical Association*, vol. 38 (1943), 178–186.

[‡]G. Tintner and O. H. Brownlee. "Production functions derived from farm records," *American Journal of Agricultural Economics*, vol. 26 (1944), 566–571.

[§]A. G. Weinstein and V. Srinivasen. "Predicting managerial success of master of business administration (MBA) graduates," *Journal of Applied Psychology*, vol. 59, no. 2 (1974), 207–212.

13. For the congressional elections of 1974, the Republican percentage, R, of the Republican-Democratic vote in a district is given (approximately) by [†]

$$R = f(E_r, E_d, I_r, I_d, N)$$

$$= 15.4725 + 2.5945E_r - .0804E_r^2 - 2.3648E_d +$$

$$.0687E_d^2 + 2.1914I_r - .0912I_r^2 - .8096I_d +$$

$$.0081I_d^2 - .0277E_rI_r + .0493E_dI_d +$$

$$.8579N - .0061N^2.$$

Here E_r and E_d are the campaign expenditures (in units of \$10,000) by Republicans and Democrats, respectively; I_r and I_d are the number of terms served in Congress, *plus* one, for the Republican and Democratic candidates, respectively; and N is the percentage of the two-party presidential vote that Richard Nixon received in the district for 1968. The variable N gives a measure of Republican strength in the district.

(a) In the Federal Election Campaign Act of 1974, Congress set a limit of \$188,000 on campaign expenditures. By analyzing $\partial R / \partial E_r$, would you have advised a Republican candidate who served nine terms in Congress to spend \$188,000 on his campaign?

(b) Determine the percentage above which the Nixon vote had a negative effect on R; that is, determine when $\partial R / \partial N < 0$. Give your answer to the nearest percent.

Suppose f defines a demand function for product A and $q_A = f(p_A, p_B)$ where q_A is the quantity of A demanded when the price per unit of A is p_A and the price per unit of product B is p_B. The partial elasticity of demand for A with respect to p_A, denoted η_{p_A}, is defined as $\eta_{p_A} = (p_A/q_A)(\partial q_A / \partial p_A)$. The partial elasticity of demand for A with respect to p_B, denoted η_{p_B}, is defined as $(p_B/q_A)(\partial q_A / \partial p_B)$. Loosely speaking, η_{p_A} is the ratio of a percentage change in the quantity of A demanded to a percentage change in the price of A when the price of B is fixed. Similarly, η_{p_B} can be loosely interpreted as the ratio of a percentage change in the quantity of A demanded to a percentage change in the price of B when the price of A is fixed. *In Problems 14–16, find η_{p_A} and η_{p_B} for the given values of p_A and p_B.*

14. $q_A = 1000 - 50p_A + 2p_B$; $p_A = 2, p_B = 10$.

15. $q_A = 100/(p_A\sqrt{p_B})$; $p_A = 1, p_B = 4$.

16. $q_A = 20 - p_A - 2p_B$; $p_A = 2, p_B = 2$.

[†]J. Silberman and G. Yochum. "The role of money in determining election outcomes," *Social Science Quarterly*, vol. 58, no. 4 (1978), 671–682.

12-4 IMPLICIT PARTIAL DIFFERENTIATION[†]

In the equation

$$z^2 - x^2 - y^2 = 0, \tag{1}$$

if $x = 1$ and $y = 1$, then $z^2 - 1 - 1 = 0$ and so $z = \pm \sqrt{2}$. Thus Eq. (1) does not define z as a function of x and y. However, solving Eq. (1) for z gives

$$z = \sqrt{x^2 + y^2} \quad \text{or} \quad z = -\sqrt{x^2 + y^2} \,,$$

each of which defines z as a function of x and y. Although Eq. (1) does not *explicitly* express z as a function of x and y, it can be thought of as expressing z *implicitly* as one of two different functions of x and y. Note that $z^2 - x^2 - y^2 = 0$ has the form $F(x, y, z) = 0$. Any equation of the form $F(x, y, z) = 0$ can be thought of as expressing z implicitly as one of a set of possible functions of x and y.

To find $\partial z / \partial x$ where

$$z^2 - x^2 - y^2 = 0, \tag{2}$$

we first differentiate both sides of Eq. (2) with respect to x while treating z as a function of x and y and treating y as a constant.

$$\frac{\partial}{\partial x}(z^2 - x^2 - y^2) = \frac{\partial}{\partial x}(0),$$

$$\frac{\partial}{\partial x}(z^2) - \frac{\partial}{\partial x}(x^2) - \frac{\partial}{\partial x}(y^2) = 0,$$

$$2z\frac{\partial z}{\partial x} - 2x - 0 = 0.$$

Solving for $\partial z / \partial x$, we obtain

$$2z\frac{\partial z}{\partial x} = 2x,$$

$$\frac{\partial z}{\partial x} = \frac{x}{z}.$$

[†]May be omitted without loss of continuity.

To find $\partial z / \partial y$ we differentiate both sides of Eq. (2) with respect to y while treating z as a function of x and y and treating x as a constant.

$$\frac{\partial}{\partial y}(z^2 - x^2 - y^2) = \frac{\partial}{\partial y}(0),$$

$$2z\frac{\partial z}{\partial y} - 0 - 2y = 0,$$

$$2z\frac{\partial z}{\partial y} = 2y.$$

Hence,

$$\frac{\partial z}{\partial y} = \frac{y}{z}.$$

The method we used to find $\partial z / \partial x$ and $\partial z / \partial y$ is called *implicit* (*partial*) *differentiation*.

EXAMPLE 1

a. If $\dfrac{xz^2}{x + y} + y^2 = 0$, *evaluate $\partial z / \partial x$ when $x = -1, y = 2$, and $z = 2$.*

We treat z as a function of x and y and differentiate both sides of the equation with respect to x.

$$\frac{\partial}{\partial x}\left(\frac{xz^2}{x + y}\right) + \frac{\partial}{\partial x}(y^2) = \frac{\partial}{\partial x}(0).$$

Using the quotient rule for the first term on the left side, we have

$$\frac{(x + y)\frac{\partial}{\partial x}(xz^2) - xz^2\frac{\partial}{\partial x}(x + y)}{(x + y)^2} + 0 = 0.$$

Using the product rule for $\frac{\partial}{\partial x}(xz^2)$ gives

$$\frac{(x + y)\left[x\left(2z\frac{\partial z}{\partial x}\right) + z^2(1)\right] - xz^2(1)}{(x + y)^2} = 0.$$

Solving for $\partial z / \partial x$, we obtain

$$2xz(x + y)\frac{\partial z}{\partial x} + z^2(x + y) - xz^2 = 0,$$

$$\frac{\partial z}{\partial x} = \frac{xz^2 - z^2(x + y)}{2xz(x + y)} = -\frac{yz}{2x(x + y)}, \qquad z \neq 0.$$

Thus,

$$\frac{\partial z}{\partial x}\bigg|_{(-1,\,2,\,2)} = 2.$$

b. *If* $se^{r^2+u^2} = u \ln (t^2 + 1)$, *determine* $\partial t / \partial u$.

We consider t as a function of r, s, and u. By differentiating both sides with respect to u while treating r and s as constants, we get

$$\frac{\partial}{\partial u}(se^{r^2+u^2}) = \frac{\partial}{\partial u}[u \ln (t^2 + 1)],$$

$$2sue^{r^2+u^2} = u\frac{\partial}{\partial u}[\ln (t^2 + 1)] + \ln (t^2 + 1)\frac{\partial}{\partial u}(u),$$

$$2sue^{r^2+u^2} = u\frac{2t}{t^2 + 1}\frac{\partial t}{\partial u} + \ln (t^2 + 1).$$

Thus,

$$\frac{\partial t}{\partial u} = \frac{(t^2 + 1)\left[2sue^{r^2+u^2} - \ln (t^2 + 1)\right]}{2ut}.$$

EXERCISE 12-4

In Problems **1–11**, *by the method of implicit partial differentiation find the indicated partial derivatives.*

1. $x^2 + y^2 + z^2 = 9$; $\partial z / \partial x$.

2. $z^2 - 3x^2 + y^2 = 0$; $\partial z / \partial x$.

3. $2z^3 - x^2 - 4y^2 = 0$; $\partial z / \partial y$.

4. $3x^2 + y^2 + 2z^3 = 9$; $\partial z / \partial y$.

5. $x^2 - 2y - z^2 + x^2yz^2 = 20$; $\partial z / \partial x$.

6. $z^3 - xz - y = 0$; $\partial z / \partial x$.

7. $e^x + e^y + e^z = 10$; $\partial z / \partial y$.

8. $xyz + 2y^2x - z^3 = 0$; $\partial z / \partial x$.

9. $\ln (z) + z - xy = 1$; $\partial z / \partial x$.

10. $\ln x + \ln y - \ln z = e^y$; $\partial z / \partial x$.

11. $(z^2 + 6xy)\sqrt{x^3 + 5} = 2$; $\partial z / \partial y$.

In Problems **12–18**, *evaluate the indicated partial derivatives for the given values of the variables.*

12. $xz + xyz - 5 = 0$; $\partial z / \partial x$, $x = 1, y = 4, z = 1$.

13. $xz^2 + yz - 12 = 0$; $\partial z / \partial x$, $x = 2, y = -2, z = 3$.

14. $e^{zx} = xyz$; $\partial z / \partial y$, $x = 1, y = -e^{-1}, z = -1$.

15. $\ln z = x + y$; $\partial z / \partial x$, $x = 5, y = -5, z = 1$.

16. $\sqrt{xz + y^2} - xy = 0$; $\partial z / \partial y$, $x = 2, y = 2, z = 6$.

17. $\dfrac{s^2 + t^2}{rs} = 10$; $\partial t / \partial r$, $r = 1, s = 2, t = 4$.

18. $\dfrac{rs}{s^2 + t^2} = t$; $\partial r / \partial t$, $r = 0, s = 1, t = 0$.

12-5 HIGHER-ORDER PARTIAL DERIVATIVES

If $z = f(x, y)$, then not only is z a function of x and y, but also f_x and f_y are each functions of x and y. Hence we may differentiate them and obtain so-called second-order partial derivatives of f. Symbolically,

$$f_{xx} \quad \text{means} \quad (f_x)_x, \qquad f_{xy} \quad \text{means} \quad (f_x)_y,$$

$$f_{yx} \quad \text{means} \quad (f_y)_x, \qquad f_{yy} \quad \text{means} \quad (f_y)_y.$$

In terms of ∂-notation,

$$\frac{\partial^2 z}{\partial x^2} \quad \text{means} \quad \frac{\partial}{\partial x}\left[\frac{\partial z}{\partial x}\right], \qquad \frac{\partial^2 z}{\partial y \, \partial x} \quad \text{means} \quad \frac{\partial}{\partial y}\left[\frac{\partial z}{\partial x}\right],$$

$$\frac{\partial^2 z}{\partial x \, \partial y} \quad \text{means} \quad \frac{\partial}{\partial x}\left[\frac{\partial z}{\partial y}\right], \qquad \frac{\partial^2 z}{\partial y^2} \quad \text{means} \quad \frac{\partial}{\partial y}\left[\frac{\partial z}{\partial y}\right].$$

Note that to find f_{xy}, first differentiate f with respect to x. For $\partial^2 z / \partial x \partial y$, first differentiate with respect to y.

We can extend our notation beyond second-order partial derivatives. For example, f_{xyx} (or $\partial^3 z / \partial x \partial y \partial x$) is a third-order partial derivative of f. It is the partial derivative of f_{xy} (or $\partial^2 z / \partial y \partial x$) with respect to x. A generalization regarding higher-order partial derivatives to functions of more than two variables should be obvious.

EXAMPLE 1

Find the four second-order partial derivatives of $f(x, y) = x^2 y + x^2 y^2$.

Since

$$f_x(x, y) = 2xy + 2xy^2,$$

then

$$f_{xx}(x, y) = \frac{\partial}{\partial x}(2xy + 2xy^2) = 2y + 2y^2$$

and

$$f_{xy}(x, y) = \frac{\partial}{\partial y}(2xy + 2xy^2) = 2x + 4xy.$$

Since

$$f_y(x, y) = x^2 + 2x^2y,$$

then

$$f_{yy}(x, y) = \frac{\partial}{\partial y}(x^2 + 2x^2y) = 2x^2$$

and

$$f_{yx}(x, y) = \frac{\partial}{\partial x}(x^2 + 2x^2y) = 2x + 4xy.$$

Observe in Example 1 that $f_{xy}(x, y) = f_{yx}(x, y)$. This equality did not occur by chance. It can be shown that for any function f, if f_{xy} and f_{yx} are both continuous, then $f_{xy} = f_{yx}$; that is, the order of differentiation is of no concern.

EXAMPLE 2

Determine the value of $\dfrac{\partial^3 w}{\partial z\,\partial y\,\partial x}\bigg|_{(1, 2, 3)}$ *if* $w = (2x + 3y + 4z)^3$.

$$\frac{\partial w}{\partial x} = 3(2x + 3y + 4z)^2 \frac{\partial}{\partial x}(2x + 3y + 4z)$$

$$= 6(2x + 3y + 4z)^2.$$

$$\frac{\partial^2 w}{\partial y\,\partial x} = 6\cdot 2(2x + 3y + 4z)\frac{\partial}{\partial y}(2x + 3y + 4z)$$

$$= 36(2x + 3y + 4z).$$

$$\frac{\partial^3 w}{\partial z\,\partial y\,\partial x} = 36\cdot 4 = 144.$$

Thus

$$\frac{\partial^3 w}{\partial z\,\partial y\,\partial x}\bigg|_{(1, 2, 3)} = 144.$$

EXAMPLE 3[†]

Determine $\partial^2 z / \partial x^2$ if $z^2 = xy$.

By implicit differentiation we first determine $\partial z / \partial x$:

$$\frac{\partial}{\partial x}(z^2) = \frac{\partial}{\partial x}(xy),$$

$$2z\frac{\partial z}{\partial x} = y,$$

$$\frac{\partial z}{\partial x} = \frac{y}{2z}, \qquad z \neq 0.$$

Differentiating both sides with respect to x, we obtain

$$\frac{\partial}{\partial x}\left[\frac{\partial z}{\partial x}\right] = \frac{\partial}{\partial x}\left[\frac{1}{2}yz^{-1}\right],$$

$$\frac{\partial^2 z}{\partial x^2} = -\frac{1}{2}yz^{-2}\frac{\partial z}{\partial x}.$$

Substituting $y/(2z)$ for $\partial z / \partial x$, we have

$$\frac{\partial^2 z}{\partial x^2} = -\frac{1}{2}yz^{-2}\left(\frac{y}{2z}\right) = -\frac{y^2}{4z^3}, \qquad z \neq 0.$$

EXERCISE 12-5

*In Problems **1–10**, find the indicated partial derivatives.*

1. $f(x, y) = 3x^2y^2;$ $f_x(x, y), f_{xy}(x, y).$

2. $f(x, y) = 3x^2y + 2xy^2 - 7y;$ $f_x(x, y), f_{xx}(x, y).$

3. $f(x, y) = e^{3xy} + 4x^2y;$ $f_y(x, y), f_{yx}(x, y), f_{yxy}(x, y).$

4. $f(x, y) = 7x^2 + 3y;$ $f_y(x, y), f_{yy}(x, y), f_{yyx}(x, y).$

5. $f(x, y) = (x^2 + xy + y^2)(x^2 + xy + 1);$ $f_x(x, y), f_{xy}(x, y).$

6. $f(x, y) = \ln(x^2 + y^2) + 2;$ $f_x(x, y), f_{xx}(x, y), f_{xy}(x, y).$

7. $f(x, y) = (x + y)^2(xy);$ $f_x(x, y), f_y(x, y), f_{xx}(x, y), f_{yy}(x, y).$

8. $f(x, y, z) = xy^2z^3;$ $f_x(x, y, z), f_{xz}(x, y, z), f_{xy}(x, y, z).$

9. $z = \sqrt{x^2 + y^2};$ $\dfrac{\partial z}{\partial x}, \dfrac{\partial^2 z}{\partial x^2}.$

[†]Omit if Sec. 12-4 was not covered.

10. $z = \dfrac{\ln(x^2 + 5)}{y};\quad \dfrac{\partial z}{\partial x},\ \dfrac{\partial^2 z}{\partial y\,\partial x}.$

11. If $f(x, y, z) = 7$, find $f_{yxx}(4, 3, -2)$.

12. If $f(x, y, z) = z^2(3x^2 - 4xy^3)$, find $f_{xyz}(1, 2, 3)$.

13. If $f(l, k) = 5l^3k^6 - lk^7$, find $f_{kkl}(2, 1)$.

14. If $f(x, y) = 2x^2y + xy^2 - x^2y^2$, find $f_{xxy}(0, 1)$.

15. If $f(x, y) = y^2e^x + \ln(xy)$, find $f_{xyy}(1, 1)$.

16. If $f(x, y) = x^3 - 3xy^2 + x^2 - y^3$, find $f_{xy}(1, -1)$.

17. For $f(x, y) = 8x^3 + 2x^2y^2 + 5y^4$, show that $f_{xy}(x, y) = f_{yx}(x, y)$.

18. For $f(x, y) = x^4y^4 + 3x^3y^2 - 7x + 4$, show that $f_{xyx}(x, y) = f_{xxy}(x, y)$.

19. For $z = \ln(x^2 + y^2)$, show that

$$\frac{\partial^2 z}{\partial x^2} + \frac{\partial^2 z}{\partial y^2} = 0.$$

†20. If $2z^2 - x^2 - 4y^2 = 0$, find $\dfrac{\partial^2 z}{\partial x^2}$.

†21. If $z^2 - 3x^2 + y^2 = 0$, find $\dfrac{\partial^2 z}{\partial y^2}$.

12-6 CHAIN RULE‡

Suppose a manufacturer of two related products A and B has a joint-cost function given by

$$c = f(q_A, q_B),$$

where c is the total cost of producing quantities q_A and q_B of A and B, respectively. Furthermore, suppose the demand functions for his products are

$$q_A = g(p_A, p_B) \quad \text{and} \quad q_B = h(p_A, p_B),$$

where p_A and p_B are the prices per unit of A and B, respectively. Since c is a function of q_A and q_B and both q_A and q_B are themselves functions of p_A and p_B, then c can be viewed as a function of p_A and p_B. (Appropriately, the variables q_A and q_B are called *intermediate variables* of c.) Consequently, we should be able

---

†Omit if Sec. 12-4 was not covered.

‡May be omitted without of loss of continuity.

to determine $\partial c / \partial p_A$, the rate of change of total cost with respect to the price of A. One way to do this is to substitute the expressions $g(p_A, p_B)$ and $h(p_A, p_B)$ for q_A and q_B, respectively, into $c = f(q_A, q_B)$. Then c is a function of p_A and p_B and we can differentiate c with respect to p_A directly. This approach has some drawbacks—especially when f, g, or h is given by a complicated expression. Another way to approach the problem would be to use the chain rule (actually *a* chain rule), which we now state without proof.

CHAIN RULE. *Let* $z = f(x, y)$, *where both x and y are functions of r and s given by* $x = x(r, s)$ *and* $y = y(r, s)$. *If f, x, and y have continuous partial derivatives, then z is a function of r and s and*

$$\frac{\partial z}{\partial r} = \frac{\partial z}{\partial x} \frac{\partial x}{\partial r} + \frac{\partial z}{\partial y} \frac{\partial y}{\partial r}$$

and

$$\frac{\partial z}{\partial s} = \frac{\partial z}{\partial x} \frac{\partial x}{\partial s} + \frac{\partial z}{\partial y} \frac{\partial y}{\partial s}.$$

Note that in the chain rule the number of intermediate variables of z (two) is the same as the number of terms that compose each of $\partial z / \partial r$ and $\partial z / \partial s$.

Returning to the original situation concerning the manufacturer, we see that if f, q_A, and q_B have continuous partial derivatives, then by the chain rule

$$\frac{\partial c}{\partial p_A} = \frac{\partial c}{\partial q_A} \frac{\partial q_A}{\partial p_A} + \frac{\partial c}{\partial q_B} \frac{\partial q_B}{\partial p_A}.$$

EXAMPLE 1

For a manufacturer of cameras and film, the total cost c of producing q_C cameras and q_F units of film is given by

$$c = 30q_C + .015q_Cq_F + q_F + 900.$$

The demand functions for the cameras and film are given by

$$q_C = \frac{9000}{p_C\sqrt{p_F}} \quad and \quad q_F = 2000 - p_C - 400p_F,$$

where p_C is the price per camera and p_F is the price per unit of film. Find the rate of change of total cost with respect to the price of the camera when $p_C = 50$ and $p_F = 2$.

We must first determine $\partial c / \partial p_C$. By the chain rule,

$$\frac{\partial c}{\partial p_C} = \frac{\partial c}{\partial q_C} \frac{\partial q_C}{\partial p_C} + \frac{\partial c}{\partial q_F} \frac{\partial q_F}{\partial p_C}$$

$$= (30 + .015q_F)\left[\frac{-9000}{p_C^2\sqrt{p_F}}\right] + (.015q_C + 1)(-1).$$

When $p_C = 50$ and $p_F = 2$, then $q_C = 90\sqrt{2}$ and $q_F = 1150$. Substituting these values into $\partial c / \partial p_C$ and simplifying, we have

$$\left. \frac{\partial c}{\partial p_C} \right|_{\substack{p_C=50 \\ p_F=2}} = -123.2 \qquad \text{(approximately)}.$$

The chain rule can be extended. For example, suppose $z = f(v, w, x, y)$ and v, w, x, and y are all functions of r, s, and t. Then, if certain conditions of continuity are assumed, z is a function of r, s, and t and

$$\frac{\partial z}{\partial r} = \frac{\partial z}{\partial v} \frac{\partial v}{\partial r} + \frac{\partial z}{\partial w} \frac{\partial w}{\partial r} + \frac{\partial z}{\partial x} \frac{\partial x}{\partial r} + \frac{\partial z}{\partial y} \frac{\partial y}{\partial r},$$

$$\frac{\partial z}{\partial s} = \frac{\partial z}{\partial v} \frac{\partial v}{\partial s} + \frac{\partial z}{\partial w} \frac{\partial w}{\partial s} + \frac{\partial z}{\partial x} \frac{\partial x}{\partial s} + \frac{\partial z}{\partial y} \frac{\partial y}{\partial s},$$

and

$$\frac{\partial z}{\partial t} = \frac{\partial z}{\partial v} \frac{\partial v}{\partial t} + \frac{\partial z}{\partial w} \frac{\partial w}{\partial t} + \frac{\partial z}{\partial x} \frac{\partial x}{\partial t} + \frac{\partial z}{\partial y} \frac{\partial y}{\partial t}.$$

Observe that the number of intermediate variables of z (four) is the same as the number of terms that form each of $\partial z / \partial r$, $\partial z / \partial s$, and $\partial z / \partial t$.

Now consider the situation where $z = f(x, y)$ and $x = x(t)$ and $y = y(t)$. Then

$$\frac{dz}{dt} = \frac{\partial z}{\partial x} \frac{dx}{dt} + \frac{\partial z}{\partial y} \frac{dy}{dt}.$$

Here we use the symbol dz/dt rather than $\partial z / \partial t$, since z can be considered a function of the *one* variable t. Likewise, the symbols dx/dt and dy/dt are used rather than $\partial x / \partial t$ and $\partial y / \partial t$. As is typical, the number of terms that compose dz/dt equals the number of intermediate variables of z. Other situations would be treated in a similar way.

EXAMPLE 2

a. *If $w = f(x, y, z) = 3x^2y + xyz - 4y^2z^3$, where $x = 2r - 3s$, $y = 6r + s$, and $z = r - s$, determine $\partial w / \partial r$ and $\partial w / \partial s$.*

Since x, y, and z are functions of r and s, then by the chain rule,

$$\frac{\partial w}{\partial r} = \frac{\partial w}{\partial x} \frac{\partial x}{\partial r} + \frac{\partial w}{\partial y} \frac{\partial y}{\partial r} + \frac{\partial w}{\partial z} \frac{\partial z}{\partial r}$$

$$= (6xy + yz)(2) + (3x^2 + xz - 8yz^3)(6) + (xy - 12y^2z^2)(1)$$

$$= x(18x + 13y + 6z) + 2yz(1 - 24z^2 - 6yz)$$

and

$$\frac{\partial w}{\partial s} = \frac{\partial w}{\partial x}\frac{\partial x}{\partial s} + \frac{\partial w}{\partial y}\frac{\partial y}{\partial s} + \frac{\partial w}{\partial z}\frac{\partial z}{\partial s}$$

$$= (6xy + yz)(-3) + (3x^2 + xz - 8yz^3)(1) + (xy - 12y^2z^2)(-1)$$

$$= x(3x - 19y + z) - yz(3 + 8z^2 - 12yz).$$

b. *If* $z = \dfrac{x + e^y}{y}$, *where* $x = rs + se^{rt}$ *and* $y = 9 + rt$, *evaluate* $\partial z / \partial s$ *when* $r = -2$, $s = 5$, *and* $t = 4$.

Since x and y are functions of r, s, and t (note that we can write $y = 9 + rt + 0 \cdot s$), by the chain rule,

$$\frac{\partial z}{\partial s} = \frac{\partial z}{\partial x}\frac{\partial x}{\partial s} + \frac{\partial z}{\partial y}\frac{\partial y}{\partial s}$$

$$= \left(\frac{1}{y}\right)(r + e^{rt}) + \frac{\partial z}{\partial y} \cdot (0) = \frac{r + e^{rt}}{y}.$$

If $r = -2$, $s = 5$, and $t = 4$, then $y = 1$. Thus,

$$\left.\frac{\partial z}{\partial s}\right|_{\substack{r = -2 \\ s = 5 \\ t = 4}} = \frac{-2 + e^{-8}}{1} = -2 + e^{-8}.$$

c. *Determine* $\partial y / \partial r$ *if* $y = x^2 \ln(x^4 + 6)$ *and* $x = (r + 3s)^6$.

By the chain rule,

$$\frac{\partial y}{\partial r} = \frac{dy}{dx}\frac{\partial x}{\partial r}$$

$$= \left[x^2 \cdot \frac{4x^3}{x^4 + 6} + 2x \cdot \ln(x^4 + 6)\right]\left[6(r + 3s)^5\right]$$

$$= 12x(r + 3s)^5\left[\frac{2x^4}{x^4 + 6} + \ln(x^4 + 6)\right].$$

EXAMPLE 3

Given that $z = e^{xy}$, $x = r - 4s$, *and* $y = r - s$, *find* $\partial z / \partial r$ *in terms of* r *and* s.

$$\frac{\partial z}{\partial r} = \frac{\partial z}{\partial x}\frac{\partial x}{\partial r} + \frac{\partial z}{\partial y}\frac{\partial y}{\partial r}$$

$$= (ye^{xy})(1) + (xe^{xy})(1)$$

$$= (x + y)e^{xy}.$$

Since $x = r - 4s$ and $y = r - s$,

$$\frac{\partial z}{\partial r} = [(r - 4s) + (r - s)]e^{(r-4s)(r-s)}$$

$$= (2r - 5s)e^{r^2 - 5rs + 4s^2}.$$

EXERCISE 12-6

In Problems 1–12, find the indicated derivatives by using the chain rule.

1. $z = 5x + 3y$, $x = 2r + 3s$, $y = r - 2s$; $\partial z / \partial r$, $\partial z / \partial s$.

2. $z = x^2 + 3xy + 7y^3$, $x = r^2 - 2s$, $y = 5s^2$; $\partial z / \partial r$, $\partial z / \partial s$.

3. $z = e^{x+y}$, $x = t^2 + 3$, $y = \sqrt{t^3}$; dz / dt.

4. $z = \sqrt{8x + y}$, $x = t^2 + 3t + 4$, $y = t^3 + 4$; dz / dt.

5. $w = x^2 z^2 + xyz + yz^2$, $x = 5t$, $y = 2t + 3$, $z = 6 - t$; dw / dt.

6. $w = \ln(x^2 + y^2 + z^2)$, $x = 2 - 3t$, $y = t^2 + 3$, $z = 4 - t$; dw / dt.

7. $z = (x^2 + xy^2)^3$, $x = r + s + t$, $y = 2r - 3s + t$; $\partial z / \partial t$.

8. $z = \sqrt{x^2 + y^2}$, $x = r^2 + s - t$, $y = r - s + t$; $\partial z / \partial r$.

9. $w = x^2 + xyz + y^3 z^2$, $x = r - s^2$, $y = rs$, $z = 2r - 5s$; $\partial w / \partial s$.

10. $w = e^{xyz}$, $x = r^2 s^3$, $y = r - s$, $z = rs^2$; $\partial w / \partial r$.

11. $y = x^2 - 7x + 5$, $x = 15rs + 2s^2 t^2$; $\partial y / \partial r$.

12. $y = 4 - x^2$, $x = 2r + 3s - 4t$; $\partial y / \partial t$.

13. If $z = (4x + 3y)^3$, where $x = r^2 s$ and $y = r - 2s$, evaluate $\partial z / \partial r$ when $r = 0$ and $s = 1$.

14. If $z = \sqrt{5x + 2y}$, where $x = 4t + 7$ and $y = t^2 - 3t + 4$, evaluate dz / dt when $t = 1$.

15. If $w = e^{3x-y}(x^2 + 4z^3)$, where $x = rs$, $y = 2s - r$, and $z = r + s$, evaluate $\partial w / \partial s$ when $r = 1$ and $s = -1$.

16. If $y = x/(x - 5)$, where $x = 2t^2 - 3rs - r^2 t$, evaluate $\partial y / \partial t$ when $r = 0$, $s = 2$, and $t = -1$.

12-7 MAXIMA AND MINIMA FOR FUNCTIONS OF TWO VARIABLES

We now extend to functions of two variables the notion of relative maxima and minima (or relative extrema), which was introduced in Chapter 9.

DEFINITION. *A function $z = f(x, y)$ is said to have a **relative maximum** at the point (x_0, y_0), that is, when $x = x_0$ and $y = y_0$, if for all points (x, y) in the plane which are sufficiently "close" to*

(x_0, y_0) *we have*

$$f(x_0, y_0) \geqslant f(x, y). \tag{1}$$

*For a **relative minimum** we replace $\geqslant$ by $\leqslant$ in (1).*

To say that $z = f(x, y)$ has a relative maximum at (x_0, y_0) means geometrically that the point (x_0, y_0, z_0) on the graph of f is higher than (or is as high as) all other points on the surface that are "near" (x_0, y_0, z_0). In Fig. 12-11(a), f has a relative maximum at (x_1, y_1). Similarly, the function f in Fig. 12-11(b) has a relative minimum when $x = y = 0$, which corresponds to a *low* point on the surface.

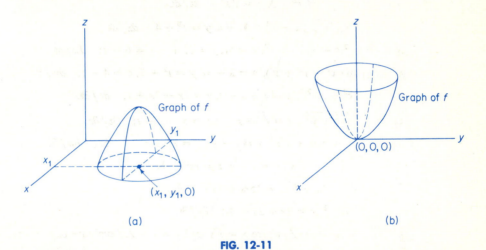

(a)   (b)

FIG. 12-11

Recall that in locating extrema for a function $y = f(x)$ of one variable, we examined those values of x for which $f'(x) = 0$ or $f'(x)$ does not exist. For functions of two (or more) variables, a similar procedure can be followed. However, for the functions that we shall work with, extrema will not occur where a derivative does not exist, and such situations will be excluded from considerations.

Suppose $z = f(x, y)$ has a relative maximum at (x_0, y_0), as indicated in Fig. 12-12(a). Then the curve where the plane $y = y_0$ intersects the surface must have a relative maximum when $x = x_0$. Hence the slope of the tangent line to the surface in the x-direction must be 0 at (x_0, y_0). Equivalently, $f_x(x, y) = 0$ at (x_0, y_0). Similarly, on the curve where the plane $x = x_0$ intersects the surface [Fig. 12-12(b)], there must be a relative maximum when $y = y_0$. Thus in the y-direction, the slope of the tangent to the surface must be 0 at (x_0, y_0). Equivalently, $f_y(x, y) = 0$ at (x_0, y_0). Since a similar discussion can be given for a relative minimum, we can combine these results as follows.

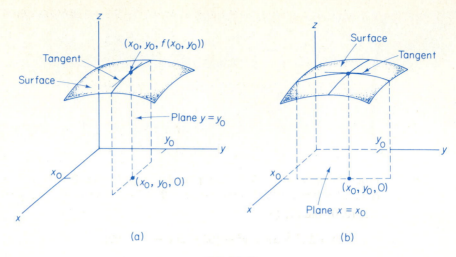

FIG. 12-12

RULE 1. *If $z = f(x, y)$ has a relative maximum or minimum at (x_0, y_0) and if both f_x and f_y are defined for all points close to (x_0, y_0), it is necessary that (x_0, y_0) be a solution of the system*

$$\begin{cases} f_x(x, y) = 0, \\ f_y(x, y) = 0. \end{cases}$$

A point (x_0, y_0) for which $f_x(x, y) = f_y(x, y) = 0$ is called a **critical point** of f. Thus from Rule 1 we infer that to locate relative extrema for a function we should examine its critical points.

PITFALL. *Rule 1 does not imply that there must be an extremum at a critical point. Just as in the case of functions of one variable, a critical point can give rise to a relative maximum, a relative minimum, or neither.*

Two additional comments: First, Rule 1, as well as the notion of a critical point, can be extended to functions of more than two variables. Thus, to locate possible extrema for $w = f(x, y, z)$ we would examine those points for which $w_x = w_y = w_z = 0$. Second, for a function whose domain is restricted, a thorough examination for absolute maxima and minima would include consideration of boundary points.

EXAMPLE 1

Examine each of the following for critical points.

a. $f(x, y) = 2x^2 + y^2 - 2xy + 5x - 3y + 1$.

Since $f_x(x, y) = 4x - 2y + 5$ and $f_y(x, y) = 2y - 2x - 3$, we solve the system

$$\begin{cases} 4x - 2y + 5 = 0, \\ -2x + 2y - 3 = 0. \end{cases}$$

This gives $x = -1$ and $y = \frac{1}{2}$. Thus, $(-1, \frac{1}{2})$ is the only critical point.

b. $f(l, k) = l^3 + k^3 - lk$.

$$\begin{cases} f_l(l, k) = 3l^2 - k = 0, & (2) \\ f_k(l, k) = 3k^2 - l = 0. & (3) \end{cases}$$

From Eq. (2), $k = 3l^2$. Substituting for k in Eq. (3) gives $0 = 27l^4 - l = l(27l^3 - 1)$. Hence, $l = 0$ or $l = \frac{1}{3}$. If $l = 0$, then $k = 0$; if $l = \frac{1}{3}$, then $k = \frac{1}{3}$. The critical points are thus $(0, 0)$ and $(\frac{1}{3}, \frac{1}{3})$.

c. $f(x, y, z) = 2x^2 + xy + y^2 + 100 - z(x + y - 100)$.

Solving the system

$$\begin{cases} f_x(x, y, z) = 4x + y - z = 0, \\ f_y(x, y, z) = x + 2y - z = 0, \\ f_z(x, y, z) = -x - y + 100 = 0 \end{cases}$$

gives the critical point $(25, 75, 175)$, as you may verify.

EXAMPLE 2

Find the critical points of $f(x, y) = x^2 - 4x + 2y^2 + 4y + 7$.

We have $f_x(x, y) = 2x - 4$ and $f_y(x, y) = 4y + 4$. The system

$$\begin{cases} 2x - 4 = 0, \\ 4y + 4 = 0 \end{cases}$$

gives the critical point $(2, -1)$. Observe that the given function can be written

$$f(x, y) = x^2 - 4x + 4 + 2(y^2 + 2y + 1) + 1$$

$$= (x - 2)^2 + 2(y + 1)^2 + 1,$$

and $f(2, -1) = 1$. Clearly, if $(x, y) \neq (2, -1)$, then $f(x, y) > 1$. Hence a relative minimum occurs at $(2, -1)$. Moreover, there is an *absolute minimum* at $(2, -1)$, since $f(x, y) > f(2, -1)$ for *all* $(x, y) \neq (2, -1)$.

Although in Example 2 we were able to show that the critical point gave rise to a relative minimum, in many cases this is not so easy to do. There is, however, a second-derivative test that gives conditions under which a critical point will be a relative maximum or minimum. We state it now, omitting the proof.

RULE 2. ***Second-Derivative Test for Functions of Two Variables.*** *Suppose* $z = f(x, y)$ *has continuous partial derivatives* $f_{xx}, f_{yy},$ *and* f_{xy} *at all points* (x, y) *near the critical point* (x_0, y_0)*. Let D be the function defined by*

$$D(x, y) = f_{xx}(x, y)f_{yy}(x, y) - [f_{xy}(x, y)]^2.$$

Then

a. *if* $D(x_0, y_0) > 0$ *and* $f_{xx}(x_0, y_0) < 0, f$ *has a relative maximum at* (x_0, y_0);

b. *if* $D(x_0, y_0) > 0$ *and* $f_{xx}(x_0, y_0) > 0, f$ *has a relative minimum at* (x_0, y_0);

c. *if* $D(x_0, y_0) < 0, f$ *has neither a relative maximum nor a relative minimum at* (x_0, y_0);

d. *if* $D(x_0, y_0) = 0,$ *no conclusion can be drawn and further analysis is required.*

EXAMPLE 3

Examine $f(x, y) = x^3 + y^3 - xy$ *for relative maxima or minima by using the second-derivative test.*

$$f_x(x, y) = 3x^2 - y, \quad f_y(x, y) = 3y^2 - x.$$

In the same manner as in Example 1(b), solving $f_x(x, y) = f_y(x, y) = 0$ gives the critical points $(0, 0)$ and $(\frac{1}{3}, \frac{1}{3})$.

At $(0, 0)$,

$$f_{xx}(x, y) = 6x = 0, \qquad f_{xy}(x, y) = -1, \qquad f_{yy}(x, y) = 6y = 0,$$

and

$$D(0, 0) = 0(0) - (-1)^2 = -1.$$

Since $D(0, 0) < 0$, there is no relative extremum at $(0, 0)$.

At $(\frac{1}{3}, \frac{1}{3})$,

$$f_{xx}(x, y) = 6\left(\tfrac{1}{3}\right) = 2, \qquad f_{xy}(x, y) = -1, \qquad f_{yy}(x, y) = 6\left(\tfrac{1}{3}\right) = 2,$$

and

$$D\left(\tfrac{1}{3}, \tfrac{1}{3}\right) = 2(2) - (-1)^2 = 3.$$

Since $D(\frac{1}{3}, \frac{1}{3}) > 0$ and $f_{xx}(\frac{1}{3}, \frac{1}{3}) > 0$, there is a relative minimum at $(\frac{1}{3}, \frac{1}{3})$. At this point the value of the given function is

$$f\left(\tfrac{1}{3}, \tfrac{1}{3}\right) = \left(\tfrac{1}{3}\right)^3 + \left(\tfrac{1}{3}\right)^3 - \left(\tfrac{1}{3}\right)\left(\tfrac{1}{3}\right) = -\tfrac{1}{27}.$$

EXAMPLE 4

Examine $f(x, y) = y^2 - x^2$ for relative extrema.

Solving

$$f_x(x, y) = -2x = 0 \quad \text{and} \quad f_y(x, y) = 2y = 0,$$

we get the critical point $(0, 0)$. Moreover, at $(0, 0)$, and indeed at any point,

$$f_{xx}(x, y) = -2. \quad f_{yy}(x, y) = 2, \quad f_{xy}(x, y) = 0.$$

Hence, $D(0, 0) = (-2)(2) - (0)^2 = -4 < 0$ and no relative extrema exist. A sketch of $z = f(x, y) = y^2 - x^2$ appears in Fig. 12-13. Note that for the surface curve cut by the plane $y = 0$, there is a *maximum* at $(0, 0)$; but for the surface curve cut by the plane $x = 0$, there is a *minimum* at $(0, 0)$. Thus, on the *surface* no relative extremum can exist at the origin, although $(0, 0)$ is a critical point. Around the origin the curve is saddle-shaped and $(0, 0)$ is called a *saddle-point* of f.

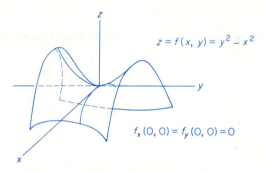

FIG. 12-13

EXAMPLE 5

Examine $f(x, y) = x^4 + (x - y)^4$ for relative extrema.

If we set

$$f_x(x, y) = 4x^3 + 4(x - y)^3 = 0, \tag{4}$$

$$f_y(x, y) = -4(x - y)^3 = 0, \tag{5}$$

then from Eq. (5) we have $x - y = 0$ or $x = y$. Substituting into Eq. (4) gives $4x^3 = 0$ or $x = 0$. Thus $x = y = 0$ and $(0, 0)$ is the only critical point. At $(0, 0)$, $f_{xx}(x, y) = 12x^2 + 12(x - y)^2 = 0$, $f_{yy}(x, y) = 12(x - y)^2 = 0$, and $f_{xy}(x, y) = -12(x - y)^2 = 0$. Hence $D(0, 0) = 0$ and the second-derivative test fails. However, for all $(x, y) \neq (0, 0)$ we have $f(x, y) > 0$ while $f(0, 0) = 0$. Hence at $(0, 0)$ the graph of f has a low point and we conclude that f has a relative (and absolute) minimum at $(0, 0)$.

In many situations involving functions of two variables, and especially in their applications, the nature of the given problem is an indicator of whether a critical point is in fact a relative (or absolute) maximum or a relative (or absolute) minimum. In such cases the second-derivative test is not needed. Often, in mathematical studies of economic phenomena the appropriate second-order conditions are assumed to hold.

EXAMPLE 6

Let P be a production function given by

$$P = f(l, k) = .54l^2 - .02l^3 + 1.89k^2 - .09k^3,$$

where l and k are the amounts of labor and capital, respectively, and P is the quantity of output produced. Find the values of l and k that maximize P.

$$P_l = 1.08l - .06l^2 \qquad P_k = 3.78k - .27k^2$$
$$= .06l(18 - l) = 0. \qquad = .27k(14 - k) = 0.$$
$$l = 0, l = 18. \qquad k = 0, k = 14.$$

The critical points are (0, 0), (0, 14), (18, 0), and (18, 14).
At (0, 0),

$$P_{ll} = 1.08 - .12l = 1.08, \qquad P_{lk} = 0, \qquad P_{kk} = 3.78 - .54k = 3.78,$$

$$D(0, 0) = 1.08(3.78) - 0^2 > 0.$$

Since $D(0, 0) > 0$ and $P_{ll} > 0$, there is a relative minimum at (0, 0).
At (0, 14),

$$P_{ll} = 1.08, \qquad P_{lk} = 0, \qquad P_{kk} = -3.78,$$

$$D(0, 14) = 1.08(-3.78) - 0^2 < 0.$$

Since $D(0, 14) < 0$, there is no relative extremum at (0, 14).
At (18, 0),

$$P_{ll} = -1.08, \qquad P_{lk} = 0, \qquad P_{kk} = 3.78,$$

$$D(18, 0) = (-1.08)(3.78) - 0 < 0.$$

Since $D(18, 0) < 0$, there is no relative extremum at (18, 0).
At (18, 14),

$$P_{ll} = -1.08, \qquad P_{lk} = 0, \qquad P_{kk} = -3.78,$$

$$D(18, 14) = (-1.08)(-3.78) - 0 > 0.$$

Since $D(18, 14) > 0$ and $P_{ll} < 0$, there is a relative maximum at (18, 14). The maximum output is obtained when $l = 18$ and $k = 14$.

EXAMPLE 7

A food manufacturer produces two types of candy, A and B, for which the average costs of production are constant at 70 and 80 cents per lb, respectively. The quantities q_A, q_B (in lb) of A and B that can be sold each week are given by the joint-demand functions

$$q_A = 240(p_B - p_A)$$

and

$$q_B = 240(150 + p_A - 2p_B),$$

where p_A and p_B are the selling prices (in cents per lb) of A and B, respectively. Determine the selling prices that will maximize the manufacturer's profit P.

For A and B the profits per lb are $p_A - 70$ and $p_B - 80$, respectively. Hence, total profit P is

$$P = (p_A - 70)q_A + (p_B - 80)q_B$$
$$= (p_A - 70)[240(p_B - p_A)] + (p_B - 80)[240(150 + p_A - 2p_B)].$$

To maximize P, we set its partial derivatives equal to 0:

$$\frac{\partial P}{\partial p_A} = (p_A - 70)[240(-1)] + [240(p_B - p_A)](1) +$$
$$(p_B - 80)[(240)(1)] = 0, \qquad (6)$$

$$\frac{\partial P}{\partial p_B} = (p_A - 70)[240(1)] + (p_B - 80)[240(-2)] +$$
$$[240(150 + p_A - 2p_B)](1) = 0. \qquad (7)$$

Simplifying the preceding two equations gives

$$\begin{cases} p_B - p_A - 5 = 0, \\ -2p_B + p_A + 120 = 0, \end{cases}$$

whose solution is $p_A = 110$ and $p_B = 115$ (both in cents). Moreover,

$$\frac{\partial^2 P}{\partial p_A^2} = -480 < 0, \qquad \frac{\partial^2 P}{\partial p_B^2} = -960, \qquad \frac{\partial^2 P}{\partial p_B \partial p_A} = 480.$$

Thus $D(110, 115) = (-480)(-960) - (480)^2 > 0$ and we indeed have a maximum. The manufacturer should sell candy A at \$1.10 per lb and B at \$1.15 per lb.

EXAMPLE 8[†]

Suppose a monopolist is practicing price discrimination by selling the same product in two separate markets at different prices. Let q_A be the number of units sold in Market A,

---

[†]Omit if Sec. 12-6 was not covered.

where the demand function is $p_A = f(q_A)$, and let q_B be the number of units sold in Market B, where the demand function is $p_B = g(q_B)$. Then the revenue functions for the two markets are

$$r_A = q_A f(q_A) \quad \text{and} \quad r_B = q_B g(q_B).$$

Assume that all units are produced at one plant and let the cost function for producing $q = q_A + q_B$ units be $c = c(q)$. Keep in mind that r_A is a function of q_A, and r_B is a function of q_B. The monopolist's profit P is

$$P = r_A + r_B - c.$$

To maximize P with respect to outputs q_A and q_B, we set its partial derivatives equal to 0. To begin with,

$$\frac{\partial P}{\partial q_A} = \frac{dr_A}{dq_A} + 0 - \frac{\partial c}{\partial q_A}$$

$$= \frac{dr_A}{dq_A} - \frac{dc}{dq}\frac{\partial q}{\partial q_A} = 0 \qquad \text{(chain rule)}.$$

But

$$\frac{\partial q}{\partial q_A} = \frac{\partial}{\partial q_A}(q_A + q_B) = 1.$$

Thus,

$$\frac{\partial P}{\partial q_A} = \frac{dr_A}{dq_A} - \frac{dc}{dq} = 0. \tag{8}$$

Similarly,

$$\frac{\partial P}{\partial q_B} = \frac{dr_B}{dq_B} - \frac{dc}{dq} = 0. \tag{9}$$

From Eqs. (8) and (9) we get

$$\frac{dr_A}{dq_A} = \frac{dc}{dq} = \frac{dr_B}{dq_B}.$$

But dr_A/dq_A and dr_B/dq_B are marginal revenues, and dc/dq is marginal cost. Hence, to maximize profit it is necessary to charge prices (and distribute output) so that the marginal revenues in both markets will be the same and, loosely speaking, will also be equal to the cost of the last unit produced in the plant.

EXERCISE 12-7

In Problems 1–6, find the critical points of the functions.

1. $f(x, y) = x^2 + y^2 - 5x + 4y + xy.$

2. $f(x, y) = x^2 + 4y^2 - 6x + 16y.$

3. $f(x, y) = 2x^3 + y^3 - 3x^2 + 1.5y^2 - 12x - 90y$.

4. $f(x, y) = xy - \dfrac{1}{x} - \dfrac{1}{y}$.

5. $f(x, y, z) = 2x^2 + xy + y^2 + 100 - z(x + y - 200)$.

6. $f(x, y, z, w) = x^2 + y^2 + z^2 - w(x - y + 2z - 6)$.

In Problems 7–18, find the critical points of the functions. Determine, by the second-derivative test, whether these points correspond to a relative maximum, to a relative minimum, to neither, or whether the test fails.

7. $f(x, y) = x^2 + 3y^2 + 4x - 9y + 3$.

8. $f(x, y) = -2x^2 + 8x - 3y^2 + 24y + 7$.

9. $f(x, y) = y - y^2 - 3x - 6x^2$.

10. $f(x, y) = x^2 + y^2 + xy - 9x + 1$.

11. $f(x, y) = x^3 - 3xy + y^2 + y - 5$.

12. $f(x, y) = \dfrac{x^3}{3} + y^2 - 2x + 2y - 2xy$.

13. $f(x, y) = \frac{1}{3}(x^3 + 8y^3) - 2(x^2 + y^2) + 1$.

14. $f(x, y) = x^2 + y^2 - xy + x^3$.

15. $f(l, k) = 2lk - l^2 + 264k - 10l - 2k^2$.

16. $f(l, k) = l^3 + k^3 - 3lk$.

17. $f(p, q) = pq - \dfrac{1}{p} - \dfrac{1}{q}$.

18. $f(x, y) = (x - 3)(y - 3)(x + y - 3)$.

In Problems 19–26, unless otherwise indicated the variables p_A and p_B denote selling prices of products A and B, respectively. Similarly, q_A and q_B denote quantities of A and B which are produced and sold during some time period. In all cases, the variables employed will be assumed to be units of output, input, money, etc.

19. Suppose $P = f(l, k) = 1.08l^2 - .03l^3 + 1.68k^2 - .08k^3$ is a production function for a firm. Find the quantities of inputs, l and k, so as to maximize output P.

20. In a certain automated manufacturing process, machines M and N are utilized for m and n hours, respectively. If daily output Q is a function of m and n, namely $Q = 4.5m + 5n - .5m^2 - n^2 - .25mn$, find the values of m and n that will maximize Q.

21. A candy company produces two varieties of candy, A and B, for which the constant average costs of production are 60 and 70 (cents per lb), respectively. The demand functions for A and B are respectively given by $q_A = 5(p_B - p_A)$ and $q_B = 500 + 5(p_A - 2p_B)$. Find the selling prices p_A and p_B that would maximize the company's profit.

22. Repeat Problem 21 if the constant costs of production of A and B are a and b (cents per lb), respectively.

23. Suppose a monopolist is practicing price discrimination in the sale of a product by charging different prices in two separate markets. In market A the demand function is $p_A = 100 - q_A$ and in B it is $p_B = 84 - q_B$, where q_A and q_B are the quantities sold per week in A and B, and p_A and p_B are the respective prices per unit. If the monopolist's cost function is $c = 600 + 4(q_A + q_B)$, how much should be sold in each market to maximize profit? What selling prices would give this maximum profit? Find the maximum profit.

24. A monopolist sells two competitive products, A and B, for which the demand functions are $q_A = 1 - 2p_A + 4p_B$ and $q_B = 11 + 2p_A - 6p_B$. If the constant average cost of producing a unit of A is 4 and for B it is 1, how many units of A and B should be sold to maximize the monopolist's profit?

25. For products A and B, the joint-cost function for a manufacturer is $c = 1.5q_A^2 + 4.5q_B^2$ and the demand functions are $p_A = 36 - q_A^2$ and $p_B = 30 - q_B^2$. Find the level of production that will maximize profit.

26. For a monopolist's products, A and B, the joint-cost function is $c = (q_A + q_B)^2$ and the demand functions are $q_A = 26 - p_A$ and $q_B = 10 - .25p_B$. Find the values of p_A and p_B that will maximize profit. What are the quantities of A and B which correspond to these prices? What is the total profit?

27. An open-top rectangular box is to have a volume of 6 cubic feet. The cost per sq ft of materials is $3 for the the bottom, $1 for the front and back, and $0.50 for the other two sides. Find the dimensions of the box so that the cost of materials is minimized. See Fig. 12-14.

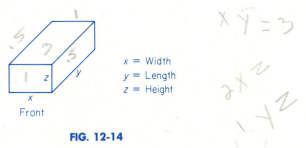

x = Width
y = Length
z = Height

Front

FIG. 12-14

28. Suppose A and B are the only two firms in the market selling the same product (we say they are *duopolists*). The industry demand function for the product is $p = 92 - q_A - q_B$ where q_A and q_B denote the output produced and sold by A and B, respectively. For A the cost function is $c_A = 10q_A$ and for B it is $c_B = .5q_B^2$. Suppose the firms decide to enter into an agreement on output and price control by jointly acting as a monopoly. In this case we say they enter into *collusion*. Show that the profit function for the monopoly is given by

$$P = pq_A - c_A + pq_B - c_B.$$

Express P as a function of q_A and q_B and determine how output should be allocated so as to maximize the profit of the monopoly.

29. Suppose $f(x, y) = -2x^2 + 5y^2 + 7$, where x and y must satisfy the equation $3x - 2y = 7$. Find the relative extrema of f subject to the given conditions on x and y by first solving the second equation for y. Substitute the result for y in the given equation. Thus f is expressed as a function of one variable for which extrema may be found in the usual way.

30. Repeat Problem 29 if $f(x, y) = x^2 + 4y^2 + 6$ subject to the condition that $2x - 8y = 20$.

12-8 LAGRANGE MULTIPLIERS

We shall now find relative maxima and minima for a function on which certain *constraints* are imposed. Such a situation could arise if a manufacturer wished to minimize the total cost of factors of input and yet obtain a particular level of output.

Suppose we wish to find the relative extrema of

$$w = x^2 + y^2 + z^2 \tag{1}$$

subject to the constraint that x, y, and z must satisfy

$$x - y + 2z = 6. \tag{2}$$

Solving Eq. (2) for x, we get

$$x = y - 2z + 6, \tag{3}$$

which when substituted for x in Eq. (1) gives

$$w = (y - 2z + 6)^2 + y^2 + z^2. \tag{4}$$

Since w in Eq. (4) is expressed as a function of two variables, to find relative extrema we follow the usual procedure of setting its partial derivatives equal to 0:

$$\frac{\partial w}{\partial y} = 2(y - 2z + 6) + 2y = 4y - 4z + 12 = 0, \tag{5}$$

$$\frac{\partial w}{\partial z} = -4(y - 2z + 6) + 2z = -4y + 10z - 24 = 0. \tag{6}$$

Solving Eqs. (5) and (6) simultaneously gives $y = -1$ and $z = 2$. Substituting in Eq. (3), we get $x = 1$. Hence, the only critical point of (1) subject to constraint (2) is $(1, -1, 2)$. Evaluating the second-order derivatives of (4) when $y = -1$ and $z = 2$ gives

$$\frac{\partial^2 w}{\partial y^2} = 4, \qquad \frac{\partial^2 w}{\partial z^2} = 10, \qquad \frac{\partial^2 w}{\partial z \, \partial y} = -4,$$

$$D(-1, 2) = 4(10) - (-4)^2 = 24 > 0.$$

Thus w, subject to the constraint, has a relative minimum at $(1, -1, 2)$.

This solution was found by using the constraint to express one of the variables in the original function in terms of the other variables. Often this is not practical, but there is another technique, called the method of **Lagrange multipliers**[†], which avoids this step and yet allows us to obtain critical points.

The method is as follows. Suppose we have a function $f(x, y, z)$ subject to the constraint $g(x, y, z) = 0$. We construct a new function F of four variables defined by the following (where λ is a Greek letter read "lambda"):

$$F(x, y, z, \lambda) = f(x, y, z) - \lambda g(x, y, z).$$

It can be shown that if (x_0, y_0, z_0) is a critical point of f subject to the constraint $g(x, y, z) = 0$, there exists a value of λ, say λ_0, such that $(x_0, y_0, z_0, \lambda_0)$ is a critical point of F. Also, if $(x_0, y_0, z_0, \lambda_0)$ is a critical point of F, then (x_0, y_0, z_0) is a critical point of f subject to the constraint. Thus to find critical points of f subject to $g(x, y, z) = 0$, we instead find critical points of F. These are obtained by solving the simultaneous equations

$$\begin{cases} F_x(x, y, z, \lambda) = 0, \\ F_y(x, y, z, \lambda) = 0, \\ F_z(x, y, z, \lambda) = 0, \\ F_\lambda(x, y, z, \lambda) = 0. \end{cases}$$

At times, ingenuity must be used to do this. Once we obtain a critical point $(x_0, y_0, z_0, \lambda_0)$ of F, we can conclude that (x_0, y_0, z_0) is a critical point of f subject to the constraint $g(x, y, z) = 0$.

Let us illustrate for the original situation:

$$f(x, y, z) = x^2 + y^2 + z^2 \quad \text{subject to} \quad x - y + 2z = 6.$$

First, we write the constraint as $g(x, y, z) = x - y + 2z - 6 = 0$. Second, we form the function

$$\begin{aligned} F(x, y, z, \lambda) &= f(x, y, z) - \lambda g(x, y, z) \\ &= x^2 + y^2 + z^2 - \lambda(x - y + 2z - 6). \end{aligned}$$

Next we set each partial derivative of F equal to 0. For convenience we shall

---

[†]After the French mathematician, Joseph-Louis Lagrange (1736–1813).

write $F_x(x, y, z, \lambda)$ as F_x, etc.

$$\begin{cases} F_x = 2x - \lambda = 0, & (7) \\ F_y = 2y + \lambda = 0, & (8) \\ F_z = 2z - 2\lambda = 0, & (9) \\ F_\lambda = -x + y - 2z + 6 = 0. & (10) \end{cases}$$

From Eqs. (7)–(9) we see immediately that

$$x = \frac{\lambda}{2}, \quad y = -\frac{\lambda}{2}, \quad \text{and} \quad z = \lambda. \qquad (11)$$

Substituting these values in Eq. (10), we obtain

$$-\frac{\lambda}{2} - \frac{\lambda}{2} - 2\lambda + 6 = 0,$$

$$\lambda = 2.$$

Thus from Eq. (11), $x = 1$, $y = -1$, and $z = 2$. Hence the only critical point of f subject to the constraint is $(1, -1, 2)$ at which there may exist a relative maximum, a relative minimum, or neither of these. The method of Lagrange multipliers does not directly indicate which of these possibilities occur, although from our previous work we saw that it is indeed a relative minimum. In applied problems, the nature of the problem itself may give a clue as to how a critical point is to be regarded. Often the existence of either a relative minimum or a relative maximum is assumed and a critical point is treated accordingly. Actually, sufficient second-order conditions for relative extrema are available, but we shall not consider them.

EXAMPLE 1

Find the critical points for $z = f(x, y) = 3x - y + 6$ subject to the constraint $x^2 + y^2 = 4$.

We write the constraint as $g(x, y) = x^2 + y^2 - 4 = 0$ and construct the function

$$F(x, y, \lambda) = f(x, y) - \lambda g(x, y) = 3x - y + 6 - \lambda(x^2 + y^2 - 4).$$

Setting $F_x = F_y = F_\lambda = 0$:

$$\begin{cases} 3 - 2x\lambda = 0, & (12) \\ -1 - 2y\lambda = 0, & (13) \\ -x^2 - y^2 + 4 = 0. & (14) \end{cases}$$

From Eqs. (12) and (13),

$$x = \frac{3}{2\lambda} \quad \text{and} \quad y = -\frac{1}{2\lambda}.$$

Substituting in Eq. (14), we obtain

$$-\frac{9}{4\lambda^2} - \frac{1}{4\lambda^2} + 4 = 0,$$

$$\lambda = \pm \frac{\sqrt{10}}{4}.$$

If $\lambda = \sqrt{10}/4$,

$$x = \frac{3}{2\left(\dfrac{\sqrt{10}}{4}\right)} = \frac{3\sqrt{10}}{5}, \qquad y = -\frac{1}{2\left(\dfrac{\sqrt{10}}{4}\right)} = -\frac{\sqrt{10}}{5}.$$

Similarly, if $\lambda = -\sqrt{10}/4$,

$$x = -\frac{3\sqrt{10}}{5}, \qquad y = \frac{\sqrt{10}}{5}.$$

Therefore the critical points of f subject to the constraint are $(3\sqrt{10}/5, -\sqrt{10}/5)$ and $(-3\sqrt{10}/5, \sqrt{10}/5)$. Note that the values of λ do not appear in the answer. They are simply a means to obtain it.

EXAMPLE 2

Find critical points for $f(x, y, z) = xyz$, *where* $xyz \neq 0$, *subject to the constraint* $x + 2y + 3z = 36$.

Set

$$F(x, y, z, \lambda) = xyz - \lambda(x + 2y + 3z - 36).$$

Then

$$\begin{cases} F_x = yz - \lambda = 0, \\ F_y = xz - 2\lambda = 0, \\ F_z = xy - 3\lambda = 0, \\ F_\lambda = -x - 2y - 3z + 36 = 0. \end{cases}$$

We can write the system as

$$\begin{cases} yz = \lambda, & (15) \\ xz = 2\lambda, & (16) \\ xy = 3\lambda, & (17) \\ x + 2y + 3z - 36 = 0. & (18) \end{cases}$$

Dividing each side of Eq. (15) by the corresponding side of Eq. (16), we get

$$\frac{yz}{xz} = \frac{\lambda}{2\lambda} \qquad \text{or} \qquad y = \frac{x}{2}.$$

This division is valid since $xyz \neq 0$. Similarly, from Eqs. (15) and (17) we get

$$z = \frac{x}{3}.$$

Substituting into Eq. (18) gives

$$x + 2\left(\frac{x}{2}\right) + 3\left(\frac{x}{3}\right) - 36 = 0,$$

$$x = 12.$$

Thus $y = 6$ and $z = 4$. Hence $(12, 6, 4)$ is the only critical point satisfying the given conditions.

EXAMPLE 3

Suppose a firm has an order for 200 units of its product and wishes to distribute their manufacture between two of its plants, Plant 1 and Plant 2. Let q_1 and q_2 denote the outputs of Plants 1 and 2, respectively, and suppose the total cost function is given by $c = f(q_1, q_2)$ $= 2q_1^2 + q_1q_2 + q_2^2 + 200$. How should the output be distributed in order to minimize costs?

We must minimize $c = f(q_1, q_2)$ subject to the constraint $q_1 + q_2 = 200$.

$$F(q_1, q_2, \lambda) = 2q_1^2 + q_1q_2 + q_2^2 + 200 - \lambda(q_1 + q_2 - 200).$$

$$\begin{cases} \dfrac{\partial F}{\partial q_1} = 4q_1 + q_2 - \lambda = 0, & (19) \\[2mm] \dfrac{\partial F}{\partial q_2} = q_1 + 2q_2 - \lambda = 0, & (20) \\[2mm] \dfrac{\partial F}{\partial \lambda} = -q_1 - q_2 + 200 = 0. & (21) \end{cases}$$

Solving Eqs. (19) and (20) gives

$$q_1 = \frac{\lambda}{7}, \qquad q_2 = \frac{3\lambda}{7}.$$

Substituting these values in Eq. (21) gives $\lambda = 350$. Thus $q_1 = 50$ and $q_2 = 150$. Plant 1 should produce 50 units and Plant 2, 150 units, in order to minimize costs.

An interesting observation can be made concerning Example 3. From Eq. (19), $\lambda = 4q_1 + q_2 = \partial c / \partial q_1$, the marginal cost of Plant 1. From Eq. (20), $\lambda = q_1 + 2q_2 = \partial c / \partial q_2$, the marginal cost of Plant 2. Hence, $\partial c / \partial q_1 = \partial c / \partial q_2$, and we conclude that to minimize cost it is necessary that the marginal costs of each plant be equal to each other.

EXAMPLE 4

Suppose a firm must produce a given quantity P_0 of output in the cheapest possible manner. If there are two input factors l and k, and their prices per unit are fixed at p_l and p_k

respectively, discuss the economic significance of combining input to achieve least cost. That is, describe the least-cost input combination.

Let $P = f(l, k)$ be the production function. Then we must minimize the cost function

$$c = lp_l + kp_k$$

subject to

$$P_0 = f(l, k).$$

We construct

$$F(l, k, \lambda) = lp_l + kp_k - \lambda[f(l, k) - P_0].$$

We have

$$\frac{\partial F}{\partial l} = p_l - \lambda \frac{\partial}{\partial l}[f(l, k)] = 0, \tag{22}$$

$$\frac{\partial F}{\partial k} = p_k - \lambda \frac{\partial}{\partial k}[f(l, k)] = 0, \tag{23}$$

$$\frac{\partial F}{\partial \lambda} = -f(l, k) + P_0 = 0.$$

From Eqs. (22) and (23),

$$\lambda = \frac{p_l}{\dfrac{\partial}{\partial l}[f(l, k)]} = \frac{p_k}{\dfrac{\partial}{\partial k}[f(l, k)]}. \tag{24}$$

Hence,

$$\frac{p_l}{p_k} = \frac{\dfrac{\partial}{\partial l}[f(l, k)]}{\dfrac{\partial}{\partial k}[f(l, k)]}.$$

We conclude that when the least-cost combination of factors is used, the ratio of the marginal products of the input factors must be equal to the ratio of their corresponding prices.

The method of Lagrange multipliers is by no means restricted to problems involving a single constraint. For example, suppose $f(x, y, z, w)$ were subject to constraints $g_1(x, y, z, w) = 0$ and $g_2(x, y, z, w) = 0$. Then there would be two Lagrange multipliers, λ_1 and λ_2 (one for each constraint), and we would construct the function $F = f - \lambda_1 g_1 - \lambda_2 g_2$. We would then solve the system $F_x = F_y = F_z = F_w = F_{\lambda_1} = F_{\lambda_2} = 0$.

EXAMPLE 5

Find critical points for $f(x, y, z) = xy + yz$ subject to the constraints $x^2 + y^2 = 8$ and $yz = 8$.

Set

$$F(x, y, z, \lambda_1, \lambda_2) = xy + yz - \lambda_1(x^2 + y^2 - 8) - \lambda_2(yz - 8).$$

Then

$$\begin{cases} F_x = y - 2x\lambda_1 = 0, \\ F_y = x + z - 2y\lambda_1 - z\lambda_2 = 0, \\ F_z = y - y\lambda_2 = 0, \\ F_{\lambda_1} = -x^2 - y^2 + 8 = 0, \\ F_{\lambda_2} = -yz + 8 = 0. \end{cases}$$

We can write the system as

$$\begin{cases} \dfrac{y}{2x} = \lambda_1, & (25) \\ x + z - 2y\lambda_1 - z\lambda_2 = 0, & (26) \\ \lambda_2 = 1, & (27) \\ x^2 + y^2 = 8, & (28) \\ z = \dfrac{8}{y}. & (29) \end{cases}$$

Substituting $\lambda_2 = 1$ from Eq. (27) into Eq. (26) and simplifying gives $x - 2y\lambda_1 = 0$, and so

$$\lambda_1 = \frac{x}{2y}.$$

Substituting into Eq. (25) gives

$$\frac{y}{2x} = \frac{x}{2y},$$

$$y^2 = x^2. \qquad (30)$$

Substituting into Eq. (28) gives $x^2 + x^2 = 8$ from which $x = \pm 2$. If $x = 2$, then from Eq. (30) we have $y = \pm 2$. Similarly, if $x = -2$, then $y = \pm 2$. Thus if $x = 2$ and $y = 2$, then from Eq. (29) we obtain $z = 4$. Continuing in this manner, we obtain four critical points:

$$(2, 2, 4), \ (2, -2, -4), \ (-2, 2, 4), \ (-2, -2, -4).$$

EXERCISE 12-8

*In Problems **1–12** find, by the method of Lagrange multipliers, the critical points of the functions subject to the given constraints.*

1. $f(x, y) = x^2 + 4y^2 + 6; \quad 2x - 8y = 20.$

2. $f(x, y) = -2x^2 + 5y^2 + 7; \quad 3x - 2y = 7.$

3. $f(x, y, z) = x^2 + y^2 + z^2$; $2x + y - z = 9$.

4. $f(x, y, z) = x + y + z$; $xyz = 27$.

5. $f(x, y, z) = x^2 + xy + 2y^2 + z^2$; $x - 3y - 4z = 16$.

6. $f(x, y, z) = xyz^2$; $x - y + z = 20$, $(xyz^2 \neq 0)$.

7. $f(x, y, z) = xyz$; $x + 2y + 3z = 18$, $(xyz \neq 0)$.

8. $f(x, y, z) = x^2 + y^2 + z^2$; $x + y + z = 1$.

9. $f(x, y, z) = x^2 + 2y - z^2$; $2x - y = 0, y + z = 0$.

10. $f(x, y, z) = x^2 + y^2 + z^2$; $x + y + z = 1, x - y + z = 1$.

11. $f(x, y, z) = xyz$; $x + y + z = 12, x + y - z = 0$, $(xyz \neq 0)$.

12. $f(x, y, z, w) = 2x^2 + 2y^2 + 3z^2 - 4w^2$; $4x - 8y + 6z + 16w = 6$.

13. To fill an order for 100 units of its product, a firm wishes to distribute the production between its two plants, Plant 1 and Plant 2. The total cost function is given by $c = f(q_1, q_2) = .1q_1^2 + 7q_1 + 15q_2 + 1000$, where q_1 and q_2 are the number of units produced at Plants 1 and 2, respectively. How should the output be distributed in order to minimize costs?

14. Repeat Problem 13 if the cost function is $c = 3q_1^2 + q_1 q_2 + 2q_2^2$ and a total of 200 units are to be produced.

15. The production function for a firm is $f(l, k) = 12l + 20k - l^2 - 2k^2$. The cost to the firm of l and k is 4 and 8 per unit, respectively. If the firm wants the total cost of input to be 88, find the greatest output possible subject to this budget constraint.

16. Repeat Problem 15, given that $f(l, k) = 60l + 30k - 2l^2 - 3k^2$ and the budget constraint is $2l + 3k = 30$.

*Problems 17–20 refer to the following definition. A **utility function** is a function that attaches a measure to the satisfaction or utility a consumer gets from the consumption of products per unit of time. Suppose $U = f(x, y)$ is such a function, where x and y are the amounts of two products, X and Y. The **marginal utility** of X is $\partial U / \partial x$ and approximately represents the change in total utility resulting from a one-unit change in consumption of product X per unit of time. We define the marginal utility of Y in similar fashion. If the prices of X and Y are p_x and p_y, respectively, and the consumer has an income or budget of I to spend, then his budget constraint is $xp_x + yp_y = I$. In the following problems you are asked to find the quantities of each product which the consumer should buy, subject to his budget, which will allow him to maximize his satisfaction. That is, you are to maximize $U = f(x, y)$ subject to $xp_x + yp_y = I$. Assume such a maximum exists.*

17. $U = x^3 y^3$; $p_x = 2, p_y = 3, I = 48$, $(x^3 y^3 \neq 0)$.

18. $U = 46x - (5x^2/2) + 34y - 2y^2$; $p_x = 5, p_y = 2, I = 30$.

19. $U = f(x, y, z) = xyz$; $p_x = 2, p_y = 1, p_z = 4, I = 60$, $(xyz \neq 0)$.

20. Let $U = f(x, y)$ be a utility function subject to the budget constraint $xp_x + yp_y = I$, where p_x, p_y, and I are constant. Show that to maximize satisfaction it is necessary

that

$$\lambda = \frac{f_x(x, y)}{p_x} = \frac{f_y(x, y)}{p_y},$$

where $f_x(x, y)$ and $f_y(x, y)$ are the marginal utilities of X and Y, respectively. Deduce that $f_x(x, y)/p_x$ is the marginal utility of one dollar's worth of X. Hence, maximum satisfaction is obtained when the consumer allocates his budget so that the marginal utility of a dollar's worth of X is equal to the marginal utility per dollar's worth of Y. Performing the same procedure as above, verify that this is true for $U = f(x, y, z, w)$ subject to the corresponding budget equation. In each case, λ is called the *marginal utility of income.*

12-9 LINES OF REGRESSION[†]

To study the influence of advertising on sales, a firm compiled the data in Table 12-4. The variable x denotes advertising expenditures in hundreds of dollars and the variable y denotes the resulting sales revenue in thousands of dollars. If each pair (x, y) of data is plotted, the result is called a *scatter diagram* [Fig. 12-15(a)].

From an observation of the distribution of the points, it is reasonable to assume that a relationship exists between x and y and that it is approximately linear. On this basis we may fit a straight line "by eye" to the data [Fig. 12-15(b)], and from this line predict a value of y for a given value of x. This line seems consistent with the trend of the data, although other lines could be drawn as well. Unfortunately, determining a line "by eye" is not very objective. We want to apply criteria in specifying what we shall call a line of "best fit." A frequently used technique is called the **method of least squares**.

To apply the method of least squares to the data in Table 12-4, we first assume that x and y are approximately linearly related and that we can fit a straight line

$$\hat{y} = \hat{a} + \hat{b}x \tag{1}$$

to the given points by a suitable objective choice of the constants $\hat{a}$ and $\hat{b}$ (read "a hat" and "b hat," respectively). For a given value of x in Eq. (1), $\hat{y}$ is the corresponding predicted value of y, and $(x, \hat{y})$ will be on the line. Our aim is that $\hat{y}$ be near y.

When $x = 2$, the observed value of y is 3. Our predicted value of y is obtained by substituting $x = 2$ in Eq. (1), which yields $\hat{y} = \hat{a} + 2\hat{b}$. The error of estimation, or vertical deviation of the point $(2, 3)$ from the line, is $\hat{y} - y$, or

$$\hat{a} + 2\hat{b} - 3.$$

[†]May be omitted without loss of continuity.

TABLE 12-4

expenditures (x)	2	3	4.5	5.5	7
revenue (y)	3	6	8	10	11

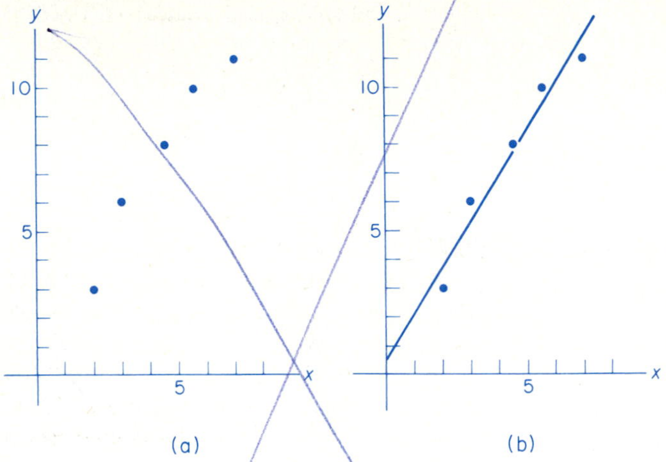

(a)   (b)

FIG. 12-15

This vertical deviation is indicated (although exaggerated for clarity) in Fig. 12-16. Similarly, the vertical deviation of (3, 6) from the line is $\hat{a} + 3\hat{b} - 6$, as is also illustrated. To avoid possible difficulties associated with positive and negative deviations, we shall consider the squares of the deviations and shall form the sum S of all such squares for the given data.

$$S = (\hat{a} + 2\hat{b} - 3)^2 + (\hat{a} + 3\hat{b} - 6)^2 + (\hat{a} + 4.5\hat{b} - 8)^2 +$$

$$(\hat{a} + 5.5\hat{b} - 10)^2 + (\hat{a} + 7\hat{b} - 11)^2.$$

The method of least squares requires that we choose as the line of "best fit" the one obtained by selecting $\hat{a}$ and $\hat{b}$ so as to minimize S. We can minimize S with respect to $\hat{a}$ and $\hat{b}$ by solving the system

$$\begin{cases} \dfrac{\partial S}{\partial \hat{a}} = 0, \\[2mm] \dfrac{\partial S}{\partial \hat{b}} = 0. \end{cases}$$

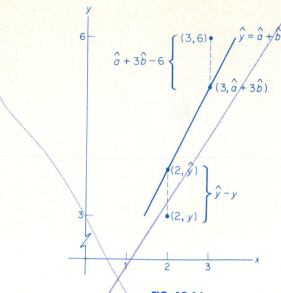

FIG. 12-16

We have

$$\frac{\partial S}{\partial \hat{a}} = 2(\hat{a} + 2\hat{b} - 3) + 2(\hat{a} + 3\hat{b} - 6) + 2(\hat{a} + 4.5\hat{b} - 8) +$$

$$2(\hat{a} + 5.5\hat{b} - 10) + 2(\hat{a} + 7\hat{b} - 11) = 0,$$

$$\frac{\partial S}{\partial \hat{b}} = 4(\hat{a} + 2\hat{b} - 3) + 6(\hat{a} + 3\hat{b} - 6) + 9(\hat{a} + 4.5\hat{b} - 8) +$$

$$11(\hat{a} + 5.5\hat{b} - 10) + 14(\hat{a} + 7\hat{b} - 11) = 0,$$

which when simplified gives

$$\begin{cases} 5\hat{a} + 22\hat{b} = 38, \\ 44\hat{a} + 225\hat{b} = 384. \end{cases}$$

Solving for $\hat{a}$ and $\hat{b}$, we obtain

$$\hat{a} = \frac{102}{157} \approx .65, \qquad \hat{b} = \frac{248}{157} \approx 1.58.$$

It can be shown that the values of $\hat{a}$ and $\hat{b}$ obtained this way always lead to a minimum value of S. Hence, in the sense of least squares, the line of best fit $\hat{y} = \hat{a} + \hat{b}x$ is

$$\hat{y} = .65 + 1.58x. \tag{2}$$

This is, in fact, the line indicated in Fig. 12-15(b). It is called the **least squares line of y on x** or the **linear regression line of y on x**. The constants $\hat{a}$ and $\hat{b}$ are called **linear regression coefficients**. With Eq. (2) we would predict that when $x = 5$, the corresponding value of y is $\hat{y} = .65 + 1.58(5) = 8.55$.

More generally, suppose we are given the following n pairs of observations:

$$(x_1, y_1), (x_2, y_2), \ldots, (x_n, y_n).$$

If we assume that x and y are approximately linearly related and that we can fit a straight line $\hat{y} = \hat{a} + \hat{b}x$ to the data, the sum of the squares of the errors $\hat{y} - y$ is

$$S = \left(\hat{a} + \hat{b}x_1 - y_1\right)^2 + \left(\hat{a} + \hat{b}x_2 - y_2\right)^2 + \ldots + \left(\hat{a} + \hat{b}x_n - y_n\right)^2.$$

Since S must be minimized with respect to $\hat{a}$ and $\hat{b}$,

$$\begin{cases} \dfrac{\partial S}{\partial \hat{a}} = 2\left(\hat{a} + \hat{b}x_1 - y_1\right) + 2\left(\hat{a} + \hat{b}x_2 - y_2\right) + \ldots + 2\left(\hat{a} + \hat{b}x_n - y_n\right) = 0, \\[2mm] \dfrac{\partial S}{\partial \hat{b}} = 2x_1\left(\hat{a} + \hat{b}x_1 - y_1\right) + 2x_2\left(\hat{a} + \hat{b}x_2 - y_2\right) + \ldots + 2x_n\left(\hat{a} + \hat{b}x_n - y_n\right) = 0. \end{cases}$$

Dividing both equations by 2 and using sigma notation, we have

$$\begin{cases} \hat{a}n + \hat{b}\displaystyle\sum_{i=1}^{n} x_i - \sum_{i=1}^{n} y_i = 0, \\[4mm] \hat{a}\displaystyle\sum_{i=1}^{n} x_i + \hat{b}\sum_{i=1}^{n} x_i^2 - \sum_{i=1}^{n} x_i y_i = 0. \end{cases}$$

Equivalently we have the system of so-called *normal equations:*

$$\begin{cases} \displaystyle\sum_{i=1}^{n} y_i = \hat{a}n + \hat{b}\sum_{i=1}^{n} x_i, & (3) \\[4mm] \displaystyle\sum_{i=1}^{n} x_i y_i = \hat{a}\sum_{i=1}^{n} x_i + \hat{b}\sum_{i=1}^{n} x_i^2. & (4) \end{cases}$$

To solve for $\hat{b}$ we first multiply Eq. (3) by $\displaystyle\sum_{i=1}^{n} x_i$ and Eq. (4) by n:

$$\begin{cases} \left(\displaystyle\sum_{i=1}^{n} x_i\right)\left(\sum_{i=1}^{n} y_i\right) = \hat{a}n \sum_{i=1}^{n} x_i + \hat{b}\left(\sum_{i=1}^{n} x_i\right)^2, & (5) \\[4mm] n\displaystyle\sum_{i=1}^{n} x_i y_i = \hat{a}n \sum_{i=1}^{n} x_i + \hat{b}n \sum_{i=1}^{n} x_i^2. & (6) \end{cases}$$

Subtracting Eq. (5) from Eq. (6), we obtain

$$n \sum_{i=1}^{n} x_i y_i - \left(\sum_{i=1}^{n} x_i \right)\left(\sum_{i=1}^{n} y_i \right) = \hat{b} n \sum_{i=1}^{n} x_i^2 - \hat{b} \left(\sum_{i=1}^{n} x_i \right)^2$$

$$= \hat{b} \left[n \sum_{i=1}^{n} x_i^2 - \left(\sum_{i=1}^{n} x_i \right)^2 \right].$$

Thus,

$$\hat{b} = \frac{n \sum_{i=1}^{n} x_i y_i - \left(\sum_{i=1}^{n} x_i \right)\left(\sum_{i=1}^{n} y_i \right)}{n \sum_{i=1}^{n} x_i^2 - \left(\sum_{i=1}^{n} x_i \right)^2}. \tag{7}$$

It can also be shown that

$$\hat{a} = \frac{\left(\sum_{i=1}^{n} x_i^2 \right)\left(\sum_{i=1}^{n} y_i \right) - \left(\sum_{i=1}^{n} x_i \right)\left(\sum_{i=1}^{n} x_i y_i \right)}{n \sum_{i=1}^{n} x_i^2 - \left(\sum_{i=1}^{n} x_i \right)^2}. \tag{8}$$

Computing the linear regression coefficients $\hat{a}$ and $\hat{b}$ by the formulas of Eqs. (7) and (8) gives the linear regression line of y on x, namely $\hat{y} = \hat{a} + \hat{b}x$, which can be used to estimate y for a given value of x.

In the next example, as well as in the exercises, you will encounter *index numbers*. They are used to relate a variable in one period of time to the same variable in another period, this latter period called the *base period*. An index number is a *relative* number which describes data that are changing over time. Such data are referred to as *time series*.

For example, consider the time series data of total production of widgets in the United States for 1975–1979 indicated in Table 12-5. If we choose 1976 as the base year and assign to it the index number 100, then the other index numbers are obtained by dividing each year's production by the 1976 production of 900 and multiplying the result by 100. We can, for example, interpret the index 106 for 1979 as meaning that production for that year was 106 percent of the production in 1976.

In time series analysis, index numbers are obviously of great advantage if the data involve numbers of great magnitude. But regardless of the magnitude of

TABLE 12-5

YEAR	PRODUCTION IN THOUSANDS	INDEX [1976 = 100]
1975	828	92
1976	900	100
1977	936	104
1978	891	99
1979	954	106

the data, index numbers simplify the task of comparing changes in data over periods of time.

EXAMPLE 1

By means of the least squares linear regression line, represent the trend for the Index of Industrial Production from 1963 to 1968 (Table 12-6 [1967 = 100]).

TABLE 12-6

Year	1963	1964	1965	1966	1967	1968
Index	79	84	91	99	100	105

Source: Economic Report of the President, 1971, U.S. Government Printing Office, Washington, D.C., 1971.

We shall let x denote time, y denote the index, and treat y as a linear function of x. Also, we shall designate 1963 by $x = 1$, 1964 by $x = 2$, etc. There are $n = 6$ pairs of measurements. To use the formulas of Eqs. (7) and (8) we first perform the arithmetic shown in Table 12-7.

TABLE 12-7

YEAR	x_i	y_i	$x_i y_i$	x_i^2
1963	1	79	79	1
1964	2	84	168	4
1965	3	91	273	9
1966	4	99	396	16
1967	5	100	500	25
1968	6	105	630	36
Total	21	558	2046	91
	$= \sum_{i=1}^{6} x_i$	$= \sum_{i=1}^{6} y_i$	$= \sum_{i=1}^{6} x_i y_i$	$= \sum_{i=1}^{6} x_i^2$

Hence by Eq. (8),

$$\hat{a} = \frac{91(558) - 21(2046)}{6(91) - (21)^2} = \frac{7812}{105} = 74.4,$$

and by Eq. (7),

$$\hat{b} = \frac{6(2046) - 21(558)}{6(91) - (21)^2} = \frac{558}{105} = 5.31 \text{ (approximately)}.$$

Thus the regression line of y on x is

$$\hat{y} = 74.4 + 5.31x,$$

whose graph, as well as a scatter diagram, appears in Fig. 12-17.

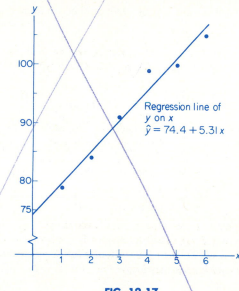

FIG. 12-17

EXERCISE 12-9

In Problems **1–4**, find an equation of the least squares linear regression line of y on x for the given data and sketch both the line and the data. Predict the value of y corresponding to $x = 3.5$.

1.

x	1	2	3	4	5	6
y	1.5	2.3	2.6	3.7	4.0	4.5

2.

x	1	2	3	4	5	6	7
y	1	1.8	2	4	4.5	7	9

3.

x	2	3	4.5	5.5	7
y	3	5	8	10	11

4.

x	2	3	4	5	6	7
y	2.4	2.9	3.3	3.8	4.3	4.9

5. A firm finds that when the price of its product is p dollars per unit, the number of units sold is q as indicated below. Find an equation of the regression line of q on p.

price (p)	10	30	40	50	60	70
demand (q)	70	68	63	50	46	32

6. On a farm the amount of water applied (in inches) and the corresponding yield of a certain crop (in tons per acre) are as given below. Find an equation of the regression line of y on x. Predict y when $x = 12$.

water (x)	8	16	24	32
yield (y)	4.1	4.5	5.1	6.1

7. The average cost $\bar{c}$, in dollars per unit, of producing q hundred units of a product is given in the table below. Find an equation of the regression line of $\bar{c}$ on q and predict $\bar{c}$ when $q = 5$.

output (q)	2	4	6	8	10
average cost ($\bar{c}$)	7.9	7	6.2	5.5	5

For the time series in Problems **8–10** *fit a linear regression line by least squares; that is, find an equation of the linear regression line of y on x. In each case let the first year in the table correspond to x = 1.*

8.

PRODUCTION OF PRODUCT A, 1975–1979
(in thousands of units)

YEAR	PRODUCTION
1975	10
1976	15
1977	16
1978	18
1979	21

9. In the following, let 1961 correspond to $x = 1$, 1963 correspond to $x = 3$, etc.

WHOLESALE PRICE INDEX—LUMBER AND WOOD PRODUCTS
[1967 = 100]

YEAR	INDEX
1961	91
1963	94
1965	96
1967	100

Source: Economic Report of the President, 1971, U.S. Government Printing Office, Washington, D.C., 1971.

10.

EQUIPMENT EXPENDITURES OF ALLIED COMPUTER COMPANY, 1973–1978
(in millions of dollars)

YEAR	EXPENDITURES
1973	15
1974	22
1975	21
1976	27
1977	26
1978	34

11. (a) By the method of least squares find an equation of the linear regression line of y on x for the following data. Refer to 1971 as year $x = 1$, etc.

OVERSEAS SHIPMENTS OF BUTTONS BY ACME BUTTON CO., INC.
(in millions)

YEAR	QUANTITY
1971	35
1972	31
1973	26
1974	24
1975	26

(b) For the data in part (a), refer to 1971 as year $x = -2$, 1972 as year $x = -1$, 1973 as year $x = 0$, etc. Then $\sum_{i=1}^{5} x_i = 0$. Fit a least squares line and observe how the calculation is simplified.

12. For the following time series, find an equation of the linear regression line that best fits the data. Refer to 1965 as year $x = -2$, 1966 as year $x = -1$, etc.

CONSUMER PRICE INDEX—MEDICAL CARE
1965–1969, [1967 = 100]

YEAR	INDEX
1965	90
1966	93
1967	100
1968	106
1969	113

Source: Economic Report of the President, 1971, U.S. Government Printing Office, Washington, D.C., 1971.

12-10 A COMMENT ON HOMOGENEOUS FUNCTIONS[†]

Many of the functions that are useful in economic analysis share the property of being homogeneous.

DEFINITION. *A function $z = f(x, y)$ is said to be **homogeneous of degree n** (n being a constant) if, for **all** positive real values of* λ,

$$f(\lambda x, \lambda y) = \lambda^n f(x, y).$$

Verbally, if both x and y are multiplied by the same positive real number, then the resulting functional value is a power of the number times the functional value $f(x, y)$. For example, if

$$f(x, y) = x^3 - 2xy^2,$$

then

$$f(\lambda x, \lambda y) = (\lambda x)^3 - 2(\lambda x)(\lambda y)^2 = \lambda^3 x^3 - 2\lambda^3 xy^2$$
$$= \lambda^3(x^3 - 2xy^2) = \lambda^3 f(x, y).$$

Thus f is homogeneous of degree three.

An important homogeneous function in economics is the Cobb-Douglas production function:

$$P = f(l, k) = Al^\alpha k^{1-\alpha} \qquad (\alpha \text{ and } A \text{ are constants}).$$

[†]This section contains material from Sec. 12-6 and may be omitted without loss of continuity.

We have

$$f(\lambda l, \lambda k) = A(\lambda l)^{\alpha}(\lambda k)^{1-\alpha} = A\lambda^{\alpha}l^{\alpha}\lambda^{1-\alpha}k^{1-\alpha}$$

$$= \lambda A l^{\alpha}k^{1-\alpha} = \lambda f(l, k).$$

Thus f is homogeneous of degree one. For example, $f(l, k) = 2l^{.3}k^{.7}$ is a homogeneous function of degree one.

Homogeneous production functions of degree one have an interesting property. If f is such a function, then

$$f(\lambda l, \lambda k) = \lambda f(l, k).$$

Thus if all inputs are doubled, then

$$f(2l, 2k) = 2f(l, k),$$

and output is doubled. Similarly, if all inputs are tripled, output is tripled, etc. In short, the same proportional change in each input factor of production results in the same proportional change in output.

By considering the partial derivatives of a homogeneous function, an important result can be obtained. Let $f(l, k)$ be a homogeneous production function of degree n. Then we have the identity

$$f(\lambda l, \lambda k) = \lambda^{n}f(l, k). \tag{1}$$

Consider the left side of Eq. (1). If we set $r = \lambda l$ and $s = \lambda k$, then Eq. (1) becomes

$$f(r, s) = \lambda^{n}f(l, k). \tag{2}$$

Now, for each side we shall take the partial with respect to λ. For the left side, $f(r, s)$, by the chain rule we have

$$\frac{\partial}{\partial \lambda}[f(r, s)] = \frac{\partial}{\partial r}[f(r, s)]\frac{\partial r}{\partial \lambda} + \frac{\partial}{\partial s}[f(r, s)]\frac{\partial s}{\partial \lambda}$$

$$= \frac{\partial}{\partial r}[f(r, s)]l + \frac{\partial}{\partial s}[f(r, s)]k. \tag{3}$$

For the right side of Eq. (2),

$$\frac{\partial}{\partial \lambda}[\lambda^{n}f(l, k)] = n\lambda^{n-1}f(l, k). \tag{4}$$

Setting (3) and (4) equal to each other, we have

$$l\frac{\partial}{\partial r}[f(r, s)] + k\frac{\partial}{\partial s}[f(r, s)] = n\lambda^{n-1}f(l, k).$$

In particular, if $\lambda = 1$, then $r = l$ and $s = k$, and so $f(r, s) = f(l, k)$. Thus, $\dfrac{\partial}{\partial r}[f(r, s)] = \dfrac{\partial}{\partial l}[f(l, k)]$ and $\dfrac{\partial}{\partial s}[f(r, s)] = \dfrac{\partial}{\partial k}[f(l, k)]$. Hence we have what is called *Euler's Theorem* for homogeneous functions:

$$l\frac{\partial}{\partial l}[f(l, k)] + k\frac{\partial}{\partial k}[f(l, k)] = nf(l, k). \tag{5}$$

Now, if f is homogeneous of degree one, such as the Cobb-Douglas function, then $n = 1$ and Eq. (5) becomes

$$l\frac{\partial}{\partial l}[f(l, k)] + k\frac{\partial}{\partial k}[f(l, k)] = f(l, k).$$

Thus if we multiply the marginal product of each input by the quantity of the input, the sum is equal to the total product.

12-11 MULTIPLE INTEGRALS

Recall that the definite integral of a function of one variable is concerned with integration over an *interval*. There are also definite integrals of functions of two variables, called (definite) *double integrals*. These involve integration over a region in the plane.

For example, the symbol

$$\int_0^1 \int_{x^3}^{x^2} (x^3 - xy)\, dy\, dx, \quad \text{or equivalently} \quad \int_0^1 \left[\int_{x^3}^{x^2} (x^3 - xy)\, dy \right] dx,$$

is the double integral of $f(x, y) = x^3 - xy$ over a region determined by the indicated limits of integration. That region is all points (x, y) in the x, y-plane such that $x^3 \leqslant y \leqslant x^2$ and $0 \leqslant x \leqslant 1$, which is shown in Fig. 12-18. Essentially,

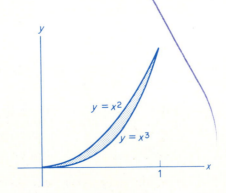

FIG. 12-18

a definite integral is a limit of a sum of the form $\Sigma f(x, y)\, \Delta y\, \Delta x$, where in our case the points (x, y) are in the shaded region. A geometric interpretation of a double integral will be given later. To evaluate

$$\int_0^1 \int_{x^3}^{x^2}(x^3 - xy)\, dy\, dx \quad \text{or} \quad \int_0^1\left[\int_{x^3}^{x^2}(x^3 - xy)\, dy\right]dx,$$

we use successive integrations starting with the innermost integral.

First, we evaluate

$$\int_{x^3}^{x^2}(x^3 - xy)\, dy$$

by treating x as a constant and integrating with respect to y between the limits x^3 and x^2:

$$\int_{x^3}^{x^2}(x^3 - xy)\, dy = \left(x^3 y - \frac{xy^2}{2}\right)\Bigg|_{x^3}^{x^2}.$$

Substituting the limits for the variable y, we have

$$\left[x^3(x^2) - \frac{x(x^2)^2}{2}\right] - \left[x^3(x^3) - \frac{x(x^3)^2}{2}\right]$$

$$= x^5 - \frac{x^5}{2} - x^6 + \frac{x^7}{2} = \frac{x^5}{2} - x^6 + \frac{x^7}{2}.$$

Now we integrate this result with respect to x between the limits 0 and 1.

$$\int_0^1\left(\frac{x^5}{2} - x^6 + \frac{x^7}{2}\right)dx = \left(\frac{x^6}{12} - \frac{x^7}{7} + \frac{x^8}{16}\right)\Bigg|_0^1 = \left(\frac{1}{12} - \frac{1}{7} + \frac{1}{16}\right) - 0 = \frac{1}{336}.$$

Thus,

$$\int_0^1\int_{x^3}^{x^2}(x^3 - xy)\, dy\, dx = \frac{1}{336}.$$

Now consider the integral

$$\int_0^2\int_3^4 xy\, dx\, dy \quad \text{or} \quad \int_0^2\left[\int_3^4 xy\, dx\right]dy.$$

The region over which the integration takes place is all points (x, y) for which $3 \leqslant x \leqslant 4$ and $0 \leqslant y \leqslant 2$. See Fig. 12-19. This double integral is evaluated by

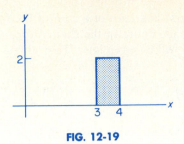

FIG. 12-19

first treating y as a constant and then integrating xy with respect to x between 3 and 4. Then we integrate the result with respect to y between 0 and 2.

$$\int_0^2 \int_3^4 xy \; dx \; dy = \int_0^2 \left[\int_3^4 xy \; dx \right] dy = \int_0^2 \left(\frac{x^2 y}{2} \right) \Big|_3^4 dy$$

$$= \int_0^2 \left(8y - \frac{9}{2}y \right) dy = \int_0^2 \left(\frac{7}{2}y \right) dy$$

$$= \frac{7y^2}{4} \Big|_0^2 = 7 - 0 = 7.$$

EXAMPLE 1

Evaluate $\int_{-1}^1 \int_0^{1-x}(2x + 1) \; dy \; dx.$

$$\int_{-1}^1 \int_0^{1-x}(2x + 1) \; dy \; dx = \int_{-1}^1 \left[\int_0^{1-x}(2x + 1) \; dy \right] dx$$

$$= \int_{-1}^1 (2xy + y) \Big|_0^{1-x} dx = \int_{-1}^1 \{ [2x(1 - x) + (1 - x)] - 0 \} \; dx$$

$$= \int_{-1}^1 (-2x^2 + x + 1) \; dx = \left(-\frac{2x^3}{3} + \frac{x^2}{2} + x \right) \Big|_{-1}^1$$

$$= \left(-\frac{2}{3} + \frac{1}{2} + 1 \right) - \left(\frac{2}{3} + \frac{1}{2} - 1 \right) = \frac{2}{3} .$$

EXAMPLE 2

Evaluate $\int_1^{\ln 2} \int_{e^y}^2 dx\, dy$.

$$\int_1^{\ln 2} \int_{e^y}^2 dx\, dy = \int_1^{\ln 2} \left[\int_{e^y}^2 dx \right] dy = \int_1^{\ln 2} x \Big|_{e^y}^2 \, dy$$

$$= \int_1^{\ln 2} (2 - e^y)\, dy = (2y - e^y) \Big|_1^{\ln 2}$$

$$= (2 \ln 2 - 2) - (2 - e) = 2 \ln 2 - 4 + e$$

$$= \ln 4 - 4 + e.$$

A double integral can be interpreted in terms of the volume of the region between the x, y-plane and a surface $z = f(x, y)$ if $z \geqslant 0$. In Fig. 12-20 is a region whose volume we shall consider. The element of volume for this region is a vertical column. It has height $z = f(x, y)$ and a base area of $\Delta y\, \Delta x$. Thus its

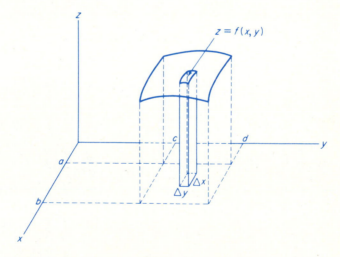

FIG. 12-20

volume is $f(x, y)\, \Delta y\, \Delta x$. The double integral sums up all such volumes for $a \leqslant x \leqslant b$ and $c \leqslant y \leqslant d$. Thus

$$\text{Volume} = \int_a^b \int_c^d f(x, y)\, dy\, dx.$$

Triple integrals are handled by successively evaluating three integrals, as the next example shows.

EXAMPLE 3

Evaluate $\int_0^1 \int_0^x \int_0^{x-y} x \, dz \, dy \, dx$.

$$\int_0^1 \int_0^x \int_0^{x-y} x \, dz \, dy \, dx = \int_0^1 \int_0^x \left[\int_0^{x-y} x \, dz \right] dy \, dx$$

$$= \int_0^1 \int_0^x (xz) \Big|_0^{x-y} dy \, dx = \int_0^1 \int_0^x [x(x-y) - 0] \, dy \, dx$$

$$= \int_0^1 \int_0^x (x^2 - xy) \, dy \, dx = \int_0^1 \left[\int_0^x (x^2 - xy) \, dy \right] dx$$

$$= \int_0^1 \left(x^2 y - \frac{xy^2}{2} \right) \Big|_0^x dx = \int_0^1 \left[\left(x^3 - \frac{x^3}{2} \right) - 0 \right] dx$$

$$= \int_0^1 \frac{x^3}{2} \, dx = \frac{x^4}{8} \Big|_0^1 = \frac{1}{8} .$$

EXERCISE 12-11

In Problems **1–22**, evaluate the given multiple integrals.

1. $\int_0^3 \int_0^4 x \, dy \, dx$.

2. $\int_0^2 \int_1^2 y \, dy \, dx$.

3. $\int_0^1 \int_0^1 xy \, dx \, dy$.

4. $\int_0^2 \int_0^3 x^2 \, dy \, dx$.

5. $\int_1^3 \int_1^2 (x^2 - y) \, dx \, dy$.

6. $\int_{-1}^2 \int_1^4 (x^2 - 2xy) \, dy \, dx$.

7. $\int_0^1 \int_0^2 (x + y) \, dy \, dx$.

8. $\int_0^3 \int_0^x (x^2 + y^2) \, dy \, dx$.

9. $\int_0^6 \int_0^{3x} y \, dy \, dx$.

10. $\int_1^2 \int_0^{x-1} y \, dy \, dx$.

11. $\int_0^1 \int_{3x}^{x^2} 2x^2 y \, dy \, dx$.

12. $\int_0^2 \int_0^{x^2} xy \, dy \, dx$.

13. $\int_0^2 \int_0^{\sqrt{4-y^2}} x \, dx \, dy$.

14. $\int_0^1 \int_y^{\sqrt{y}} y \, dx \, dy$.

15. $\int_{-1}^1 \int_x^{1-x} (x + y) \, dy \, dx$.

16. $\int_0^3 \int_{y^2}^{3y} x \, dx \, dy$.

17. $\int_0^1 \int_0^y e^{x+y} \, dx \, dy$.

18. $\int_2^3 \int_0^2 e^{x-y} \, dx \, dy$.

19. $\int_{-1}^{0}\int_{-1}^{2}\int_{1}^{2}6xy^2z^3\,dx\,dy\,dz.$ 20. $\int_{0}^{1}\int_{0}^{x}\int_{0}^{x-y}x\,dz\,dy\,dx.$

21. $\int_{0}^{1}\int_{x^2}^{x}\int_{0}^{xy}dz\,dy\,dx.$ 22. $\int_{0}^{2}\int_{y^2}^{3y}\int_{0}^{x}dz\,dx\,dy.$

23. In statistics a joint density function $z = f(x, y)$ defined on a region in the x, y-plane is represented by a surface in space. The probability that $a \leqslant x \leqslant b$ and $c \leqslant y \leqslant d$ is given by

$$P(a \leqslant x \leqslant b, c \leqslant y \leqslant d) = \int_{c}^{d}\int_{a}^{b}f(x, y)\,dx\,dy$$

and is represented by the volume between the graph of f and the rectangular region determined by $a \leqslant x \leqslant b$ and $c \leqslant y \leqslant d$. If $f(x, y) = e^{-(x+y)}$ is a joint density function, where $x \geqslant 0$ and $y \geqslant 0$, find $P(0 \leqslant x \leqslant 2, 1 \leqslant y \leqslant 2)$ and give your answer in terms of e.

24. In Problem 23, let $f(x, y) = 12e^{-4x-3y}$ for $x, y \geqslant 0$. Find $P(3 \leqslant x \leqslant 4, 2 \leqslant y \leqslant 6)$ and give your answer in terms of e.

25. In Problem 23, let $f(x, y) = \dfrac{x}{8}$, where $0 \leqslant x \leqslant 2$ and $0 \leqslant y \leqslant 4$. Find $P(x \geqslant 1, y \geqslant 2)$.

26. In Problem 23, let f be the uniform distribution $f(x, y) = 1$ defined over the unit square $0 \leqslant x \leqslant 1, 0 \leqslant y \leqslant 1$. Find the probability that $0 \leqslant x \leqslant \frac{1}{3}$ and $\frac{1}{4} \leqslant y \leqslant \frac{3}{4}$.

12-12 REVIEW

IMPORTANT TERMS AND SYMBOLS IN CHAPTER 12

3-dimensional coordinate system *(p. 510)*	x, y-plane; x, z-plane; y, z-plane *(p. 513)*
function of n variables *(p. 512)*	partial derivative *(p. 518)*
$\partial z/\partial x, f_x(x, y)$ *(p. 520)*	$\dfrac{\partial z}{\partial x}\Big|_{(x_0, y_0)}, f_x(x_0, y_0)$ *(p. 520)*
joint-cost function *(p. 524)*	production function *(p. 525)*
marginal productivity *(p. 525)*	competitive products *(p. 526)*
complementary products *(p. 526)*	implicit partial differentiation *(p. 530)*
$\dfrac{\partial^2 z}{\partial x \partial y}, \dfrac{\partial^2 z}{\partial x^2}, \dfrac{\partial^2 z}{\partial y^2}$ *(p. 533)*	f_{xy}, f_{xx}, f_{yy} *(p. 533)*
chain rule *(p. 537)*	intermediate variables *(p. 536)*
relative maxima and minima *(p. 540)*	critical point *(p. 542)*
second-derivative test *(p. 544)*	method of Lagrange multipliers *(p. 552)*

REVIEW SECTION

1. If $f(x, y) = 3x^2y^3$, then to find $f_x(x, y)$, which do we think of as a constant, x or y?

_____

Ans. y.

2. If $f(x, y) = x^2y + 3y$, then $f_x(x, y) = $ ___(a)___ and $f_y(x, y) = $ ___(b)___ .

Ans.   (a) $2xy$;   (b) $x^2 + 3$.

3. If $f(x, y) = 2$, then $f_{xy}(x, y) = $ _____ .

Ans.   0.

†4. If $w = f(x, y, z)$, $x = g(r, s)$, $y = h(r, s)$, and $z = k(r, s)$, then the number of terms in $\partial w / \partial s$ is _____ .

Ans.   3.

5. If $(1, 2)$ is a critical point of $z = f(x, y)$ and $f_{xx}(1, 2) = f_{yy}(1, 2) = 2$ and $f_{xy}(1, 2) = 1$, then f has a relative (maximum) (minimum) at $(1, 2)$.

Ans.   minimum.

6. If $f(x, y, z) = x^2 + y^2 + z^2 + 6z$, then f has a critical point at _____ .

Ans. $(0, 0, -3)$.

†7. If $yz - x + y = 0$, then $\partial z / \partial x = $ _____ .

Ans. $1/y$.

†8. Let $z = x^2 + y^2$, $x = r^2 + rs$, $y = s^2 + s$. Then $\partial z / \partial r = $ _____ .

Ans. $2x(2r + s)$.

9. If $z = xyw^2$, then $\dfrac{\partial^2 z}{\partial x \, \partial w} = $ _____ .

Ans. $2yw$.

10. True or false: If $f(x, y) = 2x^2 + xy + y^2$, then $f_{xy}(1, 2) = f_{yx}(1, 2)$. _____

Ans.   true.

11. In finding critical points of $f(x, y, z) = x^2 - xy + 2z^2$ subject to $x - y + z = 6$ by the method of Lagrange multipliers, we examine the function $F(x, y, z, \lambda) = $ _____ .

Ans. $x^2 - xy + 2z^2 - \lambda(x - y + z - 6)$ or
$x^2 - xy + 2z^2 - \lambda(6 - x + y - z)$.

---

†Refers to Secs. 12-4, 12-6, 12-9, or 12-10.

†12. If $z = f(x, y)$, $x = g(s, t)$, and $y = h(s, t)$, then by the chain rule, $\partial z / \partial s = $ _____ .

Ans. $(\partial z / \partial x)(\partial x / \partial s) + (\partial z / \partial y)(\partial y / \partial s)$.

13. A critical point of $w = f(x, y, z)$ is a point where $w_x = w_y = w_z =$ _____ .

Ans.   0.

14. In a three-dimensional coordinate system, the graph of $y = 2$ is a (a) (line) (plane) parallel to the (b) (x, y) (x, z) -plane.

Ans.   (a) plane;   (b) x, z.

15. There is a natural one-to-one correspondence between all points in space and all ordered (pairs) (triples) of real numbers.

Ans.   triples.

16. If $P = f(l, k)$ is a production function, then in terms of partial derivatives the marginal product of l is _____ .

Ans. $\partial P / \partial l$.

†17. True or false: If $z = f(x)$ and $x = g(y, w)$, then $\partial z / \partial y = (dz / dx)(\partial x / \partial y)$.

_____

Ans.   true.

†18. A function $z = f(x, y)$ is a homogeneous function of degree one if $f(\lambda x, \lambda y) = $ _____ .

Ans. $\lambda f(x, y)$.

19. In evaluating $\int_0^3 \int_0^2 xy \, dx \, dy$, we integrate first with respect to _____ .

Ans. x.

REVIEW PROBLEMS

In Problems 1–12, find the indicated partial derivatives.

1. $f(x, y) = 2x^2 + 3xy + y^2 - 1$; $f_x(x, y), f_y(x, y)$.

2. $P = l^3 + k^3 - lk$; $\partial P / \partial l, \partial P / \partial k$.

3. $z = x/(x + y)$; $\partial z / \partial x, \partial z / \partial y$.

4. $w = \dfrac{\sqrt{x^2 + y^2}}{y}$; $\partial w / \partial x$.

5. $w = e^{x^3yz}$; $w_{xy}(x, y)$.

6. $f(x, y) = xy \ln (xy)$; $f_{xy}(x, y)$.

7. $f(x, y) = \ln \sqrt{x^2 + y^2}$; $\dfrac{\partial}{\partial y}[f(x, y)]$.

----

†Refers to Secs. 12-4, 12-6, 12-9, or 12-10.

8. $f(p_A, p_B) = (p_A - 20)q_A + (p_B - 30)q_B;$ $f_{p_A}(p_A, p_B).$

9. $f(x, y, z) = (x + y)(y + z^2);$ $\dfrac{\partial^2}{\partial z^2}[f(x, y, z)].$

10. $z = (x^2 - y)(y^2 - 2xy);$ $\partial^2 z / \partial y^2.$

11. $w = xe^{yz} \ln z;$ $\partial w / \partial y, \partial^2 w / \partial x \partial z.$

12. $P = 2.4l^{.11}k^{.89};$ $\partial P / \partial k.$

†13. If $w = x^2 + 2xy + 3y^2$, $x = e^r$, and $y = \ln (r + s)$, find $\partial w / \partial r$ and $\partial w / \partial s$.

†14. If $z = \ln (x/y) + e^y - xy$, $x = r^2 s^2$, and $y = r + s$, find $\partial z / \partial s$.

†15. If $x^2 + 2xy - 2z^2 + xz + 2 = 0$, find $\partial z / \partial x$.

†16. If $z^2 - e^{yz} + \ln z + e^{xz} = 0$, find $\partial z / \partial y$.

17. Examine $f(x, y) = x^2 + 2y^2 - 2xy - 4y + 3$ for relative extrema.

18. Examine $f(w, z) = 2w^3 + 2z^3 - 6wz + 7$ for relative extrema.

19. Find all critical points of $f(x, y, z) = x^2 + y^2 + z^2$ subject to the constraint $3x + 2y + z = 14$.

20. Find all critical points of $f(x, y, z) = xyz$ subject to $3x + 2y + 4z - 120 = 0$ ($xyz \neq 0$).

†21. Find an equation of the linear regression line of y on x given the following data:

x	1	2	3	4	5
y	2.8	5.1	7.1	8.7	10

22. A manufacturer's cost for producting x units of product X and y units of product Y is given by $c = 5x + .03xy + 7y + 200$. Determine the (partial) marginal cost with respect to x when $x = 100$ and $y = 200$.

23. If a manufacturer's production function is defined by $P = 20l^{.7}k^{.3}$, determine the marginal productivity functions.

24. If $q_A = 200 - 3p_A + p_B$ and $q_B = 50 - 5p_B + p_A$, where q_A and q_B are the number of units demanded of products A and B, respectively, and p_A and p_B are their respective prices per unit, determine whether A and B are competitive or complementary products.

In Problems 25–28, evaluate the double integrals.

25. $\displaystyle \int_1^2 \int_0^y x^2 y^2 \, dx \, dy.$

26. $\displaystyle \int_0^4 \int_{y/2}^2 xy \, dx \, dy.$

27. $\displaystyle \int_0^3 \int_{y^2}^{3y} x \, dx \, dy.$

28. $\displaystyle \int_0^1 \int_{\sqrt{x}}^{x^2} (x^2 + 2xy - 3y^2) \, dy \, dx.$

---

†Refers to Secs. 12-4, 12-6, 12-9, or 12-10.

29. For industry there is a model that describes the rate α (a Greek letter read "alpha") at which a new innovation substitutes for an established process. It is given by[†]

$$\alpha = Z + .530P - .027S,$$

where Z is a constant that depends on the particular industry, P is an index of profitability of the new innovation, and S is an index of the extent of the investment necessary to make use of the innovation. Find $\partial\alpha/\partial P$ and $\partial\alpha/\partial S$.

30. A person's general status, S_g, is believed to be a function of status attributable to education, S_e, and status attributable to income, S_i, where S_g, S_e, and S_i are represented numerically. If $S_g = 7\sqrt[3]{S_e}\ \sqrt{S_i}$, determine $\dfrac{\partial S_g}{\partial S_e}$ and $\dfrac{\partial S_g}{\partial S_i}$ when $S_e = 125$ and $S_i = 100$, and interpret your results (adapted from Leik and Meeker[‡]).

In Problems 31–34, sketch the given surfaces.

31. $2x + 3y + z = 9.$ **32.** $z = x.$

33. $z = y^2.$ **34.** $x^2 + z^2 = 1.$

35. An open-top rectangular cardboard box is to have a volume of 32 cubic feet. Find the dimensions of the box so that the amount of cardboard used is minimized.

[†]A. P. Hurter, Jr., A. H. Rubenstein, et al. "Market penetration by new innovations: the technological literature," *Technological Forecasting and Social Change*, vol. 11 (1978), 197–221.

[‡]R. K. Leik and B. F. Meeker. *Mathematical Sociology.* Englewood Cliffs, N. J.: Prentice-Hall, Inc., 1975.

PROBABILITY

The term *probability* is familiar to most of us. It is not uncommon to hear such phrases as "the probability of rain," "the probability of flooding," and "the probability of receiving an A in a course." Loosely speaking, probability refers to a number that indicates the degree of likelihood that some future event will have a particular outcome. For example, before tossing a well-balanced coin, you do not know with certainty whether a head or tail will show. However, if the coin were tossed a large number of times, approximately half of the tosses would give heads. Thus we say that the probability that a head occurs on any toss is $\frac{1}{2}$ or 50 percent. In this chapter we shall be concerned with the fundamentals of probability. Section 13-1 will involve some preliminary concepts.

13-1 SAMPLE SPACES AND EVENTS

Inherent in any discussion of probability is the performance of an *experiment* whose *outcome* is determined by chance. Such an experiment is sometimes called a *random process*. For example, consider the experiment of tossing a coin and observing whether it lands heads upward (H) or lands tails upward (T). (We assume that the coin does not land on an edge.) The actual outcome is determined by chance. We can write the set of all possible outcomes as

$$\{H, T\},$$

which is called the *sample space* for this experiment. The elements H and T are called *sample points*.

> **DEFINITION.** *The **sample space** S of an experiment is the set of all possible outcomes of the experiment. The elements of the sample space are called **sample points**.*

EXAMPLE 1

A die is tossed and the number of dots appearing on the top face is observed. Determine the sample space of this experiment.

The only possible outcomes are that 1, 2, 3, 4, 5, or 6 dots show. Thus the sample space S is

$$S = \{1, 2, 3, 4, 5, 6\}.$$

EXAMPLE 2

If a coin is tossed twice in succession and the outcome (H or T) on each toss is observed, determine the sample space of this experiment.

One possible outcome is H on the first toss and H on the second toss. We can indicate this by the ordered pair (H, H), or simply by HH since no confusion arises.

When an experiment like this involves a sequence of processes, a systematic way to determine all possible outcomes is by a **tree diagram**, as illustrated in Fig. 13-1. From the starting point, there are two branches that indicate the possible results for the first toss. From each of these branches are two more branches that indicate the possible results for

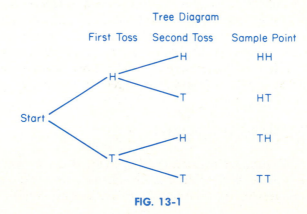

Tree Diagram

First Toss   Second Toss   Sample Point

FIG. 13-1

the second toss. This tree determines four paths, each beginning at the starting point and ending at a tip. Each path determines a sample point. Thus, the sample space S is

$$S = \{HH, HT, TH, TT\}.$$

In Example 2, the procedure of tossing a coin twice in succession involves a sequence of two processes. The first can occur in either of two ways (H or T), and then the second can occur in either of two ways (H or T). Since each of the first two ways can be paired with each of the second two ways, the total number of ways that the coins can fall is $2 \cdot 2$, or 4. Thus the sample space has 4 sample points. This multiplication procedure can be generalized into a basic counting principle.

BASIC COUNTING PRINCIPLE

Suppose that a procedure involves a sequence of k processes. Let n_1 be the number of ways the first can occur and n_2 be the number of ways the second can occur after the first process has occurred. Continuing in this way, let n_k be the number of ways the kth process can occur after the first $k - 1$ processes have occurred. Then the total number of ways the procedure can occur is $n_1 \cdot n_2 \cdots n_k$.

EXAMPLE 3

If a coin is tossed three times in succession and the result of each toss is observed, determine the number of sample points in the sample space and construct a tree diagram that gives these points.

On each toss the coin can fall two ways, H or T. The result of one toss does not affect the result of another. Since three tosses are involved, by the basic counting principle the total number of sample points is $2 \cdot 2 \cdot 2 = 8$. A tree diagram is shown in Fig. 13-2.

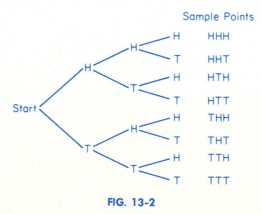

FIG. 13-2

EXAMPLE 4

An urn contains four colored marbles: one red, one white, one blue, and one yellow.

a. *A marble is selected at random and its color is noted. After the marble is replaced in the urn, a second marble is selected and its color noted. Determine the number of sample points in the sample space.*

In this experiment we say that two marbles are drawn from the urn in succession **with replacement**. Let R, W, B, and Y denote drawing red, white, blue, and yellow marbles, respectively. We can represent, for example, the selection of a red and then a white marble by RW. The sample points are RW, RB, WR, WW, and so on. For the first selection there are four possibilities: R, W, B, or Y. Since the first marble is replaced in the urn, there are also four possibilities for the second selection. By the basic counting principle, the number of sample points is $4 \cdot 4 = 16$.

b. *Determine the number of sample points in the sample space if two marbles are selected in succession **without replacement** and the colors are noted.*

The first marble drawn can be any of four colors. Since it is *not* returned to the urn, the second marble drawn can have any of the *three* colors that remain. Thus the number of sample points is $4 \cdot 3 = 12$.

At times we are concerned with whether or not the outcome of an experiment satisfies a particular relationship. For example, we may be interested in whether the outcome of tossing a die is an even number—that is 2, 4, or 6. This relationship can be considered to be the set $\{2, 4, 6\}$, which is a subset of the sample space $\{1, 2, 3, 4, 5, 6\}$. In general, any subset of a sample space is called an *event* for the experiment. Thus,

$$\{2, 4, 6\} \text{ is the event that an even number appears.}$$

Note that although an event is a set, it can be described verbally. Usually an event is denoted by E; when several events are involved in a discussion, they are denoted by E_1, E_2, E_3, and so on.

DEFINITION. *Any subset E of the sample space of an experiment is called an **event** for the experiment. If the outcome of the experiment is a sample point in E, then event E is said to **occur**.*

For the experiment of rolling a die, the sample space is $\{1, 2, 3, 4, 5, 6\}$. We saw that $\{2, 4, 6\}$ is an event. Thus, if a 2 shows, the event $\{2, 4, 6\}$ occurs. Some other events are

$$E_1 = \{1, 3, 5\} \qquad \text{an odd number shows,}$$
$$E_2 = \{3, 4, 5, 6\} \qquad \text{a number greater than or equal to 3 shows,}$$
$$E_3 = \{1\} \qquad \text{1 shows.}$$

Since S is a subset of itself, S is an event. S is called a **certain event** because it must occur no matter what the outcome. We can also consider an event such as "7 appears." Since no outcome meets this condition, this event is the empty set $\varnothing$ (the set with no elements in it). Sometimes $\varnothing$ is called an **impossible event** because it can never occur.

It is sometimes convenient to represent the sample space S and an event E for an experiment by a *Venn diagram*, as in Fig. 13-3. The rectangular region

Venn Diagram

FIG. 13-3

represents S and contains all the sample points in S. (Here the sample points are not specifically shown.) The circular region represents E and contains all sample points in E. Since E is a subset of S, the circular region cannot extend outside the rectangle.

With Venn diagrams it is easy to see how events for an experiment can be used to form other events. Figure 13-4 shows sample space S and event E. The

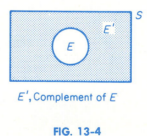

E', Complement of E

FIG. 13-4

shaded region represents the set of all sample points in S that are not in E. This set is an event called the *complement of E* and denoted by E'. Figure 13-5(a) shows two events, E_1 and E_2. The shaded region represents the set of all sample points in E_1, or E_2, or both. This set is an event called the *union of E_1 and E_2* and

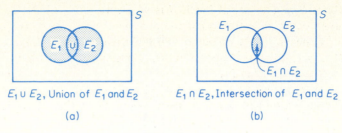

$E_1 \cup E_2$, Union of E_1 and E_2 $E_1 \cap E_2$, Intersection of E_1 and E_2

(a)   (b)

FIG. 13-5

denoted by $E_1 \cup E_2$. The shaded region in Fig. 13-5(b) represents the event that consists of all sample points that are common to both E_1 and E_2. This event is called the *intersection of E_1 and E_2* and is denoted by $E_1 \cap E_2$. In summary, we have the following definitions.

DEFINITION. *Suppose that S is the sample space of an experiment with events E, E_1, and E_2. The **complement of E**, denoted by E', is the event consisting of all sample points in S that are not in E. The **union of E_1 and E_2**, denoted by $E_1 \cup E_2$, is the event consisting of all sample points that are in E_1, or E_2, or both. The **intersection of E_1 and E_2**, denoted by $E_1 \cap E_2$, is the event consisting of all sample points that are common to both E_1 and E_2.*

EXAMPLE 5

Given the sample space $S = \{1,2,3,4,5,6\}$ for the rolling of a die, let E_1, E_2, and E_3 be the events

$$E_1 = \{1,3,5\}, \qquad E_2 = \{3,4,5,6\}, \qquad E_3 = \{1\}.$$

Determine each of the following events.

a. E_1'.

We must find those sample points in S that are not in E_1.

$$E_1' = \{2,4,6\}.$$

Thus, E_1' is the event that an even number appears.

b. $E_1 \cup E_2$.

We want the sample points in E_1, or E_2, or both.

$$E_1 \cup E_2 = \{1,3,4,5,6\}.$$

c. $E_1 \cap E_2$.

The sample points common to both E_1 and E_2 are 3 and 5. Thus,

$$E_1 \cap E_2 = \{3,5\}.$$

d. $E_2 \cap E_3$.

Since E_2 and E_3 have no sample points in common,

$$E_2 \cap E_3 = \varnothing.$$

e. $E_1 \cup E_1'$.

$$E_1 \cup E_1' = \{1,3,5\} \cup \{2,4,6\} = \{1,2,3,4,5,6\} = S.$$

f. $E_1 \cap E_1'$.

$$E_1 \cap E_1' = \{1,3,5\} \cap \{2,4,6\} = \varnothing.$$

The results of Examples 5(e) and 5(f) can be generalized.

> If E is any event for an experiment with sample space S, then
> $$E \cup E' = S \quad \text{and} \quad E \cap E' = \varnothing.$$

When two events have no sample points in common, they are called *mutually exclusive events*.

DEFINITION. *Events E_1 and E_2 for an experiment are **mutually exclusive events** if and only if $E_1 \cap E_2 = \varnothing$.*

For example, in the tossing of a die the events $\{2,4,6\}$ and $\{1\}$ are mutually exclusive. Also, an event and its complement are mutually exclusive, since $E \cap E' = \varnothing$. For mutually exclusive events, the occurrence of one event means that the other does not occur.

EXERCISE 13-1

Suppose that $S = \{1,2,3,4,5,6,7,8,9,10\}$ is the sample space for an experiment with events

$$E_1 = \{1,3,5\}, \qquad E_2 = \{3,5,7,9\}, \quad and \quad E_3 = \{2,4,6,8\}.$$

*In Problems **1–8**, determine the indicated events.*

1. $E_1 \cup E_2$.   **2.** E_3'.   **3.** $E_1 \cap E_2$.   **4.** $E_1 \cap E_3$.

5. E_2'.   **6.** $(E_1 \cup E_2)'$.   **7.** $(E_2 \cap E_3)'$.   **8.** $(E_1 \cup E_3) \cap E_2'$.

9. Of the following events, which pairs are mutually exclusive?

$$E_1 = \{1,2,3\}, \qquad E_2 = \{3,4,5\}, \qquad E_3 = \{1,2,\}, \qquad E_4 = \{5,6,7\}.$$

10. From a standard deck of 52 playing cards, 2 cards are drawn without replacement. Suppose E_A is the event that both cards are aces, E_H is the event that both cards are hearts, and E_2 is the event that both cards are 2s. Which pairs of these events are mutually exclusive?

11. An urn contains three colored marbles: one red, one white, and one blue. Determine the sample space if (a) two marbles are selected with replacement, and (b) two marbles are selected without replacement.

12. A company makes a product that goes through three processes during its manufacture. The first is an assembly line, the second is a finishing line, and the third is an inspection line. There are three assembly lines (A, B, and C), two finishing lines (D and E), and two inspection lines (F and G). For each process the company chooses a line at random. Determine the sample space.

13. A coin is tossed three times in succession and the results are observed. Determine each of the following.
    a. Sample space S.
    b. Event E_1 that at least one head occurs.
    c. Event E_2 that at least one tail occurs.
    d. $E_1 \cup E_2$.
    e. $E_1 \cap E_2$.
    f. $(E_1 \cup E_2)'$.
    g. $(E_1 \cap E_2)'$.

14. A husband and wife have two children. The outcome of the first child being a boy and the second a girl can be represented by BG. Determine each of the following.
    a. Sample space that describes all the orders of the possible sexes of the children.
    b. Event that at least one child is a girl.
    c. Event that at least one child is a boy.
    d. Is the event in (c) the complement of the event in (b)?

15. Persons A, B, and C enter a building at different times. The outcome of A arriving first, B second, and C third can be indicated by ABC. Determine each of the following.
    a. Sample space involved for the arrivals.
    b. Event that A arrives first.
    c. Event that A does not arrive first.

16. A manufacturer can order electronic components from suppliers U, V, W, or X and mechanical components from suppliers U, V, Y, or Z. The manufacturer selects one supplier for each type of component. The outcome of U being selected for electronic components and V for mechanical components can be represented by UV.
    a. Determine the sample space.
    b. Determine the event E that the suppliers are different.
    c. Determine E' and give a verbal description of this event.

17. In how many ways can a student answer a ten-question, multiple-choice examination if there are four choices for each question? Assume that the student answers each question by indicating exactly one choice.

18. The transmitter for an electric garage door opener transmits a coded signal to a receiver. The code is determined by five switches, each of which is either in an "on" or an "off" position. Determine the number of different codes that may be transmitted.

19. From a group of ten people, a president, a vice-president, a secretary, and a treasurer are to be selected. In how many ways can this be done (a) if no person can serve in two or more offices, and (b) if there is no restriction on the number of offices a person may hold?

20. From a standard deck of 52 playing cards, 2 cards are successively drawn without replacement. For this experiment, how many sample points does the sample space contain?

21. Next semester a student must enroll in four courses: mathematics, economics, English, and history. If four mathematics, two economics, two English, and four history courses are available, in how many ways can the student select the courses?

22. A company wants to fill two sales positions and has 20 qualified applicants. In how many ways can the company fill the positions?

13-2 PROBABILITY

Suppose a well-balanced die is tossed and the number of dots on the top face is observed. Then the sample space is $S = \{1, 2, 3, 4, 5, 6\}$. Before the experiment is performed, we cannot predict the outcome with certainty, but it must be one of the sample points in S. Because the die is well-balanced, each of the six sample points has the same chance of occurring. This does not mean that in six tosses each sample point must occur once. It means that if the experiment were performed a large number of times, each sample point would occur about $\frac{1}{6}$ of the time. In particular, let the experiment be performed n times. Each performance of an experiment is called a **trial**. Suppose that we are interested in the occurrence of the sample point 1. If 1 occurs k times in the n trials, then the proportion of times that 1 occurs is k/n, which is called the **relative frequency** of the outcome 1 in n trials. Since the die is well balanced, we expect that in the long run a 1 will occur $\frac{1}{6}$ of the time. That is, as n becomes very large, k/n approaches $\frac{1}{6}$. In terms of limits, we have

$$\lim_{n \to \infty} \frac{k}{n} = \frac{1}{6}.$$

This limit is defined as the probability of 1 occurring on a toss of a well-balanced die and is written $P(1)$. Thus $P(1) = \frac{1}{6}$. Similarly, $P(2) = \frac{1}{6}$, $P(3) = \frac{1}{6}, \ldots,$ and $P(6) = \frac{1}{6}$.

> **DEFINITION.** *Let s be a sample point in the sample space of an experiment and k be the number of occurrences of s in n trials. The **probability of s**, denoted by P(s), is given by*

$$P(s) = \lim_{n \to \infty} \frac{k}{n}.$$

Note that we can think of P as a function that associates with each outcome s in the sample space the probability of s, namely, $P(s)$.

Since k is the number of occurrences of s in n trials, we have $0 \leqslant k \leqslant n$. Thus, $0 \leqslant k/n \leqslant 1$ and, consequently, $0 \leqslant \lim_{n \to \infty} k/n \leqslant 1$. This gives

$$0 \leqslant P(s) \leqslant 1 \qquad \text{for any } s \text{ in the sample space.}$$

We can also define the probability of an *event*.

> **DEFINITION.** *If E is an event for an experiment with sample space S, then the **probability of E**, denoted by P(E), is*
>
> $$P(E) = \lim_{n \to \infty} \frac{k}{n},$$
>
> *where k is the number of times E occurs in n trials of the experiment.*

Thus $P(E)$ is the relative frequency of E in the long run. Since $0 \leqslant k \leqslant n$,

$$\boxed{0 \leqslant P(E) \leqslant 1.}$$

In particular, if $E = \varnothing$, then $k = 0$; thus,

$$\boxed{P(\varnothing) = 0.}$$

If $E = S$, then $k = n$, so

$$\boxed{P(S) = 1.}$$

That is, the probability of an impossible event is 0 and the probability of a certain event is 1.

In any trial, either E or E' occurs, but not both. For n trials, if E occurs k times, then E' occurs $n - k$ times. Thus,

$$P(E') = \lim_{n \to \infty} \frac{n - k}{n} = \lim_{n \to \infty} \left(1 - \frac{k}{n} \right) = 1 - \lim_{n \to \infty} \frac{k}{n} = 1 - P(E),$$

or more simply,

$$P(E') = 1 - P(E).$$

Equivalently,

$$P(E) = 1 - P(E').$$

Thus, if we know the probability of an event, then we can easily find the probability of its complement, and vice versa. For example, if $P(E) = \frac{1}{4}$, then $P(E') = 1 - \frac{1}{4} = \frac{3}{4}$.

Suppose that E contains exactly j sample points: $E = \{s_1, s_2, \ldots, s_j\}$. Also, in n trials let us assume s_1 occurs k_1 times, s_2 occurs k_2 times, $\ldots$, and s_j occurs k_j times. Since E occurs if and only if the outcome is $s_1, s_2, \ldots,$ or s_j, the relative frequency of E is

$$\frac{\text{number of occurrences of } E}{\text{number of trials}} = \frac{k_1 + k_2 + \cdots + k_j}{n}.$$

Thus,

$$P(E) = \lim_{n \to \infty} \frac{k_1 + k_2 + \cdots + k_j}{n}$$

$$= \lim_{n \to \infty} \frac{k_1}{n} + \lim_{n \to \infty} \frac{k_2}{n} + \cdots + \lim_{n \to \infty} \frac{k_j}{n}$$

$$= P(s_1) + P(s_2) + \cdots + P(s_j).$$

That is, $P(E)$ *is the sum of the probabilities of the sample points in* E.

$$\boxed{\text{If } E \text{ is an event and } E = \{s_1, s_2, \ldots, s_j\}, \text{ then}}$$
$$P(E) = P(s_1) + P(s_2) + \cdots + P(s_j).$$

When the possible outcomes of an experiment have the same probability, they are called **equally likely outcomes**. For the rolling of a well-balanced die, we know that possible outcomes have the same probability. Thus, the outcomes are equally likely, because the die is *well-balanced*. Other words and phrases used to indicate equally likely outcomes are "fair," "unbiased," and "at random." For example, we may have a *fair* coin or an *unbiased* die, or we may select a marble *at random* from an urn.

Suppose an experiment has exactly N possible outcomes, which are equally likely, and the sample space is $S = \{s_1, s_2, \ldots, s_N\}$. Let the probability of each outcome be p. Since $P(S) = 1$,

$$P(S) = P(s_1) + P(s_2) + \cdots + P(s_N) = \underbrace{p + p + \cdots + p}_{N \text{ terms}} = Np = 1.$$

Thus $p = 1/N$. If an event E for this experiment has exactly K sample points, then $P(E)$, which is the sum of the probabilities of the outcomes in E, is

$$K \cdot \frac{1}{N} = \frac{K}{N}.$$

> Suppose an experiment has exactly N possible outcomes, which are equally likely. For any outcome s, $P(s) = \dfrac{1}{N}$. If E is an event that has exactly K possible outcomes, then $P(E) = K \cdot \dfrac{1}{N} = \dfrac{K}{N}$.

EXAMPLE 1

From a production run of 5000 light bulbs, 2 percent of which are defective, 1 bulb is selected at random. What is the probability that the bulb is defective? What is the probability that it is not defective?

The sample space consists of the 5000 bulbs. Since a bulb is selected at random, the possible outcomes are equally likely. Let E be the event of selecting a defective bulb. The number of sample points in E is $.02 \cdot 5000$, or 100. Thus,

$$P(E) = \tfrac{100}{5000} = \tfrac{1}{50} = .02.$$

Alternatively, since the probability of selecting a particular bulb is $\frac{1}{5000}$ and E contains 100 sample points, by summing probabilities we have

$$P(E) = 100 \cdot \tfrac{1}{5000} = .02.$$

The event that the bulb selected is *not* defective is E'. Hence,

$$P(E') = 1 - P(E) = 1 - .02 = .98.$$

EXAMPLE 2

If two fair coins are tossed and the result on each coin is observed, determine the probability that (a) two heads occur, and (b) at least one head occurs.

Suppose we distinguish between the two coins by calling them coin 1 and coin 2. One possible result is that coin 1 shows heads and coin 2 shows tails, which we can represent by HT. Another result can be TH (coin 1 shows tails and coin 2 shows heads). Thus HT and TH are *different*, although in each case one head and one tail appear. The set of all possible results is the sample space

$$S = \{HH, HT, TH, TT\}.$$

(Note that S is the same sample space as that of tossing of one coin twice in succession.) Since both coins are fair, the outcomes are equally likely.

a. Since there are four equally likely outcomes in S, the probability of each outcome is $\frac{1}{4}$. Thus,

$$P(HH) = \frac{1}{4}.$$

b. Let E be the event that at least one head appears. Then

$$E = \{HH, HT, TH\}.$$

Since E contains three sample points,

$$P(E) = \frac{3}{4}.$$

Suppose that E_1 and E_2 are events of an experiment and both consist of a finite number of sample points. Let us find $P(E_1 \cup E_2)$, which is the sum of the probabilities of the sample points in $E_1 \cup E_2$. Now, $P(E_1)$ is the sum of the probabilities of the sample points in E_1, and a similar statement is true for $P(E_2)$. Since $E_1 \cap E_2$ is contained in both E_1 and E_2 (see Fig. 13-6), both $P(E_1)$ and

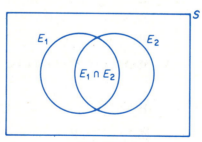

FIG. 13-6

$P(E_2)$ include $P(E_1 \cap E_2)$. If we were to add $P(E_1)$ and $P(E_2)$, this sum would include $P(E_1 \cap E_2)$ twice. By subtracting $P(E_1 \cap E_2)$ from $P(E_1) + P(E_2)$, we obtain $P(E_1 \cup E_2)$.

> If the events E_1 and E_2 each consist of a finite number of sample points, then
>
> $$P(E_1 \cup E_2) = P(E_1) + P(E_2) - P(E_1 \cap E_2).$$

For example, if a die is rolled, let E_1 be $\{1,3,5\}$ and E_2 be $\{1,2,3\}$. Then $E_1 \cap E_2 = \{1,3\}$, so

$$P(E_1 \cup E_2) = P(E_1) + P(E_2) - P(E_1 \cap E_2)$$
$$= \tfrac{3}{6} + \tfrac{3}{6} - \tfrac{2}{6} = \tfrac{2}{3}.$$

Alternatively, since $E_1 \cup E_2 = \{1,2,3,5\}$, $P(E_1 \cup E_2) = \tfrac{4}{6} = \tfrac{2}{3}$.

If E_1 and E_2 are mutually exclusive events, then $E_1 \cap E_2 = \varnothing$, so we have $P(E_1 \cap E_2) = P(\varnothing) = 0$. Thus, from Eq. (1) we can conclude the following:

> If E_1 and E_2 are *mutually exclusive* events, each consisting of a finite number of sample points, then
>
> $$P(E_1 \cup E_2) = P(E_1) + P(E_2).$$

EXAMPLE 3

A pair of well-balanced dice are rolled. Determine the probability that the sum of the numbers of dots appearing on the upward faces is (a) 7, (b) 7 *or* 11, *and* (c) *greater than* 3.

Since each die can roll any of six ways, the total number of sample points in the sample space is $6 \cdot 6 = 36$. The sample points can be considered as ordered pairs:

$(1,1)$,	$(1,2)$,	$(1,3)$,	$(1,4)$,	$(1,5)$,	$(1,6)$,
$(2,1)$,	$(2,2)$,	$(2,3)$,	$(2,4)$,	$(2,5)$,	$(2,6)$,
$(3,1)$,	$(3,2)$,	$(3,3)$,	$(3,4)$,	$(3,5)$,	$(3,6)$,
$(4,1)$,	$(4,2)$,	$(4,3)$,	$(4,4)$,	$(4,5)$,	$(4,6)$,
$(5,1)$,	$(5,2)$,	$(5,3)$,	$(5,4)$,	$(5,5)$,	$(5,6)$,
$(6,1)$,	$(6,2)$,	$(6,3)$,	$(6,4)$,	$(6,5)$,	$(6,6)$.

Since the dice are well balanced, the outcomes are equally likely. Thus, the probability of any outcome is $\tfrac{1}{36}$.

a. Let E_7 be the event that the sum of the numbers appearing is 7. Then

$$E_7 = \{(1,6), (2,5), (3,4), (4,3), (5,2), (6,1)\}.$$

Since E_7 has six sample points,

$$P(E_7) = \tfrac{6}{36} = \tfrac{1}{6}.$$

b. Let $E_{7 \text{ or } 11}$ be the event that the sum is 7 or 11. If E_{11} is the event that the sum is 11, then

$$E_{11} = \{(5,6), (6,5)\},$$

which has two sample points. Since $E_{7 \text{ or } 11} = E_7 \cup E_{11}$ and E_7 and E_{11} are mutually exclusive, we have

$$P(E_{7 \text{ or } 11}) = P(E_7) + P(E_{11}) = \tfrac{6}{36} + \tfrac{2}{36} = \tfrac{8}{36} = \tfrac{2}{9}.$$

Alternatively, we can determine $P(E_{7 \text{ or } 11})$ by counting the number of points in $E_{7 \text{ or } 11}$.

$$E_{7 \text{ or } 11} = \{(1,6), (2,5), (3,4), (4,3), (5,2), (6,1), (5,6), (6,5)\},$$

which has eight sample points. Thus,

$$P(E_{7 \text{ or } 11}) = \tfrac{8}{36} = \tfrac{2}{9}.$$

c. Let E be the event that the sum is greater than 3. The number of sample points in E is relatively large. Thus, to determine $P(E)$, it is easier to find E', rather than E, and then use the formula $P(E) = 1 - P(E')$. Here E' is the event that the sum is 2 or 3. Thus,

$$E' = \{(1,1), (1,2), (2,1)\},$$

which has three sample points. Hence,

$$P(E) = 1 - P(E') = 1 - \tfrac{3}{36} = \tfrac{11}{12}.$$

EXAMPLE 4

From a standard deck of 52 playing cards, two cards are drawn in succession and at random without replacement. Find $P(E)$, where E is the event that one card is a 2 and the other is a 3.

For drawing two cards without replacement, there are 52 possibilities for the first card and 51 for the second. Thus the sample space consists of $52 \cdot 51 = 2652$ ordered pairs, which are equally likely. For example, (5 of hearts, king of diamonds) is one such possible outcome. The event E occurs when either the first card drawn is a 2 and the second card is a 3, or vice versa. If E_1 is the event that the first card is a 2 and the second is a 3, then, in terms of sample points, E_1 occurs in $4 \cdot 4 = 16$ ways, since there are four 2s and four 3s. Similarly, if E_2 is the event that the first card is a 3 and the second is a 2, then E_2 also occurs in 16 ways. Since $E = E_1 \cup E_2$ and E_1 and E_2 are mutually exclusive,

$$P(E) = P(E_1) + P(E_2) = \frac{16}{2652} + \frac{16}{2652} = \frac{8}{663}.$$

EXERCISE 13-2

1. When a biased die is tossed, the probabilities of 1, 3, and 5 showing are the same. The probabilities of 2, 4, and 6 showing are also the same, but are twice those of 1, 3 and 5. Determine $P(1)$.

2. For the sample space $\{a, b, c, d, e\}$, suppose that the probabilities of a, b, c, and d are the same. Is it possible to determine $P(e)$?

3. A stock is selected at random from a list of 60 utility stocks, 48 of which have an annual dividend yield of 10 percent or more. Find the probability that the stock pays an annual dividend that yields (a) 10 percent or more, and (b) less than 10 percent.

4. A clothing store maintains its inventory of suits so that 25 percent are 100 percent pure wool. If a suit is selected at random, what is the probability that it is (a) 100 percent pure wool, and (b) not 100 percent pure wool?

5. On an examination given to 40 students, 10 percent received an A, 25 percent a B, 35 percent a C, 25 percent a D, and 5 percent an F. If a student is selected at random, what is the probability that the student (a) received an A, (b) received an A or B, (c) received neither a D nor an F, and (d) did not receive an F? (e) Answer questions (a)–(d) if the number of students that were given the examination is unknown.

6. A pair of fair dice are tossed. Determine the probability that at least one die shows a 2.

7. A pair of well-balanced dice are tossed. Find the probability that the sum of the number of dots showing is (a) 8, (b) 2 or 3, (c) 3, 4, or 5, (d) 12 or 13, (e) even, (f) odd, and (g) less than 10.

8. A marble is randomly drawn from an urn that contains seven red, five white, and eight blue marbles. Find the probability that the marble is (a) blue, (b) not red, (c) red or white, (d) neither red nor blue, (e) yellow, and (f) red or yellow.

9. A card is randomly selected from a standard deck of 52 playing cards. Determine the probability that the card is (a) a king of hearts, (b) a diamond, (c) a jack, (d) red, (e) a heart or a club, (f) a club and a 4, (g) a club or a 4, (h) red and a king, and (i) a spade and a heart.

10. A fair coin and a fair die are tossed. Find the probability that (a) a head and a 5 show, (b) a head shows, (c) a 3 shows, and (d) a head and an even number show.

11. A fair coin and a fair die are tossed and a card is randomly selected from a standard deck of 52 playing cards. Determine the probability that the coin, die, and card respectively show (a) a tail, a 3, the queen of hearts, (b) a tail, a 3, and a queen, (c) a head, a 2 or 3, and a queen, and (d) a head, an even number, and a diamond.

12. Three fair coins are tossed. Find the probability that (a) three heads show, (b) exactly one tail shows, (c) no more than two heads show, and (d) no more than one tail shows.

13. Assuming that the sex of a person is determined at random, determine the probability that a family with three children has (a) three girls, (b) exactly one boy, (c) no girls, and (e) at least one girl.

14. Two urns contain colored marbles. Urn 1 contains three red and two green marbles and urn 2 contains four red and five green marbles. A marble is selected at random from each urn. Find the probability that (a) both marbles are red, and (b) one marble is red and the other is green.

15. Two cards from a standard deck of 52 playing cards are successively drawn at random without replacement. Find the probability that (a) both cards are kings, and (b) one card is a diamond and the other is a heart.

16. Two cards from a standard deck of 52 playing cards are successively drawn at random with replacement. Find the probability that (a) both cards are kings, and (b) one card is a king and the other is a heart.

17. From a group of two women and three men, two persons are selected at random to form a committee. Find the probability that the committee consists of women only.

18. For the committee selection in Problem 17, find the probability that the committee consists of a man and a woman.

19. A student answers each question on a ten-question true-false examination in a random fashion. If each question is worth 10 points, what is the probability that the student scores (a) 100 points, and (b) 90 or more points.

20. On a five-question, multiple-choice examination there are four choices for each question, only one of which is correct. If a student answers each question in a random fashion, find the probability that the student answers (a) each question correctly, and (b) exactly four questions correctly.

13-3 DISCRETE RANDOM VARIABLES

With some experiments, we are interested in events associated with numbers. For example, if two coins are tossed, our interest may be in the *number* of heads that occur. Thus we can consider the events

0 heads show,   1 head shows,   2 heads show.

If we let X be a variable that represents the number of heads that occur, then the only values that X can assume are 0, 1, and 2. The value of X is determined by the outcome of the experiment, and hence by chance. In general, a variable whose values depend on the outcome of a random process is called a **random variable**. Usually, random variables are denoted by capital letters such as X, Y, or Z, and the values that these variables assume may be denoted by corresponding lower case letters (x, y, z). Thus, for the number of heads (X) that occur in the tossing of two coins, we may indicate the possible values by writing

$$X = x, \qquad \text{where } x = 0, 1, 2,$$

or, more simply,

$$X = 0, 1, 2.$$

EXAMPLE 1

a. Suppose a die is rolled and X is the number of dots that show. Then X is a random variable and $X = 1, 2, 3, 4, 5, 6$.

b. Suppose a coin is successively tossed until a head appears. If Y is the number of such tosses, then Y is a random variable and $Y = y$, where $y = 1, 2, 3, 4, \ldots$. Note that Y may assume infinitely many values.

c. A student is taking an exam with a 1-hour time limit. If X is the number of minutes it takes to complete the exam, then X is a random variable. The values that X may assume form the interval $(0, 60]$. That is, $0 < X \leqslant 60$.

A random variable is called a **discrete random variable** if it may assume only a finite number of values or if its values can be placed in one-to-one correspondence with the positive integers. In Example 1(a) and 1(b), X and Y are discrete. A random variable is called a **continuous random variable** if it may assume any value in some interval or intervals, such as X does in Example 1(c). In this section we shall be concerned with discrete random variables.

If X is a random variable, the probability of the event that X assumes the value x is denoted $P(X = x)$. Similarly, we can consider probabilities of events such as $X \leqslant x$ and $X > x$. If X is discrete, then the function f that assigns the number $P(X = x)$ to each possible value of X is called the **probability function**, the **probability distribution**, or—more simply—the **distribution** of the random variable X. Thus,

$$f(x) = P(X = x).$$

EXAMPLE 2

Suppose that X is the number of heads that appear on the toss of two well-balanced coins. Determine the distribution of X.

We must find the probabilities of the events $X = 0$, $X = 1$, and $X = 2$. The sample space is

$$S = \{HH, HT, TH, TT\},$$

where the four outcomes are equally likely.

The event $X = 0$ is $\{TT\}$.

The event $X = 1$ is $\{HT, TH\}$.

The event $X = 2$ is $\{HH\}$.

The probability for each of these events is

$$\frac{\text{number of sample points in event}}{4}$$

and is given in Table 13-1. If f is the distribution for X, that is, $f(x) = P(X = x)$, then

$$f(0) = \tfrac{1}{4}, \quad f(1) = \tfrac{1}{2}, \quad \text{and} \quad f(2) = \tfrac{1}{4}.$$

Table 13-1

x	$P(X = x)$
0	$\frac{1}{4}$
1	$\frac{2}{4} = \frac{1}{2}$
2	$\frac{1}{4}$

In Example 2, the distribution f was indicated by the listing

$$f(0) = \tfrac{1}{4}, \qquad f(1) = \tfrac{1}{2}, \quad \text{and} \quad f(2) = \tfrac{1}{4}.$$

However, Table 13-1, called the **probability table** for X, gives the same information and is an acceptable way of expressing the distribution. Another way is by its graph, as shown in Fig. 13-7. The vertical lines from the x-axis to the points on

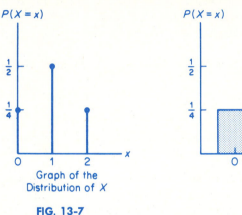

Graph of the
Distribution of X

FIG. 13-7

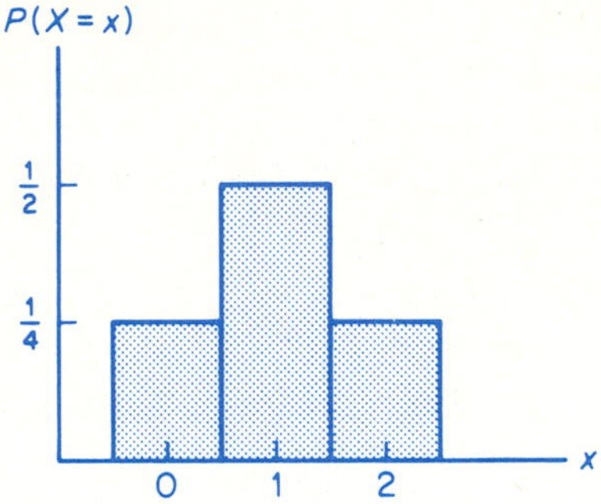

Probability
Histogram for X

FIG. 13-8

the graph merely emphasize the heights of the points. Another representation for the distribution of X is the rectangle diagram in Fig. 13-8, called the **probability histogram** for X. Here each rectangle is centered over the corresponding value of X. The rectangle above x has width 1 and height $P(X = x)$. Thus its *area* is the probability $P(X = x)$. This interpretation of area as a probability is important in later sections.

Note that in Example 2, the sum of $f(0)$, $f(1)$, and $f(2)$ is 1:

$$f(0) + f(1) + f(2) = \tfrac{1}{4} + \tfrac{1}{2} + \tfrac{1}{4} = 1.$$

This occurs because any two of the events $X = 0$, $X = 1$, and $X = 2$ are mutually exclusive and the union of all three is the sample space [and $P(S) = 1$]. In general, for any probability distribution f, the sum of all functional values is 1, which we can indicate symbolically by

$$\sum_x f(x) = 1.$$

Thus, in any probability histogram the sum of the areas of its rectangles is 1.

The probability distribution for a random variable X gives the relative frequencies of the values of X in the long run. However, it is often useful to determine the "average" value of X in the long run. In Example 2 for instance, suppose that the two coins were tossed n times, in which $X = 0$ occurred k_0 times, $X = 1$ occurred k_1 times, and $X = 2$ occurred k_2 times. Then the average value of X for these n tosses is

$$\frac{0 \cdot k_0 + 1 \cdot k_1 + 2 \cdot k_2}{n}$$

or, equivalently,

$$0 \cdot \frac{k_0}{n} + 1 \cdot \frac{k_1}{n} + 2 \cdot \frac{k_2}{n}.$$

In the long run, the average value of X is the limit of this expression as $n \to \infty$:

$$\lim_{n \to \infty} \left[0 \cdot \frac{k_0}{n} + 1 \cdot \frac{k_1}{n} + 2 \cdot \frac{k_2}{n} \right]$$

$$= 0 \cdot \lim_{n \to \infty} \frac{k_0}{n} + 1 \cdot \lim_{n \to \infty} \frac{k_1}{n} + 2 \cdot \lim_{n \to \infty} \frac{k_2}{n}.$$

But the limits of k_0/n, k_1/n, and k_2/n, which are limits of relative frequencies, are the probabilities of the events $X = 0$, $X = 1$, and $X = 2$, respectively. These probabilities are given by the distribution of X. Thus, the average value of X in the long run is

$$0 \cdot f(0) + 1 \cdot f(1) + 2 \cdot f(2) \qquad (1)$$

$$= 0 \cdot \tfrac{1}{4} + 1 \cdot \tfrac{1}{2} + 2 \cdot \tfrac{1}{4} = 1.$$

This means that if we tossed the coins many times, the average number of heads appearing per toss is very close to 1. This value is called the *mean*, *expected value*, or *expectation of X* and is denoted by μ (the Greek letter "mu") or $E(X)$. The mean does not necessarily have to be an outcome for the experiment. Note that from (1), μ has the form $\sum_x x f(x)$. In general, we have the following definition.

DEFINITION. *If X is a discrete random variable with probability distribution f, then the **mean** (or expected value or **expectation**) of X, denoted by μ, or $E(X)$, is given by*

$$\mu = E(X) = \sum_x x f(x).$$

EXAMPLE 3

An insurance company offers a $60,000 catastrophic fire insurance policy to homeowners of a certain type of house. The policy provides protection in the event that such a house is totally destroyed by fire in a 1-year period. The company has determined that the probability of such an event is .0002. If the annual policy premium is $52, find the expected gain to the company.

If an insured house does not suffer a catastrophic fire, the company gains $52. However, if there is such a fire, the company loses $60,000 − $52 (insured value of house minus premium), or $59,948. Let X be the gain (in dollars) to the company. Then X is a random variable that may assume the values 52 and −59,948 (a loss is considered a negative gain). If f is the probability function for X then

$$f(-59{,}948) = P(X = -59{,}948) = .0002$$

and

$$f(52) = P(X = 52) = 1 - .0002 = .9998.$$

The expected gain to the company is the expected value of X.

$$E(X) = \sum_x x f(x) = -59{,}948 \, f(-59{,}948) + 52 \, f(52)$$

$$= -59{,}948(.0002) + 52(.9998) = 40.$$

Thus if the company sold many policies, it could expect to gain approximately $40 per policy, which could be applied to such expenses as advertising, overhead, and profit.

Since $E(X)$ is the average value of X in the long run, it is a measure of the central tendency of X. However, $E(X)$ does not indicate the dispersion or spread of X from the mean in the long run. For example, Fig. 13-9 shows the graphs of two probability distributions, f and g, for the random variables X and Y. It can

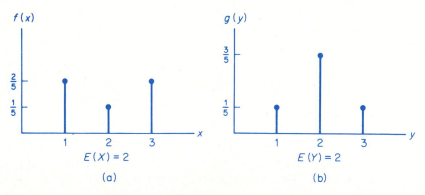

FIG. 13-9

easily be shown that both X and Y have the same mean: $E(X) = 2$ and $E(Y) = 2$. You should verify these results. But from Fig. 13-9, X is more likely to assume the values 1 or 3 than is Y, because $f(1)$ and $f(3)$ are $\frac{2}{5}$, while $g(1)$ and $g(3)$ are $\frac{1}{5}$. Thus X has more likelihood of assuming values away from the mean than does Y, so there is more dispersion for X in the long run.

There are various ways to measure dispersion for a random variable X. One way is to determine the long run average of the absolute values of the deviations from μ, that is $E(|X - \mu|)$:

$$E(|X - \mu|) = \sum_x |x - \mu| f(x).^\dagger$$

However, since $|x - \mu|$ is involved, $E(|X - \mu|)$ is a mathematically awkward expression.

Although many other measures of dispersion can be considered, two are most widely accepted. One is *variance* and the other (which is related to variance) is *standard deviation*. The **variance of** X, denoted by $\text{Var}(X)$, is the long-run average of the *squares* of the deviations of X from μ.

> **VARIANCE OF** X
>
> $$\text{Var}(X) = E[(X - \mu)^2] = \sum_x (x - \mu)^2 f(x).^{\dagger\dagger}$$

(2)

Since $(X - \mu)^2$ is involved in $\text{Var}(X)$ and both X and μ have the same units of measurement, the units for $\text{Var}(X)$ are those of X^2. For instance, in Example 3, X is in dollars; thus $\text{Var}(X)$ has units of dollars squared. It is convenient to have a measure of dispersion in the same units as X. Such a measure is $\sqrt{\text{Var}(X)}$, which is called the **standard deviation of** X and is denoted by σ (the Greek letter "sigma").

> **STANDARD DEVIATION OF** X
>
> $$\sigma = \sqrt{\text{Var}(X)}.$$

Note that σ has the property that

$$\sigma^2 = \text{Var}(X).$$

Both $\text{Var}(X)$ [or σ^2] and σ are measures of the dispersion of X. The greater the value of $\text{Var}(X)$, or σ, the greater the dispersion. In fact, one result of a famous theorem, *Chebyshev's inequality*, is that the probability of X falling within two standard deviations of the mean is at least $\frac{3}{4}$.

---

†It can be shown that if $Y = g(X)$, then $E(Y) = \sum_x g(x) f(x)$, where f is the probability function for X. For example, if $Y = |X - \mu|$ then $E(|X - \mu|) = \sum_x |x - \mu| f(x)$.

††By the previous footnote, if $Y = (X - \mu)^2$, then $E[(X - \mu)^2] = \sum_x (x - \mu)^2 f(x)$.

In addition to the formula for variance given by Eq. (2), we can derive an equivalent formula. By Eq. (2),

$$\text{Var}(X) = \sum_x (x - \mu)^2 f(x).$$

Thus,

$$\text{Var}(X) = \sum_x (x^2 - 2x\mu + \mu^2) f(x)$$

$$= \sum_x x^2 f(x) - \sum_x 2x\mu f(x) + \sum_x \mu^2 f(x)$$

$$= \sum_x x^2 f(x) - 2\mu \sum_x x f(x) + \mu^2 \sum_x f(x).$$

But $\sum_x x^2 f(x) = E(X^2)^\dagger$, $\sum_x x f(x) = \mu$, and $\sum_x f(x) = 1$. Therefore,

$$\text{Var}(X) = \sigma^2 = E(X^2) - 2\mu^2 + \mu^2,$$

or

$$\boxed{\begin{aligned} \text{Var}(X) = \sigma^2 &= E(X^2) - \mu^2 \\ &= \sum_x x^2 f(x) - \mu^2. \end{aligned}} \qquad (3)$$

This formula for variance is quite useful since it usually simplifies computations.

EXAMPLE 4

An urn contains ten marbles, each of which shows a number. Five marbles show 1, two show 2, and three show 3. A marble is drawn at random. If X is the number that shows, determine μ, Var(X), and σ.

The sample space consists of ten equally likely outcomes (the marbles). The values that X can assume are 1, 2, and 3. The events $X = 1$, $X = 2$, and $X = 3$ contain 5, 2, and 3 sample points, respectively. Thus, if f is the probability function for X,

$$f(1) = P(X = 1) = \tfrac{5}{10} = \tfrac{1}{2},$$

$$f(2) = P(X = 2) = \tfrac{2}{10} = \tfrac{1}{5},$$

$$f(3) = P(X = 3) = \tfrac{3}{10}.$$

$\dagger$ This is a consequence of a previous footnote.

Calculating the mean μ gives

$$\mu = \sum_x xf(x) = 1 \cdot f(1) + 2 \cdot f(2) + 3 \cdot f(3)$$

$$= 1 \cdot \tfrac{5}{10} + 2 \cdot \tfrac{2}{10} + 3 \cdot \tfrac{3}{10} = \tfrac{18}{10} = \tfrac{9}{5}.$$

To find Var(X), either Eq. (2) or Eq. (3) can be used. Both will be used here so that we can compare the arithmetical computations involved. By Eq. (2),

$$\text{Var}(X) = \sum_x (x - \mu)^2 f(x)$$

$$= \left(1 - \frac{9}{5}\right)^2 f(1) + \left(2 - \frac{9}{5}\right)^2 f(2) + \left(3 - \frac{9}{5}\right)^2 f(3)$$

$$= \left(-\frac{4}{5}\right)^2 \cdot \frac{5}{10} + \left(\frac{1}{5}\right)^2 \cdot \frac{2}{10} + \left(\frac{6}{5}\right)^2 \cdot \frac{3}{10}$$

$$= \frac{16}{25} \cdot \frac{5}{10} + \frac{1}{25} \cdot \frac{2}{10} + \frac{36}{25} \cdot \frac{3}{10}$$

$$= \frac{80 + 2 + 108}{250} = \frac{190}{250} = \frac{19}{25}.$$

By Eq. (3),

$$\text{Var}(X) = \sum_x x^2 f(x) - \mu^2$$

$$= 1^2 \cdot f(1) + 2^2 \cdot f(2) + 3^2 \cdot f(3) - \left(\frac{9}{5}\right)^2$$

$$= 1 \cdot \frac{5}{10} + 4 \cdot \frac{2}{10} + 9 \cdot \frac{3}{10} - \frac{81}{25}$$

$$= \frac{5 + 8 + 27}{10} - \frac{81}{25} = \frac{40}{10} - \frac{81}{25}$$

$$= 4 - \frac{81}{25} = \frac{19}{25}.$$

Notice that Eq. (2) involves $(x - \mu)^2$, but Eq. (3) involves x^2. Because of this, it is easier to compute variance by Eq. (3) than by Eq. (2).

Since $\sigma^2 = \text{Var}(X) = \tfrac{19}{25}$, the standard deviation σ is

$$\sigma = \sqrt{\text{Var}(X)} = \sqrt{\frac{19}{25}} = \frac{\sqrt{19}}{5}.$$

EXERCISE 13-3

*In Problems **1–4**, the distribution of the random variable X is given. Determine μ, Var(X), and σ. In Problem 1, construct the probability histogram. In Problem 2, graph the distribution.*

1. $f(0) = .1, f(1) = .4, f(2) = .2, f(3) = .3.$

2. $f(4) = .4, f(5) = .6.$

3.

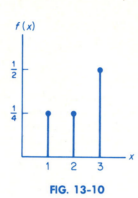

FIG. 13-10

4.

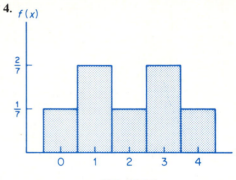

FIG. 13-11

*In Problems **5–8**, determine E(X), σ², and σ for the random variable X.*

5. Three fair coins are tossed. Let X be the number of heads that occur.

6. An urn contains six marbles, each of which shows a number. Four marbles show a 1 and two show a 2. A marble is randomly selected and the number that shows, X, is observed.

7. From a group of two women and three men, two persons are selected at random to form a committee. Let X be the number of women on the committee.

8. An urn contains two red and three green marbles. Two marbles are randomly drawn in succession with replacement and the number of red marbles, X, is observed.

9. A landscaper earns $200 per day when working and loses $30 per day when not working. If the probability of working on any day is $\frac{4}{7}$, find the landscaper's expected daily earnings.

10. A fast-food chain estimates that if it opens a restaurant in a shopping center, the probability that the restaurant is successful is .65. To the chain, a successful restaurant earns an annual profit of $75,000; one that is not loses $20,000. What is the expected gain to the chain if it opens a restaurant in a shopping center?

11. An insurance company offers a hospitalization policy to individuals in a certain group. For a 1-year period, the company will pay $100 per day, to a maximum of 5 days, for each day the policyholder is hospitalized. The company estimates that the probability that any person in this group is hospitalized for exactly one day is .001;

for exactly two days, .002; for exactly three days, .003; for exactly four days, .004; and for 5 or more days, .008. Find the expected gain per policy to the company if the annual premium is $10.

12. Table 13-2 gives the probability that X units of a company's product are demanded weekly. Determine the expected weekly demand.

Table 13-2

x	0	1	2	3	4	5
$P(X = x)$	.05	.20	.40	.24	.10	.01

13. In Example 3, if the company wants an expected gain of $50 per policy, determine the annual premium.

14. In the game of roulette, there is a wheel with 37 slots numbered with the integers from 0 to 36, inclusive. A player bets $1 (for example) and chooses a number. The wheel is spun and a ball rolls on the wheel. If the ball lands in the slot showing the chosen number, the player receives the $1 bet plus $35. Otherwise, the player loses the $1 bet. Assume that all numbers are equally likely and determine the expected gain or loss per play.

15. Suppose that you pay $1.25 to play a game in which two fair coins are tossed. You receive the number of dollars equal to the number of heads that occur. What is your expected gain (or loss) on each play? The game is said to be *fair* to you when your expected gain is $0. What should you pay to play if this is to be a fair game?

13-4 THE BINOMIAL DISTRIBUTION

Later in this section you will see that the terms in the expansion of a power of a binomial are useful in describing the probability distributions of certain random variables. It is worthwhile, therefore, to first discuss the *binomial theorem*, which is a formula for expanding $(a + b)^n$. The notation in this formula involves *factorials*.

DEFINITION. *If n is a positive integer, then **n factorial**, written n!, is defined by*

$$n! = n \cdot (n - 1) \cdot (n - 2) \cdots 3 \cdot 2 \cdot 1.$$

If n = 0, then 0! = 1.

Thus if n is a positive integer, $n!$ is simply the product of all integers from n to 1. For example,

$$3! = 3 \cdot 2 \cdot 1 = 6,$$
$$2!3! = (2 \cdot 1)(3 \cdot 2 \cdot 1) = 12,$$
$$6! = 6 \cdot 5 \cdot 4 \cdot 3 \cdot 2 \cdot 1 = 720.$$

Note that $6! = 6 \cdot 5!$. In general,

$$\boxed{n! = n \cdot (n-1)!}$$

This property is useful for simplifying fractions involving factorials. For example, watch how we cancel factorials:

$$\frac{18!}{17!} = \frac{18 \cdot 17!}{17!} = 18.$$

Keep in mind that $0! = 1$, not 0, Thus,

$$\frac{4!}{0!4!} = \frac{4!}{1(4!)} = 1.$$

PITFALL. *Although* $\dfrac{6}{3} = 2$, $\dfrac{6!}{3!} \neq 2!$:

$$\frac{6!}{3!} = \frac{6 \cdot 5 \cdot 4 \cdot 3!}{3!} = 6 \cdot 5 \cdot 4 = 120.$$

We can now state the formula for expanding $(a+b)^n$.

BINOMIAL THEOREM

If n is a positive integer, then

$$(a+b)^n = \binom{n}{0}a^n + \binom{n}{1}a^{n-1}b + \binom{n}{2}a^{n-2}b^2$$

$$+ \cdots + \binom{n}{n-1}ab^{n-1} + \binom{n}{n}b^n,$$

where the symbol $\binom{n}{r}$ is called a **binomial coefficient** and is defined by

$$\binom{n}{r} = \frac{n!}{r!\,(n-r)!}.$$

An example of a binomial coefficient is

$$\binom{5}{3} = \frac{5!}{3!\,(5-3)!} = \frac{5!}{3!2!} = \frac{5 \cdot 4 \cdot 3!}{3!\,(1 \cdot 2)} = \frac{5 \cdot 4}{1 \cdot 2} = 10.$$

EXAMPLE 1

Use the binomial theorem to expand $(q + p)^3$.

Here $n = 3$, $a = q$, and $b = p$.

$$(q + p)^3 = \binom{3}{0}q^3 + \binom{3}{1}q^{3-1}p + \binom{3}{2}q^{3-2}p^2 + \binom{3}{3}p^3$$

$$= \frac{3!}{0!3!}q^3 + \frac{3!}{1!2!}q^2p + \frac{3!}{2!1!}qp^2 + \frac{3!}{3!0!}p^3.$$

$$(q + p)^3 = q^3 + 3q^2p + 3qp^2 + p^3.$$

Before applying the binomial theorem to a distribution, we shall consider probabilities associated with repetitions, or **trials**, of an experiment such that the trials are *independent* of each other. By **independent trials** we mean that the outcome of any single trial does not effect the outcome of any other trial. An example of two independent trials of an experiment is tossing a coin twice in succession. Suppose the coin is biased so that the probability a head occurs on any toss is $\frac{1}{3}$. Thus, the probability that a tail occurs on any toss is $1 - \frac{1}{3} = \frac{2}{3}$. Let us determine $P(HT)$. If tossing the coin twice is performed n times, where n is a large number, we should expect that heads occur approximately $\frac{1}{3}n$ times on the first toss. Because what occurs on the first toss in no way influences what occurs on the second, we should expect that a tail on the second toss occurs approximately $\frac{2}{3}$ of these $\frac{1}{3}n$ times. Thus in n performances, the number of times we get HT is about $\frac{2}{3}(\frac{1}{3}n)$. Therefore, the relative frequency of HT is

$$\frac{\frac{2}{3}(\frac{1}{3}n)}{n} = \frac{2}{3} \cdot \frac{1}{3}.$$

Since n is large, we conclude that

$$P(HT) = \frac{1}{3} \cdot \frac{2}{3} = P(H) \cdot P(T),$$

where $P(H)$ is the probability of a head on the first toss and $P(T)$ is the probability of a tail on the second toss. Notice that $P(HT)$ is the product of two probabilities, $P(H)$ and $P(T)$. This result can be generalized to n independent trials of an experiment.

MULTIPLICATION RULE FOR n INDEPENDENT TRIALS

Let there be n independent trials of an experiment and let a_i be a possible outcome of the ith trial. Then

$$P(a_1, a_2, \ldots, a_n) = P(a_1)P(a_2) \cdots P(a_n).$$

EXAMPLE 2

A fair die is tossed three times. Find the probability of the event of getting 1 *on the first toss*, 3 *on the second toss, and* 5 *on the third toss.*

Here we have three independent trials of tossing a coin. The probability of a particular number showing on any toss is $\frac{1}{6}$. By the multiplication rule,

$$P(1,3,5) = P(1) \cdot P(3) \cdot P(5) = \tfrac{1}{6} \cdot \tfrac{1}{6} \cdot \tfrac{1}{6} = \tfrac{1}{216}.$$

We shall now consider independent trials of an experiment such that each trial has only *two* possible outcomes. For example, tossing a coin has two possible outcomes, H or T. For a particular coin, suppose that $P(H) = p$ and $P(T) = q = 1 - p$. Let the coin be tossed twice. The sample space is

$$\{TT, TH, HT, HH\}.$$

Using the multiplication rule, we can determine the probabilities of the outcomes.

$$P(TT) = qq = q^2,$$

$$P(TH) = qp,$$

$$P(HT) = pq,$$

$$P(HH) = pp = p^2.$$

If X is the number of heads that occur on the two trials, then $X = 0$, 1, or 2. The distribution f for X is given by

$$f(0) = P(X = 0) = P(TT) = q^2,$$

$$f(1) = P(X = 1) = P(TH) + P(HT) = qp + pq = 2qp,$$

$$f(2) = P(X = 2) = P(HH) = p^2.$$

We can generate the sample space and distribution of X in another, but interesting, way. If we think of T and H as numbers, then the terms in the expansion of the square of the binomial $T + H$ can be interpreted as the points in the sample space:

$$(T + H)^2 = (T + H)(T + H) = T(T + H) + H(T + H)$$
$$= TT + TH + HT + HH. \tag{1}$$

$$(T + H)^2 = T^2 + 2TH + H^2. \tag{2}$$

Notice that the terms in (1) indicate all possible outcomes. The terms TH and HT each refer to one head occurring. Thus in (2), the coefficient 2 of 2TH indicates that one head can occur *two* ways (TH and HT). Both TH and HT have the same probability of occurring: qp (or pq). Thus $P(X = 1) = 2(qp)$. We can obtain this probability from the term 2TH in (2) by replacing T by q and H by p. Similarly, in (2) the coefficient 1 of T^2 indicates that there is only *one* way that no heads can occur. Replacing T by q here gives $P(X = 0)$, or q^2. Also, the coefficient 1 of H^2 indicates that there is only *one* way that two heads can occur. Replacing H by p gives the corresponding probability $P(X = 2)$, which is p^2. In summary, replacing T by q and H by p in (2) completely describes the distribution of X. That is, the distribution of X (number of heads that occur) is given by the terms in the square of the binomial $q + p$.

This pattern holds for any number of trials. For example, if the coin is tossed three times, then we get the distribution for X from the terms in the *cube* of $q + p$:

$$(q + p)^3 = q^3 + 3q^2p + 3qp^2 + p^3.$$

If we think of q's as T's and p's as H's, then q^3 is the probability of three tails, that is, $P(X = 0)$. Similarly, $3q^2p$ is the probability of two tails and one head, that is, $P(X = 1)$. Thus,

$$P(X = 0) = q^3,$$

$$P(X = 1) = 3q^2p,$$

$$P(X = 2) = 3qp^2,$$

$$P(X = 3) = p^3.$$

We shall now generalize our results so that they apply to n independent trials of an experiment in which each trial has only two possible outcomes—call them *success* and *failure*—and the probability of success in each trial remains the same. Such trials are called **Bernoulli trials** and the experiment is called a **binomial experiment**.

BINOMIAL DISTRIBUTION

If X is the number of successes in n independent trials of a binomial experiment with probability p of success and q of failure on any trial, then the distribution f for X is given by

$$f(x) = P(X = x) = \binom{n}{x} p^x q^{n-x},$$

where x is an integer such that $0 \leqslant x \leqslant n$, $q = 1 - p$, and $\binom{n}{x}$ is the binomial coefficient $\dfrac{n!}{x!(n-x)!}$. Any random variable with this distribution is called a **binomial random variable** and is said to have a **binomial distribution**. The mean and standard deviation of X are given, respectively, by

$$\mu = np, \qquad \sigma = \sqrt{npq}.$$

EXAMPLE 3

Suppose X is a binomial random variable with $n = 4$ and $p = \frac{1}{3}$. Find the distribution for X.

Here $q = 1 - p = 1 - \frac{1}{3} = \frac{2}{3}$.

$$P(X = x) = \binom{n}{x} p^x q^{n-x}, \qquad x = 0, 1, 2, 3, 4.$$

Thus,

$$P(X = 0) = \binom{4}{0}\left(\frac{1}{3}\right)^0\left(\frac{2}{3}\right)^4 = \frac{4!}{0!4!}(1)\left(\frac{16}{81}\right) = 1(1)\left(\frac{16}{81}\right) = \frac{16}{81}.$$

$$P(X = 1) = \binom{4}{1}\left(\frac{1}{3}\right)^1\left(\frac{2}{3}\right)^3 = \frac{4!}{1!3!}\left(\frac{1}{3}\right)\left(\frac{8}{27}\right) = 4\left(\frac{1}{3}\right)\left(\frac{8}{27}\right) = \frac{32}{81}.$$

$$P(X = 2) = \binom{4}{2}\left(\frac{1}{3}\right)^2\left(\frac{2}{3}\right)^2 = \frac{4!}{2!2!}\left(\frac{1}{9}\right)\left(\frac{4}{9}\right) = 6\left(\frac{1}{9}\right)\left(\frac{4}{9}\right) = \frac{8}{27}.$$

$$P(X = 3) = \binom{4}{3}\left(\frac{1}{3}\right)^3\left(\frac{2}{3}\right)^1 = \frac{4!}{3!1!}\left(\frac{1}{27}\right)\left(\frac{2}{3}\right) = 4\left(\frac{1}{27}\right)\left(\frac{2}{3}\right) = \frac{8}{81}.$$

$$P(X = 4) = \binom{4}{4}\left(\frac{1}{3}\right)^4\left(\frac{2}{3}\right)^0 = \frac{4!}{4!0!}\left(\frac{1}{81}\right)(1) = 1\left(\frac{1}{81}\right)(1) = \frac{1}{81}.$$

The probability histogram for X is given in Fig. 13-12. Note that the mean μ for X is $np = 4\left(\frac{1}{3}\right) = \frac{4}{3}$, and the standard deviation σ is

$$\sqrt{npq} = \sqrt{4\left(\frac{1}{3}\right)\left(\frac{2}{3}\right)} = \sqrt{\frac{8}{9}} = \frac{2\sqrt{2}}{3}.$$

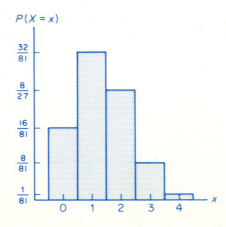

FIG. 13-12

EXAMPLE 4

A fair coin is tossed eight times. Find the probability of getting at least two heads.

If X is the number of heads that occur, then X has a binomial distribution with $n = 8, p = \frac{1}{2}$ and $q = \frac{1}{2}$. To simplify our work, we use the fact that

$$P(X \geqslant 2) = 1 - P(X < 2).$$

Now,

$$P(X < 2) = P(X = 0) + P(X = 1)$$

$$= \binom{8}{0}\left(\frac{1}{2}\right)^0\left(\frac{1}{2}\right)^8 + \binom{8}{1}\left(\frac{1}{2}\right)^1\left(\frac{1}{2}\right)^7$$

$$= \frac{8!}{0!8!}(1)\left(\frac{1}{256}\right) + \frac{8!}{1!7!}\left(\frac{1}{2}\right)\left(\frac{1}{128}\right)$$

$$= 1(1)\left(\frac{1}{256}\right) + 8\left(\frac{1}{2}\right)\left(\frac{1}{128}\right)$$

$$= \frac{9}{256}.$$

Thus,

$$P(X \geqslant 2) = 1 - \frac{9}{256} = \frac{247}{256}.$$

A probability histogram for X is given in Fig. 13-13.

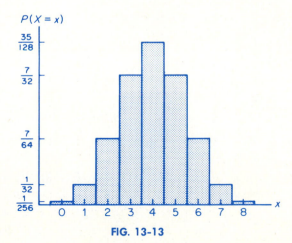

FIG. 13-13

EXAMPLE 5

For a particular group of individuals, 20 percent of their income tax returns are audited each year. Of five randomly chosen individuals, what is the probability that exactly two will have their returns audited?

We shall consider this to be a binomial experiment with 5 trials (selecting an individual). Actually, the experiment is not truly binomial because selecting an individual from this group affects the probability that another individual's return will be audited. For example, if there are 5000 individuals, then 20 percent or 1000, will be audited. The probability that the first individual selected will be audited is $\frac{1000}{5000}$. If that event occurs, the probability that the second individual selected will be audited is $\frac{999}{4999}$. Thus the trials are not independent. However, we assume that the number of individuals is large, so that for practical purposes the probability of auditing an individual remains constant from trial to trial.

For each trial, the two outcomes are *being audited* or *not being audited*. Here we shall define a success as being audited. Letting X be the number of returns audited, $p = .2$, and $q = 1 - .2 = .8$, we have

$$P(X = 2) = \binom{5}{2}(.2)^2(.8)^3 = \frac{5!}{2!3!}(.04)(.512)$$

$$= 10(.04)(.512) = .2048.$$

EXERCISE 13-4

*In Problems **1–8**, simplify the expressions.*

1. 4!0!.

2. $\dfrac{7!}{3!}$.

3. $\dfrac{6!}{2!4!}$.

4. $\binom{3}{2}$.

5. $\binom{5}{2}$.

6. $\binom{4}{4}$.

7. $\binom{4}{0}$.

8. $\binom{n}{n-1}$.

*In Problems **9** and **10**, determine the distribution f for the binomial random variable X if the number of trials is n and the probability of success on any trial is p. Also find μ and σ.*

9. $n = 2$, $p = \frac{1}{4}$.

10. $n = 3$, $p = \frac{1}{2}$.

11. A biased coin is tossed three times in succession. The probability of heads on any toss is $\frac{1}{4}$. Find the probability that (a) exactly 2 heads occur, and (b) 2 or 3 heads occur.

12. A fair coin is tossed ten times. What is the probability that exactly eight heads occur?

13. If a family has five children, find the probability that at least two are girls. (Assume the probability that a child is a girl is $\frac{1}{2}$.)

14. From a standard deck of 52 playing cards, 3 cards are randomly selected in succession with replacement. Determine the probability that exactly 2 cards are aces.

15. A manufacturer produces electrical switches, of which 2 percent are defective. From a production run of 50,000 switches, 4 are randomly selected and each is tested.

Determine, to three decimal places, the probability that the sample contains exactly 2 defective switches. Assume that the four trials are independent and that the number of defective switches in the sample has a binomial distribution.

13-5 CONTINUOUS RANDOM VARIABLES

In previous sections the random variables were primarily discrete. Now we shall concern ourselves with *continuous* random variables. Recall that a random variable is continuous if it can assume any value in some interval or intervals. A continuous random variable usually represents data that are measured, such as heights, weights, distances, and time periods.

For example, the number of hours of life of a pocket-calculator battery is a continuous random variable, X. If the maximum possible life is 1000 hours, then X can assume any value in the interval $[0, 1000]$. In a practical sense, the likelihood that X will assume a single specified value, like 764.1238, is extremely remote. It is more meaningful to consider the likelihood of X lying within an *interval*, such as that between 764 and 765 (that is, $764 < X < 765$). In general, *with a continuous random variable our concern is the likelihood that it falls within an interval, and not that it assumes a particular value.*

As another example, consider an experiment in which a number X is randomly selected from the interval $[0, 2]$. Then X is a continuous random variable. What is the probability that X lies in the interval $[0, 1]$? Because we can loosely think of $[0, 1]$ as being "half" the interval $[0, 2]$, a reasonable (and correct) answer is $\frac{1}{2}$. Similarly, if we think of the interval $\left[0, \frac{1}{2}\right]$ as being one-fourth of $[0, 2]$, then $P\left(0 \leqslant X \leqslant \frac{1}{2}\right) = \frac{1}{4}$. Actually, each one of these probabilities is simply the length of the given interval divided by the length of $[0, 2]$. For example,

$$P\left(0 \leqslant X \leqslant \frac{1}{2}\right) = \frac{\text{length of } \left[0, \frac{1}{2}\right]}{\text{length of } [0, 2]} = \frac{\frac{1}{2}}{2} = \frac{1}{4}.$$

Let us now consider a similar experiment in which X denotes a number chosen at random from the interval $[0, 1]$. As you might expect, the probability that X will assume a value in any given interval within $[0, 1]$ is equal to the length of the given interval divided by the length of $[0, 1]$. Because $[0, 1]$ has length 1, we can simply say that the probability of X falling in an interval is the length of the interval. For example $P(.2 \leqslant X \leqslant .5) = .5 - .2 = .3$, and $P(.2 \leqslant X \leqslant .2001) = .0001$. Clearly, as the length of an interval approaches 0, the probability that X assumes a value in that interval approaches 0. Keeping this in mind, we can think of a single number like .2 as the limiting case of an interval as the length of the interval approaches 0 (think of $[.2, .2 + x]$ as $x \to 0$). Thus, $P(X = .2) = 0$. In general, *the probability that a continuous random variable X assumes a particular value is* 0. As a result, **the probability that X lies in some interval is not affected by whether or not either of the endpoints of the interval are included or excluded.** For example,

$$P(X \leqslant .4) = P(X < .4) + P(X = .4)$$
$$= P(X < .4) + 0$$
$$= P(X < .4).$$

Similarly, $P(.2 \leqslant X \leqslant .5) = P(.2 < X < .5)$.

We can geometrically represent the probabilities associated with a continuous random variable X. This is done by means of the graph of a function $y = f(x)$ such that the area under this graph (and above the x-axis) between the lines $x = a$ and $x = b$ represents the probability that X assumes a value between a and b (see Fig. 13-14). Since this area is given by the definite integral $\int_a^b f(x)\, dx$, we have

$$P(a \leqslant X \leqslant b) = \int_a^b f(x)\, dx.$$

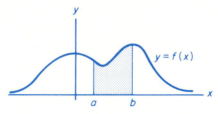

$P(a \leq X \leq b) = $ Area of Shaded Region

FIG. 13-14

We call the function f the *probability density function* for X (or simply the *density function* for X) and say that it defines the *distribution* of X. Because probabilities are always nonnegative, it is always true that $f(x) \geqslant 0$. Also, because the event $-\infty < X < \infty$ must occur, the total area under the density function curve must be 1. That is, $\int_{-\infty}^{\infty} f(x)\, dx = 1$. In summary we have the following.

DEFINITION.   *If X is a continuous random variable, then a function $y = f(x)$ is called a (**probability**) density function for X if and only if it has the following properties:*

  1. $f(x) \geqslant 0.$

  2. $\displaystyle\int_{-\infty}^{\infty} f(x)\, dx = 1.$

  3. $\displaystyle P(a \leqslant X \leqslant b) = \int_a^b f(x)\, dx.$

To illustrate a density function, we return to the previous experiment in which a number X is chosen at random from the interval $[0, 1]$. Recall that

$$P(a \leqslant X \leqslant b) = \text{length of } [a, b] = b - a. \tag{1}$$

We shall show that the function

$$f(x) = \begin{cases} 1, & \text{if } 0 \leqslant x \leqslant 1, \\ 0, & \text{otherwise,} \end{cases} \tag{2}$$

whose graph appears in Fig. 13-15(a), is a density function for X. First, since $f(x)$

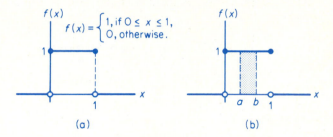

FIG. 13-15

is either 0 or 1, we have $f(x) \geqslant 0$. Next, since $f(x) = 0$ for x outside $[0, 1]$,

$$\int_{-\infty}^{\infty} f(x)\, dx = \int_{0}^{1} 1\, dx = x \Big|_{0}^{1} = 1.$$

Finally, to verify that $P(a \leqslant X \leqslant b) = \int_{a}^{b} f(x)\, dx$, we compute the area under the graph between $x = a$ and $x = b$ [Fig. 13-15(b)].

$$\int_{a}^{b} f(x)\, dx = \int_{a}^{b} 1\, dx = x \Big|_{a}^{b} = b - a,$$

which, as stated in (1), is $P(a \leqslant X \leqslant b)$.

The function in (2) is called the **uniform density function** over $[0, 1]$, and X is said to have a **uniform distribution**. The word *uniform* is meaningful in the sense that the graph of the density function is horizontal, or "flat," over $[0, 1]$. As a result, X is just as likely to assume a value in one interval within $[0, 1]$ as in another of equal length. A more general uniform distribution is given in Example 1.

EXAMPLE 1

The uniform density function over $[a, b]$ for the random variable X is given by

$$f(x) = \begin{cases} \dfrac{1}{b-a}, & \text{if } a \leqslant x \leqslant b, \\ 0, & \text{otherwise.} \end{cases}$$

See Fig. 13-16. Note that over $[a, b]$, the region under the graph is a rectangle with height

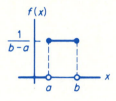

FIG. 13-16

$1/(b-a)$ and width $b-a$. Thus its area is $[1/(b-a)][b-a] = 1$; so $\int_{-\infty}^{\infty} f(x)\,dx = 1$, as must be the case for a density function. If $[c, d]$ is any interval within $[a, b]$, then

$$P(c \leqslant X \leqslant d) = \int_{c}^{d} f(x)\,dx = \int_{c}^{d} \frac{1}{b-a}\,dx$$

$$= \frac{x}{b-a}\bigg|_{c}^{d} = \frac{d-c}{b-a}.$$

For example, suppose X is uniformly distributed over the interval $[1, 4]$ and we need to find $P(2 < X < 3)$. Then $a = 1$, $b = 4$, $c = 2$, and $d = 3$. Thus,

$$P(2 < X < 3) = \frac{3-2}{4-1} = \frac{1}{3}.$$

EXAMPLE 2

The density function for a random variable X is given by

$$f(x) = \begin{cases} kx, & \text{if } 0 \leqslant x \leqslant 2, \\ 0, & \text{otherwise,} \end{cases}$$

where k is a constant.

a. *Find k.*

Since $\int_{-\infty}^{\infty} f(x)\,dx$ must be 1 and $f(x) = 0$ outside $[0, 2]$, we have

$$\int_{-\infty}^{\infty} f(x)\,dx = \int_{0}^{2} kx\,dx = \frac{kx^2}{2}\bigg|_{0}^{2} = 2k = 1.$$

Thus $k = \frac{1}{2}$, so $f(x) = \frac{1}{2}x$ on $[0, 2]$.

b. *Find* $P\left(\frac{1}{2} < X < 1\right)$.

$$P\left(\frac{1}{2} < X < 1\right) = \int_{1/2}^{1} \frac{1}{2}x \, dx = \frac{x^2}{4}\Big|_{1/2}^{1} = \frac{1}{4} - \frac{1}{16} = \frac{3}{16}.$$

c. *Find* $P(X < 1)$.

Since $f(x) = 0$ for $x < 0$, we need only compute the area under the density function between 0 and 1.

$$P(X < 1) = \int_{0}^{1} \frac{1}{2}x \, dx = \frac{x^2}{4}\Big|_{0}^{1} = \frac{1}{4}.$$

EXAMPLE 3

The **exponential density function** *is defined by*

$$f(x) = \begin{cases} ke^{-kx}, & \text{if } x \geqslant 0, \\ 0, & \text{if } x < 0, \end{cases}$$

where k is a positive constant, called a **parameter**, *whose value depends on the experiment under consideration. If X is a random variable with this density function, then X is said to have an* **exponential distribution**. *Let $k = 1$. Then $f(x) = e^{-x}$ for $x \geqslant 0$, and $f(x) = 0$ for $x < 0$ (Fig. 13-17).*

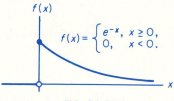

FIG. 13-17

a. *Find* $P(2 < X < 3)$.

$$P(2 < X < 3) = \int_{2}^{3} e^{-x} \, dx = -e^{-x}\Big|_{2}^{3}$$

$$= -e^{-3} - (-e^{-2}) = e^{-2} - e^{-3}$$

$$= .13534 - .04979 = .086 \quad \text{(approximately)}.$$

b. *Find* $P(X > 4)$.

$$P(X > 4) = \int_4^\infty e^{-x}\,dx = \lim_{r \to \infty} \int_4^r e^{-x}\,dx$$

$$= \lim_{r \to \infty} -e^{-x}\Big|_4^r = \lim_{r \to \infty} (-e^{-r} + e^{-4})$$

$$= \lim_{r \to \infty} \left(-\frac{1}{e^r} + e^{-4}\right) = 0 + .01832$$

$$= .018 \quad (\text{approximately}).$$

Alternatively, we can avoid an improper integral because

$$P(X > 4) = 1 - P(X \leqslant 4) = 1 - \int_0^4 e^{-x}\,dx.$$

The **cumulative distribution function** F for the continuous random variable X with density function f is defined by

$$F(x) = P(X \leqslant x) = \int_{-\infty}^x f(t)\,dt.$$

For example, $F(2)$ represents the entire area under the density curve that is to the left of the line $x = 2$ (Fig. 13-18). Under certain conditions of continuity, it can

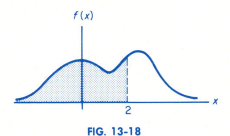

FIG. 13-18

be shown that

$$F'(x) = f(x).$$

That is, the derivative of the cumulative distribution function is the density function. Thus F is an antiderivative of f, so

$$P(a < X < b) = \int_a^b f(x)\,dx = F(b) - F(a). \tag{3}$$

This means that the area under the density curve between a and b (Fig. 13-19) is simply the area to the left of b minus the area to the left of a.

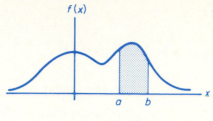

FIG. 13-19

EXAMPLE 4

Suppose X is a random variable with density function given by

$$f(x) = \begin{cases} \frac{1}{2}x, & \text{if } 0 \leqslant x \leqslant 2, \\ 0, & \text{otherwise}, \end{cases}$$

as shown in Fig. 13-20.

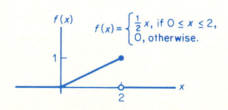

FIG. 13-20

a. *Find and sketch the cumulative distribution function.*

Because $f(x) = 0$ if $x < 0$, the area under the density curve to the left of $x = 0$ is 0. Thus $F(x) = 0$ if $x < 0$. If $0 \leqslant x \leqslant 2$, then

$$F(x) = \int_{-\infty}^{x} f(t)\, dt = \int_{0}^{x} \frac{1}{2} t\, dt = \frac{t^2}{4}\Big|_{0}^{x} = \frac{x^2}{4}.$$

Since f is a density function and $f(x) = 0$ for $x < 0$ and for $x > 2$, the area under the density curve from $x = 0$ to $x = 2$ is 1. Thus if $x > 2$, the area to the left of x is 1, so

$F(x) = 1$. Hence the cumulative distribution function is

$$F(x) = \begin{cases} 0, & \text{if } x < 0, \\ \dfrac{x^2}{4}, & \text{if } 0 \leqslant x \leqslant 2, \\ 1, & \text{if } x > 2, \end{cases}$$

which is shown in Fig. 13-21.

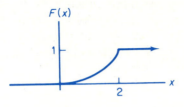

FIG. 13-21

b. *Find $P(X < 1)$ and $P(1 < X < 1.1)$.*

Using the results of (a), we have

$$P(X < 1) = F(1) = \frac{1^2}{4} = \frac{1}{4}.$$

From (3),

$$P(1 < X < 1.1) = F(1.1) - F(1) = \frac{1.1^2}{4} - \frac{1}{4} = .0525.$$

For a random variable X with density function f, the mean μ [or expectation $E(X)$] is given by

$$\mu = E(X) = \int_{-\infty}^{\infty} x f(x)\, dx,$$

and the variance σ^2 [or Var(X)] is given by

$$\sigma^2 = \text{Var}(X) = \int_{-\infty}^{\infty} (x - \mu)^2 f(x)\, dx.$$

You may have noticed that these formulas are similar to the corresponding ones for a discrete random variable. It can be shown that an alternative formula for variance is

$$\sigma^2 = \text{Var}(X) = \int_{-\infty}^{\infty} x^2 f(x)\, dx - \mu^2.$$

The standard deviation is σ, where

$$\sigma = \sqrt{\text{Var}(X)}.$$

For example, if X is exponentially distributed (see Example 3), it can be shown that $\mu = 1/k$ and $\sigma = 1/k$. As with a discrete random variable, the standard deviation of a continuous random variable X is small if X is likely to assume values close to the mean but unlikely to assume values far from the mean. It is large if the reverse is true.

EXAMPLE 5

If X is a random variable with density function given by

$$f(x) = \begin{cases} \frac{1}{2}x, & \text{if } 0 \le x \le 2, \\ 0, & \text{otherwise,} \end{cases}$$

find its mean and standard deviation.

The mean is given by

$$\mu = \int_{-\infty}^{\infty} xf(x)\, dx = \int_0^2 x \cdot \frac{1}{2}x\, dx = \frac{x^3}{6}\bigg|_0^2 = \frac{4}{3}.$$

By the alternative formula for variance, we have

$$\sigma^2 = \int_{-\infty}^{\infty} x^2 f(x)\, dx - \mu^2 = \int_0^2 x^2 \cdot \frac{1}{2}x\, dx - \left(\frac{4}{3}\right)^2$$

$$= \frac{x^4}{8}\bigg|_0^2 - \frac{16}{9} = 2 - \frac{16}{9} = \frac{2}{9}.$$

Thus the standard deviation is

$$\sigma = \sqrt{\frac{2}{9}} = \frac{\sqrt{2}}{3}.$$

We conclude this section by emphasizing that a density function for a continuous random variable must not be confused with a probability distribution function for a discrete random variable. Evaluating such a probability distribution function at a *point* gives a probability. But evaluating a density function at a point does not. Instead, the *area* under the density function curve over an *interval* is interpreted as a probability. That is, probabilities associated with a continuous random variable are given by integrals.

EXERCISE 13-5

1. Suppose X is a continuous random variable with density function given by

$$f(x) = \begin{cases} \frac{1}{6}(x+1), & \text{if } 1 < x < 3, \\ 0, & \text{otherwise.} \end{cases}$$

   a. Find $P(1 < X < 2)$.
   b. Find $P(X < 2.5)$.
   c. Find $P(X \geq \frac{3}{2})$.
   d. Find c such that $P(X < c) = \frac{1}{2}$. Give your answer in radical form.

2. Suppose X is a continuous random variable with density function given by

$$f(x) = \begin{cases} \dfrac{1000}{x^2}, & \text{if } x > 1000, \\ 0, & \text{otherwise.} \end{cases}$$

   a. Find $P(3000 < X < 4000)$.
   b. Find $P(X > 2000)$.

3. Suppose X is a continuous random variable that is uniformly distributed on $[1, 4]$.
   a. What is the formula of the density function for X? Sketch its graph.
   b. Find $P(2 < X < 3)$.   c. Find $P(0 < X < 1)$.
   d. Find $P(X \leq 3.5)$.   e. Find $P(X > 2)$.
   f. Find $P(X = 3)$.   g. Find $P(X < 5)$.
   h. Find μ.   i. Find σ.
   j. Find the cumulative distribution function F and sketch its graph. Use F to find $P(X < 2)$ and $P(1 < X < 3)$.

4. Suppose X is a continuous random variable that is uniformly distributed on $[0, 5]$.
   a. What is the formula of the density function for X? Sketch its graph.
   b. Find $P(1 < X < 3)$.   c. Find $P(4.5 \leq X < 5)$.
   d. Find $P(X = 4)$.   e. Find $P(X > 1)$.
   f. Find $P(X < 5)$.   g. Find $P(X > 5)$.
   h. Find μ.   i. Find σ.
   j. Find the cumulative distribution function F and sketch its graph. Use F to find $P(1 < X < 3.5)$.

5. Suppose X is uniformly distributed on $[a, b]$.
   a. What is the density function for X?
   b. Find μ.
   c. Find σ^2 and σ.

6. Suppose X is a continuous random variable with density function given by

$$f(x) = \begin{cases} k, & \text{if } a \leqslant x \leqslant b, \\ 0, & \text{otherwise.} \end{cases}$$

   a. Show that $k = \dfrac{1}{b - a}$ and thus X is uniformly distributed.
   b. Find the cumulative distribution function F.

7. Suppose the random variable X is exponentially distributed with $k = 2$.
   a. Find $P(1 < X < 3)$.   b. Find $P(X < 2)$.
   c. Find $P(X > 2.5)$.   d. Find $P(\mu - \sigma < X < \mu + \sigma)$.
   e. Show that the area under the density function is 1.

8. Suppose the random variable X is exponentially distributed with $k = .5$.
   a. Find $P(X > 4)$.   b. Find $P(.5 < X < 2.6)$.
   c. Find $P(X < 5)$.   d. Find $P(X = 4)$.
   e. Find c such that $P(0 < X < c) = \frac{1}{2}$.

9. The density function for a random variable X is given by

$$f(x) = \begin{cases} kx, & \text{if } 0 \leqslant x \leqslant 4, \\ 0, & \text{otherwise.} \end{cases}$$

   a. Find k.   b. Find $P(2 < X < 3)$.
   c. Find $P(X > 2.5)$.   d. Find $P(X > 0)$.
   e. Find μ.   f. Find σ.
   g. Find c such that $P(X < c) = \frac{1}{2}$.   h. Find $P(3 < X < 5)$.

10. The density function for a random variable X is given by

$$f(x) = \begin{cases} \frac{1}{2}x + k, & \text{if } 2 \leqslant x \leqslant 4, \\ 0, & \text{otherwise.} \end{cases}$$

   a. Find k.   b. Find $P(X \geqslant 2.5)$.
   c. Find μ.   d. Find $P(2 < X < \mu)$.

11. At a bus stop, the time X (in minutes) that a randomly arriving person must wait for a bus is uniformly distributed with density function given by $f(x) = \frac{1}{10}$ if $0 \leqslant x \leqslant 10$ and $f(x) = 0$ otherwise. What is the probability that such a person must wait at most 7 minutes? What is the average time that a person must wait?

12. An automatic soft-drink dispenser at a fast-food restaurant dispenses X ounces of cola in a 12-ounce drink. Suppose X is uniformly distributed over $[11.92, 12.08]$.

What is the probability that less than 12 ounces will be dispensed? What is the probability that exactly 12 ounces will be dispensed? What is the average amount dispensed?

13. At a particular hospital, the length of time X (in hours) between successive arrivals at the emergency room is exponentially distributed with $k = 3$. What is the probability that more than 1 hour passes without an arrival?

14. The length of life X (in years) of an electronic component has an exponential distribution with $k = \frac{1}{5}$. What is the probability that such a component will fail within 4 years of use? What is the probability that it will last more than 6 years?

13-6 THE NORMAL DISTRIBUTION

Quite often, measured data in nature—such as heights of individuals in a population—are represented by a random variable whose density function may be approximated by the bell-shaped curve in Fig. 13-22. The curve extends indefinitely to the right and left and never touches the x-axis, although it appears to do so at points where the ordinates are close to zero. This curve, called the **normal curve**, is the graph of the most important of all density functions, the *normal density function*.

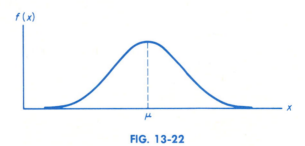

FIG. 13-22

DEFINITION. *A continuous random variable X is a **normal random variable** or has a **normal** (or **Gaussian**†) **distribution** if its density function is given by*

$$f(x) = \frac{1}{\sigma\sqrt{2\pi}} e^{-(1/2)[(x-\mu)/\sigma]^2}, \qquad -\infty < x < \infty,$$

———————

†After the German mathematician Carl Friedrich Gauss (1777–1855).

*called the **normal density function**. The parameters μ and σ are the mean and standard deviation of X, respectively.*

Observe in Fig. 13-22 that $f(x) \to 0$ as $x \to \pm\infty$. That is, the normal curve has the x-axis as a horizontal asymptote. Also note that the normal curve is symmetric about the vertical line $x = \mu$. That is, the height of a point on the curve d units to the right of $x = \mu$ is the same as the height of the point on the curve that is d units to the left of $x = \mu$. Because of this symmetry and the fact that the area under the normal curve is 1, the area to the right (left) of the mean must be $\frac{1}{2}$.

Each choice of values for μ and σ determines a different normal curve. The value of μ determines where the curve is "centered," and σ determines how "spread out" the curve is. The smaller the value of σ, the less spread out is the area near μ. For example, Fig. 13-23 shows normal curves C_1, C_2, and C_3, where C_1 has mean μ_1 and standard deviation σ_1, C_2 has mean μ_2, and so on. Here, C_1

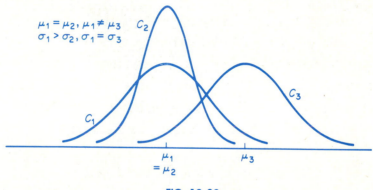

FIG. 13-23

and C_2 have the same mean but different standard deviations: $\sigma_1 > \sigma_2$. C_1 and C_3 have the same standard deviation but different means: $\mu_1 < \mu_3$. Curves C_2 and C_3 have different means and different standard deviations.

The standard deviation plays a significant role in describing probabilities associated with a normal random variable X. More precisely, the probability that X will lie within one standard deviation of the mean is approximately .68:

$$P(\mu - \sigma < X < \mu + \sigma) = .68.$$

In other words, approximately 68 percent of the area under a normal curve is within one standard deviation of the mean (Fig. 13-24). Between $\mu \pm 2\sigma$ is about

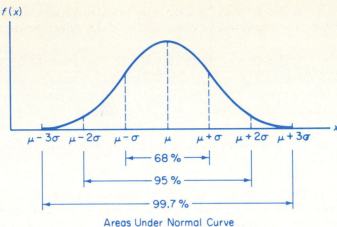

Areas Under Normal Curve

FIG. 13-24

95 percent of the area, and between $\mu \pm 3\sigma$ is about 99.7 percent:

$$P(\mu - 2\sigma < X < \mu + 2\sigma) = .95,$$
$$P(\mu - 3\sigma < X < \mu + 3\sigma) = .997.$$

Thus it is highly likely that X will lie within three standard deviations of the mean.

EXAMPLE 1

Let X be a random variable whose values are the test scores obtained on a nationwide test given to high school seniors. Suppose, for modeling purposes, X is normally distributed with mean 600 and standard deviation 90. Then the probability that X lies within $2\sigma = 2(90) = 180$ points of 600 is .95. In other words, 95 percent of the scores lie between 420 and 780. Similarly, 99.7 percent of the scores are within $3\sigma = 3(90) = 270$ points of 600—that is, between 330 and 870.

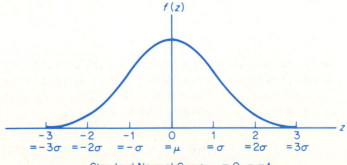

Standard Normal Curve: $\mu = 0, \sigma = 1$

FIG. 13-25

If Z is a normally distributed random variable with $\mu = 0$ and $\sigma = 1$, we obtain the normal curve shown in Fig. 13-25, called the **standard normal curve**.

DEFINITION. *A continuous random variable Z is a **standard normal random variable** (or has a **standard normal distribution**) if its density function is given by*

$$f(z) = \frac{1}{\sqrt{2\pi}} e^{-z^2/2},$$

*called the **standard normal density function**. The variable Z has mean 0 and standard deviation 1.*

Because a standard normal random variable Z has mean 0 and standard deviation 1, its values are in units of standard deviations from the mean, which are called **standard units**. For example, if $0 < Z < 2.54$, then Z lies within 2.54 standard deviations to the right of 0, the mean. That is, $0 < Z < 2.54\sigma$. To find $P(0 < Z < 2.54)$, we have

$$P(0 < Z < 2.54) = \int_0^{2.54} \frac{e^{-z^2/2}}{\sqrt{2\pi}}\, dz.$$

This integral cannot be evaluated by elementary methods. However, values for integrals of this kind have been computed and put in table form.

One such table is given in Appendix F. This table gives the area under a standard normal curve between $z = 0$ and $z = z_0$, where $z_0 \geqslant 0$. This area is shaded in Fig. 13-26 and is denoted by $A(z_0)$. In the left-hand columns of the

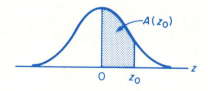

FIG. 13-26

table are z-values to the nearest tenth. The numbers across the top are the hundredths' values. For example, the entry in the row for 2.5 and column under .04 corresponds to $z = 2.54$ and is .4945. Thus the area under a standard normal curve between $z = 0$ and $z = 2.54$ is .4945:

$$P(0 < Z < 2.54) = A(2.54) = .4945.$$

Similarly, you should verify that $A(2) = .4772$ and $A(.33) = .1293$.

Using symmetry, we compute an area to the left of $z = 0$ by computing the corresponding area to the right of $z = 0$. For example,

$$P(-z_0 < Z < 0) = P(0 < Z < z_0) = A(z_0),$$

as shown in Fig. 13-27. Thus $P(-2.54 < Z < 0) = A(2.54) = .4945$.

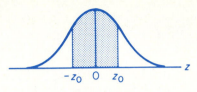

FIG. 13-27

When computing probabilities for a standard normal variable, you may have to add or subtract areas. A useful aid for doing this properly is a rough sketch of a standard normal curve in which you have shaded the entire area that you want to find, as Example 2 shows.

EXAMPLE 2

Suppose Z is a standard normal variable.

a. *Find $P(Z > 1.5)$.*

This probability is the area to the right of $z = 1.5$ (Fig. 13-28). This area is equal to the

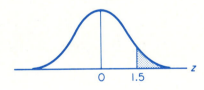

FIG. 13-28

difference between the total area to the right of $z = 0$, which is .5, and the area between $z = 0$ and $z = 1.5$, which is $A(1.5)$. Thus,

$$P(Z > 1.5) = .5 - A(1.5)$$

$$= .5 - .4332 = .0668.$$

b. *Find $P(.5 < Z < 2)$.*

This probability is the area between $z = .5$ and $z = 2$ (Fig. 13-29). This area is the

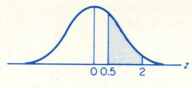

FIG. 13-29

difference of two areas. It is the area between $z = 0$ and $z = 2$, or $A(2)$, minus the area between $z = 0$ and $z = .5$, or $A(.5)$. Thus,

$$P(.5 < Z < 2) = A(2) - A(.5)$$

$$= .4772 - .1915 = .2857.$$

c. *Find $P(Z \leqslant 2)$.*

This probability is the area to the left of $z = 2$ (Fig. 13-30). This area is equal to the

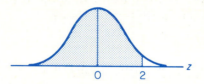

FIG. 13-30

sum of the area to the left of $z = 0$, which is .5, and the area between $z = 0$ and $z = 2$, or $A(2)$. Thus,

$$P(Z \leqslant 2) = .5 + A(2)$$

$$= .5 + .4772 = .9772.$$

d. *Find $P(-2 < Z < -.5)$.*

This probability is the area between $z = -2$ and $z = -.5$ (Fig. 13-31). By symmetry,

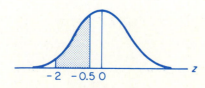

FIG. 13-31

this is equal to the area between $z = .5$ and $z = 2$, which was computed in (b). We have

$$P(-2 < Z < -.5) = P(.5 < Z < 2)$$

$$= A(2) - A(.5) = .2857.$$

e. *Find z_0 such that $P(-z_0 < Z < z_0) = .9642$.*

Figure 13-32 shows the corresponding area. Because the total area is .9642, by

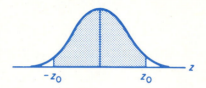

FIG. 13-32

symmetry the area between $z = 0$ and $z = z_0$ is $\frac{1}{2}(.9642) = .4821$, which is $A(z_0)$. Looking at the body of the table in Appendix F, we see that .4821 corresponds to a Z-value of 2.1. Thus $z_0 = 2.1$.

If X is normally distributed with mean μ and standard deviation σ, you might think that a table of areas is needed for each pair of values of μ and σ. Fortunately this is not the case. Appendix F is still used. But you must first express a given area as an equivalent area under a standard normal curve. This involves transforming X into a standard normal variable Z (with mean 0 and standard deviation 1) by using the following change of variable formula:

$$Z = \frac{X - \mu}{\sigma}. \qquad (1)$$

On the right side, subtracting μ from X gives the distance from μ to X. Dividing by σ expresses this distance in terms of units of standard deviation. Thus Z is the number of standard deviations that X is from μ. That is, formula (1) converts units of X to standard units (Z-values). For example, if $X = \mu$, then using formula (1) gives $Z = 0$. Hence μ is zero standard deviations from μ.

Suppose X is normally distributed with $\mu = 4$ and $\sigma = 2$. To find—for example—$P(0 < X < 6)$, we first use formula (1) to convert the X-values 0 and 6 to Z-values (standard units).

$$z_1 = \frac{x_1 - \mu}{\sigma} = \frac{0 - 4}{2} = -2,$$

$$z_2 = \frac{x_2 - \mu}{\sigma} = \frac{6 - 4}{2} = 1.$$

It can be shown that

$$P(0 < X < 6) = P(-2 < Z < 1).$$

This means that the area under a normal curve with $\mu = 4$ and $\sigma = 2$ between $x = 0$ and $x = 6$ is equal to the area under a standard normal curve between $z = -2$ and $z = 1$ (Fig. 13-33). This area is the sum of the area A_1 between

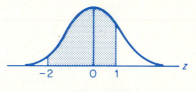

FIG. 13-33

$z = -2$ and $z = 0$ and the area A_2 between $z = 0$ and $z = 1$. Using symmetry for A_1, we have

$$P(-2 < Z < 1) = A_1 + A_2 = A(2) + A(1)$$
$$= .4772 + .3413 = .8185.$$

EXAMPLE 3

The weekly salaries of 5000 employees of a large corporation are assumed to be normally distributed with mean $300 and standard deviation $40. How many employees earn less than $250 per week?

Converting to standard units, we have

$$P(X < 250) = P\left(Z < \frac{250 - 300}{40}\right)$$

$$= P(Z < -1.25).$$

This probability is the area shown in Fig. 13-34(a). By symmetry, this is equal to the area

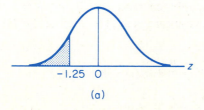

(a)   (b)

FIG. 13-34

in Fig. 13-34(b) that corresponds to $P(Z > 1.25)$. This area is the difference between the total area to the right of $z = 0$, which is .5, and the area between $z = 0$ and $z = 1.25$, which is $A(1.25)$. Thus,

$$P(X < 250) = P(Z < -1.25) = P(Z > 1.25)$$

$$= .5 - A(1.25) = .5 - .3944$$

$$= .1056.$$

This means that 10.56 percent of the employees have salaries less than \$250. This corresponds to .1056(5000) = 528 employees.

EXERCISE 13-6

1. If Z is a standard normal random variable, find each of the following probabilities.
   a. $P(0 < Z < 1.8)$.   b. $P(.45 < Z < 2.81)$.
   c. $P(Z > -1.22)$.   d. $P(Z \leqslant 2.93)$.
   e. $P(-2.61 < Z \leqslant 1.4)$ f. $P(Z > .07)$.

2. If Z is a standard normal random variable, find each of the following.
   a. $P(-1.96 < Z < 1.96)$.   b. $P(-2.11 < Z < -1.25)$.
   c. $P(Z < -1.05)$.   d. $P(Z > 3\sigma)$.
   e. $P(|Z| > 2)$.   f. $P(|Z| < \tfrac{1}{2})$.

In Problems 3–8, find z_0 such that the given statement is true. Assume that Z is a standard normal random variable.

3. $P(Z < z_0) = .5517$.   4. $P(Z < z_0) = .0668$.

5. $P(Z > z_0) = .8599$.   6. $P(Z > z_0) = .4960$.

7. $P(-z_0 < Z < z_0) = .2662$.   8. $P(|Z| > z_0) = .2186$.

9. If X is normally distributed with $\mu = 16$ and $\sigma = 4$, find each of the following probabilities.
   a. $P(X < 22)$.   b. $P(X < 10)$.   c. $P(10.8 < X < 12.4)$.

10. If X is normally distributed with $\mu = 200$ and $\sigma = 40$, find each of the following probabilities.
    a. $P(X > 150)$.   b. $P(210 < X < 250)$.

11. If X is normally distributed with $\mu = -3$ and $\sigma = 2$, find $P(X > -2)$.

12. If X is normally distributed with $\mu = 0$ and $\sigma = 1.5$, find $P(X < 3)$.

13. If X is normally distributed with $\mu = 25$ and $\sigma^2 = 9$, find $P(19 < X \leqslant 28)$.

14. If X is normally distributed with $\mu = 8$ and $\sigma = 1$, find $P(X > \mu - \sigma)$.

15. If X is normally distributed with $\mu = 40$ and $P(X > 54) = .0401$, find σ.

16. If X is normally distributed with $\mu = 16$ and $\sigma = 2.25$, find x_0 such that the probability that X is between x_0 and 16 is .4641.

17. The scores on a national achievement test are normally distributed with mean 500 and standard deviation 100. What percentage of those who took the test had a score greater than 630?

18. In a test given to a large group of people, the scores were normally distributed with mean 70 and standard deviation 10. What is the least whole-number score that a person could get and yet score in about the top 15 percent?

19. The heights (in inches) of adults in a large population are normally distributed with $\mu = 68$ and $\sigma = 3$. What percentage of the group is under 6 feet tall?

20. The yearly income for a group of 10,000 professional people is normally distributed with $\mu = \$30,000$ and $\sigma = \$2500$.
    a. What is the probability that a person from this group has a yearly income of less than $28,000?
    b. How many of these people have yearly incomes of over $35,000?

21. The I.Q.'s for a large population of children are normally distributed with mean 100.4 and standard deviation 11.6.
    a. What percentage of the children have I.Q.'s greater than 125?
    b. About 90 percent of the children have I.Q.'s greater than what value?

22. Suppose X is a random variable with $\mu = 10$ and $\sigma = 2$. If $P(4 < X < 16) = .25$, can X be normally distributed?

13-7 THE NORMAL APPROXIMATION TO THE BINOMIAL DISTRIBUTION

We conclude this chapter by bringing together the notions of a discrete random variable and a continuous random variable. To begin with, you would no doubt agree that calculating probabilities for a binomial random variable can be quite tedious when the number of trials (n) is large. For example, just imagine trying to compute $\binom{100}{40}(.3)^{40}(.7)^{60}$. To handle expressions like this, we can approximate a binomial distribution by a normal distribution and then use a table of areas to estimate a binomial probability.

To show how this is done, let us take a simple example. Figure 13-35 gives a probability histogram for a binomial experiment with $n = 10$ and $p = .5$. The

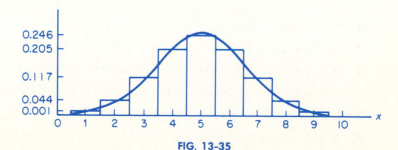

FIG. 13-35

rectangles centered at $x = 0$ and $x = 10$ are not shown because their heights are very close to 0. Superimposed on the histogram is a normal curve, which approximates it. The approximation would even be better if n were larger. That is, as n gets larger, then the width of each unit interval appears to get smaller and the outline of the histogram tends to take on the appearance of a smooth curve. In fact, *it is not unusual to think of a density curve as the limiting case of a probability histogram.* In spite of the fact that in our case n is only 10, the approximation shown does not seem too bad. The question that now arises is, "Which normal distribution approximates the binomial distribution?" Since the mean and standard deviation are measures of central tendency and dispersion of a random variable, we choose the approximating normal distribution to have the same mean and standard deviation as that of the binomial distribution. For this choice we can estimate the areas of rectangles in the histogram (that is, the binomial probabilities) by finding the corresponding area under the normal curve. In summary, we have the following:

> If X is a binomial random variable and n is sufficiently large, then the distribution of X can be approximated by a normal random variable whose mean and standard deviation are the same as for X, which are np and $\sqrt{npq}$, respectively.

Perhaps a word of explanation is appropriate concerning the phrase "n is sufficiently large." Generally speaking, a normal approximation to a binomial distribution is not good if n is small and p is near 0 or 1, because much of the area in the binomial histogram would be concentrated at one end of the distribution (that is, at 0 or n). Thus the distribution would not be fairly symmetric and a normal curve would not "fit" well. A general rule that you can follow is that the normal approximation to the binomial distribution is reasonable if np and nq are at least 5. This is the case in our example: $np = 10(.5) = 5$ and $nq = 10(.5) = 5$.

Let us now use the normal approximation to estimate a binomial probability for $n = 10$ and $p = .5$. If X denotes the number of successes, then its mean is $np = 10(.5) = 5$ and its standard deviation is $\sqrt{npq} = \sqrt{10(.5)(.5)} = 1.58$. The probability function for X is given by

$$f(x) = \binom{10}{x}(.5)^x(.5)^{10-x}.$$

We approximate this distribution by the normal distribution with $\mu = 5$ and $\sigma = 1.58$.

Suppose we estimate the probability that there are between 4 and 7 successes, inclusive, which is given by

$$P(4 \leqslant X \leqslant 7) = P(X = 4) + P(X = 5) + P(X = 6) + P(X = 7)$$

$$= \sum_{x=4}^{7} \binom{10}{x}(.5)^x(.5)^{10-x}.$$

This probability is the sum of the areas of the *rectangles* for $X = 4$, 5, 6, and 7 in Fig. 13-36. Under the normal curve we have shaded the corresponding area that

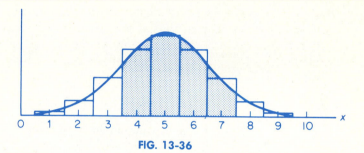

FIG. 13-36

we shall compute as an approximation to this probability. Note that the shading extends not from 4 to 7, but from $4 - \frac{1}{2}$ to $7 + \frac{1}{2}$, that is, from 3.5 to 7.5. This "continuity correction" of .5 on each end of the interval allows most of the area in the appropriate rectangles to be included in the approximation, and *such a correction must always be made*. The phrase *continuity correction* is used because X is treated as though it were a continuous random variable. We now convert the X-values 3.5 and 7.5 to Z-values.

$$z_1 = \frac{3.5 - 5}{1.58} = -.95,$$

$$z_2 = \frac{7.5 - 5}{1.58} = 1.58.$$

Thus,

$$P(4 \leqslant X \leqslant 7) \approx P(-.95 \leqslant Z \leqslant 1.58),$$

which corresponds to the area under a standard normal curve between $z = -.95$ and $z = 1.58$ (Fig. 13-37). This area is the sum of the area between $z = -.95$ and

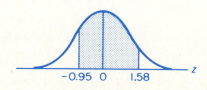

FIG. 13-37

$z = 0$, which by symmetry is $A(.95)$, and the area between $z = 0$ and $z = 1.58$, which is $A(1.58)$. Hence,

$$P(4 \leqslant X \leqslant 7) \approx P(-.95 \leqslant Z \leqslant 1.58)$$
$$= A(.95) + A(1.58)$$
$$= .3289 + .4429 = .7718.$$

This result is close to the true value .7734.

EXAMPLE 1

Suppose X is a binomial random variable with $n = 100$ and $p = .3$. Estimate $P(X = 40)$ using the normal approximation.

We have

$$P(X = 40) = \binom{100}{40}(.3)^{40}(.7)^{60},$$

which was mentioned at the beginning of this section. We use a normal distribution with $\mu = np = 100(.3) = 30$ and $\sigma = \sqrt{npq} = \sqrt{100(.3)(.7)} = 4.58$. Converting the corrected X-values 39.5 and 40.5 to Z-values gives

$$z_1 = \frac{39.5 - 30}{4.58} = 2.07,$$

$$z_2 = \frac{40.5 - 30}{4.58} = 2.29.$$

Thus,

$$P(X = 40) \approx P(2.07 \leqslant Z \leqslant 2.29).$$

This probability is the area under a standard normal curve between $z = 2.07$ and $z = 2.29$ (Fig. 13-38). This area is the difference of the area between $z = 0$ and $z = 2.29$, which is

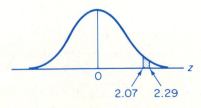

FIG. 13-38

$A(2.29)$, and the area between $z = 0$ and $z = 2.07$, which is $A(2.07)$. Thus,

$$P(X = 40) \approx P(2.07 \leqslant Z \leqslant 2.29)$$

$$= A(2.29) - A(2.07)$$

$$= .4890 - .4808 = .0082.$$

EXAMPLE 2

In a quality-control experiment, a sample of 500 items is taken from an assembly line. Customarily, 8 percent of the items produced are defective. What is the probability that more than 50 defective items appear in the sample?

If X is the number of defective items in the sample, then we shall consider X to be binomial with $n = 500$ and $p = .08$. To find $P(X \geqslant 51)$, we use the normal approximation to the binomial distribution with $\mu = np = 500(.08) = 40$ and $\sigma = \sqrt{npq} = \sqrt{500(.08)(.92)} = 6.066$. Converting the corrected value 50.5 to a Z-value gives

$$z = \frac{50.5 - 40}{6.066} = 1.73.$$

Thus,

$$P(X \geqslant 51) \approx P(Z \geqslant 1.73).$$

This probability is the area under a standard normal curve to the right of $z = 1.73$ (Fig. 13-39). This area is the difference of the area to the right of $z = 0$, which is .5, and the area

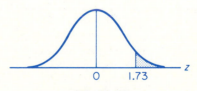

FIG. 13-39

between $z = 0$ and $z = 1.73$, which is $A(1.73)$. Thus,

$$P(X \geqslant 51) \approx P(Z \geqslant 1.73)$$

$$= .5 - A(1.73) = .5 - .4582 = .0418.$$

EXERCISE 13-7

*In Problems **1–4**, X is a binomial random variable with the given values of n and p. Calculate the indicated probabilities by using the normal approximation.*

1. $n = 150, p = .4.$ $P(X \leqslant 52), P(X \geqslant 74).$

2. $n = 50, p = .3.$ $P(X = 18), P(X \leqslant 18).$

3. $n = 200$, $p = .6$. $P(X = 125)$, $P(110 \leqslant X \leqslant 135)$.

4. $n = 25$, $p = .25$. $P(X \geqslant 5)$.

5. Suppose a fair die is tossed 300 times. What is the probability that a 5 turns up between 45 and 60 times, inclusive?

6. For a biased coin, $P(H) = .4$ and $P(T) = .6$. If the coin is tossed 200 times, what is the probability of getting between 90 and 100 heads, inclusive?

7. A delivery service has a fleet of 60 trucks. At any given time the probability of a truck being out of use due to factors such as breakdowns and maintenance is .1. What is the probability that 7 or more trucks are out of service at any time?

8. In a manufacturing plant, a sample of 100 items is taken from the assembly line. For each item in the sample, the probability of being defective is .06. What is the probability that there are 3 or more defective items in the sample?

9. In a true-false exam with 20 questions, what is the probability of getting at least 12 correct answers by just guessing on all the questions? If there are 100 questions instead of 20, what is the probability of getting at least 60 correct answers by just guessing?

10. In a multiple-choice test with 50 questions, each question has four answers, only one of which is correct. If a student guesses on the last 20 questions, what is the probability of getting at least half of them correct?

11. In a poker game, the probability of being dealt a hand consisting of three cards of one suit and two cards of another suit (in any order) is about .1. In 100 dealt hands, what is the probability that 16 or more of them will be as described above?

12. A major cola company sponsors a national taste test, in which subjects sample its cola as well as the best-selling brand. Neither cola is identified by brand. The subjects are then asked to choose the cola that tastes better. If each of the 25 subjects in a supermarket actually have no preference and arbitrarily chose one of the colas, what is the probability that 15 or more of them choose the cola from the sponsoring company?

13-8 REVIEW

IMPORTANT TERMS AND SYMBOLS IN CHAPTER 13

sample space *(p. 581)*	sample point *(p. 581)*
tree diagram *(p. 581)*	basic counting principle *(p. 582)*
event *(p. 583)*	certain event *(p. 584)*
impossible event *(p. 584)*	complement, E' *(p. 584)*
union, $\cup$ *(p. 584)*	intersection, $\cap$ *(p. 585)*
mutually exclusive events *(p. 586)*	trial *(p. 588)*
relative frequency *(p. 588)*	probability of event *(p. 589)*

equally likely outcomes *(p. 590)*   random variable *(p. 596)*

discrete random variable *(p. 597)*   continuous random variable *(p. 597)*

probability function *(p. 597)*   probability distribution *(p. 597)*

probability histogram *(p. 598)*   mean, μ *(p. 599)*

expected value, $E(X)$ *(p. 599)*   variance, $\text{Var}(X)$ *(p. 601)*

standard deviation *(p. 601)* σ, σ^2 *(p. 601)*

factorial, $n!$ *(p. 605)*   binomial theorem *(p. 606)*

binomial coefficient, $\binom{n}{r}$ *(p. 606)* independent trials *(p. 607)*

Bernoulli trials *(p. 609)*   binomial experiment *(p. 609)*

binomial distribution *(p. 609)*   density function *(p. 614)*

uniform distribution *(p. 615)*   exponential distribution *(p. 617)*

cumulative distribution function *(p. 618)*   normal curve *(p. 624)*

normal distribution *(p. 624)*   standard normal curve *(p. 627)*

standard normal distribution *(p. 627)*   continuity correction *(p. 635)*

REVIEW SECTION

1. If X has a binomial distribution, is X a discrete or a continuous random variable? _____

Ans. discrete.

2. If $S = \{1,2,3,4\}$ is a sample space and $E_1 = \{1,2\}$, then $E' =$ _____ .

Ans. $\{3,4\}$.

3. If a coin is tossed four times, the number of sample points in the sample space of this experiment is _____ .

Ans. 16.

4. From a committee of five people, in how many ways can two different persons be selected to fill the positions of chairperson and secretary? _____

Ans. 20.

5. If $P(E) = .4$, then $P(E') =$ _____ .

Ans. .6.

6. True or false: If $P(E_1) = .2$ and $P(E_2) = .3$, then $P(E_1 \cup E_2)$ must be .5. _____

Ans. False.

7. If $\text{Var}(X) = 4$, then $\sigma =$ _____ .

Ans. 2.

8. If an event is certain to occur, then its probability is __(a)__ . If an event is impossible, its probability is __(b)__ .

Ans. (a) 1; (b) 0.

9. If X is a continuous random variable, the probability that it assumes a specified value is _____ .

                                                  Ans.   0.

10. If X has a binomial distribution with $n = 10$ and $p = .3$, then $\mu = $ _____ .

                                                  Ans.   3.

11. If X is normally distributed with $\mu = 4$ and $\sigma = 2$, then the Z-value corresponding to $X = 6$ is _____ .

                                                  Ans.   1.

12. The value of 4! is _____ .

                                                  Ans.   24.

13. The value of $\dfrac{9!}{8!}$ is _____ .

                                                  Ans.   9.

14. The value of $\binom{6}{4}$ is _____ .

                                                  Ans.   15.

15. If E_1 and E_2 are events, then the event consisting of all sample points that are common to both E_1 and E_2 is called the ___(a)___ of E_1 and E_2 and is denoted by ___(b)___ .

                     Ans.   (a) intersection; (b) $E_1 \cap E_2$.

16. If X has distribution f with mean 3 and $\Sigma x^2 f(x) = 11$, then Var$(X) = $ _____ .

                                                  Ans.   2.

REVIEW PROBLEMS

1. Suppose $S = \{1, 2, 3, 4, 5, 6, 7, 8\}$ is the sample space and $E_1 = \{1, 2, 3, 4, 5, 6\}$ and $E_2 = \{4, 5, 6, 7\}$ are events for an experiment. Determine (a) $E_1 \cup E_2$, (b) $E_1 \cap E_2$, (c) $E_1' \cup E_2$, (d) $E_1 \cap E_1'$, and (e) $(E_1 \cap E_2')'$. (f) Are E_1 and E_2 mutually exclusive?

2. A die is rolled and then a coin is tossed. (a) Determine the sample space for this experiment. Determine the events that (b) a 2 shows, and (c) a head and an even number shows.

3. Three urns, labeled 1, 2, and 3, each contain two marbles, one red and the other green. A marble is selected from each urn. (a) Determine the sample space for this experiment. Determine the events that (b) exactly two marbles are red, and (c) the marbles are the same color.

4. Suppose that E_1 and E_2 are events for an experiment that has a finite number of sample points. If $P(E_1) = .6$, $P(E_1 \cup E_2) = .7$, and $P(E_1 \cap E_2) = .2$, find $P(E_2)$.

5. A manufacturer places a four-symbol serial number on a product. Each of the first three symbols is an integer between 0 and 9, inclusive; the fourth is a letter. How many different serial numbers are possible?

6. In how many ways can five different books be arranged on a bookshelf in terms of order from left to right?

7. In a restaurant, a complete dinner consists of one appetizer, one entree, and one dessert. The choices for the appetizer are soup and juice; for the entree, chicken, steak, lobster, and veal; for the dessert, ice cream, pie, and pudding. How many complete dinners are possible?

8. Each of 80 white mice was injected with one of four drugs, A, B, C, or D. Drug A was given to 25 percent, B to 20 percent, and C to 20 percent. If a mouse is chosen at random, determine the probability that it was injected with either C or D.

9. An urn contains four red and six green marbles.
    a. If two marbles are randomly selected in succession with replacement, determine the probability that both are red.
    b. If the selection is made without replacement, determine the probability that both are red.

10. A pair of fair dice are rolled. Determine the probability that the sum of the number of dots showing is (a) 4 or 5, (b) a multiple of 4, and (c) no less than 5.

11. Two cards from a standard deck of 52 playing cards are randomly drawn in succession with replacement. Determine the probability that (a) both cards are red, and (b) one card is red and the other is a club.

12. Two cards from a standard deck of 52 playing cards are randomly drawn in succession without replacement. Determine the probability that (a) both are hearts, and (b) one is an ace and the other is a red king.

*In Problems **13** and **14**, the distribution for the random variable X is given. Construct the probability histogram and determine* μ, $Var(X)$, *and* σ.

13. $f(1) = .7, f(2) = .1, f(3) = .2$.   14. $f(0) = \frac{1}{2}, f(1) = \frac{1}{8}, f(2) = \frac{3}{8}$.

15. A fair coin and a fair die are tossed. Let X be the sum of the number of heads and the number of dots that show. Determine (a) the distribution f for X, and (b) $E(X)$.

16. Two cards from a standard deck of 52 playing cards are randomly drawn in succession without replacement and the number of aces, X, is observed. Determine (a) the distribution f for X, and (b) $E(X)$.

17. In a game, a player pays \$0.25 to randomly draw two cards, with replacement, from a standard deck of 52 playing cards. For each 10 that appears, the player receives \$1. What is the player's expected gain or loss? Give your answer to the nearest cent.

18. An oil company determines that the probability that a gas station located along an interstate highway is successful is .45. A successful station earns an annual profit of \$40,000; one that is not loses \$10,000 annually. What is the expected gain to the company if it locates a station along an interstate highway?

19. A biased coin is tossed four times. The probability that a head occurs on any toss is $\frac{1}{3}$. Find the probability that at least two heads occur.

20. The probability that a certain type of seed germinates is .8. If five seeds are planted, what is the probability that none will germinate?

21. Suppose X is a continuous random variable with density function given by

$$f(x) = \begin{cases} \frac{1}{3} + kx^2, & \text{if } 0 \leqslant x \leqslant 1, \\ 0, & \text{otherwise.} \end{cases}$$

   a. Find k.   b. Find $P(\frac{1}{2} < X < \frac{3}{4})$.
   c. Find $P(X \geqslant \frac{1}{2})$.
   d. Find the cumulative distribution function.

22. Suppose X is exponentially distributed with $k = \frac{1}{4}$. Find $P(X > 1)$.

23. Suppose X is a random variable with density function given by

$$f(x) = \begin{cases} \frac{2}{9}x, & \text{if } 0 \leqslant x \leqslant 3, \\ 0, & \text{otherwise.} \end{cases}$$

   a. Find μ.   b. Find σ.

24. Suppose X is uniformly distributed over the interval $[2, 6]$. Find $P(X < 5)$.

Let X be normally distributed with mean 20 and standard deviation 4. In Problems 25–30, determine the given probabilities.

25. $P(X > 22)$.   26. $P(X < 21)$.

27. $P(12 < X < 18)$.   28. $P(X > 10)$.

29. $P(X < 16)$.   30. $P(22 < X < 32)$.

In Problems 31 and 32, X is a binomial random variable with $n = 100$ and $p = .35$. Find the given probabilities by using the normal approximation.

31. $P(25 \leqslant X \leqslant 47)$.   32. $P(X = 48)$.

33. The heights (in inches) of individuals in a certain group are normally distributed with mean 68 and standard deviation 2. Find the probability that an individual from this group is taller than 6 feet.

34. If a fair coin is tossed 400 times, use the normal approximation to the binomial distribution to estimate the probability that a head comes up at least 185 times.

MATRIX ALGEBRA

14-1 MATRICES

Finding ways to describe many situations in mathematics and economics leads to the study of rectangular arrays of numbers. Consider, for example, the system of linear equations

$$\begin{cases} 3x + 4y + 3z = 0, \\ 2x + \ y - \ z = 0, \\ 9x - 6y + 2z = 0. \end{cases}$$

The features that characterize this system are the numerical coefficients in the equations, together with their relative positions. For this reason the system can be described by the rectangular array

$$\begin{bmatrix} 3 & 4 & 3 \\ 2 & 1 & -1 \\ 9 & -6 & 2 \end{bmatrix},$$

which is called a *matrix* (plural: *matrices*, pronounced may′tri sees). We shall consider such rectangular arrays to be objects in themselves and our custom, as shown above, will be to enclose them by brackets. Parentheses ( ) are also commonly used.

PITFALL. *Do not use vertical bars, | |, instead of brackets or parentheses, for they have a different meaning.*

In symbolically representing matrices, we shall use bold capital letters such as **A, B, C,** etc.

In economics it is often convenient to use matrices in formulating problems and displaying data. For example, a manufacturer who produces products *A*, *B*, and *C* could represent the units of labor and material involved in one week's production of these items as in Table 14-1. More simply, these data can

TABLE 14-1

		PRODUCT	
	A	*B*	*C*
labor	10	12	16
material	5	9	7

be represented by the matrix

$$\mathbf{A} = \begin{bmatrix} 10 & 12 & 16 \\ 5 & 9 & 7 \end{bmatrix}.$$

The horizontal rows of a matrix are numbered consecutively from top to bottom, and the vertical columns are numbered from left to right. For matrix **A** above we have

$$\begin{array}{cccc} & \text{column 1} & \text{column 2} & \text{column 3} \\ \text{row 1} & 10 & 12 & 16 \\ \text{row 2} & 5 & 9 & 7 \end{array} \Bigg] = \mathbf{A}.$$

Since **A** has two rows and three columns, we say **A** has *order* (or *dimension*) 2×3 (read "2 by 3"), where the number of rows is specified first. Similarly, the matrices

$$\mathbf{B} = \begin{bmatrix} 1 & 6 & -2 \\ 5 & 1 & -4 \\ -3 & 5 & 0 \end{bmatrix} \quad \text{and} \quad \mathbf{C} = \begin{bmatrix} 1 & 2 \\ -3 & 4 \\ 5 & 6 \\ 7 & -8 \end{bmatrix}$$

have orders 3×3 and 4×2, respectively.

The numbers in a matrix are called its **entries** or **elements**. To denote arbitrary entries in a matrix, say one of order 2×3, there are two common methods. First, we may use different letters:

$$\begin{bmatrix} a & b & c \\ d & e & f \end{bmatrix}.$$

Second, a single letter may be used, say a, along with appropriate *double* subscripts to indicate position:

$$\begin{bmatrix} a_{11} & a_{12} & a_{13} \\ a_{21} & a_{22} & a_{23} \end{bmatrix}.$$

For the entry a_{12} (read "a sub one-two"), the first subscript 1 specifies the row and the second subscript 2, the column in which the entry appears. Similarly, the entry a_{23} (read "a sub two-three") is the entry in the second row and the third column. Generalizing, we say that the symbol a_{ij} denotes the entry in the ith row and jth column.

Our concern in this chapter is the manipulation and application of various types of matrices. For completeness, we now give a formal definition of a matrix.

DEFINITION. *A rectangular array of numbers consisting of **m** rows and **n** columns,*

$$\begin{bmatrix} a_{11} & a_{12} & \cdots & a_{1n} \\ a_{21} & a_{22} & \cdots & a_{2n} \\ \cdot & \cdot & \cdots & \cdot \\ \cdot & \cdot & \cdots & \cdot \\ \cdot & \cdot & \cdots & \cdot \\ a_{m1} & a_{m2} & \cdots & a_{mn} \end{bmatrix},$$

*is called an **m × n matrix** or a **matrix of order (dimension) m × n**. For the entry a_{ij}, i is the row subscript and j the column subscript.*

For brevity, an $m \times n$ matrix can be denoted by the symbol $[a_{ij}]_{m \times n}$ or more simply $[a_{ij}]$, where the dimension is understood to be that which is appropriate for the given context. This notation merely indicates what types of symbols we are using to denote the general entry. Thus if $[a_{ij}]$ is 2×3, then $i = 1, 2$ and $j = 1, 2, 3$ and $[a_{ij}]$ has six entries. More generally, $m \times n$ matrices have mn entries.

PITFALL. *Do not confuse the general entry a_{ij} with the matrix $[a_{ij}]$.*

A matrix that has exactly one row, such as

$$\mathbf{A} = \begin{bmatrix} 1 & 7 & 12 & 3 \end{bmatrix},$$

is called a **row matrix**. Here **A** has order 1×4. Similarly, a matrix consisting of a single column, such as the 5×1 matrix

$$\mathbf{B} = \begin{bmatrix} 1 \\ -2 \\ 15 \\ 9 \\ 16 \end{bmatrix},$$

is called a **column matrix**.

EXAMPLE 1

The matrices

$$\mathbf{A} = [1 \quad 2 \quad 0], \qquad \mathbf{B} = \begin{bmatrix} 1 & -6 \\ 5 & 1 \\ 9 & 4 \end{bmatrix}, \qquad \mathbf{C} = [7], \qquad \mathbf{D} = \begin{bmatrix} 1 & 3 & 7 & -2 & 4 \\ 9 & 11 & 5 & 6 & 8 \\ 6 & -2 & -1 & 1 & 1 \end{bmatrix}$$

have orders 1×3, 3×2, 1×1, and 3×5, respectively.

EXAMPLE 2

a. *Construct a 3-entry column matrix such that $a_{21} = 6$ and $a_{ij} = 0$ otherwise.*

The matrix is given by

$$\begin{bmatrix} 0 \\ 6 \\ 0 \end{bmatrix}.$$

b. *If $\mathbf{A} = [a_{ij}]$ has order 3×4 and $a_{ij} = i + j$, find $\mathbf{A}$.*

Here $i = 1, 2, 3$ and $j = 1, 2, 3, 4$, and $\mathbf{A}$ has $(3)(4) = 12$ entries. Since $a_{ij} = i + j$, the entry in row i and column j is obtained by adding the numbers i and j. Hence, $a_{11} = 1 + 1 = 2$, $a_{12} = 1 + 2 = 3$, $a_{13} = 1 + 3 = 4$, etc., and

$$\mathbf{A} = \begin{bmatrix} 1+1 & 1+2 & 1+3 & 1+4 \\ 2+1 & 2+2 & 2+3 & 2+4 \\ 3+1 & 3+2 & 3+3 & 3+4 \end{bmatrix} = \begin{bmatrix} 2 & 3 & 4 & 5 \\ 3 & 4 & 5 & 6 \\ 4 & 5 & 6 & 7 \end{bmatrix}.$$

c. *Construct the 3×3 matrix $\mathbf{I}$ given that $a_{11} = a_{22} = a_{33} = 1$ and $a_{ij} = 0$ otherwise.*

The matrix is given by

$$\mathbf{I} = \begin{bmatrix} 1 & 0 & 0 \\ 0 & 1 & 0 \\ 0 & 0 & 1 \end{bmatrix}.$$

DEFINITION. *Two matrices are **equal** if and only if they have the same order and corresponding entries are equal.*

For example,

$$\begin{bmatrix} 1+1 & \frac{2}{2} \\ 2 \cdot 3 & 0 \end{bmatrix} = \begin{bmatrix} 2 & 1 \\ 6 & 0 \end{bmatrix},$$

but

$$\begin{bmatrix} 1 & 1 \end{bmatrix} \neq \begin{bmatrix} 1 \\ 1 \end{bmatrix}$$

and

$$\begin{bmatrix} 1 & 1 \end{bmatrix} \neq \begin{bmatrix} 1 & 1 & 1 \end{bmatrix}.$$

By the definition of equality, for the matrix equation

$$\begin{bmatrix} x & y+1 \\ 2z & 5w \end{bmatrix} = \begin{bmatrix} 2 & 7 \\ 4 & 2 \end{bmatrix}$$

to be a true statement, it must be equivalent to the system

$$\begin{cases} x = 2, \\ y + 1 = 7, \\ 2z = 4, \\ 5w = 2. \end{cases}$$

Solving gives $x = 2$, $y = 6$, $z = 2$, and $w = \frac{2}{5}$. It is a significant fact that a matrix equation can define a system of linear equations as has been shown above.

Certain types of matrices play important roles in matrix theory. We now consider three such types.

An $m \times n$ matrix whose entries are all 0 is called the $m \times n$ **zero matrix**, denoted $\mathbf{O}_{m \times n}$ or, more simply, $\mathbf{O}$. Thus the 2×3 zero matrix is

$$\mathbf{O} = \begin{bmatrix} 0 & 0 & 0 \\ 0 & 0 & 0 \end{bmatrix}$$

and in general

$$\mathbf{O} = \begin{bmatrix} 0 & 0 & \cdots & 0 \\ 0 & 0 & \cdots & 0 \\ \cdot & \cdot & \cdots & \cdot \\ \cdot & \cdot & \cdots & \cdot \\ \cdot & \cdot & \cdots & \cdot \\ 0 & 0 & \cdots & 0 \end{bmatrix}.$$

A matrix having the same number of columns as rows, for example n rows and n columns, is called a **square matrix** of order n. That is, an $m \times n$ matrix is square if and only if $m = n$. Thus

$$\begin{bmatrix} -2 & 7 & 4 \\ 6 & 2 & 0 \\ 4 & 6 & 1 \end{bmatrix} \quad \text{and} \quad [3]$$

are square matrices of orders 3 and 1, respectively.

In a square matrix of order n, the entries $a_{11}, a_{22}, a_{33}, \ldots, a_{nn}$ which lie on the diagonal extending from the upper left corner to the lower right corner are called the **main diagonal** entries, or more simply the **main diagonal**. Thus in the matrix

$$\begin{bmatrix} 1 & 2 & 3 \\ 4 & 5 & 6 \\ 7 & 8 & 9 \end{bmatrix},$$

the main diagonal consists of $a_{11} = 1$, $a_{22} = 5$, and $a_{33} = 9$.

A square matrix is said to be an **upper (lower) triangular matrix** if all entries below (above) the main diagonal are zero. Thus

$$\begin{bmatrix} 5 & 1 & 1 \\ 0 & -3 & 7 \\ 0 & 0 & 4 \end{bmatrix} \quad \text{and} \quad \begin{bmatrix} 7 & 0 & 0 & 0 \\ -3 & 2 & 0 & 0 \\ 6 & 5 & -4 & 0 \\ 1 & 6 & 0 & 1 \end{bmatrix}$$

are upper and lower triangular matrices, respectively.

EXERCISE 14-1

1. Given the matrices

$$A = \begin{bmatrix} 1 & -6 & 2 \\ -4 & 2 & 1 \end{bmatrix}, \quad B = \begin{bmatrix} 1 & 2 & 3 \\ 4 & 5 & 6 \\ 7 & 8 & 9 \end{bmatrix}, \quad C = \begin{bmatrix} 1 & 1 \\ 2 & 2 \\ 3 & 3 \end{bmatrix},$$

$$D = \begin{bmatrix} 1 & 0 \\ 2 & 3 \end{bmatrix}, \quad E = \begin{bmatrix} 1 & 2 & 3 & 4 \\ 0 & 1 & 6 & 0 \\ 0 & 0 & 2 & 0 \\ 0 & 0 & 6 & 1 \end{bmatrix}, \quad F = [6 \quad 2],$$

$$G = \begin{bmatrix} 5 \\ 6 \\ 1 \end{bmatrix}, \quad H = \begin{bmatrix} 1 & 6 & 2 \\ 0 & 0 & 0 \\ 0 & 0 & 0 \end{bmatrix}, \quad J = [4],$$

(a) State the order of each matrix.

(b) Which matrices are square?

(c) Which matrices are upper triangular? lower triangular?

(d) Which are row matrices?

(e) Which are column matrices?

In Problems 2–9, let

$$A = [a_{ij}] = \begin{bmatrix} 7 & -2 & 14 & 6 \\ 6 & 2 & 3 & -2 \\ 5 & 4 & 1 & 0 \\ 8 & 0 & 2 & 0 \end{bmatrix}.$$

2. What is the order of **A**?

Find

3. a_{43}.   **4.** a_{12}.   **5.** a_{32}.   **6.** a_{34}.   **7.** a_{14}.   **8.** a_{55}.

9. What are the main diagonal entries?

10. Write the upper triangular matrix of order 5 given that all entries which are not required to be 0 are equal to 1.

11. Write $A = [a_{ij}]$ if **A** is 3×4 and $a_{ij} = 2i + 3j$.

12. Write $B = [b_{ij}]$ if **B** is 2×2 and $b_{ij} = (-1)^{i+j}(i^2 + j^2)$.

13. If $A = [a_{ij}]$ is 12×10, how many entries does **A** have? If $a_{ij} = 1$ for $i = j$, and $a_{ij} = 0$ for $i \neq j$, find a_{33}, a_{52}, $a_{10,10}$, and $a_{12,10}$.

14. List the main diagonal of

$$\text{(a)} \quad \begin{bmatrix} 1 & 4 & -2 & 0 \\ 7 & 0 & 4 & -1 \\ -6 & 6 & -5 & 1 \\ 2 & 1 & 7 & 2 \end{bmatrix}, \qquad \text{(b)} \quad \begin{bmatrix} x & 1 & y \\ 9 & y & 7 \\ y & 0 & z \end{bmatrix}.$$

15. Write the zero matrix of order (a) 4; (b) 6.

In Problems 16–19, solve the matrix equation.

16. $\begin{bmatrix} 2x & y \\ z & 3w \end{bmatrix} = \begin{bmatrix} 4 & 6 \\ 0 & 7 \end{bmatrix}.$

17. $\begin{bmatrix} 6 & 2 \\ x & 7 \\ 3y & 2z \end{bmatrix} = \begin{bmatrix} 6 & 2 \\ 6 & 7 \\ 2 & 7 \end{bmatrix}.$

18. $\begin{bmatrix} 4 & 2 & 1 \\ 3x & y & 3z \\ 0 & w & 7 \end{bmatrix} = \begin{bmatrix} 4 & 2 & 1 \\ 6 & 7 & 9 \\ 0 & 9 & 8 \end{bmatrix}.$

19. $\begin{bmatrix} 2x & 7 \\ 7 & 2y \end{bmatrix} = \begin{bmatrix} y & 7 \\ 7 & y \end{bmatrix}.$

20. A stock broker sold a customer 200 shares of Stock A, 300 shares of Stock B, 500 shares of Stock C, and 300 shares of Stock D. Write a row matrix that gives the number of shares of each stock sold. If the stocks sell for $20, $30, $45, and $100 per share, respectively, write this information as a column matrix.

21. The Widget Company has its monthly sales reports given by means of matrices where the rows, in order, represent the number of regular, deluxe, and super-duper models sold, and the columns, in order, give the number of red, white, blue, and purple units sold. The matrices for January (J) and February (F) are

$$J = \begin{bmatrix} 2 & 6 & 1 & 2 \\ 0 & 1 & 3 & 5 \\ 2 & 7 & 6 & 0 \end{bmatrix}; \qquad F = \begin{bmatrix} 0 & 2 & 4 & 4 \\ 2 & 3 & 3 & 2 \\ 4 & 0 & 2 & 6 \end{bmatrix}.$$

(a) How many white super-duper models were sold in January? (b) How many blue deluxe models were sold in February? (c) In which month were more purple regular models sold? (d) Which model and color sold the same number of units in both months? (e) In which month were more deluxe models sold? (f) In which month were more red widgets sold? (g) How many widgets were sold in January?

22. Input-output matrices, which were developed by W. W. Leontief, indicate the interrelationships that exist among the various sectors of an economy during some period of time. A hypothetical example for a simplified economy is given by matrix **M** below. The consuming sectors are the same as the producing sectors and can be thought of as manufacturers, government, steel, agriculture, households, etc. Each row shows how the output of a given sector is consumed by the four sectors. For example, of the total output of Industry A, 50 went to Industry A itself, 70 to B, 200 to C, and 360 to all others. The sum of the entries in row 1, namely 680, gives the total output of A for a given time period. Each column gives the output of each sector that is consumed by a given sector. For example, in producing 680 units, Industry A consumed 50 units of A, 90 of B, 120 of C, and 420 from all other producers. For each column, find the sum of the entries. Do the same for each row. What do you observe in comparing these totals? Suppose sector A increases its output by 20 percent, namely by 136 units. Assuming this results in a uniform 20 percent increase of all its inputs, by how many units will sector B have to increase its output? Answer the same question for C and D.

CONSUMERS

		Industry A	Industry B	Industry C	All Other Consumers
	PRODUCERS				
	Industry A	50	70	200	360
M =	Industry B	90	30	270	320
	Industry C	120	240	100	1,050
	All Other Producers	420	370	940	4,960

14-2 MATRIX ADDITION AND SCALAR MULTIPLICATION

Consider a snowmobile dealer who sells two models, Deluxe and Super. Each is available in one of two colors, red and blue. Suppose the sales for January and February are represented by the sales matrices

$$\begin{array}{cc} & \text{Deluxe} \quad \text{Super} \end{array}$$

$$\mathbf{J} = \begin{array}{c} \text{red} \\ \text{blue} \end{array}\begin{bmatrix} 1 & 2 \\ 3 & 5 \end{bmatrix}, \qquad \mathbf{F} = \begin{bmatrix} 3 & 1 \\ 4 & 2 \end{bmatrix}.$$

Each row of $\mathbf{J}$ and $\mathbf{F}$ gives the number of each model sold for a given color. Each column gives the number of each color sold for a given model. A matrix representing total sales for each model and color for both months can be obtained by adding the corresponding entries in $\mathbf{J}$ and $\mathbf{F}$:

$$\begin{bmatrix} 4 & 3 \\ 7 & 7 \end{bmatrix}.$$

This situation provides some motivation for introducing the operation of matrix addition for two matrices of the same order.

DEFINITION. *If $\mathbf{A}$ and $\mathbf{B}$ are both $m \times n$ matrices, then $\mathbf{A} + \mathbf{B}$ is the $m \times n$ matrix obtained by adding corresponding entries of $\mathbf{A}$ and $\mathbf{B}$.*

Thus if

$$\mathbf{A} = \begin{bmatrix} a_{11} & a_{12} & a_{13} \\ a_{21} & a_{22} & a_{23} \end{bmatrix} \quad \text{and} \quad \mathbf{B} = \begin{bmatrix} b_{11} & b_{12} & b_{13} \\ b_{21} & b_{22} & b_{23} \end{bmatrix},$$

then $\mathbf{A}$ and $\mathbf{B}$ have the same order (2×3) and

$$\mathbf{A} + \mathbf{B} = \begin{bmatrix} a_{11} + b_{11} & a_{12} + b_{12} & a_{13} + b_{13} \\ a_{21} + b_{21} & a_{22} + b_{22} & a_{23} + b_{23} \end{bmatrix}.$$

EXAMPLE 1

a. $\begin{bmatrix} 1 & 2 \\ 3 & 4 \\ 5 & 6 \end{bmatrix} + \begin{bmatrix} 7 & -2 \\ -6 & 4 \\ 3 & 0 \end{bmatrix} = \begin{bmatrix} 1+7 & 2-2 \\ 3-6 & 4+4 \\ 5+3 & 6+0 \end{bmatrix} = \begin{bmatrix} 8 & 0 \\ -3 & 8 \\ 8 & 6 \end{bmatrix}.$

b. $\begin{bmatrix} 1 & 2 \\ 3 & 4 \end{bmatrix} + \begin{bmatrix} 2 \\ 1 \end{bmatrix}$ is not defined since the matrices do not have the same order.

If **A**, **B**, **C**, and **O** have the same order, then the following properties hold for matrix addition:

a. $\mathbf{A} + \mathbf{B} = \mathbf{B} + \mathbf{A}$ Commutative Property,

b. $\mathbf{A} + (\mathbf{B} + \mathbf{C}) = (\mathbf{A} + \mathbf{B}) + \mathbf{C}$ Associative Property,

c. $\mathbf{A} + \mathbf{O} = \mathbf{O} + \mathbf{A} = \mathbf{A}$ Identity Property.

These properties are illustrated in the following example.

EXAMPLE 2

Let

$$\mathbf{A} = \begin{bmatrix} 1 & 2 & 1 \\ -2 & 0 & 1 \end{bmatrix}, \quad \mathbf{B} = \begin{bmatrix} 0 & 1 & 2 \\ 1 & -3 & 1 \end{bmatrix}, \quad \mathbf{C} = \begin{bmatrix} -2 & 1 & -1 \\ 0 & -2 & 1 \end{bmatrix}, \quad \mathbf{O} = \begin{bmatrix} 0 & 0 & 0 \\ 0 & 0 & 0 \end{bmatrix}.$$

a. $\mathbf{A} + \mathbf{B} = \begin{bmatrix} 1 & 3 & 3 \\ -1 & -3 & 2 \end{bmatrix}; \quad \mathbf{B} + \mathbf{A} = \begin{bmatrix} 1 & 3 & 3 \\ -1 & -3 & 2 \end{bmatrix}.$

b. $\mathbf{A} + (\mathbf{B} + \mathbf{C}) = \mathbf{A} + \begin{bmatrix} -2 & 2 & 1 \\ 1 & -5 & 2 \end{bmatrix} = \begin{bmatrix} -1 & 4 & 2 \\ -1 & -5 & 3 \end{bmatrix},$

$(\mathbf{A} + \mathbf{B}) + \mathbf{C} = \begin{bmatrix} 1 & 3 & 3 \\ -1 & -3 & 2 \end{bmatrix} + \mathbf{C} = \begin{bmatrix} -1 & 4 & 2 \\ -1 & -5 & 3 \end{bmatrix}.$

c. $\mathbf{A} + \mathbf{O} = \begin{bmatrix} 1 & 2 & 1 \\ -2 & 0 & 1 \end{bmatrix} + \begin{bmatrix} 0 & 0 & 0 \\ 0 & 0 & 0 \end{bmatrix} = \begin{bmatrix} 1 & 2 & 1 \\ -2 & 0 & 1 \end{bmatrix} = \mathbf{A}.$

Thus the zero matrix plays the same role in matrix addition as the number zero does in addition of real numbers.

Returning to the snowmobile dealer, recall that February sales were given by the matrix

$$\mathbf{F} = \begin{bmatrix} 3 & 1 \\ 4 & 2 \end{bmatrix}.$$

If in March the dealer doubles February's sales of each model and color of snowmobile, the sales matrix **M** for March could be obtained by multiplying each entry in **F** by 2:

$$\mathbf{M} = \begin{bmatrix} 2 \cdot 3 & 2 \cdot 1 \\ 2 \cdot 4 & 2 \cdot 2 \end{bmatrix}.$$

It seems reasonable to write this operation as

$$\mathbf{M} = 2\mathbf{F} = 2\begin{bmatrix} 3 & 1 \\ 4 & 2 \end{bmatrix} = \begin{bmatrix} 2 \cdot 3 & 2 \cdot 1 \\ 2 \cdot 4 & 2 \cdot 2 \end{bmatrix} = \begin{bmatrix} 6 & 2 \\ 8 & 4 \end{bmatrix},$$

which is thought of as multiplying a matrix by a real number. Indeed, we have the following definition.

DEFINITION. *If* **A** *is an* $m \times n$ *matrix and* k *is a real number (also called a scalar), then by* k**A** *we denote the* $m \times n$ *matrix obtained by multiplying each entry in* **A** *by* k. *This operation is called* **scalar multiplication**.

Thus if

$$\mathbf{A} = \begin{bmatrix} a_{11} & a_{12} \\ a_{21} & a_{22} \end{bmatrix},$$

then

$$k\mathbf{A} = \begin{bmatrix} ka_{11} & ka_{12} \\ ka_{21} & ka_{22} \end{bmatrix}.$$

EXAMPLE 3

Let

$$\mathbf{A} = \begin{bmatrix} 1 & 2 \\ 4 & -2 \end{bmatrix}, \quad \mathbf{B} = \begin{bmatrix} 3 & -4 \\ 7 & 1 \end{bmatrix}, \quad \mathbf{O} = \begin{bmatrix} 0 & 0 \\ 0 & 0 \end{bmatrix}.$$

Find the following matrices.

a. **4A**.

$$4\mathbf{A} = 4\begin{bmatrix} 1 & 2 \\ 4 & -2 \end{bmatrix} = \begin{bmatrix} 4 \cdot 1 & 4 \cdot 2 \\ 4 \cdot 4 & 4 \cdot (-2) \end{bmatrix} = \begin{bmatrix} 4 & 8 \\ 16 & -8 \end{bmatrix}.$$

b. $-\dfrac{2}{3}\mathbf{B}$.

$$-\frac{2}{3}\mathbf{B} = \begin{bmatrix} -\frac{2}{3}(3) & -\frac{2}{3}(-4) \\ -\frac{2}{3}(7) & -\frac{2}{3}(1) \end{bmatrix} = \begin{bmatrix} -2 & \frac{8}{3} \\ -\frac{14}{3} & -\frac{2}{3} \end{bmatrix}.$$

c. $\dfrac{1}{2}\mathbf{A} + 3\mathbf{B}$.

$$\frac{1}{2}\mathbf{A} + 3\mathbf{B} = \frac{1}{2}\begin{bmatrix} 1 & 2 \\ 4 & -2 \end{bmatrix} + 3\begin{bmatrix} 3 & -4 \\ 7 & 1 \end{bmatrix}$$

$$= \begin{bmatrix} \frac{1}{2} & 1 \\ 2 & -1 \end{bmatrix} + \begin{bmatrix} 9 & -12 \\ 21 & 3 \end{bmatrix} = \begin{bmatrix} \frac{19}{2} & -11 \\ 23 & 2 \end{bmatrix}.$$

d. **0A**.

$$0\mathbf{A} = 0\begin{bmatrix} 1 & 2 \\ 4 & -2 \end{bmatrix} = \begin{bmatrix} 0 & 0 \\ 0 & 0 \end{bmatrix} = \mathbf{O}.$$

e. $k\mathbf{O}$.

$$k\mathbf{O} = k\begin{bmatrix} 0 & 0 \\ 0 & 0 \end{bmatrix} = \begin{bmatrix} 0 & 0 \\ 0 & 0 \end{bmatrix} = \mathbf{O}.$$

If $\mathbf{A}$, $\mathbf{B}$, and $\mathbf{O}$ have the same order, then for any scalars k, k_1, and k_2 we have the following properties of scalar multiplication:

(a) $k(\mathbf{A} + \mathbf{B}) = k\mathbf{A} + k\mathbf{B}$,

(b) $(k_1 + k_2)\mathbf{A} = k_1\mathbf{A} + k_2\mathbf{A}$,

(c) $k_1(k_2\mathbf{A}) = (k_1k_2)\mathbf{A}$,

(d) $0\mathbf{A} = \mathbf{O}$,

(e) $k\mathbf{O} = \mathbf{O}$.

Properties (d) and (e) were illustrated in (d) and (e) of Example 3; the others will be illustrated in the exercises. Remember that $\mathbf{O} \neq 0$, for 0 is a *scalar* and $\mathbf{O}$ is a zero *matrix*.

For the case that $k = -1$, then $k\mathbf{A} = (-1)\mathbf{A}$, which will be denoted by simply writing $-\mathbf{A}$, called the *negative* of $\mathbf{A}$. Thus if

$$\mathbf{A} = \begin{bmatrix} 3 & 1 \\ -4 & 5 \end{bmatrix},$$

then

$$-\mathbf{A} = (-1)\begin{bmatrix} 3 & 1 \\ -4 & 5 \end{bmatrix} = \begin{bmatrix} -3 & -1 \\ 4 & -5 \end{bmatrix}.$$

Note that $-\mathbf{A}$ is the matrix obtained by multiplying each entry of $\mathbf{A}$ by -1.

Subtraction of matrices can now be defined.

DEFINITION. *If $\mathbf{A}$ and $\mathbf{B}$ have the same order, then by $\mathbf{A} - \mathbf{B}$ we mean the matrix $\mathbf{A} + (-\mathbf{B})$.*

EXAMPLE 4

a. $\begin{bmatrix} 2 & 6 \\ -4 & 1 \\ 3 & 2 \end{bmatrix} - \begin{bmatrix} 6 & -2 \\ 4 & 1 \\ 0 & 3 \end{bmatrix} = \begin{bmatrix} 2 & 6 \\ -4 & 1 \\ 3 & 2 \end{bmatrix} + (-1)\begin{bmatrix} 6 & -2 \\ 4 & 1 \\ 0 & 3 \end{bmatrix}$

$$= \begin{bmatrix} 2 & 6 \\ -4 & 1 \\ 3 & 2 \end{bmatrix} + \begin{bmatrix} -6 & 2 \\ -4 & -1 \\ 0 & -3 \end{bmatrix}$$

$$= \begin{bmatrix} 2-6 & 6+2 \\ -4-4 & 1-1 \\ 3+0 & 2-3 \end{bmatrix} = \begin{bmatrix} -4 & 8 \\ -8 & 0 \\ 3 & -1 \end{bmatrix}.$$

More simply, to find $\mathbf{A} - \mathbf{B}$ we can subtract each entry in $\mathbf{B}$ from the corresponding entry in $\mathbf{A}$.

b.
$$\begin{bmatrix} 6 & -4 & 7 & 1 \\ -1 & 6 & 0 & -4 \\ 2 & -1 & 3 & 1 \end{bmatrix} - \begin{bmatrix} 2 & -3 & 3 & 2 \\ 4 & 2 & 1 & 3 \\ 1 & 0 & -1 & -4 \end{bmatrix}$$

$$= \begin{bmatrix} 6-2 & -4+3 & 7-3 & 1-2 \\ -1-4 & 6-2 & 0-1 & -4-3 \\ 2-1 & -1-0 & 3+1 & 1+4 \end{bmatrix} = \begin{bmatrix} 4 & -1 & 4 & -1 \\ -5 & 4 & -1 & -7 \\ 1 & -1 & 4 & 5 \end{bmatrix}.$$

EXAMPLE 5

Solve $2\begin{bmatrix} x_1 \\ x_2 \end{bmatrix} - \begin{bmatrix} 3 \\ 4 \end{bmatrix} = 5\begin{bmatrix} 5 \\ -4 \end{bmatrix}$.

$$2\begin{bmatrix} x_1 \\ x_2 \end{bmatrix} - \begin{bmatrix} 3 \\ 4 \end{bmatrix} = 5\begin{bmatrix} 5 \\ -4 \end{bmatrix},$$

$$\begin{bmatrix} 2x_1 \\ 2x_2 \end{bmatrix} - \begin{bmatrix} 3 \\ 4 \end{bmatrix} = \begin{bmatrix} 25 \\ -20 \end{bmatrix},$$

$$\begin{bmatrix} 2x_1 - 3 \\ 2x_2 - 4 \end{bmatrix} = \begin{bmatrix} 25 \\ -20 \end{bmatrix}.$$

By equality of matrices we must have $2x_1 - 3 = 25$, which gives $x_1 = 14$; from $2x_2 - 4 = -20$ we get $x_2 = -8$.

EXAMPLE 6

Consider a simplified hypothetical economy having three industries, say coal, electricity, and steel, and three consumers 1, 2, and 3. Moreover, assume each consumer may use some of the output of each industry, and also that each industry uses some of the output of each other industry. The needs of each consumer and industry can be represented by a (row) demand matrix where the entries, in order, give the amount of coal, electricity, and steel needed by the consumer or industry in some convenient units. For example, the demand matrices for the consumers might be

$$\mathbf{D}_1 = [3 \quad 2 \quad 5], \qquad \mathbf{D}_2 = [0 \quad 17 \quad 1], \qquad \mathbf{D}_3 = [4 \quad 6 \quad 12],$$

and for the industries they might be

$$\mathbf{D}_C = [0 \quad 1 \quad 4], \qquad \mathbf{D}_E = [20 \quad 0 \quad 8], \qquad \mathbf{D}_S = [30 \quad 5 \quad 0],$$

where the subscripts C, E, and S stand for coal, electricity, and steel respectively. The total demand for these goods by the consumers is given by the sum

$$\mathbf{D}_1 + \mathbf{D}_2 + \mathbf{D}_3 = [3 \quad 2 \quad 5] + [0 \quad 17 \quad 1] + [4 \quad 6 \quad 12] = [7 \quad 25 \quad 18].$$

The total industrial demand is given by the sum

$$\mathbf{D}_C + \mathbf{D}_E + \mathbf{D}_S = [0 \quad 1 \quad 4] + [20 \quad 0 \quad 8] + [30 \quad 5 \quad 0] = [50 \quad 6 \quad 12].$$

Therefore, the total overall demand is given by

$$[7 \quad 25 \quad 18] + [50 \quad 6 \quad 12] = [57 \quad 31 \quad 30].$$

Thus the coal industry sells a total of 57 units, the total units of electricity sold are 31, and the total units of steel which are sold are 30.[†]

EXERCISE 14-2

In Problems 1–12 perform the indicated operations.

1. $\begin{bmatrix} 2 & 0 & -3 \\ -1 & 4 & 0 \\ 1 & -6 & 5 \end{bmatrix} + \begin{bmatrix} 2 & -3 & 4 \\ -1 & 6 & 5 \\ 9 & 11 & -2 \end{bmatrix}.$

2. $\begin{bmatrix} 2 & -7 \\ -6 & 4 \end{bmatrix} + \begin{bmatrix} 7 & -4 \\ -2 & 1 \end{bmatrix} + \begin{bmatrix} 2 & 7 \\ 7 & 2 \end{bmatrix}.$

3. $\begin{bmatrix} 1 & 4 \\ -2 & 7 \\ 6 & 9 \end{bmatrix} - \begin{bmatrix} 6 & -1 \\ 7 & 2 \\ 1 & 0 \end{bmatrix}.$ 4. $2\begin{bmatrix} 3 & -1 & 4 \\ 2 & 1 & -1 \\ 0 & 0 & 2 \end{bmatrix}.$

5. $3[1 \quad -3 \quad 2 \quad 1] + 2[-6 \quad 1 \quad 0 \quad 4] - 0[-2 \quad 7 \quad 6 \quad 4].$

6. $[7 \quad 7] + 66.$

7. $\begin{bmatrix} 1 & 2 \\ 3 & 4 \end{bmatrix} + \begin{bmatrix} 5 \\ 6 \end{bmatrix}.$ 8. $\begin{bmatrix} 2 & -1 \\ 7 & 4 \end{bmatrix} + 3\begin{bmatrix} 0 & 0 \\ 0 & 0 \end{bmatrix}.$

9. $-6\begin{bmatrix} 2 & -6 & 7 & 1 \\ 7 & 1 & 6 & -2 \end{bmatrix}.$ 10. $\begin{bmatrix} 1 & -1 \\ 2 & 0 \\ 3 & -6 \\ 4 & 9 \end{bmatrix} - 3\begin{bmatrix} -6 & 9 \\ 2 & 6 \\ 1 & -2 \\ 4 & 5 \end{bmatrix}.$

11. $\begin{bmatrix} 2 & -4 & 0 \\ 0 & 6 & -2 \\ -4 & 0 & 10 \end{bmatrix} + \frac{1}{3}\begin{bmatrix} 9 & 0 & 3 \\ 0 & 3 & 0 \\ 3 & 9 & 9 \end{bmatrix}.$

12. $2\begin{bmatrix} 1 & 0 & 0 \\ 0 & 1 & 0 \\ 0 & 0 & 1 \end{bmatrix} - 3\left(\begin{bmatrix} 2 & 1 & 0 \\ 1 & -2 & 3 \\ 1 & 0 & 0 \end{bmatrix} - \begin{bmatrix} 6 & -2 & 1 \\ -5 & 1 & -2 \\ 0 & 1 & 3 \end{bmatrix} \right).$

[†]This example, as well as some others in this chapter, are from John G. Kemeny, J. Laurie Snell, and Gerald L. Thompson, *Introduction to Finite Mathematics*, 3rd ed, © 1974. Reprinted by permission of Prentice-Hall, Inc., Englewood Cliffs, New Jersey.

In Problems **13–24**, *compute the required matrices if*

$$A = \begin{bmatrix} 2 & 1 \\ 3 & -3 \end{bmatrix}, \quad B = \begin{bmatrix} -6 & -5 \\ 2 & -3 \end{bmatrix}, \quad C = \begin{bmatrix} -2 & -1 \\ -3 & 3 \end{bmatrix}, \quad O = \begin{bmatrix} 0 & 0 \\ 0 & 0 \end{bmatrix}.$$

13. $-B$.

14. $-(A - B)$.

15. $2O$.

16. $A + B - C$.

17. $2(A - 2B)$.

18. $0(A + B)$.

19. $3(A - C) + 6$.

20. $A + (C + 2O)$.

21. $2B - 3A + 2C$.

22. $3C - 2B$.

23. $\frac{1}{2}A - 2(B + 2C)$.

24. $2A - \frac{1}{2}(B - C)$.

For matrices **A, B,** *and* **C** *above, verify that:*

25. $3(A + B) = 3A + 3B$.

26. $(2 + 3)A = 2A + 3A$.

27. $k_1(k_2A) = (k_1k_2)A$.

28. $k(A + B + C) = kA + kB + kC$.

29. Express the matrix equation

$$x\begin{bmatrix} 2 \\ 1 \end{bmatrix} - y\begin{bmatrix} -3 \\ 5 \end{bmatrix} = 2\begin{bmatrix} 8 \\ 11 \end{bmatrix}$$

as a system of linear equations and solve.

30. In the reverse of the manner used in Problem 29, write the system

$$\begin{cases} 3x + 5y = 16 \\ 2x - 6y = -4 \end{cases}$$

as a matrix equation.

In Problems **31–34**, *solve the matrix equations.*

31. $3\begin{bmatrix} x \\ y \end{bmatrix} - 3\begin{bmatrix} -2 \\ 4 \end{bmatrix} = 4\begin{bmatrix} 6 \\ -2 \end{bmatrix}$.

32. $3\begin{bmatrix} x \\ 2 \end{bmatrix} - 4\begin{bmatrix} 7 \\ -y \end{bmatrix} = \begin{bmatrix} -x \\ 2y \end{bmatrix}$.

33. $\begin{bmatrix} 2 \\ 4 \\ 6 \end{bmatrix} + 2\begin{bmatrix} x \\ y \\ 4z \end{bmatrix} = \begin{bmatrix} -10 \\ -24 \\ 14 \end{bmatrix}$.

34. $x\begin{bmatrix} 2 \\ 0 \\ 3 \end{bmatrix} + 2\begin{bmatrix} -1 \\ 0 \\ 6 \end{bmatrix} + y\begin{bmatrix} 0 \\ 2 \\ -3 \end{bmatrix} = \begin{bmatrix} 8 \\ 4 \\ 3x + 12 - 3y \end{bmatrix}$.

35. Suppose the prices of products A, B, and C are given, in that order, by the price

matrix

$$P = [p_1 \quad p_2 \quad p_3].$$

If the prices are to be increased by 10 percent, the matrix of the new prices can be obtained by multiplying P by what scalar?

14-3 MATRIX MULTIPLICATION

Besides the operations of matrix addition and scalar multiplication, the product AB of matrices A and B can be defined under certain conditions, namely that the number of columns of A is equal to the number of rows of B.

DEFINITION. *Let A be an $m \times n$ matrix and B be an $n \times p$ matrix. Then the product AB is the $m \times p$ matrix C whose entry c_{ij} in the i-th row and j-th column is obtained as follows: sum the products formed by multiplying, in order, each entry (that is, first, second, etc.) in the i-th row of A by the "corresponding" entry (that is, first, second, etc.) in the j-th column of B.*

Three points must be completely understood concerning the above definition of AB. First, the condition that A be $m \times n$ and B be $n \times p$ is equivalent to saying that the number of columns of A must be equal to the number of rows of B. Second, the product will be a matrix of order $m \times p$; it will have as many rows as A and as many columns as B. Third, the definition refers to the product AB, *in that order;* A is the left factor and B is the right factor. For AB we say B is *premultiplied* by A or A is *postmultiplied* by B.

To apply the definition, let us find the product

$$AB = \begin{bmatrix} 2 & 1 & -6 \\ 1 & -3 & 2 \end{bmatrix} \begin{bmatrix} 1 & 0 & -3 \\ 0 & 4 & 2 \\ -2 & 1 & 1 \end{bmatrix}.$$

The number of columns of A is equal to the number of rows of B and so the product is defined. Since A is 2×3 $(m \times n)$ and B is 3×3 $(n \times p)$, the product C will have order 2×3 $(m \times p)$:

$$C = \begin{bmatrix} c_{11} & c_{12} & c_{13} \\ c_{21} & c_{22} & c_{23} \end{bmatrix}.$$

The entry c_{11} is obtained by summing the products of each entry in row 1 of A by the "corresponding" entry in column 1 of B. That is,

$$\overset{\text{row 1 entries of } A}{c_{11} = (2)(1) + (1)(0) + (-6)(-2) = 14.}$$

column 1 entries of B

Similarly, for c_{21} we use the entries in row 2 of **A** and those in column 1 of **B**:

row 2 entries of **A**

$$c_{21} = (1)(1) + (-3)(0) + (2)(-2) = -3.$$

column 1 entries of **B**

Also,

$$c_{12} = (2)(0) + (1)(4) + (-6)(1) = -2,$$
$$c_{22} = (1)(0) + (-3)(4) + (2)(1) = -10,$$
$$c_{13} = (2)(-3) + (1)(2) + (-6)(1) = -10,$$
$$c_{23} = (1)(-3) + (-3)(2) + (2)(1) = -7.$$

Thus

$$\mathbf{AB} = \begin{bmatrix} 2 & 1 & -6 \\ 1 & -3 & 2 \end{bmatrix} \begin{bmatrix} 1 & 0 & -3 \\ 0 & 4 & 2 \\ -2 & 1 & 1 \end{bmatrix} = \begin{bmatrix} 14 & -2 & -10 \\ -3 & -10 & -7 \end{bmatrix}.$$

If we reverse the order of the factors, then

$$\mathbf{BA} = \begin{bmatrix} 1 & 0 & -3 \\ 0 & 4 & 2 \\ -2 & 1 & 1 \end{bmatrix} \begin{bmatrix} 2 & 1 & -6 \\ 1 & -3 & 2 \end{bmatrix}.$$

This product is *not* defined since the number of columns of **B** does *not* equal the number of rows of **A**. This shows that matrix multiplication is not commutative. That is, for any matrices **A** and **B** it is usually the case that $\mathbf{AB} \neq \mathbf{BA}$ (even if both products are defined).

EXAMPLE 1

Determine

$$\mathbf{AB} = \begin{bmatrix} 2 & -4 & 2 \\ 0 & 1 & -3 \end{bmatrix} \begin{bmatrix} 2 & 1 \\ 0 & 4 \\ 2 & 2 \end{bmatrix}.$$

Since **A** is 2×3 and **B** is 3×2, the product **AB** is defined and will have order 2×2. By simultaneously moving the index finger of your left hand along the rows of **A** and the index finger of your right hand along the columns of **B**, you can find mentally the entries of the product.

$$\begin{bmatrix} 2 & -4 & 2 \\ 0 & 1 & -3 \end{bmatrix} \begin{bmatrix} 2 & 1 \\ 0 & 4 \\ 2 & 2 \end{bmatrix} = \begin{bmatrix} 8 & -10 \\ -6 & -2 \end{bmatrix}.$$

EXAMPLE 2

Evaluate each of the following.

a. $[1 \quad 2 \quad 3]\begin{bmatrix} 4 \\ 5 \\ 6 \end{bmatrix}$.

The product has order 1×1:

$$[1 \quad 2 \quad 3]\begin{bmatrix} 4 \\ 5 \\ 6 \end{bmatrix} = [32].$$

b. $\begin{bmatrix} 1 \\ 2 \\ 3 \end{bmatrix}[1 \quad 6]$.

The product has order 3×2:

$$\begin{bmatrix} 1 \\ 2 \\ 3 \end{bmatrix}[1 \quad 6] = \begin{bmatrix} 1 & 6 \\ 2 & 12 \\ 3 & 18 \end{bmatrix}.$$

c. $\begin{bmatrix} 1 & 3 & 0 \\ -2 & 2 & 1 \\ 1 & 0 & -4 \end{bmatrix}\begin{bmatrix} 1 & 0 & 2 \\ 5 & -1 & 3 \\ 2 & 1 & -2 \end{bmatrix} = \begin{bmatrix} 16 & -3 & 11 \\ 10 & -1 & 0 \\ -7 & -4 & 10 \end{bmatrix}$.

d. $\begin{bmatrix} a_{11} & a_{12} \\ a_{21} & a_{22} \end{bmatrix}\begin{bmatrix} b_{11} & b_{12} \\ b_{21} & b_{22} \end{bmatrix} = \begin{bmatrix} a_{11}b_{11} + a_{12}b_{21} & a_{11}b_{12} + a_{12}b_{22} \\ a_{21}b_{11} + a_{22}b_{21} & a_{21}b_{12} + a_{22}b_{22} \end{bmatrix}$.

EXAMPLE 3

*Find **AB** and **BA** if*

$$\mathbf{A} = \begin{bmatrix} 2 & -1 \\ 3 & 1 \end{bmatrix} \quad and \quad \mathbf{B} = \begin{bmatrix} -2 & 1 \\ 1 & 4 \end{bmatrix}.$$

We have

$$\mathbf{AB} = \begin{bmatrix} 2 & -1 \\ 3 & 1 \end{bmatrix}\begin{bmatrix} -2 & 1 \\ 1 & 4 \end{bmatrix} = \begin{bmatrix} -5 & -2 \\ -5 & 7 \end{bmatrix},$$

$$\mathbf{BA} = \begin{bmatrix} -2 & 1 \\ 1 & 4 \end{bmatrix}\begin{bmatrix} 2 & -1 \\ 3 & 1 \end{bmatrix} = \begin{bmatrix} -1 & 3 \\ 14 & 3 \end{bmatrix}.$$

Although both **AB** and **BA** are defined, $\mathbf{AB} \neq \mathbf{BA}$.

Matrix multiplication satisfies the following properties if it is assumed that all sums and products are defined:

(1) $A(BC) = (AB)C$ Associative Property,

(2) $A(B + C) = AB + AC,$
 $\left.(A + B)C = AC + BC\right\}$ Distributive Properties.

EXAMPLE 4

If

$$A = \begin{bmatrix} 1 & -2 \\ -3 & 4 \end{bmatrix}, \quad B = \begin{bmatrix} 3 & 0 & -1 \\ 1 & 1 & 2 \end{bmatrix}, \quad and \quad C = \begin{bmatrix} 1 & 0 \\ 0 & 2 \\ 1 & 1 \end{bmatrix},$$

find **ABC** *in two ways*.

$$A(BC) = \begin{bmatrix} 1 & -2 \\ -3 & 4 \end{bmatrix} \left(\begin{bmatrix} 3 & 0 & -1 \\ 1 & 1 & 2 \end{bmatrix} \begin{bmatrix} 1 & 0 \\ 0 & 2 \\ 1 & 1 \end{bmatrix} \right)$$

$$= \begin{bmatrix} 1 & -2 \\ -3 & 4 \end{bmatrix} \begin{bmatrix} 2 & -1 \\ 3 & 4 \end{bmatrix} = \begin{bmatrix} -4 & -9 \\ 6 & 19 \end{bmatrix}.$$

$$(AB)C = \left(\begin{bmatrix} 1 & -2 \\ -3 & 4 \end{bmatrix} \begin{bmatrix} 3 & 0 & -1 \\ 1 & 1 & 2 \end{bmatrix} \right) \begin{bmatrix} 1 & 0 \\ 0 & 2 \\ 1 & 1 \end{bmatrix}$$

$$= \begin{bmatrix} 1 & -2 & -5 \\ -5 & 4 & 11 \end{bmatrix} \begin{bmatrix} 1 & 0 \\ 0 & 2 \\ 1 & 1 \end{bmatrix}$$

$$= \begin{bmatrix} -4 & -9 \\ 6 & 19 \end{bmatrix}.$$

EXAMPLE 5

Verify that **A(B + C) = AB + AC** *if*

$$A = \begin{bmatrix} 1 & 0 \\ 2 & 3 \end{bmatrix}, \quad B = \begin{bmatrix} -2 & 0 \\ 1 & 3 \end{bmatrix}, \quad and \quad C = \begin{bmatrix} -2 & 1 \\ 0 & 2 \end{bmatrix}.$$

$$A(B + C) = \begin{bmatrix} 1 & 0 \\ 2 & 3 \end{bmatrix} \left(\begin{bmatrix} -2 & 0 \\ 1 & 3 \end{bmatrix} + \begin{bmatrix} -2 & 1 \\ 0 & 2 \end{bmatrix} \right)$$

$$= \begin{bmatrix} 1 & 0 \\ 2 & 3 \end{bmatrix} \begin{bmatrix} -4 & 1 \\ 1 & 5 \end{bmatrix} = \begin{bmatrix} -4 & 1 \\ -5 & 17 \end{bmatrix}.$$

$$AB + AC = \begin{bmatrix} 1 & 0 \\ 2 & 3 \end{bmatrix} \begin{bmatrix} -2 & 0 \\ 1 & 3 \end{bmatrix} + \begin{bmatrix} 1 & 0 \\ 2 & 3 \end{bmatrix} \begin{bmatrix} -2 & 1 \\ 0 & 2 \end{bmatrix}$$

$$= \begin{bmatrix} -2 & 0 \\ -1 & 9 \end{bmatrix} + \begin{bmatrix} -2 & 1 \\ -4 & 8 \end{bmatrix} = \begin{bmatrix} -4 & 1 \\ -5 & 17 \end{bmatrix}.$$

Thus **A(B + C) = AB + AC**.

A square matrix of order n whose main diagonal entries are all 1 and all of whose other entries are 0 is called the **identity matrix** of order n. It is denoted by **I**. For example, the identity matrices of orders 3 and 4, respectively, are

$$\mathbf{I} = \begin{bmatrix} 1 & 0 & 0 \\ 0 & 1 & 0 \\ 0 & 0 & 1 \end{bmatrix} \quad \text{and} \quad \mathbf{I} = \begin{bmatrix} 1 & 0 & 0 & 0 \\ 0 & 1 & 0 & 0 \\ 0 & 0 & 1 & 0 \\ 0 & 0 & 0 & 1 \end{bmatrix}.$$

If **A** is a square matrix and both **A** and **I** have the same order, then

$$\mathbf{AI} = \mathbf{IA} = \mathbf{A}.$$

Thus the identity matrix plays the same role in matrix multiplication as does the number 1 in the multiplication of real numbers. For example,

$$\begin{bmatrix} 2 & 4 \\ 1 & 5 \end{bmatrix} \begin{bmatrix} 1 & 0 \\ 0 & 1 \end{bmatrix} = \begin{bmatrix} 2 & 4 \\ 1 & 5 \end{bmatrix}$$

and

$$\begin{bmatrix} 1 & 0 \\ 0 & 1 \end{bmatrix} \begin{bmatrix} 2 & 4 \\ 1 & 5 \end{bmatrix} = \begin{bmatrix} 2 & 4 \\ 1 & 5 \end{bmatrix}.$$

EXAMPLE 6

If

$$\mathbf{A} = \begin{bmatrix} 3 & 2 \\ 1 & 4 \end{bmatrix}, \quad \mathbf{B} = \begin{bmatrix} \frac{2}{5} & -\frac{1}{5} \\ -\frac{1}{10} & \frac{3}{10} \end{bmatrix}, \quad \mathbf{I} = \begin{bmatrix} 1 & 0 \\ 0 & 1 \end{bmatrix}, \quad \text{and} \quad \mathbf{O} = \begin{bmatrix} 0 & 0 \\ 0 & 0 \end{bmatrix},$$

determine each of the following.

a. **I** − **A**.

$$\mathbf{I} - \mathbf{A} = \begin{bmatrix} 1 & 0 \\ 0 & 1 \end{bmatrix} - \begin{bmatrix} 3 & 2 \\ 1 & 4 \end{bmatrix} = \begin{bmatrix} -2 & -2 \\ -1 & -3 \end{bmatrix}.$$

b. 3(**A** − 2**I**).

$$3(\mathbf{A} - 2\mathbf{I}) = 3\left(\begin{bmatrix} 3 & 2 \\ 1 & 4 \end{bmatrix} - 2\begin{bmatrix} 1 & 0 \\ 0 & 1 \end{bmatrix} \right)$$

$$= 3\left(\begin{bmatrix} 3 & 2 \\ 1 & 4 \end{bmatrix} - \begin{bmatrix} 2 & 0 \\ 0 & 2 \end{bmatrix} \right)$$

$$= 3\begin{bmatrix} 1 & 2 \\ 1 & 2 \end{bmatrix} = \begin{bmatrix} 3 & 6 \\ 3 & 6 \end{bmatrix}.$$

c. **AO**.

$$\mathbf{AO} = \begin{bmatrix} 3 & 2 \\ 1 & 4 \end{bmatrix} \begin{bmatrix} 0 & 0 \\ 0 & 0 \end{bmatrix} = \begin{bmatrix} 0 & 0 \\ 0 & 0 \end{bmatrix} = \mathbf{O}.$$

d. **AB**.

$$AB = \begin{bmatrix} 3 & 2 \\ 1 & 4 \end{bmatrix} \begin{bmatrix} \frac{2}{5} & -\frac{1}{5} \\ -\frac{1}{10} & \frac{3}{10} \end{bmatrix} = \begin{bmatrix} 1 & 0 \\ 0 & 1 \end{bmatrix} = I.$$

Systems of linear equations can be represented by using matrix multiplication. Consider the left side of the matrix equation .

$$\begin{bmatrix} a_{11} & a_{12} \\ a_{21} & a_{22} \end{bmatrix} \begin{bmatrix} x_1 \\ x_2 \end{bmatrix} = \begin{bmatrix} c_1 \\ c_2 \end{bmatrix}. \tag{1}$$

The product on the left side has order 2×1 and hence is a column matrix:

$$\begin{bmatrix} a_{11}x_1 + a_{12}x_2 \\ a_{21}x_1 + a_{22}x_2 \end{bmatrix} = \begin{bmatrix} c_1 \\ c_2 \end{bmatrix}.$$

By equality of matrices we must have

$$\begin{cases} a_{11}x_1 + a_{12}x_2 = c_1, \\ a_{21}x_1 + a_{22}x_2 = c_2. \end{cases}$$

Hence, a system of linear equations can be defined by a matrix equation. We usually describe Eq. (1) by saying it has the form

$$AX = C.$$

EXAMPLE 7

Represent the system

$$\begin{cases} 2x_1 + 5x_2 = 4, \\ 8x_1 + 3x_2 = 7 \end{cases}$$

in terms of matrices.

If

$$A = \begin{bmatrix} 2 & 5 \\ 8 & 3 \end{bmatrix}, \quad X = \begin{bmatrix} x_1 \\ x_2 \end{bmatrix}, \quad and \quad C = \begin{bmatrix} 4 \\ 7 \end{bmatrix},$$

then the given system is equivalent to

$$AX = C$$

or

$$\begin{bmatrix} 2 & 5 \\ 8 & 3 \end{bmatrix} \begin{bmatrix} x_1 \\ x_2 \end{bmatrix} = \begin{bmatrix} 4 \\ 7 \end{bmatrix}.$$

EXAMPLE 8

Suppose the prices (in dollars per unit) for products A, B, and C are represented by the price matrix

Price of
A B C

$\mathbf{P} = [2 \quad 3 \quad 4].$

If the quantities (in units) of A, B, and C that are purchased are given by the column matrix

$$\mathbf{Q} = \begin{bmatrix} 7 \\ 5 \\ 11 \end{bmatrix} \begin{matrix} \text{units of } A \\ \text{units of } B \\ \text{units of } C, \end{matrix}$$

then the total cost (in dollars) of the purchases is given by the entry in $\mathbf{PQ}$:

$$\mathbf{PQ} = [2 \quad 3 \quad 4] \begin{bmatrix} 7 \\ 5 \\ 11 \end{bmatrix} = [(2 \cdot 7) + (3 \cdot 5) + (4 \cdot 11)] = [73].$$

EXAMPLE 9

Suppose a building contractor has accepted orders for five ranch-style houses, seven Cape Cod-style houses, and twelve colonial-style houses. Then his orders can be represented by the row matrix

$$\mathbf{Q} = [5 \quad 7 \quad 12].$$

Furthermore, suppose the "raw materials" that go into each type of house are steel, wood, glass, paint, and labor. The entries in the matrix $\mathbf{R}$ below give the number of units of each raw material going into each type of house. (The entries are not necessarily realistic, but are chosen for convenience.)

	STEEL	WOOD	GLASS	PAINT	LABOR	
Ranch	5	20	16	7	17	
Cape Cod	7	18	12	9	21	= R.
Colonial	6	25	8	5	13	

Each row indicates the amount of each raw material needed for a given kind of house; each column indicates the amount of a given raw material needed for each type of house. Suppose now that the contractor wishes to compute the amount of each raw material needed to fulfill his contracts. Then such information is given by $\mathbf{QR}$:

$$\mathbf{QR} = [5 \quad 7 \quad 12] \begin{bmatrix} 5 & 20 & 16 & 7 & 17 \\ 7 & 18 & 12 & 9 & 21 \\ 6 & 25 & 8 & 5 & 13 \end{bmatrix}$$

$$= [146 \quad 526 \quad 260 \quad 158 \quad 388].$$

Thus, the contractor should order 146 units of steel, 526 units of wood, 260 units of glass, etc.

The contractor is also interested in the costs he will have to pay for these materials. Suppose steel costs $1500 per unit, wood costs $800 per unit, and glass, paint, and labor cost $500, $100, and $1000 per unit, respectively. These data can be written as the column cost matrix

$$\mathbf{C} = \begin{bmatrix} 1500 \\ 800 \\ 500 \\ 100 \\ 1000 \end{bmatrix}.$$

Then **RC** gives the cost of each type of house:

$$\mathbf{RC} = \begin{bmatrix} 5 & 20 & 16 & 7 & 17 \\ 7 & 18 & 12 & 9 & 21 \\ 6 & 25 & 8 & 5 & 13 \end{bmatrix} \begin{bmatrix} 1500 \\ 800 \\ 500 \\ 100 \\ 1000 \end{bmatrix} = \begin{bmatrix} 49,200 \\ 52,800 \\ 46,500 \end{bmatrix}.$$

Thus, the cost of materials for the ranch-style house is $49,200, for the Cape Cod house, $52,800, and for the colonial house, $46,500.

The total cost of raw materials for all the houses is given by

$$\mathbf{QRC} = \mathbf{Q(RC)} = \begin{bmatrix} 5 & 7 & 12 \end{bmatrix} \begin{bmatrix} 49,200 \\ 52,800 \\ 46,500 \end{bmatrix} = [1,173,600].$$

The total cost is $1,173,600.

EXAMPLE 10

In Example 6 of Sec. 14-2, suppose the price of coal is $10,000 per unit, the price of electricity is $20,000 per unit, and the price of steel is $40,000 per unit. These prices can be represented by the (column) price matrix

$$\mathbf{P} = \begin{bmatrix} 10,000 \\ 20,000 \\ 40,000 \end{bmatrix}.$$

Consider the steel industry. It sells a total of 30 units of steel at $40,000 per unit and its total income is therefore $1,200,000. Its costs for the various goods are given by the matrix product

$$D_S P = [30 \quad 5 \quad 0] \begin{bmatrix} 10{,}000 \\ 20{,}000 \\ 40{,}000 \end{bmatrix} = [400{,}000].$$

Hence the profit for the steel industry is $1,200,000 − $400,000 = $800,000.

EXERCISE 14-3

If

$$A = \begin{bmatrix} 1 & 3 & -2 \\ -2 & 1 & -1 \\ 0 & 4 & 3 \end{bmatrix}, \qquad B = \begin{bmatrix} 0 & -2 & 3 \\ -2 & 4 & -2 \\ 3 & 1 & -1 \end{bmatrix},$$

and $AB = C$, *find each of the following.*

1. c_{11}.   2. c_{23}.

3. c_{32}.   4. c_{33}.

5. c_{22}.   6. c_{13}.

If A is 2×3, B is 3×1, C is 2×5, D is 4×3, E is 3×2, and F is 2×3, find the order and number of entries of each of the following.

7. AE.   8. DE.

9. EC.   10. DB.

11. FB.   12. BA.

13. EA.   14. E(AE).

15. E(FB).   16. (F + A)B.

Write the identity matrix that has the following order.

17. 4.   18. 6.

In Problems 19–36 perform the indicated operations.

19. $\begin{bmatrix} 2 & -4 \\ 3 & 2 \end{bmatrix} \begin{bmatrix} 3 & 0 \\ -1 & 4 \end{bmatrix}$.   20. $\begin{bmatrix} -1 & 1 \\ 0 & 4 \\ 2 & 1 \end{bmatrix} \begin{bmatrix} 1 & -2 \\ 3 & 4 \end{bmatrix}$.

21. $\begin{bmatrix} 2 & 0 & 3 \\ -1 & 4 & 5 \end{bmatrix} \begin{bmatrix} 1 \\ 4 \\ 7 \end{bmatrix}$.   22. $[1 \quad 0 \quad 6 \quad 2] \begin{bmatrix} 0 \\ 1 \\ 2 \\ 3 \end{bmatrix}$.

23. $\begin{bmatrix} 1 & 4 & -1 \\ 0 & 0 & 2 \\ -2 & 1 & 1 \end{bmatrix} \begin{bmatrix} -2 & 1 & 0 \\ 0 & 1 & 1 \\ 1 & 1 & 2 \end{bmatrix}.$ **24.** $\begin{bmatrix} 3 & 2 & -1 \\ 4 & 10 & 0 \\ 0 & 1 & 2 \end{bmatrix} \begin{bmatrix} 2 & 0 & 1 & 0 \\ 0 & 1 & 0 & 0 \\ 0 & 1 & 0 & 1 \end{bmatrix}.$

25. $\begin{bmatrix} -1 & 2 & 3 \end{bmatrix} \begin{bmatrix} 3 & 1 & -1 & 2 \\ 0 & 4 & 3 & 1 \\ -1 & 3 & 1 & -2 \end{bmatrix}.$ **26.** $\begin{bmatrix} 1 & -4 \end{bmatrix} \begin{bmatrix} -2 & 1 \\ 0 & 5 \\ 1 & 0 \end{bmatrix}.$

27. $\begin{bmatrix} 2 \\ 3 \\ -4 \\ 1 \end{bmatrix} \begin{bmatrix} 2 & 3 & -2 & 3 \end{bmatrix}.$

28. $\begin{bmatrix} 0 & 1 \\ 2 & 3 \end{bmatrix} \left(\begin{bmatrix} 1 & 0 & 1 \\ 0 & 1 & 0 \end{bmatrix} + \begin{bmatrix} 0 & 1 & 0 \\ 0 & 0 & 1 \end{bmatrix} \right).$

29. $3 \left(\begin{bmatrix} -2 & 0 & 2 \\ 3 & -1 & 1 \end{bmatrix} + 2 \begin{bmatrix} -1 & 0 & 2 \\ 1 & 1 & -2 \end{bmatrix} \right) \begin{bmatrix} 1 & 2 \\ 3 & 4 \\ 5 & 6 \end{bmatrix}.$

30. $\begin{bmatrix} -1 & 3 \\ -1 & 0 \end{bmatrix} \begin{bmatrix} -1 & 0 & 2 & -1 \\ 2 & 1 & -3 & -2 \end{bmatrix}.$ **31.** $\begin{bmatrix} 1 & 2 \\ 3 & 4 \end{bmatrix} \left(\begin{bmatrix} 2 & 0 & 1 \\ 1 & 0 & -2 \end{bmatrix} \begin{bmatrix} 1 & -2 \\ 2 & 1 \\ 3 & 0 \end{bmatrix} \right).$

32. $3 \begin{bmatrix} 1 & 2 \\ -1 & 4 \end{bmatrix} - 4 \left(\begin{bmatrix} 1 & 0 \\ 0 & 1 \end{bmatrix} \begin{bmatrix} -2 & 4 \\ 6 & 1 \end{bmatrix} \right).$ **33.** $\begin{bmatrix} 1 & 0 & 0 \\ 0 & 1 & 0 \\ 0 & 0 & 1 \end{bmatrix} \begin{bmatrix} x \\ y \\ z \end{bmatrix}.$

34. $\begin{bmatrix} a_{11} & a_{12} \\ a_{21} & a_{22} \end{bmatrix} \begin{bmatrix} x_1 \\ x_2 \end{bmatrix}.$ **35.** $\begin{bmatrix} 2 & 1 & 3 \\ 4 & 9 & 7 \end{bmatrix} \begin{bmatrix} x_1 \\ x_2 \\ x_3 \end{bmatrix}.$

36. $\begin{bmatrix} 1 & -2 \\ 0 & 1 \\ 3 & 2 \end{bmatrix} \begin{bmatrix} x_1 \\ x_2 \end{bmatrix}.$

In Problems **37–51** *compute the required matrices if*

$$A = \begin{bmatrix} 1 & -2 \\ 0 & 3 \end{bmatrix}, \qquad B = \begin{bmatrix} -2 & 3 & 0 \\ 1 & -4 & 1 \end{bmatrix}, \qquad C = \begin{bmatrix} -1 & 1 \\ 0 & 3 \\ 2 & 4 \end{bmatrix},$$

$$D = \begin{bmatrix} 1 & 0 & 0 \\ 0 & 1 & 1 \\ 1 & 2 & 1 \end{bmatrix}, \qquad E = \begin{bmatrix} 1 & 2 & 4 \end{bmatrix}, \qquad F = \begin{bmatrix} 2 \\ 1 \end{bmatrix},$$

$$G = \begin{bmatrix} 3 & 0 & 0 \\ 0 & 6 & 0 \\ 0 & 0 & 3 \end{bmatrix}, \qquad H = \begin{bmatrix} \frac{1}{3} & 0 & 0 \\ 0 & \frac{1}{6} & 0 \\ 0 & 0 & \frac{1}{3} \end{bmatrix}, \qquad I = \begin{bmatrix} 1 & 0 & 0 \\ 0 & 1 & 0 \\ 0 & 0 & 1 \end{bmatrix}.$$

37. AB.   **38.** BD.   **39.** CF.

40. FE − 3B.   **41.** DG.   **42.** D² (= DD).

43. EC.

44. GC.

45. DI − $\frac{1}{3}$G.

46. B(D + G).

47. 3A − 2BC.

48. G(2D − 3I).

49. 2I − $\frac{1}{2}$GH.

50. A(BC).

51. (DC)A.

In Problems 52–55 represent the given system by using matrix multiplication.

52. $\begin{cases} 2x - y = 4, \\ 3x + y = 5. \end{cases}$

53. $\begin{cases} 3x + y = 6, \\ 7x - 2y = 5. \end{cases}$

54. $\begin{cases} x + y + z = 6, \\ x - y + z = 2, \\ 2x - y + 3z = 6. \end{cases}$

55. $\begin{cases} 4r - s + 3t = 9, \\ 3r \quad - t = 7, \\ 3s + 2t = 15. \end{cases}$

56. A stockbroker sold a customer 200 shares of Stock *A*, 300 shares of Stock *B*, 500 shares of Stock *C*, and 250 shares of Stock *D*. The prices per share of *A*, *B*, *C*, and *D* are $100, $150, $200, and $300, respectively. Write a row matrix representing the number of shares of each stock bought. Write a column matrix representing the price per share of each stock. Using matrix multiplication, find the total cost of stocks.

57. In Example 9 assume that the contractor is to build seven ranch-style, three Cape Cod, and five colonial-style houses. Compute, using matrix multiplication, the total cost of raw materials.

58. In Example 9 assume that the contractor wishes to take into account the cost of transporting raw materials to the building site as well as the purchasing cost. Suppose the costs are given in the matrix below.

$$\mathbf{C} = \begin{matrix} & \text{Purchase} & \text{Transport} \\ \begin{bmatrix} & & \\ & & \\ & & \\ & & \\ & & \end{bmatrix} & \begin{matrix} 1500 \\ 800 \\ 500 \\ 100 \\ 1000 \end{matrix} & \begin{matrix} 45 \\ 20 \\ 30 \\ 5 \\ 0 \end{matrix} \end{matrix} \begin{matrix} \text{Steel} \\ \text{Wood} \\ \text{Glass} \\ \text{Paint} \\ \text{Labor.} \end{matrix}$$

(a) By computing **RC**, find a matrix whose entries give the purchase and transportation costs of the materials for each kind of house.

(b) Find the matrix **QRC** whose first entry gives the total purchase price and whose second entry gives the total transportation cost.

(c) Let $\mathbf{Z} = \begin{bmatrix} 1 \\ 1 \end{bmatrix}$ and then compute **QRCZ**, which gives the total cost of materials and transportation for all houses being built.

59. Perform the following calculations for Example 10:

(a) Compute the amount that each industry and each consumer have to pay for the goods they receive.

(b) Compute the profit earned by each industry.

(c) Find the total amount of money that is paid out by all the industries and consumers.

(d) Find the proportion of the total amount of money found in (c) paid out by the industries. Find the proportion of the total amount of money found in (c) that is paid out by the consumers.

14-4 METHOD OF REDUCTION

In this section we shall illustrate a method by which matrices can be used to solve a system of linear equations, *the method of reduction*. In introducing the method we shall first solve a system in the usual way. Then we shall obtain the same solution by using matrices.

Let us consider the system

$$\begin{cases} 3x - y = 1, & \text{(1)} \\ x + 2y = 5 & \text{(2)} \end{cases}$$

consisting of two linear equations in two unknowns, x and y. Although this system can be solved by various algebraic methods, we shall solve it by a method which is readily adapted to matrices.

For reasons that will be obvious later, we begin by replacing Eq. (1) by Eq. (2), and Eq. (2) by Eq. (1), thus obtaining the equivalent system

$$\begin{cases} x + 2y = 5, & \text{(3)} \\ 3x - y = 1. & \text{(4)} \end{cases}$$

In $x + 2y = 5$, multiplying both sides by -3 gives $-3x - 6y = -15$. Adding the left and right sides of this equation to the corresponding sides of Eq. (4) gives the equivalent system

$$\begin{cases} x + 2y = 5, & \text{(5)} \\ 0x - 7y = -14. & \text{(6)} \end{cases}$$

Multiplying both sides of Eq. (6) by $-\frac{1}{7}$ gives the equivalent system

$$\begin{cases} x + 2y = 5, & \text{(7)} \\ 0x + y = 2. & \text{(8)} \end{cases}$$

By Eq. (8), $y = 2$ and hence $-2y = -4$. Adding the sides of $-2y = -4$ to the corresponding sides of Eq. (7), we get the equivalent system

$$\begin{cases} x + 0y = 1, \\ 0x + y = 2. \end{cases}$$

Therefore, $x = 1$ and $y = 2$, and the original system is solved.

Before showing a method of solving

$$\begin{cases} 3x - y = 1, \\ x + 2y = 5 \end{cases}$$

by matrices, we first define some terms. We say that the matrix

$$\begin{bmatrix} 3 & -1 \\ 1 & 2 \end{bmatrix}$$

is the **coefficient matrix** of this system. The entries in the first column correspond to the coefficients of the x's in the equations. For example, the entry in the first row and first column corresponds to the coefficient of x in the first equation; the entry in the second row and first column corresponds to the coefficient of x in the second equation. Similarly, the entries in the second column correspond to the coefficients of the y's.

Another matrix associated with this system is called the **augmented coefficient matrix** and is given by

$$\begin{bmatrix} 3 & -1 & \vdots & 1 \\ 1 & 2 & \vdots & 5 \end{bmatrix}.$$

The first and second columns are the first and second columns, respectively, of the coefficient matrix. The entries in the third column correspond to the constant terms in the system: the entry in the first row of this column is the constant term of the first equation, while the entry in the second row is the constant term of the second equation. Although it is not necessary to include the broken line in the augmented coefficient matrix, it serves to remind us that the 1 and the 5 are the constant terms that appear on the right sides of the equations. The augmented coefficient matrix itself completely describes the system of equations.

The procedure that was used to solve the original system involved a number of equivalent systems. With each of these systems we can associate its augmented coefficient matrix. Listed below are the systems that were involved, together with their corresponding augmented coefficient matrices, which we have labeled **A**, **B**, **C**, **D**, and **E**.

$$\begin{cases} 3x - y = 1, \\ x + 2y = 5. \end{cases} \qquad \begin{bmatrix} 3 & -1 & \vdots & 1 \\ 1 & 2 & \vdots & 5 \end{bmatrix} = \mathbf{A}.$$

$$\begin{cases} x + 2y = 5, \\ 3x - y = 1. \end{cases} \qquad \begin{bmatrix} 1 & 2 & \vdots & 5 \\ 3 & -1 & \vdots & 1 \end{bmatrix} = \mathbf{B}.$$

$$\begin{cases} x + 2y = 5, \\ 0x - 7y = -14. \end{cases} \qquad \begin{bmatrix} 1 & 2 & \vdots & 5 \\ 0 & -7 & \vdots & -14 \end{bmatrix} = \mathbf{C}.$$

$$\begin{cases} x + 2y = 5, \\ 0x + y = 2. \end{cases} \qquad \begin{bmatrix} 1 & 2 & \vdots & 5 \\ 0 & 1 & \vdots & 2 \end{bmatrix} = \mathbf{D}.$$

$$\begin{cases} x + 0y = 1, \\ 0x + y = 2. \end{cases} \qquad \begin{bmatrix} 1 & 0 & \vdots & 1 \\ 0 & 1 & \vdots & 2 \end{bmatrix} = \mathbf{E}.$$

Let us see how these matrices are related.

B can be obtained from **A** by interchanging the first and second rows of **A**. This operation corresponds to the interchanging of the two equations in the original system.

C can be obtained from **B** by adding to each entry in the second row of **B**, -3 times the corresponding entry in the first row of **B**.

$$C = \begin{bmatrix} 1 & 2 & \vdots & 5 \\ 3 + (-3)(1) & -1 + (-3)(2) & \vdots & 1 + (-3)(5) \end{bmatrix}$$

$$= \begin{bmatrix} 1 & 2 & \vdots & 5 \\ 0 & -7 & \vdots & -14 \end{bmatrix}.$$

This operation is described as the addition of -3 times the first row of **B** to the second row of **B**.

D can be obtained from **C** by multiplying each entry in the second row of **C** by $-\frac{1}{7}$. This operation is referred to as multiplying the second row of **C** by $-\frac{1}{7}$.

E can be obtained from **D** by adding -2 times the second row of **D** to the first row of **D**.

Observe that **E**, which essentially gives the solution, can be obtained from **A** by a series of operations which include:

(1) interchanging two rows of a matrix;

(2) adding a multiple of one row of a matrix to a different row of that matrix;

(3) multiplying a row of a matrix by a nonzero scalar.

We refer to these operations as **elementary row operations**. Whenever a matrix can be obtained from another by one or more elementary row operations, we say that the matrices are **equivalent**. Thus **A** is equivalent to **E**, and we write **A ~ E**.

We are now ready to describe a matrix procedure for solving a system of linear equations. First, form the augmented coefficient matrix of the system; then, by means of elementary row operations, determine an equivalent matrix that clearly indicates the solution. Let us be quite specific as to what we mean by a matrix that clearly indicates the solution. It is a matrix, called a **reduced matrix**, such that

(1) the first nonzero entry in each row is 1 while all other entries in the column in which the 1 appears are zeros,

(2) the first nonzero entry in each row is to the right of the first nonzero entry of each preceding row,

(3) **each row that consists entirely of zeros is below each row that contains a nonzero entry.**[†]

In other words, to solve the system we must find a reduced matrix such that the augmented coefficient matrix is equivalent to it. Note that **E** above,

$$E = \begin{bmatrix} 1 & 0 & \vdots & 1 \\ 0 & 1 & \vdots & 2 \end{bmatrix},$$

is a reduced matrix.

EXAMPLE 1

For each matrix below, determine whether it is reduced or not reduced.

a. $\begin{bmatrix} 1 & 0 \\ 0 & 3 \end{bmatrix}$.

b. $\begin{bmatrix} 1 & 0 & 0 \\ 0 & 1 & 0 \end{bmatrix}$.

c. $\begin{bmatrix} 0 & 1 \\ 1 & 0 \end{bmatrix}$.

d. $\begin{bmatrix} 0 & 0 & 0 \\ 0 & 0 & 0 \end{bmatrix}$.

e. $\begin{bmatrix} 1 & 0 & 0 \\ 0 & 0 & 0 \\ 0 & 1 & 0 \end{bmatrix}$.

f. $\begin{bmatrix} 0 & 1 & 0 & 3 \\ 0 & 0 & 1 & 2 \\ 0 & 0 & 0 & 0 \end{bmatrix}$.

a. Not a reduced matrix, since the first nonzero entry in the second row is not 1.

b. Reduced matrix.

c. Not a reduced matrix, since the first nonzero entry in the second row is not to the right of the first nonzero entry in the first row.

d. Reduced matrix.

e. Not a reduced matrix, since the second row, consisting entirely of zeros, is not below each row which contains nonzero entries.

f. Reduced matrix.

The method of reduction we have described for solving our original system can be generalized to systems consisting of m linear equations in n unknowns.

[†]From Paul C. Shields, *Elementary Linear Algebra* (New York: Worth Publishers, Inc., 2nd ed., 1973), p. 7.

To solve such a system as

$$
\begin{cases}
a_{11}x_1 + a_{12}x_2 + \cdots + a_{1n}x_n = c_1, \\
a_{21}x_1 + a_{22}x_2 + \cdots + a_{2n}x_n = c_2, \\
\quad\vdots \qquad \vdots \qquad\qquad \vdots \qquad \vdots \\
a_{m1}x_1 + a_{m2}x_2 + \cdots + a_{mn}x_n = c_m
\end{cases}
$$

involves

i. determining the augmented coefficient matrix of the system:

$$
\begin{bmatrix}
a_{11} & a_{12} & \cdots & a_{1n} & c_1 \\
a_{21} & a_{22} & \cdots & a_{2n} & c_2 \\
\vdots & \vdots & & \vdots & \vdots \\
a_{m1} & a_{m2} & \cdots & a_{mn} & c_m
\end{bmatrix}
$$

and

ii. determining a reduced matrix such that the augmented coefficient matrix is equivalent to it.

Frequently, step ii is called *reducing the augmented coefficient matrix.*

EXAMPLE 2

By using matrix reduction, solve the system

$$
\begin{cases}
2x + 3y = -1, \\
2x + \ y = 5, \\
\ x + \ y = 1.
\end{cases}
$$

The augmented coefficient matrix of the system is

$$
\begin{bmatrix}
2 & 3 & -1 \\
2 & 1 & 5 \\
1 & 1 & 1
\end{bmatrix}.
$$

Reducing this matrix, we have

$$\begin{bmatrix} 2 & 3 & \vdots & -1 \\ 2 & 1 & \vdots & 5 \\ 1 & 1 & \vdots & 1 \end{bmatrix}$$

$$\sim \begin{bmatrix} 1 & 1 & \vdots & 1 \\ 2 & 1 & \vdots & 5 \\ 2 & 3 & \vdots & -1 \end{bmatrix} \qquad \text{(by interchanging the first and third rows)}$$

$$\sim \begin{bmatrix} 1 & 1 & \vdots & 1 \\ 0 & -1 & \vdots & 3 \\ 2 & 3 & \vdots & -1 \end{bmatrix} \qquad \text{(by adding } -2 \text{ times the first row to the second)}$$

$$\sim \begin{bmatrix} 1 & 1 & \vdots & 1 \\ 0 & -i & \vdots & 3 \\ 0 & 1 & \vdots & -3 \end{bmatrix} \qquad \text{(by adding } -2 \text{ times the first row to the third)}$$

$$\sim \begin{bmatrix} 1 & 1 & \vdots & 1 \\ 0 & 1 & \vdots & -3 \\ 0 & 1 & \vdots & -3 \end{bmatrix} \qquad \text{(by multiplying the second row by } -1)$$

$$\sim \begin{bmatrix} 1 & 0 & \vdots & 4 \\ 0 & 1 & \vdots & -3 \\ 0 & 1 & \vdots & -3 \end{bmatrix} \qquad \text{(by adding } -1 \text{ times the second row to the first)}$$

$$\sim \begin{bmatrix} 1 & 0 & \vdots & 4 \\ 0 & 1 & \vdots & -3 \\ 0 & 0 & \vdots & 0 \end{bmatrix} \qquad \text{(by adding } -1 \text{ times the second row to the third).}$$

The last matrix is reduced and corresponds to the system

$$\begin{cases} x + 0y = 4, \\ 0x + y = -3, \\ 0x + 0y = 0. \end{cases}$$

Since the original system is equivalent to this system, it has a unique solution, namely

$$x = 4,$$

$$y = -3.$$

We point out that the sequence of steps that are used to reduce a matrix is not unique.

EXAMPLE 3

Using matrix reduction, solve

$$\begin{cases} x + 2y + 4z - 6 = 0, \\ 2z + y - 3 = 0, \\ x + y + 2z - 1 = 0. \end{cases}$$

Rewriting the system so that the variables are aligned and the constant terms appear on the right sides of the equations, we have

$$\begin{cases} x + 2y + 4z = 6, \\ y + 2z = 3, \\ x + y + 2z = 1. \end{cases}$$

Reducing the augmented coefficient matrix, we have

$$\begin{bmatrix} 1 & 2 & 4 & \vdots & 6 \\ 0 & 1 & 2 & \vdots & 3 \\ 1 & 1 & 2 & \vdots & 1 \end{bmatrix}$$

$$\sim \begin{bmatrix} 1 & 2 & 4 & \vdots & 6 \\ 0 & 1 & 2 & \vdots & 3 \\ 0 & -1 & -2 & \vdots & -5 \end{bmatrix}$$ (by adding -1 times the first row to the third)

$$\sim \begin{bmatrix} 1 & 0 & 0 & \vdots & 0 \\ 0 & 1 & 2 & \vdots & 3 \\ 0 & 0 & 0 & \vdots & -2 \end{bmatrix}$$ (by adding -2 times the second row to the first, and adding the second row to the third)

$$\sim \begin{bmatrix} 1 & 0 & 0 & \vdots & 0 \\ 0 & 1 & 2 & \vdots & 3 \\ 0 & 0 & 0 & \vdots & 1 \end{bmatrix}$$ (by multiplying the third row by $-\frac{1}{2}$)

$$\sim \begin{bmatrix} 1 & 0 & 0 & \vdots & 0 \\ 0 & 1 & 2 & \vdots & 0 \\ 0 & 0 & 0 & \vdots & 1 \end{bmatrix}$$ (by adding -3 times the third row to the second).

The last matrix is reduced and corresponds to

$$\begin{cases} x = 0, \\ y + 2z = 0, \\ 0 = 1. \end{cases}$$

Since $0 \neq 1$, there are no values of x, y, and z for which all equations are satisfied simultaneously. Thus, the original system has no solution.

EXAMPLE 4

Using matrix reduction, solve

$$\begin{cases} 2x_1 + 3x_2 + 2x_3 + 6x_4 = 10, \\ x_2 + 2x_3 + x_4 = 2, \\ 3x_1 - 3x_3 + 6x_4 = 9. \end{cases}$$

Reducing the augmented coefficient matrix, we have

$$\begin{bmatrix} 2 & 3 & 2 & 6 & \vdots & 10 \\ 0 & 1 & 2 & 1 & \vdots & 2 \\ 3 & 0 & -3 & 6 & \vdots & 9 \end{bmatrix}$$

$$\sim \begin{bmatrix} 1 & \frac{3}{2} & 1 & 3 & \vdots & 5 \\ 0 & 1 & 2 & 1 & \vdots & 2 \\ 3 & 0 & -3 & 6 & \vdots & 9 \end{bmatrix} \quad \text{(by multiplying the first row by } \tfrac{1}{2}\text{)}$$

$$\sim \begin{bmatrix} 1 & \frac{3}{2} & 1 & 3 & \vdots & 5 \\ 0 & 1 & 2 & 1 & \vdots & 2 \\ 0 & -\frac{9}{2} & -6 & -3 & \vdots & -6 \end{bmatrix} \quad \begin{array}{l} \text{(by adding } -3 \text{ times the first row to} \\ \text{the third)} \end{array}$$

$$\sim \begin{bmatrix} 1 & 0 & -2 & \frac{3}{2} & \vdots & 2 \\ 0 & 1 & 2 & 1 & \vdots & 2 \\ 0 & 0 & 3 & \frac{3}{2} & \vdots & 3 \end{bmatrix} \quad \begin{array}{l} \text{(by adding } -\frac{3}{2} \text{ times the second row to the} \\ \text{first, and adding } \frac{9}{2} \text{ times the second row to} \\ \text{the third)} \end{array}$$

$$\sim \begin{bmatrix} 1 & 0 & -2 & \frac{3}{2} & \vdots & 2 \\ 0 & 1 & 2 & 1 & \vdots & 2 \\ 0 & 0 & 1 & \frac{1}{2} & \vdots & 1 \end{bmatrix} \quad \text{(by multiplying the third row by } \tfrac{1}{3}\text{)}$$

$$\sim \begin{bmatrix} 1 & 0 & 0 & \frac{5}{2} & \vdots & 4 \\ 0 & 1 & 0 & 0 & \vdots & 0 \\ 0 & 0 & 1 & \frac{1}{2} & \vdots & 1 \end{bmatrix} \quad \begin{array}{l} \text{(by adding 2 times the third row to the first,} \\ \text{and adding } -2 \text{ times the third row to the} \\ \text{second).} \end{array}$$

The last matrix is reduced and corresponds to the system

$$\begin{cases} x_1 + \frac{5}{2}x_4 = 4, \\ x_2 = 0, \\ x_3 + \frac{1}{2}x_4 = 1. \end{cases}$$

Thus,

$$x_1 = -\frac{5}{2}x_4 + 4, \tag{9}$$

$$x_2 = 0, \tag{10}$$

$$x_3 = -\frac{1}{2}x_4 + 1, \tag{11}$$

$$x_4 = x_4. \tag{12}$$

If x_4 is any real number, then Eqs. (9)–(12) determine a particular solution to the original system. For example if $x_4 = 0$, then a *particular* solution is $x_1 = 4$, $x_2 = 0$, $x_3 = 1$, and $x_4 = 0$. If $x_4 = 2$, then $x_1 = -1$, $x_2 = 0$, $x_3 = 0$, and $x_4 = 2$ is a particular solution. The variable x_4, on which x_1 and x_3 depend, is called a *parameter*. Clearly, there are an infinite number of solutions to the system—one corresponding to each value of the parameter. We say that the *general* solution of the original system is given by Eqs. (9)–(12).

EXERCISE 14-4

In each of Problems **1–6**, *determine whether the matrix is reduced or not reduced.*

1. $\begin{bmatrix} 1 & 2 \\ 3 & 0 \end{bmatrix}$.

2. $\begin{bmatrix} 1 & 0 & 0 & 3 \\ 0 & 0 & 1 & 2 \end{bmatrix}$.

3. $\begin{bmatrix} 1 & 0 & 0 \\ 0 & 1 & 0 \\ 0 & 0 & 1 \end{bmatrix}$.

4. $\begin{bmatrix} 1 & 1 \\ 0 & 1 \\ 0 & 0 \\ 0 & 0 \end{bmatrix}$.

5. $\begin{bmatrix} 0 & 0 & 0 & 0 \\ 0 & 1 & 0 & 0 \\ 0 & 0 & 1 & 0 \\ 0 & 0 & 0 & 0 \end{bmatrix}$.

6. $\begin{bmatrix} 0 & 0 & 5 \\ 1 & 0 & 4 \\ 0 & 1 & 2 \\ 0 & 0 & 0 \end{bmatrix}$.

In each of Problems **7–12**, *reduce the given matrix.*

7. $\begin{bmatrix} 1 & 3 \\ 4 & 0 \end{bmatrix}$.

8. $\begin{bmatrix} 0 & -2 & 0 & 1 \\ 1 & 2 & 0 & 4 \end{bmatrix}$.

9. $\begin{bmatrix} 2 & 4 & 6 \\ 1 & 2 & 3 \\ 1 & 2 & 3 \end{bmatrix}$.

10. $\begin{bmatrix} 2 & 3 \\ 1 & -6 \\ 4 & 8 \\ 1 & 7 \end{bmatrix}$.

11. $\begin{bmatrix} 2 & 0 & 3 & 1 \\ 1 & 4 & 2 & 2 \\ -1 & 3 & 1 & 4 \\ 0 & 2 & 1 & 0 \end{bmatrix}$.

12. $\begin{bmatrix} 0 & 0 & 2 \\ 2 & 0 & 3 \\ 0 & -1 & 0 \\ 0 & 4 & 1 \end{bmatrix}$.

Solve Problems **13–26** *by the method of reduction.*

13. $\begin{cases} 2x + 3y = 5, \\ x - 2y = -1. \end{cases}$

14. $\begin{cases} x - 3y = -11, \\ 4x + 3y = 9. \end{cases}$

15. $\begin{cases} 3x + y = 4, \\ 12x + 4y = 2. \end{cases}$

16. $\begin{cases} x + 2y - 3z = 0, \\ -2x - 4y + 6z = 1. \end{cases}$

17. $\begin{cases} x + 2y + z - 4 = 0, \\ 3x + 2z - 5 = 0. \end{cases}$

18. $\begin{cases} x + 2y + 5z - 1 = 0, \\ x + y + 3z - 2 = 0. \end{cases}$

19. $\begin{cases} x_1 - 3x_2 = 0, \\ 2x_1 + 2x_2 = 3, \\ 5x_1 - x_2 = 1. \end{cases}$

20. $\begin{cases} x_1 + 3x_2 = 5, \\ 2x_1 + x_2 = 5, \\ x_1 + x_2 = 3. \end{cases}$

21. $\begin{cases} x - y - 3z = -4, \\ 2x - y - 4z = -7, \\ x + y - z = -2. \end{cases}$

22. $\begin{cases} x + y - z = 6, \\ 2x - 3y - 2z = 2, \\ x - y - 5z = 18. \end{cases}$

23. $\begin{cases} 2x - 4z = 8, \\ x - 2y - 2z = 14, \\ x + y - 2z = -1, \\ 3x + y + z = 0. \end{cases}$

24. $\begin{cases} x + 3z = -1, \\ 3x + 2y + 11z = 1, \\ x + y + 4z = 1, \\ 2x - 3y + 3z = -8. \end{cases}$

25. $\begin{cases} x_1 + x_2 - x_3 + x_4 + x_5 = 0, \\ x_1 + x_2 + x_3 - x_4 + x_5 = 0, \\ x_1 - x_2 - x_3 + x_4 - x_5 = 0, \\ x_1 + x_2 - x_3 - x_4 - x_5 = 0. \end{cases}$

26. $\begin{cases} x_1 + x_2 - x_3 + x_4 = 0, \\ x_1 + x_2 + x_3 - x_4 = 0, \\ x_1 - x_2 - x_3 + x_4 = 0, \\ x_1 + x_2 - x_3 - x_4 = 0. \end{cases}$

Solve Problems 27–31 by using matrix reduction.

27. A company has taxable income of $312,000. The federal tax is 25 percent of that portion which is left after the state tax has been paid. The state tax is 10 percent of that portion which is left after the federal tax has been paid. Find the federal and state taxes.

28. A manufacturer produces two products, A and B. For each unit of A sold the profit is $8, and for each unit of B sold the profit is $11. From past experience it has been found that 25 percent more of A can be sold than of B. Next year the manufacturer desires a total profit of $42,000. How many units of each product must be sold?

29. A manufacturer produces three products, A, B, and C. The profits for each unit sold of A, B, and C are $1, 2, and $3, respectively. Fixed costs are $17,000 per year and the costs of producing each unit of A, B, and C are $4, $5, and $7, respectively. Next year, a total of 11,000 units of all three products is to be produced and sold, and a total profit of $25,000 is to be realized. If total cost is to be $80,000, how many units of each of the products should be produced next year?

30. National Desk Co. has plants for producing desks on both the east and west coasts. At the east coast plant, fixed costs are $16,000 per year and the cost of producing each desk is $90. At the west coast plant, fixed costs are $20,000 per year and the cost of producing each desk is $80. Next year the company wants to produce a total of 800 desks. Determine the production order for each plant for the forthcoming year if the total cost for each plant is to be the same.

31. A person is ordered by a doctor to take 10 units of vitamin A, 9 units of vitamin D, and 19 units of vitamin E each day. The person can choose from three brands of vitamin pills. Brand X contains two units of vitamin A, three units of vitamin D, and five units of vitamin E; brand Y has 1, 3, and 4 units, respectively; and brand Z has 1 unit of vitamin A, none of vitamin D, and 1 of vitamin E.

    (a) Find all possible combinations of pills that will provide exactly the required amounts of vitamins.

(b) If brand X costs 1¢ a pill, brand Y 6¢, and brand Z 3¢, are there any combinations in part (a) costing exactly 15¢ a day?

(c) What is the least expensive combination in part (a)? The most expensive?

14-5 METHOD OF REDUCTION (CONTINUED)[†]

As we saw in Sec. 14-4, a system of linear equations may have a unique solution, no solution, or an infinite number of solutions. When there are infinitely many, the general solution is expressed in terms of at least one parameter. For example, the general solution in Example 4 was given in terms of the parameter x_4:

$$x_1 = -\tfrac{5}{2}x_4 + 4,$$
$$x_2 = 0,$$
$$x_3 = -\tfrac{1}{2}x_4 + 1,$$
$$x_4 = x_4.$$

At times, more than one parameter is necessary, as the following example shows.

EXAMPLE 1

Using matrix reduction, solve

$$\begin{cases} x_1 + 2x_2 + 5x_3 + 5x_4 = -3, \\ x_1 + \ x_2 + 3x_3 + 4x_4 = -1, \\ x_1 - \ x_2 - \ x_3 + 2x_4 = 3. \end{cases}$$

The augmented coefficient matrix is

$$\begin{bmatrix} 1 & 2 & 5 & 5 & \vdots & -3 \\ 1 & 1 & 3 & 4 & \vdots & -1 \\ 1 & -1 & -1 & 2 & \vdots & 3 \end{bmatrix},$$

which is equivalent to the reduced matrix

$$\begin{bmatrix} 1 & 0 & 1 & 3 & \vdots & 1 \\ 0 & 1 & 2 & 1 & \vdots & -2 \\ 0 & 0 & 0 & 0 & \vdots & 0 \end{bmatrix}.$$

Hence,

$$\begin{cases} x_1 + \ x_3 + 3x_4 = 1, \\ x_2 + 2x_3 + \ x_4 = -2. \end{cases}$$

[†] This section may be omitted.

Thus the general solution can be given by

$$x_1 = 1 - x_3 - 3x_4,$$

$$x_2 = -2 - 2x_3 - x_4,$$

$$x_3 = x_3,$$

$$x_4 = x_4,$$

where parameters x_3 and x_4 are involved. By assigning specific values to x_3 and x_4, we get particular solutions. For example, if $x_3 = 1$ and $x_4 = 2$, then the corresponding particular solution is $x_1 = -6$, $x_2 = -6$, $x_3 = 1$, and $x_4 = 2$.

It is customary to classify a system of linear equations as being either *homogeneous* or *nonhomogeneous*. The appropriate classification depends on the constant terms, as the following definition indicates.

DEFINITION.   *The system*

$$\begin{cases} a_{11}x_1 + a_{12}x_2 + \cdots + a_{1n}x_n = c_1, \\ a_{21}x_1 + a_{22}x_2 + \cdots + a_{2n}x_n = c_2, \\ \phantom{a_{11}x_1}\vdots \phantom{+a_{12}}\vdots \phantom{+\cdots +a_{1n}}\vdots \vdots \\ a_{m1}x_1 + a_{m2}x_2 + \cdots + a_{mn}x_n = c_m \end{cases}$$

*is a **homogeneous system** if $c_1 = c_2 = \cdots = c_m = 0$. The system is a **nonhomogeneous** system if at least one of the c's is not equal to 0.*

EXAMPLE 2

The system

$$\begin{cases} 2x + 3y = 4, \\ 3x - 4y = 0 \end{cases}$$

is nonhomogeneous due to the 4 in the top equation. The system

$$\begin{cases} 2x + 3y = 0, \\ 3x - 4y = 0 \end{cases}$$

is homogeneous.

If the homogeneous system

$$\begin{cases} 2x + 3y = 0, \\ 3x - 4y = 0 \end{cases}$$

were solved by the method of reduction, first the augmented coefficient matrix would be written:

$$\begin{bmatrix} 2 & 3 & \vdots & 0 \\ 3 & -4 & \vdots & 0 \end{bmatrix}.$$

Observe that the last column consists entirely of zeros. This is typical of the augmented coefficient matrix of any homogeneous system. We would then reduce this matrix by using elementary row operations:

$$\begin{bmatrix} 2 & 3 & \vdots & 0 \\ 3 & -4 & \vdots & 0 \end{bmatrix} \sim \cdots \sim \begin{bmatrix} 1 & 0 & \vdots & 0 \\ 0 & 1 & \vdots & 0 \end{bmatrix}.$$

The last column of the reduced matrix also consists only of zeros. This does not occur by chance. When any elementary row operation is performed on a matrix that has a column consisting entirely of zeros, the corresponding column of the resulting matrix will also be all zeros. For convenience it will be our custom when solving a homogeneous system by matrix reduction to delete the last column of the matrices involved. That is, we shall reduce only the *coefficient matrix* of the system. For the above system we would have

$$\begin{bmatrix} 2 & 3 \\ 3 & -4 \end{bmatrix} \sim \cdots \sim \begin{bmatrix} 1 & 0 \\ 0 & 1 \end{bmatrix}.$$

Here the reduced matrix, called the *reduced coefficient matrix*, corresponds to

$$\begin{cases} x + 0y = 0, \\ 0x + y = 0, \end{cases}$$

and so the solution is $x = 0$ and $y = 0$.

Let us now consider the number of solutions of the homogeneous system

$$\begin{cases} a_{11}x_1 + a_{12}x_2 + \cdots + a_{1n}x_n = 0, \\ a_{21}x_1 + a_{22}x_2 + \cdots + a_{2n}x_n = 0, \\ \vdots \quad\quad \vdots \quad\quad\quad\quad \vdots \quad\quad \vdots \\ a_{m1}x_1 + a_{m2}x_2 + \cdots + a_{mn}x_n = 0. \end{cases}$$

One solution always occurs when $x_1 = 0$, $x_2 = 0, \cdots$, and $x_n = 0$ since each equation is satisfied for these values. This solution, called the **trivial solution**, is a solution of *every* homogeneous system.

There is a theorem which allows us to determine whether a homogeneous system has a unique solution (the trivial solution only) or an infinite number of solutions. The theorem is based on the number of nonzero rows that appear in the reduced coefficient matrix of the system. A *nonzero row* is a row that does not consist entirely of zeros.

Theorem.   Let **A** be the *reduced* coefficient matrix of a homogeneous system of m linear equations in n unknowns. If **A** has exactly k nonzero rows, then $k \leqslant n$. Moreover,

a. if $k < n$, the system has an infinite number of solutions;

and

b. if $k = n$, the system has a unique solution (the trivial solution).

If a homogeneous system consists of m equations in n unknowns, then the coefficient matrix of the system has dimension $m \times n$. Thus, if $m < n$ and k is the number of nonzero rows in the reduced coefficient matrix, then $k \leqslant m$ and hence $k < n$. By the theorem, the system must have an infinite number of solutions. Consequently we have the following.

Corollary.   A homogeneous system of linear equations with fewer equations than unknowns has an infinite number of solutions.

EXAMPLE 3

Determine whether the system

$$\begin{cases} x + y - 2z = 0, \\ 2x + 2y - 4z = 0 \end{cases}$$

has a unique solution or an infinite number of solutions.

There are two equations in this homogeneous system and this number is less than the number of unknowns (three). Thus by the corollary above, the system has an infinite number of solutions.

PITFALL.   *The theorem and corollary above apply only to **homogeneous systems** of linear equations. For example, consider the system*

$$\begin{cases} x + y - 2z = 3, \\ 2x + 2y - 4z = 4, \end{cases}$$

*which consists of two linear equations in three unknowns. We **cannot** conclude that this system has an infinite number of solutions, since it is not homogeneous. Indeed, you should verify that it has no solution.*

EXAMPLE 4

Determine whether the following homogeneous systems have a unique solution or an infinite number of solutions; then solve the system.

$$a. \quad \begin{cases} x - 2y + z = 0, \\ 2x - y + 5z = 0, \\ x + y + 4z = 0. \end{cases}$$

Reducing the coefficient matrix, we have

$$\begin{bmatrix} 1 & -2 & 1 \\ 2 & -1 & 5 \\ 1 & 1 & 4 \end{bmatrix} \sim \cdots \sim \begin{bmatrix} 1 & 0 & 3 \\ 0 & 1 & 1 \\ 0 & 0 & 0 \end{bmatrix}.$$

The number of nonzero rows (2) in the reduced coefficient matrix is less than the number of unknowns (3) in the system. By the theorem above, there are an infinite number of solutions.

Since the reduced coefficient matrix corresponds to

$$\begin{cases} x + 3z = 0, \\ y + z = 0, \end{cases}$$

the solution may be given by

$$x = -3z,$$

$$y = -z,$$

$$z = z,$$

where z is any real number.

b. $\begin{cases} 3x + 4y = 0, \\ x - 2y = 0, \\ 2x + y = 0, \\ 2x + 3y = 0. \end{cases}$

Reducing the coefficient matrix, we have

$$\begin{bmatrix} 3 & 4 \\ 1 & -2 \\ 2 & 1 \\ 2 & 3 \end{bmatrix} \sim \cdots \sim \begin{bmatrix} 1 & 0 \\ 0 & 1 \\ 0 & 0 \\ 0 & 0 \end{bmatrix}.$$

The number of nonzero rows (2) in the reduced coefficient matrix equals the number of unknowns in the system. By the theorem the system must have a unique solution, namely the trivial solution $x = 0, y = 0$.

EXERCISE 14-5

In Problems **1–8**, solve the systems by using matrix reduction.

1. $\begin{cases} w - x - y + 4z = 5, \\ 2w - 3x - 4y + 9z = 13, \\ 2w + x + 4y + 5z = 1. \end{cases}$

2. $\begin{cases} 3w - x + 12y + 18z = -4, \\ w - 2x + 4y + 11z = -13, \\ w + x + 4y + 2z = 8. \end{cases}$

3. $\begin{cases} 3w - x - 3y - z = -2, \\ 2w - 2x - 6y - 6z = -4, \\ 2w - x - 3y - 2z = -2, \\ 3w + x + 3y + 7z = 2. \end{cases}$

4. $\begin{cases} w + x + 5z = 1, \\ w + y + 2z = 1, \\ w - 3x + 4y - 7z = 1, \\ x - y + 3z = 0. \end{cases}$

5. $\begin{cases} w + x + 3y - z = 2, \\ 2w + x + 5y - 2z = 0, \\ 2w - x + 3y - 2z = -8, \\ 3w + 2x + 8y - 3z = 2, \\ w + 2y - z = -2. \end{cases}$

6. $\begin{cases} w + x + y + 2z = 4, \\ 2w + x + 2y + 2z = 7, \\ w + 2x + y + 4z = 5, \\ 3w - 2x + 3y - 4z = 7, \\ 4w - 3x + 4y - 6z = 9. \end{cases}$

7. $\begin{cases} 4x_1 - 3x_2 + 5x_3 - 10x_4 + 11x_5 = -8, \\ 2x_1 + x_2 + 5x_3 + 3x_5 = 6. \end{cases}$

8. $\begin{cases} x_1 + 2x_3 + x_4 + 4x_5 = 1, \\ x_2 + x_3 - 3x_4 = -2, \\ 4x_1 - 3x_2 + 5x_3 + 13x_4 + 16x_5 = 10, \\ x_1 + 2x_2 + 4x_3 - 5x_4 + 4x_5 = -3. \end{cases}$

For each of Problems **9–14**, *determine whether the system has an infinite number of solutions or only the trivial solution. Do not solve the systems.*

9. $\begin{cases} .07x + .3y + .02z = 0, \\ .053x - .4y + .08z = 0. \end{cases}$

10. $\begin{cases} 3w + 5x - 4y + 2z = 0, \\ 7w - 2x + 9y + 3z = 0. \end{cases}$

11. $\begin{cases} 3x - 4y = 0, \\ x + 5y = 0, \\ 4x - y = 0. \end{cases}$

12. $\begin{cases} 2x + 3y + 12z = 0, \\ 3x - 2y + 5z = 0, \\ 4x + y + 14z = 0. \end{cases}$

13. $\begin{cases} x + y + z = 0, \\ x - z = 0, \\ x - 2y - 5z = 0. \end{cases}$

14. $\begin{cases} 2x + 5y = 0, \\ x + 4y = 0, \\ 3x - 2y = 0. \end{cases}$

Solve each of the following systems.

15. $\begin{cases} x + y = 0, \\ 3x - 4y = 0. \end{cases}$

16. $\begin{cases} 2x - 5y = 0, \\ 8x - 20y = 0. \end{cases}$

17. $\begin{cases} x + 6y - 2z = 0, \\ 2x - 3y + 4z = 0. \end{cases}$

18. $\begin{cases} 4x + 7y = 0. \\ 2x + 3y = 0. \end{cases}$

19. $\begin{cases} x + y = 0, \\ 3x - 4y = 0, \\ 5x - 8y = 0. \end{cases}$

20. $\begin{cases} 4x - 3y + 2z = 0, \\ x + 2y + 3z = 0, \\ x + y + z = 0. \end{cases}$

21. $\begin{cases} x + y + z = 0, \\ 5x - 2y - 9z = 0, \\ 3x + y - z = 0, \\ 3x - 2y - 7z = 0. \end{cases}$

22. $\begin{cases} x + y + 7z = 0, \\ x - y - z = 0, \\ 2x - 3y - 6z = 0, \\ 3x + y + 13z = 0. \end{cases}$

23. $\begin{cases} w + x + y + 4z = 0, \\ w + x + 5z = 0, \\ 2w + x + 3y + 4z = 0, \\ w - 3x + 2y - 9z = 0. \end{cases}$

24. $\begin{cases} w + x + 2y + 7z = 0, \\ w - 2x - y + z = 0, \\ w + 2x + 3y + 9z = 0, \\ 2w - 3x - y + 4z = 0. \end{cases}$

14-6 INVERSES

We have seen how useful the method of reduction is for solving systems of linear equations. But it is by no means the only method which uses matrices. In this section we shall discuss a different method which applies to many systems of n linear equations in n unknowns.

To introduce the general technique, we consider the system

$$\begin{cases} a_{11}x_1 + a_{12}x_2 = c_1, \\ a_{21}x_1 + a_{22}x_2 = c_2. \end{cases}$$

We know from Sec. 14-3 that this system can be represented by the matrix equation

$$\begin{bmatrix} a_{11} & a_{12} \\ a_{21} & a_{22} \end{bmatrix} \begin{bmatrix} x_1 \\ x_2 \end{bmatrix} = \begin{bmatrix} c_1 \\ c_2 \end{bmatrix}. \tag{1}$$

Note that the 2×2 matrix in Eq. (1), which we shall denote by **A**, is the coefficient matrix of the system. Let us assume that there exists a 2×2 matrix **B**,

$$\mathbf{B} = \begin{bmatrix} p & q \\ r & s \end{bmatrix},$$

such that **B** times **A** is the identity matrix:

$$\begin{bmatrix} p & q \\ r & s \end{bmatrix} \begin{bmatrix} a_{11} & a_{12} \\ a_{21} & a_{22} \end{bmatrix} = \begin{bmatrix} 1 & 0 \\ 0 & 1 \end{bmatrix}.$$

If both sides of Eq. (1) are premultiplied by **B**, we have

$$\begin{bmatrix} p & q \\ r & s \end{bmatrix} \begin{bmatrix} a_{11} & a_{12} \\ a_{21} & a_{22} \end{bmatrix} \begin{bmatrix} x_1 \\ x_2 \end{bmatrix} = \begin{bmatrix} p & q \\ r & s \end{bmatrix} \begin{bmatrix} c_1 \\ c_2 \end{bmatrix},$$

$$\begin{bmatrix} 1 & 0 \\ 0 & 1 \end{bmatrix} \begin{bmatrix} x_1 \\ x_2 \end{bmatrix} = \begin{bmatrix} p & q \\ r & s \end{bmatrix} \begin{bmatrix} c_1 \\ c_2 \end{bmatrix},$$

$$\begin{bmatrix} x_1 \\ x_2 \end{bmatrix} = \begin{bmatrix} pc_1 + qc_2 \\ rc_1 + sc_2 \end{bmatrix}.$$

Thus, $x_1 = pc_1 + qc_2$, $x_2 = rc_1 + sc_2$, and the system is solved.

Summarizing our procedure, we first express the system as a matrix equation of the form

$$\mathbf{AX} = \mathbf{C}. \tag{2}$$

Then, provided there exists a matrix **B** such that $\mathbf{BA} = \mathbf{I}$, we premultiply both sides of Eq. (2) by $\mathbf{B}$:

$$\mathbf{BAX} = \mathbf{BC}.$$

Simplifying, we have

$$\mathbf{IX} = \mathbf{BC},$$
$$\mathbf{X} = \mathbf{BC}. \tag{3}$$

Thus the solution is given by $\mathbf{X} = \mathbf{BC}$. This procedure is based on our assuming the existence of a matrix $\mathbf{B}$ such that $\mathbf{BA} = \mathbf{I}$. When such a matrix does exist, we say that it is an *inverse* matrix of $\mathbf{A}$.

DEFINITION. *If $\mathbf{A}$ and $\mathbf{B}$ are $n \times n$ matrices, then $\mathbf{B}$ is an **inverse matrix** of $\mathbf{A}$ (or $\mathbf{B}$ is an inverse of $\mathbf{A}$) if and only if $\mathbf{BA} = \mathbf{I}$.*

EXAMPLE 1

Let $\mathbf{A} = \begin{bmatrix} 1 & 2 \\ 3 & 7 \end{bmatrix}$ and $\mathbf{B} = \begin{bmatrix} 7 & -2 \\ -3 & 1 \end{bmatrix}$. Since

$$\mathbf{BA} = \begin{bmatrix} 7 & -2 \\ -3 & 1 \end{bmatrix} \begin{bmatrix} 1 & 2 \\ 3 & 7 \end{bmatrix} = \begin{bmatrix} 1 & 0 \\ 0 & 1 \end{bmatrix},$$

$\mathbf{B}$ is an inverse matrix of $\mathbf{A}$.

It can be shown that if $\mathbf{B}$ is an inverse matrix of $\mathbf{A}$, then that inverse is unique. Thus in Example 1, $\mathbf{B}$ is the *only* matrix that has the property that $\mathbf{BA} = \mathbf{I}$. In keeping with common practice, we denote *the* inverse of a matrix $\mathbf{A}$ by $\mathbf{A}^{-1}$. Hence, $\mathbf{B} = \mathbf{A}^{-1}$ and

$$\mathbf{A}^{-1}\mathbf{A} = \mathbf{I}.$$

We may now write Eq. (3) as

$$\mathbf{X} = \mathbf{A}^{-1}\mathbf{C}. \tag{4}$$

It is also true that $\mathbf{A}^{-1}\mathbf{A} = \mathbf{A}\mathbf{A}^{-1}$. When $\mathbf{A}^{-1}$ does exist, we say $\mathbf{A}$ is **invertible** (or **nonsingular**).

Not all square matrices are invertible. For example, if

$$\mathbf{A} = \begin{bmatrix} 0 & 1 \\ 0 & 1 \end{bmatrix},$$

then

$$\begin{bmatrix} a & b \\ c & d \end{bmatrix} \begin{bmatrix} 0 & 1 \\ 0 & 1 \end{bmatrix} = \begin{bmatrix} 0 & a+b \\ 0 & c+d \end{bmatrix} \neq \begin{bmatrix} 1 & 0 \\ 0 & 1 \end{bmatrix}.$$

Hence there is no matrix which when postmultiplied by **A** yields the identity matrix. Thus **A** is not invertible.

Before discussing a procedure for finding the inverse of an invertible matrix, we introduce the concept of *elementary matrices*. An $n \times n$ **elementary matrix** is a matrix obtained from the $n \times n$ identity matrix **I** by an elementary row operation. Thus there are three basic types of elementary matrices:

(1) one obtained by interchanging two rows of I;

(2) one obtained by multiplying a row of I by a nonzero scalar; and

(3) one obtained by adding a multiple of one row of I to another.

EXAMPLE 2

The matrices

$$\mathbf{E}_1 = \begin{bmatrix} 1 & 0 & 0 \\ 0 & 0 & 1 \\ 0 & 1 & 0 \end{bmatrix}, \qquad \mathbf{E}_2 = \begin{bmatrix} -4 & 0 \\ 0 & 1 \end{bmatrix}, \qquad \text{and} \qquad \mathbf{E}_3 = \begin{bmatrix} 1 & 0 \\ 3 & 1 \end{bmatrix}$$

are elementary matrices. $\mathbf{E}_1$ is obtained from the 3×3 identity matrix by interchanging the second and third rows. $\mathbf{E}_2$ is obtained from the 2×2 identity matrix by multiplying the first row by -4. $\mathbf{E}_3$ is obtained from the 2×2 identity matrix by adding 3 times the first row to the second.

Suppose **E** is an $n \times n$ elementary matrix obtained from **I** by a certain elementary row operation, and **A** is an $n \times n$ matrix. Then it can be shown that the product **EA** is equal to the matrix that is obtained from **A** by applying the same elementary row operation to **A**. For example, let

$$\mathbf{A} = \begin{bmatrix} 1 & 2 \\ 3 & 4 \end{bmatrix}, \qquad \mathbf{E}_1 = \begin{bmatrix} 0 & 1 \\ 1 & 0 \end{bmatrix}, \qquad \mathbf{E}_2 = \begin{bmatrix} 1 & 0 \\ 0 & 2 \end{bmatrix}, \qquad \text{and} \qquad \mathbf{E}_3 = \begin{bmatrix} 1 & -2 \\ 0 & 1 \end{bmatrix}.$$

$\mathbf{E}_1$, $\mathbf{E}_2$, and $\mathbf{E}_3$ are elementary matrices. $\mathbf{E}_1$ is obtained by interchanging the first and second rows of **I**. Likewise, the product

$$\mathbf{E}_1\mathbf{A} = \begin{bmatrix} 0 & 1 \\ 1 & 0 \end{bmatrix}\begin{bmatrix} 1 & 2 \\ 3 & 4 \end{bmatrix} = \begin{bmatrix} 3 & 4 \\ 1 & 2 \end{bmatrix}$$

is the matrix obtained from **A** by interchanging the first and second rows of **A**. $\mathbf{E}_2$ is obtained by multiplying the second row of **I** by 2. Accordingly, the product

$$\mathbf{E}_2\mathbf{A} = \begin{bmatrix} 1 & 0 \\ 0 & 2 \end{bmatrix}\begin{bmatrix} 1 & 2 \\ 3 & 4 \end{bmatrix} = \begin{bmatrix} 1 & 2 \\ 6 & 8 \end{bmatrix}$$

is the matrix obtained by multiplying the second row of **A** by 2. $\mathbf{E_3}$ is obtained by adding -2 times the second row of **I** to the first row. The product

$$\mathbf{E_3A} = \begin{bmatrix} 1 & -2 \\ 0 & 1 \end{bmatrix}\begin{bmatrix} 1 & 2 \\ 3 & 4 \end{bmatrix} = \begin{bmatrix} -5 & -6 \\ 3 & 4 \end{bmatrix}$$

is the matrix obtained from **A** by the same elementary row operation.

If we wanted to reduce the matrix

$$\mathbf{A} = \begin{bmatrix} 1 & 0 \\ 2 & 2 \end{bmatrix},$$

we might proceed through a sequence of steps as follows:

$$\mathbf{A} = \begin{bmatrix} 1 & 0 \\ 2 & 2 \end{bmatrix}$$

$$\sim \begin{bmatrix} 1 & 0 \\ 0 & 2 \end{bmatrix} \qquad \text{(by adding } -2 \text{ times the first row to the second)}$$

$$\sim \begin{bmatrix} 1 & 0 \\ 0 & 1 \end{bmatrix} \qquad \left(\text{by multiplying the second row by } \tfrac{1}{2}\right).$$

Since this involves elementary row operations, it seems natural that elementary matrices can be used to reduce **A**. If **A** is premultiplied by the elementary matrix $\mathbf{E_1} = \begin{bmatrix} 1 & 0 \\ -2 & 1 \end{bmatrix}$, then $\mathbf{E_1A}$ is the matrix obtained from **A** by adding -2 times the first row to the second row:

$$\mathbf{E_1A} = \begin{bmatrix} 1 & 0 \\ -2 & 1 \end{bmatrix}\begin{bmatrix} 1 & 0 \\ 2 & 2 \end{bmatrix} = \begin{bmatrix} 1 & 0 \\ 0 & 2 \end{bmatrix}.$$

Premultiplying $\mathbf{E_1A}$ by the elementary matrix $\mathbf{E_2} = \begin{bmatrix} 1 & 0 \\ 0 & \tfrac{1}{2} \end{bmatrix}$ gives the matrix obtained by multiplying the second row of $\mathbf{E_1A}$ by $\tfrac{1}{2}$:

$$\mathbf{E_2(E_1A)} = \begin{bmatrix} 1 & 0 \\ 0 & \tfrac{1}{2} \end{bmatrix}\begin{bmatrix} 1 & 0 \\ 0 & 2 \end{bmatrix} = \begin{bmatrix} 1 & 0 \\ 0 & 1 \end{bmatrix} = \mathbf{I}.$$

Thus we have reduced **A** by multiplying **A** by a product of elementary matrices.

Since $(\mathbf{E_2E_1})\mathbf{A} = \mathbf{E_2(E_1A)} = \mathbf{I}$, the product $\mathbf{E_2E_1}$ is $\mathbf{A^{-1}}$. However, $\mathbf{A^{-1}} = \mathbf{E_2E_1} = (\mathbf{E_2E_1})\mathbf{I} = \mathbf{E_2(E_1I)}$. Thus $\mathbf{A^{-1}}$ can be obtained by applying the same elementary row operations, beginning with **I**, that were used to reduce **A** to **I**.

$$\mathbf{I} = \begin{bmatrix} 1 & 0 \\ 0 & 1 \end{bmatrix}$$

$$\sim \begin{bmatrix} 1 & 0 \\ -2 & 1 \end{bmatrix} \quad \text{(by adding } -2 \text{ times the first row to the second)}$$

$$\sim \begin{bmatrix} 1 & 0 \\ -1 & \frac{1}{2} \end{bmatrix} \quad \left(\text{by multiplying the second row by } \tfrac{1}{2}\right).$$

Therefore,

$$\mathbf{A}^{-1} = \begin{bmatrix} 1 & 0 \\ -1 & \frac{1}{2} \end{bmatrix}.$$

Our result can be verified by showing $\mathbf{A}^{-1}\mathbf{A} = \mathbf{I}$:

$$\mathbf{A}^{-1}\mathbf{A} = \begin{bmatrix} 1 & 0 \\ -1 & \frac{1}{2} \end{bmatrix}\begin{bmatrix} 1 & 0 \\ 2 & 2 \end{bmatrix} = \begin{bmatrix} 1 & 0 \\ 0 & 1 \end{bmatrix} = \mathbf{I}.$$

In summary, to find $\mathbf{A}^{-1}$ we apply the identical elementary row operations, beginning with **I** and proceeding in the same order, as those that were used to reduce **A** to **I**. Finding $\mathbf{A}^{-1}$ by this technique can be done conveniently by using the following format. First, we write the matrix

$$[\mathbf{A} \mathrel{\vdots} \mathbf{I}] = \begin{bmatrix} 1 & 0 & \vdots & 1 & 0 \\ 2 & 2 & \vdots & 0 & 1 \end{bmatrix}.$$

Then we apply elementary row operations until $[\mathbf{A} \mathrel{\vdots} \mathbf{I}]$ is equivalent to a matrix which has **I** as its first two columns. The last two columns of this matrix will be $\mathbf{A}^{-1}$. Thus,

$$[\mathbf{A} \mathrel{\vdots} \mathbf{I}] = \begin{bmatrix} 1 & 0 & \vdots & 1 & 0 \\ 2 & 2 & \vdots & 0 & 1 \end{bmatrix} \sim \begin{bmatrix} 1 & 0 & \vdots & 1 & 0 \\ 0 & 2 & \vdots & -2 & 1 \end{bmatrix}$$

$$\sim \begin{bmatrix} 1 & 0 & \vdots & 1 & 0 \\ 0 & 1 & \vdots & -1 & \frac{1}{2} \end{bmatrix} = [\mathbf{I} \mathrel{\vdots} \mathbf{A}^{-1}].$$

Note that the first two columns of $[\mathbf{I} \mathrel{\vdots} \mathbf{A}^{-1}]$ form a reduced matrix.

This procedure can be extended to find the inverse of *any* invertible $n \times n$ matrix. If **M** is such a matrix, form the $n \times (2n)$ matrix $[\mathbf{M} \mathrel{\vdots} \mathbf{I}]$. Then perform elementary row operations until the first n columns form a reduced matrix equal to **I**. The last n columns will be $\mathbf{M}^{-1}$.

$$[\mathbf{M} \mathrel{\vdots} \mathbf{I}] \sim \cdots \sim [\mathbf{I} \mathrel{\vdots} \mathbf{M}^{-1}].$$

If a matrix **M** does not reduce to **I**, then $\mathbf{M}^{-1}$ does not exist.

EXAMPLE 3

Determine $\mathbf{A}^{-1}$ *if* $\mathbf{A}$ *is invertible.*

a. $\mathbf{A} = \begin{bmatrix} 1 & 0 & -2 \\ 4 & -2 & 1 \\ 1 & 2 & -10 \end{bmatrix}$.

$$[\mathbf{A} \mid \mathbf{I}] = \begin{bmatrix} 1 & 0 & -2 & \vdots & 1 & 0 & 0 \\ 4 & -2 & 1 & \vdots & 0 & 1 & 0 \\ 1 & 2 & -10 & \vdots & 0 & 0 & 1 \end{bmatrix}$$

$$\sim \cdots \sim \begin{bmatrix} 1 & 0 & 0 & \vdots & -9 & 2 & 2 \\ 0 & 1 & 0 & \vdots & -\frac{41}{2} & 4 & \frac{9}{2} \\ 0 & 0 & 1 & \vdots & -5 & 1 & 1 \end{bmatrix}.$$

The first three columns of the last matrix form $\mathbf{I}$. Thus $\mathbf{A}$ is invertible and

$$\mathbf{A}^{-1} = \begin{bmatrix} -9 & 2 & 2 \\ -\frac{41}{2} & 4 & \frac{9}{2} \\ -5 & 1 & 1 \end{bmatrix}.$$

b. $\mathbf{A} = \begin{bmatrix} 3 & 2 \\ 6 & 4 \end{bmatrix}$.

$$[\mathbf{A} \mid \mathbf{I}] = \begin{bmatrix} 3 & 2 & \vdots & 1 & 0 \\ 6 & 4 & \vdots & 0 & 1 \end{bmatrix} \sim \begin{bmatrix} 3 & 2 & \vdots & 1 & 0 \\ 0 & 0 & \vdots & -2 & 1 \end{bmatrix}$$

$$\sim \begin{bmatrix} 1 & \frac{2}{3} & \vdots & \frac{1}{3} & 0 \\ 0 & 0 & \vdots & -2 & 1 \end{bmatrix}.$$

The first two columns of the last matrix form a reduced matrix different from $\mathbf{I}$. Thus $\mathbf{A}$ is not invertible.

EXAMPLE 4

Solve each system by finding the inverse of the coefficient matrix.

a. $\begin{cases} x_1 + 2x_2 = 0, \\ 4x_1 + 9x_2 = 1. \end{cases}$

The system can be expressed as the matrix equation $\mathbf{AX} = \mathbf{C}$, where $\mathbf{A}$ is the coefficient matrix of the system.

$$\begin{bmatrix} 1 & 2 \\ 4 & 9 \end{bmatrix} \begin{bmatrix} x_1 \\ x_2 \end{bmatrix} = \begin{bmatrix} 0 \\ 1 \end{bmatrix}.$$

Since $\mathbf{AX} = \mathbf{C}$, then $\mathbf{A}^{-1}\mathbf{AX} = \mathbf{A}^{-1}\mathbf{C}$ and so our solution is given by

$$\mathbf{X} = \mathbf{A}^{-1}\mathbf{C}.$$

Thus we need to find $\mathbf{A}^{-1}$.

$$\begin{bmatrix} 1 & 2 & \vdots & 1 & 0 \\ 4 & 9 & \vdots & 0 & 1 \end{bmatrix} \sim \cdots \sim \begin{bmatrix} 1 & 0 & \vdots & 9 & -2 \\ 0 & 1 & \vdots & -4 & 1 \end{bmatrix}.$$

$$\mathbf{A}^{-1} = \begin{bmatrix} 9 & -2 \\ -4 & 1 \end{bmatrix}.$$

Hence

$$\mathbf{X} = \begin{bmatrix} x_1 \\ x_2 \end{bmatrix} = \mathbf{A}^{-1}\mathbf{C} = \begin{bmatrix} 9 & -2 \\ -4 & 1 \end{bmatrix}\begin{bmatrix} 0 \\ 1 \end{bmatrix} = \begin{bmatrix} -2 \\ 1 \end{bmatrix}.$$

Therefore, $x_1 = -2$ and $x_2 = 1$.

b. $\begin{cases} x_1 \quad\quad - 2x_3 = 1, \\ 4x_1 - 2x_2 + x_3 = 2, \\ x_1 + 2x_2 - 10x_3 = -1. \end{cases}$

The coefficient matrix of the system is

$$\mathbf{A} = \begin{bmatrix} 1 & 0 & -2 \\ 4 & -2 & 1 \\ 1 & 2 & -10 \end{bmatrix}.$$

By Example 3(a),

$$\mathbf{A}^{-1} = \begin{bmatrix} -9 & 2 & 2 \\ -\frac{41}{2} & 4 & \frac{9}{2} \\ -5 & 1 & 1 \end{bmatrix}.$$

Thus,

$$\begin{bmatrix} x_1 \\ x_2 \\ x_3 \end{bmatrix} = \begin{bmatrix} -9 & 2 & 2 \\ -\frac{41}{2} & 4 & \frac{9}{2} \\ -5 & 1 & 1 \end{bmatrix}\begin{bmatrix} 1 \\ 2 \\ -1 \end{bmatrix} = \begin{bmatrix} -7 \\ -17 \\ -4 \end{bmatrix}.$$

Consequently, $x_1 = -7$, $x_2 = -17$, and $x_3 = -4$.

It can be shown that a system of n linear equations in n unknowns has a unique solution if and only if the coefficient matrix is invertible. Indeed, in both parts of the last example the coefficient matrices were invertible and unique solutions did in fact exist. When the coefficient matrix is not invertible, the system will have either no solution or an infinite number of solutions.

EXAMPLE 5

Solve the system

$$\begin{cases} x - 2y + z = 0, \\ 2x - y + 5z = 0, \\ x + y + 4z = 0. \end{cases}$$

The coefficient matrix is

$$\begin{bmatrix} 1 & -2 & 1 \\ 2 & -1 & 5 \\ 1 & 1 & 4 \end{bmatrix}.$$

Since

$$\begin{bmatrix} 1 & -2 & 1 & \vdots & 1 & 0 & 0 \\ 2 & -1 & 5 & \vdots & 0 & 1 & 0 \\ 1 & 1 & 4 & \vdots & 0 & 0 & 1 \end{bmatrix} \sim \cdots \sim \begin{bmatrix} 1 & 0 & 3 & \vdots & -\frac{1}{3} & \frac{2}{3} & 0 \\ 0 & 1 & 1 & \vdots & -\frac{2}{3} & \frac{1}{3} & 0 \\ 0 & 0 & 0 & \vdots & 1 & -1 & 1 \end{bmatrix},$$

the coefficient matrix is not invertible. Hence, the system *cannot* be solved by inverses. Another method must be used. In Example 4(a) of Sec. 14-5, the solution was found to be $x = -3z, y = -z, z = z$.

EXERCISE 14-6

In each of Problems 1–18, if the given matrix is invertible, find its inverse.

1. $\begin{bmatrix} 6 & 1 \\ 5 & 1 \end{bmatrix}$.

2. $\begin{bmatrix} 2 & 8 \\ 3 & 12 \end{bmatrix}$.

3. $\begin{bmatrix} 1 & 1 \\ 1 & 1 \end{bmatrix}$.

4. $\begin{bmatrix} 4 & 9 \\ 0 & -6 \end{bmatrix}$.

5. $\begin{bmatrix} 1 & 0 & 0 \\ 0 & -3 & 0 \\ 0 & 0 & 4 \end{bmatrix}$.

6. $\begin{bmatrix} 2 & 0 & 8 \\ -1 & 4 & 0 \\ 2 & 1 & 0 \end{bmatrix}$.

7. $\begin{bmatrix} 1 & 2 & 3 \\ 0 & 0 & 4 \\ 0 & 0 & 5 \end{bmatrix}$.

8. $\begin{bmatrix} 2 & 0 & 0 \\ 0 & 0 & 0 \\ 0 & 0 & -4 \end{bmatrix}$.

9. $\begin{bmatrix} 2 & 4 \\ 8 & 1 \\ 6 & 3 \end{bmatrix}$.

10. $\begin{bmatrix} 0 & 0 & 0 \\ 0 & 0 & 0 \\ 0 & 0 & 0 \end{bmatrix}$.

11. $\begin{bmatrix} 1 & 1 & 1 \\ 0 & 1 & 1 \\ 0 & 0 & 1 \end{bmatrix}$.

12. $\begin{bmatrix} 1 & 2 & -1 \\ 0 & 1 & 4 \\ 1 & -1 & 2 \end{bmatrix}$.

13. $\begin{bmatrix} 7 & 0 & -2 \\ 0 & 1 & 0 \\ -3 & 0 & 1 \end{bmatrix}$.

14. $\begin{bmatrix} 7 & -8 & 5 \\ -4 & 5 & -3 \\ 1 & -1 & 1 \end{bmatrix}$.

15. $\begin{bmatrix} 2 & 1 & 0 \\ 4 & -1 & 5 \\ 1 & -1 & 2 \end{bmatrix}.$

16. $\begin{bmatrix} -5 & 4 & -3 \\ 10 & -7 & 6 \\ 8 & -6 & 5 \end{bmatrix}.$

17. $\begin{bmatrix} 1 & 2 & 3 \\ 1 & 3 & 5 \\ 1 & 5 & 12 \end{bmatrix}.$

18. $\begin{bmatrix} 2 & -1 & 3 \\ 0 & 2 & 0 \\ 2 & 1 & 1 \end{bmatrix}.$

*For each of Problems **19–32**, if the coefficient matrix of the system is invertible, solve the system by using the inverse. If not, solve the system by the method of reduction.*

19. $\begin{cases} 6x + 5y = 2, \\ x + y = -3. \end{cases}$

20. $\begin{cases} 2x + 3y = 4, \\ -x + 5y = -2. \end{cases}$

21. $\begin{cases} 2x + y = 5, \\ 3x - y = 0. \end{cases}$

22. $\begin{cases} 3x + 2y = 26, \\ 4x + 3y = 37. \end{cases}$

23. $\begin{cases} 2x + 6y = 2, \\ 3x + 9y = 3. \end{cases}$

24. $\begin{cases} 2x + 8y = 3, \\ 3x + 12y = 6. \end{cases}$

25. $\begin{cases} x + 2y + z = 4, \\ 3x + z = 2, \\ x - y + z = 1. \end{cases}$

26. $\begin{cases} x + y + z = 2, \\ x - y + z = -2, \\ x - y - z = 0. \end{cases}$

27. $\begin{cases} x + y + z = 2, \\ x - y + z = 1, \\ x - y - z = 0. \end{cases}$

28. $\begin{cases} 2x + 8z = 8, \\ -x + 4y = 36, \\ 2x + y = 9. \end{cases}$

29. $\begin{cases} x + 3y + 3z = 7, \\ 2x + y + z = 4, \\ x + y + z = 4. \end{cases}$

30. $\begin{cases} x + 3y + 3z = 7, \\ 2x + y + z = 4, \\ x + y + z = 3. \end{cases}$

31. $\begin{cases} w + 2y + z = 4, \\ w - x + 2z = 12, \\ 2w + x + z = 12, \\ w + 2x + y + z = 12. \end{cases}$

32. $\begin{cases} w + x + z = 2, \\ w + y = 0, \\ x + y + z = 4, \\ y + z = 1. \end{cases}$

Find $(\mathbf{I} - \mathbf{A})^{-1}$ for each of the following matrices $\mathbf{A}$.

33. $\begin{bmatrix} 2 & -1 \\ 1 & 3 \end{bmatrix}.$

34. $\begin{bmatrix} -3 & 2 \\ 4 & 3 \end{bmatrix}.$

35. Solve the following problems by using the inverse of the matrix involved.

(a) An automobile factory produces two models. The first requires 1 man-hour to paint and $\frac{1}{2}$ man-hour to polish; the second requires 1 man-hour for each process. During each hour that the assembly line is operating, there are 100 man-hours available for painting and 80 man-hours for polishing. How many of each model can be produced each hour if all the man-hours available are to be utilized?

(b) Suppose each car of the first type requires 10 widgets and 14 shims, and each car of the second type requires 7 widgets and 10 shims. The factory can obtain 800 widgets and 1130 shims each hour. How many cars of each model can it produce while utilizing all the parts available?

14-7 DETERMINANTS

We now introduce a new function, the *determinant function*. Here our inputs will be *square* matrices, but our outputs will be real numbers. If **A** is a square matrix, then the determinant function associates with **A** exactly one real number called the *determinant* of **A**. Denoting the determinant of **A** by $|\mathbf{A}|$ (that is, using vertical bars), we can think of the determinant function as a correspondence:

$$\begin{array}{ccc} \mathbf{A} & \longrightarrow & |\mathbf{A}| \\ \text{Square} & & \text{Real} & = & \text{Determinant} \\ \text{Matrix} & & \text{Number} & & \text{of } \mathbf{A} \end{array}$$

The use of determinants in solving systems of linear equations will be discussed later. Turning to how a real number is assigned to a square matrix, we shall first consider the special cases of matrices of orders one and two. Then we shall extend the definition to matrices of order n.

DEFINITION.   *If* $\mathbf{A} = [a_{11}]$ *is a square matrix of order one, then* $|\mathbf{A}| = a_{11}$.

That is, the determinant function assigns to the one-entry matrix $[a_{11}]$ the number a_{11}. Hence if $\mathbf{A} = [6]$, then $|\mathbf{A}| = 6$.

DEFINITION.   *If*

$$\mathbf{A} = \begin{bmatrix} a_{11} & a_{12} \\ a_{21} & a_{22} \end{bmatrix}$$

is a square matrix of order two, then

$$|\mathbf{A}| = a_{11}a_{22} - a_{12}a_{21}.$$

That is, the determinant of a 2×2 matrix is obtained by taking the product of the entries in the main diagonal and subtracting from it the product of the entries in the other diagonal. We speak of the determinant of a 2×2 matrix as a *determinant of order* 2.

EXAMPLE 1

Find $|\mathbf{A}|$ *if* $\mathbf{A} =$

$$a. \begin{bmatrix} 2 & 1 \\ 3 & -4 \end{bmatrix}, \quad b. \begin{bmatrix} -3 & -2 \\ 0 & 1 \end{bmatrix}, \quad c. \begin{bmatrix} 1 & 0 \\ 0 & 1 \end{bmatrix}, \quad d. \begin{bmatrix} x & 0 \\ y & 1 \end{bmatrix}.$$

We have

a. $|\mathbf{A}| = \begin{vmatrix} 2 & 1 \\ 3 & -4 \end{vmatrix} = (2)(-4) - (1)(3) = -8 - 3 = -11.$

b. $|\mathbf{A}| = \begin{vmatrix} -3 & -2 \\ 0 & 1 \end{vmatrix} = (-3)(1) - (-2)(0) = -3 - 0 = -3.$

c. $|\mathbf{A}| = \begin{vmatrix} 1 & 0 \\ 0 & 1 \end{vmatrix} = (1)(1) - (0)(0) = 1.$

d. $|\mathbf{A}| = \begin{vmatrix} x & 0 \\ y & 1 \end{vmatrix} = (x)(1) - (0)(y) = x.$

The determinant of a square matrix $\mathbf{A}$ of order n ($n > 2$) is defined in the following manner. With a given entry of $\mathbf{A}$ we associate the square matrix of order $n - 1$ obtained by deleting the entries in the row and column in which the given entry lies. For example, given the matrix

$$\begin{bmatrix} a_{11} & a_{12} & a_{13} \\ a_{21} & a_{22} & a_{23} \\ a_{31} & a_{32} & a_{33} \end{bmatrix},$$

with entry a_{21} we delete the entries in row 2 and column 1,

$$\begin{bmatrix} a_{11} & a_{12} & a_{13} \\ a_{21} & a_{22} & a_{23} \\ a_{31} & a_{32} & a_{33} \end{bmatrix},$$

leaving the matrix of order 2

$$\begin{bmatrix} a_{12} & a_{13} \\ a_{32} & a_{33} \end{bmatrix}.$$

The *determinant* of this matrix is called the **minor** of a_{21}. Similarly, the minor of a_{22} is

$$\begin{vmatrix} a_{11} & a_{13} \\ a_{31} & a_{33} \end{vmatrix},$$

and for a_{23} it is

$$\begin{vmatrix} a_{11} & a_{12} \\ a_{31} & a_{32} \end{vmatrix}.$$

With each entry a_{ij} we also associate a number determined by the subscript of the entry:

$$(-1)^{i+j},$$

where $i + j$ is the sum of the row number i and column number j in which the entry lies. With a_{21} we associate $(-1)^{2+1} = -1$, with a_{22} the number $(-1)^{2+2} = 1$, and with a_{23} the number $(-1)^{2+3} = -1$. The **cofactor** c_{ij} of the entry a_{ij} is the product of $(-1)^{i+j}$ and the minor of a_{ij}. For example, the cofactor of a_{21} is

$$c_{21} = (-1)^{2+1} \begin{vmatrix} a_{12} & a_{13} \\ a_{32} & a_{33} \end{vmatrix}.$$

The only difference between a cofactor and a minor is the factor $(-1)^{i+j}$.

To find the determinant of any square matrix $\mathbf{A}$ of order n, select *any* row (or column) of $\mathbf{A}$ and multiply each entry in the row (column) by its cofactor. The sum of these numbers is defined to be the determinant of $\mathbf{A}$ and is called a **determinant of order n**.

Let us find the determinant of

$$\begin{bmatrix} 2 & -1 & 3 \\ 3 & 0 & -5 \\ 2 & 1 & 1 \end{bmatrix}$$

by applying the above rule to the first row (sometimes referred to as "expanding along the first row"). For

$$a_{11} \text{ we obtain } (2)(-1)^{1+1} \begin{vmatrix} 0 & -5 \\ 1 & 1 \end{vmatrix} = (2)(1)(5) = 10,$$

$$a_{12} \text{ we obtain } (-1)(-1)^{1+2} \begin{vmatrix} 3 & -5 \\ 2 & 1 \end{vmatrix} = (-1)(-1)(13) = 13,$$

$$a_{13} \text{ we obtain } (3)(-1)^{1+3} \begin{vmatrix} 3 & 0 \\ 2 & 1 \end{vmatrix} = 3(1)(3) = 9.$$

Hence,

$$\begin{vmatrix} 2 & -1 & 3 \\ 3 & 0 & -5 \\ 2 & 1 & 1 \end{vmatrix} = 10 + 13 + 9 = 32.$$

If we had expanded along the second column, then

$$\begin{vmatrix} 2 & -1 & 3 \\ 3 & 0 & -5 \\ 2 & 1 & 1 \end{vmatrix} = (-1)(-1)^{1+2} \begin{vmatrix} 3 & -5 \\ 2 & 1 \end{vmatrix} + 0 + (1)(-1)^{3+2} \begin{vmatrix} 2 & 3 \\ 3 & -5 \end{vmatrix}$$

$$= 13 + 0 + 19 = 32 \text{ as before.}$$

It can be shown that the determinant of a matrix is unique and does not depend on the row or column chosen for its evaluation. In the above problem, the second expansion is preferable since the 0 in column 2 contributed nothing to the sum, thus simplifying the calculation.

EXAMPLE 2

Find $|\mathbf{A}|$ *if*

a. $\mathbf{A} = \begin{bmatrix} 12 & -1 & 3 \\ -3 & 1 & -1 \\ -10 & 2 & -3 \end{bmatrix}$.

Expanding along the first row, we have

$$|\mathbf{A}| = 12(-1)^{1+1}\begin{vmatrix} 1 & -1 \\ 2 & -3 \end{vmatrix} + (-1)(-1)^{1+2}\begin{vmatrix} -3 & -1 \\ -10 & -3 \end{vmatrix} + 3(-1)^{1+3}\begin{vmatrix} -3 & 1 \\ -10 & 2 \end{vmatrix}$$

$$= 12(1)(-1) + (-1)(-1)(-1) + 3(1)(4) = -1.$$

b. $\mathbf{A} = \begin{bmatrix} 0 & 1 & 1 \\ 2 & 3 & 2 \\ 0 & -1 & 3 \end{bmatrix}$.

Expanding along column 1 for convenience, we have

$$|\mathbf{A}| = 0 + 2(-1)^{2+1}\begin{vmatrix} 1 & 1 \\ -1 & 3 \end{vmatrix} + 0 = 2(-1)(4) = -8.$$

EXAMPLE 3

Evaluate

$$|\mathbf{A}| = \begin{vmatrix} 2 & 0 & 0 & 1 \\ 0 & 1 & 0 & 3 \\ 0 & 0 & 1 & 2 \\ 1 & 2 & 3 & 0 \end{vmatrix}$$

by expanding along the first row,

$$|\mathbf{A}| = 2(-1)^{1+1}\begin{vmatrix} 1 & 0 & 3 \\ 0 & 1 & 2 \\ 2 & 3 & 0 \end{vmatrix} + 1(-1)^{1+4}\begin{vmatrix} 0 & 1 & 0 \\ 0 & 0 & 1 \\ 1 & 2 & 3 \end{vmatrix}.$$

We have now expressed $|\mathbf{A}|$ in terms of determinants of order three. Expanding each of these along the first row, we have

$$|\mathbf{A}| = 2(1)\left[1(-1)^{1+1}\begin{vmatrix} 1 & 2 \\ 3 & 0 \end{vmatrix} + 3(-1)^{1+3}\begin{vmatrix} 0 & 1 \\ 2 & 3 \end{vmatrix}\right] + 1(-1)\left[1(-1)^{1+2}\begin{vmatrix} 0 & 1 \\ 1 & 3 \end{vmatrix}\right]$$

$$= 2[1(1)(-6) + 3(1)(-2)] + (-1)[(1)(-1)(-1)] = -25.$$

The evaluation of determinants is often simplified by the use of various properties, some of which we now list. In each case, **A** denotes a square matrix.

(1) If each of the entries in a row (or column) of A is 0, then |A| = 0.
Thus,

$$\begin{vmatrix} 6 & 2 & 5 \\ 7 & 1 & 4 \\ 0 & 0 & 0 \end{vmatrix} = 0.$$

(2) If two rows (or columns) of A are identical, |A| = 0.
Thus,

$$\begin{vmatrix} 2 & 5 & 2 & 1 \\ 2 & 6 & 2 & 3 \\ 2 & 4 & 2 & 1 \\ 6 & 5 & 6 & 1 \end{vmatrix} = 0, \text{ since column 1 = column 3.}$$

(3) If all the entries below (or above) the main diagonal of A are 0, then |A| is equal to the product of the main diagonal entries.
Thus,

$$\begin{vmatrix} 2 & 6 & 1 & 0 \\ 0 & 5 & 7 & 6 \\ 0 & 0 & -2 & 5 \\ 0 & 0 & 0 & 1 \end{vmatrix} = (2)(5)(-2)(1) = -20.$$

From this property we conclude that the determinant of an identity matrix is 1.

(4) If B is the matrix obtained by adding a multiple of one row (or column) of A to another row (column), then |B| = |A|.
Thus if

$$\mathbf{A} = \begin{bmatrix} 2 & 4 & 2 & 6 \\ 1 & 3 & 5 & 2 \\ 1 & 2 & 1 & 3 \\ 0 & 5 & 6 & 2 \end{bmatrix}$$

and **B** is the matrix obtained from **A** by adding −2 times row 3 to row 1, then

$$|\mathbf{A}| = \begin{vmatrix} 2 & 4 & 2 & 6 \\ 1 & 3 & 5 & 2 \\ 1 & 2 & 1 & 3 \\ 0 & 5 & 6 & 2 \end{vmatrix} = \begin{vmatrix} 0 & 0 & 0 & 0 \\ 1 & 3 & 5 & 2 \\ 1 & 2 & 1 & 3 \\ 0 & 5 & 6 & 2 \end{vmatrix} = |\mathbf{B}|.$$

By property 1, |**B**| = 0 and hence |**A**| = 0.

(5) **If B is the matrix obtained by interchanging two rows (or columns) of A, then $|A| = -|B|$.**
Thus if

$$A = \begin{bmatrix} 2 & 2 & 1 & 6 \\ 0 & 0 & 0 & 1 \\ 0 & 0 & 2 & 0 \\ 0 & 1 & -3 & 4 \end{bmatrix}$$

and **B** is obtained from **A** by interchanging rows 2 and 4, then

$$|A| = \begin{vmatrix} 2 & 2 & 1 & 6 \\ 0 & 0 & 0 & 1 \\ 0 & 0 & 2 & 0 \\ 0 & 1 & -3 & 4 \end{vmatrix} = -\begin{vmatrix} 2 & 2 & 1 & 6 \\ 0 & 1 & -3 & 4 \\ 0 & 0 & 2 & 0 \\ 0 & 0 & 0 & 1 \end{vmatrix} = -|B|.$$

By property 3, $|B| = 4$ and hence $|A| = -4$.

(6) **If B is the matrix obtained by multiplying each entry of a row (or column) of A by the same number k, then $|B| = k|A|$.**
Thus,

$$\begin{vmatrix} 2 \cdot 3 & 2 \cdot 5 & 2 \cdot 7 \\ 5 & 2 & 1 \\ 6 & 4 & 3 \end{vmatrix} = 2\begin{vmatrix} 3 & 5 & 7 \\ 5 & 2 & 1 \\ 6 & 4 & 3 \end{vmatrix}.$$

Essentially, a number can be "factored out" of one row or column.

(7) **The determinant of the product of two matrices of order n is the product of their determinants. That is, $|AB| = |A||B|$.**
Thus if

$$A = \begin{bmatrix} 1 & 2 \\ 3 & 4 \end{bmatrix} \quad \text{and} \quad B = \begin{bmatrix} 1 & 2 \\ 0 & 3 \end{bmatrix},$$

then

$$|AB| = |A| \cdot |B| = \begin{vmatrix} 1 & 2 \\ 3 & 4 \end{vmatrix} \cdot \begin{vmatrix} 1 & 2 \\ 0 & 3 \end{vmatrix} = (-2)(3) = -6.$$

EXAMPLE 4

Evaluate

$$|A| = \begin{vmatrix} 1 & 1 & 0 & 5 \\ 1 & 2 & 1 & 0 \\ 0 & 2 & 1 & 1 \\ 3 & 0 & 0 & -4 \end{vmatrix}.$$

We shall express **A** in upper triangular form (we say that we "triangulate") and then, by property 3, take the product of the main diagonal.

$$\begin{vmatrix} 1 & 1 & 0 & 5 \\ 1 & 2 & 1 & 0 \\ 0 & 2 & 1 & 1 \\ 3 & 0 & 0 & -4 \end{vmatrix} = \begin{vmatrix} 1 & 1 & 0 & 5 \\ 0 & 1 & 1 & -5 \\ 0 & 2 & 1 & 1 \\ 0 & -3 & 0 & -19 \end{vmatrix}$$
(by adding -1 times row 1 to row 2; adding -3 times row 1 to row 4)

$$= \begin{vmatrix} 1 & 1 & 0 & 5 \\ 0 & 1 & 1 & -5 \\ 0 & 0 & -1 & 11 \\ 0 & 0 & 3 & -34 \end{vmatrix}$$
(by adding -2 times row 2 to row 3; adding 3 times row 2 to row 4)

$$= \begin{vmatrix} 1 & 1 & 0 & 5 \\ 0 & 1 & 1 & -5 \\ 0 & 0 & -1 & 11 \\ 0 & 0 & 0 & -1 \end{vmatrix}$$
(by adding 3 times row 3 to row 4)

$$= (1)(1)(-1)(-1) = 1.$$

EXERCISE 14-7

Evaluate the determinants in Problems 1–6.

1. $\begin{vmatrix} 2 & 1 \\ 3 & 2 \end{vmatrix}$.

2. $\begin{vmatrix} 3 & 2 \\ -5 & -4 \end{vmatrix}$.

3. $\begin{vmatrix} -2 & -3 \\ -4 & -6 \end{vmatrix}$.

4. $\begin{vmatrix} -3 & 1 \\ -a & b \end{vmatrix}$.

5. $\begin{vmatrix} 1 & x \\ 0 & y \end{vmatrix}$.

6. $\begin{vmatrix} -2 & -a \\ -a & 2 \end{vmatrix}$.

In Problems 7 and 8, evaluate the given expressions.

7. $\begin{vmatrix} 1 & 2 \\ 3 & 4 \end{vmatrix}$
   $\begin{vmatrix} 2 & 1 \\ 5 & 6 \end{vmatrix}$.

8. $\begin{vmatrix} 6 & 2 \\ 1 & 5 \end{vmatrix}$
   $\begin{vmatrix} 2 & -6 \\ 5 & 3 \end{vmatrix}$.

9. Solve for k if

$$\begin{vmatrix} 2 & 3 \\ 4 & k \end{vmatrix} = 12.$$

If $\mathbf{A} = \begin{bmatrix} 1 & 2 & 3 \\ 4 & 5 & 6 \\ 7 & 8 & 9 \end{bmatrix}$, *determine each of the following.*

10. The minor of a_{31}.

11. The minor of a_{22}.

12. The cofactor of a_{23}.

13. The cofactor of a_{32}.

14. If $\mathbf{A} = [a_{ij}]$ is 50×50 and the minor of $a_{43,\,47}$ equals 20, what is the value of the cofactor of $a_{43,\,47}$?

If

$$A = \begin{bmatrix} a_{11} & a_{12} & a_{13} & a_{14} \\ a_{21} & a_{22} & a_{23} & a_{24} \\ a_{31} & a_{32} & a_{33} & a_{34} \\ a_{41} & a_{42} & a_{43} & a_{44} \end{bmatrix},$$

write each of the following.

15. The minor of a_{32}.

16. The minor of a_{24}.

17. The cofactor of a_{13}.

18. The cofactor of a_{43}.

*In Problems **19–34**, evaluate the determinant. Use properties of determinants if possible.*

19. $\begin{vmatrix} 2 & 1 & 3 \\ 2 & 0 & 1 \\ -4 & 0 & 6 \end{vmatrix}$.

20. $\begin{vmatrix} 3 & 2 & 1 \\ 1 & -2 & 3 \\ -1 & 3 & 2 \end{vmatrix}$.

21. $\begin{vmatrix} 1 & 2 & -3 \\ 4 & 5 & 4 \\ 3 & -2 & 1 \end{vmatrix}$.

22. $\begin{vmatrix} 1 & 0 & -1 \\ 0 & 1 & 0 \\ 1 & -1 & 1 \end{vmatrix}$.

23. $\begin{vmatrix} 2 & 1 & 5 \\ -3 & 4 & -1 \\ 0 & 6 & -1 \end{vmatrix}$.

24. $\begin{vmatrix} 1 & 2 & 3 \\ 4 & 5 & 4 \\ 3 & 2 & 1 \end{vmatrix}$.

25. $\begin{vmatrix} 2 & -1 & 3 \\ 1 & 1 & -1 \\ 1 & 2 & -3 \end{vmatrix}$.

26. $\begin{vmatrix} 1 & 2 & 3 \\ 4 & 5 & 6 \\ 7 & 8 & 9 \end{vmatrix}$.

27. $\begin{vmatrix} \frac{1}{2} & \frac{2}{3} & -\frac{1}{2} \\ -1 & \frac{1}{3} & \frac{2}{3} \\ 3 & -4 & 1 \end{vmatrix}$.

28. $\begin{vmatrix} -\frac{1}{3} & \frac{1}{4} & 4 \\ \frac{3}{2} & \frac{3}{8} & -2 \\ -\frac{1}{8} & \frac{9}{2} & 1 \end{vmatrix}$.

29. $\begin{vmatrix} 1 & 0 & 3 & 2 \\ 4 & -1 & 0 & 1 \\ 2 & 1 & 0 & 3 \\ -1 & 2 & 3 & -1 \end{vmatrix}$.

30. $\begin{vmatrix} 7 & 6 & 0 & 5 \\ -3 & 2 & 0 & 1 \\ 4 & -3 & 0 & 2 \\ 1 & 0 & 0 & 6 \end{vmatrix}$.

31. $\begin{vmatrix} 1 & 7 & -3 & 8 \\ 0 & 1 & -5 & 4 \\ 0 & 0 & 1 & 7 \\ 0 & 0 & 0 & 1 \end{vmatrix}$.

32. $\begin{vmatrix} 1 & 2 & -3 & 4 \\ 3 & -1 & 2 & 4 \\ -2 & -4 & 6 & -8 \\ 0 & 3 & -1 & 2 \end{vmatrix}$.

33. $\begin{vmatrix} 1 & 0 & 0 & 0 \\ 0 & -2 & 0 & 0 \\ 0 & 0 & 4 & 0 \\ 0 & 0 & 0 & -3 \end{vmatrix}$.

34. $\begin{vmatrix} 1 & -3 & 2 & 6 & 4 \\ 0 & 13 & 0 & 1 & 5 \\ -2 & 1 & 2 & 3 & 4 \\ 1 & 1 & 4 & 5 & 9 \end{vmatrix}$.

*In Problems **35** and **36**, solve for x.*

35. $\begin{vmatrix} x & -2 \\ 7 & 7-x \end{vmatrix} = 26$.

36. $\begin{vmatrix} 3 & x & 2x \\ 0 & x & 99 \\ 0 & 0 & x-1 \end{vmatrix} = 60$.

37. If A is of order 4×4 and $|A| = 12$, what is the value of the determinant obtained by multiplying every element in A by 2?

14-8 CRAMER'S RULE

Determinants can be applied to solving a system of n linear equations in n unknowns. In fact, it is from the analysis of such systems that the study of determinants took its origin. Although the method of reduction is more practical for systems involving a large number of unknowns, the method of solution by determinants is interesting enough to warrant some attention here and it also allows us to solve for one unknown without having to solve for the others. We shall first consider a system of two linear equations in two unknowns. Then the results will be extended to include more general situations.

Let us solve

$$\begin{cases} a_{11}x + a_{12}y = c_1, \\ a_{21}x + a_{22}y = c_2. \end{cases} \qquad (1)$$

To find an explicit formula for x, we look at $x\begin{vmatrix} a_{11} & a_{12} \\ a_{21} & a_{22} \end{vmatrix}$.

$$x\begin{vmatrix} a_{11} & a_{12} \\ a_{21} & a_{22} \end{vmatrix} = \begin{vmatrix} a_{11}x & a_{12} \\ a_{21}x & a_{22} \end{vmatrix} \qquad \text{(property 6 of Sec. 14-7)}$$

$$= \begin{vmatrix} a_{11}x + a_{12}y & a_{12} \\ a_{21}x + a_{22}y & a_{22} \end{vmatrix} \qquad \text{(adding } y \text{ times column 2 to column 1)}$$

$$= \begin{vmatrix} c_1 & a_{12} \\ c_2 & a_{22} \end{vmatrix} \qquad \text{[from Eq. (1)].}$$

Thus,

$$x\begin{vmatrix} a_{11} & a_{12} \\ a_{21} & a_{22} \end{vmatrix} = \begin{vmatrix} c_1 & a_{12} \\ c_2 & a_{22} \end{vmatrix},$$

and so

$$x = \frac{\begin{vmatrix} c_1 & a_{12} \\ c_2 & a_{22} \end{vmatrix}}{\begin{vmatrix} a_{11} & a_{12} \\ a_{21} & a_{22} \end{vmatrix}}. \qquad (2)$$

To find a formula for y, we look at $y\begin{vmatrix} a_{11} & a_{12} \\ a_{21} & a_{22} \end{vmatrix}$.

$$y\begin{vmatrix} a_{11} & a_{12} \\ a_{21} & a_{22} \end{vmatrix} = \begin{vmatrix} a_{11} & a_{12}y \\ a_{21} & a_{22}y \end{vmatrix} \qquad \text{(property 6 of Sec. 14-7)}$$

$$= \begin{vmatrix} a_{11} & a_{11}x + a_{12}y \\ a_{21} & a_{21}x + a_{22}y \end{vmatrix} \qquad \text{(adding } x \text{ times column 1 to column 2)}$$

$$= \begin{vmatrix} a_{11} & c_1 \\ a_{21} & c_2 \end{vmatrix} \qquad \text{[from Eq. (1)].}$$

Thus,

$$y\begin{vmatrix} a_{11} & a_{12} \\ a_{21} & a_{22} \end{vmatrix} = \begin{vmatrix} a_{11} & c_1 \\ a_{21} & c_2 \end{vmatrix},$$

and so

$$y = \frac{\begin{vmatrix} a_{11} & c_1 \\ a_{21} & c_2 \end{vmatrix}}{\begin{vmatrix} a_{11} & a_{12} \\ a_{21} & a_{22} \end{vmatrix}}. \qquad (3)$$

Note that in Eqs. (2) and (3) the denominators are the same, namely the determinant of the coefficient matrix of the given system. In finding x, the numerator in Eq. (2) is the determinant of the matrix obtained by replacing the "x-column" (that is, column 1) of the coefficient matrix by the column of constants $\begin{matrix} c_1 \\ c_2 \end{matrix}$. Similarly, the numerator in Eq. (3) is the determinant of the matrix obtained from the coefficient matrix when the "y-column" (that is, column 2) is replaced by $\begin{matrix} c_1 \\ c_2 \end{matrix}$. Provided that the determinant of the coefficient matrix is not zero, the original system will have a unique solution. However, if this determinant is zero, the procedure is not applicable and the system may have either no solution or an infinite number of solutions. In such cases previous methods should be used to solve the system.

We shall illustrate the above results by solving the system

$$\begin{cases} 2x + y + 5 = 0, \\ 3y + x = 6. \end{cases}$$

First, the system is written in the appropriate form:

$$\begin{cases} 2x + y = -5, \\ x + 3y = 6. \end{cases}$$

The determinant Δ of the coefficient matrix is

$$\Delta = \begin{vmatrix} 2 & 1 \\ 1 & 3 \end{vmatrix} = 2(3) - 1(1) = 5.$$

Since $\Delta \neq 0$, there is a unique solution. Solving for x, we have

$$x = \frac{\begin{vmatrix} -5 & 1 \\ 6 & 3 \end{vmatrix}}{\Delta} = \frac{-21}{5} = -\frac{21}{5}.$$

Solving for y, we obtain

$$y = \frac{\begin{vmatrix} 2 & -5 \\ 1 & 6 \end{vmatrix}}{\Delta} = \frac{17}{5}.$$

Thus the solution is $x = -\frac{21}{5}$ and $y = \frac{17}{5}$.

The method described above can be extended to systems of n linear equations in n unknowns and is referred to as *Cramer's rule*.

CRAMER'S RULE.   Let a system of n linear equations in n unknowns be given by

$$\begin{cases} a_{11}x_1 + a_{12}x_2 + \cdots + a_{1n}x_n = c_1, \\ a_{21}x_1 + a_{22}x_2 + \cdots + a_{2n}x_n = c_2, \\ \vdots \qquad\quad \vdots \qquad\qquad\quad \vdots \quad \vdots \\ a_{n1}x_1 + a_{n2}x_2 + \cdots + a_{nn}x_n = c_n. \end{cases}$$

If the determinant Δ of the coefficient matrix $\mathbf{A}$ is different from 0, then the system has a unique solution. Moreover, the solution is given by

$$x_1 = \frac{\Delta_1}{\Delta}, \qquad x_2 = \frac{\Delta_2}{\Delta}, \qquad \cdots, \qquad x_n = \frac{\Delta_n}{\Delta},$$

where Δ_k, the numerator of x_k, is the determinant of the matrix obtained by replacing the kth column of $\mathbf{A}$ by the column of constants.

EXAMPLE 1

Solve the following system by Cramer's rule.

$$\begin{cases} 2x + y + z = 0, \\ 4x + 3y + 2z = 2, \\ 2x - y - 3z = 0. \end{cases}$$

The determinant of the coefficient matrix is

$$\Delta = \begin{vmatrix} 2 & 1 & 1 \\ 4 & 3 & 2 \\ 2 & -1 & -3 \end{vmatrix} = -8.$$

Since $\Delta \neq 0$, there is a unique solution. Solving for x, we obtain

$$x = \frac{\begin{vmatrix} 0 & 1 & 1 \\ 2 & 3 & 2 \\ 0 & -1 & -3 \end{vmatrix}}{\Delta} = \frac{4}{-8} = -\frac{1}{2}.$$

Similarly,

$$y = \frac{\begin{vmatrix} 2 & 0 & 1 \\ 4 & 2 & 2 \\ 2 & 0 & -3 \end{vmatrix}}{\Delta} = \frac{-16}{-8} = 2,$$

$$z = \frac{\begin{vmatrix} 2 & 1 & 0 \\ 4 & 3 & 2 \\ 2 & -1 & 0 \end{vmatrix}}{\Delta} = \frac{8}{-8} = -1.$$

$$2(-1)^{2+2} \begin{vmatrix} 2 & 1 \\ 2 & -3 \end{vmatrix}$$

$$2(-6-2)$$

$$2(-8) = -16$$

The solution is $x = -\frac{1}{2}$, $y = 2$, and $z = -1$.

EXAMPLE 2

Solve the following system for z by using Cramer's rule.

$$\begin{cases} x + y \qquad + 5w = 6, \\ x + 2y + z \qquad = 4, \\ \qquad 2y + z + w = 6, \\ 3x \qquad - 4w = 2. \end{cases}$$

We have

$$\Delta = \begin{vmatrix} 1 & 1 & 0 & 5 \\ 1 & 2 & 1 & 0 \\ 0 & 2 & 1 & 1 \\ 3 & 0 & 0 & -4 \end{vmatrix} = \begin{vmatrix} 1 & 1 & 0 & 5 \\ 0 & 1 & 1 & -5 \\ 0 & 0 & -1 & 11 \\ 0 & 0 & 0 & -1 \end{vmatrix} = 1.$$

Here we transformed into upper-triangular form and found the product of the main diagonal entries (Sec. 14-7, Example 4). In a similar fashion we obtain

$$\Delta_z = \begin{vmatrix} 1 & 1 & 6 & 5 \\ 1 & 2 & 4 & 0 \\ 0 & 2 & 6 & 1 \\ 3 & 0 & 2 & -4 \end{vmatrix} = \begin{vmatrix} 1 & 1 & 6 & 5 \\ 0 & 1 & -2 & -5 \\ 0 & 0 & 10 & 11 \\ 0 & 0 & 0 & -\frac{49}{5} \end{vmatrix} = -98.$$

Hence $z = \Delta_z / \Delta = -98/1 = -98$.

EXERCISE 14-8

Solve each of the following. Use Cramer's rule, if possible.

1. $\begin{cases} 2x - y = 4, \\ 3x + y = 5. \end{cases}$ 2. $\begin{cases} 3x + y = 6, \\ 7x - 2y = 5. \end{cases}$

3. $\begin{cases} -2x = 4 - 3y, \\ \quad y = 6x - 1. \end{cases}$ 4. $\begin{cases} x + 2y - 6 = 0, \\ \quad y - 1 = 3x. \end{cases}$

5. $\begin{cases} 3(x + 2) = 5, \\ 6(x + y) = -8. \end{cases}$ 6. $\begin{cases} w - 2z = 4, \\ 3w - 4z = 6. \end{cases}$

7. $\begin{cases} \frac{3}{2}x - \frac{1}{4}z = 1, \\ \frac{1}{3}x + \frac{1}{2}z = 2. \end{cases}$ 8. $\begin{cases} .6x - .7y = .33, \\ 2.1x - .9y = .69. \end{cases}$

9. $\begin{cases} x + y + z = 6, \\ x - y + z = 2, \\ 2x - y + 3z = 6. \end{cases}$ 10. $\begin{cases} 2x - y + 3z = 12, \\ x + y - z = -3, \\ x + 2y - 3z = -10. \end{cases}$

11. $\begin{cases} 2x - 3y + 4z = 0, \\ x + y - 3z = 4, \\ 3x + 2y - z = 0. \end{cases}$ 12. $\begin{cases} 3r \quad - t = 7, \\ 4r - s + 3t = 9, \\ \quad 3s + 2t = 15. \end{cases}$

13. $\begin{cases} x - 2y + z = 3, \\ 2x + y + 2z = 6, \\ x + 8y + z = 3. \end{cases}$ 14. $\begin{cases} 2x + y + z = 1, \\ x - y + z = 4, \\ 5x + y + 3z = 5. \end{cases}$

15. $\begin{cases} 2x - 3y + z = -2, \\ x - 6y + 3z = -2, \\ 3x + 3y - 2z = 2. \end{cases}$ 16. $\begin{cases} x - z = 14, \\ y + z = 21, \\ x - y + z = -10. \end{cases}$

In each of the following solve for the indicated unknowns.

17. $\begin{cases} x - y + 3z + w = -14, \\ x + 2y \quad - 3w = 12, \\ 2x + 3y + 6z + w = 1, \\ x + y + z + w = 6. \end{cases}$; $y, w.$ 18. $\begin{cases} x + y + 5z = 6, \\ x + 2y + w = 4, \\ 2y + z + w = 6, \\ 3x - 4z = 2. \end{cases}$; $x, y.$

19. Show that Cramer's rule does *not* apply to

$$\begin{cases} 2 - y = x, \\ 3 + x = -y, \end{cases}$$

but that from geometrical considerations there is no solution.

14-9 INVERSES USING THE ADJOINT

Determinants and cofactors can be used to find the inverse of a matrix, if it exists. To begin we need the idea of the *transpose* of a matrix.

DEFINITION.   *The **transpose** of an m by n matrix* $\mathbf{A}$, *denoted* $\mathbf{A}^T$, *is the n by m matrix whose i-th row is the i-th column of* $\mathbf{A}$.

EXAMPLE 1

If $\mathbf{A} = \begin{bmatrix} 1 & 2 & 3 \\ 4 & 5 & 6 \\ 7 & 8 & 9 \end{bmatrix}$, *find* $\mathbf{A}^T$.

Column 1 of $\mathbf{A}$ becomes row 1 of $\mathbf{A}^T$, column 2 becomes row 2, and column 3 becomes row 3. Thus,

$$\mathbf{A}^T = \begin{bmatrix} 1 & 4 & 7 \\ 2 & 5 & 8 \\ 3 & 6 & 9 \end{bmatrix}.$$

DEFINITION.   *The **adjoint** of a square matrix* $\mathbf{A}$, *denoted adj* $\mathbf{A}$, *is the transpose of the matrix obtained by replacing each entry* a_{ij} *in* $\mathbf{A}$ *by its cofactor* c_{ij}. *That is, it is the transpose of the cofactor matrix* $[c_{ij}]$.

EXAMPLE 2

If $\mathbf{A} = \begin{bmatrix} 2 & -1 & 3 \\ 3 & 0 & -5 \\ 2 & 1 & 1 \end{bmatrix}$, *find adj* $\mathbf{A}$.

We first find the cofactor c_{ij} of each entry a_{ij} in $\mathbf{A}$.

$$c_{11} = (-1)^{1+1} \begin{vmatrix} 0 & -5 \\ 1 & 1 \end{vmatrix} = (1)(5) = 5.$$

$$c_{12} = (-1)^{1+2} \begin{vmatrix} 3 & -5 \\ 2 & 1 \end{vmatrix} = (-1)(13) = -13.$$

$$c_{13} = (-1)^{1+3} \begin{vmatrix} 3 & 0 \\ 2 & 1 \end{vmatrix} = (1)(3) = 3.$$

$$c_{21} = (-1)^{2+1} \begin{vmatrix} -1 & 3 \\ 1 & 1 \end{vmatrix} = (-1)(-4) = 4.$$

$$c_{22} = (-1)^{2+2} \begin{vmatrix} 2 & 3 \\ 2 & 1 \end{vmatrix} = (1)(-4) = -4.$$

$$c_{23} = (-1)^{2+3} \begin{vmatrix} 2 & -1 \\ 2 & 1 \end{vmatrix} = (-1)(4) = -4.$$

$$c_{31} = (-1)^{3+1} \begin{vmatrix} -1 & 3 \\ 0 & -5 \end{vmatrix} = (1)(5) = 5.$$

$$c_{32} = (-1)^{3+2} \begin{vmatrix} 2 & 3 \\ 3 & -5 \end{vmatrix} = (-1)(-19) = 19.$$

$$c_{33} = (-1)^{3+3} \begin{vmatrix} 2 & -1 \\ 3 & 0 \end{vmatrix} = (1)(3) = 3.$$

The cofactor matrix $[c_{ij}]$ is thus

$$[c_{ij}] = \begin{bmatrix} 5 & -13 & 3 \\ 4 & -4 & -4 \\ 5 & 19 & 3 \end{bmatrix}.$$

The adjoint is

$$\text{adj } \mathbf{A} = [c_{ij}]^T = \begin{bmatrix} 5 & 4 & 5 \\ -13 & -4 & 19 \\ 3 & -4 & 3 \end{bmatrix}.$$

It can be shown that if $|\mathbf{A}| \neq 0$, then $\mathbf{A}^{-1}$ exists and

$$\boxed{\mathbf{A}^{-1} = \frac{1}{|\mathbf{A}|} \text{adj } \mathbf{A}.}$$

EXAMPLE 3

If $\mathbf{A} = \begin{bmatrix} 2 & -1 & 3 \\ 3 & 0 & -5 \\ 2 & 1 & 1 \end{bmatrix}$, *find* $\mathbf{A}^{-1}$.

We find that

$$|\mathbf{A}| = 32 \neq 0.$$

Thus $\mathbf{A}^{-1}$ exists. Also, from Example 2,

$$\text{adj } \mathbf{A} = \begin{bmatrix} 5 & 4 & 5 \\ -13 & -4 & 19 \\ 3 & -4 & 3 \end{bmatrix}.$$

Thus,

$$\mathbf{A}^{-1} = \frac{1}{|\mathbf{A}|} \text{adj } \mathbf{A}$$

$$= \frac{1}{32} \begin{bmatrix} 5 & 4 & 5 \\ -13 & -4 & 19 \\ 3 & -4 & 3 \end{bmatrix}$$

$$= \begin{bmatrix} \frac{5}{32} & \frac{1}{8} & \frac{5}{32} \\ -\frac{13}{32} & -\frac{1}{8} & \frac{19}{32} \\ \frac{3}{32} & -\frac{1}{8} & \frac{3}{32} \end{bmatrix}.$$

You should verify that $\mathbf{A}^{-1}\mathbf{A} = \mathbf{I}$.

EXAMPLE 4

Find $\mathbf{A}^{-1}$ *if* $\mathbf{A} = \begin{bmatrix} 1 & 0 & -2 \\ 4 & -2 & 1 \\ 1 & 2 & -10 \end{bmatrix}$.

We first find the cofactors of $\mathbf{A}$.

$$c_{11} = (-1)^2 \begin{vmatrix} -2 & 1 \\ 2 & -10 \end{vmatrix} = 18.$$

$$c_{12} = (-1)^3 \begin{vmatrix} 4 & 1 \\ 1 & -10 \end{vmatrix} = 41.$$

$$c_{13} = (-1)^4 \begin{vmatrix} 4 & -2 \\ 1 & 2 \end{vmatrix} = 10.$$

$$c_{21} = (-1)^3 \begin{vmatrix} 0 & -2 \\ 2 & -10 \end{vmatrix} = -4.$$

$$c_{22} = (-1)^4 \begin{vmatrix} 1 & -2 \\ 1 & -10 \end{vmatrix} = -8.$$

$$c_{23} = (-1)^5 \begin{vmatrix} 1 & 0 \\ 1 & 2 \end{vmatrix} = -2.$$

$$c_{31} = (-1)^4 \begin{vmatrix} 0 & -2 \\ -2 & 1 \end{vmatrix} = -4.$$

$$c_{32} = (-1)^5 \begin{vmatrix} 1 & -2 \\ 4 & 1 \end{vmatrix} = -9.$$

$$c_{33} = (-1)^6 \begin{vmatrix} 1 & 0 \\ 4 & -2 \end{vmatrix} = -2.$$

Since the inverse of $\mathbf{A}$ involves $|\mathbf{A}|$, we compute $|\mathbf{A}|$ next. We already have the cofactors, so we shall find $|\mathbf{A}|$ by expanding along the first row.

$$|\mathbf{A}| = (1)(18) + 0 + (-2)(10) = -2.$$

The cofactor matrix is

$$[c_{ij}] = \begin{bmatrix} 18 & 41 & 10 \\ -4 & -8 & -2 \\ -4 & -9 & -2 \end{bmatrix},$$

and the adjoint is $[c_{ij}]^{\mathrm{T}}$:

$$\text{adj } \mathbf{A} = \begin{bmatrix} 18 & -4 & -4 \\ 41 & -8 & -9 \\ 10 & -2 & -2 \end{bmatrix}.$$

Thus,

$$\mathbf{A}^{-1} = \frac{1}{|\mathbf{A}|}\text{adj } \mathbf{A}$$

$$= \frac{1}{-2}\begin{bmatrix} 18 & -4 & -4 \\ 41 & -8 & -9 \\ 10 & -2 & -2 \end{bmatrix}$$

$$= \begin{bmatrix} -9 & 2 & 2 \\ -\frac{41}{2} & 4 & \frac{9}{2} \\ -5 & 1 & 1 \end{bmatrix}.$$

This result was also obtained in Example 3(a) of Sec. 14-6 by reduction.

EXAMPLE 5

Find $\mathbf{A}^{-1}$ if $\mathbf{A} = \begin{bmatrix} \frac{4}{5} & -\frac{1}{3} \\ -\frac{3}{10} & \frac{13}{15} \end{bmatrix}$.

Since $|\mathbf{A}| = (\frac{4}{5})(\frac{13}{15}) - (-\frac{1}{3})(-\frac{3}{10}) = \frac{89}{150} \neq 0$, $\mathbf{A}^{-1}$ exists. The cofactors are (the vertical bars here denote determinants, not absolute value)

$$c_{11} = (-1)^2\left|\frac{13}{15}\right| = \frac{13}{15}, \quad c_{12} = (-1)^3\left|-\frac{3}{10}\right| = \frac{3}{10},$$

$$c_{21} = (-1)^3\left|-\frac{1}{3}\right| = \frac{1}{3}, \quad c_{22} = (-1)^4\left|\frac{4}{5}\right| = \frac{4}{5}.$$

The cofactor matrix is

$$[c_{ij}] = \begin{bmatrix} \frac{13}{15} & \frac{3}{10} \\ \frac{1}{3} & \frac{4}{5} \end{bmatrix},$$

and the adjoint is

$$\text{adj } \mathbf{A} = [c_{ij}]^T = \begin{bmatrix} \frac{13}{15} & \frac{1}{3} \\ \frac{3}{10} & \frac{4}{5} \end{bmatrix}.$$

Hence,

$$\mathbf{A}^{-1} = \frac{1}{|\mathbf{A}|}\text{adj } \mathbf{A} = \frac{150}{89}\begin{bmatrix} \frac{13}{15} & \frac{1}{3} \\ \frac{3}{10} & \frac{4}{5} \end{bmatrix}$$

$$= \begin{bmatrix} \frac{130}{89} & \frac{50}{89} \\ \frac{45}{89} & \frac{120}{89} \end{bmatrix}.$$

EXERCISE 14-9

For the following, use adjoints to find the inverses.

1. $\begin{bmatrix} 3 & -2 \\ 1 & 2 \end{bmatrix}$.

2. $\begin{bmatrix} 2 & -1 \\ 1 & 3 \end{bmatrix}$.

3. $\begin{bmatrix} \frac{1}{4} & \frac{3}{8} \\ 0 & \frac{1}{6} \end{bmatrix}$.

4. $\begin{bmatrix} \frac{1}{2} & \frac{1}{4} \\ \frac{1}{3} & \frac{2}{3} \end{bmatrix}$.

5. $\begin{bmatrix} 2 & 3 & -1 \\ 1 & 2 & 1 \\ -1 & -1 & 3 \end{bmatrix}$.

6. $\begin{bmatrix} -1 & 2 & -3 \\ 2 & 1 & 0 \\ 4 & -2 & 5 \end{bmatrix}$.

7. $\begin{bmatrix} 1 & -\frac{2}{3} & \frac{5}{3} \\ -1 & \frac{4}{3} & -\frac{10}{3} \\ -1 & 1 & -2 \end{bmatrix}$.

8. $\begin{bmatrix} 1 & 2 & 3 \\ 1 & 3 & 5 \\ 1 & 5 & 12 \end{bmatrix}$.

9. $\begin{bmatrix} -\frac{1}{4} & -\frac{1}{2} & \frac{3}{4} \\ 0 & \frac{1}{2} & 0 \\ \frac{1}{2} & \frac{1}{2} & -\frac{1}{2} \end{bmatrix}$.

10. $\begin{bmatrix} \frac{11}{2} & -\frac{5}{2} & -\frac{3}{2} \\ -4 & 2 & 1 \\ -\frac{7}{2} & \frac{3}{2} & \frac{1}{2} \end{bmatrix}$.

11. $\begin{bmatrix} \frac{2}{5} & -\frac{1}{5} & \frac{9}{15} \\ \frac{4}{15} & \frac{1}{5} & -\frac{4}{15} \\ -\frac{1}{15} & \frac{1}{5} & \frac{1}{15} \end{bmatrix}$.

12. $\begin{bmatrix} 0 & -\frac{1}{9} & \frac{4}{9} \\ 0 & \frac{2}{9} & \frac{1}{9} \\ \frac{1}{8} & \frac{1}{36} & -\frac{1}{9} \end{bmatrix}$.

14-10 INPUT-OUTPUT ANALYSIS

Input-output matrices, which were developed by Wassily W. Leontief[†] of Harvard, indicate the supply and demand interrelationships that exist between the various sectors of an economy during some time period. The phrase "input-output" is used because the matrices show the values of outputs of each industry that are sold as inputs to each industry and for final use by consumers.

[†]Leontief won the 1973 Nobel prize in economic science for the development of the "input-output" method and its applications to economic problems.

A hypothetical example for an oversimplified two-industry economy is given by the matrix below. Before we explain the matrix, let us say that the *industrial* sectors can be thought of as manufacturing, steel, agriculture, coal, etc. The *other production factors* sector consists of costs to the respective industries such as labor, profits, etc. The *final demand* sector could be consumption by households, government, etc.

Consumers (input)

	Industry A	Industry B	Final Demand	Totals
Producers (output):				*Totals*
Industry A	240	500	460	1200
Industry B	360	200	940	1500
Other Production Factors	600	800	—	
Totals	1200	1500		

Each industry appears in a row and column. The row shows the purchases of an industry's output by the industrial sectors and by consumers for final use (hence the term "final demand"). The entries represent the value of the products and might be in units of millions of dollars of product. For example, of the total output of industry A, 240 went as input to industry A itself (for internal use), 500 to industry B, and 460 went direct to the final demand sector. The total output of A is the sum of industrial demand and final demand ($240 + 500 + 460 = 1200$).

Each industry column gives the value of what the industry purchased for input from each industry as well as what it spent for other costs. For example, in order to produce its 1200 units, A purchased 240 units of output from itself, 360 of B's output, and had labor and other costs of 600 units.

Note that for each industry, the sum of the entries in its row is equal to the sum of the entries in its column. That is, the value of total output of A is equal to the value of total input to A.

Input-output analysis allows us to estimate the total production of each *industrial* sector if there is a change in final demand *as long as the basic structure of the economy remains the same*. This important assumption means that for each industry, the amount spent on each input for each dollar's worth of output must remain fixed.

For example, in producing 1200 units worth of product, industry A purchases 240 units' worth from industry A, 360 units' worth from B, and spends 600 units on other costs. Thus for each dollar's worth of output, industry A spends $\frac{240}{1200} = \frac{1}{5}$ (= \$0.20) on A, $\frac{360}{1200} = \frac{3}{10}$ (= \$0.30) on B, and $\frac{600}{1200} = \frac{1}{2}$ (= \$0.50) on other costs. Combining these fixed ratios of industry A with those of industry B, we can give the input requirements per dollar of output for each industry.

$$
\begin{array}{c}
\\
A \\
\\
B \\
\\
\text{Other}
\end{array}
\begin{array}{cc}
A & B \\
\end{array}
\left[
\begin{array}{c|c}
\dfrac{240}{1200} & \dfrac{500}{1500} \\[2mm]
\dfrac{360}{1200} & \dfrac{200}{1500} \\[2mm]
\hline
\dfrac{600}{1200} & \dfrac{800}{1500}
\end{array}
\right]
=
\left[
\begin{array}{c|c}
\dfrac{1}{5} & \dfrac{1}{3} \\[2mm]
\dfrac{3}{10} & \dfrac{2}{15} \\[2mm]
\hline
\dfrac{1}{2} & \dfrac{8}{15}
\end{array}
\right]
\begin{array}{c}
A \\
\\
B \\
\\
\text{Other}
\end{array}
$$

The entries in the matrix are called *input-output coefficients*. The sum of each column is 1.

Now, suppose the value of final demand changes from 460 to 500 for industry A and from 940 to 1200 for industry B. We would like to estimate the value of *total* output that A and B must produce for both industry and final demand to meet this goal, provided the structure in the preceding matrix remains the same.

Let the new values of total outputs for industries A and B be X_A and X_B, respectively. Since

$$
\begin{array}{c}
\text{Total value of} \\
\text{output of } A
\end{array}
=
\begin{array}{c}
\text{Value consumed} \\
\text{by } A
\end{array}
+
\begin{array}{c}
\text{Value consumed} \\
\text{by } B
\end{array}
+
\begin{array}{c}
\text{Value consumed} \\
\text{by final demand}
\end{array},
$$

we have

$$
X_A = \tfrac{1}{5} X_A + \tfrac{1}{3} X_B + 500.
$$

Similarly,

$$
X_B = \tfrac{3}{10} X_A + \tfrac{2}{15} X_B + 1200.
$$

In matrix notation,

$$
\begin{bmatrix} X_A \\ X_B \end{bmatrix}
=
\begin{bmatrix} \tfrac{1}{5} & \tfrac{1}{3} \\[1mm] \tfrac{3}{10} & \tfrac{2}{15} \end{bmatrix}
\begin{bmatrix} X_A \\ X_B \end{bmatrix}
+
\begin{bmatrix} 500 \\ 1200 \end{bmatrix}.
\tag{1}
$$

Let

$$
\mathbf{X} = \begin{bmatrix} X_A \\ X_B \end{bmatrix}, \qquad
\mathbf{A} = \begin{bmatrix} \tfrac{1}{5} & \tfrac{1}{3} \\[1mm] \tfrac{3}{10} & \tfrac{2}{15} \end{bmatrix}, \qquad
\text{and} \quad
\mathbf{C} = \begin{bmatrix} 500 \\ 1200 \end{bmatrix}.
$$

We call **X** the **output matrix**, **A** the **coefficient matrix**, and **C** the **final demand**

matrix. From Eq. (1)

$$X = AX + C,$$
$$X - AX = C.$$

If **I** is the 2×2 identity matrix, then

$$IX - AX = C,$$
$$(I - A)X = C.$$

If $(I - A)^{-1}$ exists, then

$$\boxed{X = (I - A)^{-1}C.}$$

The matrix $I - A$ is called the Leontief matrix. Now,

$$I - A = \begin{bmatrix} 1 & 0 \\ 0 & 1 \end{bmatrix} - \begin{bmatrix} \frac{1}{5} & \frac{1}{3} \\ \frac{3}{10} & \frac{2}{15} \end{bmatrix}$$

$$= \begin{bmatrix} \frac{4}{5} & -\frac{1}{3} \\ -\frac{3}{10} & \frac{13}{15} \end{bmatrix}.$$

From Example 5 of Sec. 14-9,

$$(I - A)^{-1} = \begin{bmatrix} \frac{130}{89} & \frac{50}{89} \\ \frac{45}{89} & \frac{120}{89} \end{bmatrix};$$

Hence the output matrix is

$$X = (I - A)^{-1}C = \begin{bmatrix} \frac{130}{89} & \frac{50}{89} \\ \frac{45}{89} & \frac{120}{89} \end{bmatrix} \begin{bmatrix} 500 \\ 1200 \end{bmatrix}$$

$$= \begin{bmatrix} 1404.49 \\ 1870.79 \end{bmatrix}.$$

Thus to meet the goal, industry A must produce 1404.49 units of value and industry B must produce 1870.79. If we were interested in the value of other production factors for A, then

$$P_A = \tfrac{1}{2}X_A = 702.25.$$

EXAMPLE 1

Given the input-output matrix below, suppose final demand changes to 77 for A, 154 for B, and 231 for C. Find the output matrix for the economy. (The entries are in millions of dollars.)

		Industry			Final
		A	B	C	Demand
Industry:	A	240	180	144	36
	B	120	36	48	156
	C	120	72	48	240
Other		120	72	240	—

We separately add the entries in the first three rows. The total value of output for industries A, B, and C are 600, 360, and 480, respectively. To get the coefficient matrix, we divide the industry entries in each industry column by the total value of output for that industry.

$$
\mathbf{A} = \begin{bmatrix} \frac{240}{600} & \frac{180}{360} & \frac{144}{480} \\ \frac{120}{600} & \frac{36}{360} & \frac{48}{480} \\ \frac{120}{600} & \frac{72}{360} & \frac{48}{480} \end{bmatrix} = \begin{bmatrix} \frac{2}{5} & \frac{1}{2} & \frac{3}{10} \\ \frac{1}{5} & \frac{1}{10} & \frac{1}{10} \\ \frac{1}{5} & \frac{1}{5} & \frac{1}{10} \end{bmatrix}.
$$

Thus, if $\mathbf{I}$ is the 3×3 identity matrix,

$$
\mathbf{I} - \mathbf{A} = \begin{bmatrix} \frac{3}{5} & -\frac{1}{2} & -\frac{3}{10} \\ -\frac{1}{5} & \frac{9}{10} & -\frac{1}{10} \\ -\frac{1}{5} & -\frac{1}{5} & \frac{9}{10} \end{bmatrix}.
$$

We shall find $(\mathbf{I} - \mathbf{A})^{-1}$ by using the adjoint. Computing the cofactors, we have

$$
c_{11} = (-1)^{1+1} \begin{vmatrix} \frac{9}{10} & -\frac{1}{10} \\ -\frac{1}{5} & \frac{9}{10} \end{vmatrix} = \frac{79}{100},
$$

$$
c_{12} = (-1)^{1+2} \begin{vmatrix} -\frac{1}{5} & -\frac{1}{10} \\ -\frac{1}{5} & \frac{9}{10} \end{vmatrix} = \frac{1}{5},
$$

$$
c_{13} = (-1)^{1+3} \begin{vmatrix} -\frac{1}{5} & \frac{9}{10} \\ -\frac{1}{5} & -\frac{1}{5} \end{vmatrix} = \frac{11}{50}.
$$

At this point we can evaluate $|\mathbf{I} - \mathbf{A}|$ along row 1 by using cofactors.

$$|\mathbf{I} - \mathbf{A}| = \tfrac{3}{5}\left(\tfrac{79}{100}\right) - \tfrac{1}{2}\left(\tfrac{1}{5}\right) - \tfrac{3}{10}\left(\tfrac{11}{50}\right) = \tfrac{77}{250}.$$

Continuing, we obtain

$$c_{21} = (-1)^{2+1}\begin{vmatrix} -\tfrac{1}{2} & -\tfrac{3}{10} \\ -\tfrac{1}{5} & \tfrac{9}{10} \end{vmatrix} = \frac{51}{100},$$

$$c_{22} = (-1)^{2+2}\begin{vmatrix} \tfrac{3}{5} & -\tfrac{3}{10} \\ -\tfrac{1}{5} & \tfrac{9}{10} \end{vmatrix} = \frac{12}{25},$$

$$c_{23} = (-1)^{2+3}\begin{vmatrix} \tfrac{3}{5} & -\tfrac{1}{2} \\ -\tfrac{1}{5} & -\tfrac{1}{5} \end{vmatrix} = \frac{11}{50},$$

$$c_{31} = (-1)^{3+1}\begin{vmatrix} -\tfrac{1}{2} & -\tfrac{3}{10} \\ \tfrac{9}{10} & -\tfrac{1}{10} \end{vmatrix} = \frac{8}{25},$$

$$c_{32} = (-1)^{3+2}\begin{vmatrix} \tfrac{3}{5} & -\tfrac{3}{10} \\ -\tfrac{1}{5} & -\tfrac{1}{10} \end{vmatrix} = \frac{3}{25},$$

$$c_{33} = (-1)^{3+3}\begin{vmatrix} \tfrac{3}{5} & -\tfrac{1}{2} \\ -\tfrac{1}{5} & \tfrac{9}{10} \end{vmatrix} = \frac{11}{25}.$$

Thus,

$$(\mathbf{I} - \mathbf{A})^{-1} = \frac{1}{|\mathbf{I} - \mathbf{A}|}\,\mathrm{adj}(\mathbf{I} - \mathbf{A})$$

$$= \frac{250}{77}\begin{bmatrix} \tfrac{79}{100} & \tfrac{51}{100} & \tfrac{8}{25} \\ \tfrac{1}{5} & \tfrac{12}{25} & \tfrac{3}{25} \\ \tfrac{11}{50} & \tfrac{11}{50} & \tfrac{11}{25} \end{bmatrix}$$

$$= \begin{bmatrix} \tfrac{395}{154} & \tfrac{255}{154} & \tfrac{80}{77} \\ \tfrac{50}{77} & \tfrac{120}{77} & \tfrac{30}{77} \\ \tfrac{5}{7} & \tfrac{5}{7} & \tfrac{10}{7} \end{bmatrix}.$$

Hence

$$X = (I - A)^{-1}C$$

$$= \begin{bmatrix} \frac{395}{154} & \frac{255}{154} & \frac{80}{77} \\ \frac{50}{77} & \frac{120}{77} & \frac{30}{77} \\ \frac{5}{7} & \frac{5}{7} & \frac{10}{7} \end{bmatrix} \begin{bmatrix} 77 \\ 154 \\ 231 \end{bmatrix}$$

$$= \begin{bmatrix} 692.5 \\ 380 \\ 495 \end{bmatrix}.$$

EXERCISE 14-10

1. Given the input-output matrix below, find the output matrix if final demand changes to 600 for A and 805 for B. Find the total value of other production costs that this involves.

	Industry		Final
	A	B	Demand
Industry: A	200	500	500
B	400	200	900
Other	600	800	—

2. Given the input-output matrix below, find the output matrix if final demand changes to (a) 200 for A and 300 for B; (b) 64 for A and 64 for B.

	Industry		Final
	A	B	Demand
Industry: A	40	120	40
B	120	90	90
Other	40	90	—

3. Given the input-output matrix below, find the output matrix if final demand changes

to (a) 50 for A, 40 for B, and 30 for C; (b) 10 for A, 10 for B, and 24 for C.

	Industry			Final Demand
	A	B	C	
Industry: A	18	30	45	15
B	27	30	60	3
C	54	40	60	26
Other	9	20	15	—

4. Given the input-output matrix below, find the output matrix if final demand changes to 300 for A, 200 for B, and 400 for C.

	Industry			Final Demand
	A	B	C	
Industry: A	100	400	240	260
B	100	80	480	140
C	300	160	240	500
Other	500	160	240	—

14-11 REVIEW

IMPORTANT TERMS AND SYMBOLS IN CHAPTER 14

matrix *(p. 644)*

order (dimension) *(p. 644)*

entry *(p. 645)*

a_{ij} *(p. 645)*

$[a_{ij}]$ *(p. 645)*

row matrix *(p. 646)*

column matrix *(p. 646)*

zero matrix, $\mathbf{O}$ *(p. 647)*

equality of matrices *(p. 647)*

main diagonal *(p. 648)*

square matrix *(p. 648)*

upper triangular matrix *(p. 648)*

matrix addition *(p. 651)*

scalar multiplication *(p. 653)*

matrix multiplication *(p. 658)*

identity matrix, $\mathbf{I}$ *(p. 662)*

coefficient matrix *(p. 670)*

augmented coefficient matrix *(p. 670)*

elementary row operation *(p. 671)*

equivalent matrices *(p. 671)*

parameter *(p. 677)*

reduced matrix *(p. 671)*

homogeneous system   *(p. 680)*   nonhomogeneous system
                                              (p. 680)

trivial solution   *(p. 681)*   inverse matrix   *(p. 686)*

invertible   *(p. 686)*   elementary matrices   *(p. 687)*

determinant   *(p. 694)*   minor   *(p. 695)*

cofactor   *(p. 696)*   Cramer's rule   *(p. 704)*

transpose   *(p. 707)*   adjoint   *(p. 707)*

input-output matrix   *(p. 711)*

REVIEW SECTION

1. The order of the matrix given below is ___(a)___ and the entry in the third row and second column is ___(b)___.

$$\begin{bmatrix} 2 & 3 & 0 & 1 \\ 1 & 4 & 6 & 0 \\ 7 & 5 & 2 & 3 \end{bmatrix}$$

Ans.   (a) 3×4;   (b) 5.

2. A matrix of order 3×1 is a ___(a) (row) (column)___ matrix while a matrix of order 1×3 is a ___(b) (row) (column)___ matrix.

Ans.   (a) column;   (b) row.

3. If $[3x \quad 5] = [6 \quad 5]$, then $x =$ _____.

Ans.   2.

4. If $[4 \quad 1] = k[12 \quad 3]$, then $k =$ _____.

Ans. $\frac{1}{3}$.

5. The main diagonal entries of $\begin{bmatrix} 1 & 3 \\ 2 & 4 \end{bmatrix}$ are _____ and _____.

Ans.   1, 4.

6. If $2\begin{bmatrix} 3 & -4 \\ 0 & 1 \end{bmatrix} - 3\begin{bmatrix} 0 & 1 \\ -2 & 3 \end{bmatrix} = \begin{bmatrix} p & q \\ r & s \end{bmatrix}$, then $p =$ ___(a)___, $q =$ ___(b)___, $r =$ ___(c)___, and $s =$ ___(d)___.

Ans.   (a) 6;   (b) −11;   (c) 6;   (d) −7.

7. True or false: Both matrix addition and multiplication are commutative. _____

Ans.   false.

8. The entry in the first row and second column of the product

$$\begin{bmatrix} 2 & 1 \\ 3 & 4 \end{bmatrix} \cdot \begin{bmatrix} 0 & 1 \\ 2 & 3 \end{bmatrix} \text{ is } \underline{\qquad}.$$

Ans.   5.

9. True or false: In order to multiply two matrices, it is necessary that they be of the same order. _____

Ans.   false

10. True or false: The identity matrix is an upper triangular matrix. _____

Ans.   true

11. With respect to the system

$$\begin{cases} 2x + y = 4, \\ 3x + 2y = 5, \end{cases}$$

the matrix $\begin{bmatrix} 2 & 1 \\ 3 & 2 \end{bmatrix}$ is called the ___(a)___ matrix, while $\begin{bmatrix} 2 & 1 & \vdots & 4 \\ 3 & 2 & \vdots & 5 \end{bmatrix}$ is called the ___(b)___ matrix.

Ans.   (a) coefficient;   (b) augmented coefficient.

12. If **B** is obtained from **A** by an elementary row operation, then **A** is said to be _____ to **B**.

Ans.   equivalent.

13. True or false: The 3×3 identity matrix and the 3×3 zero matrix are both reduced matrices. _____

Ans.   true.

14. A system of m linear equations in n unknowns has either a unique solution, an infinite number of solutions, or _____ .

Ans.   no solution.

15. The system

$$\begin{cases} 2x - 3y + 4z = 0, \\ 3x - 4y + 2z = 0 \end{cases}$$

is ___(a) (homogeneous) (nonhomogeneous)___ and it has a(n) ___(b) (finite) (infinite)___ number of solutions.

Ans.   (a) homogeneous;   (b) infinite

16. If a homogeneous system of three linear equations in the variables x, y, and z has only the trivial solution, then $x =$ ___(a)___ , $y =$ ___(b)___ , and $z =$ ___(c)___ .

Ans.   (a) 0;   (b) 0;   (c) 0

17. The inverse of **I** is _____ .

Ans.   **I**

18. True or false: Every square matrix is invertible. ___(a)___ . If **A** is invertible, then $A^{-1}A = I$. ___(b)___

Ans.   (a) false;   (b) true

19. If the inverse of $\begin{bmatrix} 3 & 7 \\ 2 & 5 \end{bmatrix}$ is $\begin{bmatrix} 5 & -7 \\ -2 & 3 \end{bmatrix}$, and

$$\begin{cases} 3x + 7y = 0, \\ 2x + 5y = 1, \end{cases}$$

then $x = $ __(a)__ and $y = $ __(b)__ .

Ans. (a) −7; (b) 3.

20. $[4]^{-1} = $ _____.

Ans. $\left[\frac{1}{4} \right]$.

21. If $[A \vdots I] \sim \cdots \sim [I \vdots B]$, then $A^{-1} = $ _____.

Ans. **B**.

22. The determinant of a 2×2 matrix is a __(real number) (2×2 matrix)__ .

Ans. real number.

23. $\|[5]\| = $ __(a)__ and $\begin{vmatrix} 1 & 3 \\ 2 & 4 \end{vmatrix} = $ __(b)__ .

Ans. (a) 5; (b) −2.

24. The minor of the element 3 in the matrix $\begin{bmatrix} 0 & 1 & 3 \\ 2 & 4 & 0 \\ 1 & 5 & 2 \end{bmatrix}$ is __(a)__ and its cofactor is

__(b)__ .

Ans. (a) 6; (b) 6.

25. $\begin{vmatrix} 0 & 0 & 0 \\ 4 & 5 & 1 \\ 2 & 3 & 4 \end{vmatrix} = $ __(a)__ and $\begin{vmatrix} 1 & 2 & 3 \\ 0 & 5 & 6 \\ 0 & 0 & 4 \end{vmatrix} = $ __(b)__ .

Ans. (a) 0; (b) 20.

26. Cramer's rule is applicable in solving systems of n linear equations in __(a)__ unknowns provided the determinant of the coefficient matrix is __(b)__ .

Ans. (a) n; (b) different from 0.

27. $\begin{bmatrix} 1 & 1 \\ 2 & 2 \end{bmatrix}^{\mathrm{T}} = $ _____.

Ans. $\begin{bmatrix} 1 & 2 \\ 1 & 2 \end{bmatrix}$.

28. If adj $A = \begin{bmatrix} \frac{1}{2} & 0 \\ 0 & 1 \end{bmatrix}$ and $|A| = \frac{1}{2}$, then $A^{-1} = $ _____.

Ans. $\begin{bmatrix} 1 & 0 \\ 0 & 2 \end{bmatrix}$.

REVIEW PROBLEMS

In Problems 1–6, simplify.

1. $3 \begin{bmatrix} 3 & 4 \\ -5 & 1 \end{bmatrix} - 2 \begin{bmatrix} 1 & 0 \\ 2 & 4 \end{bmatrix}$.

2. $5 \begin{bmatrix} 1 & 2 \\ 7 & 0 \end{bmatrix} - 6 \begin{bmatrix} 1 & 0 \\ 0 & 1 \end{bmatrix}$.

3. $\begin{bmatrix} 1 & 7 \\ 2 & -3 \\ 1 & 0 \end{bmatrix}\begin{bmatrix} 1 & 0 & -2 \\ 0 & 5 & 1 \end{bmatrix}.$

4. $\begin{bmatrix} 1 & 4 & 5 \end{bmatrix}\begin{bmatrix} 2 & 1 \\ 0 & -1 \\ 8 & 1 \end{bmatrix}.$

5. $\begin{bmatrix} 1 & 0 \\ -1 & 4 \end{bmatrix}\left(\begin{bmatrix} 1 & 4 \\ 6 & 5 \end{bmatrix} - \begin{bmatrix} 2 & 6 \\ 5 & 0 \end{bmatrix}\right).$

6. $-\left(\begin{bmatrix} 2 & 0 \\ 7 & 8 \end{bmatrix} + 2\begin{bmatrix} 0 & -5 \\ 6 & -4 \end{bmatrix}\right).$

In Problems 7 and 8, solve for x and y.

7. $\begin{bmatrix} 5 \\ 2 \end{bmatrix}[x] = \begin{bmatrix} 15 \\ y \end{bmatrix}.$

8. $\begin{bmatrix} 1 & x \\ 2 & y \end{bmatrix}\begin{bmatrix} 2 & 1 \\ x & 3 \end{bmatrix} = \begin{bmatrix} 3 & 4 \\ 3 & y \end{bmatrix}.$

In Problems 9–12, reduce the given matrices.

9. $\begin{bmatrix} 1 & 4 \\ 5 & 8 \end{bmatrix}.$

10. $\begin{bmatrix} 0 & 0 & 4 \\ 0 & 3 & 5 \end{bmatrix}.$

11. $\begin{bmatrix} 2 & 4 & 3 \\ 1 & 2 & 3 \\ 4 & 8 & 6 \end{bmatrix}.$

12. $\begin{bmatrix} 0 & 0 & 0 & 1 \\ 0 & 0 & 0 & 0 \\ 1 & 0 & 0 & 0 \end{bmatrix}.$

In Problems 13–16, solve each of the systems by the method of reduction.

13. $\begin{cases} 2x - 5y = 0, \\ 4x + 3y = 0. \end{cases}$

14. $\begin{cases} x - y + 2z = 3, \\ 3x + y + z = 5. \end{cases}$

15. $\begin{cases} x + y + 2z = 1, \\ 3x - 2y - 4z = -7, \\ 2x - y - 2z = 2. \end{cases}$

16. $\begin{cases} x - y - z - 2 = 0, \\ x + y + 2z + 5 = 0, \\ 2x + z + 3 = 0. \end{cases}$

In Problems 17–20, find the inverses of the matrices by using reduction.

17. $\begin{bmatrix} 1 & 5 \\ 3 & 9 \end{bmatrix}.$

18. $\begin{bmatrix} 0 & 1 \\ 1 & 0 \end{bmatrix}.$

19. $\begin{bmatrix} 1 & 3 & -2 \\ 4 & 1 & 0 \\ 3 & -2 & 2 \end{bmatrix}.$

20. $\begin{bmatrix} 1 & 0 & 0 \\ 2 & 3 & -1 \\ 1 & -1 & 2 \end{bmatrix}.$

In Problems 21 and 22, solve the given system by finding the inverse of the coefficient matrix.

21. $\begin{cases} 3x + y + 4z = 1, \\ x + z = 0, \\ 2y + z = 2. \end{cases}$

22. $\begin{cases} 2x + y - z = 0, \\ 3x + z = 0, \\ x - y + z = 0. \end{cases}$

In Problems 23–28, evaluate the determinants.

23. $\begin{vmatrix} 2 & -1 \\ 4 & 7 \end{vmatrix}.$

24. $\begin{vmatrix} 5 & 8 \\ 3 & 0 \end{vmatrix}.$

25. $\begin{vmatrix} 1 & 2 & -1 \\ 0 & 1 & 4 \\ 1 & 2 & 2 \end{vmatrix}.$

26. $\begin{vmatrix} 2 & 0 & 3 \\ 1 & 4 & 6 \\ -1 & 2 & -1 \end{vmatrix}.$

27. $\begin{vmatrix} r & p & q & a \\ 0 & i & j & m \\ 0 & 0 & c & n \\ 0 & 0 & 0 & h \end{vmatrix}.$

28. $\begin{vmatrix} e & 0 & 0 & 0 \\ a & r & 0 & 0 \\ p & j & n & 0 \\ s & k & t & i \end{vmatrix}.$

Solve the systems in Problems 29 and 30 by using Cramer's rule.

29. $\begin{cases} 3x - y = 1, \\ 2x + 3y = 8. \end{cases}$

30. $\begin{cases} x + 2y - z = 0, \\ y + 4z = 0, \\ x + 2y + 2z = 0. \end{cases}$

In Problems 31 and 32, use the adjoint to find the inverse of each matrix.

31. $\begin{bmatrix} 1 & 1 & 1 \\ 0 & 2 & 1 \\ 1 & 3 & 1 \end{bmatrix}.$

32. $\begin{bmatrix} 2 & 1 & -1 \\ 1 & 1 & 3 \\ -1 & 1 & -1 \end{bmatrix}.$

33. Given the input-output matrix below, find the output matrix if final demand changes to 2 for A and 2 for B. (Data are in tens of billions of dollars.)

	Industry		Final
	A	B	Demand
Industry: A	0	2	1
B	1	0	3
Other	2	2	—

LINEAR PROGRAMMING

15

15-1 LINEAR INEQUALITIES IN TWO VARIABLES

Suppose a consumer receives a fixed income of $60 per week and uses it all to purchase products A and B. If x kilograms of A cost $2 per kg and y kg of B cost $3 per kg, then the possible combinations of A and B that can be purchased must satisfy the consumer's *budget equation*

$$2x + 3y = 60, \quad \text{where } x, y \geqslant 0.$$

The solution is represented by the *budget line* in Fig. 15-1. For example, if 15 kg of A are purchased at a total cost of $30, then 10 kg of B must be bought at a total cost of $30. Thus (15, 10) lies on the line.

On the other hand, suppose the consumer does not necessarily wish to spend his *total* income. In this case the possible combinations are described by the inequality

$$2x + 3y \leqslant 60, \quad \text{where } x, y \geqslant 0. \tag{1}$$

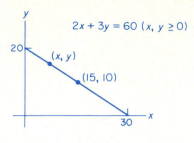

FIG. 15-1

When inequalities in one variable were discussed in Chapter 2, their solutions were represented geometrically by *intervals* on the real number line. However, for an inequality in two variables, as in (1), the solution is usually represented by a *region* in the coordinate plane. We shall find the region corresponding to (1) after considering such inequalities in general.

DEFINITION.   *A **linear inequality** in the variables x and y is an inequality that can be written in the form*

$$ax + by + c < 0 \quad (or \leqslant 0, \geqslant 0, > 0),$$

where a, b, and c are constants and a and b are not both zero.

Geometrically, the solution of a linear inequality in x and y consists of all points in the plane whose coordinates satisfy the inequality. In particular, the graph of a nonvertical line $y = mx + b$ separates the plane into three distinct parts (Fig. 15-2):

(a) the line itself, consisting of all points (x, y) whose coordinates satisfy $y = mx + b$;

(b) the region above the line, consisting of all points (x, y) which satisfy $y > mx + b$;

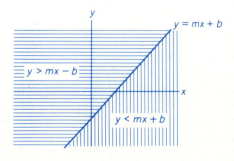

FIG. 15-2

(c) the region below the line, consisting of all points (x, y) satisfying $y < mx + b$.

For a vertical line $x = a$, we speak of regions to the right $(x > a)$ or to the left $(x < a)$ of the line (Fig. 15-3).

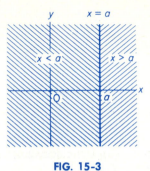

FIG. 15-3

To apply these facts we shall solve $2x + y < 5$. The corresponding *line* $2x + y = 5$ is first sketched by choosing two points on it, for instance the intercepts $(\frac{5}{2}, 0)$ and $(0, 5)$ [Fig. 15-4]. By writing the inequality in the equivalent

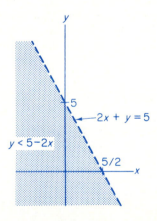

FIG. 15-4

form $y < 5 - 2x$, we conclude from (c) above that the solution consists of all points below the line. Part of this region is shaded in the diagram. Thus if (x_0, y_0) is *any* point in this region, then its ordinate y_0 is less than the number $5 - 2x_0$ (Fig. 15-5). For example, the point $(-2, -1)$ is in the region and $-1 < 5 - 2(-2)$. If we had required that $y \leqslant 5 - 2x$, the line $y = 5 - 2x$ would also have been included in the solution as indicated by the solid line in

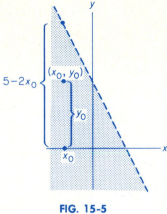

FIG. 15-5

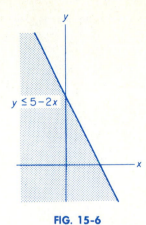

FIG. 15-6

Fig. 15-6. We shall adopt the conventions that **a solid line *is* included in the solution, and a broken line *is not*.**

EXAMPLE 1

a. *Find the region described by $y \leqslant 5$.*

Since x does not appear, the inequality is assumed to be true for all values of x. The solution consists of the line $y = 5$ *and* the region below it (see Fig. 15-7), since the y-coordinate of each point in that region is less than 5.

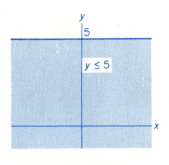

FIG. 15-7

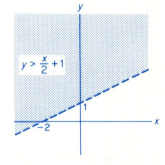

FIG. 15-8

b. *Solve $2(2x - y) < 2(x + y) - 4$.*

The inequality is equivalent to

$$4x - 2y < 2x + 2y - 4,$$
$$-4y < -2x - 4,$$
$$y > \frac{x}{2} + 1.$$

In the last step, both sides were divided by -4 and the sense of the inequality was reversed. We now sketch the line $y = (x/2) + 1$ by noting that its intercepts are $(0, 1)$ and $(-2, 0)$. Then we shade the region above it (see Fig. 15-8). Every point in this region is a solution.

The solution of a *system* of inequalities consists of all points whose coordinates simultaneously satisfy all of the given inequalities. Geometrically, it is the region which is common to all the regions determined by the given inequalities. For example, let us solve the system

$$\begin{cases} 2x + y > 3, \\ x \geqslant y, \\ 2y - 1 > 0. \end{cases}$$

This system is equivalent to

$$\begin{cases} y > -2x + 3, \\ y \leqslant x, \\ y > \tfrac{1}{2}. \end{cases}$$

Note that each inequality has been written so that y is isolated. Thus the appropriate regions with respect to the corresponding lines will be apparent. We first sketch the lines $y = -2x + 3$, $y = x$, and $y = \tfrac{1}{2}$ and then shade the region that is simultaneously above the first line, on or below the second line, and above the third line (see Fig. 15-9). This region is the solution. When one is

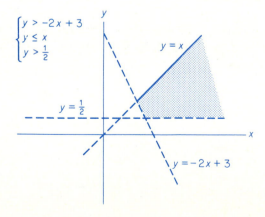

FIG. 15-9

sketching the lines, **it is best to draw broken lines everywhere until it is clear which portions are to be included in the solution.**

EXAMPLE 2

Solve

$$\begin{cases} y \geqslant -2x + 10, \\ y \geqslant x - 2. \end{cases}$$

The solution consists of all points that are simultaneously on or above $y = -2x + 10$ and on or above $y = x - 2$. It is the shaded region in Fig. 15-10.

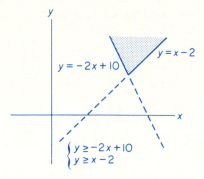

FIG. 15-10

EXAMPLE 3

Find the region described by

$$\begin{cases} 2x + 3y \leqslant 60, \\ x \geqslant 0, \\ y \geqslant 0. \end{cases}$$

This system relates to inequality (1) in the discussion of budget lines at the beginning of this section. The latter two inequalities restrict the solution to points which are both on or

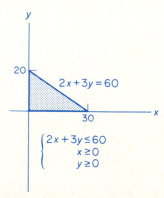

FIG. 15-11

to the right of the y-axis *and* also on or above the x-axis. The desired region is shaded in Fig. 15-11.

EXERCISE 15-1

Sketch the region described by the following inequalities.

1. $2x + 3y > 6.$

2. $3x - 2y \geqslant 12.$

3. $x + 2y \leqslant 7.$

4. $y > 6 - 2x.$

5. $-x \leqslant 2y - 4.$

6. $2x + y \geqslant 10.$

7. $3x + y < 0.$

8. $x + 5y < -5.$

9. $\begin{cases} 3x - 2y < 6, \\ x - 3y > 9. \end{cases}$

10. $\begin{cases} 2x + 3y > -6, \\ 3x - y < 6. \end{cases}$

11. $\begin{cases} 2x + 3y \leqslant 6, \\ x \geqslant 0. \end{cases}$

12. $\begin{cases} 2y - 3x < 6, \\ x < 0. \end{cases}$

13. $\begin{cases} y - 3x < 6, \\ x - y > -3. \end{cases}$

14. $\begin{cases} x - y < 1, \\ y - x < 1. \end{cases}$

15. $\begin{cases} 2x - 2 \geqslant y, \\ 2x \leqslant 3 - 2y. \end{cases}$

16. $\begin{cases} 2y < 4x + 2, \\ y < 2x + 1. \end{cases}$

17. $\begin{cases} x - y > 4, \\ x < 2, \\ y > -5. \end{cases}$

18. $\begin{cases} 2x + y < -1, \\ y > -x, \\ 2x + 6 < 0. \end{cases}$

19. $\begin{cases} y < 2x + 4, \\ x \geqslant -2, \\ y < 1. \end{cases}$

20. $\begin{cases} 4x + 3y \geqslant 12, \\ y \geqslant x, \\ 2y \leqslant 3x + 6. \end{cases}$

21. $\begin{cases} x + y > 1, \\ 3x - 5 \leqslant y, \\ y < 2x. \end{cases}$

22. $\begin{cases} 2x - 3y > -12, \\ 3x + y > -6, \\ y > x. \end{cases}$

23. $\begin{cases} 3x + y > -6, \\ x - y > -5, \\ x \geqslant 0. \end{cases}$

24. $\begin{cases} 5y - 2x \leqslant 10, \\ 4x - 6y \leqslant 12, \\ y \geqslant 0. \end{cases}$

*If a consumer wants to spend no more than P dollars to purchase quantities x and y of two products having prices of p_1 and p_2 dollars per unit, respectively, then $p_1 x + p_2 y \leqslant P$, where $x, y \geqslant 0$. In Problems **25** and **26**, find geometrically the possible combinations of purchases by determining the solution of this system for the given values of p_1, p_2, and P.*

25. $p_1 = 5, p_2 = 3, P = 15.$

26. $p_1 = 6, p_2 = 4, P = 24.$

27. If a manufacturer wishes to purchase a *total* of no more than 100 lb of Product Z from suppliers A and B, set up a system of inequalities which describes the possible combinations of quantities that can be purchased from each supplier. Sketch the solution in the plane.

15-2 LINEAR PROGRAMMING

Sometimes it is desired to maximize or minimize a function subject to certain restrictions (or *constraints*). For example, a manufacturer may want to maximize a profit function subject to production restrictions imposed by limitations on the use of machinery and labor.

We shall now consider how to solve such problems when the function to be maximized or minimized is *linear*. A **linear function in x and y** has the form

$$Z = ax + by,$$

where a and b are constants. We shall also require that the corresponding constraints be represented by a system of linear inequalities (involving " $\leqslant$ " or " $\geqslant$ ") or linear equations in x and y, and that all variables be nonnegative. A problem involving all of these conditions is called a *linear programming problem*.

Linear programming was developed by George B. Danzig in the late 1940's, and was first used by the U.S. Air Force as an aid in decision-making. Today it has wide application in industrial and economic analysis.

In a linear programming problem, the function to be maximized or minimized is called the **objective function**. Although there are usually infinitely many solutions to the system of constraints (these are called **feasible solutions** or **feasible points**), the aim is to find one such solution that is an **optimum solution** (that is, one that gives the maximum or minimum value of the objective function).

We shall now give a geometrical approach to linear programming. In Sec. 15-4 a matrix approach will be discussed that will enable us to work with more than two variables and, hence, a wider range of problems.

Suppose a company produces two types of widgets, manual and electric. Each requires in its manufacture the use of three machines: A, B, and C. A manual widget requires the use of machine A for 2 hours, machine B for 1 hour, and machine C for 1 hour. An electric widget requires 1 hour on A, 2 hours on B, and 1 hour on C. Furthermore, suppose the maximum numbers of hours available per month for the use of machines A, B, and C are 180, 160, and 100, respectively. The profit on a manual widget is \$4 and on an electric widget it is \$6. (See Table 15-1 for a summary of data.) If the company can sell all the widgets it can produce, how many of each type should it make in order to maximize the monthly profit?

TABLE 15-1

	A	B	C	PROFIT/UNIT
Manual	2 hr	1 hr	1 hr	\$4
Electric	1 hr	2 hr	1 hr	\$6
Hours available	180	160	100	

To answer the question we let x and y denote the number of manual and electric widgets, respectively, that are made in a month. Since the number of widgets made is not negative, we have

$$x \geqslant 0, \qquad y \geqslant 0.$$

For machine A, the time needed for working on x manual widgets is $2x$ hours, and the time needed for working on y electric widgets is $1y$ hours. The sum of these times cannot be greater than 180, and so we have

$$2x + y \leqslant 180.$$

Similarly, the restrictions for machines B and C give

$$x + 2y \leqslant 160 \qquad \text{and} \qquad x + y \leqslant 100.$$

The profit P is a function of x and y and is given by the *profit function*

$$P = 4x + 6y.$$

Summarizing, we want to maximize the *objective function*

$$P = 4x + 6y \tag{1}$$

subject to the condition that x and y must be a solution to the system of constraints

$$\left\{ \begin{array}{ll} x \geqslant 0, & \text{(2)} \\ y \geqslant 0, & \text{(3)} \\ 2x + y \leqslant 180, & \text{(4)} \\ x + 2y \leqslant 160, & \text{(5)} \\ x + y \leqslant 100. & \text{(6)} \end{array} \right.$$

Thus we have a linear programming problem. Constraints (2) and (3) are called **nonnegativity conditions**. The region simultaneously satisfying (2)–(6) is shaded in Fig. 15-12. Each point in this region represents a feasible solution, and the region is called the **feasible region**. Although there are infinitely many feasible solutions, we must find one that maximizes the profit function.

Since $P = 4x + 6y$ is equivalent to

$$y = -\frac{2}{3}x + \frac{P}{6},$$

it defines a so-called "family" of parallel lines, each having a slope of $-2/3$ and

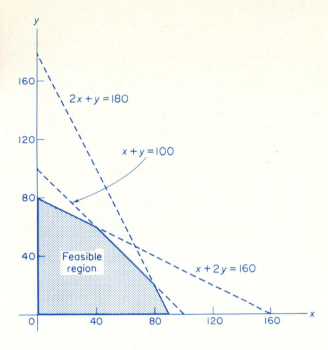

FIG. 15-12

y-intercept $(0, P/6)$. For example, if $P = 600$, then we obtain the line $y = -\frac{2}{3}x + 100$ shown in Fig. 15-13. This line, called an *isoprofit line*, gives all possible combinations of x and y that yield the same profit, \$600. Note that this isoprofit line has no point in common with the feasible region, whereas the isoprofit line for $P = 300$ has infinitely many such points. Let us look for the member of the family that contains a feasible point and whose P-value is maximum. *It will be the line whose y-intercept is furthest from the origin (this gives a maximum value of P) and which has at least one point in common with the feasible region.* It is not difficult to observe that such a line will contain the *corner point A*. Any isoprofit line with a greater profit will contain no points of the feasible region.

From Fig. 15-12, A lies on both the line $x + y = 100$ and the line $x + 2y = 160$. Thus its coordinates may be found by solving the system

$$\begin{cases} x + y = 100, \\ x + 2y = 160. \end{cases}$$

This gives $x = 40$ and $y = 60$. Substituting these values in $P = 4x + 6y$, we find

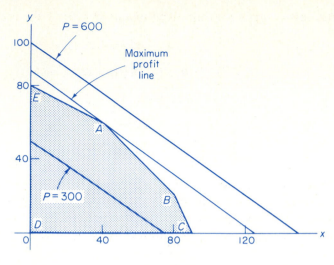

FIG. 15-13

that the maximum profit subject to the constraints is \$520, which is obtained by producing 40 manual widgets and 60 electric widgets per month.

If a feasible region can be contained within a circle, such as the region in Fig. 15-13, it is called a **bounded feasible region**. Otherwise, it is **unbounded**. When a feasible region contains at least one point, it is said to be **nonempty**. Otherwise, it is **empty**. Thus the region in Fig. 15-13 is a nonempty bounded feasible region.

It can be shown that

> **a linear function defined on a nonempty bounded feasible region has a maximum (minimum) value, and this value can be found at a corner point.**

This statement gives us a way of finding an optimum solution without drawing isoprofit lines as we did above. We could evaluate the objective function at each of the corner points of the feasible region and then choose a corner point at which the function is optimum.

For example, in Fig. 15-13 the corner points are A, B, C, D, and E. We found A before to be $(40, 60)$. To find B, we see from Fig. 15-12 that we must solve $2x + y = 180$ and $x + y = 100$ simultaneously. This gives $B = (80, 20)$. In a similar way we obtain all the corner points:

$$A = (40, 60), \quad B = (80, 20), \quad C = (90, 0), \quad D = (0, 0), \quad E = (0, 80).$$

We now evaluate the objective function $P = 4x + 6y$ at each point:

$$P(A) = 4(40) + 6(60) = 520,$$

$$P(B) = 4(80) + 6(20) = 440,$$

$$P(C) = 4(90) + 6(0) = 360,$$

$$P(D) = 4(0) + 6(0) = 0,$$

$$P(E) = 4(0) + 6(80) = 480.$$

Thus P has a maximum value of 520 at A, where $x = 40$ and $y = 60$.

The optimum solution to a linear programming problem is given by the point where the optimum value of the objective function occurs. We shall also include the optimum value of the objective function.

EXAMPLE 1

Maximize the objective function $Z = 3x + y$ subject to the constraints

$$2x + y \leqslant 8,$$
$$2x + 3y \leqslant 12,$$
$$x \geqslant 0,$$
$$y \geqslant 0.$$

In Fig. 15-14 the feasible region is nonempty and bounded. Thus Z is maximum at one of the four corner points. The coordinates of A, B, and D are obvious on inspection. To find

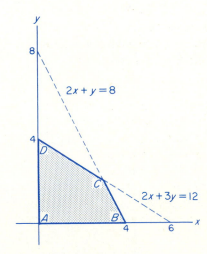

FIG. 15-14

C we solve the equations $2x + y = 8$ and $2x + 3y = 12$ simultaneously, which gives $x = 3, y = 2$. Thus,

$$A = (0, 0), \qquad B = (4, 0), \qquad C = (3, 2), \qquad D = (0, 4).$$

Evaluating Z at these points, we obtain

$$Z(A) = 3(0) + 0 = 0,$$

$$Z(B) = 3(4) + 0 = 12,$$

$$Z(C) = 3(3) + 2 = 11,$$

$$Z(D) = 3(0) + 4 = 4.$$

Hence the maximum value of Z, subject to the constraints, is 12 and it occurs when $x = 4$ and $y = 0$.

EXAMPLE 2

Minimize the objective function $Z = 8x - 3y$ subject to the constraints

$$-x + 3y = 21,$$

$$x + y \leqslant 5,$$

$$x \geqslant 0,$$

$$y \geqslant 0.$$

Notice that the first constraint $-x + 3y = 21$ is an *equality*. The portions of the lines $-x + 3y = 21$ and $x + y = 5$ for which $x \geqslant 0$ and $y \geqslant 0$ are shown in Fig. 15-15. They will remain broken lines until we determine whether or not they are to be included in the feasible region. A feasible point (x, y) must have $x \geqslant 0, y \geqslant 0$, and must lie both on the top dotted line and on or below the bottom dotted line. However, no such point exists. Hence the feasible region is *empty* and thus this problem has *no* optimum solution.

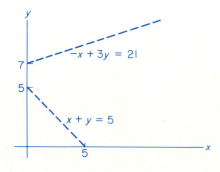

FIG. 15-15

The result in Example 2 can be made more general:

whenever the feasible region of a linear programming problem is empty, no optimum solution exists.

Suppose a feasible region is defined by

$$y = 2, \quad x \geqslant 0, \quad \text{and} \quad y \geqslant 0.$$

This region is the portion of the horizontal line $y = 2$ that is indicated in Fig. 15-16. Since the region cannot be contained within a circle, it is *unbounded*. Let

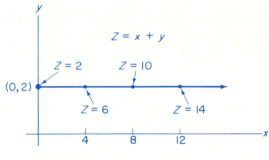

FIG. 15-16

us consider maximizing

$$Z = x + y$$

subject to the above constraints. Since $y = 2$, then $Z = x + 2$. Clearly, as x increases without bound, so does Z. Thus no feasible point maximizes Z, and so no optimum solution exists. In this case we say that the solution is "unbounded." On the other hand, suppose we want to *minimize* $Z = x + y$ over the same region. Since $Z = x + 2$, then Z is minimum when x is as small as possible, namely when $x = 0$. This gives a minimum value of $Z = x + y = 0 + 2 = 2$, and the optimum solution is the corner point $(0, 2)$.

In general, it can be shown that

if a feasible region is unbounded, and if the objective function has a maximum (or minimum) value, then that value occurs at a corner point.

EXAMPLE 3

A produce grower is purchasing fertilizer containing three nutrients, A, B, and C. The minimum needs are 160 units of A, 200 units of B, and 80 units of C. There are two popular brands of fertilizer on the market. Fast Grow, costing $4 a bag, contains 3 units of A, 5 units of B, and 1 unit of C. Easy Grow, costing $3 a bag, contains 2 units of each nutrient. If the *grower wishes to minimize cost while still maintaining the nutrients required, how many bags of each brand should be bought? The information is summarized in Table* 15-2.

TABLE 15-2

	A	B	C	COST/BAG
Fast Grow	3 units	5 units	1 unit	$4
Easy Grow	2 units	2 units	2 units	$3
Units required	160	200	80	

Let x be the number of bags of Fast Grow that are bought and y the number of bags of Easy Grow that are bought. Then we wish to *minimize* the cost function

$$C = 4x + 3y \tag{7}$$

subject to the constraints

$$x \geqslant 0, \tag{8}$$
$$y \geqslant 0, \tag{9}$$
$$3x + 2y \geqslant 160, \tag{10}$$
$$5x + 2y \geqslant 200, \tag{11}$$
$$x + 2y \geqslant 80. \tag{12}$$

The feasible region satisfying (8)–(12) is shaded in Fig. 15-17, along with *isocost lines* for

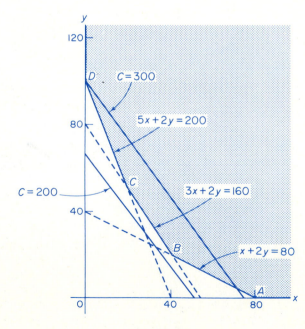

FIG. 15-17

$C = 200$ and $C = 300$. The feasible region is unbounded. The member of the family of lines $C = 4x + 3y$ which gives a minimum cost, subject to the constraints, intersects the feasible region at corner point B. Here we chose the isocost line whose y-intercept was *closest* to the origin and which had at least one point in common with the feasible region. The coordinates of B are found by solving the system

$$\begin{cases} 3x + 2y = 160, \\ x + 2y = 80. \end{cases}$$

Thus $x = 40$ and $y = 20$, which gives a minimum cost of $220. The produce grower should buy 40 bags of Fast Grow and 20 bags of Easy Grow.

In Example 3 we found that the function $C = 4x + 3y$ has a minimum value at a corner point of the unbounded feasible region. On the other hand, suppose we want to *maximize* C over that region and take the approach of evaluating C at all corner points. These points are

$$A = (80, 0), \qquad B = (40, 20), \qquad C = (20, 50), \qquad D = (0, 100),$$

from which

$$C(A) = 4(80) + 3(0) = 320,$$
$$C(B) = 4(40) + 3(20) = 220, \checkmark$$
$$C(C) = 4(20) + 3(50) = 230, \checkmark$$
$$C(D) = 4(0) + 3(100) = 300. \checkmark$$

A hasty conclusion is that the maximum value of C is 320. This is *false*! There is *no* maximum value, since isocost lines with arbitrarily large values of C intersect the feasible region.

PITFALL. *When working with an unbounded feasible region, do not simply conclude that an optimum solution exists at a corner point, since there may not be an optimum solution!*

EXERCISE 15-2

1. Maximize
   $$P = 10x + 12y$$
   subject to
   $$x + y \leqslant 60,$$
   $$x - 2y \geqslant 0,$$
   $$x, y \geqslant 0.$$

   $x \geq 2y + 0$

   $-2y \geq -x + 0$

   $y \leq \frac{1}{2}x + 0$

2. Maximize
   $$P = 5x + 6y$$
   subject to
   $$x + y \leqslant 80,$$
   $$3x + 2y \leqslant 220,$$
   $$2x + 3y \leqslant 210,$$
   $$x, y \geqslant 0.$$

3. Maximize
$$Z = 4x - 6y$$
subject to
$$y \leqslant 7,$$
$$3x - y \leqslant 3,$$
$$x + y \geqslant 5,$$
$$x, y \geqslant 0.$$

4. Minimize
$$Z = x + y$$
subject to
$$x - y \geqslant 0,$$
$$4x + 3y \geqslant 12,$$
$$9x + 11y \leqslant 99,$$
$$x \leqslant 8,$$
$$x, y \geqslant 0.$$

5. Maximize
$$Z = 4x - 10y$$
subject to
$$x - 4y \geqslant 4,$$
$$2x - y \leqslant 2,$$
$$x, y \geqslant 0.$$

6. Minimize
$$Z = 20x + 30y$$
subject to
$$2x + y \leqslant 10,$$
$$3x + 4y \leqslant 24,$$
$$8x + 7y \geqslant 56,$$
$$x, y \geqslant 0.$$

7. Minimize
$$Z = 7x + 3y$$
subject to
$$3x - y \geqslant -2,$$
$$x + y \leqslant 9,$$
$$x - y = -1,$$
$$x, y \geqslant 0.$$

8. Maximize
$$Z = .5x - .3y$$
subject to
$$x - y \geqslant -2,$$
$$2x - y \leqslant 4,$$
$$2x + y = 8,$$
$$x, y \geqslant 0.$$

9. Minimize
$$C = 2x + y$$
subject to
$$3x + y \geqslant 3,$$
$$4x + 3y \geqslant 6,$$
$$x + 2y \geqslant 2,$$
$$x, y \geqslant 0.$$

10. Minimize
$$C = 2x + 2y$$
subject to
$$x + 2y \geqslant 80,$$
$$3x + 2y \geqslant 160,$$
$$5x + 2y \geqslant 200,$$
$$x, y \geqslant 0.$$

11. Maximize
$$Z = 10x + 2y$$
subject to
$$x + 2y \geqslant 4,$$
$$x - 2y \geqslant 0,$$
$$x, y \geqslant 0.$$

12. Minimize
$$Z = y - x$$
subject to
$$x \geqslant 3,$$
$$x + 3y \geqslant 6,$$
$$x - 3y \geqslant -6,$$
$$x, y \geqslant 0.$$

13. A toy manufacturer preparing his production schedule for two new toys, widgets and wadgits, must use the information concerning their construction times given in Table 15-3. For example, each widget requires 2 hours on Machine A. The available

TABLE 15-3

	MACHINE A	MACHINE B	FINISHING
Widgets	2 hr	1 hr	1 hr
Wadgits	1 hr	1 hr	3 hr

employee hours per week are as follows: for operating machine A, 70 hours; for B, 40 hours; for finishing, 90 hours. If the profits on each widget and wadgit are $4 and $6, respectively, how many of each toy should be made per week in order to maximize profit? What would the maximum profit be?

14. A manufacturer produces two types of barbecue grills, Old Smokey and Blaze Away. During production the grills require the use of two machines, A and B. The number of hours needed on both are indicated in Table 15-4. If each machine can be used 24 hours a day, and the profits on the Old Smokey and Blaze Away models are $4 and $6, respectively, how many of each type of grill should be made per day to obtain maximum profit? What is the maximum profit?

TABLE 15-4

	MACHINE A	MACHINE B
Old Smokey	2 hr	4 hr
Blaze Away	4 hr	2 hr

15. A diet is to contain at least 16 units of carbohydrates and 20 units of protein. Food A contains 2 units of carbohydrates and 4 of protein; Food B contains 2 units of carbohydrates and 1 of protein. If Food A costs $1.20 per unit and Food B costs $0.80 per unit, how many units of each food should be purchased in order to minimize cost? What is the minimum cost?

16. A produce grower is purchasing fertilizer containing three nutrients: A, B, and C. The minimum weekly requirements are 80 units of A, 120 of B, and 240 of C. There are two popular blends of fertilizer on the market. Blend I, costing $4 a bag, contains 2 units of A, 6 of B, and 4 of C. Blend II, costing $5 a bag, contains 2 units of A, 2 of B, and 12 of C. How many bags of each blend should the grower buy each week to minimize the cost of meeting the nutrient requirements?

17. A company extracts minerals from ore. The number of pounds of minerals A and B that can be extracted from each ton of ores I and II are given in Table 15-5 together with the costs per ton of the ores. If the company must produce at least 3000 lb of A and 2500 lb of B, how many tons of each ore should be processed in order to minimize cost? What is the minimum cost?

TABLE 15-5

	ORE I	ORE II
Mineral A	100 lb	200 lb
Mineral B	200 lb	50 lb
Cost per ton	$50	$60

15-3 MULTIPLE OPTIMUM SOLUTIONS†

Sometimes an objective function attains its optimum value at more than one feasible point, in which case **multiple optimum solutions** are said to exist. Example 1 will illustrate.

EXAMPLE 1

Maximize $Z = 2x + 4y$ subject to the constraints

$$x - 4y \leqslant -8,$$
$$x + 2y \leqslant 16,$$
$$x \geqslant 0, y \geqslant 0.$$

The feasible region appears in Fig. 15-18. Since the region is nonempty and bounded, Z

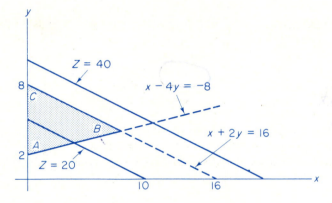

FIG. 15-18

has a maximum value at a corner point. The corner points are

$$A = (0, 2), \qquad B = (8, 4), \qquad C = (0, 8).$$

Evaluating the objective function at A, B, and C gives

$$Z(A) = 2(0) + 4(2) = 8,$$
$$Z(B) = 2(8) + 4(4) = 32,$$
$$Z(C) = 2(0) + 4(8) = 32.$$

Thus the maximum value of Z over the region is 32, and it occurs at *two* corner points, B

---

†This section can be omitted.

and C. In fact, this maximum value also occurs at *all* points on the line segment *joining B and C*, for the following reason. Each member of the family of lines $Z = 2x + 4y$ has slope $-\frac{1}{2}$. Moreover, the constraint line $x + 2y = 16$, which contains B and C, also has slope $-\frac{1}{2}$, and hence is parallel to each member of $Z = 2x + 4y$. Figure 15-18 shows lines for $Z = 20$ and $Z = 40$. Thus the member of the family that maximizes Z contains not only B and C but also all points on the line segment BC. It thus has infinitely many points in common with the feasible region. Hence this linear programming problem has infinitely many optimum solutions. In fact, it can be shown that

> **if (x_1, y_1) and (x_2, y_2) are two corner points at which an objective function is optimum, then the function will also be optimum at all points (x, y) where**
>
> $$x = (1 - t)x_1 + tx_2,$$
> $$y = (1 - t)y_1 + ty_2,$$
> **and $0 < t < 1$.**

In our case, if $(x_1, y_1) = B = (8, 4)$ and $(x_2, y_2) = C = (0, 8)$, then Z is maximum at any point (x, y) where

$$x = (1 - t)8 + t \cdot 0 = 8(1 - t),$$
$$y = (1 - t)4 + t \cdot 8 = 4(1 + t),$$
$$\text{and } 0 \leqslant t \leqslant 1.$$

These equations give the coordinates of any point on the line segment BC. In particular, if $t = 0$, then $x = 8$ and $y = 4$, which gives the corner point $B = (8, 4)$. If $t = 1$, we get the corner point $C = (0, 8)$. The value $t = \frac{1}{2}$ gives the point $(4, 6)$. Notice that at $(4, 6)$, $Z = 2(4) + 4(6) = 32$, which is the maximum value of Z.

EXERCISE 15-3

1. Minimize
   $$Z = 3x + 9y$$
   subject to
   $$y \geqslant -\tfrac{3}{2}x + 6,$$
   $$y \geqslant -\tfrac{1}{3}x + \tfrac{11}{3},$$
   $$y \geqslant x - 3,$$
   $$x, y \geqslant 0.$$

2. Maximize
   $$Z = 3x + 6y$$
   subject to
   $$x - y \geqslant -3,$$
   $$2x - y \leqslant 4,$$
   $$x + 2y = 12,$$
   $$x, y \geqslant 0.$$

3. Maximize
   $$Z = 18x + 9y$$
   subject to
   $$2x + 3y \leqslant 12,$$
   $$2x + y \leqslant 8,$$
   $$x, y \geqslant 0.$$

15-4 THE SIMPLEX METHOD

Up to now we have solved linear programming problems by a geometric method. This method will not be handy when the number of variables increases to three, and will not be possible beyond that. Now we shall look at a different technique—the *simplex method*, whose name is linked in more advanced discussions to a geometrical object called a simplex.

The simplex method begins with a feasible solution and tests whether or not it is optimum. If not optimum, the method proceeds to a *better* solution. We say "better" in the sense that the new solution brings you closer to optimization of the objective function.[†] Should this new solution not be optimum, then we repeat the procedure. Eventually the simplex method leads to an optimum solution, if one exists.

Besides being efficient, there are other advantages to the simplex method. It is completely mechanical (we use matrices, elementary row operations, and basic arithmetic). Moreover, no geometry is involved. This allows us to solve linear programming problems having any number of constraints and variables.

In this section we shall consider only so-called *standard* linear programming problems. These can be put in the form:

$$\text{Maximize } Z = c_1 x_1 + c_2 x_2 + \ldots + c_n x_n$$

such that

$$\left.\begin{aligned}
a_{11} x_1 + a_{12} x_2 + \ldots + a_{1n} x_n &\leqslant b_1, \\
a_{21} x_1 + a_{22} x_2 + \ldots + a_{2n} x_n &\leqslant b_2, \\
\vdots \qquad \vdots \qquad\qquad \vdots \quad\; \vdots \\
a_{m1} x_1 + a_{m2} x_2 + \ldots + a_{mn} x_n &\leqslant b_m,
\end{aligned}\right\} \tag{1}$$

where $x_1, x_2, \ldots, x_n$ and $b_1, b_2, \ldots, b_m$ are nonnegative.

Note that one feasible solution to a standard linear programming problem is always $x_1 = 0$, $x_2 = 0$, ..., $x_n = 0$. Other types of linear programming problems will be discussed in Sections 15-6 and 15-7.

We shall now apply the simplex method to the problem in Example 1 of Sec. 15-2 that has the form:

$$\text{Maximize } Z = 3x_1 + x_2$$

[†]In most cases this is true. In some situations, however, the new solution may be just as good as the previous one. Example 2 will illustrate this.

subject to the constraints

$$2x_1 + x_2 \leqslant 8 \tag{2}$$

$$\text{and} \quad 2x_1 + 3x_2 \leqslant 12, \tag{3}$$

where $x_1 \geqslant 0$ and $x_2 \geqslant 0$. This problem is of standard form. We begin by expressing constraints (2) and (3) as equations. In (2), $2x_1 + x_2$ will *equal* 8 if we add some nonnegative number s_1 to $2x_1 + x_2$:

$$2x_1 + x_2 + s_1 = 8, \quad \text{where } s_1 \geqslant 0.$$

We call s_1 a **slack variable** since it makes up for the "slack" on the left side of (2) so that we have equality. Similarly, inequality (3) can be written as an equation by using the slack variable s_2:

$$2x_1 + 3x_2 + s_2 = 12, \quad \text{where } s_2 \geqslant 0.$$

The variables x_1 and x_2 are called **structural variables**.

Now we can restate the problem in terms of equations:

$$\text{Maximize } Z = 3x_1 + x_2 \tag{4}$$

such that

$$2x_1 + x_2 + s_1 = 8 \tag{5}$$

$$\text{and} \quad 2x_1 + 3x_2 + s_2 = 12, \tag{6}$$

where x_1, x_2, s_1, and s_2 are nonnegative.

From Sec. 15-2 we know that the optimum solution occurs at a corner point of the feasible region in Fig. 15-19. At each of these points at least *two* of

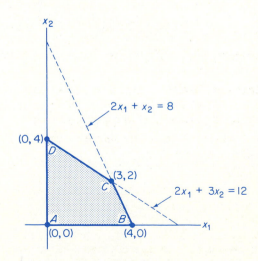

FIG. 15-19

the variables x_1, x_2, s_1, and s_2 are 0.

(a) At A, we have $x_1 = 0$ and $x_2 = 0$.

(b) At B, $x_1 = 4$ and $x_2 = 0$. But from Eq. (5), $2(4) + 0 + s_1 = 8$. Thus $s_1 = 0$.

(c) At C, $x_1 = 3$ and $x_2 = 2$. But from Eq. (5), $2(3) + 2 + s_1 = 8$. Thus $s_1 = 0$. From Eq. (6), $2(3) + 3(2) + s_2 = 12$. Thus $s_2 = 0$.

(d) At D, $x_1 = 0$ and $x_2 = 4$. From Eq. (6), $2(0) + 3(4) + s_2 = 12$ and so $s_2 = 0$.

It can also be shown that any solution to Eqs. (5) and (6), such that at least *two* of the four variables x_1, x_2, s_1, and s_2 are zero, corresponds to a corner point. Any such solution where at least two of these variables are zero is called a **basic feasible solution** (abbreviated B.F.S.). This number, 2, is determined by the expression $n - m$, where m is the number of constraints (excluding the nonnegativity conditions) and n is the number of variables that occur after these constraints are converted to equations. In our case $n = 4$ and $m = 2$. For any particular B.F.S., the two variables held at zero value are called **nonbasic variables**, while the others are called **basic variables** for that B.F.S. Thus, for the B.F.S. corresponding to (c) above, s_1 and s_2 are the nonbasic variables, but for the B.F.S. corresponding to (d) the nonbasic variables are x_1 and s_2. We eventually want to find a B.F.S. that maximizes Z.

We shall first find an initial B.F.S. and then determine whether the corresponding value of Z can be increased by a different B.F.S. Since $x_1 = 0$ and $x_2 = 0$ is a feasible solution to this standard linear programming problem, let us initially find the B.F.S. where the structural variables x_1 and x_2 are nonbasic. That is, we choose $x_1 = 0$ and $x_2 = 0$ and find the corresponding values of s_1, s_2, and Z. This can be done most conveniently by matrix techniques, based on the methods developed in the previous chapter.

If we write Eq. (4) as $-3x_1 - x_2 + Z = 0$, then Eqs. (5), (6), and (4) form the system

$$\begin{cases} 2x_1 + x_2 + s_1 & = 8, \\ 2x_1 + 3x_2 + s_2 & = 12, \\ -3x_1 - x_2 + Z = 0. \end{cases}$$

In terms of an augmented coefficient matrix (also called a **simplex tableau**), we have

$$\begin{array}{c} \\ s_1 \\ s_2 \\ \\ Z \end{array} \begin{array}{cccccc} x_1 & x_2 & s_1 & s_2 & Z & \\ \left[\begin{array}{ccccc|c} 2 & 1 & 1 & 0 & 0 & 8 \\ 2 & 3 & 0 & 1 & 0 & 12 \\ \hline -3 & -1 & 0 & 0 & 1 & 0 \end{array} \right]. \end{array}$$

nonbasic   basic

The first two rows correspond to the constraints, and the last row corresponds to the objective equation—thus the broken horizontal separating line. Notice that if $x_1 = 0$ and $x_2 = 0$, then from rows 1, 2, and 3 we can directly read off the values of s_1, s_2, and Z; $s_1 = 8$, $s_2 = 12$, and $Z = 0$. That is why we placed the letters s_1, s_2, and Z to the left of the rows. (We remind you that s_1 and s_2 are the basic variables.) Thus, our initial basic feasible solution is

$$x_1 = 0, \qquad x_2 = 0, \qquad s_1 = 8, \qquad s_2 = 12,$$

at which $Z = 0$. Let us see if we can find a B.F.S. that gives a larger value of Z.

The variables x_1 and x_2 are nonbasic in the B.F.S. above. We shall now look for a B.F.S. in which one of these variables is basic while the other remains nonbasic. Which one should we choose as the basic variable? Let us examine the possibilities. From the Z-row of the above matrix, $Z = 3x_1 + x_2$. If x_1 is allowed to become basic, then x_2 remains at 0 and $Z = 3x_1$; thus, for each one-unit increase in x_1, Z increases by three units. On the other hand, if x_2 is allowed to become basic, then x_1 remains at 0 and $Z = x_2$; thus, for each one-unit increase in x_2, Z increases by one unit. Hence, we get a *greater* increase in the value of Z if x_1, rather than x_2, enters the basic variable category. In this case we call x_1 an **entering variable**. Thus, in terms of the simplex tableau below (which is the same as the matrix above except for some additional labeling) the entering variable can be found by looking at the "most negative" of the numbers enclosed by the brace in the Z-row. Since that number is -3 and appears in the x_1-column, x_1 is the entering variable. The numbers in the brace are sometimes called **indicators**.

$$
\begin{array}{c}
\quad\;\; x_1 \quad x_2 \quad s_1 \quad s_2 \quad Z \\
\begin{array}{c} s_1 \\ s_2 \\ \\ Z \end{array}
\left[
\begin{array}{ccccc|c}
2 & 1 & 1 & 0 & 0 & 8 \\
2 & 3 & 0 & 1 & 0 & 12 \\
\hline
-3 & -1 & 0 & 0 & 1 & 0
\end{array}
\right].
\end{array}
$$

$$\underbrace{\qquad\qquad}\quad \text{indicators}$$

$$\uparrow$$
entering
variable

Let us summarize the information that can be obtained from this tableau. It gives a B.F.S. where s_1 and s_2 are the basic variables and x_1 and x_2 are nonbasic. This B.F.S. is $s_1 = 8$ (= the right-hand side of the s_1-row), $s_2 = 12$ (= the right-hand side of the s_2-row), $x_1 = 0$, and $x_2 = 0$. The -3 in the x_1-column of the Z-row indicates that if x_2 remains 0, then Z increases three units for each one-unit increase in x_1. The -1 in the x_2-column of the Z-row indicates that if x_1 remains 0, then Z increases one unit for each one-unit increase in x_2. The column in which the most negative indicator -3 lies gives the entering variable x_1, that is, the variable that should become basic in the next B.F.S.

In our new B.F.S., the larger the increase in x_1 (from $x_1 = 0$), the larger the increase in Z. Now, by how much can we increase x_1? Since x_2 is still held at 0, from rows 1 and 2 of the simplex tableau above it follows that

$$s_1 = 8 - 2x_1$$

$$\text{and} \quad s_2 = 12 - 2x_1.$$

Since s_1 and s_2 are nonnegative, we have

$$8 - 2x_1 \geqslant 0$$

$$\text{and} \quad 12 - 2x_1 \geqslant 0.$$

From the first inequality, $x_1 \leqslant \frac{8}{2} = 4$; from the second, $x_1 \leqslant \frac{12}{2} = 6$. Thus x_1 must be less than or equal to the smaller of the quotients $\frac{8}{2}$ and $\frac{12}{2}$, which is $\frac{8}{2}$. Hence x_1 can increase at most by 4. However, in a B.F.S. two variables must be 0. We already have $x_2 = 0$. Since $s_1 = 8 - 2x_1$, s_1 must be 0 for $x_1 = 4$. Thus we have a new B.F.S. with x_1 replacing s_1 as a basic variable. That is, s_1 will *depart* from the category of basic variables in the previous B.F.S. and will be nonbasic in the new B.F.S. We say that s_1 is the **departing variable** for the previous B.F.S. In summary, for our new B.F.S. we want x_1 and s_2 as basic variables with $x_1 = 4$, and x_2 and s_1 as nonbasic variables ($x_2 = 0$, $s_1 = 0$).

Before proceeding, let us update our tableau. To the right of the tableau below, the quotients $\frac{8}{2}$ and $\frac{12}{2}$ are indicated. They are obtained by dividing each entry in the first two rows of the b-column by the entry in the corresponding row of the entering variable column. Notice that the departing variable is in the same row as the *smaller* quotient $8 \div 2$.

$$
\begin{array}{c}
\text{departing} \\
\text{variable} \rightarrow
\end{array}
\begin{array}{c}
s_1 \\
s_2 \\
Z
\end{array}
\begin{array}{c}
\begin{array}{cccccc}
x_1 & x_2 & s_1 & s_2 & Z & b
\end{array} \\
\left[
\begin{array}{cccccc}
2 & 1 & 1 & 0 & 0 & 8 \\
2 & 3 & 0 & 1 & 0 & 12 \\
-3 & -1 & 0 & 0 & 1 & 0
\end{array}
\right]
\end{array}
\begin{array}{l}
Quotients \\
8 \div 2 = 4. \\
12 \div 2 = 6.
\end{array}
$$

$$\uparrow$$
$$\text{entering variable}$$

Since x_1 and s_2 will be basic variables in our new B.F.S., it would be convenient to change our previous tableau by elementary row operations into a form where the values of x_1, s_2, and Z can be read off with ease (just as we were able to do with the solution corresponding to $x_1 = 0$ and $x_2 = 0$). To do this we want to find a matrix which is equivalent to the tableau above but which has the form

$$
\begin{array}{ccccc}
x_1 & x_2 & s_1 & s_2 & Z \\
\end{array}
$$

$$
\left[
\begin{array}{ccccc|c}
1 & ? & ? & 0 & 0 & ? \\
0 & ? & ? & 1 & 0 & ? \\
\hline
0 & ? & ? & 0 & 1 & ? \\
\end{array}
\right]
$$

where the question marks represent numbers to be determined. Notice here that if $x_2 = 0$ and $s_1 = 0$, then x_1 equals the number in row 1 of the last column, s_2 equals the number in row 2, and Z is the number in row 3. Thus we must transform the tableau

$$
\begin{array}{cccccc}
 & x_1 & x_2 & s_1 & s_2 & Z \\
\end{array}
$$

$$
\begin{array}{c}
\text{departing} \rightarrow s_1 \\
\text{variable} \quad\quad s_2 \\
Z
\end{array}
\left[
\begin{array}{ccccc|c}
② & 1 & 1 & 0 & 0 & 8 \\
2 & 3 & 0 & 1 & 0 & 12 \\
\hline
-3 & -1 & 0 & 0 & 1 & 0 \\
\end{array}
\right] \quad (7)
$$

$$
\underset{\text{entering variable}}{\uparrow}
$$

into an equivalent matrix that has a 1 where the circle appears and 0's elsewhere in the x_1-column. The entry in the circle is called the **pivot entry**—it is in the column of the entering variable and the row of the departing variable. By elementary row operations, we have

$$
\begin{array}{ccccc}
x_1 & x_2 & s_1 & s_2 & Z \\
\end{array}
$$

$$
\left[
\begin{array}{ccccc|c}
② & 1 & 1 & 0 & 0 & 8 \\
2 & 3 & 0 & 1 & 0 & 12 \\
\hline
-3 & -1 & 0 & 0 & 1 & 0 \\
\end{array}
\right]
$$

$$
\sim
\left[
\begin{array}{ccccc|c}
1 & \frac{1}{2} & \frac{1}{2} & 0 & 0 & 4 \\
2 & 3 & 0 & 1 & 0 & 12 \\
\hline
-3 & -1 & 0 & 0 & 1 & 0 \\
\end{array}
\right]
\quad
\begin{array}{l}
\text{(by multiplying} \\
\text{first row by } \frac{1}{2})
\end{array}
$$

$$
\sim
\left[
\begin{array}{ccccc|c}
1 & \frac{1}{2} & \frac{1}{2} & 0 & 0 & 4 \\
0 & 2 & -1 & 1 & 0 & 4 \\
\hline
0 & \frac{1}{2} & \frac{3}{2} & 0 & 1 & 12 \\
\end{array}
\right]
\quad
\begin{array}{l}
\text{(by adding } -2 \text{ times first row} \\
\text{to the second, and adding 3} \\
\text{times first row to the third).}
\end{array}
$$

Thus we have a new simplex tableau:

$$
\begin{array}{c}
\begin{array}{ccccc} x_1 & x_2 & s_1 & s_2 & Z \end{array} \\
\begin{array}{c} x_1 \\ s_2 \\ \\ Z \end{array}
\left[
\begin{array}{ccccc|c}
1 & \frac{1}{2} & \frac{1}{2} & 0 & 0 & 4 \\
0 & 2 & -1 & 1 & 0 & 4 \\
\hline
0 & \frac{1}{2} & \frac{3}{2} & 0 & 1 & 12
\end{array}
\right]
\end{array}
\qquad (8)
$$

$$\underbrace{}_{\text{indicators}}$$

For $x_2 = 0$ and $s_1 = 0$, then from the first row we have $x_1 = 4$; from the second, $s_2 = 4$. These values give us the new B.F.S. Note that we replaced the s_1 located to the left of the initial tableau in (7) by x_1 in our new tableau (8)—thus s_1 *departed* and x_1 *entered*. From row 3, for $x_2 = 0$ and $s_1 = 0$ we get $Z = 12$, which is a larger value than we had before (it was $Z = 0$).

In our present B.F.S., x_2 and s_1 are nonbasic variables ($x_2 = 0$, $s_1 = 0$). Suppose we look for another B.F.S. that gives a larger value of Z and such that one of x_2 or s_1 is basic. The equation corresponding to the Z-row is given by $\frac{1}{2}x_2 + \frac{3}{2}s_1 + Z = 12$ or

$$Z = 12 - \tfrac{1}{2}x_2 - \tfrac{3}{2}s_1. \qquad (9)$$

If x_2 becomes basic and therefore s_1 remains nonbasic, then

$$Z = 12 - \tfrac{1}{2}x_2 \qquad (\text{since } s_1 = 0).$$

Here, each one-unit increase in x_2 *decreases* Z by $\frac{1}{2}$ unit. Thus any increase in x_2 would make Z smaller than before. On the other hand, if s_1 becomes basic and x_2 remains nonbasic, then from Eq. (9),

$$Z = 12 - \tfrac{3}{2}s_1 \qquad (\text{since } x_2 = 0).$$

Here each one-unit increase in s_1 *decreases* Z by $\frac{3}{2}$ units. Thus any increase in s_1 would make Z smaller than before. We cannot move to a better B.F.S. In short, no B.F.S. gives a larger value of Z than the B.F.S. $x_1 = 4$, $s_2 = 4$, $x_2 = 0$, $s_1 = 0$ (which gives $Z = 12$).

In fact, since $x_2 \geqslant 0$ and $s_1 \geqslant 0$ and the coefficients of x_2 and s_1 in Eq. (9) are negative, then Z is maximum when $x_2 = 0$ and $s_1 = 0$. That is, in (8), *having all nonnegative indicators means that we have an optimum solution*.

In terms of our original problem, if

$$Z = 3x_1 + x_2,$$

such that

$$2x_1 + x_2 \leqslant 8, \qquad 2x_1 + 3x_2 \leqslant 12, \qquad x_1 \geqslant 0, \quad \text{and} \quad x_2 \geqslant 0,$$

then Z is maximum when $x_1 = 4$ and $x_2 = 0$, and the maximum value of Z is 12 (this confirms our result in Example 1 of Sec. 15-2). Note that the values of s_1 and s_2 do not have to appear here.

Let us outline the simplex method for a standard linear programming problem with three structural variables and four constraints not counting nonnegativity conditions. This is to imply how the simplex method works for any number of structural variables and constraints.

SIMPLEX METHOD

Problem:

$$\text{Maximize } Z = c_1 x_1 + c_2 x_2 + c_3 x_3$$

such that

$$a_{11}x_1 + a_{12}x_2 + a_{13}x_3 \leqslant b_1,$$
$$a_{21}x_1 + a_{22}x_2 + a_{23}x_3 \leqslant b_2,$$
$$a_{31}x_1 + a_{32}x_2 + a_{33}x_3 \leqslant b_3,$$
$$a_{41}x_1 + a_{42}x_2 + a_{43}x_3 \leqslant b_4,$$

where x_1, x_2, x_3 and b_1, b_2, b_3, b_4 are nonnegative.

Method:

(1) Set up the initial simplex tableau.

	x_1	x_2	x_3	s_1	s_2	s_3	s_4	Z	b
s_1	a_{11}	a_{12}	a_{13}	1	0	0	0	0	b_1
s_2	a_{21}	a_{22}	a_{23}	0	1	0	0	0	b_2
s_3	a_{31}	a_{32}	a_{33}	0	0	1	0	0	b_3
s_4	a_{41}	a_{42}	a_{43}	0	0	0	1	0	b_4
Z	$-c_1$	$-c_2$	$-c_3$	0	0	0	0	1	0

indicators

There are four slack variables, s_1, s_2, s_3, and s_4—one for each constraint.

(2) If all the indicators in the last row are nonnegative, then Z has a maximum when $x_1 = 0$, $x_2 = 0$, and $x_3 = 0$. The maximum value is 0.

If there are any negative indicators, locate the column in which the most negative indicator appears. This column gives the entering variable.

(3) Divide each *positive*[†] entry above the broken line in the entering variable column *into* the corresponding value of b.

(4) Place a circle around the entry in the entering variable column that corresponds to the smallest quotient in step (3). This is the pivot entry. The departing variable is the one to the left of the pivot entry row.

(5) Use elementary row operations to transform the tableau into a new equivalent tableau that has a 1 where the pivot entry was and 0's elsewhere in that column.

(6) On the left side of this tableau the entering variable replaces the departing variable.

(7) If the indicators of the new tableau are all nonnegative, you have an optimum solution. The maximum value of Z is the entry in the last row and last column. It occurs when the variables to the left of the tableau are equal to the corresponding entries in the last column. All other variables are 0.

If at least one of the indicators is negative, repeat the process beginning with step (2) applied to the new tableau.

As an aid in understanding the simplex method, you should be able to interpret certain entries in a tableau. Suppose that we obtain a tableau where the last row is indicated below.

$$
\begin{array}{c}
\begin{array}{ccccccccc} x_1 & x_2 & x_3 & s_1 & s_2 & s_3 & s_4 & Z \end{array} \\
Z \begin{bmatrix} \vdots & \vdots & \vdots & \vdots & \vdots & \vdots & \vdots & \vdots & \vdots \\ a & b & c & d & e & f & g & 1 & h \end{bmatrix}.
\end{array}
$$

We can interpret the entry b, for example, as follows. If x_2 is nonbasic and were to become basic, then for each one-unit increase in x_2,

if $b < 0$, Z *increases* by $|b|$ units;

if $b > 0$, Z *decreases* by b units;

if $b = 0$, there is no change in Z.

[†]This will be discussed after Example 1.

EXAMPLE 1

Maximize $Z = 5x_1 + 4x_2$ subject to

$$x_1 + x_2 \leqslant 20,$$

$$2x_1 + x_2 \leqslant 35,$$

$$-3x_1 + x_2 \leqslant 12,$$

and $x_1 \geqslant 0$, $x_2 \geqslant 0$.

This linear programming problem fits the standard form. The initial simplex tableau is

	x_1	x_2	s_1	s_2	s_3	Z	b	Quotients
s_1	1	1	1	0	0	0	20	$20 \div 1 = 20.$
$\rightarrow s_2$	②	1	0	1	0	0	35	$35 \div 2 = \frac{35}{2}.$
s_3	-3	1	0	0	1	0	12	no quotient since -3 is not positive.
Z	-5	-4	0	0	0	1	0	

departing variable $\rightarrow s_2$

↑   indicators

entering variable

The most negative indicator, -5, occurs in the x_1-column. Thus x_1 is the entering variable. The smaller quotient is $\frac{35}{2}$, and so s_2 is the departing variable. The pivot entry is 2. Using elementary row operations to get a 1 in the pivot position and 0's elsewhere in its column, we have

$$\begin{array}{ccccccc}
x_1 & x_2 & s_1 & s_2 & s_3 & Z & b \\
\end{array}$$

$$\begin{bmatrix}
1 & 1 & 1 & 0 & 0 & 0 & 20 \\
② & 1 & 0 & 1 & 0 & 0 & 35 \\
-3 & 1 & 0 & 0 & 1 & 0 & 12 \\
\hline
-5 & -4 & 0 & 0 & 0 & 1 & 0
\end{bmatrix}$$

$$\sim \begin{bmatrix}
1 & 1 & 1 & 0 & 0 & 0 & 20 \\
1 & \frac{1}{2} & 0 & \frac{1}{2} & 0 & 0 & \frac{35}{2} \\
-3 & 1 & 0 & 0 & 1 & 0 & 12 \\
\hline
-5 & -4 & 0 & 0 & 0 & 1 & 0
\end{bmatrix}$$ (by multiplying row two by $\frac{1}{2}$)

$$\sim \begin{bmatrix}
0 & \frac{1}{2} & 1 & -\frac{1}{2} & 0 & 0 & \frac{5}{2} \\
1 & \frac{1}{2} & 0 & \frac{1}{2} & 0 & 0 & \frac{35}{2} \\
0 & \frac{5}{2} & 0 & \frac{3}{2} & 1 & 0 & \frac{129}{2} \\
\hline
0 & -\frac{3}{2} & 0 & \frac{5}{2} & 0 & 1 & \frac{175}{2}
\end{bmatrix}$$ (by adding -1 times row two to row one; adding 3 times row two to row three; adding 5 times row two to row four).

Our new tableau is

$$
\begin{array}{c}
\phantom{\text{departing}} \\
\phantom{\text{variable}}
\end{array}
\begin{array}{c}
s_1 \\ x_1 \\ s_3 \\ \\ Z
\end{array}
\begin{array}{c}
x_1 \;\;\; x_2 \;\;\; s_1 \;\;\; s_2 \;\;\; s_3 \;\;\; Z \;\;\; b \\
\left[\begin{array}{cccccc|c}
0 & \tfrac{1}{2} & 1 & -\tfrac{1}{2} & 0 & 0 & \tfrac{5}{2} \\
1 & \tfrac{1}{2} & 0 & \tfrac{1}{2} & 0 & 0 & \tfrac{35}{2} \\
0 & \tfrac{5}{2} & 0 & \tfrac{3}{2} & 1 & 0 & \tfrac{129}{2} \\
\hline
0 & -\tfrac{3}{2} & 0 & \tfrac{5}{2} & 0 & 1 & \tfrac{175}{2}
\end{array}\right]
\end{array}
$$

departing variable → s_1

Quotients:
$\tfrac{5}{2} \div \tfrac{1}{2} = 5.$
$\tfrac{35}{2} \div \tfrac{1}{2} = 35.$
$\tfrac{129}{2} \div \tfrac{5}{2} = 25\tfrac{4}{5}.$

indicators ↑ entering variable

Note that on the left side, x_1 replaced s_2. Since $-\tfrac{3}{2}$ is the most negative indicator, we must continue our process. The entering variable is now x_2. The smallest quotient is 5. Thus s_1 is the departing variable and $\tfrac{1}{2}$ is the pivot entry. Using elementary row operations, we have

$$
\begin{array}{c}
x_1 \;\; x_2 \;\; s_1 \;\; s_2 \; s_3 \; Z \;\; b \\
\left[\begin{array}{cccccc|c}
0 & \tfrac{1}{2} & 1 & -\tfrac{1}{2} & 0 & 0 & \tfrac{5}{2} \\
1 & \tfrac{1}{2} & 0 & \tfrac{1}{2} & 0 & 0 & \tfrac{35}{2} \\
0 & \tfrac{5}{2} & 0 & \tfrac{3}{2} & 1 & 0 & \tfrac{129}{2} \\
\hline
0 & -\tfrac{3}{2} & 0 & \tfrac{5}{2} & 0 & 1 & \tfrac{175}{2}
\end{array}\right]
\end{array}
$$

$$
\sim \left[\begin{array}{cccccc|c}
0 & \tfrac{1}{2} & 1 & -\tfrac{1}{2} & 0 & 0 & \tfrac{5}{2} \\
1 & 0 & -1 & 1 & 0 & 0 & 15 \\
0 & 0 & -5 & 4 & 1 & 0 & 52 \\
\hline
0 & 0 & 3 & 1 & 0 & 1 & 95
\end{array}\right]
$$

(by adding -1 times row one to row two; adding -5 times row one to row three; adding 3 times row one to row four)

$$
\sim \left[\begin{array}{cccccc|c}
0 & 1 & 2 & -1 & 0 & 0 & 5 \\
1 & 0 & -1 & 1 & 0 & 0 & 15 \\
0 & 0 & -5 & 4 & 1 & 0 & 52 \\
\hline
0 & 0 & 3 & 1 & 0 & 1 & 95
\end{array}\right]
$$

(by multiplying row one by 2).

Our new tableau is

$$
\begin{array}{c}
x_2 \\ x_1 \\ s_3 \\ \\ Z
\end{array}
\begin{array}{c}
x_1 \;\;\; x_2 \;\;\; s_1 \;\;\; s_2 \;\;\; s_3 \;\;\; Z \;\;\; b \\
\left[\begin{array}{cccccc|c}
0 & 1 & 2 & -1 & 0 & 0 & 5 \\
1 & 0 & -1 & 1 & 0 & 0 & 15 \\
0 & 0 & -5 & 4 & 1 & 0 & 52 \\
\hline
0 & 0 & 3 & 1 & 0 & 1 & 95
\end{array}\right],
\end{array}
$$

indicators

where x_2 replaced s_1 on the left side. Since all indicators are nonnegative, the maximum value of Z is 95 and occurs when $x_2 = 5$ and $x_1 = 15$ (and $s_3 = 52$, $s_1 = 0$, and $s_2 = 0$).

It is interesting to see how the values of Z got progressively "better" in successive tableaus in Example 1. These are the entries in the last row and column of each tableau. In the initial tableau we had $Z = 0$. From then on we obtained $Z = \frac{175}{2} = 87\frac{1}{2}$ and then $Z = 95$, the maximum.

In Example 1, you may wonder why no quotient is considered in the third row of the initial tableau. The B.F.S. for this tableau is

$$s_1 = 20, \qquad s_2 = 35, \qquad s_3 = 12, \qquad x_1 = 0, \qquad x_2 = 0,$$

where x_1 is the entering variable. The quotients 20 and $\frac{35}{2}$ reflect that for the next B.F.S., we have $x_1 \leqslant 20$ and $x_1 \leqslant \frac{35}{2}$. Since the third row represents the equation $s_3 = 12 + 3x_1 - x_2$, and $x_2 = 0$, then $s_3 = 12 + 3x_1$. But $s_3 \geqslant 0$, and so $12 + 3x_1 \geqslant 0$, which implies $x_1 \geqslant -\frac{12}{3} = -4$. Thus we have

$$x_1 \leqslant 20, \qquad x_1 \leqslant \frac{35}{2}, \quad \text{and} \quad x_1 \geqslant -4.$$

Hence x_1 can increase at most by $\frac{35}{2}$. The condition $x_1 \geqslant -4$ has no influence in determining the maximum increase in x_1. That is why the quotient $12/(-3) = -4$ is not considered in row 3. In general, *no quotient is considered for a row if the entry in the entering variable column is negative* (or, of course, 0).

Although the simplex procedure that has been developed in this section applies only to linear programming problems of standard form, other forms may be adapted to fit this form. Suppose that a constraint has the form

$$a_1x_1 + a_2x_2 + \ldots + a_nx_n \geqslant -b,$$

where $b > 0$. Here the inequality symbol is " $\geqslant$ " and the constant on the right side is *negative*. Thus the constraint is not in standard form. However, multiplying both sides by -1 gives

$$-a_1x_1 - a_2x_2 - \ldots - a_nx_n \leqslant b,$$

which *does* have the proper form. Thus, it may be necessary to rewrite a constraint before proceeding with the simplex method.

In a simplex tableau, several indicators may "tie" for being most negative. In this case, choose any one of these indicators to give the column for the entering variable. Likewise, there may be several quotients that "tie" for being the smallest. You may choose any one of these quotients to give you the departing variable and pivot entry. Example 2 will illustrate this. When a tie for the smallest quotient exists, then along with the nonbasic variables a B.F.S. will have a basic variable that is 0. In this case we say that the B.F.S. is *degenerate* or that the linear programming problem has a *degeneracy*. More will be said about this in Sec. 15-5.

EXAMPLE 2

Maximize $Z = 3x_1 + 4x_2 + \frac{3}{2}x_3$ subject to

$$-x_1 - 2x_2 \qquad\ \geqslant -10, \tag{10}$$
$$2x_1 + 2x_2 + x_3 \leqslant \ \ \ 10,$$

and $x_1, x_2, x_3 \geqslant 0$.

Constraint (10) does not fit the standard form. However, multiplying both sides of (10) by -1 gives

$$x_1 + 2x_2 \leqslant 10,$$

which *does* have the proper form. Thus our initial simplex tableau is Tableau I.

SIMPLEX TABLEAU I

		x_1	x_2	x_3	s_1	s_2	Z	b		*Quotients*
departing variable $\rightarrow$	s_1	1	②	0	1	0	0	10		$10 \div 2 = 5.$
	s_2	2	2	1	0	1	0	10		$10 \div 2 = 5.$
	Z	-3	-4	$-\frac{3}{2}$	0	0	1	0		

indicators

↑
entering
variable

The entering variable is x_2. Since there is a tie for the smallest quotient, we can choose either s_1 or s_2 as the departing variable. Let us choose s_1. The pivot entry is circled. Using elementary row operations, we get Tableau II.

SIMPLEX TABLEAU II

		x_1	x_2	x_3	s_1	s_2	Z	b		*Quotients*
	x_2	$\frac{1}{2}$	1	0	$\frac{1}{2}$	0	0	5		no quotient since 0 is not positive.
departing variable $\rightarrow$	s_2	1	0	①	-1	1	0	0		$0 \div 1 = 0.$
	Z	-1	0	$-\frac{3}{2}$	2	0	1	20		

↗ indicators

entering
variable

Tableau II corresponds to a B.F.S. where a basic variable, s_2, is zero. Thus the B.F.S. is degenerate. Since there are negative indicators, we continue. The entering variable is now x_3, the departing variable is s_2, and the pivot is circled. Using elementary row operations, we get Tableau III.

SIMPLEX TABLEAU III

$$
\begin{array}{c}
\begin{array}{ccccccc}
x_1 & x_2 & x_3 & s_1 & s_2 & Z & b
\end{array} \\
\begin{array}{c}
x_2 \\
x_3 \\
\\
Z
\end{array}
\left[
\begin{array}{cccccc|c}
\frac{1}{2} & 1 & 0 & \frac{1}{2} & 0 & 0 & 5 \\
1 & 0 & 1 & -1 & 1 & 0 & 0 \\
\hline
\frac{1}{2} & 0 & 0 & \frac{1}{2} & \frac{3}{2} & 1 & 20
\end{array}
\right]
\end{array}
$$

$$\underbrace{\qquad\qquad\qquad}_{\text{indicators}}$$

Since all indicators are nonnegative, Z is maximum when $x_2 = 5$ and $x_3 = 0$, and $x_1 = s_1 = s_2 = 0$. The maximum value is $Z = 20$. Note that this value is the same as that value of Z corresponding to Tableau II. In degenerate problems it is possible to arrive at the same value of Z at various stages of the simplex process. In Exercise 15-4 you are asked to solve this example problem by using s_2 as the departing variable in the initial tableau.

Because of its mechanical nature, the simplex procedure is readily adaptable to computers to solve linear programming problems involving many variables and constraints.

EXERCISE 15-4

Use the simplex method to solve the following problems.

1. Maximize
    $$Z = x_1 + 2x_2$$
    subject to
    $$2x_1 + x_2 \leqslant 8,$$
    $$2x_1 + 3x_2 \leqslant 12,$$
    $$x_1, x_2 \geqslant 0.$$

2. Maximize
    $$Z = 2x_1 + x_2$$
    subject to
    $$-x_1 + x_2 \leqslant 4,$$
    $$x_1 + x_2 \leqslant 6,$$
    $$x_1, x_2 \geqslant 0.$$

3. Maximize
    $$Z = -x_1 + 3x_2$$
    subject to
    $$x_1 + x_2 \leqslant 6,$$
    $$-x_1 + x_2 \leqslant 4,$$
    $$x_1, x_2 \geqslant 0.$$

4. Maximize
    $$Z = 3x_1 + 8x_2$$
    subject to
    $$x_1 + 2x_2 \leqslant 8,$$
    $$x_1 + 6x_2 \leqslant 12,$$
    $$x_1, x_2 \geqslant 0.$$

5. Maximize
$$Z = 8x_1 + 2x_2$$
subject to
$$x_1 - x_2 \leqslant 1,$$
$$x_1 + 2x_2 \leqslant 8,$$
$$x_1 + x_2 \leqslant 5,$$
$$x_1, x_2 \geqslant 0.$$

6. Maximize
$$Z = 2x_1 - 6x_2$$
subject to
$$x_1 - x_2 \leqslant 4,$$
$$-x_1 + x_2 \leqslant 4,$$
$$x_1 + x_2 \leqslant 6,$$
$$x_1, x_2 \geqslant 0.$$

7. Solve the problem in Example 2 by choosing s_2 as the departing variable in Tableau I.

8. Maximize
$$Z = 2x_1 - x_2 + x_3$$
subject to
$$2x_1 + x_2 - x_3 \leqslant 4,$$
$$x_1 + x_2 + x_3 \leqslant 2,$$
$$x_1, x_2, x_3 \geqslant 0.$$

9. Maximize
$$Z = 2x_1 + x_2 - x_3$$
subject to
$$x_1 + x_2 \qquad \leqslant 1,$$
$$x_1 - 2x_2 - x_3 \geqslant -2,$$
$$x_1, x_2, x_3 \geqslant 0.$$

10. Maximize
$$Z = -x_1 + 2x_2$$
subject to
$$x_1 + x_2 \leqslant 1,$$
$$x_1 - x_2 \leqslant 1,$$
$$x_1 - x_2 \geqslant -2,$$
$$x_1 \leqslant 2,$$
$$x_1, x_2 \leqslant 0.$$

11. Maximize
$$Z = x_1 + x_2$$
subject to
$$x_1 - x_2 \leqslant 4,$$
$$-x_1 + x_2 \leqslant 4,$$
$$8x_1 + 5x_2 \leqslant 40,$$
$$2x_1 + x_2 \leqslant 6,$$
$$x_1, x_2 \geqslant 0.$$

12. Maximize
$$W = 2x_1 + x_2 - 2x_3$$
subject to
$$-2x_1 + x_2 + x_3 \geqslant -2,$$
$$x_1 - x_2 + x_3 \leqslant 4,$$
$$x_1 + x_2 + 2x_3 \leqslant 6,$$
$$x_1, x_2, x_3 \geqslant 0.$$

13. Maximize
$$W = x_1 - 12x_2 + 4x_3$$
subject to
$$4x_1 + 3x_2 - x_3 \leqslant 1,$$
$$x_1 + x_2 - x_3 \geqslant -2,$$
$$-x_1 + x_2 + x_3 \geqslant -1,$$
$$x_1, x_2, x_3 \geqslant 0.$$

14. Maximize
$$W = 4x_1 + 0x_2 - x_3$$
subject to
$$x_1 + x_2 + x_3 \leqslant 6,$$
$$x_1 - x_2 + x_3 \leqslant 10,$$
$$x_1 - x_2 - x_3 \leqslant 4,$$
$$x_1, x_2, x_3 \geqslant 0.$$

15. Maximize
$$Z = 60x_1 + 0x_2 + 90x_3 + 0x_4$$
subject to
$$x_1 - 2x_2 \qquad\qquad \leqslant 2,$$
$$x_1 + x_2 \qquad\qquad \leqslant 5,$$
$$x_3 + x_4 \leqslant 4,$$
$$x_3 - 2x_4 \leqslant 7,$$
$$x_1, x_2, x_3, x_4 \geqslant 0.$$

16. Maximize
$$Z = 4x_1 + 10x_2 - 6x_3 - x_4$$
subject to
$$x_1 \qquad + x_3 - x_4 \leqslant 1,$$
$$x_1 - x_2 \qquad + x_4 \leqslant 2,$$
$$x_1 + x_2 - x_3 + x_4 \leqslant 4,$$
$$x_1, x_2, x_3, x_4 \geqslant 0.$$

17. A freight company handles shipments by two corporations, A and B, that are located in the same city. Corporation A ships boxes that each weigh 3 lb and have a volume of 2 cu ft; B ships 1 cu ft boxes that weigh 5 lb each. Both A and B ship to the same destination. The transportation cost for each box from A is \$0.75, and from B it is \$0.50. The freight company has a truck with 2400 cu ft of cargo space and a maximum capacity of 9200 lb. In one haul, how many boxes from each corporation should be transported by this truck so that the freight company receives maximum revenue? What is the maximum revenue?

18. A company manufactures three products: X, Y, and Z. Each product requires machine time and finishing time as given in Table 15-6. The numbers of hours of machine time and finishing time available per month are 900 and 5000, respectively. The unit profit on X, Y, and Z is \$3, \$4, and \$6, respectively. What is the maximum profit per month that can be obtained?

TABLE 15-6

	MACHINE TIME	FINISHING TIME
X	1 hr	4 hr
Y	2 hr	4 hr
Z	3 hr	8 hr

19. A company manufactures three types of patio furniture: chairs, rockers, and chaise lounges. Each requires wood, plastic, and aluminum as given in Table 15-7. The company has available 400 units of wood, 500 units of plastic, and 1450 units of aluminum. Each chair, rocker, and chaise lounge sells at \$7, \$8, and \$12, respectively. Assuming that all furniture can be sold, determine a production order so that total revenue will be maximum. What is the maximum revenue?

TABLE 15-7

	WOOD	PLASTIC	ALUMINUM
Chair	1 unit	1 unit	2 units
Rocker	1 unit	1 unit	3 units
Chaise lounge	1 unit	2 units	5 units

15-5 DEGENERACY, UNBOUNDED SOLUTIONS, MULTIPLE OPTIMUM SOLUTIONS[†]

In the last section we stated that a basic feasible solution is **degenerate** if along with one of the nonbasic variables one of the basic variables is 0. Suppose x_1, x_2, x_3, and x_4 are the variables in a degenerate B.F.S., where x_1 and

---

[†]This section may be omitted.

x_2 are basic with $x_1 = 0$, and x_3 and x_4 are nonbasic, and x_3 is the entering variable. The corresponding simplex tableau has the form

$$
\begin{array}{c}
\text{departing} \\
\text{variable}
\end{array} \rightarrow x_1
\quad
\begin{array}{c}
 \\ \\ x_2 \\ \\ Z
\end{array}
\begin{bmatrix}
\begin{array}{cccccc|c}
x_1 & x_2 & x_3 & x_4 & Z & & b \\
1 & 0 & \boxed{a_{13}} & a_{14} & 0 & & 0 \\
0 & 1 & a_{23} & a_{24} & 0 & & a \\
\hline
0 & 0 & d_1 & d_2 & 1 & & d_3
\end{array}
\end{bmatrix}
\quad 0 \div a_{13} = 0.
$$

indicators   ↖ entering variable

Thus the B.F.S. is

$$x_1 = 0, \qquad x_2 = a, \qquad x_3 = 0, \qquad x_4 = 0.$$

Suppose $a_{13} > 0$. Then the smaller quotient is 0 and we can choose a_{13} as the pivot entry. Thus x_1 is the departing variable. Elementary row operations give the following tableau, where the question marks represent numbers to be determined.

$$
\begin{array}{c}
x_3 \\ x_2 \\ Z
\end{array}
\begin{bmatrix}
\begin{array}{ccccc|c}
x_1 & x_2 & x_3 & x_4 & Z & b \\
? & 0 & 1 & ? & 0 & 0 \\
? & 1 & 0 & ? & 0 & a \\
\hline
? & 0 & 0 & ? & 1 & d_3
\end{array}
\end{bmatrix}
$$

For the B.F.S. corresponding to this tableau, x_3 and x_2 are basic variables, and x_1 and x_4 are nonbasic. The B.F.S. is

$$x_3 = 0, \qquad x_2 = a, \qquad x_1 = 0, \qquad x_4 = 0,$$

which is the same B.F.S. as before. Actually, these are usually considered different B.F.S.'s, where the only distinction is that x_1 is basic in the first B.F.S., while in the second it is nonbasic. The value of Z for both B.F.S.'s is the same, d_3. Thus, no "improvement" in Z is obtained.

In a degenerate situation, some problems may develop in the simplex procedure. It is possible to obtain a sequence of tableaus that correspond to B.F.S.'s which give the same Z value. Moreover, we may eventually return to the first tableau in the sequence. In Fig. 15-20 we arrive at B.F.S.$_1$, proceed to B.F.S.$_2$, then B.F.S.$_3$, and finally return to B.F.S.$_1$. This is called *cycling*. When cycling occurs, it is possible that we may never obtain the optimum value of Z. This phenomenon rarely is encountered in practical linear programming problems. However, there are techniques (which will not be considered in this text) for resolving such difficulties.

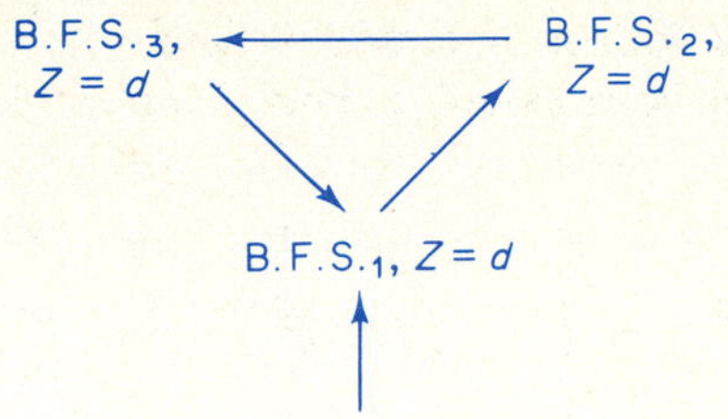

$$\text{B.F.S.}_3,\ Z = d \longleftarrow \text{B.F.S.}_2,\ Z = d$$

$$\text{B.F.S.}_1,\ Z = d$$

FIG. 15-20

A degenerate B.F.S. will occur when two quotients in a simplex tableau tie for being the smallest. For example, consider the following (partial) tableau:

$$
\begin{array}{c}
 & x_3 & & & Quotients \\
x_1 & \left[\ \textcircled{q_1} \quad \vdots \quad p_1 \right] & p_1/q_1. \\
x_2 & \left[\ q_2 \quad\quad \vdots \quad p_2 \right] & p_2/q_2.
\end{array}
$$

Here x_1 and x_2 are basic variables. Suppose x_3 is nonbasic and entering, and p_1/q_1 and p_2/q_2 are equal and also the smallest quotients involved. Choosing q_1 as the pivot entry, by elementary row operations we obtain

$$
\begin{array}{c}
 & x_3 & \\
x_3 & \left[\ 1 \quad\quad \vdots \quad p_1/q_1 \right] \\
x_2 & \left[\ 0 \quad\quad \vdots \quad p_2 - q_2\dfrac{p_1}{q_1} \right]
\end{array}
$$

Since $p_1/q_1 = p_2/q_2$, then $p_2 - q_2(p_1/q_1) = 0$. Thus the B.F.S. corresponding to this tableau has $x_2 = 0$, which gives a *degenerate* B.F.S. Although such a B.F.S. may produce cycling, we shall not encounter such situations in this book.

We now turn our attention to "unbounded problems." In Sec. 15-2 you saw that a linear programming problem may have no maximum value because the feasible region is such that the objective function may become arbitrarily large therein. In this case the problem is said to have an **unbounded solution**. This is a way of saying specifically that no optimum solution exists. Such a situation occurs when no quotients are possible in a simplex tableau for an entering variable. For example, consider the following tableau:

$$
\begin{array}{c}
 \\
x_1 \\
x_3 \\
Z
\end{array}
\begin{array}{c}
\begin{array}{cccccc}
x_1 & x_2 & x_3 & x_4 & Z & b
\end{array} \\
\left[
\begin{array}{cccccc|c}
1 & -3 & 0 & 2 & 0 & 5 \\
0 & 0 & 1 & 4 & 0 & 1 \\
\hline
0 & -5 & 0 & -2 & 1 & 10
\end{array}
\right]
\end{array}
\quad
\begin{array}{l}
\text{no quotient.} \\[1.2em]
\text{no quotient.}
\end{array}
$$

$$\underbrace{}$$
$\uparrow$ indicators
entering
variable

Here x_2 is the entering variable and for each one-unit increase in x_2, Z increases by 5. Since there are no positive entries in the first two rows of the x_2 column, no quotients exist. From rows 1 and 2 we get

$$x_1 = 5 + 3x_2 - 2x_4$$
$$\text{and} \quad x_3 = 1 - 4x_4.$$

In the B.F.S. for this tableau, $x_4 = 0$. Thus $x_1 = 5 + 3x_2$ and $x_3 = 1$. Since $x_1 \geqslant 0$, then $x_2 \geqslant -\frac{5}{3}$. Thus there is no upper bound on x_2. Hence Z can be arbitrarily large and we have an unbounded solution. In general,

if no quotients exist in a simplex tableau, then the linear
programming problem has an unbounded solution.

EXAMPLE 1

Maximize $Z = x_1 + 4x_2 - x_3$ subject to

$$-5x_1 + 6x_2 - 2x_3 \leqslant 30,$$
$$-x_1 + 3x_2 + 6x_3 \leqslant 12,$$

and $x_1, x_2, x_3 \geqslant 0$.

The initial simplex tableau is

$$
\begin{array}{c}
 \\
s_1 \\
\text{departing} \to s_2 \\
\text{variable} \\
Z
\end{array}
\begin{array}{c}
\begin{array}{ccccccc}
x_1 & x_2 & x_3 & s_1 & s_2 & Z & b
\end{array} \\
\left[
\begin{array}{ccccccc|c}
-5 & 6 & -2 & 1 & 0 & 0 & 30 \\
-1 & ③ & 6 & 0 & 1 & 0 & 12 \\
\hline
-1 & -4 & 1 & 0 & 0 & 1 & 0
\end{array}
\right]
\end{array}
\quad
\begin{array}{l}
\textit{Quotients} \\[0.3em]
30 \div 6 = 5. \\[1em]
12 \div 3 = 4.
\end{array}
$$

$$\underbrace{}$$
indicators
$\uparrow$
entering variable

The second tableau is

$$
\begin{array}{c c}
 & \begin{array}{c c c c c c c}
\ \ x_1 & x_2 & x_3 & s_1 & s_2 & Z & b
\end{array} \\
\begin{array}{c}
s_1 \\[18pt]
x_2 \\[24pt]
Z
\end{array} &
\left[
\begin{array}{c c c c c c | c}
-3 & 0 & -14 & 1 & -2 & 0 & 6 \\[8pt]
-\frac{1}{3} & 1 & 2 & 0 & \frac{1}{3} & 0 & 4 \\[4pt]
\hline
-\frac{7}{3} & 0 & 9 & 0 & \frac{4}{3} & 1 & 16
\end{array}
\right]
\end{array}
\qquad
\begin{array}{l}
\text{no quotient.} \\[18pt]
\text{no quotient.}
\end{array}
$$

$$
\begin{array}{c}
\uparrow \qquad\qquad \text{indicators} \\
\text{entering variable}
\end{array}
$$

Here the entering variable is x_1. Since the entries in the first two rows of the x_1-column are negative, no quotients exist. Hence the problem has an unbounded solution.

We conclude this section with a discussion of "multiple optimum solutions." Suppose that

$$
x_1 = a_1, \qquad x_2 = a_2, \qquad \ldots, \qquad x_n = a_n
$$
$$
\text{and} \qquad x_1 = b_1, \qquad x_2 = b_2, \qquad \ldots, \qquad x_n = b_n
$$

are two *different* B.F.S.'s for which a linear programming problem is optimum. By "different B.F.S.'s" we mean that $a_i \neq b_i$ for some i, where $1 \leqslant i \leqslant n$. It can be shown that the values

$$
\begin{aligned}
x_1 &= (1 - t)a_1 + tb_1, \\
x_2 &= (1 - t)a_2 + tb_2, \\
&\ \vdots \\
x_n &= (1 - t)a_n + tb_n, \\
&\text{for any } t \text{ where } 0 \leqslant t \leqslant 1,
\end{aligned}
\tag{1}
$$

also give an optimum solution (although it may not necessarily be a B.F.S.). Thus there are *multiple (optimum) solutions* to the problem.

We can determine the possibility of multiple optimum solutions from a simplex tableau that gives an optimum solution, such as the (partial) tableau below:

$$
\begin{array}{c c}
 & \begin{array}{c c c c c}
\ \ x_1 & x_2 & x_3 & x_4 & Z
\end{array} \\
\begin{array}{c}
x_1 \\[14pt]
x_2 \\[18pt]
Z
\end{array} &
\left[
\begin{array}{c c c c | c}
 & & & & p_1 \\[8pt]
 & & & & q_1 \\[4pt]
\hline
0 & 0 & a & 0 & 1 \ \ r
\end{array}
\right]
\end{array}
\cdot
$$

$$
\text{indicators}
$$

Here a must be nonnegative. The corresponding B.F.S. is

$$x_1 = p_1, \qquad x_2 = q_1, \qquad x_3 = 0, \qquad x_4 = 0,$$

and the maximum value of Z is r. If x_4 were to become basic, the indicator 0 in the x_4-column means that for each one-unit increase in x_4, Z does not change. Thus we can find a B.F.S. in which x_4 is basic and the corresponding Z-value is the same as before. This is done by treating x_4 as an entering variable in the tableau above. If, for instance, x_1 is the departing variable, the new B.F.S. has the form

$$x_1 = 0, \qquad x_2 = q_2, \qquad x_3 = 0, \qquad x_4 = p_2.$$

If this B.F.S. is different from the previous one, multiple solutions exist. In fact, from (1) an optimum solution is given by any values of x_1, x_2, x_3, and x_4 such that

$$
\begin{aligned}
x_1 &= (1 - t)p_1 + t{\cdot}0 = (1 - t)p_1, \\
x_2 &= (1 - t)q_1 + tq_2, \\
x_3 &= (1 - t){\cdot}0 + t{\cdot}0 = 0, \\
x_4 &= (1 - t){\cdot}0 + tp_2 = tp_2, \\
&\text{where } 0 \leqslant t \leqslant 1.
\end{aligned}
$$

Note that when $t = 0$ we get the first optimum B.F.S.; when $t = 1$ we get the second. Of course, it may be possible to repeat the procedure by using the tableau corresponding to the last B.F.S. and obtain more optimum solutions by using (1).

In general,

> **in a tableau that gives an optimum solution, a zero indicator for a nonbasic variable suggests the possibility of multiple optimum solutions.**

EXAMPLE 2

Maximize $Z = -x_1 + 4x_2 + 6x_3$ subject to

$$x_1 + 2x_2 + 3x_3 \leqslant 6,$$
$$-2x_1 - 5x_2 + x_3 \leqslant 10,$$

and $x_1, x_2, x_3 \geqslant 0.$

Our initial simplex tableau is

		x_1	x_2	x_3	s_1	s_2	Z	b	Quotients
departing →	s_1	1	2	③	1	0	0	6	$6 \div 3 = 2.$
variable	s_2	-2	-5	1	0	1	0	10	$10 \div 1 = 10.$
	Z	1	-4	-6	0	0	1	0	

↗ indicators
entering variable

Since there is a negative indicator, we continue.

		x_1	x_2	x_2	s_1	s_2	Z	b	Quotients
departing →	x_3	$\frac{1}{3}$	$\boxed{\frac{2}{3}}$	1	$\frac{1}{3}$	0	0	2	$2 \div \frac{2}{3} = 3.$
variable	s_2	$-\frac{7}{3}$	$-\frac{17}{3}$	0	$-\frac{1}{3}$	1	0	8	no quotient.
	Z	3	0	0	2	0	1	12	

↑ indicators
entering variable

[handwritten: even s_1, s_2 slack variables. can be nonbasics = 0. If so row reduce those also]

All indicators are nonnegative and hence an optimum solution occurs for the B.F.S.

$$x_3 = 2, \qquad s_2 = 8, \qquad x_1 = 0, \qquad x_2 = 0, \qquad s_1 = 0,$$

and the maximum value of Z is 12. However, since x_2 is a nonbasic variable and its indicator is 0, we shall check for multiple solutions. Treating x_2 as an entering variable, we obtain the following tableau:

	x_1	x_2	x_3	s_1	s_2	Z	b
x_2	$\frac{1}{2}$	1	$\frac{3}{2}$	$\frac{1}{2}$	0	0	3
s_2	$\frac{1}{2}$	0	$\frac{17}{2}$	$\frac{5}{2}$	1	0	25
Z	3	0	0	2	0	1	12

The B.F.S. here is

$$x_2 = 3, \qquad s_2 = 25, \qquad x_1 = 0, \qquad x_3 = 0, \qquad s_1 = 0$$

(for which $Z = 12$, as before) and is different from the previous one. Thus multiple solutions exist. Since we are concerned only with values of the structural variables, we have an optimum solution

$$x_1 = (1 - t)\cdot 0 + t\cdot 0 = 0,$$

$$x_2 = (1 - t)\cdot 0 + t\cdot 3 = 3t,$$

$$x_3 = (1 - t)\cdot 2 + t\cdot 0 = 2(1 - t)$$

for each value of t where $0 \leqslant t \leqslant 1$. (For example, if $t = 1/2$, then $x_1 = 0$, $x_2 = 3/2$, and $x_3 = 1$ is an optimum solution.)

In the last B.F.S., x_3 is nonbasic and its indicator is 0. However, if we repeated the process for determining other optimum solutions, we would return to the second tableau. Thus our procedure gives no other optimum solutions.

EXERCISE 15-5

In each of Problems 1 and 2, does the linear programming problem associated with the given tableau have a degeneracy? If so, why?

1.

$$
\begin{array}{c}
\quad\quad\; x_1 \quad x_2 \quad s_1 \quad s_2 \quad Z \\
\begin{array}{c} x_1 \\ s_2 \\ Z \end{array}
\left[
\begin{array}{ccccc|c}
1 & 2 & 4 & 0 & 0 & 6 \\
0 & 1 & 1 & 1 & 0 & 3 \\
\hline
0 & -3 & -2 & 0 & 1 & 10
\end{array}
\right].
\end{array}
$$

$$\underbrace{\qquad\qquad\qquad}_{\text{indicators}}$$

2.

$$
\begin{array}{c}
\quad\quad\; x_1 \quad x_2 \quad x_3 \quad s_1 \quad s_2 \quad Z \\
\begin{array}{c} s_1 \\ x_2 \\ Z \end{array}
\left[
\begin{array}{cccccc|c}
2 & 0 & 2 & 1 & 1 & 0 & 4 \\
3 & 1 & 1 & 0 & 1 & 0 & 0 \\
\hline
-5 & 0 & 1 & 0 & -3 & 1 & 2
\end{array}
\right].
\end{array}
$$

$$\underbrace{\qquad\qquad\qquad}_{\text{indicators}}$$

In Problems 3–11, use the simplex method.

3. Maximize

$$Z = 2x_1 + 7x_2$$

subject to

$$4x_1 - 3x_2 \leqslant 4,$$
$$3x_1 - x_2 \leqslant 6,$$
$$5x_1 \quad\quad \leqslant 8,$$
$$x_1, x_2 \geqslant 0.$$

4. Maximize

$$Z = x_1 + x_2$$

subject to

$$x_1 - x_2 \leqslant 4,$$
$$-x_1 + x_2 \leqslant 4,$$
$$8x_1 + 5x_2 \leqslant 40,$$
$$x_1 + x_2 \leqslant 6,$$
$$x_1, x_2 \geqslant 0.$$

5. Maximize

$$Z = 3x_1 - 3x_2$$

subject to

$$x_1 - x_2 \leqslant 4,$$
$$-x_1 + x_2 \leqslant 4,$$
$$x_1 + x_2 \leqslant 6,$$
$$x_1, x_2 \geqslant 0.$$

6. Maximize

$$Z = 4x_1 + x_2 + 2x_3$$

subject to

$$x_1 - x_2 + 4x_3 \leqslant 6,$$
$$x_1 - x_2 - x_3 \geqslant -4,$$
$$x_1 - 6x_2 + x_3 \leqslant 8,$$
$$x_1, x_2, x_3 \geqslant 0.$$

7. Maximize

$$Z = 5x_1 + 6x_2 + x_3$$

subject to

$$9x_1 + 3x_2 - 2x_3 \leqslant 5,$$
$$4x_1 + 2x_2 - x_3 \leqslant 2,$$
$$x_1 - 4x_2 + x_3 \leqslant 3,$$
$$x_1, x_2, x_3 \geqslant 0.$$

8. Maximize

$$Z = 2x_1 + x_2 - 4x_3$$

subject to

$$6x_1 + 3x_2 - 3x_3 \leqslant 10,$$
$$x_1 - x_2 + x_3 \leqslant 1,$$
$$2x_1 - x_2 + 2x_3 \leqslant 12,$$
$$x_1, x_2, x_3 \geqslant 0.$$

9. Maximize
    $$Z = 6x_1 + 2x_2 + x_3$$
    subject to
    $$2x_1 + x_2 + x_3 \leqslant 7,$$
    $$-4x_1 - x_2 \qquad \geqslant -6,$$
    $$x_1, x_2, x_3 \geqslant 0.$$

10. Maximize
    $$P = 4x_1 + 3x_2 + 2x_3 + x_4$$
    subject to
    $$x_1 - x_2 \qquad\qquad \leqslant 5,$$
    $$x_2 - x_3 \qquad \leqslant 2,$$
    $$x_2 - 2x_3 + x_4 \leqslant 4,$$
    $$x_1, x_2, x_3, x_4 \geqslant 0.$$

11. A company manufacturers three types of patio furniture: chairs, rockers, and chaise
    lounges. Each requires wood, plastic, and aluminum as given in Table 15-8. The
    company has available 400 units of wood, 600 units of plastic, and 1500 units of
    aluminum. Each chair, rocker, and chaise lounge sells at $6, $8, and $12, respec-
    tively. Assuming that all furniture can be sold, what is the maximum total revenue
    that can be obtained? Determine the possible production orders that will generate
    this revenue.

TABLE 15-8

	WOOD	PLASTIC	ALUMINUM
Chair	1 unit	1 unit	2 units
Rocker	1 unit	1 unit	3 units
Chaise lounge	1 unit	2 units	5 units

15-6 ARTIFICIAL VARIABLES

To initiate the simplex method, a basic feasible solution is required. For a
standard linear programming problem, we begin with the B.F.S. in which all
structural variables are zero. However, for a maximization problem that is not of
standard form, such a B.F.S. may not exist. In this section you will see how the
simplex method is used in such situations.

Let us consider the following problem:

$$\text{Maximize } Z = x_1 + 2x_2$$

subject to

$$x_1 + x_2 \leqslant 9, \qquad\qquad (1)$$
$$x_1 - x_2 \geqslant 1, \qquad\qquad (2)$$

and $x_1, x_2 \geqslant 0$. Since constraint (2) can not be written as $a_1x_1 + a_2x_2 \leqslant b$,
where b is nonnegative, this problem cannot be put into standard form. Note
that $(0, 0)$ is not a feasible point. To solve this problem, we begin by writing
constraints (1) and (2) as equations. Constraint (1) becomes

$$x + x_2 + s_1 = 9, \qquad\qquad (3)$$

where s_1 is a slack variable and $s_1 \geqslant 0$. For constraint (2), $x_1 - x_2$ will equal 1 if we *subtract* a nonnegative slack variable s_2 from $x_1 - x_2$. That is, by subtracting s_2 we are making up for the "surplus" on the left side of (2) so that we have equality. Thus

$$x_1 - x_2 - s_2 = 1, \tag{4}$$

where $s_2 \geqslant 0$. We can now restate the problem.

$$\text{Maximize } Z = x_1 + 2x_2 \tag{5}$$

subject to

$$x_1 + x_2 + s_1 = 9, \tag{6}$$
$$x_1 - x_2 - s_2 = 1, \tag{7}$$

and $x_1, x_2, s_1, s_2 \geqslant 0$.

Since $(0, 0)$ is not in the feasible region, we do not have a B.F.S. in which $x_1 = x_2 = 0$. In fact, if $x_1 = 0$ and $x_2 = 0$ are substituted into Eq. (7), then $0 - 0 - s_2 = 1$, which gives $s_2 = -1$. But this contradicts the condition that $s_2 \geqslant 0$.

To get the simplex method started, we need an initial B.F.S. Although none is obvious, there is an ingenious method to arrive at one *artificially*. It requires that we consider a related linear programming problem called the *artificial problem*. First, a new equation is formed by adding a nonnegative variable t to the left side of the equation in which the coefficient of the slack variable is -1. The variable t is called an **artificial variable**. In our case, we replace Eq. (7) by $x_1 - x_2 - s_2 + t = 1$. Thus, Eqs. (6) and (7) become

$$x_1 + x_2 + s_1 = 9, \tag{8}$$
$$x_1 - x_2 - s_2 + t = 1, \tag{9}$$

where $x_1, x_2, s_1, s_2, t \geqslant 0$.

An obvious solution to Eqs. (8) and (9) is found by setting x_1, x_2, and s_2 equal to 0. This gives

$$x_1 = x_2 = s_2 = 0, \qquad s_1 = 9, \qquad t = 1.$$

Note that these values do not satisfy Eqs. (6) and (7). However, it is clear that any solution of Eqs. (8) and (9) for which $t = 0$ will give a solution to Eqs. (6) and (7), and conversely.

We can eventually force t to be 0 if we alter the original objective function. We define the **artificial objective function** to be

$$W = Z - Mt = x_1 + 2x_2 - Mt, \tag{10}$$

where the constant M is a large positive number. We shall not worry about the particular value of M and shall proceed to maximize W by the simplex method. Since there are $m = 2$ constraints (excluding the nonnegativity conditions) and

$n = 5$ variables in Eqs. (8) and (9), any B.F.S. must have at least $n - m = 3$ variables equal to zero. We start with the following B.F.S.:

$$x_1 = x_2 = s_2 = 0, \qquad s_1 = 9, \qquad t = 1. \tag{11}$$

In this initial B.F.S., the nonbasic variables are the structural variables and the slack variable with coefficient -1 in Eqs. (8) and (9). The corresponding value of W is $W = x_1 + 2x_2 - Mt = -M$, which is "extremely" negative. A significant improvement in W will occur if we can find another B.F.S. for which $t = 0$. Since the simplex method seeks better values of W at each stage, we shall apply it until we reach such a B.F.S., if possible. That solution will be an initial B.F.S. for the original problem.

To apply the simplex method to the artificial problem, we first write Eq. (10) as

$$-x_1 - 2x_2 + Mt + W = 0. \tag{12}$$

The augmented coefficient matrix of Eqs. (8), (9), and (12) is

$$
\begin{array}{c}
\\
s_1 \\
t \\
\\
\end{array}
\begin{array}{cccccc}
x_1 & x_2 & s_1 & s_2 & t & W \\
\left[\begin{array}{cccccc|c}
1 & 1 & 1 & 0 & 0 & 0 & 9 \\
1 & -1 & 0 & -1 & 1 & 0 & 1 \\
\hline
-1 & -2 & 0 & 0 & M & 1 & 0
\end{array}\right].
\end{array}
\tag{13}
$$

An initial B.F.S. is given by (11). Notice that from row 1, when $x_1 = x_2 = s_2 = 0$, we can directly read the value of s_1, namely $s_1 = 9$. From row 2 we get $t = 1$. From row 3, $Mt + W = 0$. Since $t = 1$, then $W = -M$. But in a simplex tableau we want the value of W to appear in the last row and last column. This is not so in (13), and thus we modify that matrix.

To do this, we transform (13) into an equivalent matrix whose last row has the form

$$
\begin{array}{cccccc}
x_1 & x_2 & s_1 & s_2 & t & W \\
? & ? & 0 & ? & 0 & 1 \mid ?
\end{array}
$$

That is, the M in the t-column is replaced by 0. As a result, if $x_1 = x_2 = s_2 = 0$, then W equals the last entry. Proceeding to obtain such a matrix, we have

$$
\begin{array}{cccccc}
x_1 & x_2 & s_1 & s_2 & t & W \\
\left[\begin{array}{cccccc|c}
1 & 1 & 1 & 0 & 0 & 0 & 9 \\
1 & -1 & 0 & -1 & 1 & 0 & 1 \\
\hline
-1 & -2 & 0 & 0 & M & 1 & 0
\end{array}\right]
\end{array}
$$

$$
\sim
\begin{array}{cccccc}
x_1 & x_2 & s_1 & s_2 & t & W \\
\left[\begin{array}{cccccc|c}
1 & 1 & 1 & 0 & 0 & 0 & 9 \\
1 & -1 & 0 & -1 & 1 & 0 & 1 \\
\hline
-1-M & -2+M & 0 & M & 0 & 1 & -M
\end{array}\right]
\end{array}
\quad
\begin{array}{l}
\text{(by adding } -M \text{ times} \\
\text{row 2 to row 3).}
\end{array}
$$

Let us now check things out. If $x_1 = 0$, $x_2 = 0$, and $s_2 = 0$, then from row 1 we get $s_1 = 9$; from row 2, $t = 1$; and from row 3, $W = -M$. Thus we now have initial simplex Tableau I.

SIMPLEX TABLEAU I

		x_1	x_2	s_1	s_2	t	W			Quotients
	s_1	1	1	1	0	0	0	9		$9 \div 1 = 9.$
departing → variable	t	①	-1	0	-1	1	0	1		$1 \div 1 = 1.$
	W	$-1-M$	$-2+M$	0	M	0	1	$-M$		

↑ indicators
entering
variable

From this point we can use the procedures of Sec. 15-4. Since M is a large positive number, the most negative indicator is $-1 - M$. Thus the entering variable is x_1. From the quotients we choose t as the departing variable. The pivot entry is circled. Using elementary row operations to get 1 in the pivot position and 0's elsewhere in that column, we get Tableau II.

SIMPLEX TABLEAU II

		x_1	x_2	s_1	s_2	t	W			Quotients
departing → variable	s_1	0	②	1	1	-1	0	8		$8 \div 2.$
	x_1	1	-1	0	-1	1	0	1		(no quotient, since -1 is not positive).
	W	0	-3	0	-1	$M+1$	1	1		

↑ indicators
entering
variable

From Tableau II, we have the following B.F.S.:

$$s_1 = 8, \qquad x_1 = 1, \qquad x_2 = 0, \qquad s_2 = 0, \qquad t = 0.$$

Since $t = 0$, the values $s_1 = 8$, $x_1 = 1$, $x_2 = 0$, and $s_2 = 0$ form an initial B.F.S. for the *original* problem! The artificial variable has served its purpose. For succeeding tableaus we shall delete the t-column (since we want to solve the original problem) and change the W's to Z's (since $W = Z$ for $t = 0$). From Tableau II, the entering variable is x_2, the departing variable is s_1, and the pivot entry is circled. Using elementary row operations (omitting the t-column), we get Tableau III.

SIMPLEX TABLEAU III

$$
\begin{array}{c}
\begin{array}{ccccc}
x_1 & x_2 & s_1 & s_2 & Z
\end{array} \\
\begin{array}{c}
x_2 \\
x_1 \\
\\
Z
\end{array}
\left[
\begin{array}{ccccc|c}
0 & 1 & \frac{1}{2} & \frac{1}{2} & 0 & 4 \\
1 & 0 & \frac{1}{2} & -\frac{1}{2} & 0 & 5 \\
\hline
0 & 0 & \frac{3}{2} & \frac{1}{2} & 1 & 13
\end{array}
\right] \\
\underbrace{\hspace{5cm}}_{\text{indicators}}
\end{array}
$$

Since all the indicators are nonnegative, the maximum value of Z is 13. It occurs when $x_1 = 5$ and $x_2 = 4$.

It is worthwhile to review the steps we performed to solve our problem.

$$\text{Maximize } Z = x_1 + 2x_2$$

subject to

$$x_1 + x_2 \leqslant 9, \tag{14}$$

$$x_1 - x_2 \geqslant 1, \tag{15}$$

and $x_1 \geqslant 0$, $x_2 \geqslant 0$. We write (14) as

$$x_1 + x_2 + s_1 = 9. \tag{16}$$

Since (15) involves the symbol $\geqslant$ and the constant on the right side is nonnegative, we write (15) in a form having both a slack variable (with coefficient -1) and an artificial variable.

$$x_1 - x_2 - s_2 + t = 1. \tag{17}$$

The artificial objective equation to consider is $W = x_1 + 2x_2 - Mt$, or equivalently,

$$-x_1 - 2x_2 + Mt + W = 0. \tag{18}$$

The augmented coefficient matrix of the system formed by Eqs. (16)–(18) is

$$
\begin{array}{c}
\begin{array}{cccccc}
x_1 & x_2 & s_1 & s_2 & t & W
\end{array} \\
\left[
\begin{array}{cccccc|c}
1 & 1 & 1 & 0 & 0 & 0 & 9 \\
1 & -1 & 0 & -1 & 1 & 0 & 1 \\
\hline
-1 & -2 & 0 & 0 & M & 1 & 0
\end{array}
\right].
\end{array}
$$

Next we remove the M from the artificial variable column and replace it by 0 by using elementary row operations. The resulting simplex Tableau I corresponds to the initial B.F.S. of the aritificial problem in which the structural variables, x_1 and x_2, and the slack variable s_2 (the one associated with the constraint involving the symbol $\geqslant$) are each 0.

SIMPLEX TABLEAU I

$$
\begin{array}{c}
\\
s_1 \\
t \\
\\
W
\end{array}
\begin{array}{c}
x_1 \quad\quad x_2 \quad\quad s_1 \quad s_2 \quad t \quad W \\
\left[\begin{array}{cccccc|c}
1 & 1 & 1 & 0 & 0 & 0 & 9 \\
1 & -1 & 0 & -1 & 1 & 0 & 1 \\
\hline
-1-M & -2+M & 0 & M & 0 & 1 & -M
\end{array}\right].
\end{array}
$$

The basic variables s_1 and t on the left side of the tableau correspond to the nonstructural variables in Eqs. (16) and (17) that have positive coefficients. We now apply the simplex method until we obtain a B.F.S. in which the artificial variable t equals 0. Then we can delete the artificial variable column, change the W's to Z's, and continue the procedure until the maximum value Z is obtained.

EXAMPLE 1

Use the simplex method to maximize $Z = 2x_1 + x_2$ subject to

$$x_1 + x_2 \leqslant 12, \tag{19}$$

$$x_1 + 2x_2 \leqslant 20, \tag{20}$$

$$-x_1 + x_2 \geqslant 2, \tag{21}$$

and $x_1 \geqslant 0$, $x_2 \geqslant 0$.

The equations for (19)–(21) will involve a total of three slack variables: s_1, s_2, and s_3. Since (21) contains the symbol $\geqslant$ and the constant on the right side is nonnegative, its equation will also involve an artificial variable t, and the coefficient of its slack variable s_3 will be -1.

$$x_1 + x_2 + s_1 \qquad\qquad\qquad = 12, \tag{22}$$

$$x_1 + 2x_2 \qquad + s_2 \qquad\qquad = 20, \tag{23}$$

$$-x_1 + x_2 \qquad\qquad - s_3 + t = 2. \tag{24}$$

We consider $W = Z - Mt = 2x_1 + x_2 - Mt$ as the artificial objective equation, or equivalently,

$$-2x_1 - x_2 + Mt + W = 0, \tag{25}$$

where M is a large positive number. Now we construct the augmented coefficient matrix of Eqs. (22)–(25).

$$\begin{array}{ccccccc} x_1 & x_2 & s_1 & s_2 & s_3 & t & W \\ \end{array}$$

$$\left[\begin{array}{ccccccc|c} 1 & 1 & 1 & 0 & 0 & 0 & 0 & 12 \\ 1 & 2 & 0 & 1 & 0 & 0 & 0 & 20 \\ -1 & 1 & 0 & 0 & -1 & 1 & 0 & 2 \\ \hline -2 & -1 & 0 & 0 & 0 & M & 1 & 0 \end{array}\right].$$

To get intial simplex Tableau I, we replace the M in the artificial variable column by zero by adding $-M$ times row 3 to row 4.

SIMPLEX TABLEAU I

	x_1	x_2	s_1	s_2	s_3	t	W		Quotients
s_1	1	1	1	0	0	0	0	12	$12 \div 1 = 12.$
s_2	1	2	0	1	0	0	0	20	$20 \div 2 = 10.$
departing → t	-1	①	0	0	-1	1	0	2	$2 \div 1 = 2.$
variable									
W	$-2+M$	$-1-M$	0	0	M	0	1	$-2M$	

↑ entering variable   indicators

The variables s_1, s_2, and t on the left side of Tableau I are the nonstructural variables with positive coefficients in Eqs. (22)–(24). Since M is a large positive number, $-1 - M$ is the most negative indicator. The entering variable is x_2, the departing variable is t, and the pivot entry is circled. Proceeding, we get Tableau II.

SIMPLEX TABLEAU II

	x_1	x_2	s_1	s_2	s_3	t	W		Quotients
departing → s_1	②	0	1	0	1	-1	0	10	$10 \div 2 = 5.$
variable									
s_2	3	0	0	1	2	-2	0	16	$16 \div 3 = 5\frac{1}{3}.$
x_2	-1	1	0	0	-1	1	0	2	
W	-3	0	0	0	-1	$1+M$	1	2	

↑ entering variable   indicators

The B.F.S. corresponding to Tableau II has $t = 0$. Thus we shall delete the t-column and change W's to Z's in succeeding tableaus. Continuing, we obtain Tableau III.

SIMPLEX TABLEAU III

$$
\begin{array}{c}
\begin{array}{cccccccc}
x_1 & x_2 & s_1 & s_2 & s_3 & Z & \\
\end{array} \\
\begin{array}{c}
x_1 \\
s_2 \\
x_2 \\
Z
\end{array}
\left[
\begin{array}{cccccc|c}
1 & 0 & \frac{1}{2} & 0 & \frac{1}{2} & 0 & 5 \\
0 & 0 & -\frac{3}{2} & 1 & \frac{1}{2} & 0 & 1 \\
0 & 1 & \frac{1}{2} & 0 & -\frac{1}{2} & 0 & 7 \\
\hline
0 & 0 & \frac{3}{2} & 0 & \frac{1}{2} & 1 & 17
\end{array}
\right].
\end{array}
$$

$$\underbrace{\qquad\qquad\qquad}_{\text{indicators}}$$

All indicators are nonnegative. Thus the maximum value of Z is 17. It occurs when $x_1 = 5$ and $x_2 = 7$.

When an *equality* constraint of the form

$$a_1 x_1 + a_2 x_2 + \ldots + a_n x_n = b, \quad \text{where } b \geqslant 0,$$

occurs in a linear programming problem, artificial variables are used in the simplex method. To illustrate, consider the following problem.

$$\text{Maximize } Z = x_1 + 3x_2 - 2x_3$$

subject to

$$x_1 + x_2 - x_3 = 6, \tag{26}$$

and $x_1, x_2, x_3 \geqslant 0$. Constraint (26) is already expressed as an equation, and so no slack variable is necessary. Since $x_1 = x_2 = x_3 = 0$ is not a feasible solution, we do not have an obvious starting point for the simplex procedure. Thus we create an aritificial problem by first adding an artificial variable t to the left side of Eq. (26):

$$x_1 + x_2 - x_3 + t = 6.$$

Here an obvious B.F.S. is $x_1 = x_2 = x_3 = 0$, $t = 6$. The artificial objective function is

$$W = Z - Mt = x_1 + 3x_2 - 2x_3 - Mt,$$

where M is a large positive number. The simplex procedure is applied to this artificial problem until we obtain a B.F.S. in which $t = 0$. This solution will give an initial B.F.S. for the original problem and we then proceed as before.

In general, the simplex method may be used to

$$\text{maximize } Z = c_1x_1 + c_2x_2 + \ldots + c_nx_n$$

subject to

$$\left.\begin{array}{l}a_{11}x_1 + a_{12}x_2 + \ldots + a_{1n}x_n \; \{\leqslant, \geqslant, =\} \; b_1, \\ a_{21}x_1 + a_{22}x_2 + \ldots + a_{2n}x_n \; \{\leqslant, \geqslant, =\} \; b_2, \\ \vdots \qquad \vdots \qquad\qquad \vdots \qquad\qquad\qquad \vdots \\ a_{m1}x_1 + a_{m2}x_2 + \ldots + a_{mn}x_n \; \{\leqslant, \geqslant, =\} \; b_m,\end{array}\right\} \quad (27)$$

where $x_1, x_2, \ldots, x_n$ and $b_1, b_2, \ldots, b_m$ are nonnegative. The symbolism $\{\leqslant, \geqslant, =\}$ means that one of the relations "$\leqslant$," "$\geqslant$," or "$=$" exists for a constraint. If all constraints involve "$\leqslant$," the problem is of standard form and the simplex techniques of the previous sections apply. If any constraint involves "$\geqslant$" or "$=$," we begin with an artificial problem, which is obtained as follows.

Each constraint that contains "$\leqslant$" is written as an equation involving a slack variable s_i with coefficient $+1$:

$$a_{i1}x_1 + a_{i2}x_2 + \ldots + a_{in}x_n + s_i = b_i.$$

Each constraint that contains "$\geqslant$" is written as an equation involving a slack variable s_j with coefficient -1 and an artificial variable t_j:

$$a_{j1}x_1 + a_{j2}x_2 + \ldots + a_{jn}x_n - s_j + t_j = b_j.$$

A nonnegative artificial variable t_k is inserted into each equality constraint:

$$a_{k1}x_1 + a_{k2}x_2 + \ldots + a_{kn}x_n + t_k = b_k.$$

Should the artificial variables involved in this problem be, for example, t_1, t_2, t_3, then the artificial objective function is

$$W = Z - Mt_1 - Mt_2 - Mt_3,$$

where M is a large positive number. An initial B.F.S. occurs when $x_1 = x_2 = \ldots = x_n = 0$ and each slack variable with a coefficient of -1 equals 0. After obtaining an initial simplex tableau, we apply the simplex procedure until we arrive at a tableau that corresponds to a B.F.S. in which *all* artificial variables are 0. We then delete the artificial variable columns, change W's to Z's, and continue by using the procedures of the previous sections.

EXAMPLE 2

Use the simplex method to maximize $Z = x_1 + 3x_2 - 2x_3$ *subject to*

$$-x_1 - 2x_2 - 2x_3 = -6, \tag{28}$$

$$-x_1 - x_2 + x_3 \leqslant -2, \tag{29}$$

and $x_1, x_2, x_3 \geqslant 0$.

Constraints (28) and (29) will have the forms indicated in (27) [that is, b's positive] if we multiply both sides of each constraint by -1:

$$x_1 + 2x_2 + 2x_3 = 6, \tag{30}$$

$$x_1 + x_2 - x_3 \geqslant 2. \tag{31}$$

Since constraints (30) and (31) involve "$=$" and "$\geqslant$," two artificial variables, t_1 and t_2, will occur. The equations for the artificial problem are

$$x_1 + 2x_2 + 2x_3 \quad + t_1 \quad = 6 \tag{32}$$

$$\text{and} \quad x_1 + x_2 - x_3 - s_2 \quad + t_2 = 2. \tag{33}$$

Here the subscript 2 on s_2 reflects the order of the equations. The artificial objective function is $W = Z - Mt_1 - Mt_2$, or equivalently,

$$-x_1 - 3x_2 + 2x_3 + Mt_1 + Mt_2 + W = 0, \tag{34}$$

where M is a large positive number. The augmented coefficient matrix of Eqs. (32)–(34) is

$$
\begin{array}{ccccccc}
x_1 & x_2 & x_3 & s_2 & t_1 & t_2 & W \\
\end{array}
$$
$$
\left[
\begin{array}{ccccccc|c}
1 & 2 & 2 & 0 & 1 & 0 & 0 & 6 \\
1 & 1 & -1 & -1 & 0 & 1 & 0 & 2 \\
\hline
-1 & -3 & 2 & 0 & M & M & 1 & 0 \\
\end{array}
\right].
$$

We now use elementary row operations to remove the M's from *all* the artificial variable columns. By adding $-M$ times row 1 to row 3 and adding $-M$ times row 2 to row 3, we get initial Simplex Tableau I.

SIMPLEX TABLEAU I

	x_1	x_2	x_3	s_2	t_1	t_2	W		Quotients
t_1	1	2	2	0	1	0	0	6	$6 \div 2 = 3.$
departing variable $\to t_2$	1	①	-1	-1	0	1	0	2	$2 \div 1 = 2.$
W	$-1-2M$	$-3-3M$	$2-M$	M	0	0	1	$-8M$	

↑ indicators
entering
variable

Proceeding, we obtain Simplex Tableaus II and III.

SIMPLEX TABLEAU II

$$
\begin{array}{c}
\text{departing}\\
\text{variable} \to t_1
\end{array}
\quad
\begin{array}{c}
\begin{array}{ccccccc}
x_1 & x_2 & x_3 & s_2 & t_1 & t_2 & W \\
\end{array}\\
\left[
\begin{array}{ccccccc|c}
-1 & 0 & ④ & 2 & 1 & -2 & 0 & 2 \\
1 & 1 & -1 & -1 & 0 & 1 & 0 & 2 \\
\hline
2+M & 0 & -1-4M & -3-2M & 0 & 3+3M & 1 & 6-2M
\end{array}
\right]
\end{array}
\quad
\begin{array}{l}
\textit{Quotients}\\[2ex]
2 \div 4 = \tfrac{1}{2}.
\end{array}
$$

indicators
↗
entering variable

SIMPLEX TABLEAU III

$$
\begin{array}{c}
\text{departing}\\
\text{variable} \to x_3
\end{array}
\quad
\begin{array}{c}
\begin{array}{ccccccc}
x_1 & x_2 & x_3 & s_2 & t_1 & t_2 & W \\
\end{array}\\
\left[
\begin{array}{ccccccc|c}
-\tfrac{1}{4} & 0 & 1 & \tfrac{1}{2} & \tfrac{1}{4} & -\tfrac{1}{2} & 0 & \tfrac{1}{2} \\
\tfrac{3}{4} & 1 & 0 & -\tfrac{1}{2} & \tfrac{1}{4} & \tfrac{1}{2} & 0 & \tfrac{5}{2} \\
\hline
\tfrac{7}{4} & 0 & 0 & -\tfrac{5}{2} & \tfrac{1}{4}+M & \tfrac{5}{2}+M & 1 & \tfrac{13}{2}
\end{array}
\right]
\end{array}
\quad
\begin{array}{l}
\textit{Quotients}\\[2ex]
\tfrac{1}{2} \div \tfrac{1}{2} = 1.
\end{array}
$$

indicators ↑
entering variable

For the B.F.S. corresponding to Tableau III, the artificial variables t_1 and t_2 are both 0. We now can delete the t_1- and t_2-columns and change W's to Z's. Continuing, we obtain Simplex Tableau IV.

SIMPLEX TABLEAU IV

$$
\begin{array}{c}
\begin{array}{c}
s_2\\[1ex]
x_2\\[1ex]
Z
\end{array}
\quad
\begin{array}{c}
\begin{array}{ccccc}
x_1 & x_2 & x_3 & s_2 & Z \\
\end{array}\\
\left[
\begin{array}{ccccc|c}
-\tfrac{1}{2} & 0 & 2 & 1 & 0 & 1 \\
\tfrac{1}{2} & 1 & 1 & 0 & 0 & 3 \\
\hline
\tfrac{1}{2} & 0 & 5 & 0 & 1 & 9
\end{array}
\right]
\end{array}
\end{array}.
$$

indicators

Since all indicators are nonnegative, we have reached the final tableau. The maximum value of Z is 9 and it occurs when $x_1 = 0$, $x_2 = 3$, and $x_3 = 0$.

It is possible that the simplex procedure terminates and not all artificial variables are 0. It can be shown that in this situation *the feasible region of the original problem is empty* and hence there is *no optimum solution*. The following example will illustrate.

EXAMPLE 3

Use the simplex method to maximize $Z = 2x_1 + x_2$ subject to

$$-x_1 + x_2 \geqslant 2, \tag{35}$$

$$x_1 + x_2 \leqslant 1,$$

and $x_1, x_2 \geqslant 0$.

Since constraint (35) is of the form $a_{11}x_1 + a_{12}x_2 \geqslant b_1$ where $b_1 \geqslant 0$, an artificial variable will occur. The equations to consider are

$$-x_1 + x_2 - s_1 \qquad + t_1 = 2 \tag{36}$$

and $\qquad x_1 + x_2 \qquad + s_2 \qquad = 1, \tag{37}$

where s_1 and s_2 are slack variables and t_1 is artificial. The artificial objective function is $W = Z - Mt_1$ or, equivalently,

$$-2x_1 - x_2 + Mt_1 + W = 0. \tag{38}$$

The augmented coefficient matrix of Eqs. (36)–(38) is

$$
\begin{array}{ccccccc}
x_1 & x_2 & s_1 & s_2 & t_1 & W & \\
\left[\begin{array}{cccccc|c}
-1 & 1 & -1 & 0 & 1 & 0 & 2 \\
1 & 1 & 0 & 1 & 0 & 0 & 1 \\
\hline
-2 & -1 & 0 & 0 & M & 1 & 0
\end{array}\right]
\end{array}
$$

The simplex tableaus appear below.

SIMPLEX TABLEAU I

$$
\begin{array}{c}
\begin{array}{cccccccc}
 & x_1 & x_2 & s_1 & s_2 & t_1 & W & \\
\text{departing } t_1 & -1 & 1 & -1 & 0 & 1 & 0 & 2 \\
\text{variable} \to s_2 & 1 & \textcircled{1} & 0 & 1 & 0 & 0 & 1 \\
\hline
W & -2+M & -1-M & M & 0 & 0 & 1 & -2M
\end{array}
\end{array}
$$

$$
\begin{array}{c}
\textit{Quotients} \\
2 \div 1 = 2. \\
1 \div 1 = 1.
\end{array}
$$

$\underbrace{\qquad\qquad\qquad\qquad}_{\nearrow \text{ indicators}}$

entering variable

SIMPLEX TABLEAU II

$$
\begin{array}{ccccccc}
 & x_1 & x_2 & s_1 & s_2 & t_1 & W \\
t_1 & -2 & 0 & -1 & -1 & 1 & 0 \quad 1 \\
x_2 & 1 & 1 & 0 & 1 & 0 & 0 \quad 1 \\
\hline
W & -1+2M & 0 & M & 1+M & 0 & 1 \quad 1-M
\end{array}
$$

$$\underbrace{\qquad\qquad\qquad\qquad}_{\text{indicators}}$$

Since M is a large positive number, the indicators in Simplex Tableau II are nonnegative and so the simplex procedure terminates. The value of the artificial variable t_1 is 1. Therefore, as previously stated the feasible region of the original problem is empty and hence no solution exists. This result can be obtained geometrically. Figure 15-21 shows the graphs of $-x_1 + x_2 = 2$ and $x_1 + x_2 = 1$ for $x_1, x_2 \geqslant 0$. Since there is no point (x_1, x_2) that simultaneously lies above $-x_1 + x_2 = 2$ and below $x_1 + x_2 = 1$ such that $x_1, x_2 \geqslant 0$, the feasible region is empty and thus no solution exists.

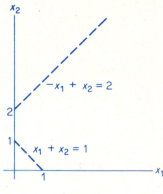

FIG. 15-21

In the next section we shall use the simplex method on minimization problems.

EXERCISE 15-6

Use the simplex method to solve the following problems.

1. Maximize
$$Z = 2x_1 + x_2$$
subject to
$$x_1 + x_2 \leqslant 6,$$
$$-x_1 + x_2 \geqslant 4,$$
$$x_1, x_2 \geqslant 0.$$

2. Maximize
$$Z = 3x_1 + 4x_2$$
subject to
$$x_1 + 2x_2 \leqslant 8,$$
$$x_1 + 6x_2 \geqslant 12,$$
$$x_1, x_2 \geqslant 0.$$

3. Maximize
$$Z = 2x_1 + x_2 - x_3$$
subject to
$$x_1 + 2x_2 + x_3 \leqslant 5,$$
$$-x_1 + x_2 + x_3 \geqslant 1,$$
$$x_1, x_2, x_3 \geqslant 0.$$

4. Maximize
$$Z = x_1 - x_2 + 4x_3$$
subject to
$$x_1 + x_2 + x_3 \leqslant 9,$$
$$x_1 - 2x_2 + x_3 \geqslant 6,$$
$$x_1, x_2, x_3 \geqslant 0.$$

5. Maximize
    $$Z = 4x_1 + x_2 + 2x_3$$
    subject to
    $$2x_1 + x_2 + 3x_3 \leqslant 10,$$
    $$x_1 - x_2 + x_3 = 4,$$
    $$x_1, x_2, x_3 \geqslant 0.$$

6. Maximize
    $$Z = x_1 + 2x_2 + 3x_3$$
    subject to
    $$x_2 - 2x_3 \geqslant 5,$$
    $$x_1 + x_2 + x_3 = 8,$$
    $$x_1, x_2, x_3 \geqslant 0.$$

7. Maximize
    $$Z = x_1 - 10x_2$$
    subject to
    $$x_1 - x_2 \leqslant 1,$$
    $$x_1 + 2x_2 \leqslant 8,$$
    $$x_1 + x_2 \geqslant 5,$$
    $$x_1, x_2 \geqslant 0.$$

8. Maximize
    $$Z = x_1 + 4x_2 - x_3$$
    subject to
    $$x_1 + x_2 - x_3 \geqslant 5,$$
    $$x_1 + x_2 + x_3 \leqslant 3,$$
    $$x_1 - x_2 + x_3 = 7,$$
    $$x_1, x_2, x_3 \geqslant 0.$$

9. Maximize
    $$Z = 3x_1 - 2x_2 + x_3$$
    subject to
    $$x_1 + x_2 + x_3 \leqslant 1,$$
    $$x_1 - x_2 + x_3 \geqslant 2,$$
    $$x_1 - x_2 - x_3 \leqslant -6,$$
    $$x_1, x_2, x_3 \geqslant 0.$$

10. Maximize
    $$Z = x_1 + 4x_2$$
    subject to
    $$x_1 + 2x_2 \leqslant 8,$$
    $$x_1 + 6x_2 \geqslant 12,$$
    $$x_2 \geqslant 2,$$
    $$x_1, x_2 \geqslant 0.$$

11. Maximize
    $$Z = -3x_1 + 2x_2$$
    subject to
    $$x_1 - x_2 \leqslant 4,$$
    $$-x_1 + x_2 = 4,$$
    $$x_1 \geqslant 6,$$
    $$x_1, x_2 \geqslant 0.$$

12. Maximize
    $$Z = x_1 - 5x_2$$
    subject to
    $$x_1 - 2x_2 \geqslant -13,$$
    $$-x_1 + x_2 \geqslant 3,$$
    $$x_1 + x_2 \geqslant 11,$$
    $$x_1, x_2 \geqslant 0.$$

13. A company manufactures two models of kitchen tables: Contemporary and Traditional. Each model requires assembly and finishing times as given in Table 15-9. The profit on each set is also indicated. The number of hours available per week in the assembly department is 400, and in the finishing department it is 510. Because of a union contract, the finishing department is guaranteed at least 240 hours of work per week. How many tables of each model should the company produce each week to maximize profit?

TABLE 15-9

	ASSEMBLY TIME	FINISHING TIME	PROFIT PER SET
Contemporary	1 hr	2 hr	$10
Traditional	2 hr	3 hr	$12

14. A company manufactures three products: X, Y, and Z. Each product requires the use of machine time on machines A and B as given in Table 15-10. The numbers of

hours per week that *A* and *B* are available for production are 40 and 30, respectively. The profit per unit on *X*, *Y*, and *Z* is $50, $60, and $75, respectively. At least five units of *Z* must be produced next week. What should be the production order for that period if maximum profit is to be achieved? What is the maximum profit?

TABLE 15-10

	MACHINE A	MACHINE B
Product X	1 hr	1 hr
Product Y	2 hr	1 hr
Product Z	2 hr	2 hr

15. The prospectus of an investment fund states that all money is invested in bonds that are rated A, AA, and AAA; no more than 30% of the total investment is in A and AA bonds, and at least 50% is in AA and AAA bonds. The A, AA, and AAA bonds respectively yield 8%, 7%, and 6% annually. Determine the percentages of the total investment that should be committed to each type of bond so that the fund maximizes its annual yield. What is this yield?

15-7 MINIMIZATION

So far we have used the simplex method to *maximize* objective functions. In general, to *minimize* a function it suffices to maximize the negative of the function. To understand why, consider the function $f(x) = x^2 - 4$. In Fig. 15-22(a) observe that the minimum value of f is -4 and it occurs when $x = 0$. Figure 15-22(b) shows the graph of $g(x) = -f(x) = -(x^2 - 4)$. This graph is the reflection through the *x*-axis of the graph of *f*. Notice that the maximum value of *g* is 4 and occurs also when $x = 0$. Thus the minimum value of $x^2 - 4$ is the negative of the maximum value of $-(x^2 - 4)$. That is,

$$\text{Min } f = -\text{Max}(-f).$$

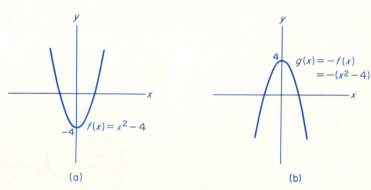

FIG. 15-22

EXAMPLE 1

Use the simplex method to minimize $Z = x_1 + 2x_2$ *subject to*

$$-2x_1 + x_2 \geqslant 1, \tag{1}$$

$$-x_1 + x_2 \geqslant 2, \tag{2}$$

and $x \geqslant 0$, $x_2 \geqslant 0$.

To minimize Z we can maximize $-Z = -x_1 - 2x_2$. Note that constraints (1) and (2) each have the form $a_1x_1 + a_2x_2 \geqslant b$, where $b \geqslant 0$. Thus their equations involve two slack variables s_1 and s_2, each with coefficient -1, and two artificial variables t_1 and t_2.

$$-2x_1 + x_2 - s_1 + t_1 = 1, \tag{3}$$

$$-x_1 + x_2 - s_2 + t_2 = 2. \tag{4}$$

Since there are *two* artificial variables, we maximize the objective function $W = (-Z) - Mt_1 - Mt_2$, where M is a large positive number. Equivalently,

$$x_1 + 2x_2 + Mt_1 + Mt_2 + W = 0. \tag{5}$$

The augmented coefficient matrix of Eqs. (3)–(5) is

$$
\begin{array}{ccccccc}
x_1 & x_2 & s_1 & s_2 & t_1 & t_2 & W \\
\end{array}
$$
$$
\left[
\begin{array}{ccccccc|c}
-2 & 1 & -1 & 0 & 1 & 0 & 0 & 1 \\
-1 & 1 & 0 & -1 & 0 & 1 & 0 & 2 \\
\hline
1 & 2 & 0 & 0 & M & M & 1 & 0
\end{array}
\right].
$$

Proceeding, we obtain Tableaus I, II, and III.

SIMPLEX TABLEAU I

		x_1	x_2	s_1	s_2	t_1	t_2	W		Quotients
departing variable $\rightarrow$	t_1	-2	①	-1	0	1	0	0	1	$1 \div 1 = 1.$
	t_2	-1	1	0	-1	0	1	0	2	$2 \div 1 = 2.$
	W	$1+3M$	$2-2M$	M	M	0	0	1	$-3M$	

$\uparrow$ indicators

entering variable

SIMPLEX TABLEAU II

	x_1	x_2	s_1	s_2	t_1	t_2	W		Quotients
x_2	-2	1	-1	0	1	0	0	1	
departing → t_2	1	0	①	-1	-1	1	0	1	$1 + 1 = 1.$
W	$5-M$	0	$2-M$	M	$-2+2M$	0	1	$-2-M$	

departing
variable

indicators

entering
variable

SIMPLEX TABLEAU III

	x_1	x_2	s_1	s_2	t_1	t_2	W	
x_2	-1	1	0	-1	0	1	0	2
s_1	1	0	1	-1	-1	1	0	1
W	3	0	0	2	M	$-2+M$	1	-4

indicators

The B.F.S. corresponding to Tableau III has both artificial variables equal to 0. Thus the t_1- and t_2-columns are no longer needed. However, the indicators in the x_1-, x_2-, s_1-, and s_2-columns are nonnegative and hence an optimum solution has been reached. Since $W = -Z$ when $t_1 = t_2 = 0$, the maximum value of $-Z$ is -4. Thus the *minimum* value of Z is $-(-4)$ or 4. It occurs when $x_1 = 0$ and $x_2 = 2$.

EXAMPLE 2

A cement plant produces 2,500,000 barrels of cement per year. The kilns emit 2 lb of dust for each barrel produced. A governmental agency dealing with environmental protection requires that the plant reduce its dust emissions to no more than 800,000 lb per year. There are two emission control devices available, A and B. Device A reduces emissions to $\frac{1}{2}$ lb per barrel and its cost is $0.20 per barrel of cement produced. For device B, emissions are reduced to $\frac{1}{5}$ lb per barrel and the cost is $0.25 per barrel of cement produced. Determine the most economical course of action that the plant should take so that it complies with the agency's requirement and also maintains its annual production of 2,500,000 barrels of cement.[†]

We must minimize the annual cost of emission control. Let x_1, x_2, and x_3 be the annual number of barrels of cement produced in kilns that use device A, device B, and no device, respectively, Then $x_1, x_2, x_3 \geq 0$ and the annual emission control cost C (in dollars) is

$$C = \tfrac{1}{5}x_1 + \tfrac{1}{4}x_2 + 0x_3. \tag{6}$$

[†]This example is adapted from Robert E. Kohn, "A Mathematical Model for Air Pollution Control," *School Science and Mathematics*, vol. 69 (1969), 487–494.

Since 2,500,000 barrels of cement are produced each year,

$$x_1 + x_2 + x_3 = 2{,}500{,}000. \tag{7}$$

The number of pounds of dust emitted annually by the kilns that use device A, device B, and no device is $\frac{1}{2}x_1$, $\frac{1}{5}x_2$, and $2x_3$, respectively. Since the total number of pounds of dust emissions is to be no more than 800,000,

$$\tfrac{1}{2}x_1 + \tfrac{1}{5}x_2 + 2x_3 \leqslant 800{,}000. \tag{8}$$

To minimize C subject to constraints (7) and (8) where $x_1, x_2, x_3 \geqslant 0$, we shall first maximize $-C$ by using the simplex method. The equations to consider are

$$x_1 + x_2 + x_3 + t_1 = 2{,}500{,}000 \tag{9}$$

and

$$\tfrac{1}{2}x_1 + \tfrac{1}{5}x_2 + 2x_3 + s_2 = 800{,}000, \tag{10}$$

where t_1 and s_2 are artificial and slack variables, respectively. The artificial objective equation is $W = (-C) - Mt_1$, or equivalently,

$$\tfrac{1}{5}x_1 + \tfrac{1}{4}x_2 + 0x_3 + Mt_1 + W = 0, \tag{11}$$

where M is a large positive number. The augmented coefficient matrix of Eqs. (9)–(11) is

$$
\begin{array}{cccccc|c}
x_1 & x_2 & x_3 & s_2 & t_1 & W & \\
1 & 1 & 1 & 0 & 1 & 0 & 2{,}500{,}000 \\
\frac{1}{2} & \frac{1}{5} & 2 & 1 & 0 & 0 & 800{,}000 \\
\hline
\frac{1}{5} & \frac{1}{4} & 0 & 0 & M & 1 & 0
\end{array}.
$$

After determining our initial simplex tableau, we proceed and obtain (after three additional tableaus) our final tableau:

$$
\begin{array}{c}
 \\
x_2 \\
x_1 \\
\\
-C
\end{array}
\begin{array}{cccccc|c}
x_1 & x_2 & x_3 & s_2 & -C & \\
0 & 1 & -5 & -\frac{10}{3} & 0 & 1{,}500{,}000 \\
1 & 0 & 6 & \frac{10}{3} & 0 & 1{,}000{,}000 \\
\hline
0 & 0 & \frac{1}{20} & \frac{1}{6} & 1 & -575{,}000
\end{array}.
$$

$$\underbrace{}_{\text{indicators}}$$

Notice that W is replaced by $-C$ when $t_1 = 0$. The maximum value of $-C$ is $-575{,}000$ and occurs when $x_1 = 1{,}000{,}000$, $x_2 = 1{,}500{,}000$, and $x_3 = 0$. Thus the *minimum* annual cost of emission control is $-(-575{,}000) = \$575{,}000$. Device A should be installed on kilns producing 1,000,000 barrels of cement annually, and device B should be installed on kilns producing 1,500,000 annually.

EXERCISE 15-7

Use the simplex method to solve the following problems.

1. Minimize
$$Z = 3x_1 + 6x_2$$
subject to
$$-x_1 + x_2 \geqslant 6,$$
$$x_1 + x_2 \geqslant 10,$$
$$x_1, x_2 \geqslant 0.$$

2. Minimize
$$Z = 8x_1 + 12x_2$$
subject to
$$2x_1 + 2x_2 \geqslant 1,$$
$$x_1 + 3x_2 \geqslant 2,$$
$$x_1, x_2 \geqslant 0.$$

3. Minimize
$$Z = 4x_1 + 2x_2 + x_3$$
subject to
$$x_1 - x_2 - x_3 \geqslant 9,$$
$$x_1, x_2, x_3 \geqslant 0.$$

4. Minimize
$$Z = x_1 + x_2 + 2x_3$$
subject to
$$x_1 + 2x_2 - x_3 \geqslant 4,$$
$$x_1, x_2, x_3 \geqslant 0.$$

5. Minimize
$$Z = 2x_1 + 3x_2 + x_3$$
subject to
$$x_1 + x_2 + x_3 \leqslant 6,$$
$$x_1 \qquad - x_3 \leqslant -4,$$
$$x_2 + x_3 \leqslant 5,$$
$$x_1, x_2, x_3 \geqslant 0.$$

6. Minimize
$$Z = 4x_1 + x_2 + 2x_3$$
subject to
$$4x_1 + x_2 - x_3 \leqslant 3,$$
$$x_1 \qquad + x_3 \leqslant 4,$$
$$x_1 + x_2 + x_3 \geqslant 1,$$
$$x_1, x_2, x_3 \geqslant 0.$$

7. Minimize
$$Z = x_1 - x_2 - 3x_3$$
subject to
$$x_1 + 2x_2 + x_3 = 4,$$
$$x_2 + x_3 = 1,$$
$$x_1 + x_2 \qquad \leqslant 6,$$
$$x_1, x_2, x_3 \geqslant 0.$$

8. Minimize
$$Z = x_1 + x_2 - 2x_3$$
subject to
$$x_1 - x_2 + x_3 \leqslant 4,$$
$$2x_1 + x_2 - 3x_3 \geqslant 6,$$
$$x_1 - x_2 - 2x_3 = 2,$$
$$x_1, x_2, x_3 \geqslant 0.$$

9. Minimize
$$Z = x_1 + 8x_2 + 5x_3$$
subject to
$$x_1 + x_2 + x_3 \geqslant 8,$$
$$-x_1 + 2x_2 + x_3 \geqslant 2,$$
$$x_1, x_2, x_3 \geqslant 0.$$

10. Minimize
$$Z = 4x_1 + 4x_2 + 6x_3$$
subject to
$$x_1 - x_2 - x_3 \leqslant 3,$$
$$x_1 - x_2 + x_3 \geqslant 3,$$
$$x_1, x_2, x_3 \geqslant 0.$$

11. A cement plant produces 3,300,000 barrels of cement per year. The kilns emit 2 lb of dust for each barrel produced. The plant must reduce its dust emissions to no more than 1,000,000 lb per year. There are two devices available, *A* and *B*, that will control emissions. Device *A* will reduce emissions to $\frac{1}{2}$ lb per barrel and the cost is $0.25 per barrel of cement produced. For device *B*, emissions are reduced to $\frac{1}{4}$ lb per barrel and the cost is $0.40 per barrel of cement produced. Determine the most economical course of action that the plant should take so that it maintains an annual production of exactly 3,300,000 barrels of cement.

12. Because of increased business, a catering service finds that it must rent additional

delivery trucks. The minimum needs are 12 units each of refrigerated and nonre-frigerated space. Two standard types of trucks are available in the rental market. Type A has 2 units of refrigerated space and 1 unit of nonrefrigerated space. Type B has 2 units of refrigerated space and 3 units of nonrefrigerated space. The costs per mile are \$0.40 for A and \$0.60 for B. How many of each type of truck should be rented so as to minimize total cost per mile? What is the minimum total cost per mile?

13.   A retailer has stores in Exton and Whyton, and has warehouses A and B in two other cities. Each store requires delivery of exactly 30 refrigerators. In warehouse A there are 50 refrigerators, and in B there are 20. The transportation costs to ship refrigerators from the warehouses to the stores are given in Table 15-11. For example, the cost to ship a refrigerator from A to the Exton store is \$15. How should the retailer order the refrigerators so that the requirements of the stores are met and the total transportation costs are minimized? What is the minimum transportation cost?

TABLE 15-11

	EXTON	WHYTON
Warehouse A	\$15	\$13
Warehouse B	\$11	\$12

14.   An auto manufacturer purchases batteries from two suppliers, X and Y. The manufacturer has two plants, A and B, and requires delivery of exactly 6000 batteries to plant A and exactly 4000 to plant B. Supplier X charges \$30 and \$32 per battery (including transportation cost) to A and B, respectively. For these prices, X requires that the auto manufacturer order at least a total of 2000 batteries. However, X can supply no more than 4000 batteries. Supplier Y charges \$34 and \$28 per battery to A and B, respectively, and requires a minimum order of 6000 batteries. Determine how the auto manufacturer should order the necessary batteries so that their total cost is a minimum. What is this minimum cost?

15.   A paper company stocks its holiday wrapping paper in 48-in. wide rolls, called stock rolls, and cuts such rolls into smaller widths depending on customers' orders. Suppose that an order for 50 rolls of 15-in. wide paper and 60 rolls of 10-in. wide paper is received. From a stock roll the company can cut three 15-in. wide rolls and one 3-in. wide roll. See Fig. 15-23. Since the 3-in. wide roll cannot be used in this

FIG. 15-23

order, 3 inches is called the trim loss for this roll. Similarly, from a stock roll, two 15-in. wide rolls, one 10-in. wide roll, and one 8-in. wide roll could be cut. Here the trim loss would be 8 inches. Table 15-12 indicates the number of 15-in. and 10-in. rolls, together with trim loss, that can be cut from a stock roll. (a) Complete the last two columns of Table 15-12. (b) Assume that the company has a sufficient number of stock rolls to fill the order and that *at least* 50 rolls of 15-in. wide and 60 rolls of 10-in. wide wrapping paper will be cut. If x_1, x_2, x_3, and x_4 are the numbers of stock rolls that are cut in a manner described by columns 1–4 of Table 15-12, respectively, determine the values of the x's so that the total trim loss is minimized. (c) What is the minimum amount of total trim loss?

TABLE 15-12

Roll width	15"	3	2	1	—
	10"	0	1	—	—
Trim loss		3	8	—	—

15-8 THE DUAL

There is a fundamental principle, called *duality*, that allows us to solve a maximization problem by solving a related minimization problem. Let us illustrate.

Suppose that a company produces two types of widgets, manual and electric, and each requires the use of machines A and B in its production. Table 15-13 indicates that a manual widget requires the use of A for 1 hour and B for 1 hour. An electric widget requires A for 2 hours and B for 4 hours. The maximum

TABLE 15-13

	MACHINE A	MACHINE B	PROFIT/UNIT
Manual	1 hr	1 hr	$10
Electric	2 hr	4 hr	$24
Hours available	120	180	

number of hours available per month for machines A and B are 120 and 180, respectively. The profit on a manual widget is $10 and on an electric widget it is $24. Assuming that the company can sell all the widgets it can produce, we shall determine the maximum monthly profit. If x_1 and x_2 are the number of manual and electric widgets produced per month, respectively, then we want to maximize the monthly profit function

$$P = 10x_1 + 24x_2,$$

subject to

$$x_1 + 2x_2 \leqslant 120, \qquad (1)$$
$$x_1 + 4x_2 \leqslant 180, \qquad (2)$$

and $x_1, x_2 \geqslant 0$. Writing constraints (1) and (2) as equations, we have

$$x_1 + 2x_2 + s_1 = 120 \qquad (3)$$
$$\text{and} \qquad x_1 + 4x_2 + s_2 = 180,$$

where s_1 and s_2 are slack variables. In Eq. (3), $x_1 + 2x_2$ is the number of hours that machine A is used. Since 120 hours on A are available, then s_1 is the number of available hours that are *not* used for production. That is, s_1 represents unused capacity (in hours) for A. Similarly, s_2 represents unused capacity for B. Solving this problem by the simplex method, we find that the final tableau is

$$
\begin{array}{c}
\begin{array}{ccccc}
x_1 & x_2 & s_1 & s_2 & P
\end{array} \\
\begin{array}{c}
x_1 \\ x_2 \\ P
\end{array}
\left[
\begin{array}{ccccc|c}
1 & 0 & 2 & -1 & 0 & 60 \\
0 & 1 & -\frac{1}{2} & \frac{1}{2} & 0 & 30 \\
\hline
0 & 0 & 8 & 2 & 1 & 1320
\end{array}
\right].
\end{array}
\qquad (4)
$$

$$\underbrace{\qquad\qquad\qquad}_{\text{indicators}}$$

Thus, the maximum profit per month is \$1320, which occurs when $x_1 = 60$ and $x_2 = 30$.

Now, let us look at the situation from a different point of view. Suppose that the company wishes to rent out machines A and B. What is the minimum monthly rental fee they should charge? Certainly if the charge is too high, no one would rent the machines. On the other hand, if the charge is too low, it may not pay the company to rent them at all. Obviously, the minimum rent should be \$1320. That is, the minimum the company should charge is the profit it could make by using the machines itself. We can arrive at this minimum rental fee directly by solving a linear programming problem.

Let R be the total monthly rental fee. To determine R, suppose the company assigns values or "worths" to each hour of capacity on machines A and B. Let these worths be y_1 and y_2 dollars, respectively, where $y_1, y_2 \geqslant 0$. Then the monthly worth of machine A is $120y_1$, and for B it is $180y_2$. Thus,

$$R = 120y_1 + 180y_2.$$

The total worth of machine time to produce a manual widget is $1y_1 + 1y_2$. This should be at least equal to the \$10 profit the company can earn by producing a manual widget. If not, the company would make more money by using the machine time to produce a manual widget. Thus,

$$1y_1 + 1y_2 \geqslant 10.$$

Similarly, the total worth of machine time to produce an electric widget should be at least $24:

$$2y_1 + 4y_2 \geqslant 24.$$

Therefore the company wants to

$$\text{minimize } R = 120y_1 + 180y_2$$

subject to

$$y_1 + y_2 \geqslant 10, \tag{5}$$
$$2y_1 + 4y_2 \geqslant 24, \tag{6}$$

and $y_1, y_2 \geqslant 0$.

To minimize R, we shall maximize $-R$. Since constraints (5) and (6) have the form $a_1y_1 + a_2y_2 \geqslant b$, where $b \geqslant 0$, we consider an artificial problem. If r_1 and r_2 are slack variables, and t_1 and t_2 are artificial variables, then we want to maximize $W = (-R) - Mt_1 - Mt_2$, where M is a large positive number, such that $y_1 + y_2 - r_1 + t_1 = 10$, $2y_1 + 4y_2 - r_2 + t_2 = 24$, and the y's, r's, and t's are nonnegative. The final simplex tableau for this problem (with the artificial variable columns deleted and W changed to $-R$) is

$$
\begin{array}{c}
\begin{array}{ccccc}
y_1 & y_2 & r_1 & r_2 & -R
\end{array} \\
\begin{array}{c}
y_1 \\
y_2 \\
-R
\end{array}
\left[
\begin{array}{ccccc|c}
1 & 0 & -2 & \frac{1}{2} & 0 & 8 \\
0 & 1 & 1 & -\frac{1}{2} & 0 & 2 \\
\hline
0 & 0 & 60 & 30 & 1 & -1320
\end{array}
\right].
\end{array}
$$

$$\underbrace{}_{\text{indicators}}$$

Since the maximum value of $-R$ is -1320, the *minimum* value of R is $-(-1320) = \$1320$ (as anticipated). It occurs when $y_1 = 8$ and $y_2 = 2$. We have therefore determined the optimum value of one linear programming problem (maximizing profit) by finding the optimum value of another problem (minimizing rental fee).

The values $y_1 = 8$ and $y_2 = 2$ could have been anticipated from the final tableau of the maximization problem. In (4), the indicator 8 in the s_1-column means that at the optimum level of production, if s_1 increases by one unit, then the profit P *decreases* by 8. That is, one unused hour of capacity on A decreases the maximum profit by \$8. Thus one hour of capacity on A is worth \$8. We say that the *shadow price* or *accounting price* of one hour of capacity on A is \$8.

Now, recall that y_1 in the rental problem is the worth of one hour of capacity on A. Thus y_1 must equal 8 in the optimum solution for that problem. Similarly, since the indicator in the s_2-column is 2, the shadow price of one hour of capacity on B is \$2, which is the value of y_2 in the optimum solution of the rental problem.

Let us now analyze the structure of our two linear programming problems:

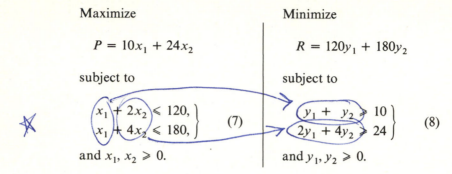

Maximize

$$P = 10x_1 + 24x_2$$

subject to

$$\left.\begin{array}{l} x_1 + 2x_2 \leq 120, \\ x_1 + 4x_2 \leq 180, \end{array}\right\} \quad (7)$$

and $x_1, x_2 \geq 0$.

Minimize

$$R = 120y_1 + 180y_2$$

subject to

$$\left.\begin{array}{l} y_1 + y_2 \geq 10 \\ 2y_1 + 4y_2 \geq 24 \end{array}\right\} \quad (8)$$

and $y_1, y_2 \geq 0$.

Note that in (7) the inequalities are all $\leq$, but in (8) they are all $\geq$. The coefficients of the objective function in the minimization problem are the constant terms in (7). The constant terms in (8) are the coefficients of the objective function of the maximization problem. The coefficients of the y_1's in (8) are the coefficients of x_1 and x_2 in the first constraint of (7); the coefficients of the y_2's in (8) are the coefficients of x_1 and x_2 in the second constraint of (7). The minimization problem is called the _dual_ of the maximization problem and vice versa.

In general, with any given linear programming problem we can associate another linear programming problem called its _dual_. The given problem is called _primal_. If the primal is a maximization problem, then its dual is a minimization. Similarly, if the primal involves minimization, then the dual involves maximization.

Any primal maximization problem can be written in the form indicated in Table 15-14. Note that there are no restrictions on the b's.[†] The corresponding dual minimization problem can be written in the form indicated in Table 15-15. Similarly, any primal minimization problem can be put in the form of Table 15-15 and its dual is the maximization problem in Table 15-14.

Let us compare the primal and its dual in Tables 15-14 and 15-15. For convenience, when we refer to constraints we shall mean those in (9) or (10); we shall not include the nonnegativity conditions. Observe that if all the constraints in the primal involve $\leq$ ($\geq$), then all the constraints in its dual involve $\geq$ ($\leq$).

[†] If an inequality constraint involves $\geq$, multiplying both sides by -1 yields an inequality involving $\leq$. If a constraint is an equality, it can be written in terms of two inequalities: one involving $\leq$ and one involving $\geq$.

TABLE 15-14   PRIMAL (DUAL)

Maximize $Z = c_1x_1 + c_2x_2 + \ldots + c_nx_n$

subject to

$$\left.\begin{array}{l} a_{11}x_1 + a_{12}x_2 + \ldots + a_{1n}x_n \leqslant b_1, \\ a_{21}x_1 + a_{22}x_2 + \ldots + a_{2n}x_n \leqslant b_2, \\ \quad\vdots \qquad\quad \vdots \qquad\qquad\quad \vdots \\ a_{m1}x_1 + a_{m2}x_2 + \ldots + a_{mn}x_n \leqslant b_m, \end{array}\right\} \quad (9)$$

and $x_1, x_2, \ldots, x_n \geqslant 0$.

TABLE 15-15   DUAL (PRIMAL)

Minimize $W = b_1y_1 + b_2y_2 + \ldots + b_my_m$

subject to

$$\left.\begin{array}{l} a_{11}y_1 + a_{21}y_2 + \ldots + a_{m1}y_m \geqslant c_1, \\ a_{12}y_1 + a_{22}y_2 + \ldots + a_{m2}y_m \geqslant c_2, \\ \quad\vdots \qquad\quad \vdots \qquad\qquad\quad \vdots \\ a_{1n}y_1 + a_{2n}y_2 + \ldots + a_{mn}y_m \geqslant c_n, \end{array}\right\} \quad (10)$$

and $y_1, y_2, \ldots, y_m \geqslant 0$.

The coefficients in the dual's objective function are the constant terms in the primal's constraints. Similarly, the constant terms in the dual's constraints are the coefficients of the primal's objective function. The coefficient matrix of the left sides of the dual's constraints is the *transpose* of the coefficient matrix of the left sides of the primal's constraints. That is, for example,

$$\begin{bmatrix} a_{11} & a_{12} & \cdots & a_{1n} \\ a_{21} & a_{22} & \cdots & a_{2n} \\ \vdots & \vdots & & \vdots \\ a_{m1} & a_{m2} & \cdots & a_{mn} \end{bmatrix}^{T} = \begin{bmatrix} a_{11} & a_{21} & \cdots & a_{m1} \\ a_{12} & a_{22} & \cdots & a_{m2} \\ \vdots & \vdots & & \vdots \\ a_{1n} & a_{2n} & \cdots & a_{mn} \end{bmatrix}.$$

If the primal involves n structural variables and m slack variables, then the dual involves m structural variables and n slack variables. It should be noted that the dual of the *dual* is the primal.

There is an important relationship between the primal and its dual:

If the primal has an optimum solution, then so does the dual and the optimum value of the primal's objective function is the *same* as that of its dual.

Moreover, suppose the primal's objective function is $Z = c_1 x_1 + c_2 x_2 + \ldots + c_n x_n$.

> **If s_i is the slack variable associated with the ith constraint in the dual, then the indicator in the s_i-column of the final simplex tableau of the dual is the value of x_i in the optimum solution of the primal.**

Thus we can solve the primal by merely solving its dual. At times this is more convenient than solving the primal directly.

EXAMPLE 1

Find the dual of the following:

$$\text{maximize } Z = 3x_1 + 4x_2 + 2x_3$$

subject to

$$x_1 + 2x_2 + 0x_3 \leqslant 10,$$
$$2x_1 + 2x_2 + x_3 \leqslant 10,$$

and $x_1, x_2, x_3 \geqslant 0$.

The primal is of the form of Table 15-14. Thus the dual is

$$\text{minimize } W = 10y_1 + 10y_2$$

subject to

$$y_1 + 2y_2 \geqslant 3,$$
$$2y_1 + 2y_2 \geqslant 4,$$
$$0y_1 + y_2 \geqslant 2,$$

and $y_1, y_2 \geqslant 0$.

EXAMPLE 2

Find the dual of the following:

$$\text{minimize } Z = 4x_1 + 3x_2$$

subject to

$$3x_1 - x_2 \geqslant 2, \tag{11}$$
$$x_1 + x_2 \leqslant 1, \tag{12}$$
$$-4x_1 + x_2 \leqslant 3, \tag{13}$$

and $x_1, x_2 \geqslant 0$.

Since the primal is a minimization problem, we want constraints (12) and (13) to involve $\geqslant$ (see Table 15-15). Multiplying both sides of (12) and (13) by -1, we get $-x_1 - x_2 \geqslant -1$ and $4x_1 - x_2 \geqslant -3$. Thus constraints (11)–(13) become

$$3x_1 - x_2 \geqslant 2,$$

$$-x_1 - x_2 \geqslant -1,$$

$$4x_1 - x_2 \geqslant -3.$$

The dual is

$$\text{maximize } W = 2y_1 - y_2 - 3y_3$$

subject to

$$3y_1 - y_2 + 4y_3 \leqslant 4,$$

$$-y_1 - y_2 - y_3 \leqslant 3,$$

and $y_1, y_2, y_3 \geqslant 0$.

EXAMPLE 3

Use the dual and the simplex method to maximize $Z = 4x_1 - x_2 - x_3$ subject to

$$3x_1 + x_2 - x_3 \leqslant 4,$$

$$x_1 + x_2 + x_3 \leqslant 2,$$

and $x_1, x_2, x_3 \geqslant 0$.

The dual is

$$\text{minimize } W = 4y_1 + 2y_2$$

subject to

$$3y_1 + y_2 \geqslant 4, \tag{14}$$

$$y_1 + y_2 \geqslant -1, \tag{15}$$

$$-y_1 + y_2 \geqslant -1, \tag{16}$$

and $y_1, y_2 \geqslant 0$. To use the simplex method we must get nonnegative constants in (15) and (16). Multiplying both sides of (15) and (16) by -1 gives

$$-y_1 - y_2 \leqslant 1, \tag{17}$$

$$y_1 - y_2 \leqslant 1. \tag{18}$$

Since (14) involves $\geqslant$, an artificial variable is required. The corresponding equations of

(14), (17), and (18) are, respectively,

$$3y_1 + y_2 - s_1 + t_1 = 4,$$

$$-y_1 - y_2 + s_2 = 1,$$

and $\quad y_1 - y_2 + s_3 = 1,$

where t_1 is an artificial variable and s_1, s_2, and s_3 are slack variables. To minimize W, we maximize $-W$. The artificial objective function is $U = (-W) - Mt$, where M is a large positive number. After computations we find that the final simplex tableau is

$$
\begin{array}{c}
y_2 \\
s_2 \\
y_1 \\
\\
-W
\end{array}
\begin{array}{c}

\end{array}
\underbrace{
\left[
\begin{array}{cccccc|c}
y_1 & y_2 & s_1 & s_2 & s_3 & -W & \\
0 & 1 & -\frac{1}{4} & 0 & -\frac{3}{4} & 0 & \frac{1}{4} \\
0 & 0 & -\frac{1}{2} & 1 & -\frac{1}{2} & 0 & \frac{5}{2} \\
1 & 0 & -\frac{1}{4} & 0 & \frac{1}{4} & 0 & \frac{5}{4} \\
\hline
0 & 0 & \frac{3}{2} & 0 & \frac{1}{2} & 1 & -\frac{11}{2}
\end{array}
\right]
}_{\text{indicators}}.
$$

The maximum value of $-W$ is $-\frac{11}{2}$, and so the *minimum* value of W is $\frac{11}{2}$. Hence the maximum value of Z is also $\frac{11}{2}$. Note that the indicators in the s_1-, s_2-, and s_3-columns are $\frac{3}{2}$, 0, and $\frac{1}{2}$, respectively. Thus the maximum value of Z occurs when $x_1 = \frac{3}{2}$, $x_2 = 0$, and $x_3 = \frac{1}{2}$.

In Example 1 of Sec. 15-7 we used the simplex method to minimize $Z = x_1 + 2x_2$ such that

$$-2x_1 + x_2 \geqslant 1,$$
$$-x_1 + x_2 \geqslant 2,$$

and $x_1, x_2 \geqslant 0$. The initial simplex tableau had 24 entries and involved two artificial variables. The tableau of the dual has only 18 entries and *no artificial variables*, and is easier to handle, as Example 4 will show. Thus, there may be a distinct advantage in solving the dual to determine the solution of the primal.

EXAMPLE 4

Use the dual and the simplex method to minimize $Z = x_1 + 2x_2$ subject to

$$-2x_1 + x_2 \geqslant 1,$$
$$-x_1 + x_2 \geqslant 2,$$

and $x_1, x_2 \geqslant 0$.

The dual is

$$\text{maximize } W = y_1 + 2y_2$$

subject to

$$-2y_1 - y_2 \leqslant 1,$$
$$y_1 + y_2 \leqslant 2,$$

and $y_1, y_2 \geqslant 0$. The initial simplex tableau is Tableau I.

SIMPLEX TABLEAU I

	y_1	y_2	s_1	s_2	W		Quotients
s_1	-2	-1	1	0	0	1	
s_2	1	①	0	1	0	2	$2 \div 1 = 2.$
W	-1	-2	0	0	1	0	

departing → variable (s_2)

↗ indicators

entering variable

Continuing, we get Tableau II.

SIMPLEX TABLEAU II

	y_1	y_2	s_1	s_2	W	
s_1	-1	0	1	1	0	3
y_2	1	1	0	1	0	2
W	1	0	0	2	1	4

indicators

Since all indicators are nonnegative in Tableau II, the maximum value of W is 4. Hence the minimum value of Z is also 4. The indicators 0 and 2 in the s_1- and s_2-columns of Tableau II mean that the minimum value of Z occurs when $x_1 = 0$ and $x_2 = 2$.

EXERCISE 15-8

In Problems **1–8,** *find the duals. Do not solve.*

1. Maximize
   $$Z = 2x_1 + 3x_2$$
   subject to
   $$x_1 + x_2 \leqslant 6,$$
   $$-x_1 + x_2 \leqslant 4,$$
   $$x_1, x_2 \geqslant 0.$$

2. Maximize
   $$Z = 2x_1 + x_2 - x_3$$
   subject to
   $$x_1 + x_2 \qquad \leqslant 1,$$
   $$-x_1 + 2x_2 + x_3 \leqslant 2,$$
   $$x_1, x_2, x_3 \geqslant 0.$$

3. Minimize
$$Z = x_1 + 8x_2 + 5x_3$$
subject to
$$x_1 + x_2 + x_3 \geqslant 8,$$
$$-x_1 + 2x_2 + x_3 \geqslant 2,$$
$$x_1, x_2, x_3 \geqslant 0.$$

4. Minimize
$$Z = 8x_1 + 12x_2$$
subject to
$$2x_1 + 2x_2 \geqslant 1,$$
$$x_1 + 3x_2 \geqslant 2,$$
$$x_1, x_2 \geqslant 0.$$

5. Maximize
$$Z = x_1 - x_2$$
subject to
$$-x_1 + 2x_2 \leqslant 13,$$
$$-x_1 + x_2 \geqslant 3,$$
$$x_1 + x_2 \geqslant 11,$$
$$x_1, x_2 \geqslant 0.$$

6. Maximize
$$Z = x_1 - x_2 + 4x_3$$
subject to
$$x_1 + x_2 + x_3 \leqslant 9,$$
$$x_1 - 2x_2 + x_3 \geqslant 6,$$
$$x_1, x_2, x_3 \geqslant 0.$$

7. Minimize
$$Z = 4x_1 + 4x_2 + 6x_3$$
subject to
$$x_1 - x_2 - x_3 \leqslant 3,$$
$$x_1 - x_2 + x_3 \geqslant 3,$$
$$x_1, x_2, x_3 \geqslant 0.$$

8. Minimize
$$Z = 6x_1 + 3x_2$$
subject to
$$-3x_1 + 4x_2 \geqslant -12,$$
$$13x_1 - 8x_2 \leqslant 80,$$
$$x_1, x_2 \geqslant 0.$$

In Problems 9–14, solve by using duals and the simplex method.

9. Minimize
$$Z = 4x_1 + 4x_2 + 6x_3$$
subject to
$$x_1 - x_2 + x_3 \geqslant 1,$$
$$-x_1 + x_2 + x_3 \geqslant 2,$$
$$x_1, x_2, x_3 \geqslant 0.$$

10. Minimize
$$Z = x_1 + x_2$$
subject to
$$x_1 + 4x_2 \geqslant 28,$$
$$2x_1 - x_2 \geqslant 2,$$
$$-3x_1 + 8x_2 \geqslant 16,$$
$$x_1, x_2 \geqslant 0.$$

11. Maximize
$$Z = 3x_1 + 8x_2$$
subject to
$$x_1 + 2x_2 \leqslant 8,$$
$$x_1 + 6x_2 \leqslant 12,$$
$$x_1, x_2 \geqslant 0.$$

12. Maximize
$$Z = 2x_1 + 6x_2$$
subject to
$$3x_1 + x_2 \leqslant 12,$$
$$x_1 + x_2 \leqslant 8,$$
$$x_1, x_2 \geqslant 0.$$

13. Minimize
$$Z = 6x_1 + 4x_2$$
subject to
$$-x_1 + x_2 \leqslant 1,$$
$$x_1 + x_2 \geqslant 3,$$
$$x_1, x_2 \geqslant 0.$$

14. Minimize
$$Z = x_1 + x_2 + 2x_3$$
subject to
$$-x_1 - x_2 + x_3 \leqslant 1,$$
$$x_1 - x_2 + x_3 \geqslant 2,$$
$$x_1, x_2, x_3 \geqslant 0.$$

15. A firm is comparing the costs of advertising in two media: newspaper and radio. For every dollar's worth of advertising, Table 15-16 gives the number of people, by income group, reached by these media. The firm wants to reach at least 8000

TABLE 15-16

	UNDER $20,000	OVER $20,000
Newspaper	40	100
Radio	50	25

persons earning under $20,000 and at least 6000 earning over $20,000. Use the dual and the simplex method to find the amounts that the firm should spend on newspaper and radio advertising so as to reach these numbers of people at a minimum total advertising cost. What is the minimum total advertising cost?

16. Use the dual and the simplex method to find the minimum total cost per mile in Problem 12 of Exercise 15-7.

17. A company pays skilled and semiskilled workers in its assembly department $7 and $4 per hour, respectively. In the shipping department, shipping clerks are paid $5 per hour and shipping clerk apprentices are paid $2 per hour. The company requires at least 90 workers in the assembly department and at least 60 in the shipping department. Because of union agreements, at least twice as many semiskilled workers must be employed as skilled workers. Also, at least twice as many shipping clerks must be employed as shipping clerk apprentices. Use the dual and the simplex method to find the number of each type of worker that the company must employ so that the total hourly wage paid to these employees is a minimum. What is the minimum total hourly wage?

15-9 REVIEW

IMPORTANT TERMS AND SYMBOLS IN CHAPTER 15

linear inequality (p. 725)

system of inequalities (p. 728)

constraints (p. 731)

linear function in x and y (p. 731)

linear programming problem (p. 731)

objective function (p. 731)

feasible solution (p. 732)

nonnegativity conditions (p. 732)

feasible region (p. 732)

isoprofit line (p. 733)

corner point (p. 733)

bounded feasible region (p. 734)

unbounded feasible region (p. 734)

nonempty feasible region (p. 734)

empty feasible region (p. 734)

isocost line (p. 738)

unbounded solution (p. 737)

multiple optimum solutions (p. 742)

standard linear programming
    problem (p. 744)

slack variable (p. 745)

structural variable (p. 745)

basic feasible solution (p. 746)

nonbasic variable *(p. 746)*   basic variable *(p. 746)*

simplex tableau *(p. 746)*   entering variable *(p. 747)*

indicator *(p. 747)*   departing variable *(p. 748)*

pivot entry *(p. 749)*   simplex method *(p. 751)*

degeneracy *(p. 755)*   artificial problem *(p. 768)*

artificial variable *(p. 768)*   artificial objective function *(p. 768)*

shadow price *(p. 789)*   dual, primal *(p. 790)*

REVIEW SECTION

1. If $y_1 \leqslant m_1 x_1 + b_1$ and $y_2 \geqslant m_2 x + b_2$ is a system of inequalities, which of the regions 1, 2, 3, and 4 in Fig. 15-24 would correspond to the solution? _____

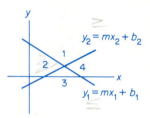

FIG. 15-24

Ans. 2.

2. The solution of $x - y > 3$ consists of all points (above) (below) the line $y = x - 3$.

Ans. below.

3. True or false: Given a system of linear inequalities in x and y, a solution of any two of the inequalities always gives a feasible point or a corner point. _____

Ans. false.

4. True or false: If $Z = Z(x, y)$ is an objective function and the feasible region is nonempty and bounded, then Z has not only a maximum value but also a minimum value. _____

Ans. true.

†5. True or false: In a linear programming problem, if an optimum solution exists, then the optimum value of the objective function may occur at more than one point in the feasible region. _____

Ans. true.

_____

†Refers to Sec. 15-3.

6. If a linear programming problem in x and y has three inequality constraints (excluding nonnegativity conditions), then there will correspond how many slack variables? _____

    Ans. three.

7. In the simplex method, an optimum solution is reached when all of the indicators are _____ .

    Ans. nonnegative.

8. Suppose that a standard linear programming problem has two constraints (excluding nonnegativity conditions) and that the constraints involve five variables when converted to equations. Then a basic feasible solution must have __(a)__ of these variables equal to 0. Such variables are called __(b)__ variables for the B.F.S.

    Ans. (a) three;   (b) nonbasic.

9. Suppose that a B.F.S. corresponding to a simplex tableau has x_i as a nonbasic variable and that the indicator in the x_i column is -4. If x_i increases by one unit, then the value of the objective function _(increases) (decreases)_ by four units.

    Ans. increases.

10. To write $3x + 4y \geqslant 8$ as an equation, we _(add) (subtract)_ a nonnegative slack variable _(to) (from)_ the left side.

    Ans. subtract; from.

11. Suppose the simplex method is applied to the following problem:

    $$\text{maximize } Z = 3x_1 + 4x_2 + 2x_3$$

    subject to

    $$2x_1 + 2x_2 + x_3 \leqslant 10,$$
    $$x_1 + 2x_2 \qquad \geqslant 2,$$

    and $x_1, x_2, x_3 \geqslant 0$. Then there will be __(a)__ slack variables and __(b)__ artificial variables. The initial simplex tableau will have __(c)__ rows. The entering variable corresponds to the most __(d)__ indicator.

    Ans. (a) 2;   (b) 1;   (c) 3;   (d) negative.

12. If Z is an objective function, to minimize Z it is sufficient to maximize _____ .

    Ans. $-Z$.

13. If the simplex method terminates and not all artificial variables are 0, then the original problem _(does) (does not)_ have an optimum solution.

    Ans. does not.

14. If a primal problem involves minimization, then its dual involves _____ .

    Ans. maximization.

15. If a primal problem has 4 structural variables and 2 slack variables, then the dual has __(a)__ structural variables and __(b)__ slack variables.

    Ans. (a) 2;   (b) 4.

16. True or false: The optimum value of the primal's objective function is equal to the optimum value of the dual's objective function. _____

Ans. true.

REVIEW PROBLEMS

In Problems **1–10** *solve the given inequality or system of inequalities.*

1. $-3x + 2y > -6$.

2. $x - 2y + 6 \geqslant 0$.

3. $2y \leqslant -3$.

4. $-x < 2$.

5. $\begin{cases} y - 3x < 6, \\ x - y > -3. \end{cases}$

6. $\begin{cases} x - 2y > 4, \\ x + y > 1. \end{cases}$

7. $\begin{cases} x - y < 4, \\ y - x < 4. \end{cases}$

8. $\begin{cases} x > y, \\ x + y < 0. \end{cases}$

9. $\begin{cases} 3x + y > -4, \\ x - y > -5, \\ x \geqslant 0. \end{cases}$

10. $\begin{cases} x - y > 4, \\ x < 2, \\ y < -4. \end{cases}$

In Problems **11–18,** *do not use the simplex method.*

11. Maximize
$$Z = x - 2y$$
subject to
$$y - x \leqslant 2,$$
$$x + y \leqslant 4,$$
$$x \leqslant 3,$$
$$x, y \geqslant 0.$$

12. Maximize
$$Z = 4x + 2y$$
subject to
$$x + 2y \leqslant 10,$$
$$x \leqslant 4,$$
$$y \geqslant 1,$$
$$x, y \geqslant 0.$$

13. Minimize
$$Z = 2x - y$$
subject to
$$x - y \geqslant -2,$$
$$x + y \geqslant 1,$$
$$x - 2y \leqslant 2,$$
$$x, y \geqslant 0.$$

14. Minimize
$$Z = x + y$$
subject to
$$x + 3y \leqslant 15,$$
$$3x + 2y \leqslant 17,$$
$$x - 5y \leqslant 0,$$
$$x, y \geqslant 0.$$

15. Minimize
$$Z = 4x - 3y$$
subject to
$$x + y \leqslant 3,$$
$$2x + 3y \leqslant 12,$$
$$5x + 8y \geqslant 40,$$
$$x, y \geqslant 0.$$

†16. Minimize
$$Z = 2x + 2y$$
subject to
$$x + y \geqslant 4,$$
$$-x + 3y \leqslant 18,$$
$$x \leqslant 6,$$
$$x, y \geqslant 0.$$

---

†Refers to Sec. 15-3.

†17. Maximize
$$Z = 9x + 6y$$
subject to
$$x + 2y \leqslant 8,$$
$$3x + 2y \leqslant 12,$$
$$x, y \geqslant 0.$$

18. Maximize
$$Z = 4x + y$$
subject to
$$x + 2y \geqslant 8,$$
$$3x + 2y \geqslant 12,$$
$$x, y \geqslant 0.$$

In Problems **19–28**, *use the simplex method.*

19. Maximize
$$Z = 4x_1 + 5x_2$$
subject to
$$x_1 + 6x_2 \leqslant 12,$$
$$x_1 + 2x_2 \leqslant 8,$$
$$x_1, x_2 \geqslant 0.$$

20. Maximize
$$Z = 18x_1 + 20x_2$$
subject to
$$2x_1 + 3x_2 \leqslant 18,$$
$$4x_1 + 3x_2 \leqslant 24,$$
$$x_2 \leqslant 5,$$
$$x_1, x_2 \geqslant 0.$$

21. Minimize
$$Z = 2x_1 + 3x_2 + x_3$$
subject to
$$x_1 + 2x_2 + 3x_3 \geqslant 6,$$
$$x_1, x_2, x_3 \geqslant 0.$$

22. Minimize
$$Z = x_1 + x_2$$
subject to
$$3x_1 + 4x_2 \geqslant 24,$$
$$x_2 \geqslant 3,$$
$$x_1, x_2 \geqslant 0.$$

23. Maximize
$$Z = x_1 + 2x_2$$
subject to
$$x_1 + x_2 \leqslant 12,$$
$$x_1 + x_2 \geqslant 5,$$
$$x_1 \qquad \leqslant 10,$$
$$x_1, x_2 \geqslant 0.$$

24. Minimize
$$Z = 2x_1 + x_2$$
subject to
$$x_1 + 2x_2 \leqslant 6,$$
$$x_1 + x_2 \geqslant 1,$$
$$x_1, x_2 \geqslant 0.$$

25. Minimize
$$Z = x_1 + 2x_2 + x_3$$
subject to
$$x_1 - x_2 - x_3 \leqslant -1,$$
$$6x_1 + 3x_2 + 2x_3 = 12,$$
$$x_1, x_2, x_3 \geqslant 0.$$

26. Maximize
$$Z = x_1 + 3x_2 + 2x_3$$
subject to
$$x_1 + x_2 + 4x_3 \geqslant 6,$$
$$2x_1 + x_2 + 3x_3 \leqslant 4,$$
$$x_1, x_2, x_3 \geqslant 0.$$

†27. Maximize
$$Z = x_1 + 4x_2 + 2x_3$$
subject to
$$4x_1 - x_2 \qquad \leqslant 2,$$
$$-10x_1 + x_2 + 3x_3 \leqslant 1,$$
$$x_1, x_2, x_3 \geqslant 0.$$

†28. Minimize
$$Z = x_1 + x_2$$
subject to
$$x_1 + x_2 + 2x_3 \leqslant 4,$$
$$x_3 \geqslant 1,$$
$$x_1, x_2, x_3 \geqslant 0.$$

---

† Refers to Sec. 15-3 or Sec. 15-5.

In Problems 29 and 30, solve by using duals and the simplex method.

29. Minimize
$$Z = 2x_1 + 7x_2 + 8x_3$$
subject to
$$x_1 + 2x_2 + 3x_3 \geqslant 35,$$
$$x_1 + x_2 + x_3 \geqslant 25,$$
$$x_1, x_2, x_3 \geqslant 0.$$

30. Maximize
$$Z = x_1 - 2x_2$$
subject to
$$x_1 - x_2 \leqslant 3,$$
$$x_1 + 2x_2 \leqslant 4,$$
$$4x_1 + x_2 \geqslant 2,$$
$$x_1, x_2, \geqslant 0.$$

31. A company manufactures three products: X, Y, and Z. Each product requires the use of machine time on machines A and B as given in Table 15-17. The number of hours per week that A and B are available for production are 40 and 34, respectively. The profit per unit on X, Y, and Z is \$10, \$15, and \$22, respectively. What should be the weekly production order if maximum profit is to be obtained? What is the maximum profit?

TABLE 15-17

	MACHINE A	MACHINE B
Product X	1 hr	1 hr
Product Y	2 hr	1 hr
Product Z	2 hr	2 hr

32. Repeat Problem 32 if the company must produce at least a total of 24 units per week.

33. An oil company has storage facilities for heating-fuel in cities A, B, C, and D. Cities C and D are each in need of exactly 500,000 gal of fuel. The company determines that A and B can each sacrifice at most 600,000 gal to satisfy the needs of C and D. Table 15-18 gives the costs per gallon to transport fuel between the cities. How should the company distribute the fuel in order to minimize the total transportation cost? What is the minimum transportation cost?

TABLE 15-18

FROM \ TO	C	D
A	\$0.01	\$0.02
B	\$0.02	\$0.04

TABLES OF POWERS–ROOTS–RECIPROCALS

n	n^2	$\sqrt{n}$	$\sqrt{10n}$	n^3	$\sqrt[3]{n}$	$\sqrt[3]{10n}$	$\sqrt[3]{100n}$	$1/n$
1.0	1.0000	1.0000	3.1623	1.0000	1.0000	2.1544	4.6416	1.0000
1.1	1.2100	1.0488	3.3166	1.3310	1.0323	2.2240	4.7914	0.9091
1.2	1.4400	1.0954	3.4641	1.7280	1.0627	2.2894	4.9324	0.8333
1.3	1.6900	1.1402	3.6056	2.1970	1.0914	2.3513	5.0658	0.7692
1.4	1.9600	1.1832	3.7417	2.7440	1.1187	2.4101	5.1925	0.7143
1.5	2.2500	1.2247	3.8730	3.3750	1.1447	2.4662	5.3133	0.6667
1.6	2.5600	1.2649	4.0000	4.0960	1.1696	2.5198	5.4288	0.6250
1.7	2.8900	1.3038	4.1231	4.9130	1.1935	2.5713	5.5397	0.5882
1.8	3.2400	1.3416	4.2426	5.8320	1.2164	2.6207	5.6462	0.5556
1.9	3.6100	1.3784	4.3589	6.8590	1.2386	2.6684	5.7489	0.5263
2.0	4.0000	1.4142	4.4721	8.0000	1.2599	2.7144	5.8480	0.5000
2.1	4.4100	1.4491	4.5826	9.2610	1.2806	2.7589	5.9439	0.4762
2.2	4.8400	1.4832	4.6904	10.6480	1.3006	2.8020	6.0368	0.4545
2.3	5.2900	1.5166	4.7958	12.1670	1.3200	2.8439	6.1269	0.4348

n	n^2	$\sqrt{n}$	$\sqrt{10n}$	n^3	$\sqrt[3]{n}$	$\sqrt[3]{10n}$	$\sqrt[3]{100n}$	$1/n$
2.4	5.7600	1.5492	4.8990	13.8240	1.3389	2.8845	6.2145	0.4167
2.5	6.2500	1.5811	5.0000	15.6250	1.3572	2.9240	6.2996	0.4000
2.6	6.7600	1.6125	5.0990	17.5760	1.3751	2.9625	6.3825	0.3846
2.7	7.2900	1.6432	5.1962	19.6830	1.3925	3.0000	6.4633	0.3704
2.8	7.8400	1.6733	5.2915	21.9520	1.4095	3.0366	6.5421	0.3571
2.9	8.4100	1.7029	5.3852	24.3890	1.4260	3.0723	6.6191	0.3448
3.0	9.0000	1.7321	5.4772	27.0000	1.4422	3.1072	6.6943	0.3333
3.1	9.6100	1.7607	5.5678	29.7910	1.4581	3.1414	6.7679	0.3226
3.2	10.2400	1.7889	5.6569	32.7680	1.4736	3.1748	6.8399	0.3125
3.3	10.8900	1.8166	5.7446	35.9370	1.4888	3.2075	6.9104	0.3030
3.4	11.5600	1.8439	5.8310	39.3040	1.5037	3.2396	6.9795	0.2941
3.5	12.2500	1.8708	5.9161	42.8750	1.5183	3.2711	7.0473	0.2857
3.6	12.9600	1.8974	6.0000	46.6560	1.5326	3.3019	7.1138	0.2778
3.7	13.6900	1.9235	6.0828	50.6530	1.5467	3.3322	7.1791	0.2703
3.8	14.4400	1.9494	6.1644	54.8720	1.5605	3.3620	7.2432	0.2632
3.9	15.2100	1.9748	6.2450	59.3190	1.5741	3.3912	7.3061	0.2564
4.0	16.0000	2.0000	6.3246	64.0000	1.5874	3.4200	7.3681	0.2500
4.1	16.8100	2.0248	6.4031	68.9210	1.6005	3.4482	7.4290	0.2439
4.2	17.6400	2.0494	6.4807	74.0880	1.6134	3.4760	7.4889	0.2381
4.3	18.4900	2.0736	6.5574	79.5070	1.6261	3.5034	7.5478	0.2326
4.4	19.3600	2.0976	6.6333	85.1840	1.6386	3.5303	7.6059	0.2273
4.5	20.2500	2.1213	6.7082	91.1250	1.6510	3.5569	7.6631	0.2222
4.6	21.1600	2.1448	6.7823	97.3360	1.6631	3.5830	7.7194	0.2174
4.7	22.0900	2.1679	6.8557	103.823	1.6751	3.6088	7.7750	0.2128
4.8	23.0400	2.1909	6.9282	110.592	1.6869	3.6342	7.8297	0.2083
4.9	24.0100	2.2136	7.0000	117.649	1.6985	3.6593	7.8837	0.2041
5.0	25.0000	2.2361	7.0711	125.000	1.7100	3.6840	7.9370	0.2000
5.1	26.0100	2.2583	7.1414	132.651	1.7213	3.7084	7.9896	0.1961
5.2	27.0400	2.2804	7.2111	140.608	1.7325	3.7325	8.0415	0.1923
5.3	28.0900	2.3022	7.2801	148.877	1.7435	3.7563	8.0927	0.1887
5.4	29.1600	2.3238	7.3485	157.464	1.7544	3.7798	8.1433	0.1852
5.5	30.2500	2.3452	7.4162	166.375	1.7652	3.8030	8.1932	0.1818
5.6	31.3600	2.3664	7.4833	175.616	1.7758	3.8259	8.2426	0.1786
5.7	32.4900	2.3875	7.5498	185.193	1.7863	3.8485	8.2913	0.1754
5.8	33.6400	2.4083	7.6158	195.112	1.7967	3.8709	8.3396	0.1724
5.9	34.8100	2.4290	7.6811	205.379	1.8070	3.8930	8.3872	0.1695
6.0	36.0000	2.4495	7.7460	216.000	1.8171	3.9149	8.4343	0.1667
6.1	37.2100	2.4698	7.8102	226.981	1.8272	3.9365	8.4809	0.1639
6.2	38.4400	2.4900	7.8740	238.328	1.8371	3.9579	8.5270	0.1613
6.3	39.6900	2.5100	7.9372	250.047	1.8469	3.9791	8.5726	0.1587
6.4	40.9600	2.5298	8.0000	262.144	1.8566	4.0000	8.6177	0.1563
6.5	42.2500	2.5495	8.0623	274.625	1.8663	4.0207	8.6624	0.1538
6.6	43.5600	2.5690	8.1240	287.496	1.8758	4.0412	8.7066	0.1515

n	n^2	$\sqrt{n}$	$\sqrt{10n}$	n^3	$\sqrt[3]{n}$	$\sqrt[3]{10n}$	$\sqrt[3]{100n}$	$1/n$
6.7	44.8900	2.5884	8.1854	300.763	1.8852	4.0615	8.7503	0.1493
6.8	46.2400	2.6077	8.2462	314.432	1.8945	4.0817	8.7937	0.1471
6.9	47.6100	2.6268	8.3066	328.509	1.9038	4.1016	8.8366	0.1449
7.0	49.0000	2.6458	8.3666	343.000	1.9129	4.1213	8.8790	0.1429
7.1	50.4100	2.6646	8.4261	357.911	1.9220	4.1408	8.9211	0.1408
7.2	51.8400	2.6833	8.4853	373.248	1.9310	4.1602	8.9628	0.1389
7.3	53.2900	2.7019	8.5440	389.017	1.9399	4.1793	9.0041	0.1370
7.4	54.7600	2.7203	8.6023	405.224	1.9487	4.1983	9.0450	0.1351
7.5	56.2500	2.7386	8.6603	421.875	1.9574	4.2172	9.0856	0.1333
7.6	57.7600	2.7568	8.7178	438.976	1.9661	4.2358	9.1258	0.1316
7.7	59.2900	2.7749	8.7750	456.533	1.9747	4.2543	9.1657	0.1299
7.8	60.8400	2.7928	8.8318	474.552	1.9832	4.2727	9.2052	0.1282
7.9	62.4100	2.8107	8.8882	493.039	1.9916	4.2908	9.2443	0.1266
8.0	64.0000	2.8284	8.9443	512.000	2.0000	4.3089	9.2832	0.1250
8.1	65.6100	2.8460	9.0000	531.441	2.0083	4.3267	9.3217	0.1235
8.2	67.2400	2.8636	9.0554	551.368	2.0165	4.3445	9.3599	0.1220
8.3	68.8900	2.8810	9.1104	571.787	2.0247	4.3621	9.3978	0.1205
8.4	70.5600	2.8983	9.1652	592.704	2.0328	4.3795	9.4354	0.1190
8.5	72.2500	2.9155	9.2195	614.125	2.0408	4.3968	9.4727	0.1176
8.6	73.9600	2.9326	9.2736	636.056	2.0488	4.4140	9.5097	0.1163
8.7	75.6900	2.9496	9.3274	658.503	2.0567	4.4310	9.5464	0.1149
8.8	77.4400	2.9665	9.3808	681.472	2.0646	4.4480	9.5828	0.1136
8.9	79.2100	2.9833	9.4340	704.969	2.0723	4.4647	9.6190	0.1124
9.0	81.000	3.0000	9.4868	729.000	2.0801	4.4814	9.6549	0.1111
9.1	82.8100	3.0166	9.5394	753.571	2.0878	4.4979	9.6905	0.1099
9.2	84.6400	3.0332	9.5917	778.688	2.0954	4.5144	9.7259	0.1087
9.3	86.4900	3.0496	9.6436	804.357	2.1029	4.5307	9.7610	0.1075
9.4	88.3600	3.0659	9.6954	830.584	2.1105	4.5468	9.7959	0.1064
9.5	90.2500	3.0822	9.7468	857.375	2.1179	4.5629	9.8305	0.1053
9.6	92.1600	3.0984	9.7980	884.736	2.1253	4.5789	9.8648	0.1042
9.7	94.0900	3.1145	9.8489	912.673	2.1327	4.5947	9.8990	0.1031
9.8	96.0400	3.1305	9.8995	941.192	2.1400	4.6104	9.9329	0.1020
9.9	98.0100	3.1464	9.9499	970.299	2.1472	4.6261	9.9666	0.1010
10.0	100.000	3.1623	10.000	1000.00	2.1544	4.6416	10.0000	0.1000

TABLE OF e^X AND e^-X

LN = e

this table used when need to e# find = X

x	e^x	e^{-x}	x	e^x	e^{-x}
0.00	1.0000	1.00000	0.15	1.1618	.86071
0.01	1.0101	0.99005	0.16	1.1735	.85214
0.02	1.0202	.98020	0.17	1.1853	.84366
0.03	1.0305	.97045	0.18	1.1972	.83527
0.04	1.0408	.96079	0.19	1.2092	.82696
0.05	1.0513	.95123	0.20	1.2214	.81873
0.06	1.0618	.94176	0.21	1.2337	.81058
0.07	1.0725	.93239	0.22	1.2461	.80252
0.08	1.0833	.92312	0.23	1.2586	.79453
0.09	1.0942	.91393	0.24	1.2712	.78663
0.10	1.1052	.90484	0.25	1.2840	.77880
0.11	1.1163	.89583	0.26	1.2969	.77105
0.12	1.1275	.88692	0.27	1.3100	.76338
0.13	1.1388	.87809	0.28	1.3231	.75578
0.14	1.1503	.86936	0.29	1.3364	.74826

x	e^x	e^{-x}	x	e^x	e^{-x}
0.30	1.3499	.74082	0.70	2.0138	.49659
0.31	1.3634	.73345	0.71	2.0340	.49164
0.32	1.3771	.72615	0.72	2.0544	.48675
0.33	1.3910	.71892	0.73	2.0751	.48191
0.34	1.4049	.71177	0.74	2.0959	.47711
0.35	1.4191	.70469	0.75	2.1170	.47237
0.36	1.4333	.69768	0.76	2.1383	.46767
0.37	1.4477	.69073	0.77	2.1598	.46301
0.38	1.4623	.68386	0.78	2.1815	.45841
0.39	1.4770	.67706	0.79	2.2034	.45384
0.40	1.4918	.67032	0.80	2.2255	.44933
0.41	1.5068	.66365	0.81	2.2479	.44486
0.42	1.5220	.65705	0.82	2.2705	.44043
0.43	1.5373	.65051	0.83	2.2933	.43605
0.44	1.5527	.64404	0.84	2.3164	.43171
0.45	1.5683	.63763	0.85	2.3396	.42741
0.46	1.5841	.63128	0.86	2.3632	.42316
0.47	1.6000	.62500	0.87	2.3869	.41895
0.48	1.6161	.61878	0.88	2.4109	.41478
0.49	1.6323	.61263	0.89	2.4351	.41066
0.50	1.6487	.60653	0.90	2.4596	.40657
0.51	1.6653	.60050	0.91	2.4843	.40252
0.52	1.6820	.59452	0.92	2.5093	.39852
0.53	1.6989	.58860	0.93	2.5345	.39455
0.54	1.7160	.58275	0.94	2.5600	.39063
0.55	1.7333	.57695	0.95	2.5857	.38674
0.56	1.7507	.57121	0.96	2.6117	.38289
0.57	1.7683	.56553	0.97	2.6379	.37908
0.58	1.7860	.55990	0.98	2.6645	.37531
0.59	1.8040	.55433	0.99	2.6912	.37158
0.60	1.8221	.54881	1.00	2.7183	.36788
0.61	1.8404	.54335	1.10	3.0042	.33287
0.62	1.8589	.53794	1.20	3.3201	.30119
0.63	1.8776	.53259	1.30	3.6693	.27253
0.64	1.8965	.52729	1.40	4.0552	.24660
0.65	1.9155	.52205	1.50	4.4817	.22313
0.66	1.9348	.51685	1.60	4.9530	.20190
0.67	1.9542	.51171	1.70	5.4739	.18268
0.68	1.9739	.50662	1.80	6.0496	.16530
0.69	1.9937	.50158	1.90	6.6859	.14957

x	e^x	e^{-x}	x	e^x	e^{-x}
2.00	7.3891	.13534	4.50	90.017	.01111
2.10	8.1662	.12246	4.60	99.484	.01005
2.20	9.0250	.11080	4.70	109.95	.00910
2.30	9.9742	.10026	4.80	121.51	.00823
2.40	11.023	.09072	4.90	134.29	.00745
2.50	12.182	.08208	5.00	148.41	.00674
2.60	13.464	.07427	5.10	164.02	.00610
2.70	14.880	.06721	5.20	181.27	.00552
2.80	16.445	.06081	5.30	200.34	.00499
2.90	18.174	.05502	5.40	221.41	.00452
3.00	20.086	.04979	5.50	244.69	.00409
3.10	22.198	.04505	5.60	270.43	.00370
3.20	24.533	.04076	5.70	298.87	.00335
3.30	27.113	.03688	5.80	330.30	.00303
3.40	29.964	.03337	5.90	365.04	.00274
3.50	33.115	.03020	6.00	403.43	.00248
3.60	36.598	.02732	6.25	518.01	.00193
3.70	40.447	.02472	6.50	665.14	.00150
3.80	44.701	.02237	6.75	854.06	.00117
3.90	49.402	.02024	7.00	1096.6	.00091
4.00	54.598	.01832	7.50	1808.0	.00055
4.10	60.340	.01657	8.00	2981.0	.00034
4.20	66.686	.01500	8.50	4914.8	.00020
4.30	73.700	.01357	9.00	8103.1	.00012
4.40	81.451	.01227	9.50	13360.	.00007
			10.00	22026.	.00005

TABLE OF NATURAL LOGARITHMS

LN = e → use when need to find the power

In the body of the table the first two digits (and decimal point) of most entries are carried over from a preceding entry in the first column. For example, $\ln 3.32 \approx 1.19996$. However, an asterisk (*) indicates that the first two digits are those of a following entry in the first column. For example, $\ln 3.33 \approx 1.20297$.

To extend this table for a number less than 1.0 or greater than 10.09, write the number in the form $x = y \cdot 10^n$ where $1.0 \leqslant y < 10$ and use the fact that $\ln x = \ln y + n \ln 10$. Some values of $n \ln 10$ are

$$1 \ln 10 \approx 2.30259, \qquad 6 \ln 10 \approx 13.81551,$$

$$2 \ln 10 \approx 4.60517, \qquad 7 \ln 10 \approx 16.11810,$$

$$3 \ln 10 \approx 6.90776, \qquad 8 \ln 10 \approx 18.42068,$$

$$4 \ln 10 \approx 9.21034, \qquad 9 \ln 10 \approx 20.72327,$$

$$5 \ln 10 \approx 11.51293, \qquad 10 \ln 10 \approx 23.02585.$$

For example,

$$\ln 332 = \ln\left[(3.32)(10^2)\right] = \ln 3.32 + 2 \ln 10$$
$$\approx 1.19996 + 4.60517 = 5.80513$$

and $$\ln .0332 = \ln\left[(3.32)(10^{-2})\right] = \ln 3.32 - 2 \ln 10$$
$$\approx 1.19996 - 4.60517 = -3.40521.$$

Properties of logarithms may be used to find the logarithm of a number such as $\frac{3}{8}$:

$$\ln \frac{3}{8} = \ln 3 - \ln 8 \approx 1.09861 - 2.07944$$
$$= -.98083.$$

N	0	1	2	3	4	5	6	7	8	9
1.0	0.0 0000	0995	1980	2956	3922	4879	5827	6766	7696	8618
1.1	9531	*0436	*1333	*2222	*3103	*3976	*4842	*5700	*6551	*7395
1.2	0.1 8232	9062	9885	*0701	*1511	*2314	*3111	*3902	*4686	*5464
1.3	0.2 6236	7003	7763	8518	9267	*0010	*0748	*1481	*2208	*2930
1.4	0.3 3647	4359	5066	5767	6464	7156	7844	8526	9204	9878
1.5	0.4 0547	1211	1871	2527	3178	3825	4469	5108	5742	6373
1.6	7000	7623	8243	8858	9470	*0078	*0672	*1282	*1879	*2473
1.7	0.5 3063	3649	4232	4812	5389	5962	6531	7098	7661	8222
1.8	8779	9333	9884	*0432	*0977	*1519	*2058	*2594	*3127	*3658
1.9	0.6 4185	4710	5233	5752	6269	6783	7294	7803	8310	8813
2.0	9315	9813	*0310	*0804	*1295	*1784	*2271	*2755	*3237	*3716
2.1	0.7 4194	4669	5142	5612	6081	6547	7011	7473	7932	8390
2.2	8846	9299	9751	*0200	*0648	*1093	*1536	*1978	*2418	*2855
2.3	0.8 3291	3725	4157	4587	5015	5442	5866	6289	6710	7129
2.4	7547	7963	8377	8789	9200	9609	*0016	*0422	*0826	*1228
2.5	0.9 1629	2028	2426	2822	3216	3609	4001	4391	4779	5166
2.6	5551	5935	6317	6698	7078	7456	7833	8208	8582	8954
2.7	9325	9695	*0063	*0430	*0796	*1160	*1523	*1885	*2245	*2604
2.8	1.0 2962	3318	3674	4028	4380	4732	5082	5431	5779	6126
2.9	6471	6815	7158	7500	7841	8181	8519	8856	9192	9527
3.0	9861	*0194	*0526	*0856	*1186	*1514	*1841	*2168	*2493	*2817
3.1	1.1 3140	3462	3783	4103	4422	4740	5057	5373	5688	6002
3.2	6315	6627	6938	7248	7557	7865	8173	8479	8784	9089
3.3	9392	9695	9996	*0297	*0597	*0896	*1194	*1491	*1788	*2083
3.4	1.2 2378	2671	2964	3256	3547	3837	4127	4415	4703	4990
3.5	5276	5562	5846	6130	6413	6695	6976	7257	7536	7815
3.6	8093	8371	8647	8923	9198	9473	9746	*0019	*0291	*0563
3.7	1.3 0833	1103	1372	1641	1909	2176	2442	2708	2972	3237
3.8	3500	3763	4025	4286	4547	4807	5067	5325	5584	5841
3.9	6098	6354	6609	6864	7118	7372	7624	7877	8128	8379
4.0	8629	8879	9128	9377	9624	9872	*0118	*0364	*0610	*0854
4.1	1.4 1099	1342	1585	1828	2070	2311	2552	2792	3031	3270
4.2	3508	3746	3984	4220	4456	4692	4927	5161	5395	5629
4.3	5862	6094	6326	6557	6787	7018	7247	7476	7705	7933
4.4	8160	8387	8614	8840	9065	9290	9515	9739	9962	*0185
4.5	1.5 0408	0630	0851	1072	1293	1513	1732	1951	2170	2388
4.6	2606	2823	3039	3256	3471	3687	3902	4116	4330	4543
4.7	4756	4969	5181	5393	5604	5814	6025	6235	6444	6653
4.8	6862	7070	7277	7485	7691	7898	8104	8309	8515	8719
4.9	8924	9127	9331	9534	9737	9939	*0141	*0342	*0543	*0744
5.0	1.6 0944	1144	1343	1542	1741	1939	2137	2334	2531	2728
5.1	2924	3120	3315	3511	3705	3900	4094	4287	4481	4673
5.2	4866	5058	5250	5441	5632	5823	6013	6203	6393	6582
5.3	6771	6959	7147	7335	7523	7710	7896	8083	8269	8455
5.4	8640	8825	9010	9194	9378	9562	9745	9928	*0111	*0293
N	0	1	2	3	4	5	6	7	8	9

N	0	1	2	3	4	5	6	7	8	9
5.5	1.7 0475	0656	0838	1019	1199	1380	1560	1740	1919	2098
5.6	2277	2455	2633	2811	2988	3166	3342	3519	3695	3871
5.7	4047	4222	4397	4572	4746	4920	5094	5267	5440	5613
5.8	5786	5958	6130	6302	6473	6644	6815	6985	7156	7326
5.9	7495	7665	7834	8002	8171	8339	8507	8675	8842	9009
6.0	1.7 9176	9342	9509	9675	9840	*0006	*0171	*0336	*0500	*0665
6.1	1.8 0829	0993	1156	1319	1482	1645	1808	1970	2132	2294
6.2	2455	2616	2777	2938	3098	3258	3418	3578	3737	3896
6.3	4055	4214	4372	4530	4688	4845	5003	5160	5317	5473
6.4	5630	5786	5942	6097	6253	6408	6563	6718	6872	7026
6.5	7180	7334	7487	7641	7794	7947	8099	8251	8403	8555
6.6	8707	8858	9010	9160	9311	9462	9612	9762	9912	*0061
6.7	1.9 0211	0360	0509	0658	0806	0954	1102	1250	1398	1545
6.8	1692	1839	1986	2132	2279	2425	2571	2716	2862	3007
6.9	3152	3297	3442	3586	3730	3874	4018	4162	4305	4448
7.0	4591	4734	4876	5019	5161	5303	5445	5586	5727	5869
7.1	6009	6150	6291	6431	6571	6711	6851	6991	7130	7269
7.2	7408	7547	7685	7824	7962	8100	8238	8376	8513	8650
7.3	8787	8924	9061	9198	9334	9470	9606	9742	9877	*0013
7.4	2.0 0148	0283	0418	0553	0687	0821	0956	1089	1223	1357
7.5	1490	1624	1757	1890	2022	2155	2287	2419	2551	2683
7.6	2815	2946	3078	3209	3340	3471	3601	3732	3862	3992
7.7	4122	4252	4381	4511	4640	4769	4898	5027	5156	5284
7.8	5412	5540	5668	5796	5924	6051	6179	6306	6433	6560
7.9	6686	6813	6939	7065	7191	7317	7443	7568	7694	7819
8.0	7944	8069	8194	8318	8443	8567	8691	8815	8939	9063
8.1	9186	9310	9433	9556	9679	9802	9924	*0047	*0169	*0291
8.2	2.1 0413	0535	0657	0779	0900	1021	1142	1263	1384	1505
8.3	1626	1746	1866	1986	2106	2226	2346	2465	2585	2704
8.4	2823	2942	3061	3180	3298	3417	3535	3653	3771	3889
8.5	4007	4124	4242	4359	4476	4593	4710	4827	4943	5060
8.6	5176	5292	5409	5524	5640	5756	5871	5987	6102	6217
8.7	6332	6447	6562	6677	6791	6905	7020	7134	7248	7361
8.8	7475	7589	7702	7816	7929	8042	8155	8267	8380	8493
8.9	8605	8717	8830	8942	9054	9165	9277	9389	9500	9611
9.0	9722	9834	9944	*0055	*0166	*0276	*0387	*0497	*0607	*0717
9.1	2.2 0827	0937	1047	1157	1266	1375	1485	1594	1703	1812
9.2	1920	2029	2138	2246	2354	2462	2570	2678	2786	2894
9.3	3001	3109	3216	3324	3431	3538	3645	3751	3858	3965
9.4	4071	4177	4284	4390	4496	4601	4707	4813	4918	5024
9.5	5129	5234	5339	5444	5549	5654	5759	5863	5968	6072
9.6	6176	6280	6384	6488	6592	6696	6799	6903	7006	7109
9.7	7213	7316	7419	7521	7624	7727	7829	7932	8034	8136
9.8	8238	8340	8442	8544	8646	8747	8849	8950	9051	9152
9.9	9253	9354	9455	9556	9657	9757	9858	9958	*0058	*0158
10.0	2.3 0259	0358	0458	0558	0658	0757	0857	0956	1055	1154
N	0	1	2	3	4	5	6	7	8	9

COMPOUND
INTEREST TABLES

D

n	$(1 + r)^n$	$(1 + r)^{-n}$	$a_{\overline{n}|r}$	$s_{\overline{n}|r}$
1	1.005000	0.995025	0.995025	1.000000
2	1.010025	0.990075	1.985099	2.005000
3	1.015075	0.985149	2.970248	3.015025
4	1.020151	0.980248	3.950496	4.030100
5	1.025251	0.975371	4.925866	5.050251
6	1.030378	0.970518	5.896384	6.075502
7	1.035529	0.965690	6.862074	7.105879
8	1.040707	0.960885	7.822959	8.141409
9	1.045911	0.956105	8.779064	9.182116
10	1.051140	0.951348	9.730412	10.228026
11	1.056396	0.946615	10.677027	11.279167
12	1.061678	0.941905	11.618932	12.335562
13	1.066986	0.937219	12.556151	13.397240
14	1.072321	0.932556	13.488708	14.464226
15	1.077683	0.927917	14.416625	15.536548
16	1.083071	0.923300	15.339925	16.614230
17	1.088487	0.918707	16.258632	17.697301
18	1.093929	0.914136	17.172768	18.785788
19	1.099399	0.909588	18.082356	19.879717
20	1.104896	0.905063	18.987419	20.979115
21	1.110420	0.900560	19.887979	22.084011
22	1.115972	0.896080	20.784059	23.194431
23	1.121552	0.891622	21.675681	24.310403
24	1.127160	0.887186	22.562866	25.431955
25	1.132796	0.882772	23.445638	26.559115
26	1.138460	0.878380	24.324018	27.691911
27	1.144152	0.874010	25.198028	28.830370
28	1.149873	0.869662	26.067689	29.974522
29	1.155622	0.865335	26.933024	31.124395
30	1.161400	0.861030	27.794054	32.280017
31	1.167207	0.856746	28.650800	33.441417
32	1.173043	0.852484	29.503284	34.608624
33	1.178908	0.848242	30.351526	35.781667
34	1.184803	0.844022	31.195548	36.960575
35	1.190727	0.839823	32.035371	38.145378
36	1.196681	0.835645	32.871016	39.336105
37	1.202664	0.831487	33.702504	40.532785
38	1.208677	0.827351	34.529854	41.735449
39	1.214721	0.823235	35.353089	42.944127
40	1.220794	0.819139	36.172228	44.158847
41	1.226898	0.815064	36.987291	45.379642
42	1.233033	0.811009	37.798300	46.606540
43	1.239198	0.806974	38.605274	47.839572
44	1.245394	0.802959	39.408232	49.078770
45	1.251621	0.798964	40.207196	50.324164
46	1.257879	0.794989	41.002185	51.575785
47	1.264168	0.791034	41.793219	52.833664
48	1.270489	0.787098	42.580318	54.097832
49	1.276842	0.783182	43.363500	55.368321
50	1.283226	0.779286	44.142786	56.645163

n	$(1 + r)^n$	$(1 + r)^{-n}$	$a_{\overline{n}\rceil r}$	$s_{\overline{n}\rceil r}$
1	1.007500	0.992556	0.992556	1.000000
2	1.015056	0.985167	1.977723	2.007500
3	1.022669	0.977833	2.955556	3.022556
4	1.030339	0.970554	3.926110	4.045225
5	1.038067	0.963329	4.889440	5.075565
6	1.045852	0.956158	5.845598	6.113631
7	1.053696	0.949040	6.794638	7.159484
8	1.061599	0.941975	7.736613	8.213180
9	1.069561	0.934963	8.671576	9.274779
10	1.077583	0.928003	9.599580	10.344339
11	1.085664	0.921095	10.520675	11.421922
12	1.093807	0.914238	11.434913	12.507586
13	1.102010	0.907432	12.342345	13.601393
14	1.110276	0.900677	13.243022	14.703404
15	1.118603	0.893973	14.136995	15.813679
16	1.126992	0.887318	15.024313	16.932282
17	1.135445	0.880712	15.905025	18.059274
18	1.143960	0.874156	16.779181	19.194718
19	1.152540	0.867649	17.646830	20.338679
20	1.161184	0.861190	18.508020	21.491219
21	1.169893	0.854779	19.362799	22.652403
22	1.178667	0.848416	20.211215	23.822296
23	1.187507	0.842100	21.053315	25.000963
24	1.196414	0.835831	21.889146	26.188471
25	1.205387	0.829609	22.718755	27.384884
26	1.214427	0.823434	23.542189	28.590271
27	1.223535	0.817304	24.359493	29.804698
28	1.232712	0.811220	25.170713	31.028233
29	1.241957	0.805181	25.975893	32.260945
30	1.251272	0.799187	26.775080	33.502902
31	1.260656	0.793238	27.568318	34.754174
32	1.270111	0.787333	28.355650	36.014830
33	1.279637	0.781472	29.137122	37.284941
34	1.289234	0.775654	29.912776	38.564578
35	1.298904	0.769880	30.682656	39.853813
36	1.308645	0.764149	31.446805	41.152716
37	1.318460	0.758461	32.205266	42.461361
38	1.328349	0.752814	32.958080	43.779822
39	1.338311	0.747210	33.705290	45.108170
40	1.348349	0.741648	34.446938	46.446482
41	1.358461	0.736127	35.183065	47.794830
42	1.368650	0.730647	35.913713	49.153291
43	1.378915	0.725208	36.638921	50.521941
44	1.389256	0.719810	37.358730	51.900856
45	1.399676	0.714451	38.073181	53.290112
46	1.410173	0.709133	38.782314	54.689788
47	1.420750	0.703854	39.486168	56.099961
48	1.431405	0.698614	40.184782	57.520711
49	1.442141	0.693414	40.878195	58.952116
50	1.452957	0.688252	41.566447	60.394257

n	$(1 + r)^n$	$(1 + r)^{-n}$	$a_{\overline{n}|r}$	$s_{\overline{n}|r}$
1	1.010000	0.990099	0.990099	1.000000
2	1.020100	0.980296	1.970395	2.010000
3	1.030301	0.970590	2.940985	3.030100
4	1.040604	0.960980	3.901966	4.060401
5	1.051010	0.951466	4.853431	5.101005
6	1.061520	0.942045	5.795476	6.152015
7	1.072135	0.932718	6.728195	7.213535
8	1.082857	0.923483	7.651678	8.285671
9	1.093685	0.914340	8.566018	9.368527
10	1.104622	0.905287	9.471305	10.462213
11	1.115668	0.896324	10.367628	11.566835
12	1.126825	0.887449	11.255077	12.682503
13	1.138093	0.878663	12.133740	13.809328
14	1.149474	0.869963	13.003703	14.947421
15	1.160969	0.861349	13.865053	16.096896
16	1.172579	0.852821	14.717874	17.257864
17	1.184304	0.844377	15.562251	18.430443
18	1.196147	0.836017	16.398269	19.614748
19	1.208109	0.827740	17.226008	20.810895
20	1.220190	0.819544	18.045553	22.019004
21	1.232392	0.811430	18.856983	23.239194
22	1.244716	0.803396	19.660379	24.471586
23	1.257163	0.795442	20.455821	25.716302
24	1.269735	0.787566	21.243387	26.973465
25	1.282432	0.779768	22.023156	28.243200
26	1.295256	0.772048	22.795204	29.525631
27	1.308209	0.764404	23.559608	30.820888
28	1.321291	0.756836	24.316443	32.129097
29	1.334504	0.749342	25.065785	33.450388
30	1.347849	0.741923	25.807708	34.784892
31	1.361327	0.734577	26.542285	36.132740
32	1.374941	0.727304	27.269589	37.494068
33	1.388690	0.720103	27.989693	38.869009
34	1.402577	0.712973	28.702666	40.257699
35	1.416603	0.705914	29.408580	41.660276
36	1.430769	0.698925	30.107505	43.076878
37	1.445076	0.692005	30.799510	44.507647
38	1.459527	0.685153	31.484663	45.952724
39	1.474123	0.678370	32.163033	47.412251
40	1.488864	0.671653	32.834686	48.886373
41	1.503752	0.665003	33.499689	50.375237
42	1.518790	0.658419	34.158108	51.878989
43	1.533978	0.651900	34.810008	53.397779
44	1.549318	0.645445	35.455454	54.931757
45	1.564811	0.639055	36.094508	56.481075
46	1.580459	0.632728	36.727236	58.045885
47	1.596263	0.626463	37.353699	59.626344
48	1.612226	0.620260	37.973959	61.222608
49	1.628348	0.614119	38.588079	62.834834
50	1.644632	0.608039	39.196118	64.463182

n	$(1 + r)^n$	$(1 + r)^{-n}$	$a_{\overline{n}|r}$	$s_{\overline{n}|r}$
1	1.012500	0.987654	0.987654	1.000000
2	1.025156	0.975461	1.963115	2.012500
3	1.037971	0.963418	2.926534	3.037656
4	1.050945	0.951524	3.878058	4.075627
5	1.064082	0.939777	4.817835	5.126572
6	1.077383	0.928175	5.746010	6.190654
7	1.090850	0.916716	6.662726	7.268038
8	1.104486	0.905398	7.568124	8.358888
9	1.118292	0.894221	8.462345	9.463374
10	1.132271	0.883181	9.345526	10.581666
11	1.146424	0.872277	10.217803	11.713937
12	1.160755	0.861509	11.079312	12.860361
13	1.175264	0.850873	11.930185	14.021116
14	1.189955	0.840368	12.770553	15.196380
15	1.204829	0.829993	13.600546	16.386335
16	1.219890	0.819746	14.420292	17.591164
17	1.235138	0.809626	15.229918	18.811053
18	1.250577	0.799631	16.029549	20.046192
19	1.266210	0.789759	16.819308	21.296769
20	1.282037	0.780009	17.599316	22.562979
21	1.298063	0.770379	18.369695	23.845016
22	1.314288	0.760868	19.130563	25.143078
23	1.330717	0.751475	19.882037	26.457367
24	1.347351	0.742197	20.624235	27.788084
25	1.364193	0.733034	21.357269	29.135435
26	1.381245	0.723984	22.081253	30.499628
27	1.398511	0.715046	22.796299	31.880873
28	1.415992	0.706219	23.502518	33.279384
29	1.433692	0.697500	24.200018	34.695377
30	1.451613	0.688889	24.888906	36.129069
31	1.469759	0.680384	25.569290	37.580682
32	1.488131	0.671984	26.241274	39.050441
33	1.506732	0.663688	26.904962	40.538571
34	1.525566	0.655494	27.560456	42.045303
35	1.544636	0.647402	28.207858	43.570870
36	1.563944	0.639409	28.847267	45.115505
37	1.583493	0.631515	29.478783	46.679449
38	1.603287	0.623719	30.102501	48.262942
39	1.623328	0.616019	30.718520	49.866229
40	1.643619	0.608413	31.326933	51.489557
41	1.664165	0.600902	31.927835	53.133177
42	1.684967	0.593484	32.521319	54.797341
43	1.706029	0.586157	33.107475	56.482308
44	1.727354	0.578920	33.686395	58.188337
45	1.748946	0.571773	34.258168	59.915691
46	1.770808	0.564714	34.822882	61.664637
47	1.792943	0.557742	35.380624	63.435445
48	1.815355	0.550856	35.931481	65.228388
49	1.838047	0.544056	36.475537	67.043743
50	1.861022	0.537339	37.012876	68.881790

n	$(1 + r)^n$	$(1 + r)^{-n}$	$a_{\overline{n}|r}$	$s_{\overline{n}|r}$
1	1.015000	0.985222	0.985222	1.000000
2	1.030225	0.970662	1.955883	2.015000
3	1.045678	0.956317	2.912200	3.045225
4	1.061364	0.942184	3.854385	4.090903
5	1.077284	0.928260	4.782645	5.152267
6	1.093443	0.914542	5.697187	6.229551
7	1.109845	0.901027	6.598214	7.322994
8	1.126493	0.887711	7.485925	8.432839
9	1.143390	0.874592	8.360517	9.559332
10	1.160541	0.861667	9.222185	10.702722
11	1.177949	0.848933	10.071118	11.863262
12	1.195618	0.836387	10.907505	13.041211
13	1.213552	0.824027	11.731532	14.236830
14	1.231756	0.811849	12.543382	15.450382
15	1.250232	0.799852	13.343233	16.682138
16	1.268986	0.788031	14.131264	17.932370
17	1.288020	0.776385	14.907649	19.201355
18	1.307341	0.764912	15.672561	20.489376
19	1.326951	0.753607	16.426168	21.796716
20	1.346855	0.742470	17.168639	23.123667
21	1.367058	0.731498	17.900137	24.470522
22	1.387564	0.720688	18.620824	25.837580
23	1.408377	0.710037	19.330861	27.225144
24	1.429503	0.699544	20.030405	28.633521
25	1.450945	0.689206	20.719611	30.063024
26	1.472710	0.679021	21.398632	31.513969
27	1.494800	0.668986	22.067617	32.986678
28	1.517222	0.659099	22.726717	34.481479
29	1.539981	0.649359	23.376076	35.998701
30	1.563080	0.639762	24.015838	37.538681
31	1.586526	0.630308	24.646146	39.101762
32	1.610324	0.620993	25.267139	40.688288
33	1.634479	0.611816	25.878954	42.298612
34	1.658996	0.602774	26.481728	43.933092
35	1.683881	0.593866	27.075595	45.592088
36	1.709140	0.585090	27.660684	47.275969
37	1.734777	0.576443	28.237127	48.985109
38	1.760798	0.567924	28.805052	50.719885
39	1.787210	0.559531	29.364583	52.480684
40	1.814018	0.551262	29.915845	54.267894
41	1.841229	0.543116	30.458961	56.081912
42	1.868847	0.535089	30.994050	57.923141
43	1.896880	0.527182	31.521232	59.791988
44	1.925333	0.519391	32.040622	61.688868
45	1.954213	0.511715	32.552337	63.614201
46	1.983526	0.504153	33.056490	65.568414
47	2.013279	0.496702	33.553192	67.551940
48	2.043478	0.489362	34.042554	69.565219
49	2.074130	0.482130	34.524683	71.608698
50	2.105242	0.475005	34.999688	73.682828

n	$(1 + r)^n$	$(1 + r)^{-n}$	$a_{\overline{n}|r}$	$s_{\overline{n}|r}$
1	1.020000	0.980392	0.980392	1.000000
2	1.040400	0.961169	1.941561	2.020000
3	1.061208	0.942322	2.883883	3.060400
4	1.082432	0.923845	3.807729	4.121608
5	1.104081	0.905731	4.713460	5.204040
6	1.126162	0.887971	5.601431	6.308121
7	1.148686	0.870560	6.471991	7.434283
8	1.171659	0.853490	7.325481	8.582969
9	1.195093	0.836755	8.162237	9.754628
10	1.218994	0.820348	8.982585	10.949721
11	1.243374	0.804263	9.786848	12.168715
12	1.268242	0.788493	10.575341	13.412090
13	1.293607	0.773033	11.348374	14.680332
14	1.319479	0.757875	12.106249	15.973938
15	1.345868	0.743015	12.849264	17.293417
16	1.372786	0.728446	13.577709	18.639285
17	1.400241	0.714163	14.291872	20.012071
18	1.428246	0.700159	14.992031	21.412312
19	1.456811	0.686431	15.678462	22.840559
20	1.485947	0.672971	16.351433	24.297370
21	1.515666	0.659776	17.011209	25.783317
22	1.545980	0.646839	17.658048	27.298984
23	1.576899	0.634156	18.292204	28.844963
24	1.608437	0.621721	18.913926	30.421862
25	1.640606	0.609531	19.523456	32.030300
26	1.673418	0.597579	20.121036	33.670906
27	1.706886	0.585862	20.706898	35.344324
28	1.741024	0.574375	21.281272	37.051210
29	1.775845	0.563112	21.844385	38.792235
30	1.811362	0.552071	22.396456	40.568079
31	1.847589	0.541246	22.937702	42.379441
32	1.884541	0.530633	23.468335	44.227030
33	1.922231	0.520229	23.988564	46.111570
34	1.960676	0.510028	24.498592	48.033802
35	1.999890	0.500028	24.998619	49.994478
36	2.039887	0.490223	25.488842	51.994367
37	2.080685	0.480611	25.969453	54.034255
38	2.122299	0.471187	26.440641	56.114940
39	2.164745	0.461948	26.902589	58.237238
40	2.208040	0.452890	27.355479	60.401983
41	2.252200	0.444010	27.799489	62.610023
42	2.297244	0.435304	28.234794	64.862223
43	2.343189	0.426769	28.661562	67.159468
44	2.390053	0.418401	29.079963	69.502657
45	2.437854	0.410197	29.490160	71.892710
46	2.486611	0.402154	29.892314	74.330564
47	2.536344	0.394268	30.286582	76.817176
48	2.587070	0.386538	30.673120	79.353519
49	2.638812	0.378958	31.052078	81.940590
50	2.691588	0.371528	31.423606	84.579401

n	$(1 + r)^n$	$(1 + r)^{-n}$	$a_{\overline{n}|r}$	$s_{\overline{n}|r}$
1	1.025000	0.975610	0.975610	1.000000
2	1.050625	0.951814	1.927424	2.025000
3	1.076891	0.928599	2.856024	3.075625
4	1.103813	0.905951	3.761974	4.152516
5	1.131408	0.883854	4.645828	5.256329
6	1.159693	0.862297	5.508125	6.387737
7	1.188686	0.841265	6.349391	7.547430
8	1.218403	0.820747	7.170137	8.736116
9	1.248863	0.800728	7.970866	9.954519
10	1.280085	0.781198	8.752064	11.203382
11	1.312087	0.762145	9.514209	12.483466
12	1.344889	0.743556	10.257765	13.795553
13	1.378511	0.725420	10.983185	15.140442
14	1.412974	0.707727	11.690912	16.518953
15	1.448298	0.690466	12.381378	17.931927
16	1.484506	0.673625	13.055003	19.380225
17	1.521618	0.657195	13.712198	20.864730
18	1.559659	0.641166	14.353364	22.386349
19	1.598650	0.625528	14.978891	23.946007
20	1.638616	0.610271	15.589162	25.544658
21	1.679582	0.595386	16.184549	27.183274
22	1.721571	0.580865	16.765413	28.862856
23	1.764611	0.566697	17.332110	30.584427
24	1.808726	0.552875	17.884986	32.349038
25	1.853944	0.539391	18.424376	34.157764
26	1.900293	0.526235	18.950611	36.011708
27	1.947800	0.513400	19.464011	37.912001
28	1.996495	0.500878	19.964889	39.859801
29	2.046407	0.488661	20.453550	41.856296
30	2.097568	0.476743	20.930293	43.902703
31	2.150007	0.465115	21.395407	46.000271
32	2.203757	0.453771	21.849178	48.150278
33	2.258851	0.442703	22.291881	50.354034
34	2.315322	0.431905	22.723786	52.612885
35	2.373205	0.421371	23.145157	54.928207
36	2.432535	0.411094	23.556251	57.301413
37	2.493349	0.401067	23.957318	59.733948
38	2.555682	0.391285	24.348603	62.227297
39	2.619574	0.381741	24.730344	64.782979
40	2.685064	0.372431	25.102775	67.402554
41	2.752190	0.363347	25.466122	70.087617
42	2.820995	0.354485	25.820607	72.839808
43	2.891520	0.345839	26.166446	75.660803
44	2.963808	0.337404	26.503849	78.552323
45	3.037903	0.329174	26.833024	81.516131
46	3.113851	0.321146	27.154170	84.554034
47	3.191697	0.313313	27.467483	87.667885
48	3.271490	0.305671	27.773154	90.859582
49	3.353277	0.298216	28.071369	94.131072
50	3.437109	0.290942	28.362312	97.484349

n	$(1 + r)^n$	$(1 + r)^{-n}$	$a_{\overline{n}|r}$	$s_{\overline{n}|r}$
1	1.030000	0.970874	0.970874	1.000000
2	1.060900	0.942596	1.913470	2.030000
3	1.092727	0.915142	2.828611	3.090900
4	1.125509	0.888487	3.717098	4.183627
5	1.159274	0.862609	4.579707	5.309136
6	1.194052	0.837484	5.417191	6.468410
7	1.229874	0.813092	6.230283	7.662462
8	1.266770	0.789409	7.019692	8.892336
9	1.304773	0.766417	7.786109	10.159106
10	1.343916	0.744094	8.530203	11.463879
11	1.384234	0.722421	9.252624	12.807796
12	1.425761	0.701380	9.954004	14.192030
13	1.468534	0.680951	10.634955	15.617790
14	1.512590	0.661118	11.296073	17.086324
15	1.557967	0.641862	11.937935	18.598914
16	1.604706	0.623167	12.561102	20.156881
17	1.652848	0.605016	13.166118	21.761588
18	1.702433	0.587395	13.753513	23.414435
19	1.753506	0.570286	14.323799	25.116868
20	1.806111	0.553676	14.877475	26.870374
21	1.860295	0.537549	15.415024	28.676486
22	1.916103	0.521893	15.936917	30.536780
23	1.973587	0.506692	16.443608	32.452884
24	2.032794	0.491934	16.935542	34.426470
25	2.093778	0.477606	17.413148	36.459264
26	2.156591	0.463695	17.876842	38.553042
27	2.221289	0.450189	18.327031	40.709634
28	2.287928	0.437077	18.764108	42.930923
29	2.356566	0.424346	19.188455	45.218850
30	2.427262	0.411987	19.600441	47.575416
31	2.500080	0.399987	20.000428	50.002678
32	2.575083	0.388337	20.388766	52.502759
33	2.652335	0.377026	20.765792	55.077841
34	2.731905	0.366045	21.131837	57.730177
35	2.813862	0.355383	21.487220	60.462082
36	2.898278	0.345032	21.832252	63.275944
37	2.985227	0.334983	22.167235	66.174223
38	3.074783	0.325226	22.492462	69.159449
39	3.167027	0.315754	22.808215	72.234233
40	3.262038	0.306557	23.114772	75.401260
41	3.359899	0.297628	23.412400	78.663298
42	3.460696	0.288959	23.701359	82.023196
43	3.564517	0.280543	23.981902	85.483892
44	3.671452	0.272372	24.254274	89.048409
45	3.781596	0.264439	24.518713	92.719861
46	3.895044	0.256737	24.775449	96.501457
47	4.011895	0.249259	25.024708	100.396501
48	4.132252	0.241999	25.266707	104.408396
49	4.256219	0.234950	25.501657	108.540648
50	4.383906	0.228107	25.729764	112.796867

n	$(1 + r)^n$	$(1 + r)^{-n}$	$a_{\overline{n}\rvert r}$	$s_{\overline{n}\rvert r}$
1	1.035000	0.966184	0.966184	1.000000
2	1.071225	0.933511	1.899694	2.035000
3	1.108718	0.901943	2.801637	3.106225
4	1.147523	0.871442	3.673079	4.214943
5	1.187686	0.841973	4.515052	5.362466
6	1.229255	0.813501	5.328553	6.550152
7	1.272279	0.785991	6.114544	7.779408
8	1.316809	0.759412	6.873956	9.051687
9	1.362897	0.733731	7.607687	10.368496
10	1.410599	0.708919	8.316605	11.731393
11	1.459970	0.684946	9.001551	13.141992
12	1.511069	0.661783	9.663334	14.601962
13	1.563956	0.639404	10.302738	16.113030
14	1.618695	0.617782	10.920520	17.676986
15	1.675349	0.596891	11.517411	19.295681
16	1.733986	0.576706	12.094117	20.971030
17	1.794676	0.557204	12.651321	22.705016
18	1.857489	0.538361	13.189682	24.499691
19	1.922501	0.520156	13.709837	26.357180
20	1.989789	0.502566	14.212403	28.279682
21	2.059431	0.485571	14.697974	30.269471
22	2.131512	0.469151	15.167125	32.328902
23	2.206114	0.453286	15.620410	34.460414
24	2.283328	0.437957	16.058368	36.666528
25	2.363245	0.423147	16.481515	38.949857
26	2.445959	0.408838	16.890352	41.313102
27	2.531567	0.395012	17.285365	43.759060
28	2.620172	0.381654	17.667019	46.290627
29	2.711878	0.368748	18.035767	48.910799
30	2.806794	0.356278	18.392045	51.622677
31	2.905031	0.344230	18.736276	54.429471
32	3.006708	0.332590	19.068865	57.334502
33	3.111942	0.321343	19.390208	60.341210
34	3.220860	0.310476	19.700684	63.453152
35	3.333590	0.299977	20.000661	66.674013
36	3.450266	0.289833	20.290494	70.007603
37	3.571025	0.280032	20.570525	73.457869
38	3.696011	0.270562	20.841087	77.028895
39	3.825372	0.261413	21.102500	80.724906
40	3.959260	0.252572	21.355072	84.550278
41	4.097834	0.244031	21.599104	88.509537
42	4.241258	0.235779	21.834883	92.607371
43	4.389702	0.227806	22.062689	96.848629
44	4.543342	0.220102	22.282791	101.238331
45	4.702359	0.212659	22.495450	105.781673
46	4.866941	0.205468	22.700918	110.484031
47	5.037284	0.198520	22.899438	115.350973
48	5.213589	0.191806	23.091244	120.388257
49	5.396065	0.185320	23.276564	125.601846
50	5.584927	0.179053	23.455618	130.997910

n	$(1 + r)^n$	$(1 + r)^{-n}$	$a_{\overline{n}\rceil r}$	$s_{\overline{n}\rceil r}$
1	1.040000	0.961538	0.961538	1.000000
2	1.081600	0.924556	1.886095	2.040000
3	1.124864	0.888996	2.775091	3.121600
4	1.169859	0.854804	3.629895	4.246464
5	4.216653	0.821927	4.451822	5.416323
6	1.265319	0.790315	5.242137	6.632975
7	1.315932	0.759918	6.002055	7.898294
8	1.368569	0.730690	6.732745	9.214226
9	1.423312	0.702587	7.435332	10.582795
10	1.480244	0.675564	8.110896	12.006107
11	1.539454	0.649581	8.760477	13.486351
12	1.601032	0.624597	9.385074	15.025805
13	1.665074	0.600574	9.985648	16.626838
14	1.731676	0.577475	10.563123	18.291911
15	1.800944	0.555265	11.118387	20.023588
16	1.872981	0.533908	11.652296	21.824531
17	1.947900	0.513373	12.165669	23.697512
18	2.025817	0.493628	12.659297	25.645413
19	2.106849	0.474642	13.133939	27.671229
20	2.191123	0.456387	13.590326	29.778079
21	2.278768	0.438834	14.029160	31.969202
22	2.369919	0.421955	14.451115	34.247970
23	2.464716	0.405726	14.856842	36.617889
24	2.563304	0.390121	15.246963	39.082604
25	2.665836	0.375117	15.622080	41.645908
26	2.772470	0.360689	15.982769	44.311745
27	2.883369	0.346817	16.329586	47.084214
28	2.998703	0.333477	16.663063	49.967583
29	3.118651	0.320651	16.983715	52.966286
30	3.243398	0.308319	17.292033	56.084938
31	3.373133	0.296460	17.588494	59.328335
32	3.508059	0.285058	17.873551	62.701469
33	3.648381	0.274094	18.147646	66.209527
34	3.794316	0.263552	18.411198	69.857909
35	3.946089	0.253415	18.664613	73.652225
36	4.103933	0.243669	18.908282	77.598314
37	4.268090	0.234297	19.142579	81.702246
38	4.438813	0.225285	19.367864	85.970336
39	4.616366	0.216621	19.584485	90.409150
40	4.801021	0.208289	19.792774	95.025516
41	4.993061	0.200278	19.993052	99.826536
42	5.192784	0.192575	20.185627	104.819598
43	5.400495	0.185168	20.370795	110.012382
44	5.616515	0.178046	20.548841	115.412877
45	5.841176	0.171198	20.720040	121.029392
46	6.074823	0.164614	20.884654	126.870568
47	6.317816	0.158283	21.042936	132.945390
48	6.570528	0.152195	21.195131	139.263206
49	6.833349	0.146341	21.341472	145.833734
50	7.106683	0.140713	21.482185	152.667084

n	$(1 + r)^n$	$(1 + r)^{-n}$	$a_{\overline{n}\rceil r}$	$s_{\overline{n}\rceil r}$
1	1.050000	0.952381	0.952381	1.000000
2	1.102500	0.907029	1.859410	2.050000
3	1.157625	0.863838	2.723248	3.152500
4	1.215506	0.822702	3.545951	4.310125
5	1.276282	0.783526	4.329477	5.525631
6	1.340096	0.746215	5.075692	6.801913
7	1.407100	0.710681	5.786373	8.142008
8	1.477455	0.676839	6.463213	3.549109
9	1.551328	0.644609	7.107822	11.026564
10	1.628895	0.613913	7.721735	12.577893
11	1.710339	0.584679	8.306414	14.206787
12	1.795856	0.556837	8.863252	15.917127
13	1.885649	0.530321	9.393573	17.712983
14	1.979932	0.505068	9.898641	19.598632
15	2.078928	0.481017	10.379658	21.578564
16	2.182875	0.458112	10.837770	23.657492
17	2.292018	0.436297	11.274066	25.840366
18	2.406619	0.415521	11.689587	28.132385
19	2.526950	0.395734	12.085321	30.539004
20	2.653298	0.376889	12.462210	33.065954
21	2.785963	0.358942	12.821153	35.719252
22	2.925261	0.341850	13.163003	38.505214
23	3.071524	0.325571	13.488574	41.430475
24	3.225100	0.310068	13.798642	44.501999
25	3.386355	0.295303	14.093945	47.727099
26	3.555673	0.281241	14.375185	51.113454
27	3.733456	0.267848	14.643034	54.669126
28	3.920129	0.255094	14.898127	58.402583
29	4.116136	0.242946	15.141074	62.322712
30	4.321942	0.231377	15.372451	66.438848
31	4.538039	0.220359	15.592811	70.760790
32	4.764941	0.209866	15.802677	75.298829
33	5.003189	0.199873	16.002549	80.063771
34	5.253348	0.190355	16.192904	85.066959
35	5.516015	0.181290	16.374194	90.320307
36	5.791816	0.172657	16.546852	95.836323
37	6.081407	0.164436	16.711287	101.628139
38	6.385477	0.156605	16.867893	107.709546
39	6.704751	0.149148	17.017041	114.095023
40	7.039989	0.142046	17.159086	120.799774
41	7.391988	0.135282	17.294368	127.839763
42	7.761588	0.128840	17.423208	135.231751
43	8.149667	0.122704	17.545912	142.993339
44	8.557150	0.116861	17.662773	151.143006
45	8.985008	0.111297	17.774070	159.700156
46	9.434258	0.105997	17.880066	168.685164
47	9.905971	0.100949	17.981016	178.119422
48	10.401270	0.096142	18.077158	188.025393
49	10.921333	0.091564	18.168722	198.426663
50	11.467400	0.087204	18.255925	209.347996

n	$(1 + r)^n$	$(1 + r)^{-n}$	$a_{\overline{n}|\,r}$	$s_{\overline{n}|\,r}$
1	1.060000	0.943396	0.943396	1.000000
2	1.123600	0.889996	1.833393	2.060000
3	1.191016	0.839619	2.673012	3.183600
4	1.262477	0.792094	3.465106	4.374616
5	1.338226	0.747258	4.212364	5.637093
6	1.418519	0.704961	4.917324	6.975319
7	1.503630	0.665057	5.582381	8.393838
8	1.593848	0.627412	6.209794	9.897468
9	1.689479	0.591898	6.801692	11.491316
10	1.790848	0.558395	7.360087	13.180795
11	1.898299	0.526788	7.886875	14.971643
12	2.012196	0.496969	8.383844	16.869941
13	2.132928	0.468839	8.852683	18.882138
14	2.260904	0.442301	9.294984	21.015066
15	2.396558	0.417265	9.712249	23.275970
16	2.540352	0.393646	10.105895	25.672528
17	2.692773	0.371364	10.477260	28.212880
18	2.854339	0.350344	10.827603	30.905653
19	3.025600	0.330513	11.158116	33.759992
20	3.207135	0.311805	11.469921	36.785591
21	3.399564	0.294155	11.764077	39.992727
22	3.603537	0.277505	12.041582	43.392290
23	3.819750	0.261797	12.303379	46.995828
24	4.048935	0.246979	12.550358	50.815577
25	4.291871	0.232999	12.783356	54.864512
26	4.549383	0.219810	13.003166	59.156383
27	4.822346	0.207368	13.210534	63.705766
28	5.111687	0.195630	13.406164	68.528112
29	5.418388	0.184557	13.590721	73.639798
30	5.743491	0.174110	13.764831	79.058186
31	6.088101	0.164255	13.929086	84.801677
32	6.453387	0.154957	14.084043	90.889778
33	6.840590	0.146186	14.230230	97.343165
34	7.251025	0.137912	14.368141	104.183755
35	7.686087	0.130105	14.498246	111.434780
36	8.147252	0.122741	14.620987	119.120867
37	8.636087	0.115793	14.736780	127.268119
38	9.154252	0.109239	14.846019	135.904206
39	9.703507	0.103056	14.949075	145.058458
40	10.285718	0.097222	15.046297	154.761966
41	10.902861	0.091719	15.138016	165.047684
42	11.557033	0.086527	15.224543	175.950545
43	12.250455	0.081630	15.306173	187.507577
44	12.985482	0.077009	15.383182	199.758032
45	13.764611	0.072650	15.455832	212.743514
46	14.590487	0.068538	15.524370	226.508125
47	15.465917	0.064658	15.589028	241.098612
48	16.393872	0.060998	15.650027	256.564529
49	17.377504	0.057546	15.707572	272.958401
50	18.420154	0.054288	15.761861	290.335905

n	$(1 + r)^n$	$(1 + r)^{-n}$	$a_{\overline{n}|r}$	$s_{\overline{n}|r}$
1	1.070000	0.934579	0.934579	1.000000
2	1.144900	0.873439	1.808018	2.070000
3	1.225043	0.816298	2.624316	3.214900
4	1.310796	0.762895	3.387211	4.439943
5	1.402552	0.712986	4.100197	5.750739
6	1.500730	0.666342	4.766540	7.153291
7	1.605781	0.622750	5.389289	8.654021
8	1.718186	0.582009	5.971299	10.259803
9	1.838459	0.543934	6.515232	11.977989
10	1.967151	0.508349	7.023582	13.816448
11	2.104852	0.475093	7.498674	15.783599
12	2.252192	0.444012	7.942686	17.888451
13	2.409845	0.414964	8.357651	20.140643
14	2.578534	0.387817	8.745468	22.550488
15	2.759032	0.362446	9.107914	25.129022
16	2.952164	0.338735	9.446649	27.888054
17	3.158815	0.316574	9.763223	30.840217
18	3.379932	0.295864	10.059087	33.999033
19	3.616528	0.276508	10.335595	37.378965
20	3.869684	0.258419	10.594014	40.995492
21	4.140562	0.241513	10.835527	44.865177
22	4.430402	0.225713	11.061240	49.005739
23	4.740530	0.210947	11.272187	53.436141
24	5.072367	0.197147	11.469334	58.176671
25	5.427433	0.184249	11.653583	63.249038
26	5.807353	0.172195	11.825779	68.676470
27	6.213868	0.160930	11.986709	74.483823
28	6.648838	0.150402	12.137111	80.697691
29	7.114257	0.140563	12.277674	87.346529
30	7.612255	0.131367	12.409041	94.460786
31	8.145113	0.122773	12.531814	102.073041
32	8.715271	0.114741	12.646555	110.218154
33	9.325340	0.107235	12.753790	118.933425
34	9.978114	0.100219	12.854009	128.258765
35	10.676581	0.093663	12.947672	138.236878
36	11.423942	0.087535	13.035208	148.913460
37	12.223618	0.081809	13.117017	160.337402
38	13.079271	0.076457	13.193473	172.561020
39	13.994820	0.071455	13.264928	185.640292
40	14.974458	0.066780	13.331709	199.635112
41	16.022670	0.062412	13.394120	214.609570
42	17.144257	0.058329	13.452449	230.632240
43	18.344355	0.054513	13.506962	247.776496
44	19.628460	0.050946	13.557908	266.120851
45	21.002452	0.047613	13.605522	285.749311
46	22.472623	0.044499	13.650020	306.751763
47	24.045707	0.041587	13.691608	329.224386
48	25.728907	0.038867	13.730474	353.270093
49	27.529930	0.036324	13.766799	378.999000
50	29.457025	0.033948	13.800746	406.528929

n	$(1 + r)^n$	$(1 + r)^{-n}$	$a_{\overline{n}\rceil r}$	$s_{\overline{n}\rceil r}$
1	1.080000	0.925926	0.925926	1.000000
2	1.166400	0.857339	1.783265	2.080000
3	1.259712	0.793832	2.577097	3.246400
4	1.360489	0.735030	3.312127	4.506112
5	1.469328	0.680583	3.992710	5.866601
6	1.586874	0.630170	4.622880	7.335929
7	1.713824	0.583490	5.206370	8.922803
8	1.850930	0.540269	5.746639	10.636628
9	1.999005	0.500249	6.246888	12.487558
10	2.158925	0.463193	6.710081	14.486562
11	2.331639	0.428883	7.138964	16.645487
12	2.518170	0.397114	7.536078	18.977126
13	2.719624	0.367698	7.903776	21.495297
14	2.937194	0.340461	8.244237	24.214920
15	3.172169	0.315242	8.559479	27.152114
16	3.425943	0.291890	8.851369	30.324283
17	3.700018	0.270269	9.121638	33.750226
18	3.996019	0.250249	9.371887	37.450244
19	4.315701	0.231712	9.603599	41.446263
20	4.660957	0.214548	9.818147	45.761964
21	5.033834	0.198656	10.016803	50.422921
22	5.436540	0.183941	10.200744	55.456755
23	5.871464	0.170315	10.371059	60.893296
24	6.341181	0.157699	10.528758	66.764759
25	6.848475	0.146018	10.674776	73.105940
26	7.396353	0.135202	10.809978	79.954415
27	7.988061	0.125187	10.935165	87.350768
28	8.627106	0.115914	11.051078	95.338830
29	9.317275	0.107328	11.158406	103.965936
30	10.062657	0.099377	11.257783	113.283211
31	10.867669	0.092016	11.349799	123.345868
32	11.737083	0.085200	11.434999	134.213537
33	12.676050	0.078889	11.513888	145.950620
34	13.690134	0.073045	11.586934	158.626670
35	14.785344	0.067635	11.654568	172.316804
36	15.968172	0.062625	11.717193	187.102148
37	17.245626	0.057986	11.775179	203.070320
38	18.625276	0.053690	11.828869	220.315945
39	20.115298	0.049713	11.878582	238.941221
40	21.724521	0.046031	11.924613	259.056519
41	23.462483	0.042621	11.967235	280.781040
42	25.339482	0.039464	12.006699	304.243523
43	27.366640	0.036541	12.043240	329.583005
44	29.555972	0.033834	12.077074	356.949646
45	31.920449	0.031328	12.108402	386.505617
46	34.474085	0.029007	12.137409	418.426067
47	37.232012	0.026859	12.164267	452.900152
48	40.210573	0.024869	12.189136	490.132164
49	43.427419	0.023027	12.212163	530.342737
50	46.901613	0.021321	12.233485	573.770156

TABLE OF SELECTED INTEGRALS

RATIONAL FORMS CONTAINING $(a + bu)$.

1. $\displaystyle\int u^n \, du = \frac{u^{n+1}}{n+1} + C, \; n \neq = -1.$

2. $\displaystyle\int \frac{du}{a + bu} = \frac{1}{b} \ln |a + bu| + C.$

3. $\displaystyle\int \frac{u \, du}{a + bu} = \frac{u}{b} - \frac{a}{b^2} \ln |a + bu| + C.$

4. $\displaystyle\int \frac{u^2 \, du}{a + bu} = \frac{u^2}{2b} - \frac{au}{b^2} + \frac{a^2}{b^3} \ln |a + bu| + C.$

5. $\displaystyle\int \frac{du}{u(a + bu)} = \frac{1}{a} \ln \left| \frac{u}{a + bu} \right| + C.$

6. $\displaystyle\int \frac{du}{u^2(a + bu)} = -\frac{1}{au} + \frac{b}{a^2} \ln \left| \frac{a + bu}{u} \right| + C.$

7. $\int \dfrac{u\,du}{(a + bu)^2} = \dfrac{1}{b^2}\left(\ln|a + bu| + \dfrac{a}{a + bu}\right) + C.$

8. $\int \dfrac{u^2\,du}{(a + bu)^2} = \dfrac{u}{b^2} - \dfrac{a^2}{b^3(a + bu)} - \dfrac{2a}{b^3}\ln|a + bu| + C.$

9. $\int \dfrac{du}{u(a + bu)^2} = \dfrac{1}{a(a + bu)} + \dfrac{1}{a^2}\ln\left|\dfrac{u}{a + bu}\right| + C.$

10. $\int \dfrac{du}{u^2(a + bu)^2} = -\dfrac{a + 2bu}{a^2 u(a + bu)} + \dfrac{2b}{a^3}\ln\left|\dfrac{a + bu}{u}\right| + C.$

11. $\int \dfrac{du}{(a + bu)(c + ku)} = \dfrac{1}{bc - ak}\ln\left|\dfrac{a + bu}{c + ku}\right| + C.$

12. $\int \dfrac{u\,du}{(a + bu)(c + ku)} = \dfrac{1}{bc - ak}\left[\dfrac{c}{k}\ln|c + ku| - \dfrac{a}{b}\ln|a + bu|\right] + C.$

FORMS CONTAINING $\sqrt{a + bu}$.

13. $\int u\sqrt{a + bu}\,du = \dfrac{2(3bu - 2a)(a + bu)^{3/2}}{15b^2} + C.$

14. $\int u^2\sqrt{a + bu}\,du = \dfrac{2(8a^2 - 12abu + 15b^2u^2)(a + bu)^{3/2}}{105b^3} + C.$

15. $\int \dfrac{u\,du}{\sqrt{a + bu}} = \dfrac{2(bu - 2a)\sqrt{a + bu}}{3b^2} + C.$

16. $\int \dfrac{u^2\,du}{\sqrt{a + bu}} = \dfrac{2(3b^2u^2 - 4abu + 8a^2)\sqrt{a + bu}}{15b^3} + C.$

17. $\int \dfrac{du}{u\sqrt{a + bu}} = \dfrac{1}{\sqrt{a}}\ln\left|\dfrac{\sqrt{a + bu} - \sqrt{a}}{\sqrt{a + bu} + \sqrt{a}}\right| + C,\ a > 0.$

18. $\int \dfrac{\sqrt{a + bu}\,du}{u} = 2\sqrt{a + bu} + a\int \dfrac{du}{u\sqrt{a + bu}}.$

FORMS CONTAINING $\sqrt{a^2 - u^2}$.

19. $\int \dfrac{du}{(a^2 - u^2)^{3/2}} = \dfrac{u}{a^2\sqrt{a^2 - u^2}} + C.$

20. $\displaystyle\int \frac{du}{u\sqrt{a^2 - u^2}} = -\frac{1}{a}\ln\left|\frac{a + \sqrt{a^2 - u^2}}{u}\right| + C.$

21. $\displaystyle\int \frac{du}{u^2\sqrt{a^2 - u^2}} = -\frac{\sqrt{a^2 - u^2}}{a^2 u} + C.$

22. $\displaystyle\int \frac{\sqrt{a^2 - u^2}\ du}{u} = \sqrt{a^2 - u^2} - a\ln\left|\frac{a + \sqrt{a^2 - u^2}}{u}\right| + C,\ a > 0.$

FORMS CONTAINING $\sqrt{u^2 \pm a^2}$.

23. $\displaystyle\int \sqrt{u^2 \pm a^2}\ du = \frac{1}{2}\left(u\sqrt{u^2 \pm a^2} \pm a^2\ln|u + \sqrt{u^2 \pm a^2}|\right) + C.$

24. $\displaystyle\int u^2\sqrt{u^2 \pm a^2}\ du = \frac{u}{8}(2u^2 \pm a^2)\sqrt{u^2 \pm a^2} - \frac{a^4}{8}\ln|u + \sqrt{u^2 \pm a^2}| + C.$

25. $\displaystyle\int \frac{\sqrt{u^2 + a^2}\ du}{u} = \sqrt{u^2 + a^2} - a\ln\left|\frac{a + \sqrt{u^2 + a^2}}{u}\right| + C.$

26. $\displaystyle\int \frac{\sqrt{u^2 \pm a^2}\ du}{u^2} = -\frac{\sqrt{u^2 \pm a^2}}{u} + \ln|u + \sqrt{u^2 \pm a^2}| + C.$

27. $\displaystyle\int \frac{du}{\sqrt{u^2 \pm a^2}} = \ln|u + \sqrt{u^2 \pm a^2}| + C.$

28. $\displaystyle\int \frac{du}{u\sqrt{u^2 + a^2}} = \frac{1}{a}\ln\left|\frac{\sqrt{u^2 + a^2} - a}{u}\right| + C.$

29. $\displaystyle\int \frac{u^2\ du}{\sqrt{u^2 \pm a^2}} = \frac{1}{2}\left(u\sqrt{u^2 \pm a^2} \mp a^2\ln|u + \sqrt{u^2 \pm a^2}|\right) + C.$

30. $\displaystyle\int \frac{du}{u^2\sqrt{u^2 \pm a^2}} = -\frac{\pm\sqrt{u^2 \pm a^2}}{a^2 u} + C.$

31. $\displaystyle\int (u^2 \pm a^2)^{3/2}\ du = \frac{u}{8}(2u^2 \pm 5a^2)\sqrt{u^2 \pm a^2} + \frac{3a^4}{8}\ln|u + \sqrt{u^2 \pm a^2}| + C.$

32. $\displaystyle\int \frac{du}{(u^2 \pm a^2)^{3/2}} = \frac{\pm u}{a^2\sqrt{u^2 \pm a^2}} + C.$

33. $\displaystyle \int \frac{u^2\, du}{(u^2 \pm a^2)^{3/2}} = \frac{-u}{\sqrt{u^2 \pm a^2}} + \ln \left| u + \sqrt{u^2 \pm a^2} \right| + C.$

RATIONAL FORMS CONTAINING $a^2 - u^2$ AND $u^2 - a^2$.

34. $\displaystyle \int \frac{du}{a^2 - u^2} = \frac{1}{2a} \ln \left| \frac{a + u}{a - u} \right| + C.$

35. $\displaystyle \int \frac{du}{u^2 - a^2} = \frac{1}{2a} \ln \left| \frac{u - a}{u + a} \right| + C.$

EXPONENTIAL AND LOGARITHMIC FORMS.

36. $\displaystyle \int e^u\, du = e^u + C.$

37. $\displaystyle \int a^u\, du = \frac{a^u}{\ln a} + C, \, a > 0, \, a \neq 1.$

38. $\displaystyle \int ue^{au}\, du = \frac{e^{au}}{a^2}(au - 1) + C.$

39. $\displaystyle \int u^n e^{au}\, du = \frac{u^n e^{au}}{a} - \frac{n}{a} \int u^{n-1} e^{au}\, du.$

40. $\displaystyle \int \frac{e^{au}\, du}{u^n} = -\frac{e^{au}}{(n-1)u^{n-1}} + \frac{a}{n-1} \int \frac{e^{au}\, du}{u^{n-1}}.$

41. $\displaystyle \int \ln u\, du = u \ln u - u + C.$

42. $\displaystyle \int u^n \ln u\, du = \frac{u^{n+1} \ln u}{n + 1} - \frac{u^{n+1}}{(n+1)^2} + C, \, n \neq -1.$

43. $\displaystyle \int u^n \ln^m u\, du = \frac{u^{n+1}}{n+1} \ln^m u - \frac{m}{n+1} \int u^n \ln^{m-1} u\, du, \quad m, n \neq -1.$

44. $\displaystyle \int \frac{du}{u \ln u} = \ln |\ln u| + C.$

45. $\displaystyle \int \frac{du}{a + be^{cu}} = \frac{1}{ac}(cu - \ln |a + be^{cu}|) + C.$

MISCELLANEOUS FORMS.

46. $\int \sqrt{\dfrac{a + u}{b + u}} \; du = \sqrt{(a + u)(b + u)} + (a - b) \ln \left(\sqrt{a + u} + \sqrt{b + u} \right) + C.$

47. $\int \dfrac{du}{\sqrt{(a + u)(b + u)}} = \ln \left| \dfrac{a + b}{2} + u + \sqrt{(a + u)(b + u)} \right| + C.$

48. $\int \sqrt{a + bu + cu^2} \; du = \dfrac{2cu + b}{4c} \sqrt{a + bu + cu^2} -$

$$\dfrac{b^2 - 4ac}{8c^{3/2}} \ln \left| 2cu + b + 2\sqrt{c} \sqrt{a + bu + cu^2} \right| + C, \; c > 0.$$

AREAS UNDER THE STANDARD NORMAL CURVE

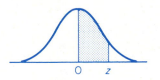

$$A(z) = \int_0^z \frac{1}{\sqrt{2\pi}} e^{-x^2/2}\, dx$$

$$A(-z) = A(z)$$

z	.00	.01	.02	.03	.04	.05	.06	.07	.08	.09
0.0	.0000	.0040	.0080	.0120	.0160	.0199	.0239	.0279	.0319	.0359
0.1	.0398	.0438	.0478	.0517	.0557	.0596	.0636	.0675	.0714	.0754
0.2	.0793	.0832	.0871	.0910	.0948	.0987	.1026	.1064	.1103	.1141
0.3	.1179	.1217	.1255	.1293	.1331	.1368	.1406	.1443	.1480	.1517
0.4	.1554	.1591	.1628	.1664	.1700	.1736	.1772	.1808	.1844	.1879
0.5	.1915	.1950	.1985	.2019	.2054	.2088	.2123	.2157	.2190	.2224
0.6	.2258	.2291	.2324	.2357	.2389	.2422	.2454	.2486	.2518	.2549
0.7	.2580	.2612	.2642	.2673	.2704	.2734	.2764	.2794	.2823	.2852
0.8	.2881	.2910	.2939	.2967	.2996	.3023	.3051	.3078	.3106	.3133
0.9	.3159	.3186	.3212	.3238	.3264	.3289	.3315	.3340	.3365	.3389
1.0	.3413	.3438	.3461	.3485	.3508	.3531	.3554	.3577	.3599	.3621
1.1	.3643	.3665	.3686	.3708	.3729	.3749	.3770	.3790	.3810	.3820
1.2	.3849	.3869	.3888	.3907	.3925	.3944	.3962	.3980	.3997	.4015
1.3	.4032	.4049	.4066	.4082	.4099	.4115	.4131	.4147	.4162	.4177
1.4	.4192	.4207	.4222	.4236	.4251	.4265	.4279	.4292	.4306	.4319
1.5	.4332	.4345	.4357	.4370	.4382	.4394	.4406	.4418	.4429	.4441
1.6	.4452	.4463	.4474	.4484	.4495	.4505	.4515	.4525	.4535	.4545
1.7	.4554	.4564	.4573	.4582	.4591	.4599	.4608	.4616	.4625	.4633
1.8	.4641	.4649	.4656	.4664	.4671	.4678	.4686	.4693	.4699	.4706
1.9	.4713	.4719	.4726	.4732	.4738	.4744	.4750	.4756	.4761	.4767
2.0	.4772	.4778	.4783	.4788	.4793	.4798	.4803	.4808	.4812	.4817
2.1	.4821	.4826	.4830	.4834	.4838	.4842	.4846	.4850	.4854	.4857
2.2	.4861	.4864	.4868	.4871	.4875	.4878	.4881	.4884	.4887	.4890
2.3	.4893	.4896	.4898	.4901	.4904	.4906	.4909	.4911	.4913	.4916
2.4	.4918	.4920	.4922	.4925	.4927	.4929	.4931	.4932	.4934	.4936
2.5	.4938	.4940	.4941	.4943	.4945	.4946	.4948	.4949	.4951	.4952
2.6	.4953	.4955	.4956	.4957	.4959	.4960	.4961	.4962	.4963	.4964
2.7	.4965	.4966	.4967	.4968	.4969	.4970	.4971	.4972	.4973	.4974
2.8	.4974	.4975	.4976	.4977	.4977	.4978	.4979	.4979	.4980	.4981
2.9	.4981	.4982	.4982	.4983	.4984	.4984	.4985	.4985	.4986	.4986
3.0	.4987	.4987	.4987	.4988	.4988	.4989	.4989	.4989	.4990	.4990
3.1	.4990	.4991	.4991	.4991	.4992	.4992	.4992	.4992	.4993	.4993
3.2	.4993	.4993	.4994	.4994	.4994	.4994	.4994	.4995	.4995	.4995
3.3	.4995	.4995	.4995	.4996	.4996	.4996	.4996	.4996	.4996	.4997
3.4	.4997	.4997	.4997	.4997	.4997	.4997	.4997	.4997	.4997	.4998
3.5	.4998	.4998	.4998	.4998	.4998	.4998	.4998	.4998	.4998	.4998

ANSWERS TO ODD-NUMBERED PROBLEMS

1. True.   **3.** False; the natural numbers are 1, 2, 3, etc.   **5.** True.

7. False; $\frac{4}{2} = 2$, a positive integer.   **9.** True.   **11.** True.

1. False.   **3.** False.   **5.** True.   **7.** True.   **9.** False.   **11.** Distributive.

13. Associative.   **15.** Commutative.   **17.** Definition of subtraction.

19. Distributive.

1. -6.   **3.** 2.   **5.** 11.   **7.** -2.   **9.** -63.   **11.** -6.   **13.** $6 - x$.

15. $-12x + 12y$ (or $12y - 12x$).   **17.** $-\dfrac{1}{3}$.   **19.** -2.   **21.** 18.   **23.** 25.

25. $3x - 12$.   **27.** $-x + 2$.   **29.** $\dfrac{8}{11}$.   **31.** $-\dfrac{5x}{7y}$.   **33.** $\dfrac{2}{3x}$.   **35.** 3.

37. $\dfrac{7}{xy}$.   **39.** $\dfrac{5}{6}$.   **41.** $-\dfrac{1}{6}$.   **43.** $\dfrac{x-y}{9}$.   **45.** $\dfrac{1}{24}$.   **47.** $\dfrac{x}{6y}$.

49. Not defined.   **51.** Not defined.

EXERCISE 0-5

1. 2^5 ($= 32$).   **3.** x^{15}.   **5.** w^{12}.   **7.** x^{13}.   **9.** $\dfrac{x^8}{y^{17}}$.   **11.** x^{48}.

13. $\dfrac{x^{10}}{y^{50}}$.   **15.** $\dfrac{x^{10}}{y^{15}}$.   **17.** $8x^6y^9$.   **19.** $\dfrac{w^4s^6}{y^4}$.   **21.** x^6.   **23.** $\dfrac{1}{8x^6}$.

25. x^{14}.   **27.** 1.   **29.** $\dfrac{z^2}{x^3y^5}$.   **31.** $\dfrac{wxz}{y^4}$.   **33.** $\dfrac{2x^4z^2}{3y^{11}}$.   **35.** $\dfrac{512x^{12}}{y^{18}}$.

37. $\sqrt[5]{(8x - y)^4}$.   **39.** $\dfrac{1}{\sqrt[5]{x^4}}$.   **41.** $\dfrac{2}{\sqrt[5]{x^2}} - \dfrac{1}{\sqrt[5]{4x^2}}$.   **43.** $\dfrac{x^3}{y^2z^2}$.   **45.** $\dfrac{2}{x^4}$.

47. $\dfrac{1}{9t^2}$.   **49.** $7^{1/3}s^{2/3}$.   **51.** $x^{1/2} - y^{1/2}$.   **53.** $\dfrac{x^{9/4}z^{3/4}}{y^{1/2}}$.

55. 5.   **57.** -2.   **59.** $\dfrac{1}{2}$.   **61.** 10.   **63.** 8.   **65.** $\dfrac{1}{4}$.   **67.** $\dfrac{1}{32}$.

69. $4\sqrt{2}$.   **71.** $x\sqrt[3]{2}$.   **73.** $4x^2$.   **75.** $3z^2$.   **77.** $\dfrac{9t^2}{4}$.   **79.** $\dfrac{3\sqrt{7}}{7}$.

81. $\dfrac{2\sqrt{2x}}{x}$.   **83.** $\dfrac{\sqrt[3]{9x^2}}{3x}$.   **85.** 4.   **87.** $\dfrac{\sqrt[12]{8x^8y^4}}{xy}$.   **89.** $\dfrac{2x^6}{y^3}$.

91. $t^{2/3}$.   **93.** $\dfrac{64y^6x^{1/2}}{x^2}$.   **95.** xyz.   **97.** $\dfrac{1}{9}$.   **99.** $\dfrac{4y^4}{x^2}$.   **101.** $x^3y^{5/2}$.

103. $\dfrac{y^{10}}{z^2}$.   **105.** $\dfrac{1}{x^4}$.   **107.** $-\dfrac{4}{s^5}$.   **109.** $\dfrac{4x^4z^4}{9y^4}$.

111. Rationalization can simplify the approximation of certain irrational numbers.

EXERCISE 0-6

1. $11x - 2y - 3$.   **3.** $6t^2 - 2s^2 + 6$.   **5.** $2\sqrt{x} + \sqrt{2y} + \sqrt{3z}$.

7. $6x^2 - 9xy - 2z + \sqrt{2} - 4$.   **9.** $\sqrt{2y} - \sqrt{3z}$.   **11.** $-7x + 14y - 19$.

13. $x^2 + 9y^2 + xy$.   **15.** $6x^2 + 96$.   **17.** $-6x^2 - 18x - 18$.   **19.** $x^2 + 9x + 20$.

21. $x^2 + x - 6$.   **23.** $10x^2 + 19x + 6$.   **25.** $x^2 + 6x + 9$.   **27.** $x^2 - 10x + 25$.

29. $2y + 6\sqrt{2y} + 9$.   **31.** $4s^2 - 1$.   **33.** $x^3 + 4x^2 - 3x - 12$.

35. $2x^4 + 2x^3 - 5x^2 - 2x + 3$.   **37.** $5x^3 + 5x^2 + 6x$.

39. $3x^2 + 2y^2 + 5xy + 2x - 8$.   **41.** $x^3 + 15x^2 + 75x + 125$.

43. $8x^3 - 36x^2 + 54x - 27$.   **45.** $z - 4$.   **47.** $3x^3 + 2x - \dfrac{1}{2x^2}$.

49. $x + \dfrac{-1}{x + 3}$.   **51.** $3x^2 - 8x + 17 + \dfrac{-37}{x + 2}$.   **53.** $t + 8 + \dfrac{64}{t - 8}$.

55. $x - 2 + \dfrac{7}{3x + 2}$.

EXERCISE 0-7

1. $2(3x + 2)$.   **3.** $5x(2y + z)$.   **5.** $4bc(2a^3 - 3ab^2d + b^3cd^2)$.

7. $(x - 5)(x + 5)$.   **9.** $(p + 3)(p + 1)$.   **11.** $(4x - 3)(4x + 3)$.

13. $(z + 4)(z + 2)$.   **15.** $(x + 3)^2$.   **17.** $2(x + 4)(x + 2)$.

19. $3(x - 1)(x + 1)$.   **21.** $(6y + 1)(y + 2)$.   **23.** $2s(3s + 4)(2s - 1)$.

25. $x^{2/3}y(1 - 2xy)(1 + 2xy)$.   **27.** $2x(x + 3)(x - 2)$.   **29.** $4(2x + 1)^2$.

31. $x(xy - 5)^2$.   **33.** $(x - 2)^2(x + 2)$.   **35.** $(y - 1)(y + 1)(y^4 + 4)^2$.

37. $(x + 2)(x^2 - 2x + 4)$.   **39.** $(x + 1)(x^2 - x + 1)(x - 1)(x^2 + x + 1)$.

41. $2(x + 3)^2(x + 1)(x - 1)$.   **43.** $2(x + 4)(x + 1)$.   **45.** $(x^2 + 4)(x + 2)(x - 2)$.

47. $(y^4 + 1)(y^2 + 1)(y + 1)(y - 1)$.   **49.** $(x^2 + 2)(x + 1)(x - 1)$.

51. $x(x + 1)^2(x - 1)^2$.

EXERCISE 0-8

1. $-\dfrac{y^2}{(y - 3)(y + 2)}$.   **3.** $\dfrac{3 - 2x}{3 + 2x}$.   **5.** $\dfrac{2(x + 4)}{(x - 4)(x + 2)}$.   **7.** $\dfrac{x}{2}$.

9. $\dfrac{1}{2n}$.   **11.** $\dfrac{2}{3}$.   **13.** $-27x^2$.   **15.** 1.   **17.** $\dfrac{2x^2}{x - 1}$.   **19.** 1.

21. $-\dfrac{(2x + 3)(1 + x)}{x + 4}$.   **23.** $x + 2$.   **25.** $\dfrac{5}{3t}$.   **27.** $-\dfrac{1}{p^2 - 1}$.

29. $\dfrac{2x^2 + 3x + 12}{(2x - 1)(x + 3)}$.   **31.** $\dfrac{2x - 3}{(x - 2)(x + 1)(x - 1)}$.   **33.** $\dfrac{35 - 8x}{(x - 1)(x + 5)}$.

35. $\dfrac{x^2 + 2x + 1}{x^2}$.   **37.** $\dfrac{x}{1 - xy}$.   **39.** $\dfrac{x + 1}{3x}$.

41. $\dfrac{(x+2)(6x-1)}{2x^2(x+3)}$. **43.** $\dfrac{2\sqrt{x}-2\sqrt{x+h}}{\sqrt{x}\,\sqrt{x+h}}$. **45.** $2-\sqrt{3}$.

47. $-\dfrac{\sqrt{6}+2\sqrt{3}}{3}$. **49.** $-4-2\sqrt{6}$. **51.** $\dfrac{x-\sqrt{5}}{x^2-5}$. **53.** $\dfrac{5\sqrt{3}-4\sqrt{2}-13}{2}$.

EXERCISE 1-1

1. 0. **3.** $\dfrac{10}{3}$. **5.** Does not satisfy equation.

7. Adding 5 to both sides; equivalence guaranteed.

9. Squaring both sides; equivalence *not* guaranteed.

11. Dividing both sides by x; equivalence *not* guaranteed.

13. Multiplying both sides by $x-1$; equivalence *not* guaranteed.

15. Multiplying both sides by $(x-5)/x$; equivalence *not* guaranteed.

17. $\dfrac{5}{2}$. **19.** 0. **21.** 1. **23.** $\dfrac{12}{5}$. **25.** -1. **27.** 2. **29.** $\dfrac{10}{3}$. **31.** 90.

33. 8. **35.** $-\dfrac{26}{9}$. **37.** $-\dfrac{37}{18}$. **39.** $\dfrac{60}{17}$. **41.** $\dfrac{14}{3}$. **43.** 3. **45.** $\dfrac{7}{8}$.

47. $P=\dfrac{I}{rt}$. **49.** $q=\dfrac{p+1}{8}$. **51.** $r=\dfrac{S-P}{Pt}$. **53.** $a_1=\dfrac{2S-na_n}{n}$.

EXERCISE 1-2

1. $\dfrac{1}{5}$. **3.** $\varnothing$. **5.** $\dfrac{8}{3}$. **7.** $\dfrac{3}{2}$. **9.** 0. **11.** $\dfrac{5}{3}$. **13.** $\dfrac{1}{8}$. **15.** 3.

17. $\dfrac{5}{13}$. **19.** $\varnothing$. **21.** 3. **23.** $\dfrac{262}{5}$. **25.** $-\dfrac{10}{9}$. **27.** 2. **29.** 7.

31. $\dfrac{49}{36}$. **33.** $-\dfrac{9}{4}$. **35.** $d=\dfrac{r}{1+rt}$. **37.** $n=\dfrac{2mI}{rB}-1$.

EXERCISE 1-3

1. 2. **3.** 4, 3. **5.** 3, -1. **7.** 6. **9.** ± 2. **11.** 0, 8.

13. $\dfrac{1}{2}$. **15.** 1, $-\dfrac{5}{2}$. **17.** 5, -2. **19.** 0, $\dfrac{3}{2}$. **21.** 0, 1, -2.

23. 0, ± 8. **25.** 0, $\dfrac{1}{2}$, $-\dfrac{4}{3}$. **27.** -3, -1, 2. **29.** 3, 4. **31.** 4, -6.

33. $\dfrac{3}{2}$. **35.** $\dfrac{5\pm\sqrt{13}}{2}$. **37.** No real roots. **39.** $\dfrac{1}{2}$, $-\dfrac{5}{3}$. **41.** 4, $-\dfrac{5}{2}$.

43. $\dfrac{-2 \pm \sqrt{14}}{2}$.   **45.** $\dfrac{3}{2}$, -1.   **47.** 6, -2.   **49.** $\dfrac{1}{2}$.   **51.** 5, -2.   **53.** $\dfrac{3}{2}$.

55. -2.   **57.** 7.   **59.** 4, 8.   **61.** 2.   **63.** 0, 4.   **65.** 4.

REVIEW PROBLEMS—CHAPTER 1

1. $\dfrac{1}{4}$.   **3.** $-\dfrac{2}{15}$.   **5.** $-\dfrac{1}{2}$.   **7.** $\varnothing$.   **9.** $\dfrac{5}{2}$.   **11.** $\dfrac{1}{3}$.   **13.** $-\dfrac{9}{7}$.

15. $-\dfrac{5}{3}$, 1.   **17.** 0, $\dfrac{7}{5}$.   **19.** 5.   **21.** $\pm\dfrac{\sqrt{15}}{3}$.   **23.** $\dfrac{5}{8}$, -3.

25. $\dfrac{5 \pm \sqrt{13}}{6}$.   **27.** 4, ± 3.   **29.** $\dfrac{1}{2}$.   **31.** $\dfrac{4 \pm \sqrt{13}}{3}$.   **33.** 4, 8.

EXERCISE 2-1

1. 181,250.   **3.** \$4000 at 6 percent, \$16,000 at $7\frac{1}{2}$ percent.   **5.** \$4.25.

7. 4 percent.   **9.** 40.   **11.** 46,000 units.   **13.** Either \$440 or \$460.   **15.** \$100.

17. 77.   **19.** 80 ft by 140 ft.   **21.** 9 cm long, 4 cm wide.   **23.** \$112,000.   **25.** 60.

27. Either 125 units of A and 100 units of B, or 150 units of A and 125 units of B.

EXERCISE 2-2

1. $x > 4$.   **3.** $x \leqslant 5$.   **5.** $x \leqslant -\dfrac{1}{2}$.   **7.** $s < -\dfrac{2}{5}$.   **9.** $y > 0$.

11. $x \geqslant -\dfrac{7}{5}$.   **13.** $x > -\dfrac{2}{7}$.   **15.** $\varnothing$.   **17.** $x < \dfrac{\sqrt{3} - 2}{2}$.

19. $x < 6$.   **21.** $y \leqslant -5$.   **23.** $-\infty < x < \infty$.   **25.** $t > \dfrac{17}{9}$.

27. $x \geqslant -\frac{14}{3}$.   **29.** $r > 0$.   **31.** $y < 0$.   **33.** $x \leqslant -2$.

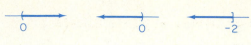

35. $444{,}000 < S < 636{,}000$.

EXERCISE 2-3

1. At least 120,001.   **3.** 12,400.   **5.** 60,000.   **7.** \$25,714.29.   **9.** 1000.

EXERCISE 2-4

1. 13.   **3.** 6.   **5.** 5.   **7.** $-3 < x < 3$.   **9.** $\sqrt{5} - 2$.

11. (a) $|x - 7| < 3$,   (b) $|x - 2| < 3$,   (c) $|x - 7| \leqslant 5$,   (d) $|x - 7| = 4$,

(e) $|x + 4| < 2$,   (f) $|x| < 3$,   (g) $|x| > 6$,   (h) $|x - 6| > 4$,

(i) $|x - 105| < 3$,   (j) $|x - 850| < 100$.

13. $|p_1 - p_2| \leqslant 2$.   **15.** ± 7.   **17.** ± 6.   **19.** $-3, 13$.   **21.** $\frac{2}{5}$.   **23.** $\frac{1}{2}, 3$.

25. $-4 < x < 4$.   **27.** $x < -8, x > 8$.   **29.** $-9 < x < -5$.

31. $x < 0, x > 1$.   **33.** $2 \leqslant x \leqslant 3$.

35. $x \leqslant 0, x \geqslant \frac{16}{3}$.   **37.** $x < \mu - h\sigma, x > \mu + h\sigma$.

REVIEW PROBLEMS—CHAPTER 2

1. $x \leqslant 0$.   **3.** $x > \frac{2}{3}$.   **5.** $\varnothing$.   **7.** $x \leqslant \frac{13}{2}$.   **9.** $-\infty < s < \infty$.

11. $-2, 5$.   **13.** $0 < t < \frac{1}{2}$.   **15.** $x \leqslant -\frac{1}{2}, x \geqslant \frac{7}{2}$.   **17.** 542.   **19.** 6000.

EXERCISE 3-1

1. x; all real numbers; $-3, 0, -\frac{5}{2}, q - 3$.

3. x; all real numbers; $1, 1 + 2u, -13, 1 + 4x, 1 - 2(x + h) = 1 - 2x - 2h$.

5. t; all real numbers; 8.73, 8.73, 8.73, 8.73.

7. p; all real numbers; $\frac{15}{2}, \frac{3(2p + 1)}{2}, \frac{3(4 + p)}{2p}$.

9. p; all real numbers; 1, 9, $x_1^2 + 2x_1 + 1$, $w^2 + 2w + 1$, $p^2 + 2ph + h^2 + 2p + 2h + 1$.

11. t; all real numbers; 16, 36, $h^2 + 12h + 36$, $h + 12$.

13. q; all real numbers; 5, 3, 0, $|2x^2 + 3| = 2x^2 + 3$.

15. x; all $x \geqslant -4$; 2, 0, 1, $\sqrt{5 + x} - \sqrt{4 + x}$.

17. s; all $s \neq \pm 3$; $-\dfrac{1}{2}$, $-\dfrac{1}{2s}$, $\dfrac{4}{4w^2 - 9}$, $\dfrac{4}{s^2 - 2s - 8}$, $\dfrac{13 - s^2}{s^2 - 9}$.

19. x; all real numbers; 4, 3, 4, 3.

21. r; all r such that $r < -2$ or $r > 2$; 8, 28, 14, 52.

23. x; all real numbers; 0, 256, $1/16$, $32\sqrt[3]{2}$.

25. t; all $t \neq \dfrac{5}{2}$, -1; $-\dfrac{1}{9}$, $\dfrac{2}{27}$, $\dfrac{2}{3(2t^2 + t - 6)}$, $\dfrac{2}{3(8t^2 - 6t - 5)}$.

27. Yes; no, since if $z = 4$, then $x = \pm 1$.

29. 3.   31. $2x + h + 2$.   33. $V = f(t) = 10,000 + 400t$.

35. Yes; P; q.   37. (a) $C = 850 + 3q$; (b) 250.

39. (a) 3000, 2900, 2300, 2000; 12, 10.   (b) 10, 12, 17, 20; 3000, 2300.

EXERCISE 3-2

1. (a) $2x + 9$,   (b) -3,   (c) $x^2 + 9x + 18$,   (d) $\dfrac{x + 3}{x + 6}$,   (e) $x + 9$,   (f) $x + 9$.

3. (a) $x^2 + 2x + 4$,   (b) $2x + 6 - x^2$,   (c) $2x^3 + 5x^2 - 2x - 5$,

   (d) $\dfrac{2x + 5}{x^2 - 1}$,   (e) $2x^2 + 3$,   (f) $4x^2 + 20x + 24$.

5. (a) $3x^2 + 3x - 11$,   (b) $-x^2 + 3x + 3$,   (c) $2x^4 + 6x^3 - 15x^2 - 21x + 28$,

   (d) $\dfrac{x^2 + 3x - 4}{2x^2 - 7}$,   (e) $4x^4 - 22x^2 + 24$,   (f) $2x^4 + 12x^3 + 2x^2 - 48x + 25$.

7. (a) 40,   (b) $30x - 5$,   (c) -12,   (d) $-3w - 9$,   (e) -16,

   (f) $54x^2 + 84x + 16$,   (g) $-\dfrac{7}{2}$,   (h) $\dfrac{9t^2 - 7}{6t^2 + 2}$.

9. (a) 18,   (b) $2x^2 + 4xh + 2h^2 - 3x - 3h + 4$,   (c) 4,

   (d) $2x^2 + 4xh + 2h^2 + 3x + 3h + 2$,   (e) 3,   (f) $-6x^{3/2} + 2x - 9x^{1/2} + 3$,

   (g) $-5/2$,   (h) $\dfrac{8p^2 + 27}{9(1 - 2p)}$.   11. 53; -32.

13. $\dfrac{4}{(t - 1)^2} + \dfrac{6}{t - 1} + 1$; $\dfrac{2}{t^2 + 3t}$.   15. $\dfrac{1}{v + 3}$; $\sqrt{\dfrac{2w^2 + 3}{w^2 + 1}}$.

17. (a) 14; all real numbers; 14.   (b) 2; all real numbers; 2.

19. $f(x) = \sqrt{x}$, $g(x) = x - 2$.   **21.** $f(x) = x^3 - x^2 + 7$, $g(x) = 3x^3 - 2x$.

23. $f(x) = \dfrac{x}{x^2 + 2}$, $g(x) = x + 1$.

25. $400m - 10m^2$; the total revenue received when the total output of m employees is sold.

EXERCISE 3-3

1.

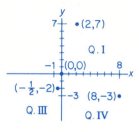

3. (a) 1, 2, 3, 0;
    (b) All real numbers;
    (c) All real numbers.

5. (a) 0, −1, −1;
    (b) All real numbers;
    (c) All nonpositive reals.

7. Function; all real numbers; all real numbers.

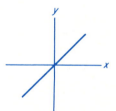

9. Function; all real numbers; all real numbers.

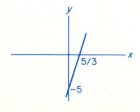

11. Function; all real numbers; all nonnegative real numbers.

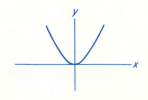

13. Not a function of x.

15. Function; all real numbers; all real numbers.

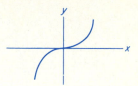

17. Not a function of x.

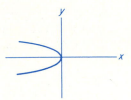

19. Function; all real numbers; all real numbers.

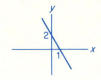

21. All real numbers; 2.

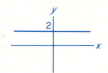

23. All real numbers; all real numbers.

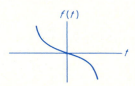

25. All real numbers $\geqslant 5$; all nonnegative reals.

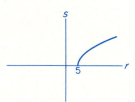

27. All real numbers; all nonnegative reals.

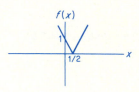

29. All real numbers; all reals $\geqslant -3$.

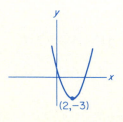

31. All nonzero real numbers; all positive real numbers.

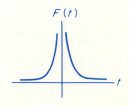

33. All real numbers;
all reals $\leqslant 4$.

35. All real numbers;
all nonnegative reals.

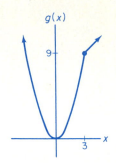

37. All positive real numbers;
all reals > 1.

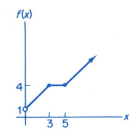

39.

As price decreases,
quantity increases;
p is a function
of q

REVIEW PROBLEMS—CHAPTER 3

1. $7, 46, 62, 3x^4 - 4x^2 + 7$; all real numbers.

3. $\dfrac{3}{2}, 18, 6\sqrt{u}, 12uh + 6h^2$; all real numbers.

5. $2, \dfrac{1}{2}, 8, \dfrac{1}{2\sqrt[3]{100}}$; all nonzero real numbers.

7. $-8, 4, 4, -92$; all real numbers $\neq 2$.

9. (a) $-x - 4$,   (b) $12 - 5x$,   (c) $-6x^2 + 32x - 32$,   (d) $\dfrac{4 - 3x}{2x - 8}$,

(e) $28 - 6x$,   (f) $-6x$.

11. $20 + 5\sqrt{23}$; 2.

13. All real numbers;
all reals $\leqslant 9$.

15. All real numbers;
all reals $\geqslant 1$.

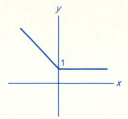

17. All $t \neq 4$; all nonzero
real numbers.

19. \$165,000; yes.

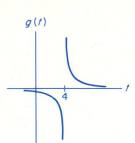

EXERCISE 4-1

1. 2. **3.** $-\dfrac{8}{13}$. **5.** Not defined. **7.** 0. **9.** $6x - y - 4 = 0$.

11. $x + 4y - 18 = 0$. **13.** $3x - 7y + 25 = 0$. **15.** $8x - 5y - 29 = 0$.

17. $y - 5 = 0$. **19.** $x - 2 = 0$. **21.** $4x - y + 7 = 0$. **23.** 2; $(0, -1)$.

25. $\dfrac{3}{8}$; $(0, -1)$. **27.** $-\dfrac{1}{2}$; $\left(0, \dfrac{3}{2}\right)$. **29.** No slope; no y-intercept.

31. 3; $(0, 0)$. **33.** 0; $(0, 1)$. **35.** $\dfrac{1}{40}$; $\left(0, -\dfrac{1}{2}\right)$.

37. $x + 2y - 4 = 0$; $y = -\dfrac{1}{2}x + 2$. **39.** $4x + 9y - 5 = 0$; $y = -\dfrac{4}{9}x + \dfrac{5}{9}$.

41. $9x - 28y - 3 = 0$; $y = \dfrac{9}{28}x - \dfrac{3}{28}$. **43.** $3x - 2y + 24 = 0$; $y = \dfrac{3}{2}x + 12$.

45. $x + y - 1 = 0$; $y = -x + 1$. **47.** 1; $(0, 1)$. **49.** -3; $(0, 5)$.

51. $f(x) = 5x - 14$. **53.** $f(x) = -3x + 9$. **55.** $(5, -4)$.

57. $p = -\dfrac{2}{5}q + 28$; 16. **59.** $x + 10y = 100$.

61. (b) $-\dfrac{2}{3}$; (c) $-\dfrac{2}{3}$; (d) they are parallel.

EXERCISE 4-2

1. Not quadratic.   **3.** Quadratic.   **5.** Not quadratic.   **7.** Quadratic.

9. (a) $(1, 11)$; (b) Highest.   **11.** (a) $(0, -8)$; (b) $(-4, 0)$, $(2, 0)$; (c) $(-1, -9)$.

13. Vertex: $(3, -4)$; intercepts: $(1, 0)$, $(5, 0)$, $(0, 5)$; range: all $y \geqslant -4$.

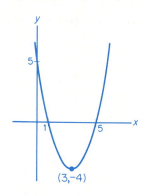

15. Vertex: $\left(-\dfrac{3}{2}, \dfrac{9}{2}\right)$; intercepts: $(0, 0)$, $(-3, 0)$; range: all $y \leqslant \dfrac{9}{2}$.

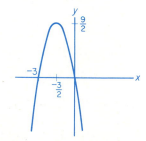

17. Vertex: $(-1, 0)$; intercepts: $(-1, 0)$, $(0, 1)$; range: all $s \geqslant 0$.

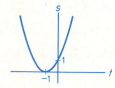

19. Vertex: $(2, -1)$; intercept: $(0, -9)$; range: all $y \leqslant -1$.

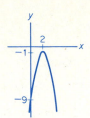

21. Vertex: $(4, -3)$; intercepts: $(4 + \sqrt{3}, 0)$, $(4 - \sqrt{3}, 0)$, $(0, 13)$; range: all $t \geqslant -3$.

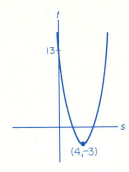

23. 24.   **25.** $q = 200$; $r = \$120{,}000$.

EXERCISE 4-3

1. $x = 4, y = -5$.   **3.** $x = 3, y = -1$.   **5.** $v = 0, w = 18$.

7. No solution.   **9.** $x = 12, y = -12$.

11. The coordinates of any point on the line $q = -\dfrac{1}{3}p + \dfrac{1}{2}$.

13. $x = \dfrac{1}{2}, y = \dfrac{1}{2}, z = \dfrac{1}{4}$.   **15.** $x = 1, y = 1, z = 1$.

17. 420 gal of 20 percent solution, 280 gal of 30 percent solution.

19. 240 units (Early American), 200 units (Contemporary).

21. 800 calculators from Exton plant, 700 from Whyton plant.

23. 4 percent on first \$100,000, 6 percent on remainder.

25. 100 chairs, 100 rockers, 200 chaise lounges.

27. 40 semiskilled workers, 20 skilled workers, 10 shipping clerks.

EXERCISE 4-4

1. $x = 4, y = -12; x = -1, y = 3.$ **3.** $p = -3, q = -5; p = 2, q = 0.$

5. $x = 0, y = 0; x = 1, y = 1.$ **7.** $x = 4, y = 8; x = -1, y = 3.$

9. $p = 0, q = 0; p = 1, q = 1.$

11. $x = 3\sqrt{2}, y = 2; x = -3\sqrt{2}, y = 2; x = \sqrt{15}, y = -1;$
$x = -\sqrt{15}, y = -1.$

13. $x = -2, y = -\dfrac{1}{3}.$

EXERCISE 4-5

1. $p = \dfrac{1}{3}q + \dfrac{55}{3}.$ **3.**

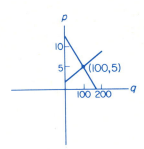

5. $(5, 212.50).$ **7.** $(9, 38).$ **9.** $(15, 5).$ **11.**

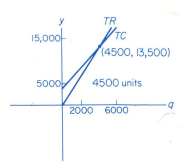

13. Cannot break even at any level of production.   **15.** 10 units or 40 units.

17. (a) \$12; (b) \$12.18.   **19.** 5840 units; 840 units; 1840 units.   **21.** \$4.

23. Total cost always exceeds total revenue—no break-even point.

25. Decreases by \$0.70.   **27.** $p_A = 5; p_B = 10.$

REVIEW PROBLEMS—CHAPTER 4

1. 9.   **3.** $y = -x + 1; x + y - 1 = 0.$ **5.** $y = \dfrac{1}{2}x - 1; x - 2y - 2 = 0.$

7. $y = 4; y - 4 = 0.$ **9.** $y = \dfrac{3}{2}x - 2; \dfrac{3}{2}.$ **11.** $y = \dfrac{4}{3}; 0.$

13. All real numbers;
all real numbers;
$-2; (0, 4).$

15. All real numbers;
all reals $\leqslant 9$; (3, 0),
$(-3, 0), (0, 9); (0, 9).$

17. All real numbers;
all reals $\geqslant -9$; (5, 0),
$(-1, 0), (0, -5); (2, -9).$

19. All real numbers;
all real numbers;
$3; (0, 0).$

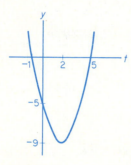

21. All real numbers;
all reals $\leqslant -2$;
$(0, -3); (-1, -2).$

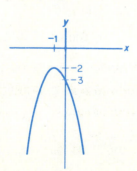

23. $x = \dfrac{17}{7}, y = -\dfrac{8}{7}.$ **25.** $x = 2, y = -1.$ **27.** $x = 8, y = 4.$

29. $x = 0, y = 1, z = 0.$ **31.** $x = -3, y = -4; x = 2, y = 1.$

33. $f(x) = -\dfrac{4}{3}x + \dfrac{19}{3}.$ **35.** 50 units; $5000.   **37.** 6.

EXERCISE 5-1

1.
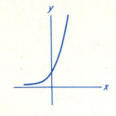

3.

5.
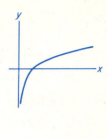

7. 4.4817.   **9.** .67032.   **11.** 1.60944.   **13.** 2.00013.   **15.** $\log_{25} 5 = \dfrac{1}{2}.$

17. $\log 10{,}000 = 4.$ **19.** $2^6 = 64.$ **21.** $2^{14} = x.$ **23.** $\ln 7.3891 = 2.$

25. $e^{1.0986} = 3.$ **27.** 9.   **29.** 125.   **31.** $\dfrac{1}{10}.$ **33.** $e^2.$ **35.** 2.   **37.** 6.

39. 2.   **41.** 4.   **43.** $\dfrac{1}{2}.$ **45.** $\dfrac{1}{81}.$ **47.** 2.   **49.** $\dfrac{5}{3}.$ **51.** $\log_2 5.$ **53.** $\dfrac{\ln 2}{3}.$

55. $\log_3 8.$ **57.** $\dfrac{5 + \ln 3}{2}.$ **59.** 140,000.   **61.** .2241.   **63.** $1491.80.

65. $26,960.   **67.** 41.50.   **69.** .399; .242; .242.   **71.** (a) 91; (b) 432; (c) 8.

EXERCISE 5-2

1. 1.1761.   **3.** .4260.   **5.** 1.5564.   **7.** 3.3010.   **9.** 48.   **11.** −4.

13. 1.5850.   **15.** $\dfrac{1}{3}.$ **17.** 4.   **19.** $\log 28.$ **21.** $\log_2 \dfrac{2x}{x + 1}.$ **23.** $\log [7^9(23)^5].$

25. $\log [100(1.05)^{10}].$ **27.** $\log x + \log (x + 2) + \log (x - 3).$

29. $\dfrac{1}{2}\log x - \log (x + 2) - 2 \log (x - 3).$

31. $\dfrac{1}{2}[2 \log x + 3 \log (x - 3) - \log (x + 2)].$ **33.** 2.   **35.** $\dfrac{1}{12}.$ **37.** $\dfrac{5}{2}.$

39. $\pm 2.$ **41.** 5.   **43.** $\dfrac{4}{3}.$ **45.** 8.   **47.** $\dfrac{\ln (x + 8)}{\ln 10}.$ **49.** (a) 3; (b) 3.78.

51. $p = \dfrac{\log (80 - q)}{\log 2}$; 4.32.

REVIEW PROBLEMS—CHAPTER 5

1. $\log_3 81 = 4$.   **3.** 3.   **5.** $\dfrac{1}{100}$.   **7.** 3.   **9.** 3.   **11.** 1.2620.

13. $2y + \dfrac{1}{2}x$.   **15.** $2x$.   **17.** $y = e^{x^2+2}$.   **19.** 134,064; 109,762.

EXERCISE 6-1

1. (a) $2676.45;   (b) $676.45.   **3.** (a) $1964.76;   (b) $1264.76.

5. (a) $11,105.58;   (b) $5105.58.   **7.** (a) $6256.36;   (b) $1256.36.   **9.** 8.24%.

11. 9.0 years.   **13.** $10,446.14.   **15.** (a) 18%; (b) 19.56%.

17. $3198.54.   **19.** 8% compounded annually.   **21.** (a) 5.47%; (b) 5.39%.

EXERCISE 6-2

1. $2261.33.   **3.** $1751.83.   **5.** $1598.37.   **7.** $14,091.11.

9. $1238.58.   **11.** $1963.28.   **13.** (a) $515.63; (b) profitable.

15. savings account.   **17.** $2550.80.

EXERCISE 6-3

1. 422/243.   **3.** 18.664613.   **5.** 8.213180.   **7.** (a) $1997.13. (b) $3325.37.

9. $1937.14.   **11.** $458.40.   **13.** (a) $3048.85; (b) $648.85.   **15.** $4149.21.

17. $3474.12.   **19.** $1598.44.   **21.** $1725.   **23.** 102.913050.   **25.** $131.34.

EXERCISE 6-4

1. (a) $249.11; (b) $75; (c) $174.11.

3.

PERIOD	PRIN. OUTS. AT BEGINNING	INTEREST FOR PERIOD	PMT. AT END	PRIN. REPAID AT END
1	5000.00	350.00	1476.14	1126.14
2	3873.86	271.17	1476.14	1204.97
3	2668.89	186.82	1476.14	1289.32
4	1379.57	96.57	1476.14	1379.57
		904.56	5904.56	5000.00

5.

PERIOD	PRIN. OUTS. AT BEGINNING	INTEREST FOR PERIOD	PMT. AT END	PRIN. REPAID AT END
1	900.00	22.50	193.72	171.22
2	728.78	18.22	193.72	175.50
3	553.28	13.83	193.72	179.89
4	373.39	9.33	193.72	184.39
5	189.00	4.72	193.72	189.00
		68.60	968.60	900.00

7. 11.   **9.** (a) \$415.28; (b) \$382.50; (c) \$32.78; (d) \$79,584.   **11.** 23.

REVIEW PROBLEMS—CHAPTER 6

1. (a) \$3829.05; (b) \$1229.05.   **3.** 8.5% compounded annually.

5. \$586.60.   **7.** \$1036.85.   **9.** \$886.98.   **11.** \$314.00.

13.

PERIOD	PRIN. OUTS. AT BEGINNING	INTEREST FOR PERIOD	PMT. AT END	PRIN. REPAID AT END
1	15,000.00	112.50	3067.84	2955.34
2	12,044.66	90.33	3067.84	2977.51
3	9,067.15	68.00	3067.84	2999.84
4	6,067.31	45.50	3067.84	3022.34
5	3,044.97	22.84	3067.84	3045.00
		339.17	15,339.20	15,000.03

EXERCISE 7-1

1. 16.   **3.** 7.   **5.** 20.   **7.** 2.82.   **9.** 0.   **11.** -1.   **13.** $-\dfrac{5}{2}$.   **15.** 0.

17. 5.   **19.** 3.   **21.** $-\dfrac{5\sqrt{3}}{6}$.   **23.** 1.   **25.** 0.   **27.** $\dfrac{1}{6}$.   **29.** $-\dfrac{1}{5}$.

31. $\dfrac{7}{2}$.   **33.** $\dfrac{11}{9}$.   **35.** 4.   **37.** $2x$.

EXERCISE 7-2

1. (a) 2;   (b) 3;   (c) Does not exist;   (d) $-\infty$;
(e) ∞;   (f) ∞;   (g) ∞;   (h) 0;   (i) 1;   (j) 1;   (k) 1.

3. 1.   **5.** $-\infty$.   **7.** $-\infty$.   **9.** ∞.   **11.** 0.   **13.** Does not exist.

15. 0.   **17.** 0.   **19.** 1.   **21.** 0.   **23.** ∞.   **25.** $-\dfrac{2}{5}$.   **27.** $-\infty$.

29. $\dfrac{2}{5}$.   **31.** $\dfrac{11}{5}$.   **33.** $-\dfrac{1}{2}$.   **35.** ∞.   **37.** ∞.

39. Does not exist.   **41.** $-\infty$.   **43.** 0.   **45.** 1.

47. (a) 1;   (b) 2;   (c) Does not exist;   (d) 1;   (e) 2.

49. (a) 0;   (b) 0;   (c) 0;   (d) $-\infty$;   (e) $-\infty$.

51.   **53.** 20,000.   **55.** -1.   **57.** $2x$.

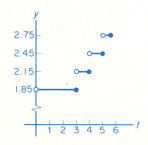

59. 1, .5, .525, .631, .912, .986, .998; conclude limit is 1.

EXERCISE 7-3

1. $5564; $1564.   **3.** $1456.88.   **5.** 4.08%.   **7.** 10.52%.   **9.** $111.63.

11. $740,820.   **13.** 4.88%.   **15.** $1377.   **17.** 22 years.

EXERCISE 7-4

7. Continuous at -2 and 0.   **9.** Discontinuous at ± 3.

11. Continuous at 2 and 0.   **13.** f is a polynomial function.

15. f is a quotient of polynomials and the denominator is never zero.

17. None.   **19.** $x = 4$.   **21.** None.   **23.** $x = -5, 3$.   **25.** $x = 0, \pm 1$.

27. None.   **29.** $x = 2$.   **31.** None.   **33.** $x = 0, 2$.

35. Discontinuities at $t = 3, 4, 5$.   **37.**

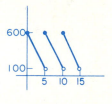

EXERCISE 7-5

1. $x < -1, x > 4$.   **3.** $2 \leqslant x \leqslant 3$.   **5.** $-\dfrac{7}{2} < x < -2$.

7. No solution.   **9.** $x \leqslant -6$, $-2 \leqslant x \leqslant 3$.   **11.** $x < -4$, $0 < x < 5$.

13. $x \geqslant 0$.   **15.** $-3 < x < 0$, $x > 1$.   **17.** $x < -1$, $0 < x < 1$.

19. $x > 1$.   **21.** $x < -5$, $-2 \leqslant x < 1$, $x \geqslant 3$.   **23.** $-4 < x < -2$.

25. $x \leqslant -1 - \sqrt{3}$, $x \geqslant -1 + \sqrt{3}$.   **27.** Between 50 and 150, inclusive.

29. 17 in. by 17 in.

REVIEW PROBLEMS—CHAPTER 7

1. -5.   **3.** 2.   **5.** x.   **7.** 0.   **9.** $\dfrac{3}{5}$.   **11.** Does not exist.   **13.** -1.

15. 0.   **17.** $\dfrac{1}{9}$.   **19.** $-\infty$.   **21.** (a) \$6661.25; (b) \$938.28.   **23.** 6.18%.

27. Continuous everywhere.   **29.** $x = -3$.   **31.** $x = -2$.

33. $x < -6$, $x > 2$.   **35.** $x \geqslant 2$, $x = 0$.   **37.** $x < -5$, $-1 < x < 1$.

39. $x < -4$, $-3 \leqslant x \leqslant 0$, $x > 2$.

EXERCISE 8-1

1. 1.   **3.** 2.   **5.** -2.   **7.** 0.   **9.** $2x + 4$.   **11.** $4p + 5$.   **13.** $-\dfrac{1}{x^2}$.

15. $\dfrac{1}{2\sqrt{x + 2}}$.   **17.** -4.   **19.** 0.   **21.** $y = x + 4$.   **23.** $y = -3x - 7$.

25. $y = -\dfrac{1}{3}x + \dfrac{5}{3}$.

EXERCISE 8-2

1. 0.   **3.** $8x^7$.   **5.** $45x^4$.   **7.** $-7w^{-8}$.   **9.** $-\dfrac{56}{5}x^{-19/5}$.

11. 3.   **13.** $\dfrac{13}{5}$.   **15.** $6x - 5$.   **17.** $-26t + 14 = 2(7 - 13t)$.

19. $42x^2 - 12x + 7$.   **21.** $-9q^2 + 9q + 9 = 9(1 + q - q^2)$.

23. $8x^7 - 42x^5 + 6x = 2x(4x^6 - 21x^4 + 3)$.

25. $1002x^{500} - 12{,}500x^{99} + .68x^{2.4}$.   **27.** $-8x^3$.   **29.** $-\dfrac{4}{3}x^3$.

31. $-4x^{-5} - 3x^{-2/3} - 2x^{-7/5}$.   **33.** $-2(27 - 70x^4) = 2(70x^4 - 27)$.

35. $-4x + \dfrac{3}{2} + x^3$.   **37.** $-x^{-2} = -\dfrac{1}{x^2}$.   **39.** $-\dfrac{5}{4}s^{-6}$.

41. $2t^{-1/2} = \dfrac{2}{\sqrt{t}}$. **43.** $-\dfrac{1}{5}x^{-6/5}$. **45.** $9x^2 - 14x + 7$.

47. $t + 4t^{-3}$. **49.** $45x^4$. **51.** $\dfrac{1}{3}x^{-2/3} - \dfrac{10}{3}x^{-5/3} = \dfrac{1}{3}x^{-5/3}(x - 10)$.

53. $8q + \dfrac{4}{q^2}$. **55.** $2(x + 2)$. **57.** 1. **59.** 4, 16, −14.

61. 0, 0, 0. **63.** $y = 13x - 2$. **65.** $y = x + 3$. **67.** $(0, 0)$, $(2, -\frac{4}{3})$.

EXERCISE 8-3

1. (a) 1; (b) $\dfrac{1}{x + 4}$; (c) 1; (d) $\dfrac{1}{9} \approx .111$; (e) 11.1%.

3. (a) $6x$; (b) $\dfrac{2x}{x^2 + 2}$; (c) 12; (d) $\dfrac{2}{3} \approx .667$; (e) 66.7%.

5. (a) $-3x^2$; (b) $-\dfrac{3x^2}{8 - x^3}$; (c) −3; (d) $-\dfrac{3}{7} \approx -.429$; (e) −42.9%.

7. $\dfrac{dc}{dq} = 10$; 10. **9.** $\dfrac{dc}{dq} = .6q + 2$; 3.8. **11.** $\dfrac{dc}{dq} = 2q + 50$; 80, 82, 84.

13. $\dfrac{dc}{dq} = .02q + 5$; 6, 7. **15.** $\dfrac{dc}{dq} = .00006q^2 - .02q + 6$; 4.6, 11.

17. $\dfrac{dr}{dq} = .7$; .7, .7, .7. **19.** $\dfrac{dr}{dq} = 250 + 90q - 3q^2$; 625, 850, 625.

21. 3.2; 21.3%. **23.** $\dfrac{dc}{dq} = 6.750 - .000656q$; 3.47.

25. (a) $\dfrac{dr}{dq} = 30 - .6q$; (b) $\dfrac{4}{45} \approx .089$; (c) 9%. **27.** $\dfrac{dP}{dR} = -4,650,000R^{-1.93}$.

EXERCISE 8-5

1. $(4x + 1)(6) + (6x + 3)(4) = 48x + 18 = 6(8x + 3)$.

3. $(8 - 7t)(2t) + (t^2 - 2)(-7) = 14 + 16t - 21t^2$.

5. $(3r^2 - 4)(2r - 5) + (r^2 - 5r + 1)(6r) = 12r^3 - 45r^2 - 2r + 20$.

7. $(x^2 + 3x - 2)(4x - 1) + (2x^2 - x - 3)(2x + 3) = 8x^3 + 15x^2 - 20x - 7$.

9. $(8w^2 + 2w - 3)(15w^2) + (5w^3 + 2)(16w + 2) = 200w^4 + 40w^3 - 45w^2 + 32w + 4$.

11. $3[(x^3 - 2x^2 + 5x - 4)(4x^3 - 6x^2 + 7) + (x^4 - 2x^3 + 7x + 1)(3x^2 - 4x + 5)]$
$= 3(7x^6 - 24x^5 + 45x^4 - 28x^3 - 15x^2 + 66x - 23)$.

13. $(x^2 - 1)(9x^2 - 6) + (3x^3 - 6x + 5)(2x) - [(x + 4)(8x + 2) + (4x^2 + 2x + 1)(1)]$
$= 15x^4 - 39x^2 - 26x - 3$.

15. $\frac{3}{2}\left[(p^{1/2} - 4)(4) + (4p - 5)\left(\frac{1}{2}p^{-1/2}\right)\right] = \frac{3}{4}(12p^{1/2} - 5p^{-1/2} - 32).$

17. $(2x^{.45} - 3)(1.3x^{.3} - 7) + (x^{1.3} - 7x)(.9x^{-.55}) = 3.5x^{.75} - 20.3x^{.45} - 3.9x^{.3} + 21.$

19. $18x^2 + 94x + 31.$ **21.** $0.$ **23.** $\dfrac{(x - 1)(1) - (x)(1)}{(x - 1)^2} = -\dfrac{1}{(x - 1)^2}.$

25. $\dfrac{(x - 1)(1) - (x + 2)(1)}{(x - 1)^2} = \dfrac{-3}{(x - 1)^2}.$

27. $\dfrac{(z^2 - 4)(-2) - (5 - 2z)(2z)}{(z^2 - 4)^2} = \dfrac{2(z - 4)(z - 1)}{(z^2 - 4)^2}.$

29. $\dfrac{(x^2 - 5x)(16x - 2) - (8x^2 - 2x + 1)(2x - 5)}{(x^2 - 5x)^2} = \dfrac{-38x^2 - 2x + 5}{(x^2 - 5x)^2}.$

31. $\dfrac{(2x^2 - 3x + 2)(2x - 4) - (x^2 - 4x + 3)(4x - 3)}{(2x^2 - 3x + 2)^2} = \dfrac{5x^2 - 8x + 1}{(2x^2 - 3x + 2)^2}.$

33. $\dfrac{-100x^{99}}{(x^{100} + 1)^2}.$ **35.** $\dfrac{4(v^5 + 2)}{v^2}.$ **37.** $\dfrac{15x^2 - 2x + 1}{3x^{4/3}}.$

39. $\dfrac{4}{(x - 8)^2} + \dfrac{2}{(3x + 1)^2}.$

41. $\dfrac{(s - 5)(2s - 2) - [(s + 2)(s - 4)](1)}{(s - 5)^2} = \dfrac{s^2 - 10s + 18}{(s - 5)^2}.$

43. $\dfrac{[(x + 2)(x - 4)](1) - (x - 5)(2x - 2)}{[(x + 2)(x - 4)]^2} = \dfrac{-(x^2 - 10x + 18)}{[(x + 2)(x - 4)]^2}.$

45. $\dfrac{[(t^2 - 1)(t^3 + 7)](2t + 3) - (t^2 + 3t)(5t^4 - 3t^2 + 14t)}{[(t^2 - 1)(t^3 + 7)]^2}$

$= \dfrac{-3t^6 - 12t^5 + t^4 + 6t^3 - 21t^2 - 14t - 21}{[(t^2 - 1)(t^3 + 7)]^2}.$

47. $\dfrac{(x^2 - 7x + 12)(2x - 3) - (x^2 - 3x + 2)(2x - 7)}{[(x - 3)(x - 4)]^2} = \dfrac{-2(2x^2 - 10x + 11)}{[(x - 3)(x - 4)]^2}.$

49. $3 - \dfrac{2x^3 + 3x^2 - 12x + 4}{[x(x - 1)(x - 2)]^2}.$ **51.** $-6.$ **53.** $y = -\dfrac{3}{2}x + \dfrac{15}{2}.$

55. $y = 16x + 24.$ **57.** $1.5.$ **59.** $\dfrac{dC}{dI} = .672.$ **61.** $\dfrac{1}{3}; \dfrac{2}{3}.$

63. $.615; .385.$ **65.** $\dfrac{dr}{dq} = 25 - .04q.$ **67.** $\dfrac{dr}{dq} = \dfrac{216}{(q + 2)^2} - 3.$

69. $\dfrac{dc}{dq} = \dfrac{5q(q + 6)}{(q + 3)^2}.$

EXERCISE 8-6

1. $(2u - 2)(2x - 1) = 4x^3 - 6x^2 - 2x + 2.$ 3. $\left(-\dfrac{2}{w^3}\right)(-1) = \dfrac{2}{(2 - x)^3}.$

5. $-2.$ 7. $0.$ 9. $56(7x + 4)^7.$ 11. $-56p(3 - 2p^2)^{13}.$

13. $\dfrac{40(3x^2 - 2)(4x^3 - 8x + 2)^9}{3}.$ 15. $-30(4r - 5)(4r^2 - 10r + 3)^{-16}.$

17. $-49x(3x - 2)(x^3 - x^2 + 2)^{-8}.$ 19. $6z(6z - 1)(4z^3 - z^2 + 2)^{-4/5}.$

21. $\dfrac{1}{2}(4x - 1)(2x^2 - x + 3)^{-1/2}.$ 23. $\dfrac{6}{5}x(x^2 + 1)^{-2/5}.$

25. $10\left(\dfrac{x - 7}{x + 4}\right)^9\left[\dfrac{(x + 4)(1) - (x - 7)(1)}{(x + 4)^2}\right] = \dfrac{110(x - 7)^9}{(x + 4)^{11}}.$

27. $10\left(\dfrac{q^3 - 2q + 4}{5q^2 + 1}\right)^4\left[\dfrac{5q^4 + 13q^2 - 40q - 2}{(5q^2 + 1)^2}\right]$

$\qquad = \dfrac{10(5q^4 + 13q^2 - 40q - 2)(q^3 - 2q + 4)^4}{(5q^2 + 1)^6}.$

29. $\dfrac{5}{2(x + 3)^2}\left(\dfrac{x - 2}{x + 3}\right)^{-1/2} = \dfrac{5}{2(x + 3)^2}\sqrt{\dfrac{x + 3}{x - 2}}.$

31. $(x^2 + 2x - 1)^3(5) + (5x + 7)[3(2x + 2)(x^2 + 2x - 1)^2]$

$\qquad = (x^2 + 2x - 1)^2(35x^2 + 82x + 37).$

33. $8[(4x + 3)(6x^2 + x + 8)]^7[(4x + 3)(12x + 1) + (6x^2 + x + 8)(4)]$

$\qquad = 8(72x^2 + 44x + 35)[(4x + 3)(6x^2 + x + 8)]^7.$

35. $\dfrac{(w^2 + 4)\left[6(2w + 3)^2\right] - (2w + 3)^3(2w)}{(w^2 + 4)^2} = \dfrac{2(w^2 - 3w + 12)(2w + 3)^2}{(w^2 + 4)^2}.$

37. $6\{(5x^2 + 2)[2x^3(x^4 + 5)^{-1/2}] + (x^4 + 5)^{1/2}(10x)\}$

$\qquad = 12x(x^4 + 5)^{-1/2}(10x^4 + 2x^2 + 25).$

39. $(4 - 3x^2)^2[3(2 - 3x)^2(-3)] + (2 - 3x)^3[2(4 - 3x^2)(-6x)]$

$\qquad = 3(4 - 3x^2)(2 - 3x)^2(21x^2 - 8x - 12).$

41. $8 + \dfrac{5}{(t + 4)^2} - (8t - 7) = 15 - 8t + \dfrac{5}{(t + 4)^2}.$

43. $\dfrac{(3x - 1)^3\left[40(8x - 1)^4\right] - (8x - 1)^5\left[9(3x - 1)^2\right]}{(3x - 1)^6} = \dfrac{(8x - 1)^4(48x - 31)}{(3x - 1)^4}.$

45. $\dfrac{(x^2 - 7)^4\left[(2x + 1)(2)(3x - 5)(3) + (3x - 5)^2(2)\right] - (2x + 1)(3x - 5)^2\left[4(x^2 - 7)^3(2x)\right]}{(x^2 - 7)^8}.$

47. 0.　　**49.** 0.　　**51.** $y = 4x - 11$.　　**53.** $y = -\frac{1}{6}x + \frac{5}{3}$.　　**55.** 96%.

57. 20.　　**59.** 13.99.

61. (a) $-\dfrac{q}{\sqrt{q^2 + 20}}$;　　(b) $-\dfrac{q}{100\sqrt{q^2 + 20} - q^2 - 20}$;

(c) $100 - \dfrac{q^2}{\sqrt{q^2 + 20}} - \sqrt{q^2 + 20}$.

63. -325.　　**65.** .456; .544.　　**67.** $\dfrac{dc}{dq} = \dfrac{5q(q^2 + 6)}{(q^2 + 3)^{3/2}}$.

EXERCISE 8-7

1. $\dfrac{3}{3x - 4}$.　　**3.** $\dfrac{2}{x}$.　　**5.** $-\dfrac{2x}{1 - x^2}$.　　**7.** $\dfrac{6p^2 + 3}{2p^3 + 3p} = \dfrac{3(2p^2 + 1)}{p(2p^2 + 3)}$.

9. $\dfrac{4 \ln^3(ax)}{x}$.　　**11.** $\dfrac{2x + 4}{x^2 + 4x + 5} = \dfrac{2(x + 2)}{x^2 + 4x + 5}$.　　**13.** $t\left(\dfrac{1}{t}\right) + (\ln t)(1) = 1 + \ln t$.

15. $\dfrac{2 \log_3 e}{2x - 1}$.　　**17.** $\dfrac{2(x^2 + 1)}{2x + 1} + 2x \ln (2x + 1)$.　　**19.** $\dfrac{4x}{x^2 + 2} + \dfrac{3x^2 + 1}{x^3 + x - 1}$.

21. $\dfrac{2}{1 - l^2}$.　　**23.** $\dfrac{x}{1 + x^2}$.　　**25.** $\dfrac{5(x + 1)^4 + 4(x + 2)^3 + 8x^7}{(x + 1)^5 + (x + 2)^4 + x^8}$.　　**27.** $\dfrac{x}{1 - x^4}$.

29. $\dfrac{z\left(\dfrac{1}{z}\right) - (\ln z)(1)}{z^2} = \dfrac{1 - \ln z}{z^2}$.　　**31.** $\dfrac{x}{2(x - 1)} + \ln\sqrt{x - 1}$.

33. $\dfrac{1}{2x\sqrt{4 + \ln x}}$.　　**35.** $y = 4x - 12$.　　**37.** $\dfrac{dr}{dq} = 25\dfrac{(q + 2) \ln (q + 2) - q}{(q + 2) \ln^2 (q + 2)}$.

EXERCISE 8-8

1. $2xe^{x^2 + 1}$.　　**3.** $-5e^{3 - 5x}$.　　**5.** $(6r + 4)e^{3r^2 + 4r + 4} = 2(3r + 2)e^{3r^2 + 4r + 4}$.

7. $x(e^x) + e^x(1) = e^x(x + 1)$.　　**9.** $2xe^{-x^2}(1 - x^2)$.　　**11.** $\dfrac{e^x - e^{-x}}{2}$.

13. $(6x)4^{3x^2} \ln 4$.　　**15.** $\dfrac{2e^{2w}(w - 1)}{w^3}$.　　**17.** $\dfrac{e^{1 + \sqrt{x}}}{2\sqrt{x}}$.

19. $3x^2 - 3^x \ln 3$.　　**21.** $\dfrac{2e^x}{(e^x + 1)^2}$.　　**23.** $e^{e^x}e^x = e^{e^x + x}$.

25. 1.　　**27.** $e^{x \ln x}(1 + \ln x)$.　　**29.** $(\log 2)^x \ln (\log 2)$.

31. $y - e^2 = e^2(x - 2)$ or $y = e^2x - e^2$.   **33.** $\dfrac{dp}{dq} = -.015e^{-.001q}$; $-.015e^{-.5}$.

37. $\dfrac{dc}{dq} = 10e^{(q+3)/400}$; $10e^{.25}$, $10e^{.5}$.   **39.** 0.

EXERCISE 8-9

1. $-\dfrac{x}{4y}$.   **3.** $-\dfrac{y}{x}$.   **5.** $\dfrac{4-y}{x-1}$.   **7.** $\dfrac{4y-x^2}{y^2-4x}$.   **9.** $-\dfrac{y^{1/4}}{x^{1/4}} = -\left(\dfrac{y}{x}\right)^{1/4}$.

11. $\dfrac{5}{12y^3}$.   **13.** $-\dfrac{\sqrt{y}}{\sqrt{x}} = -\sqrt{\dfrac{y}{x}}$.   **15.** $\dfrac{6y^{2/3}}{3y^{1/6}+2}$.   **17.** $\dfrac{1-6xy^3}{1+9x^2y^2}$.

19. $\dfrac{xe^y - y}{x(\ln x - xe^y)}$.   **21.** $-\dfrac{e^y}{xe^y + 1}$.   **23.** $-\dfrac{3}{5}$.   **25.** $y = -\dfrac{3}{4}x + \dfrac{5}{4}$.

27. $\dfrac{dq}{dp} = -\dfrac{1}{2q}$.   **29.** $\dfrac{dq}{dp} = -\dfrac{(q+5)^3}{40}$.

EXERCISE 8-10

1. $y\left[\dfrac{2}{x+1} + \dfrac{1}{x-1} + \dfrac{2x}{x^2+3}\right]$.   **3.** $y\left[\dfrac{18x^2}{3x^3-1} + \dfrac{6}{2x+5}\right]$.

5. $\dfrac{y}{2}\left[\dfrac{1}{x+1} + \dfrac{2x}{x^2-2} + \dfrac{1}{x+4}\right]$.   **7.** $y\left[\dfrac{4x}{x^2+1} - \dfrac{2}{x+1} - \dfrac{3}{3x+2}\right]$.

9. $y\left[\dfrac{4}{8x+3} + \dfrac{2x}{3(x^2+2)} - \dfrac{1}{2(1+2x)}\right]$.   **11.** $y\left[\dfrac{x}{x^2-1} + \dfrac{2}{1-2x}\right]$.

13. $x^{2x+1}\left(\dfrac{2x+1}{x} + 2\ln x\right)$.   **15.** $\dfrac{x^{1/x}(1-\ln x)}{x^2}$.

17. $xy(1+2\ln x) = x^{x^2+1}(1+2\ln x)$.   **19.** $e^x x^{3x}(4+3\ln x)$.

21. $y = 96x + 36$.   **23.** $x^x(1+\ln x)$.

EXERCISE 8-11

1. 24.   **3.** 0.   **5.** e^x.   **7.** $3 + 2\ln x$.   **9.** $-\dfrac{10}{p^6}$.   **11.** $-\dfrac{1}{4(1-r)^{3/2}}$.

13. $\dfrac{50}{(5x-6)^3}$.   **15.** $\dfrac{4}{(x-1)^3}$.   **17.** $-\left[\dfrac{1}{x^2} + \dfrac{1}{(x+1)^2}\right]$.   **19.** $e^z(z^2+4z+2)$.

21. $-\dfrac{1}{y^3}$.   **23.** $-\dfrac{4}{y^3}$.   **25.** $\dfrac{1}{8x^{3/2}}$.   **27.** $\dfrac{2(y-1)}{(1+x)^2}$.   **29.** $\dfrac{2y}{(2-y)^3}$.

31. $300(5x-3)^2$.   **33.** .6.   **35.** ± 1.

REVIEW PROBLEMS—CHAPTER 8

1. 0.   **3.** $28x^3 - 18x^2 + 10x = 2x(14x^2 - 9x + 5)$.

5. $2e^x + e^{x^2}(2x) = 2(e^x + xe^{x^2})$.   **7.** $\dfrac{2x}{5}$.   **9.** $\dfrac{1}{r^2 + 5r}(2r + 5) = \dfrac{2r + 5}{r(r + 5)}$.

11. $(x^2 + 6x)(3x^2 - 12x) + (x^3 - 6x^2 + 4)(2x + 6) = 5x^4 - 108x^2 + 8x + 24$.

13. $100(2x^2 + 4x)^{99}(4x + 4) = 400(x + 1)[(2x)(x + 2)]^{99}$.

15. $(8 + 2x)(4)(x^2 + 1)^3(2x) + (x^2 + 1)^4(2) = 2(x^2 + 1)^3(9x^2 + 32x + 1)$.

17. $\dfrac{4}{3}(4x - 1)^{-2/3}$.   **19.** $e^x(2x) + (x^2 + 2)e^x = e^x(x^2 + 2x + 2)$.

21. $\dfrac{(z^2 + 1)(2z) - (z^2 - 1)(2z)}{(z^2 + 1)^2} = \dfrac{4z}{(z^2 + 1)^2}$.

23. $\dfrac{e^x\left(\dfrac{1}{x}\right) - (\ln x)(e^x)}{e^{2x}} = \dfrac{1 - x \ln x}{xe^x}$.

25. $e^{x^2 + 4x + 5}(2x + 4) = 2(x + 2)e^{x^2 + 4x + 5}$.

27. $\dfrac{(2x^2 + 3)(2x + 1) - (x^2 + x)(4x)}{(2x^2 + 3)^2} = \dfrac{-2x^2 + 6x + 3}{(2x^2 + 3)^2}$.   **29.** $-\dfrac{y}{x + y}$.

31. $\dfrac{y}{2}\left[\dfrac{1}{x - 6} + \dfrac{1}{x + 5} - \dfrac{1}{9 - x}\right] = \dfrac{y}{2}\left[\dfrac{1}{x - 6} + \dfrac{1}{x + 5} + \dfrac{1}{x - 9}\right]$.

33. $\dfrac{2}{q + 1} + \dfrac{3}{q + 2}$.   **35.** $-\dfrac{1}{2}(1 - x)^{-3/2}(-1) = \dfrac{1}{2}(1 - x)^{-3/2}$.

37. $\dfrac{16 \log_2 e}{8x + 5}$.   **39.** $y[1 + \ln (x + 1)]$.

41. $\dfrac{\sqrt{x^2 + 5}\,(2x) - (x^2 + 6)(1/2)(x^2 + 5)^{-1/2}(2x)}{x^2 + 5} = \dfrac{x(x^2 + 4)}{(x^2 + 5)^{3/2}}$.   **43.** $-\dfrac{y}{x}$.

45. $2\left(-\dfrac{3}{8}\right)x^{-11/8} + \left(-\dfrac{3}{8}\right)(2x)^{-11/8}(2) = -\dfrac{3}{4}(1 + 2^{-11/8})x^{-11/8}$.

47. $\dfrac{1 + 2l + 3l^2}{1 + l + l^2 + l^3}$.

49. $\left(\dfrac{3}{5}\right)(x^3 + 6x^2 + 9)^{-2/5}(3x^2 + 12x) = \dfrac{9}{5}x(x + 4)(x^3 + 6x + 9)^{-2/5}$.

51. $2\left(\dfrac{1}{u}\right) + \dfrac{1}{2}\left(\dfrac{1}{1 - u}\right)(-1) = \dfrac{5u - 4}{2u(u - 1)}$.

53. $y\left[\dfrac{3}{2}\left(\dfrac{1}{x^2 + 2}\right)(2x) + \dfrac{4}{9}\left(\dfrac{1}{x^2 + 9}\right)(2x) - \dfrac{4}{11}\left(\dfrac{1}{x^3 + 6x}\right)(3x^2 + 6)\right]$

$= y\left[\dfrac{3x}{x^2 + 2} + \dfrac{8x}{9(x^2 + 9)} - \dfrac{12(x^2 + 2)}{11x(x^2 + 6)}\right]$.

55. 12. **57.** $-\dfrac{1}{128}$. **59.** $\dfrac{4}{9}$. **61.** $-\dfrac{1}{4}$. **63.** $y = -4x + 3$.

65. $y = 2x + 2(1 - \ln 2)$ or $y = 2x + 2 - \ln 4$. **67.** $7x - 2\sqrt{10}\, y - 9 = 0$.

69. $\dfrac{5}{7} \approx .714$; 71.4%. **71.** $\dfrac{dr}{dq} = 20 - .2q$. **73.** .569, .431.

75. $\dfrac{dr}{dq} = 450 - q$. **77.** $\dfrac{dc}{dq} = .125 + .00878q$; .7396.

EXERCISE 9-1

1. $(0, 0)$; sym. to origin. **3.** $(\pm 2, 0)$, $(0, 8)$; sym. to y-axis.

5. $(\pm 3, 0)$; sym. to x-axis, y-axis, origin. **7.** $(-2, 0)$; sym. to x-axis.

9. Sym. to x-axis. **11.** $(-21, 0)$, $(0, -7)$, $(0, 3)$. **13.** $(0, 0)$; sym. to origin.

15. $(\pm \sqrt{\ln 5}, 0)$, $(0, \pm \sqrt{\ln 5})$; sym. to x-axis, y-axis, origin.

17. $(0, 0)$; sym. to x-axis, y-axis, origin. **19.** $(2, 0)$, $(0, \pm 2)$; sym. to x-axis.

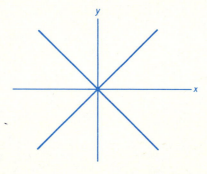 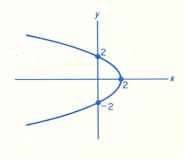

21. $(\pm 2, 0)$, $(0, 0)$; sym. to origin.

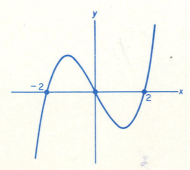

23. $(\pm 2, 0)$, $(0, \pm 4)$; sym. to x-axis, y-axis, origin.

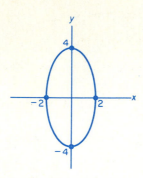

EXERCISE 9-2

1. $y = 0$, $x = 0$.   **3.** $y = 4$, $x = 6$.   **5.** None.   **7.** $y = \dfrac{1}{2}$, $x = -\dfrac{3}{2}$.

9. $y = 2$, $x = -3$, $x = 2$.   **11.** None.   **13.** $y = 4$.

15. $(0, -3)$; $y = 0$, $x = 1$.   **17.** Sym. to origin;   **19.** $(0, -1)$; sym. to y-axis;
                                              $y = 0$, $x = 0$. $y = 0$, $x = 1$, $x = -1$.

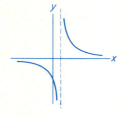
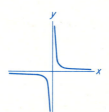
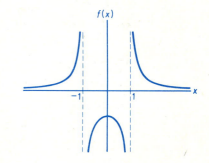

21. $(\pm 1, 0)$, $(0, \frac{1}{4})$;   **23.** $(\pm 3, 0)$; sym. to y-axis.   **25.** $((\ln 3)/2, 0)$, $(0, 2)$; $y = 3$.
sym. to y-axis;
$y = 1$, $x = 2$, $x = -2$.

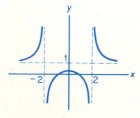
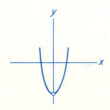

1. Dec. on $(-\infty, 0)$; inc. on $(0, \infty)$; rel. min. when $x = 0$.

3. Inc. on $\left(-\infty, \dfrac{1}{2}\right)$; dec. on $\left(\dfrac{1}{2}, \infty\right)$; rel. max. when $x = \dfrac{1}{2}$.

5. Dec. on $(-\infty, -5)$ and $(1, \infty)$; inc. on $(-5, 1)$; rel. min. when $x = -5$; rel. max. when $x = 1$.

7. Dec. on $(-\infty, -1)$ and $(0, 1)$; inc. on $(-1, 0)$ and $(1, \infty)$; rel. max. when $x = 0$; rel. min. when $x = \pm 1$.

9. Inc. on $(-\infty, 1)$ and $(3, \infty)$; dec. on $(1, 3)$; rel. max. when $x = 1$; rel. min. when $x = 3$.

11. Inc. on $(-\infty, -1)$ and $(1, \infty)$; dec. on $(-1, 0)$ and $(0, 1)$; rel. max. when $x = -1$; rel. min. when $x = 1$.

13. Dec. on $(-\infty, -4)$ and $(0, \infty)$; inc. on $(-4, 0)$; rel. min. when $x = -4$; rel. max. when $x = 0$.

15. Dec. on $(-\infty, 1)$ and $(1, \infty)$; no rel. max. or min.

17. Dec. on $(0, \infty)$; no rel. max. or min.

19. Dec. on $(-\infty, 0)$ and $(2, \infty)$; inc. on $(0, 1)$ and $(1, 2)$; rel. min. when $x = 0$; rel. max. when $x = 2$.

21. Inc. on $(-\infty, -2)$, $(-2, 11/5)$, and $(5, \infty)$; dec. on $(11/5, 5)$; rel. max. when $x = 11/5$; rel. min. when $x = 5$.

23. Dec. on $(-\infty, \infty)$; no rel. max. or min.

25. Dec. on $(0, 1)$; inc. on $(1, \infty)$; rel. min. when $x = 1$.

27. Dec. on $(-\infty, 0)$; inc. on $(0, \infty)$; rel. min. when $x = 0$.

29. Dec. on $(-\infty, 3)$; inc. on $(3, \infty)$; rel. min. when $x = 3$; intercepts: $(7, 0)$, $(-1, 0)$, $(0, -7)$.

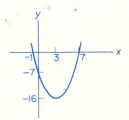

31. Dec. on $(-\infty, -1)$ and $(1, \infty)$; inc. on $(-1, 1)$; rel. min. when $x = -1$; rel. max. when $x = 1$; sym. to origin; intercepts: $(\pm \sqrt{3}, 0)$, $(0, 0)$.

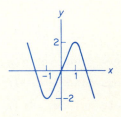

33. Inc. on $(-\infty, 1)$ and $(2, \infty)$; dec. on $(1, 2)$; rel. max. when $x = 1$; rel. min. when $x = 2$; intercept: $(0, 0)$.

35. Inc. on $(-2, -1)$ and $(0, \infty)$; dec. on $(-\infty, -2)$ and $(-1, 0)$; rel. max. when $x = -1$; rel. min. when $x = -2, 0$; intercepts $(0, 0)$, $(-2, 0)$.

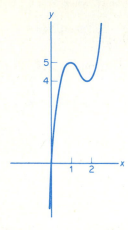

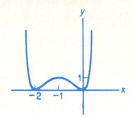

37. Dec. on $(-\infty, 1)$ and $(1, \infty)$; asym. $y = 1$, $x = 1$; intercepts: $(0, -1)$, $(-1, 0)$.

39. Inc. on $(-\infty, -6)$ and $(0, \infty)$; dec. on $(-6, -3)$ and $(-3, 0)$; rel. max. when $x = -6$; rel. min. when $x = 0$; asym. $x = -3$; intercept: $(0, 0)$.

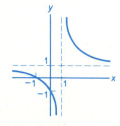

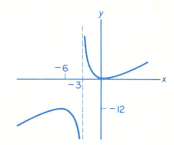

41. Ab. max. when $x = -1$; ab. min. when $x = 1$.

43. Ab. max. when $x = 0$; ab. min. when $x = 2$.

45. Ab. max. when $x = 3$; ab. min. when $x = 1$.   **49.** Never.   **51.** 40.

EXERCISE 9-4

1. Conc. down $(-\infty, \infty)$.

3. Conc. down $(-\infty, -1)$; conc. up $(-1, \infty)$; inf. pt. when $x = -1$.

5. Conc. up $(-\infty, -1)$, $(1, \infty)$; conc. down $(-1, 1)$; inf. pt. when $x = \pm 1$.

7. Conc. down $(-\infty, 1)$; conc. up $(1, \infty)$.

9. Conc. down $\left(-\infty, -\dfrac{1}{\sqrt{3}}\right)$, $\left(\dfrac{1}{\sqrt{3}}, \infty\right)$; conc. up $\left(-\dfrac{1}{\sqrt{3}}, \dfrac{1}{\sqrt{3}}\right)$;

inf. pt. when $x = \pm \dfrac{1}{\sqrt{3}}$.

11. Conc. up $(-\infty, \infty)$.

13. Conc. down $(-\infty, -2)$; conc. up $(-2, \infty)$; inf. pt. when $x = -2$.

15. Int. $(-3, 0)$, $(-1, 0)$, $(0, 3)$; dec. $(-\infty, -2)$; inc. $(-2, \infty)$; rel. min. when $x = -2$; conc. up $(-\infty, \infty)$.

17. Int. $(0, 0)$, $(4, 0)$; inc. $(-\infty, 2)$; dec. $(2, \infty)$; rel. max. when $x = 2$; conc. down $(-\infty, \infty)$.

19. Int. $(-3/2, 0)$, $(4, 0)$, $(0, -12)$; dec. $(-\infty, 5/4)$; inc. $(5/4, \infty)$; rel. min. when $x = 5/4$; conc. up $(-\infty, \infty)$.

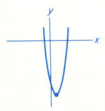

21. Int. $(0, -19)$; inc. $(-\infty, 2)$, $(4, \infty)$; dec. $(2, 4)$; rel. max. when $x = 2$; rel. min. when $x = 4$; conc. down $(-\infty, 3)$; conc. up $(3, \infty)$; inf. pt. when $x = 3$.

23. Int. $(0, 0)$, $(\pm 3, 0)$; inc. $(-\infty, -\sqrt{3})$, $(\sqrt{3}, \infty)$; dec. $(-\sqrt{3}, \sqrt{3})$; rel. max. when $x = -\sqrt{3}$; rel. min. when $x = \sqrt{3}$; conc. down $(-\infty, 0)$; conc. up $(0, \infty)$; inf. pt. when $x = 0$; sym. to origin.

25. Int. $(0, -3)$; inc. $(-\infty, 1)$, $(1, \infty)$; no rel. max. or min.; conc. down $(-\infty, 1)$; conc. up $(1, \infty)$; inf. pt. when $x = 1$.

27. Int. $(0, 0)$, $(\pm 2, 0)$; inc. $(-\infty, -\sqrt{2})$, $(0, \sqrt{2})$; dec. $(-\sqrt{2}, 0)$, $(\sqrt{2}, \infty)$; rel. max. when $x = \pm\sqrt{2}$; rel. min. when $x = 0$; conc. down $(-\infty, -\sqrt{2/3})$, $(\sqrt{2/3}, \infty)$; conc. up $(-\sqrt{2/3}, \sqrt{2/3})$; inf. pt. when $x = \pm\sqrt{2/3}$; sym. to y-axis.

29. Int. $(0, 0)$, $(4/3, 0)$; inc. $(-\infty, 0)$, $(0, 1)$; dec. $(1, \infty)$; rel. max. when $x = 1$; conc. up $(0, 2/3)$; conc. down $(-\infty, 0)$, $(2/3, \infty)$; inf. pt. when $x = 0$, $x = 2/3$.

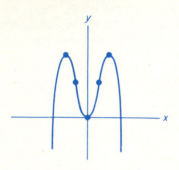

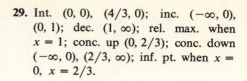

31. Int. $(0, -2)$; dec. $(-\infty, -2)$, $(2, \infty)$; inc. $(-2, 2)$; rel. min. when $x = -2$; rel. max. when $x = 2$; conc. up $(-\infty, 0)$; conc. down $(0, \infty)$; inf. pt. when $x = 0$.

33. Int. $(0, -6)$; inc. $(-\infty, 2)$, $(2, \infty)$; conc. down $(-\infty, 2)$; conc. up $(2, \infty)$; inf. pt. when $x = 2$.

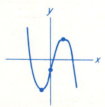

35. Int. $(0, 0)$, $(\pm\sqrt[4]{5}, 0)$; dec. $(-\infty, -1)$, $(1, \infty)$; inc. $(-1, 1)$; rel. min. when $x = -1$; rel. max. when $x = 1$; conc. up $(-\infty, 0)$; conc. down $(0, \infty)$; inf. pt. when $x = 0$; sym. to origin.

37. Int. $(0, 1)$, $(1, 0)$; dec. $(-\infty, 0)$, $(0, 1)$; inc. $(1, \infty)$; rel. min. when $x = 1$; conc. up $(-\infty, 0)$, $(2/3, \infty)$; conc. down $(0, 2/3)$; inf. pt. when $x = 0$, $x = 2/3$.

39. Dec. $(-\infty, 0)$, $(0, \infty)$; conc. down $(-\infty, 0)$; conc. up $(0, \infty)$; sym. to origin; asymptotes $x = 0$, $y = 0$.

41. Int. $(0, 0)$; inc. $(-\infty, -1)$, $(-1, \infty)$; conc. up $(-\infty, -1)$; conc. down $(-1, \infty)$; asymptotes $x = -1$, $y = 1$.

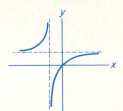

43. Dec. $(-\infty, -1)$, $(0, 1)$; inc. $(-1, 0)$, $(1, \infty)$; rel. min. when $x = \pm 1$; conc. up $(-\infty, 0)$, $(0, \infty)$; sym. to y-axis; asymptote $x = 0$.

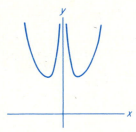

EXERCISE 9-5

1. Rel. min. when $x = \dfrac{5}{2}$; abs. min.   **3.** Rel max. when $x = \dfrac{1}{4}$; abs. max.

5. Rel. max. when $x = -3$, rel. min. when $x = 3$.

7. Rel. min. when $x = 0$, rel. max. when $x = 2$.

9. Test fails, when $x = 0$ there is rel. min. by first-deriv. test.

EXERCISE 9-6

1. 100.   **3.** \$15.   **5.** 525, \$51, \$10,525.   **7.** \$22.   **9.** 625, \$4.

11. 750.   **13.** 20 and 20.   **15.** 300 ft by 250 ft.   **17.** 4 ft by 4 ft by 2 ft.

19. 21 in. by 14 in.   **23.** 130, $p = 340$, $P = 36,980$; 125, $p = 350$, $P = 34,175$.

25. 250 per lot (4 lots).   **27.** 35.   **29.** 5000.   **31.** $5 - \sqrt{3}$ tons, $5 - \sqrt{3}$ tons.

EXERCISE 9-7

1. $3\,dx$.   **3.** $\dfrac{2x^3}{\sqrt{x^4+2}}\,dx$.   **5.** $-\dfrac{2}{x^3}\,dx$.   **7.** $\dfrac{2x}{x^2+7}\,dx$.

9. $4e^{2x^2+3}(4x^2+3x+1)\,dx$.   **11.** $-.14$.   **13.** $\dfrac{2}{15}$.   **15.** 10.05.

17. $3\frac{47}{48}$.   **19.** $-.03$.   **21.** 1.01.   **23.** $\dfrac{1}{2}$.   **25.** $\dfrac{1}{6p(p^2+5)^2}$.

27. $-p^2$.   **29.** $-\dfrac{4}{5}$.   **31.** $44;\ 41.80$.   **33.** 2.04.   **35.** $.7$.

EXERCISE 9-8

1. -3, elastic.   **3.** -1, unit elasticity.   **5.** -1, unit elasticity.

7. -1.02, elastic.   **9.** $-\dfrac{9}{32}$, inelastic.   **11.** $-\dfrac{1}{2}$, inelastic.

13. $-(150e^{-1}-1)$, elastic.

15. $|\eta|=\dfrac{10}{3}$ when $p=10$, $|\eta|=\dfrac{3}{10}$ when $p=3$, $|\eta|=1$ when $p=6.50$.

17. -1.2, $.6$ percent decrease.   **23.** $5,\ 30$.

REVIEW PROBLEMS—CHAPTER 9

1. Int. $(-4, 0)$, $(6, 0)$, $(0, -24)$; inc. $(1, \infty)$; dec. $(-\infty, 1)$; rel. min. when $x=1$; conc. up $(-\infty, \infty)$.

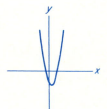

3. Int. $(0, 20)$; inc. $(-\infty, -2)$, $(2, \infty)$; dec. $(-2, 2)$; rel. max. when $x=-2$; rel. min. when $x=2$; conc. up $(0, \infty)$; conc. down $(-\infty, 0)$; inf. pt. when $x=0$.

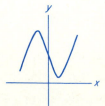

5. Int. $(0, 0)$; inc. $(-\infty, \infty)$; conc. down $(-\infty, 0)$; conc. up $(0, \infty)$; inf. pt. when $x = 0$; sym. to origin.

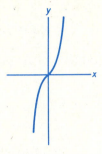

7. Int. $(-5, 0)$; inc. $(-10, 0)$; dec. $(-\infty, -10)$, $(0, \infty)$; rel. min. when $x = -10$; conc. up $(-15, 0)$, $(0, \infty)$; conc. down $(-\infty, -15)$; inf. pt. when $x = -15$; horiz. asym. $y = 0$; vert. asym. $x = 0$.

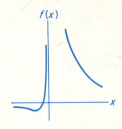

9. Int. $(0, 1)$; inc. $(0, \infty)$; dec. $(-\infty, 0)$; rel. min. when $x = 0$; conc. up $(-\infty, \infty)$; sym. to y-axis.

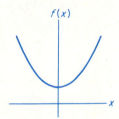

11. 20.   **13.** \$2800.   **15.** $\left[\dfrac{x^2}{x + 5} + 2x \ln (x + 5) \right] dx$.   **17.** .99.

19. Inelastic.   **21.** Elastic.   **23.** 2 in.   **25.** 20 in. by 25 in.

EXERCISE 10-1

1. $5x + C$.   **3.** $\dfrac{x^9}{9} + C$.   **5.** $\dfrac{2}{15} t^{15/2} + C$.   **7.** $-\dfrac{1}{6x^6} + C$.

9. $-\dfrac{1}{9x^9} + C$.   **11.** $-\dfrac{5}{6y^{6/5}} + C$.   **13.** $\dfrac{5\sqrt[5]{x^{11}}}{11} + C = \dfrac{5x^2\sqrt[5]{x}}{11} + C$.

15. $8\sqrt[8]{x} + C$.   **17.** $(7 + e)x + C$.   **19.** $\dfrac{x^{\sqrt{2} + 1}}{\sqrt{2} + 1} + C$.   **21.** $\dfrac{3x^8}{8} + C$.

23. $8u + \dfrac{u^2}{2} + C.$ **25.** $\dfrac{y^6}{6} + \dfrac{5y^2}{2} + C.$ **27.** $t^3 - 2t^2 - 5t + C.$

29. $\dfrac{x^2}{14} - \dfrac{3x^5}{20} + C.$ **31.** $\dfrac{4x^{3/2}}{9} + C.$ **33.** $3e^x + C.$ **35.** $6z - \dfrac{5z^3}{12} + 2e^z + C.$

37. $\dfrac{e^u}{4} + u + C.$ **39.** $-\dfrac{1}{x} + \dfrac{5}{2x^2} - \dfrac{2}{3x^3} + C.$

41. $\dfrac{x^{9.3}}{9.3} - \dfrac{9x^7}{7} - \dfrac{1}{x^3} - \dfrac{1}{2x^2} + C.$ **43.** $\dfrac{x^4}{12} + \dfrac{3}{2x^2} + C.$

45. $\dfrac{w^3}{2} + \dfrac{2}{3w} + C.$ **47.** $\dfrac{3x^{4/3}}{4} - \dfrac{4x^{5/4}}{5} + \dfrac{5x^{6/5}}{6} + C.$ **49.** $\dfrac{4x^{3/2}}{3} - \dfrac{12x^{5/4}}{5} + C.$

51. $-\dfrac{10}{x^{1/5}} + C.$ **53.** $-\dfrac{3x^{5/3}}{25} - 7x^{1/2} + 3x^2 + C.$

55. $\dfrac{x^4}{4} - x^3 + \dfrac{5x^2}{2} - 15x + C.$ **57.** $\dfrac{2x^{5/2}}{5} + 2x^{3/2} + C.$

59. $\dfrac{4u^3}{3} + 2u^2 + u + C.$ **61.** $\dfrac{2v^3}{3} + 3v + \dfrac{1}{2v^4} + C.$ **63.** $\dfrac{1}{2}(e^6x + e^x) + C.$

65. $y = \dfrac{3x^2}{2} - 4x + 1.$ **67.** $y = -\dfrac{x^4}{12} - \dfrac{x^3}{3} + \dfrac{4x}{3} + \dfrac{1}{12}.$

69. $y = \dfrac{x^4}{12} + x^2 - 5x + 2.$ **71.** $c = 1.35q + 200.$ **73.** 7715   **75.** $p = .7.$

77. $p = 275 - .5q - .1q^2.$

79. \$80. $(dc/dq = 27.50$ when $q = 50$ is not relevant to problem.)

EXERCISE 10-2

1. $\dfrac{(x + 4)^9}{9} + C.$ **3.** $\dfrac{(x^2 + 16)^4}{4} + C.$ **5.** $\dfrac{3}{5}(y^3 + 3y^2 + 1)^{5/3} + C.$

7. $-\dfrac{(3x - 1)^{-2}}{2} + C.$ **9.** $e^{3x} + C.$ **11.** $e^{t^2 + t} + C.$ **13.** $\dfrac{2(x + 10)^{3/2}}{3} + C.$

15. $\dfrac{(7x - 6)^5}{35} + C.$ **17.** $\dfrac{(x^2 + 3)^{13}}{26} + C.$ **19.** $\dfrac{3}{20}(27 + x^5)^{4/3} + C.$

21. $\dfrac{1}{10}e^{5x^2} + C.$ **23.** $-3e^{-2x} + C.$ **25.** $\ln|x + 5| + C.$

27. $\ln|x^3 + x^4| + C.$ **29.** $-\dfrac{3}{4}(z^2 - 6)^{-4} + C.$ **31.** $4\ln|x| + C.$ **33.** $\dfrac{1}{3}\ln|s^3 + 5| + C.$

35. $-\dfrac{7}{3}\ln|5 - 3x| + C.$ **37.** $\dfrac{2}{15}(5x)^{3/2} + C = \dfrac{2}{3}x\sqrt{5x} + C.$ **39.** $\sqrt{x^2 - 4} + C.$

41. $\frac{1}{2}e^{y^4+1} + C.$ **43.** $-\frac{1}{6}e^{-2v^3+1} + C.$ **45.** $-\frac{1}{5}e^{-5x} + 2e^x + C.$

47. $-\frac{1}{24}(3 - 3x^2 - 6x)^4 + C.$ **49.** $\frac{1}{3}\ln|x^3 + 6x| + C.$ **51.** $2\ln|3 - 2s + 4s^2| + C.$

53. $\frac{1}{4}\ln(2x^2 + 1) + C.$ **55.** $\frac{1}{27}(x^3 - x^6)^{-9} + C.$ **57.** $\frac{1}{4}(x^4 + x^2)^2 + C.$

59. $\frac{1}{2}(4 - 9x - 3x^2)^{-4} + C.$ **61.** $\frac{1}{6}e^{4x^3+3x^2-4} + C.$ **63.** $-\frac{1}{25}(7 - 5x^2)^{5/2} + C.$

65. $\sqrt{2x} + C.$ **67.** $\frac{x^5}{5} + \frac{2x^3}{3} + x + C.$ **69.** $\frac{1}{2}\ln(x^2 + 1) - \frac{1}{6(x^6 + 1)} + C.$

71. $\frac{1}{3}\ln|3x - 5| + \frac{1}{27}(x^3 - x^6)^{-9} + C.$ **73.** $\frac{1}{3}(2x + 3)^{3/2} - \ln\sqrt{x^2 + 3} + C.$

75. $2e^{\sqrt{x}} + C.$ **77.** $y = -\frac{1}{6}(3 - 2x)^3 + \frac{11}{2}.$ **79.** $y = -\ln|x| = \ln|1/x|.$

EXERCISE 10-3

1. $\frac{3}{2}x^2 + x - \ln|x| + C.$ **3.** $\frac{1}{3}(2x^3 + 4x + 1)^{3/2} + C.$

5. $-\frac{8}{3}\sqrt{2 - 3x} + C.$ **7.** $\frac{4^{7x}}{7\ln 4} + C.$ **9.** $7x^2 - 4e^{x^2/4} + C.$

11. $\frac{3}{2}\ln(e^{2x} + 1) + C.$ **13.** $-\frac{1}{7}e^{7/x} + C.$ **15.** $\frac{2}{9}(\sqrt{x} + 2)^3 + C.$

17. $\frac{1}{2}(\ln^2 x) + C.$ **19.** $\frac{1}{3}\ln^3(r + 1) + C.$ **21.** $e^{(x^2+3)/2} + C.$

23. $\ln|\ln(x + 3)| + C.$ **25.** $\frac{1}{2}\sqrt{x^4 - 1} - (\ln 4)x + C.$

27. $x^2 - 8x - 6\ln|x| - \frac{2}{x^2} + C.$ **29.** $x^2 - 3x + \frac{2}{3}\ln|3x - 1| + C.$

31. $x + \ln|x - 1| + C.$ **33.** $\sqrt{e^{x^2} + 2} + C.$ **35.** $-\frac{(e^{-x} + 6)^3}{3} + C.$

37. $\frac{1}{36\sqrt{2}}[(8x)^{3/2} + 3]^{3/2} + C.$ **39.** $-\frac{2}{3}e^{-\sqrt{s^3}} + C.$ **41.** $\frac{x^2}{2} + 2x + C.$

43. $p = \frac{100}{q + 2}.$ **45.** $c = 20\ln|(q + 5)/5| + 2000.$ **47.** $C = 2(\sqrt{I} + 1).$

49. $C = \frac{3}{4}I - \frac{1}{3}\sqrt{I} + \frac{71}{12}.$

EXERCISE 10-4

1. 35.   **3.** 0.   **5.** 25.   **7.** $-\dfrac{3}{16}$.   **9.** $-\dfrac{7}{6}$.   **11.** $\displaystyle\sum_{k=1}^{15} k$.   **13.** $\displaystyle\sum_{k=1}^{4} (2k-1)$.

15. $\displaystyle\sum_{k=1}^{12} k^2$.   **17.** 101,475.   **19.** 84.   **21.** 273.   **23.** 8; $850.

EXERCISE 10-5

1. $\dfrac{2}{3}$ sq unit.   **3.** $\dfrac{14}{27}$ sq unit.   **5.** $\dfrac{1}{2}$ sq unit.   **7.** $\dfrac{1}{3}$ sq unit.

9. $\dfrac{16}{3}$ sq units.   **11.** 6.   **13.** -18.   **15.** $\dfrac{5}{6}$.

EXERCISE 10-6

1. 12.   **3.** $\dfrac{9}{2}$.   **5.** $\dfrac{100}{3}$.   **7.** -24.   **9.** $\dfrac{14}{3}$.   **11.** $\dfrac{7}{3}$.   **13.** $\dfrac{15}{2}$.

15. $-\dfrac{7}{6}$.   **17.** 0.   **19.** $\dfrac{5}{3}$.   **21.** $\dfrac{32}{3}$.   **23.** $-\dfrac{1}{6}$.   **25.** 4 ln 8.

27. $\dfrac{1}{3}(e^8 - 1)$.   **29.** $\dfrac{3}{4}$.   **31.** $\dfrac{38}{9}$.   **33.** $\dfrac{15}{28}$.   **35.** $\dfrac{1}{2}$ ln 3.

37. $e + \dfrac{1}{2e^2} - \dfrac{3}{2}$.   **39.** $\dfrac{3}{2} - \dfrac{1}{e} + \dfrac{1}{2e^2}$.   **41.** $\dfrac{e^3}{2}(e^{12} - 1)$.   **43.** $\dfrac{1}{2}$; $\dfrac{1}{12}$.

45. $\displaystyle\int_{b}^{a} -Ax^{-B}\, dx$.   **47.** $160.   **49.** $2000.   **51.** $8639.   **53.** 1,973,333.

55. $\dfrac{i}{k}(1 - e^{-2kR})$.

EXERCISE 10-7

In Problems 1–33, answers are assumed to be expressed in square units.

1. 8.   **3.** $\dfrac{19}{2}$.   **5.** 8.   **7.** $\dfrac{19}{3}$.   **9.** 9.   **11.** $\dfrac{50}{3}$.   **13.** 36.

15. 8.   **17.** $\dfrac{32}{3}$.   **19.** 1.   **21.** 18.   **23.** $\dfrac{26}{3}$.   **25.** $\dfrac{3}{2}\sqrt[3]{2}$.   **27.** $e^2 - 1$.

29. $\dfrac{3}{2} + 2\ln 2 = \dfrac{3}{2} + \ln 4$.   **31.** 68.   **33.** 2.   **35.** 19 sq units.

37. (a) $\dfrac{1}{16}$; (b) $\dfrac{3}{4}$; (c) $\dfrac{7}{16}$.   **39.** (a) $\ln\dfrac{5}{3}$; (b) $\ln(4) - 1$; (c) $2 - \ln 3$.

EXERCISE 10-8

In Problems **1–21**, *the answers are assumed to be expressed in square units.*

1. $\dfrac{4}{3}$.   **3.** $\dfrac{16}{3}$.   **5.** $8\sqrt{6}$.   **7.** 40.   **9.** $\dfrac{125}{6}$.   **11.** $\dfrac{32}{81}$.

13. $\dfrac{125}{12}$.   **15.** $\dfrac{9}{2}$.   **17.** $\dfrac{44}{3}$.   **19.** $\dfrac{4}{3}(5\sqrt{5} - 2\sqrt{2})$.   **21.** $\dfrac{1}{2}$.   **23.** $\dfrac{20}{63}$.

EXERCISE 10-9

1. $CS = 25.6$, $PS = 38.4$.   **3.** $CS = 50 \ln(2) - 25$, $PS = 1.25$.

5. $CS = 800$, $PS = 1000$.

REVIEW PROBLEMS—CHAPTER 10

1. $\dfrac{x^4}{4} + x^2 - 7x + C$.   **3.** $\dfrac{117}{2}$.   **5.** $-(x + 5)^{-2} + C$.

7. $2 \ln |x^3 - 6x + 1| + C$.   **9.** $\dfrac{11\sqrt[3]{11}}{4} - 4$.   **11.** $\dfrac{y^4}{4} + \dfrac{2y^3}{3} + \dfrac{y^2}{2} + C$.

13. $\dfrac{4z^{3/4}}{3} - \dfrac{6z^{5/6}}{5} + C$.   **15.** $\dfrac{1}{3} \ln \dfrac{10}{3}$.   **17.** $\dfrac{2}{27}(3x^3 + 2)^{3/2} + C$.

19. $\dfrac{1}{2}(e^{2y} + e^{-2y}) + C$.   **21.** $\ln |x| - \dfrac{2}{x} + C$.   **23.** 11.1.   **25.** $\dfrac{7}{3}$.

27. $4 - 3\sqrt[3]{2}$.   **29.** $\dfrac{3}{t} - \dfrac{2}{\sqrt{t}} + C$.   **31.** $\dfrac{3}{2} - 5 \ln 2$.

In Problems **33–45**, *answers are assumed to be expressed in square units.*

33. $\dfrac{4}{3}$.   **35.** $\dfrac{16}{3}$.   **37.** $\dfrac{125}{6}$.   **39.** $6 + \ln 3$.   **41.** $\dfrac{2}{3}$.   **43.** 36.   **45.** $\dfrac{125}{3}$.

47. $p = 100 - \sqrt{2q}$.   **49.** \$1900.   **51.** $CS = 166\dfrac{2}{3}$, $PS = 53\dfrac{1}{3}$.

EXERCISE 11-1

1. $-e^{-x}(x + 1) + C$.   **3.** $\dfrac{y^4}{4}\left[\ln(y) - \dfrac{1}{4}\right] + C$.   **5.** $x[\ln(4x) - 1] + C$.

7. $\dfrac{2x}{3}(x + 1)^{3/2} - \dfrac{4}{15}(x + 1)^{5/2} + C = \dfrac{2}{15}(x + 1)^{3/2}(3x - 2) + C$.

9. $\dfrac{2}{3}x^{3/2} \ln(x) - \dfrac{4}{9}x^{3/2} + C$.   **11.** $\dfrac{1}{4}e^2(3e^2 - 1)$.

13. $\frac{1}{2}(1 - e^{-1})$, parts not needed.   **15.** $\frac{2}{3}(9\sqrt{3} - 10\sqrt{2})$.   **17.** $e^x(x^2 - 2x + 2) + C.$

19. $\frac{x^3}{3} + 2e^{-x}(x + 1) - \frac{e^{-2x}}{2} + C.$ **21.** $2e^3 + 1$ sq units.

EXERCISE 11-2

1. $2 \ln |x| + 3 \ln |x - 1| + C = \ln |x^2(x - 1)^3| + C.$

3. $-3 \ln |x + 1| + 4 \ln |x - 2| + C = \ln |(x - 2)^4/(x + 1)^3| + C.$

5. $\frac{1}{4}\left[\frac{3x^2}{2} + 2 \ln |x - 1| - 2 \ln |x + 1| \right] + C = \frac{1}{4}\left(\frac{3x^2}{2} + \ln\left[\frac{x - 1}{x + 1} \right]^2 \right) + C.$

7. $\ln |x| + 2 \ln |x - 4| - 3 \ln |x + 3| + C = \ln |x(x - 4)^2/(x + 3)^3| + C.$

9. $\frac{1}{2} \ln |x^6 + 2x^4 - x^2 - 2| + C$, partial fractions not required.

11. $[4/(x - 2)] - 5 \ln |x - 1| + 7 \ln |x - 2| + C.$

    $= [4/(x - 2)] + \ln|(x - 2)^7/(x - 1)^5| + C.$

13. $2 \ln |x| - \frac{1}{2} \ln (x^2 + 4) + C = \frac{1}{2} \ln \left[\frac{x^4}{x^2 + 4} \right] + C.$

15. $-\frac{1}{2} \ln (x^2 + 1) - \frac{2}{x - 3} + C.$

17. $5 \ln (x^2 + 1) + 2 \ln (x^2 + 2) + C = \ln [(x^2 + 1)^5(x^2 + 2)^2] + C.$

19. $\frac{3}{2} \ln (x^2 + 1) + \frac{1}{x^2 + 1} + C.$ **21.** $18 \ln (4) - 10 \ln (5) - 8 \ln (3).$

23. $\frac{11}{6} + 4 \ln \frac{2}{3}$ sq units.

EXERCISE 11-3

1. $\frac{1}{6} \ln \left| \frac{x}{6 + 7x} \right| + C.$ **3.** $\frac{1}{3} \ln \left| \frac{\sqrt{x^2 + 9} - 3}{x} \right| + C.$

5. $\frac{1}{2}\left[\frac{4}{5} \ln |4 + 5x| - \frac{2}{3} \ln |2 + 3x| \right] + C.$

7. $\frac{1}{8}(2x - \ln [4 + 3e^{2x}]) + C.$ **9.** $2\left[\frac{1}{1 + x} + \ln \left| \frac{x}{1 + x} \right| \right] + C.$ **11.** $1 + \ln \frac{4}{9}.$

13. $\frac{1}{2}(x\sqrt{x^2 - 3} - 3 \ln |x + \sqrt{x^2 - 3}|) + C.$ **15.** $\frac{1}{144}.$

17. $e^x(x^2 - 2x + 2) + C.$ **19.** $2\left(-\frac{\sqrt{4x^2 + 1}}{2x} + \ln |2x + \sqrt{4x^2 + 1}| \right) + C.$

21. $\frac{1}{9}\left(\ln|1 + 3x| + \frac{1}{1 + 3x}\right) + C.$ **23.** $\frac{1}{\sqrt{5}}\left(\frac{1}{2\sqrt{7}}\ln\left|\frac{\sqrt{7} + \sqrt{5}\,x}{\sqrt{7} - \sqrt{5}\,x}\right|\right) + C.$

25. $\frac{1}{3^6}\left[\frac{(3x)^6\ln(3x)}{6} - \frac{(3x)^6}{36}\right] + C = \frac{x^6}{36}[6\ln(3x) - 1] + C.$

27. $\frac{4(9x - 2)(1 + 3x)^{3/2}}{135} + C.$ **29.** $\frac{1}{2}\ln|2x + \sqrt{4x^2 - 13}| + C.$

31. $\frac{1}{4}[2x^2\ln(2x) - x^2] + C.$ **33.** $-\frac{\sqrt{9 - 4x^2}}{9x} + C.$ **35.** $\frac{1}{2}\ln(x^2 + 1) + C.$

37. $\frac{1}{6}(2x^2 + 1)^{3/2} + C.$ **39.** $\ln\left|\frac{x - 3}{x - 2}\right| + C.$ **41.** $\frac{x^4}{4}\left[\ln(x) - \frac{1}{4}\right] + C.$

43. $\frac{e^{2x}}{4}(2x - 1) + C.$ **45.** $x(\ln x)^2 - 2x\ln(x) + 2x + C.$

47. $\frac{2}{3}(9\sqrt{3} - 10\sqrt{2}).$ **49.** $2(2\sqrt{2} - \sqrt{7}).$

51. $\frac{7}{2}\ln(2) - \frac{3}{4}.$ **53.** (a) \$37,599; (b) \$4924.   **55.** (a) \$5481; (b) \$535.

EXERCISE 11-4

1. $\frac{16}{3}.$ **3.** $-1.$ **5.** $0.$ **7.** $\frac{13}{6}.$ **9.** \$12,400.   **11.** \$3156.

EXERCISE 11-5

1. .340; .333.   **3.** 1.388; 1.386.   **5.** .883.   **7.** 2,361,375.   **9.** 2.967.   **11.** .771.

EXERCISE 11-6

1. $\frac{1}{3}.$ **3.** Div.   **5.** $\frac{1}{e}.$ **7.** Div.   **9.** $-\frac{1}{2}.$ **11.** 0.   **13.** 6.

15. 3.   **17.** Div.   **19.** $\ln\frac{8}{7}$, integral is not improper.   **21.** Div.   **23.** 6.

25. Div.   **27.** (a) 800; (b) 2/3.   **29.** 4,000,000.   **31.** 1/2 sq unit.

EXERCISE 11-7

1. $y = -\frac{1}{x^2 + C}.$ **3.** $y = Ce^x, C > 0.$ **5.** $y = Cx, C > 0.$

7. $y = \frac{1}{3}(x^2 + 1)^{3/2} + C.$ **9.** $y = \sqrt{2x}.$ **11.** $y = \ln\frac{x^3 + 3}{3}.$

13. $y = \dfrac{4x^2 + 3}{2(x^2 + 1)}.$ **15.** $N = 20{,}000e^{.018t}$; $N = 20{,}000(1.2)^{t/10}$; 28,800.

17. $2e^{.946}$ billion.   **19.** $N = N_0 e^{k(t - t_0)}$, $t \geqslant t_0$.   **21.** \$62,500.

23. $N = M - (M - N_0)e^{-kt}.$

REVIEW PROBLEMS—CHAPTER 11

1. $\dfrac{x^2}{4}[2 \ln (x) - 1] + C.$ **3.** $5 + \dfrac{9}{4} \ln 3.$ **5.** $\dfrac{1}{21}(9 \ln |3 + x| - 2 \ln |2 + 3x|) + C.$

7. $\dfrac{1}{2(x + 2)} + \dfrac{1}{4} \ln \left| \dfrac{x}{x + 2} \right| + C.$ **9.** $-\dfrac{\sqrt{9 - 16x^2}}{9x} + C.$

11. $\dfrac{3}{2} \ln \left| \dfrac{x - 3}{x + 3} \right| + C.$ **13.** $\dfrac{e^{7x}}{49}(7x - 1) + C.$ **15.** $\dfrac{1}{2} \ln |\ln 2x| + C.$

17. $x - \dfrac{3}{2} \ln |3 + 2x| + C.$ **19.** 34.   **21.** (a) 1.405; (b) 1.388.   **23.** $\dfrac{1}{18}.$

25. Div.   **27.** Div.   **29.** $y = Ce^{x^3 + x^2}$, $C > 0.$

31. 144,000.   **33.** \$1443.

EXERCISE 12-1

1. 3.   **3.** 6.   **5.** 6.   **7.** 88.   **9.** 3.   **11.** $2x_0 + 2h - 5y_0 + 4.$

13. $y = -4.$ **15.** $z = 6.$

17.

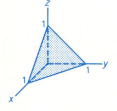

19.

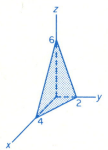

21.

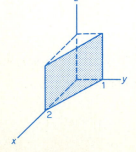

23.

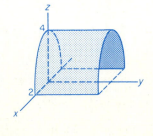

25. Surface is a sphere (top hemisphere shown).

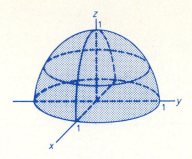

EXERCISE 12-2

1. $f_x(x, y) = 1$; $f_y(x, y) = -5$.   **3.** $f_x(x, y) = 3$; $f_y(x, y) = 0$.

5. $g_x(x, y) = 5x^4y^4 - 12x^3y^3 + 21x^2 - 3y$; $g_y(x, y) = 4x^5y^3 - 9x^4y^2 + 4y - 3x$.

7. $g_p(p, q) = \dfrac{q}{2\sqrt{pq}}$; $g_q(p, q) = \dfrac{p}{2\sqrt{pq}}$.

9. $h_s(s, t) = \dfrac{2s}{t-3}$; $h_t(s, t) = -\dfrac{s^2 + 4}{(t-3)^2}$.

11. $u_{q_1}(q_1, q_2) = \dfrac{3}{4q_1}$; $u_{q_2}(q_1, q_2) = \dfrac{1}{4q_2}$.

13. $h_x(x, y) = (x^3 + xy^2 + 3y^3)(x^2 + y^2)^{-3/2}$;

$h_y(x, y) = (3x^3 + x^2y + y^3)(x^2 + y^2)^{-3/2}$.

15. $\dfrac{\partial z}{\partial x} = 5ye^{5xy}$; $\dfrac{\partial z}{\partial y} = 5xe^{5xy}$.

17. $\dfrac{\partial z}{\partial x} = 5\left[\dfrac{2x^2}{x^2 + y} + \ln(x^2 + y)\right]$; $\dfrac{\partial z}{\partial y} = \dfrac{5x}{x^2 + y}$.

19. $f_r(r, s) = \sqrt{r + 2s}\,(3r^2 - 2s) + \dfrac{r^3 - 2rs + s^2}{2\sqrt{r + 2s}}$;

$f_s(r, s) = 2(s - r)\sqrt{r + 2s} + \dfrac{r^3 - 2rs + s^2}{\sqrt{r + 2s}}$.

21. $f_r(r, s) = -e^{3-r}\ln(7 - s)$; $f_s(r, s) = \dfrac{-e^{3-r}}{7 - s}$.

23. $g_x(x, y, z) = 6xy + 2y^2z$; $g_y(x, y, z) = 3x^2 + 4xyz$; $g_z(x, y, z) = 2xy^2 + 9z^2$.

25. $g_r(r, s, t) = 2re^{s+t}$; $g_s(r, s, t) = (7s^3 + 21s^2 + r^2)e^{s+t}$;

$g_t(r, s, t) = e^{s+t}(r^2 + 7s^3)$.

27. 50.   **29.** $\dfrac{1}{3}$.   **31.** 0.

EXERCISE 12-3

1. 20.   **3.** 784.5.   **5.** $\dfrac{\partial P}{\partial k} = 1.208648 l^{.192} k^{-.236}$; $\dfrac{\partial P}{\partial l} = .303744 l^{-.808} k^{.764}$.

7. $\dfrac{\partial q_A}{\partial p_A} = -50$; $\dfrac{\partial q_A}{\partial p_B} = 2$; $\dfrac{\partial q_B}{\partial p_A} = 4$; $\dfrac{\partial q_B}{\partial p_B} = -20$;   competitive.

9. $\dfrac{\partial q_A}{\partial p_A} = -\dfrac{100}{p_A^2 p_B^{1/2}}$; $\dfrac{\partial q_A}{\partial p_B} = -\dfrac{50}{p_A p_B^{3/2}}$; $\dfrac{\partial q_B}{\partial p_A} = -\dfrac{500}{3 p_B p_A^{4/3}}$;

$\dfrac{\partial q_B}{\partial p_B} = -\dfrac{500}{p_B^2 p_A^{1/3}}$;   complementary.

11. $\dfrac{\partial P}{\partial B} = .01 A^{.27} B^{-.99} C^{.01} D^{.23} E^{.09} F^{.27}$; $\dfrac{\partial P}{\partial C} = .01 A^{.27} B^{.01} C^{-.99} D^{.23} E^{.09} F^{.27}$.

13. (a) no;   (b) 70%.   **15.** $\eta_{p_A} = -1$, $\eta_{p_B} = -\dfrac{1}{2}$.

EXERCISE 12-4

1. $-\dfrac{x}{z}$.   **3.** $\dfrac{4y}{3z^2}$.   **5.** $\dfrac{x(yz^2 + 1)}{z(1 - x^2 y)}$.   **7.** $-e^{y-z}$.   **9.** $\dfrac{yz}{1 + z}$.

11. $-\dfrac{3x}{z}$.   **13.** $-\dfrac{9}{10}$.   **15.** 1.   **17.** $\dfrac{5}{2}$.

EXERCISE 12-5

1. $6xy^2$; $12xy$.   **3.** $3xe^{3xy} + 4x^2$; $9xye^{3xy} + 3e^{3xy} + 8x$; $9x(3xy + 2)e^{3xy}$.

5. $(2x + y)(2x^2 + 2xy + y^2 + 1)$; $6x^2 + 8xy + 3y^2 + 1$.

7. $3x^2 y + 4xy^2 + y^3$; $3xy^2 + 4x^2 y + x^3$; $6xy + 4y^2$; $6xy + 4x^2$.

9. $x(x^2 + y^2)^{-1/2}$; $y^2(x^2 + y^2)^{-3/2}$.   **11.** 0.   **13.** 1758.   **15.** $2e$.

21. $-\dfrac{y^2 + z^2}{z^3} = -\dfrac{3x^2}{z^3}$.

EXERCISE 12-6

1. $\dfrac{\partial z}{\partial r} = 13$; $\dfrac{\partial z}{\partial s} = 9$.   **3.** $\left[2t + \dfrac{3\sqrt{t}}{2} \right] e^{x+y}$.

5. $5(2xz^2 + yz) + 2(xz + z^2) - (2x^2 z + xy + 2yz)$.

7. $3(x^2 + xy^2)^2(2x + y^2 + 2xy)$.   **9.** $-2s(2x + yz) + r(xz + 3y^2 z^2) - 5(xy + 2y^3 z)$.

11. $15s(2x - 7)$.   **13.** 324.   **15.** -1.

EXERCISE 12-7

1. $\left(\dfrac{14}{3}, -\dfrac{13}{3} \right)$.   3. $(2, 5), (2, -6), (-1, 5), (-1, -6)$.   5. $(50, 150, 350)$.

7. $\left(-2, \dfrac{3}{2} \right)$, rel. min.   9. $\left(-\dfrac{1}{4}, \dfrac{1}{2} \right)$, rel. max.

11. $(1, 1)$, rel. min.; $\left(\dfrac{1}{2}, \dfrac{1}{4} \right)$, neither.

13. $(0, 0)$, rel. max.; $\left(4, \dfrac{1}{2} \right)$, rel. min.; $\left(0, \dfrac{1}{2} \right)$, $(4, 0)$, neither.

15. $(122, 127)$, rel. max.   17. $(-1, -1)$, rel. min.   19. $l = 24$, $k = 14$.

21. $p_A = 80$, $p_B = 85$.   23. $q_A = 48$, $q_B = 40$, $p_A = 52$, $p_B = 44$, profit $= 3304$.

25. $q_A = 3$, $q_B = 2$.   27. 1 ft by 2 ft by 3 ft.   29. $\left(\dfrac{105}{37}, \dfrac{28}{37} \right)$, rel. min.

EXERCISE 12-8

1. $(2, -2)$.   3. $\left(3, \dfrac{3}{2}, -\dfrac{3}{2} \right)$.   5. $\left(\dfrac{4}{3}, -\dfrac{4}{3}, -\dfrac{8}{3} \right)$.   7. $(6, 3, 2)$.

9. $\left(\dfrac{2}{3}, \dfrac{4}{3}, -\dfrac{4}{3} \right)$.   11. $(3, 3, 6)$.   13. Plant 1, 40 units;   Plant 2, 60 units.

15. 74 units (when $l = 8$, $k = 7$).   17. $x = 12$, $y = 8$.   19. $x = 10$, $y = 20$, $z = 5$.

EXERCISE 12-9

1. $\hat{y} = .98 + .61x$;   3.12.   3. $\hat{y} = .057 + 1.67x$;   5.90.

5. $\hat{q} = 82.6 - .641p$.   7. $\hat{c} = 8.51 - .365q$;   6.685.

9. $\hat{y} = 89.45 + 1.45x$.   11. (a) $\hat{y} = 35.9 - 2.5x$;   (b) $\hat{y} = 28.4 - 2.5x$.

EXERCISE 12-11

1. 18.   3. $\dfrac{1}{4}$.   5. $\dfrac{2}{3}$.   7. 3.   9. 324.   11. $-\dfrac{58}{35}$.

13. $\dfrac{8}{3}$.   15. $-\dfrac{1}{3}$.   17. $\dfrac{e^2}{2} - e + \dfrac{1}{2}$.   19. $-\dfrac{27}{4}$.   21. $\dfrac{1}{24}$.

23. $e^{-4} - e^{-2} - e^{-3} + e^{-1}$.   25. $\dfrac{3}{8}$.

REVIEW PROBLEMS—CHAPTER 12

1. $4x + 3y$; $3x + 2y$.   3. $\dfrac{y}{(x + y)^2}$; $-\dfrac{x}{(x + y)^2}$.   5. $2xze^{x^3yz}(1 + x^3yz)$.

7. $\dfrac{y}{x^2 + y^2}$.   **9.** $2(x + y)$.

11. $xze^{yz} \ln z$; $\dfrac{e^{yz}}{z} + ye^{yz} \ln z = e^{yz}\left(\dfrac{1}{z} + y \ln z\right)$.

13. $2(x + y)e^r + 2\left(\dfrac{x + 3y}{r + s}\right)$; $2\left(\dfrac{x + 3y}{r + s}\right)$.   **15.** $\dfrac{2x + 2y + z}{4z - x}$.

17. $(2, 2)$, rel. min.   **19.** $(3, 2, 1)$.   **21.** $\hat{y} = 1.34 + 1.8x$.

23. $\dfrac{\partial P}{\partial l} = 14l^{-.3}k^{.3}$; $\dfrac{\partial P}{\partial k} = 6l^{.7}k^{-.7}$.   **25.** $\dfrac{7}{2}$.   **27.** $\dfrac{81}{5}$.   **29.** $.530$; $-.027$.

31.

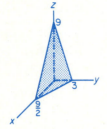

33.

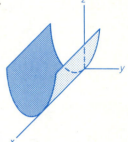

35. 4 ft by 4 ft by 2 ft.

EXERCISE 13-1

1. $\{1, 3, 5, 7, 9\}$.   **3.** $\{3, 5\}$.   **5.** $\{1, 2, 4, 6, 8, 10\}$.   **7.** S.

9. E_1 and E_4, E_2 and E_3, E_3 and E_4.

11. (a) $\{RR, RW, RB, WR, WW, WB, BR, BW, BB\}$;

   (b) $\{RW, RB, WR, WB, BR, BW\}$.

13. (a) $\{HHH, HHT, HTH, HTT, THH, THT, TTH, TTT\}$;

   (b) $\{HHH, HHT, HTH, HTT, THH, THT, TTH\}$;

   (c) $\{HHT, HTH, HTT, THH, THT, TTH, TTT\}$; (d) S;

   (e) $\{HHT, HTH, HTT, THH, THT, TTH\}$; (f) $\varnothing$;

   (g) $\{HHH, TTT\}$.

15. (a) $\{ABC, ACB, BAC, BCA, CAB, CBA\}$;   (b) $\{ABC, ACB\}$;

   (c) $\{BAC, BCA, CAB, CBA\}$.

17. $4^{10} = 1,048,576$.   **19.** (a) $10 \cdot 9 \cdot 8 \cdot 7 = 5040$;   (b) $10^4 = 10,000$.

21. $4 \cdot 2 \cdot 2 \cdot 4 = 64$.

EXERCISE 13-2

1. $\dfrac{1}{9}$.   **3.** (a) $\dfrac{4}{5}$;   (b) $\dfrac{1}{5}$.

5. (a) .1;   (b) .35;   (c) .7;   (d) .95;   (e) .1, .35, .7, .95.

7. (a) $\dfrac{5}{36}$;   (b) $\dfrac{1}{12}$;   (c) $\dfrac{1}{4}$;   (d) $\dfrac{1}{36}$;   (e) $\dfrac{1}{2}$;   (f) $\dfrac{1}{2}$;   (g) $\dfrac{5}{6}$.

9. (a) $\dfrac{1}{52}$;   (b) $\dfrac{1}{4}$;   (c) $\dfrac{1}{13}$;   (d) $\dfrac{1}{2}$;   (e) $\dfrac{1}{2}$;   (f) $\dfrac{1}{52}$;   (g) $\dfrac{4}{13}$;

(h) $\dfrac{1}{26}$;   (i) 0.

11. (a) $\dfrac{1}{624}$;   (b) $\dfrac{4}{624} = \dfrac{1}{156}$;   (c) $\dfrac{8}{624} = \dfrac{1}{78}$;   (d) $\dfrac{39}{624} = \dfrac{1}{16}$.

13. (a) $\dfrac{1}{8}$;   (b) $\dfrac{3}{8}$;   (c) $\dfrac{1}{8}$;   (d) $\dfrac{7}{8}$.

15. (a) $\dfrac{12}{2652} = \dfrac{1}{221}$;   (b) $\dfrac{338}{2652} = \dfrac{13}{102}$.   **17.** $\dfrac{2}{20} = \dfrac{1}{10}$.

19. (a) $\dfrac{1}{2^{10}} = \dfrac{1}{1024}$;   (b) $\dfrac{11}{1024}$.

EXERCISE 13-3

1. $\mu = 1.7$; $\text{Var}(X) = 1.01$; $\sigma = 1.00$.

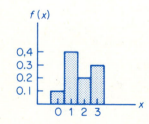

3. $\mu = \dfrac{9}{4} = 2.25$; $\text{Var}(X) = \dfrac{11}{16} = .6875$; $\sigma = .83$.

5. $E(X) = \dfrac{3}{2} = 1.5$; $\sigma^2 = \dfrac{3}{4} = .75$; $\sigma = .87$.

7. $E(X) = \dfrac{4}{5} = .8$; $\sigma^2 = \dfrac{9}{25} = .36$; $\sigma = .6$

9. $101.43.   **11.** $3.00.   **13.** $62.   **15.** Loss of $.25; $1.

EXERCISE 13-4

1. 24.   **3.** 15.   **5.** 10.   **7.** 1.

9. $f(0) = \dfrac{9}{16}$, $f(1) = \dfrac{3}{8}$, $f(2) = \dfrac{1}{16}$; $\mu = \dfrac{1}{2}$; $\sigma = \dfrac{\sqrt{6}}{4}$.   **11.** (a) $\dfrac{9}{64}$;   (b) $\dfrac{5}{32}$.

13. $\dfrac{13}{16}$.   **15.** .002.

EXERCISE 13-5

1. (a) $\dfrac{5}{12}$;   (b) $\dfrac{11}{16} = .6875$;   (c) $\dfrac{13}{16} = .8125$;   (d) $-1 + \sqrt{10}$.

3. (a) $f(x) = \begin{cases} \dfrac{1}{3}, & \text{if } 1 \leqslant x \leqslant 4, \\ 0, & \text{otherwise;} \end{cases}$

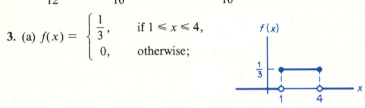

(b) $\dfrac{1}{3}$;   (c) 0;   (d) $\dfrac{5}{6}$;   (e) $\dfrac{2}{3}$;   (f) 0;   (g) 1   (h) $\dfrac{5}{2}$;   (i) $\dfrac{\sqrt{3}}{2}$;

(j) $F(x) = \begin{cases} 0, & \text{if } x < 1, \\ \dfrac{x-1}{3}, & \text{if } 1 \leqslant x \leqslant 4, \\ 1, & \text{if } x > 4, \end{cases}$

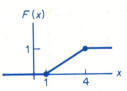

$P(X < 2) = \dfrac{1}{3}$, $P(1 < X < 3) = \dfrac{2}{3}$.

5. (a) $f(x) = \begin{cases} \dfrac{1}{b-a}, & \text{if } a \leqslant x \leqslant b, \\ 0, & \text{otherwise;} \end{cases}$ (b) $\dfrac{a+b}{2}$;

(c) $\sigma^2 = \dfrac{(b-a)^{12}}{12}$, $\sigma = \dfrac{b-a}{\sqrt{12}}$.

7. (a) .133;   (b) .982;   (c) .007;   (d) .865.   **9.** (a) $\dfrac{1}{8}$;   (b) $\dfrac{5}{16}$;

(c) $\dfrac{39}{64} = .609$;   (d) 1;   (e) $\dfrac{8}{3}$;   (f) $\dfrac{2\sqrt{2}}{3}$;   (g) $2\sqrt{2}$;   (h) $\dfrac{7}{16}$.

11. $\dfrac{7}{10}$; 5 min.   **13.** .050.

EXERCISE 13-6

1. (a) .4641;   (b) .3239;   (c) .8888;   (d) .9983;   (e) .9147;

(f) .4721.   **3.** .13.   **5.** -1.08.   **7.** .34

9. (a) .9332;   (b) .0668;   (c) .0873.   **11.** .3085.   **13.** .8185.

15. 8.   **17.** 9.68%.   **19.** 90.82%.   **21.** (a) 1.7%;   (b) 85.6.

EXERCISE 13-7

1. .1056; .0122.   **3.** .0430; .9232.   **5.** .7507.   **7.** .4129.

9. .2514; .0287.   **11.** .0336.

REVIEW PROBLEMS — CHAPTER 13

1. (a) $\{1, 2, 3, 4, 5, 6, 7\}$;   (b) $\{4, 5, 6\}$;   (c) $\{4, 5, 6, 7, 8\}$;   (d) $\varnothing$;

(e) $\{4, 5, 6, 7, 8\}$;   (f) no.

3. (a) $\{R_1R_2R_3, R_1R_2G_3, R_1G_2R_3, R_1G_2G_3, G_1R_2R_3, G_1R_2G_3, G_1G_2R_3, G_1G_2G_3\}$;

(b) $\{R_1R_2G_3, R_1G_2R_3, G_1R_2R_3\}$;   (c) $\{R_1R_2R_3, G_1G_2G_3\}$.

5. 26,000.   **7.** 24.   **9.** (a) $\dfrac{4}{25}$;   (b) $\dfrac{2}{15}$.

11. (a) $\dfrac{1}{4}$;   (b) $\dfrac{1}{4}$.   **13.** $f(x)$ $\mu = 1.5, \text{Var}(X) = .65, \sigma = .81.$

15. (a) $f(1) = \dfrac{1}{12}, f(2) = f(3) = f(4) = f(5) = f(6) = \dfrac{1}{6}, f(7) = \dfrac{1}{12}$ (b) 4.

17. Loss of $.10 per play.   **19.** $\dfrac{33}{81}$.   **21.** (a) 2;   (b) $\dfrac{9}{32}$;   (c) $\dfrac{3}{4}$;

(d) $F(x) = \begin{cases} 0, & \text{if } x < 0, \\ \dfrac{x}{3} + \dfrac{2x^3}{3}, & \text{if } 0 \leq x \leq 1, \\ 1, & \text{if } x > 1. \end{cases}$ **23.** (a) 2;   (b) $\dfrac{1}{\sqrt{2}} = .71$.

25. .3085.   **27.** .2857.   **29.** .1587.   **31.** .9817.   **33.** .0228.

EXERCISE 14-1

1. (a) $2 \times 3, 3 \times 3, 3 \times 2, 2 \times 2, 4 \times 4, 1 \times 2, 3 \times 1, 3 \times 3, 1 \times 1$;

(b) **B, D, E, H, J;**   (c) **H, J** upper triangular;   **D, J** lower triangular;

(d) **F, J;**   (e) **G, J.**

3. 2.   **5.** 4.   **7.** 6.   **9.** 7, 2, 1, 0.   **11.** $\begin{bmatrix} 5 & 8 & 11 & 14 \\ 7 & 10 & 13 & 16 \\ 9 & 12 & 15 & 18 \end{bmatrix}$.

13. 120 entries, 1, 0, 1, 0.   **15.** (a) $\begin{bmatrix} 0 & 0 & 0 & 0 \\ 0 & 0 & 0 & 0 \\ 0 & 0 & 0 & 0 \\ 0 & 0 & 0 & 0 \end{bmatrix}$; (b) $\begin{bmatrix} 0 & 0 & 0 & 0 & 0 & 0 \\ 0 & 0 & 0 & 0 & 0 & 0 \\ 0 & 0 & 0 & 0 & 0 & 0 \\ 0 & 0 & 0 & 0 & 0 & 0 \\ 0 & 0 & 0 & 0 & 0 & 0 \\ 0 & 0 & 0 & 0 & 0 & 0 \end{bmatrix}$.

17. $x = 6, y = \dfrac{2}{3}, z = \dfrac{7}{2}$.   **19.** $x = 0, y = 0$.

21. (a) 7; (b) 3; (c) February; (d) deluxe blue; (e) February; (f) February; (g) 35.

EXERCISE 14-2

1. $\begin{bmatrix} 4 & -3 & 1 \\ -2 & 10 & 5 \\ 10 & 5 & 3 \end{bmatrix}$.   **3.** $\begin{bmatrix} -5 & 5 \\ -9 & 5 \\ 5 & 9 \end{bmatrix}$.   **5.** $[-9 \quad -7 \quad 6 \quad 11]$.   **7.** Not defined.

9. $\begin{bmatrix} -12 & 36 & -42 & -6 \\ -42 & -6 & -36 & 12 \end{bmatrix}$.   **11.** $\begin{bmatrix} 5 & -4 & 1 \\ 0 & 7 & -2 \\ -3 & 3 & 13 \end{bmatrix}$.   **13.** $\begin{bmatrix} 6 & 5 \\ -2 & 3 \end{bmatrix}$.

15. O.   **17.** $\begin{bmatrix} 28 & 22 \\ -2 & 6 \end{bmatrix}$.   **19.** Not defined.   **21.** $\begin{bmatrix} -22 & -15 \\ -11 & 9 \end{bmatrix}$.

23. $\begin{bmatrix} 21 & \dfrac{29}{2} \\ \dfrac{19}{2} & -\dfrac{15}{2} \end{bmatrix}$.   **29.** $x = \dfrac{146}{13}, y = -\dfrac{28}{13}$.   **31.** $x = 6, y = \dfrac{4}{3}$.

33. $x = -6, y = -14, z = 1$.   **35.** 1.1.

EXERCISE 14-3

1. -12.   **3.** 19.   **5.** 7.   **7.** 2×2, 4.   **9.** 3×5, 15.   **11.** 2×1, 2.

13. 3×3, 9.   **15.** 3×1, 3.   **17.** $\begin{bmatrix} 1 & 0 & 0 & 0 \\ 0 & 1 & 0 & 0 \\ 0 & 0 & 1 & 0 \\ 0 & 0 & 0 & 1 \end{bmatrix}$.   **19.** $\begin{bmatrix} 10 & -16 \\ 7 & 8 \end{bmatrix}$.

21. $\begin{bmatrix} 23 \\ 50 \end{bmatrix}$.   **23.** $\begin{bmatrix} -3 & 4 & 2 \\ 2 & 2 & 4 \\ 5 & 0 & 3 \end{bmatrix}$.   **25.** $[-6 \quad 16 \quad 10 \quad -6]$.

27. $\begin{bmatrix} 4 & 6 & -4 & 6 \\ 6 & 9 & -6 & 9 \\ -8 & -12 & 8 & -12 \\ 2 & 3 & -2 & 3 \end{bmatrix}$.   **29.** $\begin{bmatrix} 78 & 84 \\ -21 & -12 \end{bmatrix}$.   **31.** $\begin{bmatrix} -5 & -8 \\ -5 & -20 \end{bmatrix}$.

33. $\begin{bmatrix} x \\ y \\ z \end{bmatrix}$.   **35.** $\begin{bmatrix} 2x_1 + x_2 + 3x_3 \\ 4x_1 + 9x_2 + 7x_3 \end{bmatrix}$.   **37.** $\begin{bmatrix} -4 & 11 & -2 \\ 3 & -12 & 3 \end{bmatrix}$.

39. $\begin{bmatrix} -1 \\ 3 \\ 8 \end{bmatrix}$.   **41.** $\begin{bmatrix} 3 & 0 & 0 \\ 0 & 6 & 3 \\ 3 & 12 & 3 \end{bmatrix}$.   **43.** $[7 \quad 23]$.   **45.** $\begin{bmatrix} 0 & 0 & 0 \\ 0 & -1 & 1 \\ 1 & 2 & 0 \end{bmatrix}$.

47. $\begin{bmatrix} -1 & -20 \\ -2 & 23 \end{bmatrix}$.   **49.** $\begin{bmatrix} \frac{3}{2} & 0 & 0 \\ 0 & \frac{3}{2} & 0 \\ 0 & 0 & \frac{3}{2} \end{bmatrix}$.   **51.** $\begin{bmatrix} -1 & 5 \\ 2 & 17 \\ 1 & 31 \end{bmatrix}$.

53. $\begin{bmatrix} 3 & 1 \\ 7 & -2 \end{bmatrix}\begin{bmatrix} x \\ y \end{bmatrix} = \begin{bmatrix} 6 \\ 5 \end{bmatrix}$.   **55.** $\begin{bmatrix} 4 & -1 & 3 \\ 3 & 0 & -1 \\ 0 & 3 & 2 \end{bmatrix}\begin{bmatrix} r \\ s \\ t \end{bmatrix} = \begin{bmatrix} 9 \\ 7 \\ 15 \end{bmatrix}$.   **57.** $735,300.

59. (a) $180,000, $520,000, $400,000, $270,000, $380,000, $640,000;

(b) $390,000, $100,000, $800,000; (c) $2,390,000; (d) $\frac{110}{239}$, $\frac{129}{239}$.

EXERCISE 14-4

1. Not reduced.   **3.** Reduced.   **5.** Not reduced.   **7.** $\begin{bmatrix} 1 & 0 \\ 0 & 1 \end{bmatrix}$.

9. $\begin{bmatrix} 1 & 2 & 3 \\ 0 & 0 & 0 \\ 0 & 0 & 0 \end{bmatrix}$.   **11.** $\begin{bmatrix} 1 & 0 & 0 & 0 \\ 0 & 1 & 0 & 0 \\ 0 & 0 & 1 & 0 \\ 0 & 0 & 0 & 1 \end{bmatrix}$.   **13.** $x = 1, y = 1$.   **15.** No solution.

17. $x = -\frac{2}{3}z + \frac{5}{3}, y = -\frac{1}{6}z + \frac{7}{6}, z = z$.   **19.** No solution.

21. $x = -3, y = 1, z = 0$.   **23.** $x = 2, y = -5, z = -1$.

25. $x_1 = 0, x_2 = -x_5, x_3 = -x_5, x_4 = -x_5, x_5 = x_5$.

27. Federal, $72,000;   state, $24,000.   **29.** A, 2000; B, 4000; C, 5000.

31. (a) 3 of X, 4 of Z;   2 of X, 1 of Y, 5 of Z;   1 of X, 2 of Y, 6 of Z;   3 of Y, 7 of Z. (b) 3 of X, 4 of Z (c) 3 of X, 4 of Z;   3 of Y, 7 of Z.

EXERCISE 14-5

1. $w = -y - 3z + 2, x = -2y + z - 3, y = y, z = z$.

3. $w = -z, x = -3y - 4z + 2, y = y, z = z$.

5. $w = -2y + z - 2, x = -y + 4, y = y, z = z$.

7. $x_1 = -2x_3 + x_4 - 2x_5 + 1$, $x_2 = -x_3 - 2x_4 + x_5 + 4$, $x_3 = x_3$, $x_4 = x_4$, $x_5 = x_5$.

9. Infinite.   **11.** Trivial.   **13.** Infinite.   **15.** $x = 0, y = 0$.

17. $x = -\dfrac{6}{5}z, y = \dfrac{8}{15}z, z = z$.   **19.** $x = 0, y = 0$.   **21.** $x = z, y = -2z, z = z$.

23. $w = -2z, x = -3z, y = z, z = z$.

EXERCISE 14-6

1. $\begin{bmatrix} 1 & -1 \\ -5 & 6 \end{bmatrix}$.   **3.** Not invertible.   **5.** $\begin{bmatrix} 1 & 0 & 0 \\ 0 & -\frac{1}{3} & 0 \\ 0 & 0 & \frac{1}{4} \end{bmatrix}$.   **7.** Not invertible.

9. Not invertible (not a square matrix).   **11.** $\begin{bmatrix} 1 & -1 & 0 \\ 0 & 1 & -1 \\ 0 & 0 & 1 \end{bmatrix}$.   **13.** $\begin{bmatrix} 1 & 0 & 2 \\ 0 & 1 & 0 \\ 3 & 0 & 7 \end{bmatrix}$.

15. $\begin{bmatrix} 1 & -\frac{2}{3} & \frac{5}{3} \\ -1 & \frac{4}{3} & -\frac{10}{3} \\ -1 & 1 & -2 \end{bmatrix}$.   **17.** $\begin{bmatrix} \frac{11}{3} & -3 & \frac{1}{3} \\ -\frac{7}{3} & 3 & -\frac{2}{3} \\ \frac{2}{3} & -1 & \frac{1}{3} \end{bmatrix}$.   **19.** $x = 17, y = -20$.

21. $x = 1, y = 3$.   **23.** $x = -3y + 1, y = y$.   **25.** $x = 0, y = 1, z = 2$.

27. $x = 1, y = \dfrac{1}{2}, z = \dfrac{1}{2}$.   **29.** No solution.   **31.** $w = 1, x = 3, y = -2, z = 7$.

33. $\begin{bmatrix} -\frac{2}{3} & -\frac{1}{3} \\ \frac{1}{3} & -\frac{1}{3} \end{bmatrix}$.

35. (a) 40 of first model, 60 of second model;   (b) 45 of first model, 50 of second model.

EXERCISE 14-7

1. 1.   **3.** 0.   **5.** y.   **7.** $-\dfrac{2}{7}$.   **9.** 12.   **11.** -12.   **13.** 6.

15. $\begin{vmatrix} a_{11} & a_{13} & a_{14} \\ a_{21} & a_{23} & a_{24} \\ a_{41} & a_{43} & a_{44} \end{vmatrix}$.   **17.** $\begin{vmatrix} a_{21} & a_{22} & a_{24} \\ a_{31} & a_{32} & a_{34} \\ a_{41} & a_{42} & a_{44} \end{vmatrix}$.   **19.** -16.   **21.** 98.   **23.** -89.

25. -1.   **27.** 2.   **29.** -90.   **31.** 1.   **33.** 24.   **35.** 3, 4.   **37.** 192.

EXERCISE 14-8

1. $x = \dfrac{9}{5}, y = -\dfrac{2}{5}$.   **3.** $x = \dfrac{7}{16}, y = \dfrac{13}{8}$.   **5.** $x = -\dfrac{1}{3}, y = -1$.

7. $x = \dfrac{6}{5}, z = \dfrac{16}{5}.$ **9.** $x = 4, y = 2, z = 0.$

11. $x = \dfrac{2}{3}, y = -\dfrac{28}{15}, z = -\dfrac{26}{15}.$ **13.** $x = 3 - z, y = 0, z = z.$

15. $x = 1, y = 3, z = 5.$ **17.** $y = 6, w = 1.$

19. Since $\Delta = \begin{vmatrix} 1 & 1 \\ 1 & 1 \end{vmatrix} = 0$, Cramer's rule does not apply.

But the equations in

$$\begin{cases} x + y = 2, \\ x + y = -3 \end{cases}$$

represent distinct parallel lines and hence no solution exists.

EXERCISE 14-9

1. $\begin{bmatrix} \frac{1}{4} & \frac{1}{4} \\ -\frac{1}{8} & \frac{3}{8} \end{bmatrix}.$ **3.** $\begin{bmatrix} 4 & -9 \\ 0 & 6 \end{bmatrix}.$ **5.** $\begin{bmatrix} 7 & -8 & 5 \\ -4 & 5 & -3 \\ 1 & -1 & 1 \end{bmatrix}.$ **7.** $\begin{bmatrix} 2 & 1 & 0 \\ 4 & -1 & 5 \\ 1 & -1 & 2 \end{bmatrix}.$

9. $\begin{bmatrix} 2 & -1 & 3 \\ 0 & 2 & 0 \\ 2 & 1 & 1 \end{bmatrix}.$ **11.** $\begin{bmatrix} 1 & 2 & -1 \\ 0 & 1 & 4 \\ 1 & -1 & 2 \end{bmatrix}.$

EXERCISE 14-10

1. $\begin{bmatrix} 1290 \\ 1425 \end{bmatrix}$; 1405.   **3.** (a) $\begin{bmatrix} 297.80 \\ 349.54 \\ 443.12 \end{bmatrix}$;   (b) $\begin{bmatrix} 102.17 \\ 125.28 \\ 175.27 \end{bmatrix}.$

REVIEW PROBLEMS — CHAPTER 14

1. $\begin{bmatrix} 7 & 12 \\ -19 & -5 \end{bmatrix}.$ **3.** $\begin{bmatrix} 1 & 35 & 5 \\ 2 & -15 & -7 \\ 1 & 0 & -2 \end{bmatrix}.$ **5.** $\begin{bmatrix} -1 & -2 \\ 5 & 22 \end{bmatrix}.$ **7.** $x = 3, y = 6.$

9. $\begin{bmatrix} 1 & 0 \\ 0 & 1 \end{bmatrix}.$ **11.** $\begin{bmatrix} 1 & 2 & 0 \\ 0 & 0 & 1 \\ 0 & 0 & 0 \end{bmatrix}.$ **13.** $x = 0, y = 0.$ **15.** No solution.

17. $\begin{bmatrix} -\frac{3}{2} & \frac{5}{6} \\ \frac{1}{2} & -\frac{1}{6} \end{bmatrix}.$ **19.** No inverse exists.   **21.** $x = 0, y = 1, z = 0.$ **23.** 18.

25. 3.   **27.** rich.   **29.** $x = 1, y = 2.$ **31.** $\begin{bmatrix} \frac{1}{2} & -1 & \frac{1}{2} \\ -\frac{1}{2} & 0 & \frac{1}{2} \\ 1 & 1 & -1 \end{bmatrix}.$ **33.** $\begin{bmatrix} 3.6 \\ 3.2 \end{bmatrix}.$

EXERCISE 15-1

1.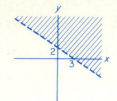

3.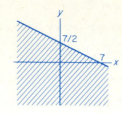

5.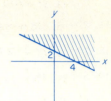

7.

9.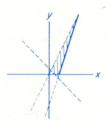

11.

13.

15.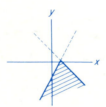

17.

19.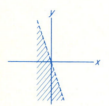

21.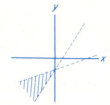

23.

25.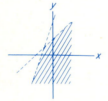

27.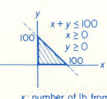

$x + y \leq 100$
$x \geq 0$
$y \geq 0$

x : number of lb. from A
y : number of lb. from B

EXERCISE 15-2

1. $P = 640$ when $x = 40$, $y = 20$.   **3.** $Z = -10$ when $x = 2$, $y = 3$.

5. No optimum solution (empty feasible region).   **7.** $Z = 3$ when $x = 0$, $y = 1$.

9. $C = 2.4$ when $x = \frac{3}{5}$, $y = \frac{6}{5}$.   **11.** No optimum solution (unbounded).

13. 15 widgets, 25 wadgits; $210.   **15.** 4 units of Food A, 4 units of Food B; $8.

17. 10 tons of ore I, 10 tons of ore II; $1100.

EXERCISE 15-3

1. $Z = 33$ when $x = (1 - t)(2) + 5t = 2 + 3t$, $y = (1 - t)(3) + 2t = 3 - t$, and $0 \leqslant t \leqslant 1$.

3. $Z = 72$ when $x = (1 - t)(3) + 4t = 3 + t$, $y = (1 - t)(2) + 0t = 2 - 2t$, and $0 \leqslant t \leqslant 1$.

EXERCISE 15-4

1. $Z = 8$ when $x_1 = 0$, $x_2 = 4$.   **3.** $Z = 14$ when $x_1 = 1$, $x_2 = 5$.

5. $Z = 28$ when $x_1 = 3$, $x_2 = 2$.   **7.** $Z = 20$ when $x_1 = 0$, $x_2 = 5$, $x_3 = 0$.

9. $Z = 2$ when $x_1 = 1$, $x_2 = 0$, $x_3 = 0$.   **11.** $Z = \frac{16}{3}$ when $x_1 = \frac{2}{3}$, $x_2 = \frac{14}{3}$.

13. $W = 13$ when $x_1 = 1$, $x_2 = 0$, $x_3 = 3$.

15. $Z = 600$ when $x_1 = 4$, $x_2 = 1$, $x_3 = 4$, $x_4 = 0$.

17. 400 from A, 1600 from B; $1100.

19. 0 chairs, 300 rockers, 100 chaise lounges; $3600.

EXERCISE 15-5

1. Yes; for the tableau, x_2 is the entering variable and the quotients $\frac{6}{2}$ and $\frac{3}{1}$ tie for being the smallest.

3. No optimum solution (unbounded).

5. $Z = 12$ when $x_1 = (1 - t)\cdot 4 + 5t = 4 + t$, $x_2 = (1 - t)\cdot 0 + 1\cdot t = t$, and $0 \leqslant t \leqslant 1$.

7. No optimum solution (unbounded).

9. $Z = 13$ when $x_1 = (1 - t)\frac{3}{2} + 0t = \frac{3}{2} - \frac{3}{2}t$, $x_2 = (1 - t)0 + 6t = 6t$, $x_3 = (1 - t)4 + 1\cdot t = 4 - 3t$, and $0 \leqslant t \leqslant 1$.

11. $3800. If x_1, x_2, and x_3 denote the number of chairs, rockers, and chaise lounges produced, respectively, then $x_1 = (1 - t)100 + 0t = 100 - 100t$, $x_2 = (1 - t)100 + 250t = 100 + 150t$, $x_3 = (1 - t)200 + 150t = 200 - 50t$, and $0 \leqslant t \leqslant 1$.

EXERCISE 15-6

1. $Z = 7$ when $x_1 = 1$, $x_2 = 5$.   **3.** $Z = 4$ when $x_1 = 1$, $x_2 = 2$, $x_3 = 0$.

5. $Z = \dfrac{58}{3}$ when $x_1 = \dfrac{14}{3}$, $x_2 = \dfrac{2}{3}$, $x_3 = 0$.   **7.** $Z = -17$ when $x_1 = 3$, $x_2 = 2$.

9. No optimum solution (empty feasible region).   **11.** $Z = 2$ when $x_1 = 6$, $x_2 = 10$.

13. 255 contemporary tables, 0 traditional tables.

15. 30% in A, 0% in AA, 70% in AAA; 6.6%.

EXERCISE 15-7

1. $Z = 54$ when $x_1 = 2$, $x_2 = 8$.   **3.** $Z = 36$ when $x_1 = 9$, $x_2 = 0$, $x_3 = 0$.

5. $Z = 4$ when $x_1 = 0$, $x_2 = 0$, $x_3 = 4$.   **7.** $Z = 0$ when $x_1 = 3$, $x_2 = 0$, $x_3 = 1$.

9. $Z = 28$ when $x_1 = 3$, $x_2 = 0$, $x_3 = 5$.

11. Install device A on kilns producing 700,000 barrels annually, and device B on kilns producing 2,600,000 barrels annually.

13. To Exton, 10 from A and 20 from B; to Whyton, 30 from A; $760.

15. (a) Column 3: 1, 3, 3; column 4: 0, 4, 8.   (b) $x_1 = 10$, $x_2 = 0$, $x_3 = 20$, $x_4 = 0$.
(c) 90 in.

EXERCISE 15-8

1. Minimize
$$W = 6y_1 + 4y_2$$
subject to
$$y_1 - y_2 \geqslant 2,$$
$$y_1 + y_2 \geqslant 3,$$
$$y_1, y_2 \geqslant 0.$$

3. Maximize
$$W = 8y_1 + 2y_2$$
subject to
$$y_1 - y_2 \leqslant 1,$$
$$y_1 + 2y_2 \leqslant 8,$$
$$y_1 + y_2 \leqslant 5,$$
$$y_1, y_2 \geqslant 0.$$

5. Minimize
$$W = 13y_1 - 3y_2 - 11y_3$$
subject to
$$-y_1 + y_2 - y_3 \geqslant 1,$$
$$2y_1 - y_2 - y_3 \geqslant -1,$$
$$y_1, y_2, y_3 \geqslant 0.$$

7. Maximize
$$W = -3y_1 + 3y_2$$
subject to
$$-y_1 + y_2 \leqslant 4,$$
$$y_1 - y_2 \leqslant 4,$$
$$y_1 + y_2 \leqslant 6,$$
$$y_1, y_2 \geqslant 0.$$

9. $Z = 11$ when $x_1 = 0$, $x_2 = \frac{1}{2}$, $x_3 = \frac{3}{2}$.   **11.** $Z = 26$ when $x_1 = 6$, $x_2 = 1$.

13. $Z = 14$ when $x_1 = 1$, $x_2 = 2$.

15. $25 on newspaper advertising, $140 on radio advertising; $165.

17. 20 shipping clerk apprentices, 40 shipping clerks, 90 semiskilled workers, 0 skilled workers; $600.

REVIEW PROBLEMS—CHAPTER 15

1.

3.

5.

7.

9.

11. $Z = 3$ when $x = 3$, $y = 0$.   **13.** $Z = -2$ when $x = 0$, $y = 2$.

15. No optimum solution (empty feasible region).

17. $Z = 36$ when $x = (1 - t)(2) + 4t = 2 + 2t$, $y = (1 - t)(3) + 0t = 3 - 3t$, and $0 \leqslant t \leqslant 1$.

19. $Z = 32$ when $x_1 = 8$, $x_2 = 0$.   **21.** $Z = 2$ when $x_1 = 0$, $x_2 = 0$, $x_3 = 2$.

23. $Z = 24$ when $x_1 = 0$, $x_2 = 12$.   **25.** $Z = \frac{7}{2}$ when $x_1 = \frac{5}{4}$, $x_2 = 0$, $x_3 = \frac{9}{4}$.

27. No optimum solution (unbounded).   **29.** $Z = 70$ when $x_1 = 35$, $x_2 = 0$, $x_3 = 0$.

31. 0 units of X, 6 units of Y, 14 units of Z; $398.

33. 500,000 gal from A to D, 100,000 gal from A to C, 400,000 gal from B to C; $19,000.

INDEX

A

Abscissa, 98–99

Absolute extrema, 336, 341, 355. *See also* Absolute maximum; Absolute minimum

Absolute maximum, 335–336, 341, 355–367

Absolute minimum, 335–336, 341, 355–367

Absolute value, 75–80, 89

Accounting price, 789

Accumulated amount. *See* Compound amount

Addition:
  fractions, 33
  matrix, 651, 658

Additive inverse, 4–5

Adjoint, 706–711

Algebraic expressions, 20–26

Amortization of loans, 201–205

Amortization schedule, 202

Annuity, 192–201, 473–475

Annuity due, 196, 198

Antiderivative, 385

Approximate integration, 480–485

Area, determination of:
  between curves, 442–447
  by definite integration, 414–423, 435–440

Artificial objective function, 768, 774, 776, 778, 794

Artificial problem, 768–778

Artificial variable, 767–781, 784, 793–794